# Chemical Signals in Vertebrates 5

## Chemical Signals in Vertebrates

*Chemical signals in vertebrates 1* (1977) (ed. D. Müller-Schwarze and M. M. Mozell). Plenum Press, New York.

*Chemical signals in vertebrates 2: vertebrates and aquatic invertebrates* (1980) (ed. D. Müller-Schwarze and R. M. Silverstein). Plenum Press, New York.

*Chemical signals in vertebrates 3* (1983) (ed. R. M. Silverstein and D. Müller-Schwarze). Plenum Press, New York.

*Chemical signals in vertebrates 4: ecology, evolution and comparative biology* (1986) (ed. D. Duvall, D. Müller-Schwarze, and R. M. Silverstein). Plenum Press, New York.

# Chemical Signals in Vertebrates 5

Edited by

D. W. MACDONALD

*Department of Zoology, University of Oxford*

D. MÜLLER-SCHWARZE

*State University of New York*
*College of Environmental Science and Forestry*

and

S. E. NATYNCZUK

*The Science School, Clifton College, Bristol*

Oxford New York Tokyo

OXFORD UNIVERSITY PRESS

1990

*Oxford University Press, Walton Street, Oxford OX2 6DP*
*Oxford New York Toronto*
*Delhi Bombay Calcutta Madras Karachi*
*Petaling Jaya Singapore Hong Kong Tokyo*
*Nairobi Dar es Salaam Cape Town*
*Melbourne Auckland*
*and associated companies in*
*Berlin Ibadan*

*Oxford is a trade mark of Oxford University Press*

*Published in the United States*
*by Oxford University Press, New York*



*British Library Cataloguing in Publication Data*
*Chemical signals in vertebrates*
*1. Vertebrates. Communication. Role of pheromones.*
*I Macdonald, D. W. (David W.) II Müller-Schwarze, Dietland. III Natynczuk, S. E.*
*596.059*
*ISBN 0–19–857731–1*

*Library of Congress Cataloging in Publication Data*
*Chemical signals in vertebrates / edited by D. W. Macdonald, D. Müller-Schwarze, and S. E. Natynczuk.*
*Includes bibliographical references and index.*
*1. Vertebrates—Behavior—Congresses. 2. Chemical senses—Congresses. 3. Pheromones—Congresses. 4. Animal communication—Congresses. 5. Semiochemicals—Congresses. 6. Vertebrates—Physiology—Congresses. I. Macdonald, David W. (David Whytè) II. Müller-Schwarze, Dietland. III. Natynczuk, S. E.*
*QL750.C48 1990 596'.051—dc20 90–14153*
*ISBN 0–19–857731–1*

*Printed in Great Britain by*
*Biddles Ltd, Guildford and King's Lynn*

# Preface

Chemical signals are a fundamental aspect of vertebrate biology. The social odours of mammals, for example, are produced by skin glands that are among the hallmarks of the order. As papers in this book demonstrate, the characteristics of these odours may be traced so close to the roots of individuality that they are coded by single genes. Mammals, of course, are notorious for their odours, but in the chapters that follow we can see that these signals are not only a uniting force in the lives of creatures from mice to men, but also for fish and reptiles. The problem is that although we may be affected by social odours, we are equipped with neither the senses to perceive them adequately nor the breadth of knowledge to interpret them. The papers in this book show how the first of these shortcomings is being remedied by technical and experimental ingenuity, and the second by a burgeoning inter-disciplinary synergism.

Scientists often aspire to inter-disciplinary exchange as an important goal. For the study of chemical signals it is not merely desirable; it is a necessity. The topic is inescapably both chemical and biological. Furthermore, it is a biological topic at an awesome variety of levels, embracing the physiological milieu within individuals (and between them, through the almost eerie actions of primer effects), the behavioural relationships between them, and the ecological traits of their populations. Questions about the causation and function of chemical signals might be answered, with equal validity, in terms of molecules or populations, physics or evolution. In the end, satisfying answers will synthesize all these levels. It was to embrace this diversity of disciplines, and the technologies and ideas of their practitioners that we arranged the meeting which spawned this book. In August 1988, and in the tradition of four similar gatherings initiated in 1976 by Dietland Müller-Schwarze, biologists and biochemists of every persuasion congregated at the Department of Zoology in Oxford. Anatomists and neurophysiologists shed their disciplinary girdles as they mixed with behaviourists. Chemists swapped stories with ecologists, while agriculturalists and medics mingled with evolutionists. The result is this book.

Our hope is that the papers herein catch the essence of an extraordinarily exciting field, at a moment when it seems poised for major intellectual advances. These advances concern not only fundamental science, but the crucial leap from pure to applied biology. The metamorphosis of the ideas presented at the Oxford conference into what we have affectionately known as The Smells Book has, like the research described, required the cooperation of many different people with varied skills. We gratefully acknowledge all of them, and most especially the authors who have tolerated our editorial whims. Our plans were greatly helped by the ideas of the six chairmen who helped organize parts of the original meeting (Drs E. Albone, D. Müller-Schwarze, B. Keverne, R. Brown, D. Duvall and D. Macdonald). We are pleased to acknowledge the helpfulness of the Zoology Department as host to the meeting, and the staff of Oxford University Press for their guidance. In addition, we acknowledge especially the editorial assistance of Jennifer Baines, Catherine Griffin, Jonathan Livingstone, Liz Bradshaw, and Rosie Woodroffe, and the meticulous work of the indexer, Frank Wallis.

*Oxford*
July, 1990

D.W.M.
D.M.-S.
S.E.N.

# Contents

SECTION FIVE PREDATORS AND PREY

# *Contributors*

D. H. ABBOT

MRC/AFRC Comparative Physiology Research Group, Institute of Zoology, Regent's Park, London NW1 4RY, UK

E. S. ALBONE

Clifton College, Bristol BSH 3JH; *and* School of Chemistry, University of Bristol, Bristol BS8 1TS, UK

P. ALDHOUS

Animal Behaviour Research Group, Department of Zoology, University of Nottingham, University Park, Nottingham NG7 2RD, UK

P. APPS

Institute for Chromatography, University of Pretoria, Pretoria 0002, South Africa; *and* Food Hygiene Section, Veterinary Research Institute, Onderstepoort 0110, South Africa

D. A. BAINES

Dalgety p.l.c., Group Research Laboratory, Station Road, Cambridge CB1 2JN, UK

J. BARD

Memorial Chemical Senses Center, University of Pennsylvania, 3500 Market Street, Philadelphia, PA 19104-3308, USA

J. BARRETT

MRC/AFRC Comparative Physiology Research Group, Institute of Zoology, Regent's Park, London NW1 4RY, UK

L. BARTOS

Research Institute of Animal Production, CS–104 00 Praha 10, Czechoslovakia

G. K. BEAUCHAMP

Monell Chemical Senses Center, University of Pennsylvania, 3500 Market Street, Philadelphia, Pa 19104-3308, USA

D. A. BOOTH

Department of Psychology, University of Birmingham, PO Box 368, Birmingham B15 2TT, UK

E. A. BOYSE

Memorial Chemical Senses Center, University of Pennsylvania, 3500 Market Street, Philadelphia, PA 19104-3308, USA

J. W. S. BRADSHAW
Department of Biology, Building 62, The University, Southampton SO9 3TU, UK

I. BROUETTE-LAHLOU
Laboratoire de Physiologie Neurosensorielle, Université Lyon I.F. 69622, Villeurbanne Cedex, France

R. E. BROWN
Department of Psychology, Dalhousie University, Halifax, Nova Scotia B3H 6JI, Canada

A. J. BUGLASS
Cambridgeshire College of Arts and Technology, East Road, Cambridge CB1 1PT, UK

G. M. BURGHARDT
Departments of Psychology and Zoology, Graduate Program in Ethology, University of Tennessee, Knoxville, TN 37996–0900, USA

D. CHISZAR
Department of Psychology, University of Colorado, Boulder, CO 80309, USA

J. R. CLARKE
Department of Zoology, University of Oxford, South Parks Road, Oxford OX1 3PS, UK

R. COCKE
Department of Psychology, University of Texas, Austin, TX 78712, USA

W. E. COOPER, JR.
Department of Biology, Auburn University at Montgomery, Montgomery, AL 36192, USA

J. J. COWLEY
Hatfield Polytechnic, College Lane, Hatfield AL10 9AB, UK

F. M. C. DARLING
Cambridgeshire College of Arts and Technology, East Road, Cambridge CB1 1PT, UK

D. A. DEWSBURY
Department of Psychology, University of Florida, Gainesville, FL 32611, USA

B. E. DIAL
Department of Biology, Chapman College, Orange, CA 92666, USA

R. L. DOTY
Smell and Taste Center, Hospital of the University of Pennsylvania, 3400 Spruce Street, Philadelphia, PA 19104, USA

K. B. DØVING
Department of Biology, University of Oslo, PO Box 1051, 0316 Oslo 3, Norway

S. J. DUGMORE
Biological Sciences, Glasgow College, Glasgow G4 0BA, Scotland

H. J. DUNCAN
University of Cincinnati, Cincinnati, O 45221, USA

D. DUVALL
Department of Zoology, University of Wyoming, Lararmie, WY 82071, USA

C. S. EVANS
Biological Sciences, Glasgow College, Glasgow G4 0BA, Scotland

C. C. FAULKES
MRC/AFRC Comparative Physiology Research Group, Institute of Zoology, Regent's Park, London NW1 4RY, UK

N. B. FORD
Department of Biology, The University of Texas at Tyler, Tyler, TX 75701, USA

D. J. GEE
Department of Forensic Medicine, University of Leeds, C.S.B., St James's University Hospital, Beckett Street, Leeds LS9 7TF, UK

K. GERBER
Department of Biology, Princeton University, Princeton, NJ 08544, USA

J. C. GILLINGHAM
Department of Biology, Central Michigan University, Mt. Pleasant, MI 48859, USA

M. L. GORMAN
University of Aberdeen, Culterty Field Station, Newburgh AB4 0AA, Scotland

L. M. GOSLING
Coypu Research Laboratory, MAFF, Jupiter Road, Norwich NR6 6SP, UK

D. B. GOWER
Division of Biochemistry, United Medical and Dental Schools, Guy's Hospital, London SE1 9RT, UK

M. A. GREEN
Department of Forensic Medicine, University of Leeds, C.S.B., St. James's University Hospital, Beckett Street, Leeds LS9 7TF, UK

N. HACKMANN
Department of Veterinary Science, University of Idaho, Moscow, ID 83843, USA

A. A. HAFER
Department of Zoology, University of Oxford, South Parks Road, Oxford OX1 3PS, UK

D. R. HALL
Overseas Development Natural Resources Institute, Central Avenue, Chatham Maritime, Chatham, Kent ME4 4TB, UK

S. P. HARDING
Department of Zoology, University of Oxford, South Parks Road, Oxford OX1 3PS, UK

D. W. HARWOOD
Department of Physiological Sciences, University of Manchester, Manchester M13 9PT, UK

G. I. HATTON
Neuroscience Program, Michigan State University, East Lansing, MI 48824, USA

P. G. HEPPER
Laboratory for Recognition Research, Department of Psychology, The Queen's University of Belfast, Belfast BT7 1NN, Northern Ireland

D. HOLLAND
Department of Biology, University of Southwestern Louisiana, Lafayette, LA 70504, USA

K. T. HOLLAND
Department of Microbiology, University of Leeds, Leeds LS2 9JT, UK

D. J. HOLMES
Department of Psychiatry, University of Medicine and Dentistry, Newark, NJ 07103, USA

B. HULTGREN
Department of Veterinary Pathobiology, College of Veterinary Medicine, University of Minnesota, St. Paul, MN 55108, USA

J. L. HURST
Animal Behaviour Research Group, Department of Zoology, University of Nottingham, University Park, Nottingham NG7 2RD, UK

F. JANNETT
The Science Museum of Minnesota, 30E Tenth Street, St. Paul, MN 55101, USA

B. JEMIOLO
Department of Chemistry, Indiana University, Bloomington, IN 47405, USA

K. L. JOHANSEN
School of Forest Resources, University of Georgia, Athens, GA 30602, USA

E. B. KEVERNE
Sub-Department of Animal Behaviour, Madingley, Cambridge CB3 8AA, UK

S. KHAN
Department of Physiological Sciences, University of Manchester, Manchester M13 9PT, UK

M. D. KIRK-SMITH
School of Management, University of Sheffield, Sheffield S10 2TN, UK

W. R. KLEMM
Department of Veterinary Anatomy, Texas A and M University, College Station, TX 77843, USA

T. W. KWAN
Department of Biochemistry, Faculty of Medicine, University of Malaya, 51900 Kuala Lumpur, Malaysia

J. J. LEPRI
Department of Biology, University of North Carolina at Greensboro, Greensboro, NC 27412, USA

F. LEVY
Institut National de la Recherche Agronomique, Nouzilly, 37380 Monnaie, France

E. LEWIS
Dalgety plc, Group Research Laboratory, Station Road, Cambridge CB1 2JN, UK

R. D. LISK
Department of Biology, Princeton University, Princeton NJ 08544, USA

D. W. MACDONALD
Department of Zoology, University of Oxford, South Parks Road, Oxford OX1 3PS, UK

J. W. MAKIN
George Peabody College of Vanderbilt University, Nashville TN 37203, USA

A. I. MALLET
Institute of Dermatology and Division of Biochemistry, United Medical and Dental Schools of Guy's and St. Thomas's Hospitals, London SE1 9RT, UK

L. MARCHINTON
School of Forest Resources, University of Georgia, Athens, GA 30602, USA

A. MARCHLEWSKA-KOJ
Jagiellonian University, 30–060 Krakow, Poland

C. A. MEANEY
Denver Museum of Natural History, City Park, Denver, CO 80205, USA

S. MCDONALD
Hatfield Polytechnic, College Lane, Hatfield AL10 9AB, UK

K. J. MERRITT
Coypu Research Laboratory, MAFF, Jupiter Road, Norwich, NR6 6SP, UK

K. V. MILLER
School of Forest Resources, University of Georgia, Athens, GA 30602, USA

D. MÜLLER-SCHWARZE
State University of New York, College of Environmental Science and Forestry, Syracuse, NY 13210, USA

S. NATYNCZUK
The Science School, Clifton College, Bristol BS8 3JH, UK; *and* School of Chemistry, University of Bristol, Bristol BS8 1TS, UK

M. V. NOVOTNY
Department of Chemistry, Indiana University, Bloomington, IN 47405, USA

P. C. PAQUET
Department of Zoology, University of Alberta, Alberta T6G 2J9, Canada

W. C. PASSANISI
Department of Zoology, University of Oxford, South Parks Road, Oxford OX1 3PS, UK

R. PEWTRESS
Hatfield Polytechnic, College Lane, Hatfield AL10 9AB, UK

V. PIKETTY
Institut National de la Recherche Agronomique, Nouzilly, 37380 Monnaie, France

P. POINDRON
Institut National de la Recherche Agronomique, Nouzilly, 37380 Monnaie, France

R. H. PORTER
George Peabody College of Vanderbilt University, Nashville TN 37203, USA

B. PROBST
Department of Animal Physiology, University of Bayreuth, POB 101251, D-8580 Bayreuth, West Germany

L. E. RASMUSSEN
Department of Chemical and Biological Sciences, Oregon Graduate Center, Beaverton, OR 97006, USA; *and* Washington Park Zoo, Portland, OR 7221, USA

C. REECE ENGEL
School of Biological Sciences, University of East Anglia, Norwich NR4 7TJ, UK

B. G. REHNBURG
Department of Biostructure and Function, University of Connecticut Health Center, Farmington, CT 06032, USA

P. J. RENNIE
Department of Microbiology, University of Leeds, Leeds LS2 9JT, UK

P. R. K. RICHARDSON
Mammal Research Unit, University of Pretoria, South Africa 0002

G. RIVARD
Department of Veterinary Anatomy, Texas A and M University, College Station, TX 77843, USA

I. ROBINSON
The Waltham Centre For Pet Nutrition, Freeby Lane, Waltham-on-the-Wolds, Melton Mowbray, Leicestershire LE14 4RT, UK

P. RODL
Institute of Hygiene and Epidemiology, CS–100 42 Praha 10, Czechoslovakia

B. ROSER
Dalhousie University, Halifax, Nova Scotia BRH 6J1, Canada

A. B. RYLANDS
Departmento de Zoologia, Instituto de Ciencias Biologicas, Universidade Federal de Minas Gerais, 31.270 Belo Horizonte, Minas Gerais, Brazil

J. RYON
Psychology Department, Dalhousie University, Halifax, Nova Scotia B3H 6JI, Canada

N. K. SANDNABBA
Department of Psychology, Abo Akademie, Vardsbergsgaten 1, SF-20700 Abo, Finland

B. SCHAAL
Laboratoire de Psycho-Biologie de l'Enfant, EPHE-CNRS UA 315, Paris, France

J. P. SIGNORET
Institut National de la Recherche Agronomique, Station de Physiologie de la Réproduction, Nouzilly 37380, Monnaie, France

P. B. SINGH
Dalhousie University, Halifax, Nova Scotia B3H 6JI, Canada

R. J. F. SMITH
Department of Biology, University of Saskatchewan, Saskatoon, Saskatchewan S7N 0WO, Canada

B. A. SOMMERVILLE
Department of Forensic Medicine, University of Leeds, C.S.B., St. James's University Hospital, Beckett Street, Leeds LS9 7TF, UK

P. W. SORENSEN
College of Forestry, Department of Fisheries and Wildlife, 200 Hodson Hall, 1980 Folwell Avenue, St. Paul MN 55108, USA

N. E. STACEY
Department of Zoology, University of Alberta, Edmonton, Alberta TG6 2E9, Canada

E. STAUBER
Department of Veterinary Science, University of Idaho, Moscow, Idaho 83843, USA

M. STAVY
Department of Zoology, Tel-Aviv University, Ramat-Aviv, 69978 Tel-Aviv, Israel

E. STEEL
MRC Unit on the Development and Integration of Behaviour, Sub-Department of Animal Behaviour, Madingley, Cambridge CB3 8AA, UK

R. D. STONE
WWF Avenue de Mont Blanc, CH–1196 Gland, Geneva, Switzerland

A. E. STOREY
Department of Psychology, Memorial University, St. John's Newfoundland A1B 3X9, Canada

S. A. TAYLOR
Department of Psychology, University of Florida, Gainesville FL 32611, USA

J. TERKEL
Department of Zoology, Tel-Aviv University, Ramat-Aviv, 69978 Tel-Aviv, Israel

D. THIESSEN
Department of Psychology, University of Texas, Austin, TX 78712, USA

E. VERNET-MAURY
Laboratoire de Physiologie Neuronsensorielle, Université Lyon I.F. 69622, Villeurbanne Cedex, France

J. S. WATERHOUSE
Cambridgeshire College of Arts and Technology, East Road, Cambridge CBI IPT, UK

W. J. WATKINS
Institute of Dermatology and Division of Biochemistry, United Medical and Dental Schools of Guy's and St. Thomas's Hospitals, London SE1 9RT, UK

P. J. WELDON
Department of Veterinary Anatomy, Texas A and M University, College Station, Texas 77843, USA

I. WERNER
Department of Zoology, University of Oxford, South Parks Road, Oxford OX1 3PS, UK

C. J. WYSOCKI
Monell Chemical Senses Center and Chemistry Department, University of Pennsylvania, 3500 Market Street, Philadelphia, PA 19104–3308, USA

K. YAMAZAKI
Monell Chemical Senses Center and Chemistry Department, University of Pennsylvania, 3500 Market Street, Philadelphia, PA 19104–3308, USA

C. S. ZAMORA
Department of Veterinary Science, Washington State University, Pullman, WA 99164, USA

# SECTION I

## Semiochemistry

# I.1 Chemistry of Rodent Pheromones: Molecular Insights into Chemical Signalling in Mammals

MILOS NOVOTNY, BOZENA JEMIOLO AND SCOTT HARVEY

## Introduction

It is now becoming clear that olfactory communication is a truly interdisciplinary endeavour where some of the best forefront skills of both biology and chemistry must be employed. Chemical studies are relatively rare compared to the rapidly growing body of biological knowledge. Structural identification of the active chemosignals is obviously a key to the elucidation of the complex physiological phenomena that are so essential for the reproduction and behaviour of various animals.

As pointed out by Albone (1984), there are two extreme views on the composition, chemical complexity, and mode of action of mammalian chemosignals which affect the choice of approach to identifying the responsible components. In the so-called response-guided strategy, the pheromone-containing material is subjected to successively refined fractionations, such as solvent-solvent extractions, purge-and-trap enrichment, or various forms of chromatography, while activity is followed by the use of a bioassay. Eventually, a pure active material should be obtained, and structure determination may ensue. The response-guided strategy clearly has problems in cases involving even a moderate degree of complexity, as this approach is destined to yield material that loses its activity as the fractionations become more exacting. The second approach, termed the chemical image strategy, assumes that the perception of the *entire matrix* is important in mammalian chemical communication. This school of thought borrows heavily from the ideas of Gestalt psychology, and assumes that the perceptual qualities of the entire matrix (a urine mark, glandular secretion, etc.) take on properties above those that can reasonably be explained by a description of the amount and identity of each individual component. While useful in predicting synergetic actions of compounds, the chemical image strategy presents a daunting challenge when applied to pheromone systems that contain infinite levels of complexity.

The chemical image strategy does have its attractive features; however, when carried to the extreme this approach is somewhat counter-productive in supporting certain pessimistic views that favour the infinite chemical complexity of communication systems and discourage enquiries into the mechanisms. Examples will be given in this article to warn against a total adherence to either strategy.

Biological studies provide a good starting point for chemical inquiries into the molecular aspects of mammalian communication. Endocrinologically altered animals often lose the ability to excrete a pheromone, while hormonal supplementation may renew its production. Consequently, endocrinological and genetic manipulations can be followed biochemically, both qualitatively and quantitatively. A fruitful bio-analytical procedure is the acquisition of volatile substance profiles by capillary gas chromatography (GC) and combined gas chromatography/mass spectometry (CG/MS). In a typical investigation, the substance profiles of intact animals are quantitatively compared with those of endocrinologically altered animals, and subsequently with hormone-treated individuals. Substantial changes in the areas of some recorded chromatographic peaks, which coincide with an increase or decrease of biological activity, direct our attention to potential pheromone candidates. Due to the extreme complexity of most mammalian samples, capillary GC, with its high resolving power, is essential in recording spectra from the peaks of interest. Positive structural elucidation may be based on a variety of spectral methods (most typically, electron impact and chemical ionization mass spectrometry in the low- or high-resolution mode, Fourier-transform infra-red spectroscopy, etc.) and chemical approaches. Structural identifications are verified through organic synthesis and comparison of the natural material with a man-made chemical. The pure chemicals (pheromone candidates) may be tested in plain water, a fixative, or in a previously inactive biological matrix. Thus, in the overall effort, three types of expertise are utilized drawing on resources from the disciplines of biology, bio-analytical chemistry, and structural and synthetic chemistry.

The primary purpose of this contribution is to summarize the recent advances made in our laboratory on the mouse primer and signalling pheromones. We will emphasize the research strategies that led to the identification of male chemosignals promoting inter-male aggression and the Whitten effect in females, sexual attraction, and dominance signalling. In the area of chemosignals of female origin, special attention will be given to puberty advancement, puberty delay, and oestrus extension. Our contributions to genetic signalling in the mouse will be briefly reviewed. Finally, we will attempt to summarize research on other mammals, done by our laboratory in definitive chemical investigations. Projections for future research will also be given.

## Studies on *Mus domesticus*: a model of chemical inquiry into the nature of olfactory communication

### MALE-ORIGINATED PHEROMONES

A growing number of chemosignals produced by intact male mice are known to be under the control of testosterone, the levels of which are frequently influenced by urinary signals from both male and female conspecifics. Bronson (1971, 1979, 1984) has elegantly summarized the basic role of the brain-pituitary-gonadal system in the production of pheromones, and explained its broader ecological significance for the house mouse, in particular, and for other mammalian species, in general. While castrated males lose their ability to excrete male-to-male and male-to-female chemosignals, treatment with testosterone often renews the relevant biological activity.

Using endocrinological manipulation (castration and testosterone supplementation), we observed that two major volatile urinary constituents, identified as 2-*sec*-butyl-4,5-dihydrothiazole and 3,4-dehydro-*exo*-brevicomin (Table 1), have clear dependency on testosterone level (Novotny *et al.* 1980, 1984; Schwende *et al.* 1986). Both compounds were synthesized as racemic mixtures (Prodan 1976; Wiesler *et al.* 1984) and utilized in behavioural and endocrinological investigations. In various experimental arrangements (Novotny *et al.* 1985) intact male urine and water were used to establish baseline control levels of the extreme situations, and compared to the castrated male urine (an otherwise inactive stimulus) which when spiked with synthetic 2-*sec*-butyl-4,5-dihydrothiazole and 3,4-dehydro-*exo*-brevicomin, had an aggression-eliciting quality similar to that of intact male urine. Neither castrated male urine spiked with either compound alone, or pure water spiked with both compounds significantly promoted aggressive behaviour. The two compounds are effective only if perceived in the general odour of mouse urine. While this is consistent with the view of Evans *et al.* (1978), that the Gestalt of the whole urine is responsible for the phenomenon, we still cannot rule out that a finite (and, perhaps, small) number of additional synthetic compounds may be needed for the full response.

Slow motion videotapes of fighting induced by intact urine and spiked castrate urine showed that the synthesized stimulus elicited fully elaborated sequences of aggressive behaviour. The synthetic stimulus appeared to communicate an olfactory message similar to the biological stimulus.

It is generally believed that the same chemical signals, when perceived in a different context, or by an animal of different sex, can cause vastly different behaviour in the recipient. With this possibility in mind, we tested 2-*sec*-butyl-4,5-dihydrothiazole and 3,4-dehydro-*exo*-brevicomin for their attractiveness to female mice. Female mice are known to become more active in the prescence of odours from males (Ropartz 1968), and exhibit preferences for intact male odour over the odour of castrated males (Scott and Pfaff 1970). Work in our laboratory (Jemiolo *et al.* 1985) was undertaken to determine whether females display preferences for our uniquely male compounds.

A series of two-choice preference tests was employed to quantify the relative attraction of female mice to whole, intact male urine (the natural biological stimulus) compared with a variety of other complex odours: castrate urine, castrate urine with the two synthetic substances, and castrate urine with each individual substance. Females were considerably more interested in intact male urine over castrate urine; however, the castrate urine spiked with 2-*sec*-butyl-4,5-dihydrothiazole and 3,4-dehydro-*exo*-brevicomin was indistinguishable from the intact male stimulus (Jemiolo *et al.* 1985). Once again, neither compound alone, in the castrate urine matrix, created an olfactory signal that was attractive to female mice.

The Whitten effect, that is, induction and synchronization of oestrus among unisexually grouped females in the presence of an intact male, is one of the best known examples of male-to-female pheromonal effects in mice (Whitten 1958, 1966). The male-originated substances act with a high degree of specificity in altering the secretion of luteinizing hormone (LH) and the corresponding steroid hormones (Bronson and Macmillan 1983), with consequent regulation of ovulatory function. The results of Whitten *et al.* (1968), Monder *et al.* (1978), and Gangrade and Dominic (1984) all suggest that the male signal involved in oestrus induction is volatile.

In agreement with Whitten *et al.* (1968), Champlin (1971) and Bronson and Whitten (1968), we found that housing conditions (singly caged vs. crowded females) had a significant effect on the number and length of oestrous cycles. Test females housed at different densities were exposed to the test stimuli (placed in a Petri dish under a wire-mesh hemisphere) so that physical contact with the sample was prevented; only airborne olfactory cues were permitted. Castrate urine spiked with 2-*sec*-butyl-4,5-dihydrothiazole and 3,4-dehydro-*exo*-brevicomin was as effective as intact male urine in increasing the frequency of oestrous cycles in crowded females (Jemiolo *et al.* 1986). The length of oestrous cycles is significantly shortened by both natural male urinary signals and their synthetic analogues. Interestingly, the two synthetics spiked into plain water appear to have properties similar to the natural signal in eliminating the suppressive effect of high population density in females (Jemiolo *et al.* 1986). However, they are somewhat less active, when spiked into plain water, in inducing oestrus on the third or fourth day after exposure than intact male urine or castrate urine containing the synthetics (Jemiolo *et al.* 1986).

These studies support the suggestion of Bronson and Macmillan (1983) that, under different conditions of perception, the same chemosignal may cause different biological responses in the recipient. In the overall context of male urine, the same chemicals were responsible for increased inter-male aggression and for attracting females. Their ability to induce the Whitten effect and to maintain a regular oestrous cycle in grouped females closely parallels the effectiveness of the natural stimulus, intact male urine. The Gestalt-type interactions appear non-essential for the Whitten effect. Obviously, in different situations, these two compounds act as either primer or releaser pheromones.

It is believed by numerous investigators that the pheromones causing the Whitten effect (Whitten 1958, 1966) and puberty acceleration in young females (Vandenbergh 1967) are identical. However, our results do not support this contention: 2-*sec*-butyl-4,5-dihydrothiazole and 3,4-dehydro-*exo*brevicomin,

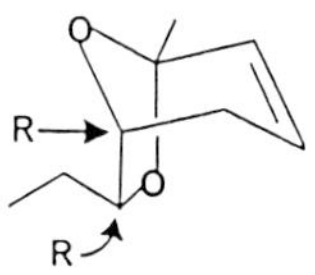

(R,R)-3,4-dehydro-*exo*-brevicomin

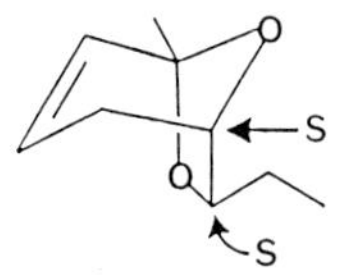

(S,S)-3,4-dehydro-*exo*-brevicomin

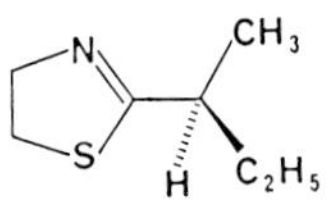

(S)-2-*sec*-butyl-4,5-dihydrothiazole

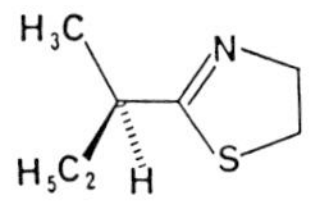

(R)-2-*sec*-butyl-4,5-dihydrothiazole

either in plain water or spiked into the castrate male urine, do not seem to promote uterine growth in prepubertal females. A partial chemical characterization (Novotny *et al.* 1980; Harvey 1988) of the puberty-accelerating pheromone points to very different chemical principles being involved.

A point of some interest is that both 3,4-dehydro-*exo*-brevicomin and 2-*sec*-butyl-4,5-dihydrothiazole can be produced in two optically active forms: What is the relation, if any, of this type of isomerism to biological activity? This could be a very important question since stereospecificity of various receptor proteins is known, including the ability of humans to distinguish the odour of different enantiomers of some compounds (Friedman and Miller 1971). Furthermore, it is well known that optical activity can be very important in insect pheromone perception (Silverstein 1981; Lanier *et al.* 1980; Riley *et al.* 1974).

Through the use of chelation capillary GC, we were able to resolve the two enantiomeric forms of 3,4-dehydro-*exo*-brevicomin (Harvey 1988). The order of their elution was verified with synthetically prepared enantiomers, and a comparison of retention for the substance isolated from the normal male urine established that it is (R,R)-3,4-dehydro-*exo*-brevicomin. This is also the isomer that is predicted biogenetically. While we have been able to resolve the racemic mixture of 2-*sec*-butyl-4,5-dihydrothiazole by both chelation chromatography and an NMR-based technique (Harvey 1988), it has been exceedingly difficult to synthesize the individual isomers. The isolated natural thiazoline produced two peaks on chromatography. Whereas it is expected that the male mouse produces only *one* antipode (most likely the S-isomer), a very facile racemization may take place on chromatography through the tautomerization process. Biological tests for the Whitten effect of the thiazoline racemic mixture with the individual optical isomers of 3,4-dehydro-*exo*-brevicomin are being pursued.

The last case of male-derived chemosignals that we wish to report for the house mouse pertains to dominance signalling. Selective advantages of dominance-associated traits, such as aggression and territoriality, have been discussed extensively (e.g. DeFries and McClearn 1970; Oakeshott 1974; Bronson 1966; Huck *et al.* 1981). Social subordination is known to suppress gonadal function in male mice; dominant males are more effective than subordinates in promoting puberty acceleration in young female mice (Lombardi and Vandenbergh 1977; Labov 1981). Laboratory evidence also indicates that male mouse urine produces an odour indicative of social status; the territorial marker produced by dominant males may be identical to the observed (Jones and Nowell 1973*a*, *b*, 1974) aversion pheromone.

Encouraged by the biological evidence of differences between dominant and subordinate males, we decided to screen the respective urines for qualitative or quantitative differences. First, dominance and subordinance were established by repeatedly pairing mice. Dominant and subordinate males that displayed a clear social relationship were chosen to serve as urine donors. As a further point of reference, urine samples were also collected from immature and castrate males. For all individuals, urinary volatile profiles were recorded and subjected to statistical analysis. The following observations were made. (i) In general, volatile profiles from dominant males tended to be more abundant than those from subordinates, a result that is supported by a very recent observation by Apps *et al.* (1988) who analysed the whole-body volatiles. (ii) Among sixteen volatile constituents exhibiting statistically significant variation, 2-*sec*-butyl-4,5 dihydrothiazole displayed substantially higher levels in dominant males. (iii) Two previously unnoticed terpenic components showed a dramatic increase in dominant urine compared to that of subordinates; these components were identified as α-farnesene and β-farnesene (Table 1). (iv) Immature animals had no detectable levels of 3,4-dehydro-*exo*-brevicomin, 2-*sec*-butyl-4,5-dihydrothiazole, or the farnesenes, whereas the castrates exhibited low residual levels of the farnesenes and no 3,4-dehydro-*exo*-brevicomin or 2-*sec*-butyl-4,5-dihydrothiazole. Detailed results on the composition of dominant/subordinate urines will be reported elsewhere (Harvey *et al.* 1989).

We were intrigued by the occurrence and endocrine dependencies of the farnesene isomers in male mice since, separately, such compounds have been implicated as insect pheromones (Cavill *et al.* 1967; Pickett and Griffiths 1980; Bowers *et al.* 1972). It is also interesting to note that farnesyl pyrophosphate is a key intermediate in steriod biosynthesis (Lehninger 1975). Because the preputial glands have previously been implicated in the production of androgen-dependent, sex-related signals, we were curious about whether any of our androgen-dependent substances originate from this sex-accessory gland. The farnesenes are absent from bladder urine; however, they are the major volatile components of the oily matrix of the preputial gland (Fig. 1). Conversely, 3,4-dehydro-*exo*-brevicomin and 2-*sec*-butyl-4,5-dihydrothiazole are undetectable in the preputial glands, and originate from bladder urine. A more detailed analysis of the preputial gland sebum by capillary GC/MS also revealed (Harvey 1988) the presence of trace amounts of myrcene and ocimene, which are chemically related to the farnesenes.

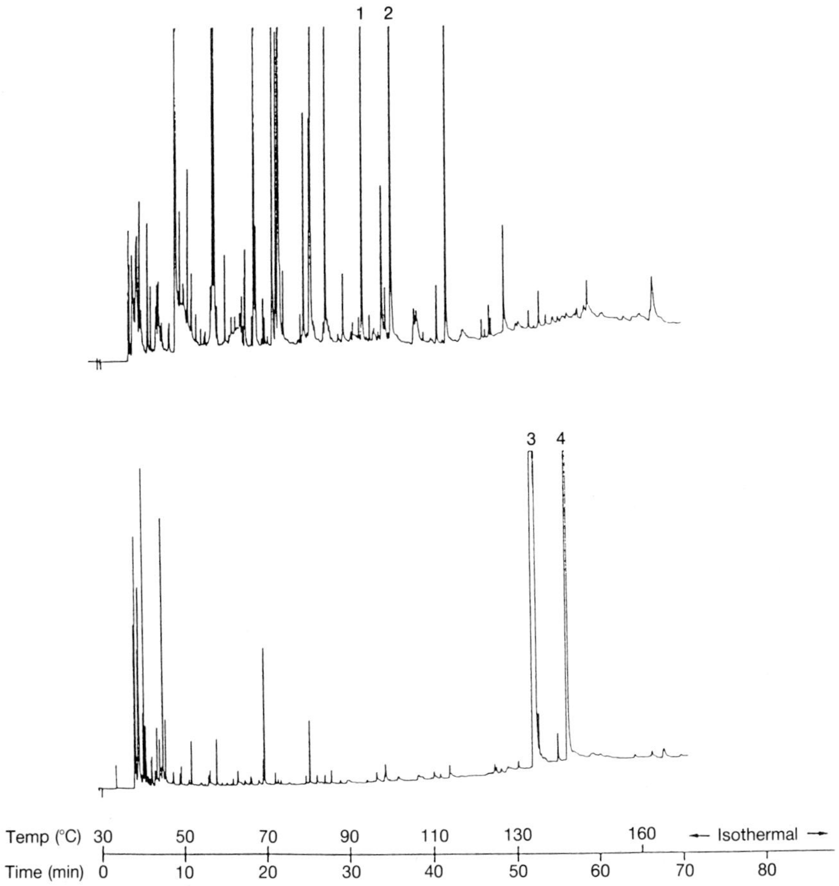


Fig. 1. Representative chromatograms of the volatiles from bladder urine of subordinate mice (upper) and from the preputial gland of dominant mice (lower). The numbers represent: (1) 3,4-dehydro-*exo*-brevocomin; (2) 2-*sec*-butyl-4,5-dihydrothiazole; (3) E-beta-farnesene; (4) E,E-alpha-farnesene.

We tested various stimuli containing α- and β-farnesene in a two-choice experimental setup (Novotny *et al.* 1989). The following urine samples were tested relative to water: (1) female urine; (2) female urine containing α- and β-farnesene at their natural concentrations in the male urine; (3) urine of dominant males; (4) urine of subordinate males; (5) bladder urine; and (6) bladder urine spiked with α and β-farnesene. Additional experiments tested bladder urine and bladder urine containing both sesquiterpenes relative to water containing α and β-farnesene. The test males were 4–6 months old and housed in groups of four. The dominant male of each cage was identified and excluded from testing. The results of tests gave a strong indication that the sesquiterpenes introduced into the previously attractive stimuli discouraged prolonged investigation. In either a urine matrix or water, the terpenes rendered the stimulus behaviourally similar to the urine of dominant males; it therefore appears that the farnesenes may be responsible for the previously described aversion signals produced by dominant males. This phenomenon could potentially be utilized in pest control.

## FEMALE-ORIGINATED PHEROMONES

A major factor that appears to decrease the mating potential of females in crowded mouse populations is the presence of one or more chemicals that delay sexual maturation (Christian 1956; Drickamer *et al.* 1978; Drickamer and McIntosh 1980; Massey and Vandenbergh 1980). The puberty delaying female-to-female chemosignal(s) is present in the urine of females, regardless of their age, only during crowded conditions (Drickamer 1983).

Intrigued by the female-to-female master pheromone and its dependence on the adrenal function of female mice (Drickamer and McIntosh 1980), we analytically compared the urine from intact and adrenalectomized females. While some individual variations occur, six urinary volatiles were consistently decreased in concentration in the urine of the adrenalectomized animals. These were subsequently identified as 2-heptanone, *trans*-5-heptene-2-one, and *trans*-4-heptene-2-one (ketones), *n*-pentyl acetate and *cis*-2-penten-1-yl acetate (acetate esters), and 2,5-dimethylpryazine (Novotny *et al.* 1986). When such synthetic compounds are mixed, in their natural concentrations, into the previously inactive adrenalectomy urine, or even into plain water, the puberty delaying activity was fully restored. Our findings indicate that it is mainly associated with *n*-pentyl acetate, *cis*-2-penten-1-yl acetate, and 2,5-dimethylpryazine (see Table 1).

Recent studies by Drickamer (1982, 1986) and co-workers (Drickamer and Hoover 1979; Hoover and Drickamer 1979) suggest that additional female signals may have biological significance. Such chemosignals may indicate the adequacy of social and environmental conditions necessary for successful reproduction. For example, the urine from oestrous, pregnant, and lactating animals appears to accelerate the sexual development of young females, and promotes oestrus in adult mice (Drickamer 1986). Both bladder and excreted urine seem to be biologically

**Table 1.** Mouse urinary compounds with biological activity

| Name | Formula | Origin | Biological Activity |
|---|---|---|---|
| 3,4-dehydro-*exo*-brevicomin | 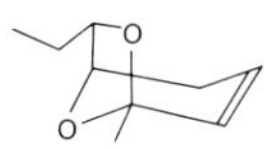 | Male | Act synergistically to:<br>(a) elicit inter-male aggression<br>(b) attract females<br>(c) promote oestrous synchrony (Whitten effect) |
| 2-*sec*-butyl-4,5-dihydrothiazole | 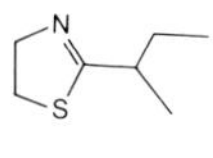 | Male | |
| E,E-alpha-farnesene | 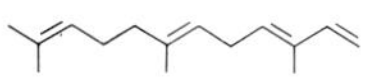 | Male | Discourage olfactory investigations and slow motor activity of submissive mice |
| E-beta-farnesone | 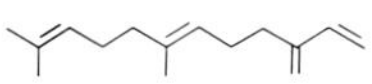 | Male | |
| *n*-pentyl acetate | 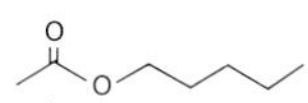 | Female | Delay puberty |
| *cis*-2-penten-1-yl acetate | 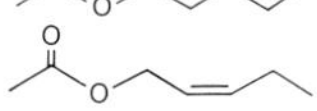 | Female | |
| 2,5-dimethylpyrazine | 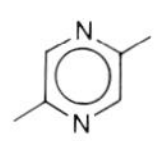 | Female | |
| 2-hexanone | 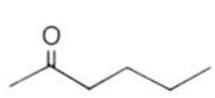 | Female | Accelerate puberty and prolong duration of first oestrus |
| 2-heptanone | 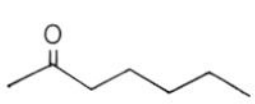 | Female | |
| 4-heptanone | 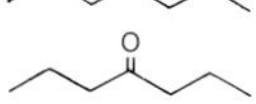 | Female | |

**Table 2.** Mean age of puberty of females exposed to various synthetic compounds.

| Stimulus | Mean age of puberty[b] (days) | Number of tested females |
|---|---|---|
| **Controls:** | | |
| Water | 30.3 (0.2) | 24 |
| Poly(ethylene glycol)[a] | 30.3 (0.6) | 30 |
| Urine of lactating females | 28.3 (0.4)* | 18 |
| **Synthetic mixtures:** | | |
| Mixture of | | |
| 5,5 dimethyl-2-ethyl-4,5 dihydrofuran | | |
| E-5,5 dimethyl-2-ethylenetetrahydrofuran | | |
| Z-5,5 dimethyl-2-ethylenetetrahydrofuran | 30.0 (0.3) | 30 |
| Mixture of | | |
| 2-hexanone | | |
| 2-heptanone | | |
| 4-heptanone | 28.3 (0.3)* | 43 |
| Mixture of | | |
| *trans*-5-hepten-2-one | | |
| *trans*-4-hepten-2-one | | |
| 6-methyl-6-hepten-2-one | | |
| 6-methyl-5-hepten-2-one | 30.0 (0.4) | 46 |
| Mixture of | | |
| 4-penten-1-ol | | |
| 1-octen-3-ol | 30.9 (0.3) | 31 |

[a] A mixture of synthetics dissolved in glycol was placed on a piece of cotton in a perforated capsule in the middle of a cage.
[b] S.E.M. values given in parenthesis.
* Significantly different at $p \leqslant 0.02$ (Anova).

active (Drickamer 1986). Moreover, oestrus-related signals are known to increase activity and the phase advance of circadian rhythm (Cushing 1985). Male mice are sexually attracted by oestrous odours (Davies and Bellamy 1972).

The volatiles excreted by female mice were followed during the natural cycle (Andreolini *et al.* 1987) and during various stages of pregnancy and lactation (Jemiolo *et al.* 1987). Statistically significant alterations in the excretion of urinary volatiles were correlated with the biologically characterized periods.

During the ovarian cycle, a very substantial endocrine dependency was noted for eleven compounds (Andreolini *et al.* 1987), of which several (Table 1) were tested for activity via acceleration of sexual development. The four ketones, 2-hexanone, 2-heptanone, *trans*-5-heptene-2-one, and *trans*-4-heptene-2-one, were significantly elevated during oestrus; however, the two acetate esters previously implicated in the delay of puberty (Novotny *et al.* 1986) also showed an

**Table 3** Percentage of females exhibiting different lengths of oestrus depending on exposure to synthetic mixtures.

| Stimulus | Length of oestrus in day intervals | | |
|---|---|---|---|
| | 1—3 | 4–6 | 7–9 |
| **Controls:** | | | |
| Water | 84 | 16 | — |
| Poly(ethylene glycol)[a] | 70 | 30 | — |
| Urine of lactating females | 38 | 62 | — |
| **Synthetic mixtures:** | | | |
| Mixture of<br>5,5 dimethyl-2-ethyl-4,5 dihydrofuran<br>E-5,5 dimethyl-2-ethylenetetrahydrofuran<br>Z-5,5 dimethyl-2-ethylenetetrahydrofuran | 33 | 57 | 10 |
| Mixture of<br>2-hexanone<br>2-heptanone<br>4-heptanone | 30 | 55 | 15 |
| Mixture of<br>*trans*-5-hepten-2-one<br>*trans*-4-hepten-2-one<br>6-methyl-6-hepten-3-one<br>6-methyl-5-hepten-3-one | 51 | 44 | 5 |
| Mixture of<br>4-penten-1-ol<br>1-octen-3-ol | 71 | 29 | — |
| Mixture of<br>*n*-pentyl acetate<br>*cis*-2-penten-1-yl acetate<br>2,5 dimethylpyrazine | 95 | 5 | - |

[a] A mixture of synthetics dissolved in glycol was placed on a piece of cotton in a perforated capsule in the middle of a cage.

enhanced concentration. Conversely, 2,5-dimethylpryazine, a compound with a clear puberty-inhibiting effect (Novotny *et al.* 1986), showed its lowest level during the peak of female oestrogen secretion. The results obtained with the ketones and acetate esters during the natural cycle parallel qualitative changes described previously in animals implanted with oestrogen pellets (Schwende *et al.* 1984*a*).

During our studies of urinary volatiles at different stages of pregnancy and lactation, considerable variations were seen in the levels of several ketones, unsaturated alcohols, esters, and cyclic vinyl ethers (Jemiolo *et al.* 1987). The urinary compounds excreted by pregnant/lactating and oestrous females, exhibiting obvious endocrinological dependencies, were tested for biological activity

(Jemiolo *et al.* 1989). Daily exposure of 21-day-old females to various mixtures of synthetic compounds revealed that 2-hexanone, 2-heptanone, and 4-heptanone could promote puberty in young female mice, while other urinary ketones, dihydrofurans, and alcohols did not seem to be effective (see Table 2). However, and extension of oestrus was also observable after exposure to other urinary compounds as well (Table 3).

### GENETIC SIGNALLING

Some strains of laboratory rodents are more convenient than others for demonstrating certain pheromonal effects. These genetic differences may concern both the pheromone producers and the responding animals.

Inter-strain differences in hormonal levels and reproductive potential are widely appreciated. As an example, genetic variations of testosterone levels in male mice (Bartke 1974) could be related to the biosynthesis of typical male chemosignals, such as those causing aggression and dominance. In highly aggressive strains developed by selective breeding (Sandnabba 1986*a*, *b*), behavioural studies indicated the occurrence of increased pheromone production. Yet caution must be exercised in evaluating such phenomena, since aggressiveness may be reflected not only in the levels of androgen-dependent aggression signals, such as 2-*sec*-butyl-4,5-dihydrothiazole and 2,3-dehydro-*exo*-brevicomin (Novotny *et al.* 1985), but also in the total urine volume (Harvey 1988) and/or general marking behaviour.

Certain strains appear to have enhanced sensitivity towards pregnancy disruption while different males are endowed with lesser or greater ability to produce this effect (Hoppe 1975). Chapman and Whitten (1969) have studied the genetics of the pregnancy-disrupting pheromone effect and found it to be a recessive trait, apparently controlled by a single gene in both males and females. Moreover, the genetic component of the phenomenon of male mouse pheromonal facilitation of gonadotrophin-induced ovulation has also been investigated (Zarrow *et al.* 1971).

While the genetic differences in pheromone excretion, discussed above, are both evident and generally understandable, some recent research has concentrated on more subtle forms of olfactory communication to which our term genetic signalling can truly be applied. The possibility of chemical communication related to symbiosis, immunology, and recognition of self from non-self postulated by Thomas (1975) has significantly influenced the basis of our perception of the overall field. Since then, the extent to which the major histocompatibility gene complex (H-2) exerts influence over mate selection under laboratory conditions or in nature has been the subject of numerous studies and discussions. Males of a given H-2 genotype show discrimination for or against females of their own type, depending on the haplotypes investigated (Yamazaki *et al.* 1976). The phenomenon was demonstrated for different H-2 mice within a particular parent strain

(C57BL, AKR, BALB, etc.). Further experiments with a Y maze (Yamazaki *et al.* 1979) strongly indicated that both male and female mice are able to discriminate between different H-2 odour sources; the tests using recombinant strains (Yamazaki *et al.* 1978) suggest that several sites at the H-2 locus could be responsible for production of the recognized olfactants. Both transmission (by females) and reception (by males) of the chemical signals appear to be under some influence of the major histocompatibility complex.

Our chemical investigations of the profiles of urinary volatiles from different parent strains indicate profound *quantitative* differences in concentration, while qualitative differences are rare (Novotny *et al.* 1980; Schwende *et al.* 1984*b*), However, mice with common histocompatibility alleles are, at first glance, hardly distinguishable. Against the odds and experimental difficulties, Schwende *et al.* (1984*b*) demonstrated that relatively simple statistical methods, when applied to the complex urinary profiles of different H-2 mice, can yield discrimination of different haplotypes. Furthermore, the extent of this discrimination appears to agree tentatively with H-2 antigenic reactivities, giving some support to a chemical basis for histocompatibility-related mating preferences in *Mus domesticus*.

More recently, the role of the T-locus on the seventeenth chromosome in aspects of chemical communication of the house mouse has also been investigated (Lenington 1983; Lenington and Egid 1985). The T-locus in the mouse is a set of dominant (T) and recessive (t) mutations, some of which have profound effects on embryonic development, sperm production and function, and genetic recombination over a large region of the chromosome (Klein 1975; Klein and Hammerberg 1977). Approximately 20–25 % of wild mice are heterozygous (+/t) for the various recessive t-alleles, and 75–80 % are homozygous (+/+) for the wild-type allele (Bennett 1978). Despite an obvious strong selection against t-alleles, they are maintained with high frequency because of the association with a strong distortion of segregation in males (Bruck 1957). When a heterozygous male reproduces, about 90–95 % of his offspring will carry his t-allele. The frequency of t-alleles in the wild population is considerably lower than would be predicted on the basis of selection against homozygotes, counteracted by segregation distortion (Bruck 1957). The factors controlling the frequency of t-alleles in a wild population are not known. One possibility explored recently by Lenington and Egid (1985) and Egid and Lenington (1985), is that social behaviour associated with this polymorphism may select against mice bearing t-alleles. It was found that both male and female mice discriminated +/+ from +/t individuals of the opposite sex on the basis of their odour alone, and preferred the odour of +/+ animals. This odour preference for +/+ over +/t individuals could provide a cue for selective mating which might reduce the frequency of lethal t-alleles in the wild population (Lenington 1983; Lenington and Egid 1985).

In our preliminary experiments, males of four genotpyes (T/t, T/+, +/t, and +/+) served as urine donors. Because breeding was necessary to determine a genotype, all animals studied were over 6 months old, with sexual experience. The males of T/t and T/+ genotypes produced significantly greater amounts of 2-heptanone, *trans*-4-heptene-2-one, *trans*-3-heptene-2-one (the three dihydro-furans) and 2,3-dehydro-*exo*-brevicomin and 2-*sec*-butyl-4, 5-dihydrothiazole than +/+ and +/t genotypes. Acetophenone seems to be an additional urinary component distinguishing the genotpyes.

The most obvious biological significance of the histocompatibility mating systems is the endowment of future generations with a wide range of immunological responses. Because of its relevance and the olfactory discrimination observed against the lethal t-alleles in wild mouse populations, these forms of genetic signalling are likely to receive considerable attention. So far as we have seen, this form of signalling fits the chemical image concept, in that relatively small concentration changes or component ratios are all that may be needed for odour discrimination. This is reasonable given the enormous olfactory capability of mice (Bowers and Alexander 1967). Any odour information presented to a mammal is likely to be interpreted in a hierarchical scale of significance: (i) species identity; (ii) sex; (iii) endocrine status; (iv) genetic information. The subtlety of some odour differences may well be beyond the capability of currently available measurement techniques.

## Chemosignals in other mammals

During recent years, interesting chemical leads have been reported in the literature on wild carnivores. Canids, in particular, attract attention because of their social interactions, ranging from the relatively solitary red fox (*Vulpes vulpes*) to a much more gregarious animal like the wolf (*Canis lupus*). Canids utilize urine and secretions of the anal sac as major sources of chemosignals.

While the initial inquiries into the chemical composition of the fox anal sac secretions revealed mostly the presence of common volatile constituents, such as acids and amines (most probably produced by the action of micro-organisms), the supracaudal (tail) gland has been further investigated (Albone 1975; Albone and Flood 1976); several terpene-like substances were found. However, one should not necessarily conclude that trivial substances are unimportant in chemical communication. In studies of this laboratory (Jorgenson *et al.* 1978), unique sulphur-containing urinary substances of the red fox, *iso*-pent-3-enyl methyl sulphide and phenylethyl methyl sulphide, were identified. Bailey *et al.* (1980), working with captive foxes, further described seasonal variations in the urinary excretion of 3-methyl-butyl methyl sulphide.

Studies in *Canis lupus* and domestic dogs (*Canis familiaris*) also produced interesting results. Once again, common odorants were initially identified in the anal sac (Preti *et al.* 1976), and subsequently a multitude of volatile constituents of this gland were followed under endocrinologically different situations and seasonal variation (Raymer *et al.* 1985). This study gives tentative support to the biological role of the anal sac as discussed extensively elsewhere (Preti *et al.* 1976).

Treatment of the anal sac with an antibiotic and subsequent analyses suggested that some of the volatile components were produced by bacteria. Volatile urinary constituents that appear to be important in the wolf include sulphides and uniquely alkylated 2-ketones (Raymer *et al.* 1984), which were shown to be under endocrinological control (Raymer *et al.* 1986).

It is noteworthy that the chemical identifications carried out in the canids appear to aid certain biologically oriented investigations. Two unique fox substances, isopentenyl methyl sulphide and 2-phenylethyl methyl sulphide, were successfully tested for their attractiveness to foxes in their natural environment (Whitten *et al.* 1980; Wilson *et al.* 1980). Further studies involved a possible predator-induced fear response in the prey species: both rats (Vernet-Maury *et al.* 1984; Fombon and Polak 1987) and snowshoe hares (Sullivan and Crump 1986) seemed to respond to the fox compounds and stress-inducing odorants.

A preliminary investigation of urine samples from the arctic fox (*Alopex lagopus*), red fox (*Vulpes vulpes*), bat-eared fox (*Octocyon megatolis*), raccoon-like dog (*Nycterentes procyonoides*), coyote (*Canis latrans*), wolf (*Canis lupus*) and dog (*C. familiaris*) indicate certain genetic similarities in the excretion of various sulphur compounds, terpenes (unique to the fox species), methyl-substituted 2-alkenones, and pyrazine derivatives. In general, some unique sulphur-containing metabolites (with wide structural variations) seem to distinguish various canids (Raymer 1984). The importance of these metabolites to the species in geographically overlapping areas and to predator–prey relationships is open to speculation.

## Conclusions

During the last decade, advances in the instrumental aspects of mammalian semiochemistry have been relatively slow. This is despite the enormous expansion of bioanalytical techniques and the capability for small-scale chemical investigations. As pointed out by Albone (1984), there are no quick rewards in this field where scientists must increasingly learn the general methods and skills of interdisciplinary research.

We believe that pursuing the current leads of rodent olfactory communication is a prudent direction for acquiring more general information on mammalian olfaction. With several pheromone systems having been identified in *Mus domesticus*, studies of the pheromone/receptor protein interactions now seem logical. Recent progress in biochemical characterization of the chemosensory membranes (for a review, see Anholt 1987) provides the opportunity for such investigations. Studies with isotopically labelled pheromone molecules are a possible first approach.

Further pheromone structures in both rodents and higher mammals will undoubtedly be reported in the future. It is likely that each of these discoveries, significant accomplishments in themselves, will lead to subsequent investigations pertaining to additional questions, including biogenesis, endocrinology and behaviour, genetics, and matters of broader ecological significance.

The possible role of substances of high molecular weight in chemical communication should not be overlooked in view of the current success with biological activity of smaller semiochemicals. The major urinary proteins of the mouse and their possible biological significance have baffled molecular biologists and chemical ecologists alike. The occurrence of non-volatile chemosignals in other mammals and the proven ability of the vomeronasal organ to accept large molecules (Wysocki *et al.* 1985) warrant further investigations.

## References


Albone, E.S. (1975). Dihydroactinidiolide in the supracaudal scent gland secretions of the red fox. *Nature*, 256, 575.

Albone, E.S. (1984). *Mammalian semiochemistry. The investigation of chemical signals between mammals.* A Wiley-Interscience Publication, N.Y.

Albone, E.S. and Flood P.F. (1976). The supracaudal scent gland of the red fox, *Vulpes vulpes*. *J. Chem. Ecol.*, 2, 167–75.

Andreoline, F., Jemiolo, B. and Novotny, M. (1987). Dynamics of excretion of urinary chemosignals in the house mouse (*Mus musculus*) during the natural estrous cycle. *Experientia*, 43, 998–1002.

Anholt, R.R. (1987). Primary events in olfactory reception. *Trends Biochem. Sci.*, 12, 58–62.

Apps, P.J., Rasa, A. and Viljoen, H.W. (1988). Quantitative chromatographic profiling of odours associated with dominance in male laboratory mice. *Aggr. Behav.*, 14(6), 451–61.

Bailey, S., Bunyan, P.J. and Page, J.M.J. (1980). Variation in the levels of some components of the volatile fraction of urine from captive red foxes (*Vulpes vulpes*) and its relationships to the state of the animals. In *Chemical signals in vertebrates and aquatic invertebrates* (ed. D. Müller-Schwarze and R.M. Silverstein), pp. 391–403. Plenum, New York.

Bartke, A. (1974). Increased sensitivity of seminal vesicles to testosterone in a mouse. *J. Endocrinol.*, 60, 145–8.

Bennett, O. (1978). Population genetics of T/t complex mutations. In *Origins of inbred mice* (ed. H. Morse), pp. 615–32. Academic Press, New York.

Bowers, J.M. and Alexander, B.K. (1967). Mice: individual recognition by olfactory cues. *Science*, 158, 1208–10.

Bowers, S.W., Nault, L.R., Webb, R.E. and Dutky, S.R. (1972). Aphid alarm pheromone: isolation, identification, synthesis. *Science*, 177, 1121–2.

Bronson, F.H. (1966). A sex attractant function for mouse preputial glands. *Bull. Ecol. Soc. Amer.*, 47, 182.

Bronson, F.H. and Whitten W.K. (1968). Oestrus-accelerating pheromones of mice: assay, androgen-dependency, and presence in bladder urine. *J. Reprod. Fertil.*, 15, 131–4.

Bronson, F.H. (1971). Rodent pheromones. *Biol. Reprod.*, 3, 344–57.

Bronson, F.H. (1979). Reproductive ecology of the house mouse. *Quart. Rev. Biol.*, 54, 265–99.

Bronson, F.H. and Macmillan B. (1983). Hormonal responses to primer pheromones. In *Pheromones and reproduction in mammals* (ed. J.G. Vandenbergh), pp. 175–97. Academic Press, New York.

Bronson, F.H. (1984). The adaptability of the house mouse. *Sci. Amer.*, 250 (3), 116–25.

Bruck, D. (1957). Male segregation ratio advantage as a factor in maintaining lethal alleles in wild population of house mice. *Proc. Natl. Acad. Sci. USA*, 43, 152–8.

Cavill, G.W.K., Williams, P.J. and Whitfield, F.B. (1967). α-Farnesene, Durour's gland secretion in the ant *Alphaenogaster longiceps* (F.S.M.). *Tetrahedron Lett.*, 23, 2201–5.

Champlin, A.K. (1971). Suppression of oestrus in grouped mice: the effects of various densities and the possible nature of the stimulus. *J. Reprod. Fertil.*, 27, 233–41.

Chapman, V.M. and Whitten, W.K. (1969). The occurrence and inheritance of pregnancy block in inbred mice. *Genetics, Princeton*, 61, Abstracts 9.

Christian, J.J. (1956). Adrenal and reproductive responses to population size in mice from freely-growing population. *Ecology*, 37, 258–73.

Cushing, B.S. (1985). A comparison of activity patterns of estrous and diestrous prairie deer mice. *Peromyscus maniculotus bairdi. J. Mammal.*, 66, 136–9.

Davies, J.V. and Bellamy, D. (1972). the olfactory response of mice to urine and effects of gonadectomy. *J. Endocrinol.*, 55, 11–20.

DeFries, J.C. and McClearn, G.E. (1970). Social dominance and Darwinian fitness in the laboratory mouse. *Amer. Nat.*, 104, 408–11.

Drickamer, L.C. (1982). Delay and acceleration of puberty in female mice by urinary chemosignals from other females. *Dev. Psychobiol.*, 15, 433–42.

Drickamer, L.C. (1983). Effect of period of grouping of donors and duration of stimulus exposure on delay of puberty in female mice by a urinary chemosignal from grouped females. *J. Reprod. Fertil.*, 69, 723–7.

Drickamer, L.C. (1986). Puberty influencing chemosignals in house mice: ecological and evolutionary considerations. In *Chemical signals in vertebrates* (ed. D. Duvall, D. Müller-Schwarze and M.R. Silverstein), pp. 441–55. Plenum, New York.

Drickamer, L.C. and Hoover, J.E. (1979). Effect of urine from pregnant and lactating female house mice on sexual maturation of juvenile females. *Dev. Psychobiol.*, 12, 545–55.

Drickamer, L.C. and McIntosh, T.K. (1980). Effect of adrenalectomy on the presence of a maturation-delaying pheromone in the urine of female mice. *Horm. Behav.*, 14, 146–52.

Drickamer, L.C., McIntosh, T.K. and Rose, E.A. (1978). Effect of ovariectomy on the presence of a maturation-delaying pheromone in the urine of female mice. *Horm. Behav*, 11, 131–7.

Egid, K. and Lenington, S. (1985). Responses of male mice to odor of female: effects of T- and H-2 locus genotype. *Behav. Genet.*, 15, 287–95.

Evans, C.M., Mackintosh, J.H., Kennedy, J.F. and Robertson, S.M. (1978). Attempts to characterize and isolate aggression-reducing olfactory signals from the urine of female mice, *Mus musculus*. *Physiol. Behav.*, 20, 129–34.

Fombon, A.M. and Polak, E.H. (1987). Odor similarity between stress-inducing odorants in Wistar rats. *J. Chem. Ecol.*, 12, 153–66.

Friedman L. and Miller, J.G. (1971). Odor incongruity and chirality. *Science*, 172, 1044–6.

Gangrade, B.K. and Dominic, C.J. (1984). Studies of the male-originating pheromones involved in the Whitten effect and the Bruce effect in mice. *Biol. Reprod.*, 31, 89–96.

Harvey, S.D. (1988). Chemical investigations of the pheromone systems in the house mouse. Ph.D. thesis, Indiana University.

Harvey, S., Jemiolo, B. and Novotny, M. (1989). Pattern of volatile compounds in dominant and subordinate male urine. *J. Chem. Ecol.*, 15(7): 2061–72.

Hoover, J.E. and Drickamer, L.C. (1979). Effect of urine from pregnant and lactating female house mice on oestrus cycle of adult females. *J. Reprod. Fertil.*, 55, 297–307.

Hoppe, P.C. (1975). Genetic and endocrine studies of the pregnancy-blocking pheromone of mice. *J. Reprod. Fertil.*, 45, 109–15.

Huck, U.W., Banks, E.M. and Wang, S.C. (1981). Olfactory discrimination of social status in the brown lemming. *Behav. Neural Biol.*, 33, 364–71.

Jemiolo, B., Alberts, J., Sochinski-Wiggins, S., Harvey, S. and Novotny, M. (1985). Behavioural and endocrine responses of female mice to synthetic analogues of volatile compounds in male urine. *Anim. Behav.*, 33, 114–8.

Jemiolo, B., Harvey, S. and Novotny, M. (1986). Promotion of the Whitten effect in female mice by synthetic analogs of male urinary constituents. *Proc. Natl. Acad. Sci. USA*, 83, 4576–9.

Jemiolo, B., Andreolini, F., Wiesler, D. and Novotny, M. (1987). Variations in mouse (*Mus musculus*) urinary volatiles during different periods of pregnancy and lactation. *J. Chem. Ecol.*, 13, 1941–56.

Jemiolo, B., Andreolini, F., Xie, T.-M., Wiesler, D. and Novotny, M. (1989). Puberty-affecting synthetic analogs of urinary chemosignals in the house mouse, *Mus domesticus*. *Physiol. Behav.*, in press.

Jones, R.B. and Nowell, N.W. (1973*a*). The effect of urine on the investigatory behavior of male albino mice. *Physiol. Behav.*, 11, 35–8.

Jones, R.B. and Nowell, N.W. (1973*b*). Effects of preputial and coagulating gland secretions upon aggressive behavior in male mice: a confirmation. *J. Endocrinol.*, 59, 203–4.

Jones, R.B. and Nowell, N.W. (1974). Effects of androgen on the aversive properties of male urine. *J. Endocrinol.*, 60, 19–25.

Jorgenson, J.W., Novotny, M., Carmack, M., Copland, G.B., Wilson, S.R., Katona, S. and Whitten, W.K. (1978). Chemical scent constituents in the urine of the red fox (*Vulpes vulpes*) during the winter season. *Science*, 199, 796–8.

Klein, J. (1975). *Biology of the mouse histocompatibility-2 complex*. Springer, New York.

Klein, J. and Hammerberg, C. (1977). The control of differentiation by the T-complex. *Immunol. Rev.*, 33, 71–104.

Labov, J.B. (1981). Male social status, physiology, and ability to block pregnancies in female house mouse (*Mus musculus*). *Behav. Ecol. Sociobiol.*, 8, 287.

Lanier, G.N., Classon, A., Stewart, T., Piston, J.J. and Silverstein, R.M. (1980). Ipspini: the basis for interpopulational differences in pheromone biology. *J. Chem. Ecol.*, 6, 677–87,

Lehninger, A.L. (1975). *Biochemistry*, 2nd Edn. Worth Publishers, New York.

Lenington, S. (1983). Social preferences for partners carrying 'good genes' in wild house mice. *Anim. Behav.*, 31, 325–33.

Lenington, S. and Egid, K. (1985). Female discrimination of male odors correlated with male genotype at the T locus: a response to T-locus or H-2 locus variability? *Behav. Genet.*, 15, 53–67.

Lombardi, J.R. and Vandenbergh, J.G. (1977). Pheromonally induced sexual maturation in females: regulation by the social environment of the male. *Science*, 196, 545–6.

Massey, A. and Vandenbergh, J.G. (1980). Puberty delay by a urinary cue from female house mice in feral populations. *Science*, 209, 821–2.

Monder, H., Lee, C., Donovick, P.J. and Burright, R.G. (1978). Male mouse urine extract effects on pheromonally mediated reproductive functions of female mice. *Physiol. Behav.*, 20, 447–52.

Novotny, M., Jorgenson, J.W., Carmack, M., Wilson, S.R., Boyse, E.A., Yamazaki, K., Wilson, M., Beamer, W. and Whitten, W.K. (1980). Chemical studies of the primer mouse pheromones. In *Chemical signals in vertebrates and aquatic invertebrates* (ed. D. Müller-Schwarze and R.M. Silverstein), pp. 377–90. Plenum, New York.

Novotny, M., Schwende, F., Wiesler, D., Jorgenson, J.W. and Carmack, M. (1984). Identification of a testosterone-dependent unique volatile constituent of male mouse urine: 7-*exo*-ethyl-5-methyl-6,8-dioxaloicyclo[3.2.1]-3-octane. *Experientia*, 40, 217–9.

Novotny, M., Harvey, S., Jemiolo, B. and Alberts, J. (1985). Synthetic pheromones that produce inter-male aggression in mice. *Proc. Natl. Acad. Sci. USA*, 82, 2059–61.

Novotny, M., Jemiolo, B., Harvey, S., Wiesler, D. and Marchlewska-Koj, A. (1986). Adrenal-mediated endogenous metabolites inhibit puberty in female mice. *Science*, 231, 722.

Novotny, M., Harvey, S.D. and Jemiolo, B. (1989). Chemistry of male dominance in the house mouse, *Mus musculus*. *Experientia*, in press.

Oakeshott, J.G. (1974). Social dominance, aggressiveness, and mating success among male house mice (*Mus musculus*). *Oecologia*, 15, 143–58.

Pickett, J.A. and Griffiths, D.C. (1980). Composition of aphid alarm pheromones. *J. Chem. Ecol.*, 6, 349–60.

Preti, G., Nuetherties, E.J., Furman, J.M., Kennelly, J.J. and Johns, B.E. (1976). Volatile constituents of the dog (*Canis familiaris*) and coyote (*Canis latrans*) anal sacs. *J. Chem. Ecol.*, 2, 177–86.

Prodan, K.A. (1976). I. Synthesis of cascarillic acid, II. Synthesis of potential mouse pheromones. M.S. thesis, Indiana University.

Raymer, J. (1984). Investigations into the chemical nature of chemo-olfactory communication in the wolf (*canis lupus*). Ph.D. thesis, Indiana University.

Raymer, J., Wiesler, D., Novotny, M., Asa, C., Seal, U.S. and Mech, L.D. (1984). Volatile constituents of wolf (*Canis lupus*) urine as related to gender and season. *Experientia*, 40, 707–9.

Raymer, J., Wiesler, D., Novotny, M., Asa, C., Seal, U.S. and Mech, L.D. (1985). Chemical investigations of wolf (*Canis lupus*) anal-sac secretion in relation to breeding season. *J. Chem. Ecol.*, 11, 593–608.

Raymer, J., Wiesler, D., Novotny, M., Asa, C., Seal, U.S. and Mech, L.D. (1986). Chemical scent constituents in urine of wolf (*Canis lupus*) and their dependence on reproductive hormones. *J. Chem. Ecol.*, 12, 297–314.

Riley, R.G., Silverstein, R.M. and Moser, J.C. (1974). Biological response of *Atta texance* to its alarm pheromone. *Science*, 183, 760–2.

Ropartz, P. (1968). Mise en évidence d'une augmentation de l'activité locomotrice des groupes de souris femelles en reponse à l'odeur d'un group de males étranges. *CR Acad. Sci. Ser. D*, 267, 2341–3.

Sandnabba, N.K. (1986*a*). Changes in male odours and urinary marking patterns due to inhibition of aggression in male mice. *Behav. Processes*, 12, 349.

Sandnabba, N.K. (1986*b*). Effects of selective breeding for high and low aggressiveness and of fighting experience on odor discrimination in mice. *Aggressive Behav.*, 12, 359–66.

Schwende, F.J., Wiesler, D. and Novotny, M. (1984*a*). Volatile compounds associated with estrus in mouse urine. Potential pheromones. *Experientia*, 40, 213–5.

Schwende, F.J., Jorgenson, J.W. and Novotny, M. (1984*b*). Possible chemical basis for histocompatibility-related mating preference in mice. *J. Chem. Ecol.*, 11, 1603–15.

Schwende, F.J., Wiesler, D., Jorgenson, J.W. and Carmack, M. (1986). Urinary volatile constituents of the house mouse (*Mus musculus*) and their endocrine dependency. *J. Chem. Ecol.*, 12, 277-96.

Scott, J.W. and Pfaff, D.W. (1970). Behavioral and electrophysiological responses of female mice to male urine odors. *Physiol. Behav.*, 5, 407–11.

Silverstein, R.M. (1981). Pheromones: background and potential for use in insects. *Science*, 213, 1326–32.

Sullivan, T.P. and Crump, D.R. (1986). Feeding responses of snowshoe hares *Lepus americanus*) to volatile constituents of red fox (*Vulpes vulpes*) urine. *J. Chem. Ecol.*, 12, 729–39.

Thomas, L. (1975). Symbiosis as an immunological problem. In *The immune system and infectious diseases* (ed. E. Neter and F. Milgram), pp. 2–11. S. Karger, Basel, New York.

Vandenbergh, J.G. (1967). Effect of the presence of a male on the social maturation of female mice. *Endocrinology*, 81, 345–9.

Vernet-Maury, E., Polak, E.H. and Demael, A. (1984). Structure-activity relationship of stress-inducing odorants in the rat. *J. Chem. Ecol.*, 10, 1007–18.

Whitten, W.K. (1958). Modification of the oestrus cycle of the mouse by external stimuli associated with the male: changes in the oestrus cycle determined by vaginal smears. *J. Endocrinol.*, 17, 307–13.

Whitten, W.K. (1966). Pheromones and mammalian reproduction. *Adv. Reprod. Physiol.*, 1, 155–77.

Whitten, W.K., Wilson, M.C., Wilson, S.R., Jorgenson, J.W., Novotny, M. and Carmack, M. (1980). Induction of marking behaviour in wild red foxes (*Vulpes vulpes*) by synthetic urinary constituents. *J. Chem. Ecol.*, 6, 49–51.

Whitten, W.K., Bronson, F.H. and Greenstein, J.A. (1968). Estrus-inducing pheromone of male mice: transport by movement of air. *Science*, 161, 584–5.

Wiesler, D., Schwende, F.J., Carmack, M. and Novotny, M. (1984). Structural determination and synthesis of a chemical signal of the male state and a potential multipurpose pheromone of the mouse, *Mus musculus*. *J. Org. Chem.*, 49, 882–4.

Wilson, M.C., Whitten, W.K., Wilson, S.R., Jorgenson, J.W., Novotny, M. and Carmack, M. (1980). Marking behavior in wild red foxes in response to synthetic volatile urinary compounds. In *Chemical signals—vertebrates and aquatic invertebrates* (ed. D. Müller-Schwarze and R.M. Silverstein), pp. 29–38, Plenum. New York.

Wysocki, C.J., Beauchamp, G.K., Reidinger, R.R. and Wellington, J.L. (1985). Access of large and nonvolatile molecules to the vomeronasal organ of mammals during social and feeding behaviors. *J. Chem. Ecol.*, 11, 1147–59.

Yamazaki, K., Boyse, E.A., Mike, V., Thaler, H.T., Mathieson, B.J., Abbot, J., Boyse, J., Zayas, A.A. and Thomas, L. (1976). Control of mating preferences in mice by genes in the major histocompatibility complex. *J. Exp. Med.*, 144, 1324–35.

Yamazaki, K., Yamaguchi, J. Andrews, P.W., Peake, V. and Boyse, E.A. (1978). Mating preferences of $F_2$ segregants of crosses between MHC-congenic mouse strains. *Immunogenetics*, 6, 253–9.

Yamazaki, K., Yamaguchi, J., Andrews, P.W., Peake, V., Boyse, E.A. and Thomas, L. (1979). Recognition among mice: evidence from the use of a

Y-maze differentially scented by congenic mice of different major histocompatibility types. *J. Exp. Med.*, 150, 755–60.

Zarrow, M.X., Christensen, C.M. and Eleftheriou, B.E. (1971). Strain differences in the ovulatory response of immature mice to PMS and to the pheromonal facilitation of PMS-induced ovulation. *Biol. Reprod.*, 4, 52–6.

# I.2 Dynamic Solvent Effect Sampling for Quantitative Analysis of Mammalian Semiochemicals

PETER APPS

## Introduction

The use of the dynamic solvent effect for sampling large, dilute, gaseous or liquid specimens for analysis by capillary gas chromatography was introduced in 1984 (Apps *et al.* 1984). The mechanism of the effect and the sampling hardware and procedures have subsequently been described (Apps *et al.* 1987). Here the suitability of the technique is considered for the analysis of mammalian social odours. Further details can be found in Apps (1988).

## Analytical challenges presented by mammalian odours

Mammalian odours pose more challenging analytical problems than do other materials. In general they are complex mixtures (which is not to say that the semiochemicals within them may not be simple). The number of identified compounds reaches 63 in wolf (*Canis lupus*) anal sac secretion (Raymer *et al.* 1985), 61 in mouse urine (Schwende *et al.* 1986), and 133 in human whole-body volatiles (Ellin *et al.* 1974). Reported concentrations of volatiles in mammalian secretions vary from the parts per million to the percentage level, and quantities per gland or per scent mark cover the nano- to milligram range (Müller-Schwarze *et al.* 1974; Preti *et al.* 1976; Singer *et al.* 1976; Crewe *et al.* 1979; Wellington *et al.* 1979; Albone 1984, p. 85). Undoubtedly these data are biased towards more abundant components, the minor compounds being undetected or non-quantifiable.

Volatiles from mammals include a wide range of chemical types (Wheeler 1977; Albone 1984, pp. 32–3, 74–134) and it cannot be assumed that any class of organic volatile will be absent from a specimen. Some compounds in mammalian secretions are thermolabile, e.g. dimethyl trisulphide (Singer *et al.* 1976) and hydroxyketones/lactols (Schwende *et al.* 1986). Probably there are other, even more thermolabile, compounds that have escaped detection altogether. In some mammalian odours it is the minor components that are active. For example, semiochemically active dimethyl disulphide is a minor volatile among those from hamsters (*Mesocricetus auratus*) (Singer *et al.* 1976). Squalene and butyrate esters make up 96% of *Saguinus fuscicollis* scent marks but do not account for their biological activity (Epple *et al.* 1979).

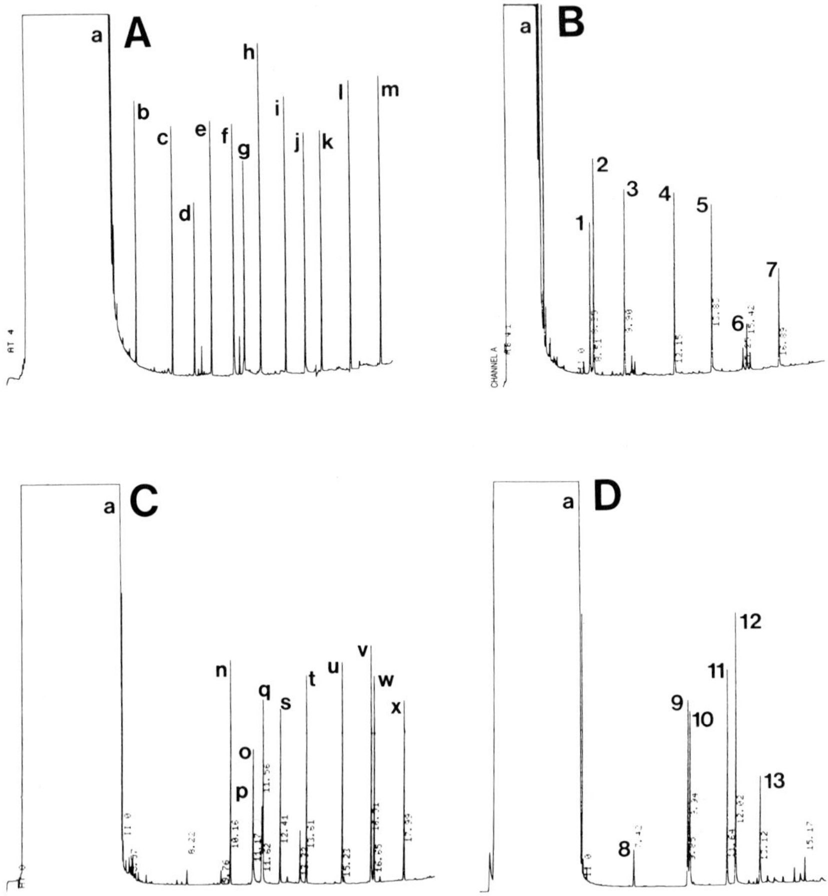


Fig. 1. Tests of chromatographic system adsorptive activity when sampling was carried out by the dynamic solvent effect. *A* Diagnostic test mix, approximately 3 ng per component. *B* Aldehydes and ketones sampled by purging from an aqueous solution, 0.9–3.5 ng per component. *C* Grob test mixture, approximately 10 ng per component. *D* Phenols, up to 10 ng per component. Peak identities are as follows: a: *n*-hexane solvent; b: *n*-octane; c: *n*-nonane; d: 2,6-dimethylheptan-2-one; e: *n*-decane; f: *p*-cresol; g: linalool; h: 2,6-dimethylaniline; i: *n*-dodecane; j: *n*-decanol; k: methyl-*n*-decanoate; l: *n*-tetradecane; m: *n*-pentadecane; n: *n*-decane; o: *n*-octanol, p: *n*-nonanal, q: 2,6-dimethylphenol, s: 2,6-dimethylaniline, t: *n*-dodecane, u: methyl-*n*-decanoate, v: dicyclohexylamine, w: methyl-*n*-undecanoate, x: methyl-*n*-dodecanoate, 1: *n*-heptanone; 2: *n*-heptanal; 3: 2,6-dimethylheptan-4-one; 4: *n*-nonanal; 5: *n*-decanal; 6: *n*- undecanal; 7: *n*-dodecanal, 8: phenol, 9: 2-chlorophenol, 10: *p*-cresol, 11: 2,4-dichlorophenol, 12: 2,6-dimethylphenol, 13: 4-nitrophenol. Column 25 m $\times$ 0.3 mm $\times$ x0.4 $\mu$m methyl silicone, programmed at 10 °C min$^{-1}$ after 6 mins.

In the analysis of mammalian semiochemicals and other odours it is desirable that the sensitivity of the analysis approaches that of the mammalian nose. Although in some cases mammals are extraordinarily sensitive to odours (Moulton 1977; O'Connell *et al.* 1979), it is encouraging that detection thresholds of around $1{:}10^9$ have been reported (Davis 1973; Amoore 1982; Punter 1983). Also, there is no reason to believe that a signalling system based on odours would normally operate near the limits of detection of its signals. This is fortunate, for were it necessary to identify by mass spectrometry a compound occurring at a concentration of $1{:}10^{14}$ in the urine of a single mouse, the collection of a sufficiently large specimen would take nearly 140 years.

Much semiochemically significant material is available in severely restricted quantities. For example, aardwolf (*Proteles cristatus*) scent marks weigh about 5 mg, one 'female equivalent' of hamster vaginal discharge is 20 mg of secretion (O'Connell *et al.* 1979), and tree shrews (*Tupaia belangeri*) deposit scent marks containing 2–20 mg of secretion (Von Holst 1985). By comparison, work with plant volatiles, even on a 'micro' scale, deals with specimens of 200–500 mg (Bicchi and Sandra 1987).

## Capabilities and performance of dynamic solvent effect sampling

When sampled and transferred to a capillary column using the dynamic solvent effect a wide variety of chromatographically intractable adsorptively active test compounds are eluted as peaks which are as sharp as the column will allow (Fig. 1).

Dynamic solvent effect sampling allows precise quantification of low nanogram quantities of a range of compounds from both gaseous and liquid specimens. For a mixture of aldehydes and ketones purged by gas stripping from 5 $cm^3$ of $8\text{–}48{:}10^9$ aqueous solution (Fig. 1B) the coefficients of variation of peak area, peak percentage area, and peak area ratio varied between 2.4% and 16.2%, 1.3% and 11.4%, and 1.1% and 15.8% respectively. For 0.6–2.0 ng of complex mixtures of airborne hydrocarbons sampled from 100 $cm^3$ of $0.7\text{–}3.0{:}10^9$ gas phase specimen, the coefficients of variation of peak area were between 0.6% and 9.3%, of peak percentage area between 1.2% and 7.4%, and of peak area ratio between 0.5% and 7.9%. For 3–4 ng of each component of a $1{:}10^8$ test mix in hexane including *p*-cresol, linalool, and dimethylaniline (Fig. 1A), the coefficients of variation of peak area were 0.05% to 3.9%, of peak percentage area 0.3% to 3.1%, and of peak area ratio 0.3% to 4.5%. For specimens in a solvent matrix the relationship of peak area to specimen concentration was almost perfectly linear; the regression coefficients were in all cases above 0.99 and in most cases above 0.999.

The dynamic solvent effect inlet is effectively a 'cold' one; the temperature at which compounds leave the inlet are similar to, or lower than, those at which they elute from the column. If a compound can be chromatographed without thermal degradation on the column it can safely be introduced to the column by a dynamic solvent effect inlet.

## Applications to mammalian odours

The performance of the dynamic solvent effect with synthetic specimens and the known properties of mammalian odours suggest that DSE sampling should make an important contribution to the analysis of these odours. Nevertheless, too little is known about mammalian semiochemicals for the suitability of an analytical system to be assessed a priori. Therefore DSE sampling has been tested in a variety of biological situations, of which some examples follow.

1 When the whole-body odours of individual male mice (*Mus musculus*) are sampled by the dynamic solvent effect over a period of 20 mins there are clear differences in the chromatographic odour profiles of mice of different social rank (Fig. 2). When odour profile and behavioural ranks are compared, the closest agreement is with a behavioural rank based on the number of aggressive encounters initiated (Apps *et al.* 1988*a*). Two of the compounds shown to increase in the odour of aggressive individuals, 2-*sec*-butyl dihydrothiazole (Fig. 2B) and dehydro-*exo*-brevicomin (Fig. 2A) are known to be semiochemicals from analyses of urine samples pooled from 4–5 individuals (Novotny *et al.*1984; Jemiolo *et al.* 1986).

2 *Mastomys coucha* and *M. natalensis* are cryptic species of small rodent. Although the two species are widely sympatric and produce viable, though infertile, hybrids when made to interbreed in captivity, no hybrids have been found in the wild. Conspecific mates can be recognized by odour; in a 2-choice olfactorium females of both species show conspecific preferences but *M. coucha* is much less discriminating than *M. natalensis*. Whole-body volatiles were sampled from males of both species. Among the complex mixture of volatiles emitted by the mice there was one compound, identified by GC-MS as 3-nonen-2-one which was always present, sometimes as the major component, in chromatograms from *M. coucha* and absent or present only in small quantities, in chromatograms from *M. natalensis* (Fig. 3). The abundance of 3-nonen-2-one in the odour profile was actually a more reliable indicator of a male's species than were the preferences of females in the 2-choice olfactorium.

3 One of the most important features of scent marking is that messages are transmitted for some time after an animal has deposited a mark. To understand how scent marks function in mammalian societies we need to be able to follow through time the changes in emission of volatiles from real scent marks.

Volatiles emitted by scent marks of the brown hyaena (*Hyaena brunnea*) (Mills *et al.* 1980) were sampled by the dynamic solvent effect over periods of up to nine days. The sampling time was increased as the marks' emission rates decreased with age.

The changes with time in the emission rates of the volatiles were unexpectedly complex and variable, and require further investigation. One aspect of the marks' behaviour illustrates both this complexity and the quantitative performance of

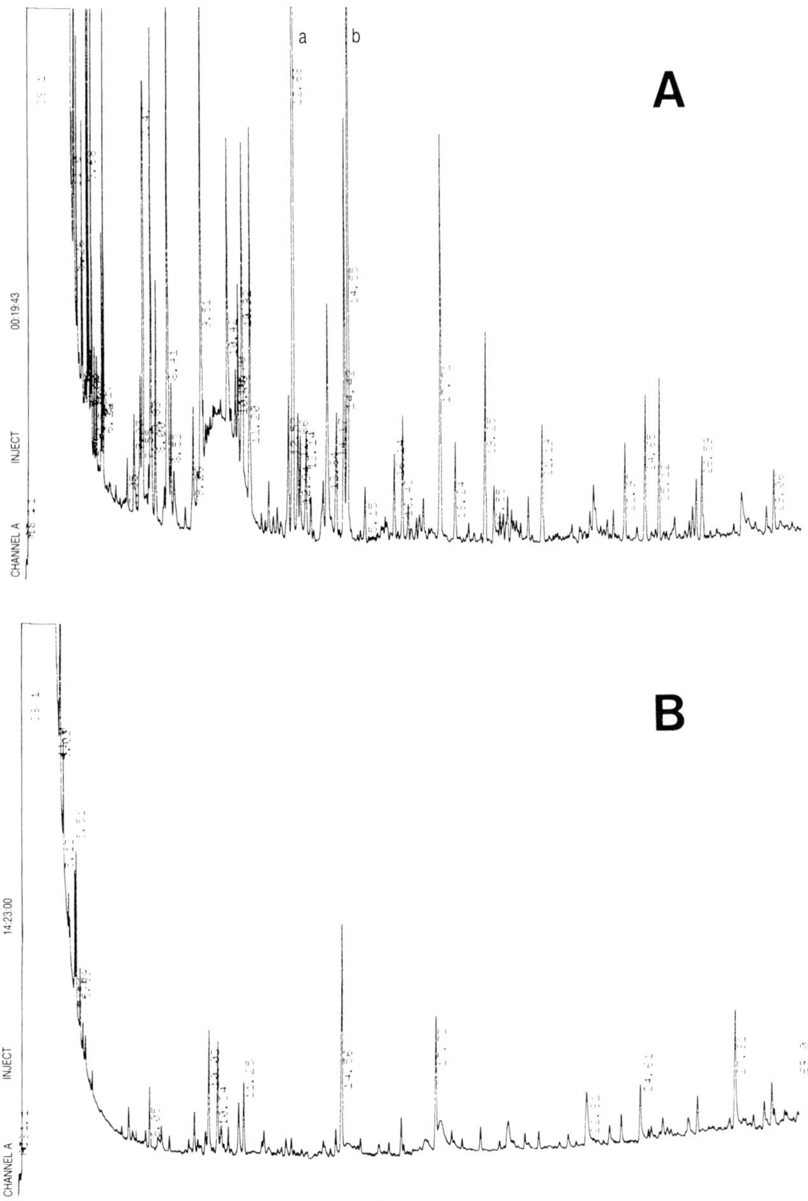


Fig. 2. Chromatographic odour profiles from individual male mice sampled by the dynamic solvent effect from a 20 $cm^3$ $min^{-1}$ flow of air for 20 min. Profile *A* is from an aggressive, dominant individual and *B* is from a non-aggressive, subordinate one. Separation on a methyl silicone column programmed at 5 °C $min^{-1}$ after 6 mins. Flame ionization detector sensitivity 1 x $10^{-11}$. a: dehydro-exo-brevicomin; b: 2-*sec*-butyl-4,5-dihydrothiazole.

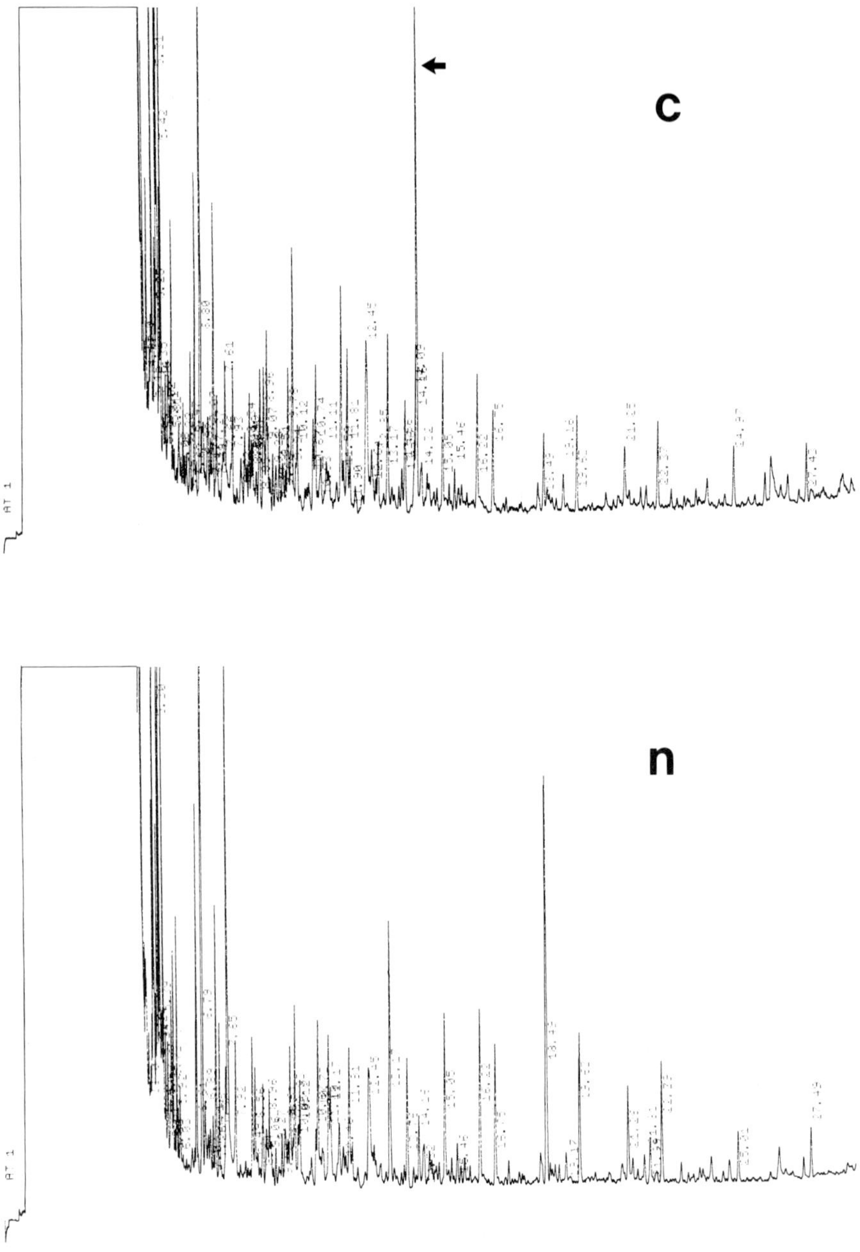


Fig. 3. Chromatographic odour profiles from male *Mastomys coucha* (*c*) and *M. natalensis* (*n*) sampled by the dynamic solvent effect. The peak due to the species-specific compound 3-nonen-2-one is arrowed. Separation conditions as in Fig. 2.

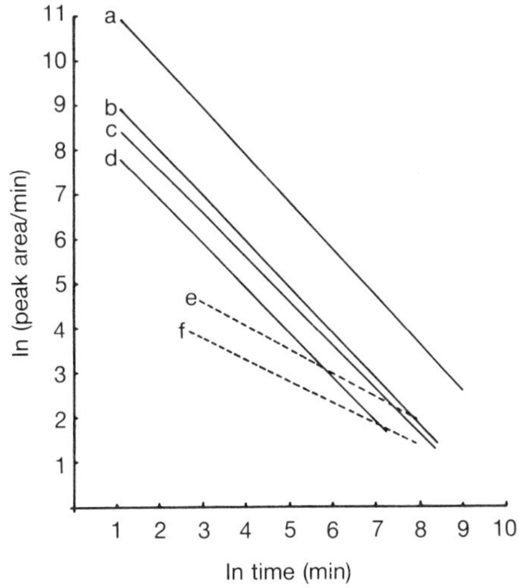


Fig. 4. Lines of best least squares fit to ln-ln transformations of emission rate vs. time for volatiles from the white paste of a brown hyaena (*Hyaena brunnea*) scent mark. Lines that are parallel are from components that maintain constant ratios of emission rates. Samples were analysed under the same separation conditions as those described for Fig. 2.

DSE sampling from biological specimens. Among the volatiles emitted by the white paste of one mark, there were four components whose emission rates declined approximately exponentially with time but whose emission rates relative to one another remained constant within narrow limits. At least one other pair of component peaks maintained emission rates that were constant relative to each other but varied in relation to the group of four (Fig. 4). Constant ratios of emission rate would be expected if scent marks signal the identity of their producer, as required by Gosling's (1982) scent matching hypothesis. Despite an increase in sampling time from 4 mins to 1 h, a 300-fold decrease in emission rates, and rates as low as $5 \times 10^{-12}$ g min$^{-1}$, the coefficients of variation of the relative peak areas within the group of four compounds were less than 8%.

4 Analysis of the anal gland secretion and scent marks of aardwolves illustrates the compatibility of the dynamic solvent effect with well-deactivated, high-efficiency capillary columns. Even free fatty acids were eluted as sharp peaks, and esters whose Kovats Indices differ by only about 4 units (Jennings and Shibamoto 1980) were resolved (Fig. 5) (Apps *et al.* 1988*b*).

5 Using DSE sampling the thermally labile compound dimethyl trisulphide has been found in anal gland secretions of the water mongoose (*Atilax paludinosus*), the yellow mongoose (*Cynictis penicillata*), and the porcupine (*Hystrix africae-australis*).

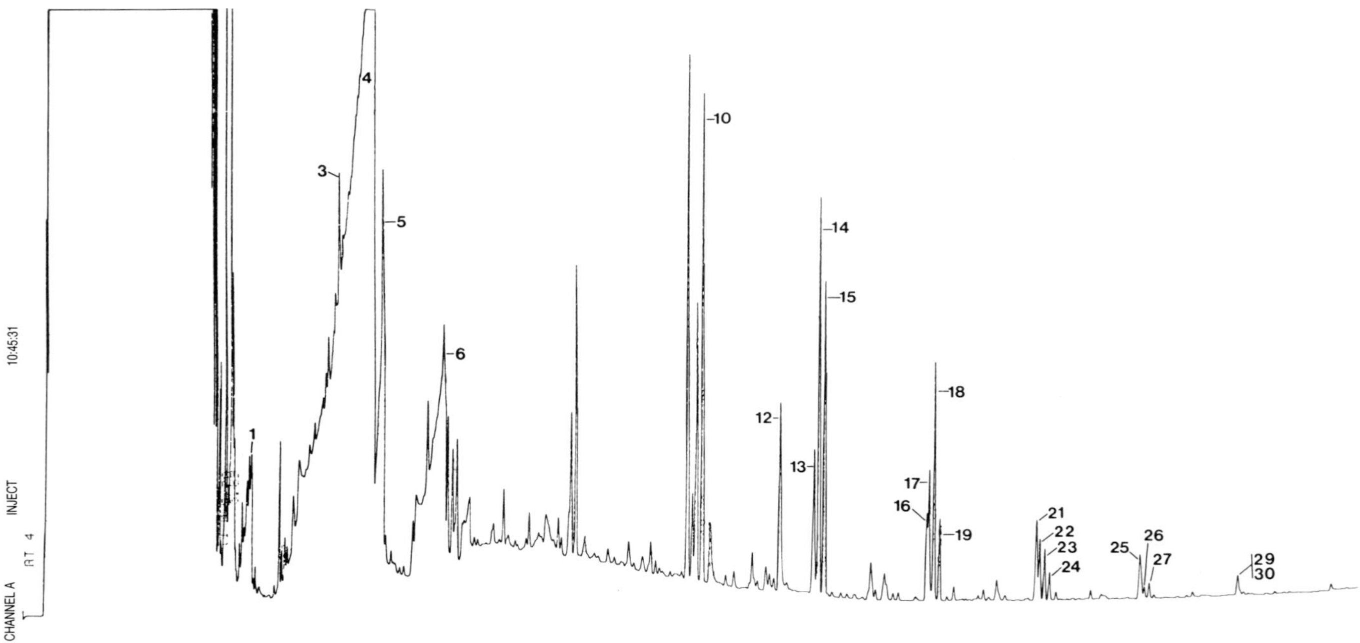


Fig. 5. Chromatogram of volatiles from the anal gland secretion of a female aardwolf (*Proteles cristatus*). Peaks 13–30 are a series of straight chain esters differing only in the position of the ester linkage within groups and total chain length between groups, the chain lengths of the five groups are 11, 12, 13, 14, and 15 carbon atoms (see Apps *et al.* 1988*b* for details).

## Conclusion

The dynamic solvent effect offers a sampling technique for high-resolution gas chromatography, which complements the rest of the analytical system in terms of quantification, sensitivity, resolution, and freedom from artefacts. By allowing precise quantification of nanogram amounts of a range of compounds sampled at low concentrations from small specimens it allows the analysis of mammalian odour signals at the level of the individual.

## Acknowledgements

I should like to thank H.W. Viljoen, P.R.K. Richardson, O.A.E. Rasa, D. Gordon and Professors V. Pretorius and J.D. Skinner. My attendance at the conference was made possible by funds from the University of Pretoria, the Institute for Chromatography and the Mammal Research Institute.

## References

Albone, E.S. (1984). *Mammalian semiochemistry. The investigation of chemical signals between mammals.* Wiley, Chichester.

Amoore, J.E. (1982). Odor theory and odor classification. In *Fragrance chemistry the science of the sense of smell* (ed. E.T. Theimer), pp. 27–76. Academic Press, New York.

Apps, P.J. (1988). Quantitative analysis of semiochemicals. Ph.D. thesis, University of Pretoria.

Apps, P.J., Pretorius, V., Rohwer, E. and Lawson, K.H. (1984). The dynamic film concentrator—a new approach to large and dilute samples in GLC. *J. High Resolut. Chromatogr. Chromatogr. Commun.*, 7, 212–4.

Apps, P.J., Lawson, K.H., Rohwer, E.R., Centner, M.R., Viljoen, H.W. and Hulse, G. (1987). Trace analysis of complex organic mixtures using capillary gas-liquid chromatography and the dynamic solvent effect. *J. High Resolut. Chromatogr. Chromatogr. Commun*, 10, 122–7.

Apps, P.J., Rasa, A. and Viljoen, H.W. (1988*a*). Quantitative chromatographic profiling of odours associated with dominance in male laboratory mice. *Agress. Behav.*, 14, 451–61.

Apps, P.J., Viljoen, H.W., Richardson, P.R.K. and Pretorius, V. (1988*b*). Volatile components of the anal gland secretion of the aardwolf (*Proteles cristatus*). *J.Chem. Ecol.* (in press).

Bicchi, C. and Sandra, P. (1987). Microtechniques in essential oil analysis. In *Capillary gas chromatography in essential oil analysis* (ed. P. Sandra and C. Bicchi), pp. 85–122. Hüthig, Heidelberg.

Crewe, R.M., Burger, B.V., Le Roux, M. and Katsir, Z. (1979). Chemical constituents of the chest gland secretion of the thick tailed galago (*Galago crassicaudatus*). *J. Chem. Ecol.*, 5, 861–8.

Davis, R.G. (1973). Olfactory psychophysical parameters in man, rat, dog and pigeon. *J. Comp. Physiol. Psychol.*, 85, 221–32.

Ellin, R.I., Farrand, R.L., Oberst, F.W., Crouse, C.L., Billups, N.B., Koon, W.S., Musselman, N.P. and Sidell, F.R. (1974). An apparatus for the detection and quantitation of volatile human effluents. *J. Chromatogr.*, 100, 137–52.

Epple, G., Golob, N.F. and Smith III, A.B. (1979). Odor communication in the tamarin *Saguinus fuscicollis* (Callitrichidae): behavioral and chemical studies. In *Chemical ecology: odour communication in animals* (ed. F.J. Ritter), pp. 117–30. Elsevier/North Holland, Amsterdam.

Gosling, L.M. (1982). A reassessment of the function of scent marking in territories. *Z. Tierpsychol.*, 60, 89–118.

Jemiolo, B., Harvey, S. and Novotny, M. (1986). Promotion of the Whitten effect in female mice by synthetic analogs of male urinary constituents. *Proc. Nat. Acad. Sci. USA*, 83, 4576–9.

Jennings, W. and Shibamoto, T. (1980). *Qualitative analysis of flavor and fragrance volatiles by glass capillary gas chromatography*. Academic Press, New York.

Mills, M.G.L., Gorman, M.L. and Mills, M.E.J. (1980). The scent marking behaviour of the brown hyaena *Hyaena brunnea*. *S. Afr. J. Zool.*, 15, 240–8.

Moulton, D.G. (1977). Minimum odorant concentrations detectable by the dog and their implications for olfactory receptor sensitivity. In *Chemical signals in vertebrates* (ed. D. Müller-Schwarze and M.M. Mozell), pp. 455–64. Plenum, New York.

Müller-Schwarze, D., Müller-Schwarze, C., Singer, A.G. and Silverstein, R.M. (1974). Mammalian pheromone: identification of active component in the subauricular scent of the male pronghorn. *Science*, 183, 860–2.

Novotny, M., Schwende, F.J., Wiesler, D., Jorgenson, J.W. and Carmack, M. (1984). Identification of a testosterone-dependent unique volatile constituent of male mouse urine: 7-*exo*-ethyl-5-methyl-6,8-dioxabicyclo[3.2.1]-3-octene. *Experientia*, 40, 217–9.

O'Connell, R.J., Singer, A.G., Pfaffman, C. and Agosta, W.C. (1979). Pheromones of hamster vaginal discharge: attraction to femtogram amounts of dimethyl disulfide and to mixtures of volatile components. *J. Chem. Ecol.*, 5, 575–85.

Preti, G., Muetterties, E.L., Furman, J.M., Kennelly, J.J. and Johns, B.E. (1976). Volatile constituents of dog (*Canis familiaris*) and coyote (*Canis latrans*) anal sacs. *J. Chem. Ecol.*, 2, 177–86.

Punter, P.H. (1983). Measurement of human olfactory thresholds for several groups of structurally related compounds. *Chem. Senses*, 7, 215–35.

Raymer, J., Wiesler, D., Novotny, M., Asa, C., Seal, U.S. and Mech, L.D. (1985). Chemical investigations of wolf (*Canis lupus*) anal-sac secretion in relation to breeding season. *J. Chem. Ecol.*, 11, 593–608.

Schwende, F.J., Wiesler, D., Jorgenson, J.W., Carmack, M. and Novotny, M. (1986). Urinary volatile constituents of the house mouse, *Mus musculus*, and their endocrine dependency. *J. Chem. Ecol.*, 12, 277–96.

Singer, A.G., Agosta, W.C., O'Connell, R.J., Pfaffman, C., Bowen, D.V. and Field, F.H. (1976). Dimethyl disulfide: an attractant pheromone in hamster vaginal secretion. *Science*, 191, 948–50.

Von Holst, D. (1985). The primitive eutherians II: a case study of the tree shrew, *Tupaia belangeri*. In *Social odours in mammals*, Vol I (ed. R.E. Brown and D.W. Macdonald), pp. 155–216. Oxford University Press, Oxford.

Wellington, Judith, L., Byrne, K.J., Preti, G., Beauchamp, G.K. and Smith III, A.B. (1979). Perineal scent gland of wild and domestic guinea pigs. A comparative chemical and behavioral study. *J. Chem. Ecol.*, 5, 737–51.

Wheeler, J.W. (1977). Properties of compounds used as chemical signals. In *Chemical signals in vertebrates* (ed. D. Müller-Schwarze and M.M. Mozell), pp. 61–70. Plenum, New York.

# I.3 Quantification of Odorous 16-androstene Steroids in Vertebrates

D.B. GOWER

## Introduction

For about 40 years, there has been considerable interest in the odorous 16-androstene steroids for two reasons: they have a pronounced odour and very large amounts of these substances are produced by porcine testes (see Gower 1984 for review). Since the early 1970s, interest has increased because two of these steroids have been identified as primer and releaser pheromones in pigs. The urinous 5α-androst-16-en-3-one (5α-androstenone) and musky 5α-androst-16-en-3α-ol (3α-androstenol) both elicit mating behaviour in an oestrous sow (Reed *et al.* 1974). In addition, Booth (1984) has shown that these steroids are involved in the acceleration of puberty in immature pigs.

In this chapter, I review the evidence that 16-androstenes may be involved in human olfactory communication; since these steroids are found in the axillae of men. Methods currently available for the estimation and biosynthesis of the odorous 16-androstenes are also surveyed.

## Structural characteristics of the 16-androstenes

The five 16-androstenes that have been studied intensively are 5α-androstenone, 4,16-androstadien-3-one (androstadienone), 3α-androstenol, 5α-androst-16-en-3β-ol (3β-androstenol) and 5,16-androstadien-3β-ol (androstadienol) (Fig. 1). Although their structures are similar to those of other $C_{19}$ steroids, they lack any substituent at C-17 and are characterized by unsaturation at C-16,17. With only one oxygen function at C-3, the 16-androstenes are very non-polar, requiring relatively non-polar solvent systems for separation by thin layer chromatography (Gower 1964). Further, as the compounds are relatively volatile (Kaufman *et al.* 1976), great care must be taken in evaporation of solutions in organic solvent so that losses of the steroids are not excessive (see Gower 1981 for review).

Ohloff *et al.* (1983) showed that the character and intensity of the odour of the 16-androstenes using human subjects varies with the structural characteristics. The two ketones, 5α-androstenone and androstadienone have quite pungent

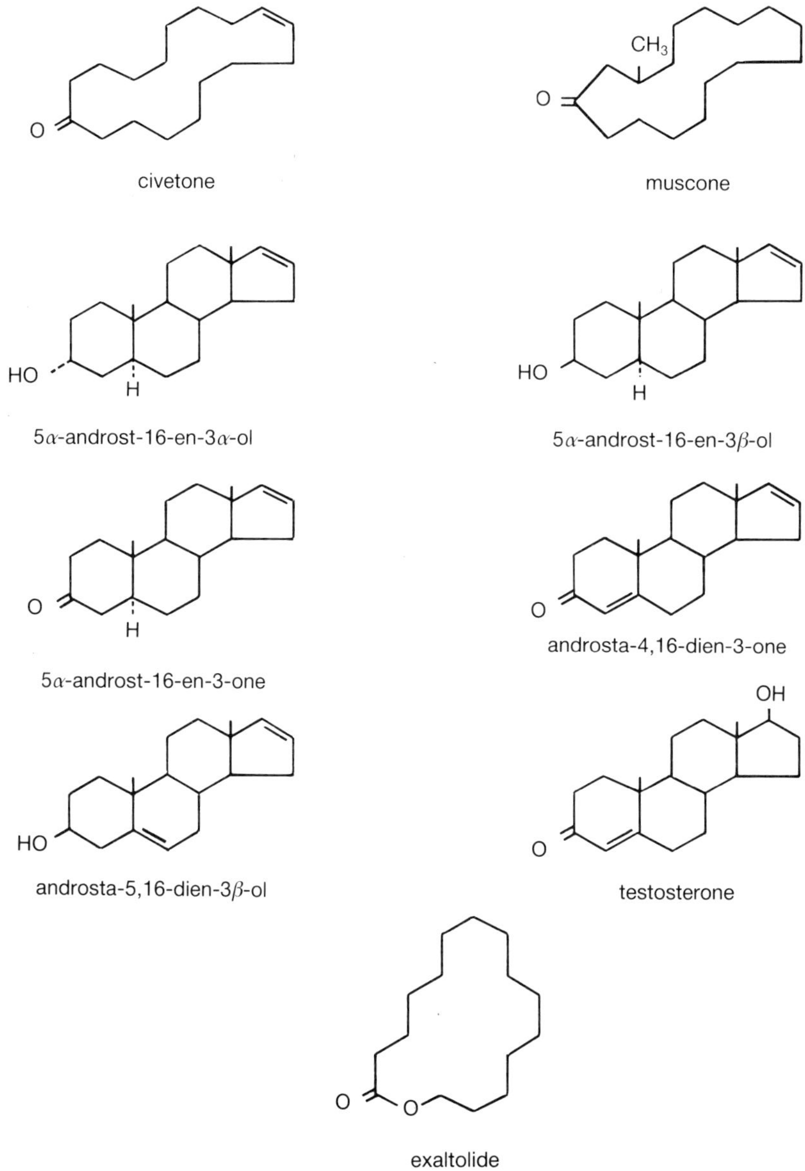


Fig. 1. Structures of some 16-androstenes. Testosterone and some cyclic odorants are shown for comparison.

'urinous' odours, the olfactory thresholds being quite low (about 0.2 and 1.8 parts per billion, respectively (Amoore *et al.* 1977; Ohloff *et al.* 1983). 3α-Androstenol has a more musk-like smell with an olfactory threshold of approximately 6.2 parts per billion. Changing the configuration at C-3 (3β-androstenol), or introducing another double bond at C-5,6 (androstadienol), results in less intense musk-like odours. The saturated ketone, 5α-androstan-3-one (5α-androstanone), has a similar odour, as perceived in man, to that of 5α-androstenone, indicating that the double bond at C-16,17 is not important with respect to odour quality.

## Possible role of 16-androstenes as human pheromones

The notion that human pheromones may exist is not new and, consistent with this, there has been accumulating evidence during the past decade to suggest that 3α-androstenol and 5α-androstenone may influence men and women as primer and releaser pheromones. These studies, some of which have been criticized because of inadequate controls, have been reviewed (Doty 1981; Gower 1981; Gower *et al.* 1988*a*; Labows 1988). The results of some of these experiments are shown in Table 1. In 1977, Cowley *et al.* published evidence that the judgements of female

**Table 1.** Possible effects associated with odorous steroids*

| |
|---|
| Alteration of judgement |
| Change of location in others' presence |
| Alteration of mood during menstrual cycle |
| Changes in cycle length |
| Changes in skin conductance |

* Kirk-Smith *et al.* (1983). *Biol. Psychol.*, 17, 221.

interviewers of men applying for a position of responsibility were modified while being exposed to the odour of 3α-androstenol. The odour also seemed to influence the opinions of men and women of photographs of normally clothed people, animals and buildings. When exposed to 3α-androstenol, both men and women judged the photographed women as more attractive, although the effect of the odour on judgements of photographed men was weaker (Kirk-Smith *et al.* 1978). It is also possible that the odour of 3α-androstenol may affect mood in women. When women were exposed to the odour throughout a menstrual cycle, they rated their moods as more submissive at mid-cycle when compared with no exposure to odour (Benton 1982). In contrast, however, Black and Biron (1982) found no differences in physical attractiveness of women to men or men to women with a musk perfume when compared with a control odour.

The more pungent, urinous smell of 5α-androstenone is also suggested as having behavioural effects in humans. Kirk-Smith and Booth (1980) noted that chairs in waiting-room situations sprayed with 5α-androstenone (3.2 μg) were more often occupied by women than by men; at a high level of 5α-androstenone (32 μg), the number of men who occupied the sprayed seats was less than at the lower concentration. Similar results were obtained in an experiment when certain theatre seats (all unreserved) were sprayed with 5α-androstenone; these seats were occupied mainly by women (Clark 1978). Although some women do not find the odour of 5α-androstenone pleasant at high concentrations (see Gower *et al.* 1988*a* for review), they may conceivably associate the odour with men (see below) and, as a result, be attracted to the seats. In contrast, men may associate the odour with the presence of other men and possibly feel uncomfortable, as suggested by Kirk-Smith and Booth (1980).

A third interesting possibility is that 16-androstenes may affect menstrual cycle length. In the 'theatre' experiment of Clark (1978), the periods of girls employed there were disrupted, presumably as a result of the odour of 5α-androstenone; previously, their cycles had been synchronized. This example of menstrual synchrony has been studied by several workers and reviewed by McClintock (1983); it occurs, probably as a pheromonal phenomenon, in women who live together in halls of residence, for example. The work of Russell *et al.* (1980), Preti *et al.* (1987) and Cutler *et al.* (1987) is of particular significance in this context. Axillary sweat extracts from a man or woman are known to have effects on menstrual cycle lengths of female subjects who are subjected to the odour. The male odour was shown to decrease the number of 'aberrant' cycles (i.e. of length significantly greater or less than the mean) whereas the female odour caused synchronization of the cycles of recipient women with that of the donor of the axillary extract. To my knowledge, the active principle in axillary sweat has not been identified, although one (or more) of the odorous 16-androstenes are likely candidates.

## 16-androstenes in the human axilla

The significance of the human axilla in sexual relationships has been discussed by Brody (1975). Clearly the axilla is an area at which odorous substances could be produced and from which they could be dispersed subsequently. The presence of the 'urinous' 5α-androstenone in the axillae of men and, to a lesser extent, of women has been known for some years (Claus and Alsing 1976; Bird and Gower 1981); the geometric mean value for secretion in men is 51 pmol/24 h and 9 pmol/24 h in women (Gower *et al.* 1985). There is some evidence that 5α-androstenone is formed by microbial action on precursors, presumably in apocrine secretion. The microbiology of the human axilla has been studied by Leyden *et al.* (1981) and by Rennie *et al.* (1990). These workers have shown that individuals with a pungent odour have a high axillary coryneform bacterial count. Almost certainly, the odorous 16-androstenes make a contribution to this odour because 5α-androstenone, androstadienone, 3α(β)-androstenol and androstadienol have been detected in the axillary hair of men (Nixon *et al.* 1988; Gower *et al.* 1988*b*).

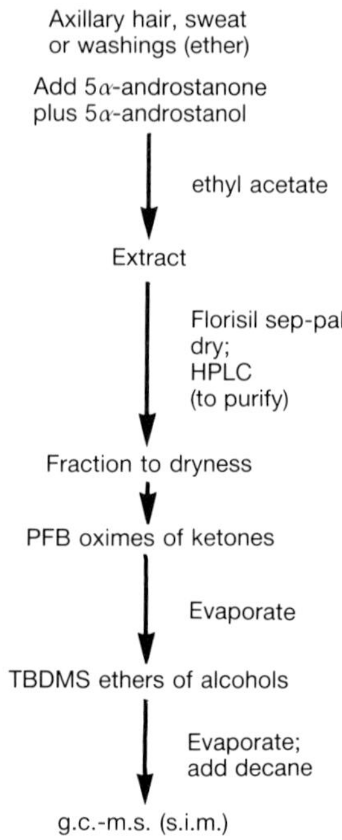


Fig. 2. Flow diagram used in the separation and quantification of 16-androstenes.

## Quantification of 16-androstenes in axillary sweat and hair

Current work in this laboratory is concerned with devising methods for the quantification of nanogram amounts of 16-androstenes in the axillary sweat and hair of men. There are inherent problems in such studies because the extracts obtained with ethyl acetate also contain vary large quantities of complex mixtures of fatty acids (Preti *et al.* 1987), triacyl glycerols, cholesterol (both free and esterified), as well as other steroids and their sulphates (see Gower *et al.* 1988*a* for review). Thus, it has been necessary to purify the ethyl acetate or ether extracts extensively (Fig. 2). Prior to any extraction of the biological material, 5α-androstanone and 5α-androstan-3β-ol were added as internal standards. Derivatization of the purified extracts was performed by methods described by Nixon *et al.* (1988) and Gower *et al.* (1988*b*) and was achieved using pentafluorobenzyl (PFB) oximes for the ketones and *t*-butyldimethylsilyl (TBDMS) ethers for the alcohols. In earlier experiments, the trimethylsilyl (TMS) ethers were used but these were found to give lower signal-to-noise ratios than the TBDMS ethers. Capillary gas chromatography–mass spectrometry was performed with a Varian 3400 gas chromatograph (Varian Associates, Walton-on-Thames, UK) with an on-column injection system. A CB sil 5 column (25 m × 0.33 mm i.d.) was directly interfaced

**Table 2.** Summary of the ions used in the estimation of 16-androstenes by selected ion monitoring

| Steroid analysed | Derivative | m/z |
|---|---|---|
| 5α-Androstenone | Pentafluorobenzyloxime | 467 |
| 4,16-Androstadienone | Pentafluorobenzyloxime | 465 |
| 5α-Androstanone (internal standard) | Pentafluorobenzyloxime | 469 |
| 3α(β)-Androstenols | Trimethylsilyl ether | 346 |
| 5,16-Androstadienol | Trimethylsilyl ether | 344 |
| 5α-Androstan-3β-ol (internal standard) | Trimethylsilyl ether | 348 |
| 3α(β)-Androstenols | *t*-Butyldimethylsilyl ether | 331* |
| 5,16-Androstadienol | *t*-Butyldimethylsilyl ether | 329* |
| 5α-Androstan-3β-ol (internal standard) | *t*-Butyldimethylsilyl ether | 333* |

* [M 57]

with the mass spectrometer (MM-305 mass spectrometer, V.G. Analytical Ltd, Manchester, UK, and Technivent data system). The ions used for monitoring the PFB oximes, TMS ethers and TBDMS ethers are shown in Table 2. Marked differences were noted in the mass spectra of the PFB oximes of 5α-androstenone and 5α-androstanone. The former showed a strong peak at *m/z* 288, corresponding to $[M-181]^+$, as a result of the loss of the pentafluorobenzyl group. In contrast, the corresponding derivative of 5α-androstanone showed an intense $[M]^+$ at *m/z* 467, with additionally [M-15] (see also Nixon *et al.* 1988). The reason why 5α-androstenone PFB oxime loses the pentafluorobenzyl group is unknown at present but may be due to the presence in the molecule of the C16-17 double bond. The selective ion monitoring (s.i.m.) trace for a skin extract from a healthy male subject is shown in Fig. 3. The PFB oxime of androstadienone is separated from those of 5α-androstenone and 5α-androstanone; in addition, the *syn*- and *anti*-form of each derivative are well resolved. The behaviour of the TMS ethers of the 16-androstenols has been described recently (Nixon *et al.* 1988) and will not be discussed here. The corresponding TBDMS ethers gave strong peaks corresponding to $[M-57]^+$, as the result of the loss of a *t*-butyl group (Fig. 4). The TBDMS ethers of 3α- androstenol and 5α-androstanol (the internal standard) were easily resolved (as were the corresponding TMS ethers; Nixon *et al.* 1988). However, the TBDMS ethers of 3β-androstenol and androstadienol were only just resolved but this was sufficient to eliminate errors due to cross-contamination between the *m/z* 329 and 331 channels. A typical s.i.m. trace for an axillary skin extract from a healthy man is shown in Fig. 5.

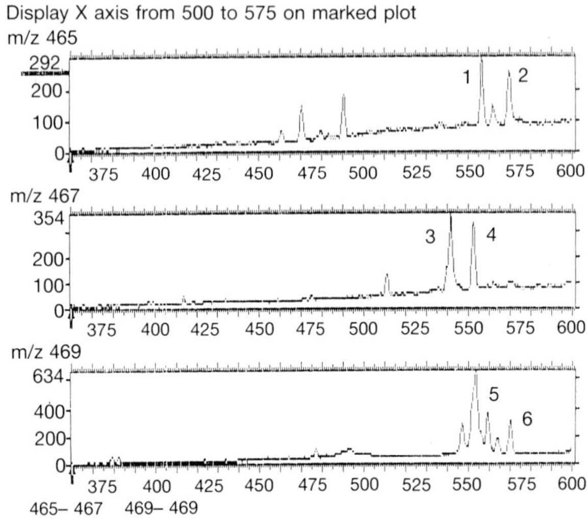


Fig. 3. S.i.m. trace of 4,16-androstadienone (1,2) and 5α-androstenone (3,4) (PFB oximes) obtained from an axillary skin extract of a healthy man. 5α-Androstan-3-one (5,6) was the internal standard. Paired peaks arise from *syn*- and *anti*-forms of derivatives.

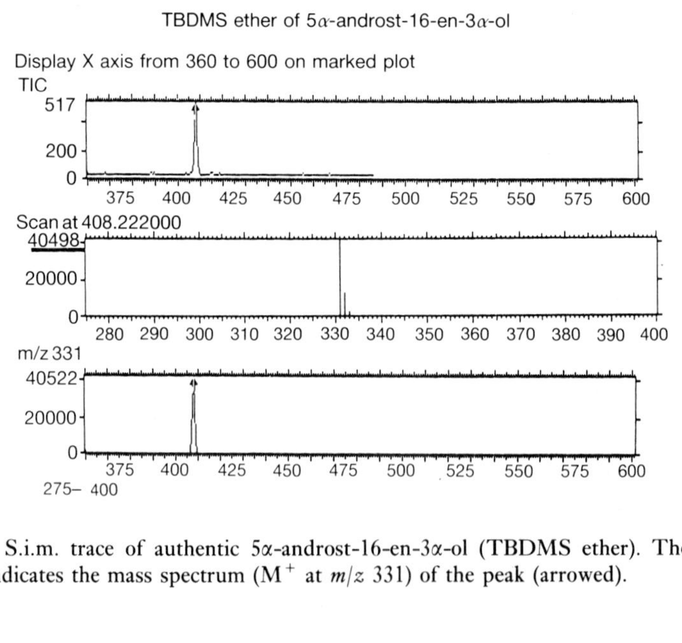


Fig. 4. S.i.m. trace of authentic 5α-androst-16-en-3α-ol (TBDMS ether). The middle trace indicates the mass spectrum ($M^+$ at $m/z$ 331) of the peak (arrowed).

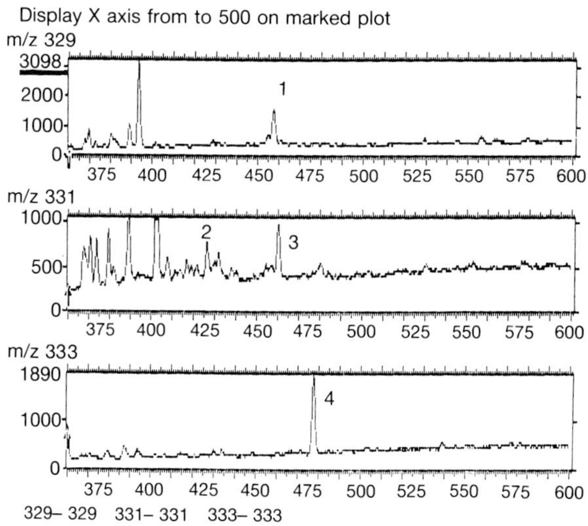


Fig. 5. S.i.m. trace of 5,16-androstadienol (1) and 3α-(2)-and 3β-(3)-androstenols (TBDMS ethers) obtained from an axillary skin extract of a healthy man. 5α-Androstan-3β-ol (4) was the internal standard.

**Table 3.** 16-Androstenes in axillary hair of men

| | Range (ng $mg^{-1}$) | |
|---|---|---|
| | Normal | T.A.* |
| 4,16-Androstadienone | 0–4.6 | 1.5 |
| 5α-Androstenone | 0–8.5 | 37.9 |
| 5,16-Androstadienol | 0–0.9 | 0.1 |
| 5α-Androst-16-en-3α-ol | 0–2.9 | 63.5 |
| 5α-Androst-16-en-3β-ol | 0–2.0 | 35.4 |

* T.A., trichomycosis axillaris.

Using these techniques, Nixon *et al.* (1988) and Gower *et al.* (1988*b*) have quantified 16-androstenes in the axillary hair of 28 healthy men (Table 3). Large ranges were recorded in each subject but the levels found in a man who had the minor dermatological condition, trichomycosis axillaris, had greatly increased quantities of 5α-androstenone, 3α- and 3β-androstenols. He also had a total axillary bacterial count about sixfold greater than that of most other subjects which is in keeping with the notion that 16-androstenes may be products of bacterial action in the axilla (Bird and Gower 1982). Recent evidence from this laboratory indicates that coryneform bacteria have the ability to metabolize testosterone (Nixon *et al.* 1986, 1987) and also androstadienol (Gower *et al.* 1988*c*). With respect to androstadienol, the 16-androstene is transformed into androstadienone, 5α-androstenone and 3α(β)-androstenols, although there is also some evidence of reversals of all these reactions (Gower *et al.* 1988*c*). As yet, the source of androstadienol in axillary secretions or hair is unknown, although androstadienone occurs in peripheral blood plasma (Brooksbank *et al.* 1972). Bearing in mind the relatively low olfactory thresholds of some 16-androstenes (see above), it can be assumed that these odorous compounds make some contribution to axillary odour.

## Testicular biosynthesis of 16-androstenes

The first report of the presence of these compounds in boar testes was by Prelog and Ruzicka (1944). These workers were attempting to study androgens, such as testosterone, but found that the 16-androstenes could be detected and identified more easily because they were present in such large quantities. Subsequent studies both *in vitro* and *in vivo* (see Gower 1984 for review) have confirmed that, in boar testes, the 16-androstenes are more important quantitatively than the androgens. Pathways of biosynthesis have been shown from the $C_{21}$ steroids, pregnenolene and progesterone, and appear to be independent of androgen biosynthesis (Fig. 6).

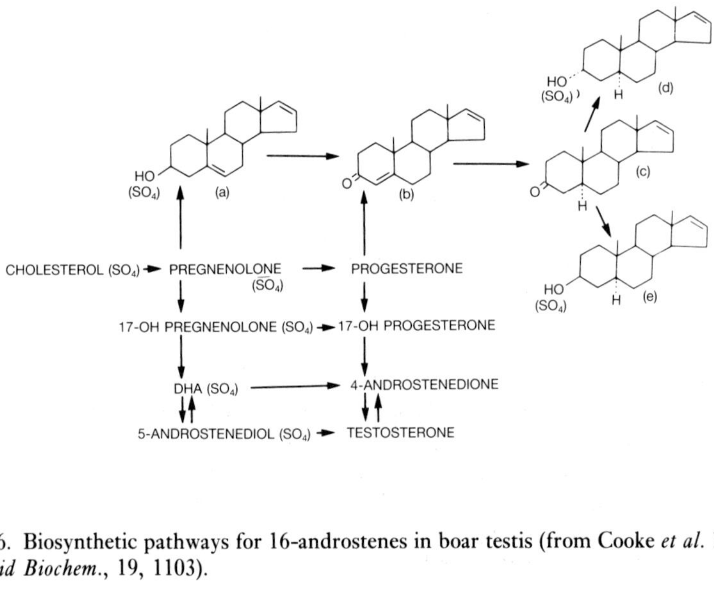


Fig. 6. Biosynthetic pathways for 16-androstenes in boar testis (from Cooke *et al.* 1983, *J. Steroid Biochem.*, 19, 1103).

Immature (three-week-old) porcine testis also had the capacity for 16-androstene biosynthesis but at a much lower level than that in the mature animals (Kwan *et al.* 1985*a*; Kwan *et al.* 1985*b*). As 3α-androstenol and other 16-androstenes are excreted in human urine (Brooksbank and Gower 1970), studies were undertaken in this laboratory with human testis to see if this tissue was also a source of 16-androstenes. Gower and Bicknell (1972) were able to show that these steroids were formed *in vitro* in histologically normal testis from a pseudo-hermaphrodite, but at a level that was about six times less than that in the boar. However, Weusten *et al.* (1987) found that, in human testis, pregnenolone was rapidly converted to androstadienol, the yield being comparable with that in boar testis (e.g. Gower and Ahmad 1967; Shimizu and Nakada 1976; Shimizu 1979).

Following the discovery of 16-androstenes in human testis, it seemed logical to detect and, if possible to quantify, these steroids in semen. Preliminary studies, using capillary gas chromatography have shown the presence of 3α-androstenol and androstadienone; 5α-androstenone was also present, but only in some samples (Kwan *et al.* 1987). In further recent work using gas chromatography–mass spectrometry, Kwan *et al.* (1989) have provided preliminary evidence (see above, for methods) for the presence of 3α- and 3β-androstenols and androstadienol in semen. The two ketones, 5α-androstenone and androstadienone, were probably also present but their presence could not be proved unequivocally by s.i.m. because of high levels of background noise. It is hoped that further purification of derivatized samples will obviate this problem.

Behavioural studies with rhesus monkeys were published many years ago, and mixtures of fatty acids present in the vagina of the females were identified as being involved as sex attractants (Michael *et al.* 1972). In present studies in this laboratory (Kwan and Gower, in preparation), homogenates of testes from *Macaca fascicularis* have been incubated with pregnenolene (0.22 mmol $l^{-1}$) for 60 mins in the presence of NADPH. After addition of $^3$H-androstenedione to allow for correction for analytical losses, the metabolites have been extracted with ethyl acetate. The metabolite mixture was then purified, derivatized as *O*-methyl oxime/TMS ethers and, after purification, subjected to capillary gas chromatography using appropriate temperature programming, so that a large range of polarities of steroids could be eluted. Using this steroid 'profiling' method, some preliminary evidence has been obtained for the formation of androstadienol, 3α- and, especially, 3β-androstenols from pregnenolone. Similar studies have been performed with rat testis subcellular fractions (Kwan and Gower 1988) and have provided evidence for the formation of androstadienol, androstadienone and 3β-androstenol. Again, the latter was the most important 16-androstene quantitatively (e.g. Ruokonen *et al.* 1973).

## Conclusions

It was thought at one time that 16-androstenes were exclusively 'boar steroids'. Comparative studies have now revealed that these compounds are also produced in human, rat and monkey testis. Further work, using more sensitive techniques, may indicate that this intriguing group of odorous steroids is present in many other animals. Whether they act as pheromones in humans, even at a subliminal level, is still an open question but their presence in the axillae of men and, as has been shown very recently, in semen may be consistent with the view that they have a role as chemical signals.

## Acknowledgements

I am most grateful to Drs T.K. Kwan, A.I. Mallet, A. Nixon, P.J. Rennie, and W.J. Watkins for their assistance and collaboration. The Dunhill Medical Trust provided financial assistance.

## References

Amoore, J.E., Pelosi, P. and Forrester, L.J. (1977). Specific anosmias to 5-α-androst-16-en-3-one and w-pentadecalactone: The urinous and musky primary odours. *Chem. Sens. Flav.*, 2, 401–25.

Benton, D. (1982). The influence of androstenol—a putative human pheromone —on mood throughout the menstrual cycle. *Biol. Psychol.*, 15, 249–56.

Bird, S. and Gower, D.B. (1981). The validation and use of a radioimmunoassay for 5-α-androst-16-en-3-one in human axillary collections. *J. Steroid Biochem.*, 14, 213–9.

Bird, S. and Gower, D.B. (1982). Axillary 5-α-androst-16-en-3-one, cholesterol and squalene in man; preliminary evidence for 5-α-androst-16-en-3-one being a product of bacterial action. *J. Steroid Biochem.*, 17, 517–22.

Black, S.L. and Biron, C. (1982). Androstenol as a human pheromone: no effect on perceived physical attractiveness. *Behav. Neurol. Biol.*, 34, 326–30.

Booth, W.D. (1984). A note on the significance of boar salivary pheromones to the male effect on puberty attainment in gilts. *Anim. Prod.*, 39, 149–52.

Brody, B. (1975). The sexual significance of the axillae. *Psychiatry*, 38, 278–89.

Brooksbank, B.W.L. and Gower, D.B. (1970). The estimation of 3α-hydroxy-5α-androst-16-ene and other C19-$\Delta^{16}$ steroids in urine by gas-liquid chromatography. *Acta Endocr. (Kbh)*, 63, 79–90.

Brooksbank, B.W.L., Wilson, D.A. and MacSweeney, D. (1972). Fate of androst-4, 16-dien-3-one and the origin of 3α-hydroxy-5α-androst-16-ene in man. *J. Endocrin.*, 52, 239–51.

Clark, T. (1978). Whose pheromone are you? *World Med.*, 26 July, pp. 21–3.

Claus, R. and Alsing, W. (1976). Occurrence of 5-α-androst-16-en-3-one. a boar pheromone in man and its relationship to testerone. *J. Endocrin.*, 68, 483–4.

Cowley, J.J., Johnson, A. and Brooksbank, B.W.L. (1977). The effect of two odorous compounds on performance in an assessment-of-people test. *Psycho-neuroendocrinology*, 2, 159–79.

Cutler, W.B., Preti, G., Krieger, A., Huggins, G.R., Garcia, C.R. and Lawley, H.J. (1986). Human axillary secretions influence women's menstrual cycles: the role of donor extract from men. *Horm. Behav.*, 20, 463–73.

Doty, R.L. (1981). Olfactory communication in humans. *Chem. Sens.*, 6, 351–76.

Gower, D.B. (1964). Chromatographic separation of $C_{19}$-16-dehydrosteroids. *J. Chromatog.*, 14, 424–31.

Gower, D.B. and Ahmad, N. (1967). Studies on the biosynthesis of 16-dehydro steroids. The metabolism of [4-$^{14}$C]pregnenolone by boar adrenal and testis tissue *in vitro*. *Biochem. J.*, 104, 550–7.

Gower, D.B. and Bicknell, D.C. (1972). Steroid excretion and biosynthesis, with special reference to the 16-unsaturated C19 steroids in cases of testicular feminization and in a male pseudohermaphrodite. *Acta Endocr. (Kbh).*, 70, 567–81.

Gower, D.B. (1981). The biosynthesis and occurrence of 16-androstenes in man. In *Hormones in normal and abnormal human tissues*, Vol. 1, (ed. K. Fotherby and S.B. Pal), pp. 1–27. W. de Gruyter, New York.

Gower, D.B. (1984). Biosynthesis of the androgens and other $C_{19}$ steroids. In *Biochemistry of steroid hormones*, 2nd edn (ed. H.L.J. Makin), pp. 170–206. Blackwell Scientific Publications, Oxford.

Gower, D.B., Bird, S., Sharma, P. and House, F.R. (1985). Axillary 5α-androst-16-en-3-one in men and women: relationships with olfactory acuity to odorous 16-androstenes. *Experentia*, 41, 1134–6.

Gower, D.B., Nixon, A. and Mallet, A.I. (1988). The significance of odorous steroids in axillary odour. In *Perfumery: the psychology and biology of fragrance* (ed. S. van Toller and G.H. Dodd), pp. 47–76. Chapman Hall, London.

Gower, D.B., Nixon, A., Mallet, A.I. and Watkins, W.J. (1988*b*). Gas chromatography-mass spectrometry (g.c.-m.s.) as a method for simultaneous measurement of odorous steroids in the axilla in health and disease. *Chem. Sens.* (in press *a*).

Gower, D.B., Watkins, W.J., Mallet, A.I., Rennie, P.J. and Holland, K. (1988*c*). Transformations of odorous steroids by human axillary coryneform bacteria. *Chem. Sens.* (in press *b*).

Kaufman, G., Ritter, K. and Schubert, K. (1976). Quantitative determination of the boar taint substance 5α-androst-16-en-3-one in fat. *J. Steroid Biochem.*, 7, 593–7.

Kirk-Smith, M.D., Booth, D.A., Carroll, D. and Davies, P. (1978). Human social attitudes affected by androstenol. *Res. Commun. in Psychol. Psychistr. and Behav.*, 3, 379–84.

Kirk-Smith, M.D. and Booth, D.A. (1980). The effect of androstenone on choice of location in others' presence. In *Olfaction and taste*, Vol. VII (ed. H. Van der Starre), pp. 397–400. Information Retrieval Ltd, London.

Kwan, T.K., Orengo, C. and Gower, D.B. (1985*a*). Biosynthesis of androgens and pheromonal steroids in neonatal porcine testicular preparations. *FEBS Lett.*, 183, 359–64.

Kwan, T.K., Taylor, N.F., Watson, D. and Gower, D.B. (1985*b*). Gas chromatographic-mass spectometric study of metabolites of $C_{21}$ and $C_{19}$ steroids in neonatal porcine testicular microsomes. *Biochem. J.*, 227, 909–16.

Kwan, T.K., Tan, G.J.S. and Gower, D.B. (1987). 16-androstenes, putative pheromones in human semen. *Med. Sci. Res.*, 15, 1443–4.

Kwan, T.K. and Gower, D.B. (1988). Inhibition of rat testicular microsomal steroidogenesis by oxytocin and metyrapone. *Biochem. Intern.*, 16, 629–37.

Labows, J.N. (1988). Odour detection, generation and etiology in the axilla. In *Antiperspirants and deodorants* (ed. C. Felgen and K. Laden), pp. 321–43. Marcel Dekker, New York.

Leyden, J.J., McGinley, K.J., Hoelzle, E., Labows, J.N. and Kligman, A.M. (1981). The microbiology of the human axilla and its relationship to axillary odor. *J. Invest. Dermatol.*, 77, 413–6.

Michael, R.P., Bonsall, R.W. and Zumpe, D. (1976). Evidence for chemical communication in primates. *Vitamins and Hormones.*, 34, 137–186.

McClintock, M.K. (1983). Synchronizing ovarian and birth cycles in female pheromones. In *Chemical signals by vertebrates*, Vol. 3 (ed. D. Müller-Schwarze and R.M. Silverstein), pp. 159–78. Plenum, New York and London.

Nixon, A., Mallet, A.I., Jackman, P.J.H. and Gower, D.B. (1986). Testosterone metabolism by isolated human axillary Corynebacterium spp: a gas-chromatographic mass-spectrometric study. *J. Steroid Biochem.*, 24, 887–93.

Nixon, A., Jackman, P.J.H., Mallet, A.I. and Gower, D.B. (1987). Testosterone metabolism by pure and mixed cultures of human corynebacteria. *FEMS Microbiol. Lett.*, 41, 53–8.

Nixon, A., Mallet, A.I. and Gower, D.B. (1988). Simultaneous quantification of five odorous steroids (16-androstenes) in the axillary hair of men. *J. Steroid Biochem.*, 29, 505–10.

Ohloff, G., Maurer, B., Winter, B. and Giersch, W. (1983). Structural and configurational dependence of the sensory process in steroids. *Helv. Chim. Acta*, 66, 192–217.

Prelog, R. and Ruzicka, L. (1944). Untersuchungen über organextraktes mitteilung. Über zwei mosschusortig riechende steroide aus schweinentestes-extracten. *Helv. Chim. Acta*, 27, 61–6.

Preti, G., Cutler, W.B., Christensen, C.M., Lawley, H., Huggins, G.R. and Garcia, C.R. (1987). Human axillary extracts: an analysis of compounds from samples which influence menstrual timing. *J. Chem. Ecol.*, 13, 717–31.

Reed, H.C.B., Melrose, D.R. and Patterson, R.L.S. (1974). Androgen steroids as an aid to the detection of oestrus in pig artificial insemination. *Br. Vet. J.*, 130, 61–7.

Rennie, P.J., Holland, K., Mallet, A.I., Watkins, W.J. and Gower, D.B. (this volume). 16-androstene steroid content of apocrine sweat and microbiology of the human axilla. In *Chemical signals in vertebrates*, Vol. 5/9 (ed. D. MacDonald), Oxford University Press, Oxford.

Ruokonen, A., Vihko, R. and Niemi, M. (1973). Steroid metabolism in testis tissue. Concentrations of testosterone, pregnenolone, and 5α-androst-16-en-3-β-ol in normal and crytorchid rat testis and in isolated interstitial and tubular tissue. *FEBS Lett.*, 31, 321–3.

Russell, M.J., Switz, G.M. and Thompson, K. (1980). Olfactory influence on the human menstrual cycle. *Pharm. Biochem. Behav.*, 12, 737–8.

Shimizu, K. (1979). Metabolism of [17-$^{2}$H]pregnenolone into 5-[17β-$^{2}$H, 17α-$^{18}$O]androstene-3β,17α-diol and other products by incubation with the microsomal fraction of boar testis under $^{18}O_2$ atmosphere. *Biochim. Biophys. Acta*, 575, 37–45.

Shimizu, K. and Nakada, F. (1976). Formation of [17-$^{2}$H]androsta-5,16-dien-3β-ol from [17,21,21,21-$^{2}$H]pregnenolone by the microsomal fraction of boar testis. *Biochim. Biophys. Acta*, 450, 441–9.

Weusten, J.J.A.M., Smals, A.G.H., Hofman, J.A., Kloppenborg, P.W.C. and Benraad, Th.J. (1987). The sex pheromone precursor androsta-5, 16-dien-3β-ol is a major early metabolite in *in vitro* pregnenolone metabolism in human testicular homogenates. *J. Clin. Endocr. Metab.*, 65, 753–6.

# I.4 Effects of Five Odorants on Mood and Assessment of Other People

M.D. KIRK-SMITH AND D.A. BOOTH

## Introduction

It has been proposed that androstenone, 1-pyrroline, and the short chain fatty acids may be human pheromones (Amoore 1974). They are secreted by the human body from sites similar to those involved in chemical communication in other mammals (Comfort 1974), and the existence of specific anosmias for the same compounds suggests that they may have specific receptor sites on the olfactory epithelium (Amoore 1974). Androstenol and androstenone, boar pheromones secreted in human male axillary sweat and urine, have been shown to influence the ratings of others by mood and behaviour, respectively (Kirk-Smith *et al.* 1978; Filsinger *et al.* 1984; Kirk-Smith and Booth 1980).

Olfaction's strong limbic and lesser neocortical projections suggest that odorants can acquire low-awareness associations with affective states (Heimer 1975), an effect that has been experimentally demonstrated using a synthetic odour (Kirk-Smith *et al.* 1984).

Androstenone, 1-pyrroline, the short chain fatty acids and perfumes have low thresholds and may be present in sexual situations. These are conditions in which such low-level or unconscious conditioning could occur. The studies reported here investigate the elicitation of changes in mood and attitude towards others by these compounds, i.e. whether they have signalling pheromone-like properties.

## Materials and methods

The following stimuli were tested at the concentrations given:

1. Androstenone was used at a concentration of 0.25 mg
2. 1-pyrroline is a monoamine whose smell occurs in semen and pubic sweat (Amoore *et al.* 1975) (12.5 $\mu$g, prepared by enzymic oxidation of putrescine)
3. Fatty acids are found in human and primate vaginal secretions (Petri and Huggins 1975; Curtis *et al.* 1971) (0.002 ml in ratios of acetic 13, propionic 20, butyric 36, iso-butyric 8, iso-valeric 20 and iso-caproic 3)
4. Shalimar (Guerlain, Paris) was chosen as an appropriate perfume since it has been rated as similar to an 'ideal erogenous perfume' (Paukner 1965) ($0.625 \times 10^{-6}$ ml).

5 Banana essence was chosen as a control odorant since it is colourless with weak associations ($0.5 \times 10^{-5}$ ml).

A surgical mask was impregnated with each odour, made up with spectroscopic ethanol or distilled water to appropriate amounts for pipetting.

Eight men and eight women tested each odorant. For half the men and half the women an odorant-impregnated surgical mask was worn at a first session. During the second session, one week later, the mask was clean. The other half of the subjects wore an impregnated mask during their second session. At each session subjects judged 16 photographs of half-torso clothed men and women counter-balanced between subjects. Mood scales were completed before and after assessing the photographs at each session.

The photographs were rated by the following terms: sensitive, confident, good, attractive, warm, cautious, femininity, masculinity, intelligent, aggressive, interesting, sexy, soft, emotional, active and friendly. These were derived from the Semantic Differential (Osgood *et al.* 1957) and scaled using a form of magnitude estimation (Stevens 1966), where average was given the value 100, with scores of 50 meaning half and 200 twice as much as average.

Subjects' mood was rated by the terms alert/drowsy, happy/sad, aggressive/defensive, self-centered/outward-going, calm/excited, confident/unconfident, strong/weak, antagonistic/friendly, tense/relaxed, sexy/unsexy, interested/bored, withdrawn/sociable, good/bad, depressed/elated and rash/cautious. These were derived from Norris' mood scales (Norris 1971; Bond and Lader 1974) and rated using 100 mm lines.

At the end of the second session subjects' attention was drawn to the odorant, which they then characterized. Specifically anosmic subjects were screened out. The intensity and pleasantness of the odour were rated. Subjects' recognition of the source of the odour was rated on terms chosen to reflect possible odour sources, associated objects, associated emotional states and control items, e.g. concrete, perspiration, perfume, semen, musk, B.O., mustiness, man, woman.

## Results

The following significant differences observed between conditions ($p < 0.05$) for mood and photograph judgements due to the odours are given in Tables 1 and 2, respectively. The ratings and descriptions of the odour, with scores out of 100 are given in Table 3.

### MOOD

A factor analysis of the mood scores identified three main factors: sociable/friendly, strong/alert and excited. Shalimar and 1-pyrroline increased subjects' ratings of sexiness and sociability. Androstenone and the fatty acid mixture

**Table 1.** Mood changes

| Raters | Sequence[a] | Mood | Odour Present | | Odour Absent |
|---|---|---|---|---|---|
| Shalimar: | | | | | |
| All | | Sexy | | | |
| | Before | | 61.1 | * | 52.7 |
| | | | * | | |
| | After | | 52.1 | | 58.0 |
| All | | Outward-going | 64.6 | ** | 50.4 |
| Men | | Excited | 38.9 | * | 52.6 |
| Women | Friendly | | 56.8 | ** | 69.6 |
| 1-Pyrroline: | | | | | |
| All | | Sociable | | | |
| | Before | | 72.6 | ** | 60.1 |
| | | | ** | | |
| | After | | 57.8 | | 59.6 |
| Men | | Outward-going | | | |
| | Before | | 61.8 | * | 53.5 |
| | After | | 47.8 | | 50.3 |
| Androstenone: | | | | | |
| All | | Strong | 40.9 | * | 50.1 |
| Men | | Alert | | | |
| | Before | | 50.5 | ⇆ | 42.3 |
| | After | | 26.5 | | 49.8 |
| Men | | Good | 35.1 | ⇆ | 44.0 |
| Women | | Good | 47.3 | | 37.4 |
| Fatty acids: | | | | | |
| All | | Excited | 45.5 | * | 36.6 |
| Banana essence | | | | | |
| All | | Friendly | 52.9 | * | 59.9 |

[a]Mood before or after judging the photographs.
* $p<0.05$
** $p<0.01$
⇆ interaction significant—no individual means significantly different.

increased the strength/alertness factor and excitability, respectively. Banana essence reduced friendliness, its only effect.

### JUDGEMENT OF PHOTOGRAPHS

A factor analysis of the photograph judgements identified a specific sexually attractive factor and a more general emotional/soft factor. Shalimar and 1-pyrroline increased the sexually attractive factor of judgement. Androstenone

**Table 2.** Ratings of people

| Judges | Photos | Assessment | Odour Present | | Absent |
|---|---|---|---|---|---|
| **Shalimar:** | | | | | |
| All | All | Sexy | | | |
| 88.5 | * | 83.7 | | | |
| All | All | Soft | 93.4 | * | 90.4 |
| All | All | Defensive | 90.9 | * | 85.0 |
| **1-Pyrroline:** | | | | | |
| Men | Women | Attractive | 117.8 | ** | 93.1 |
| Women | Men | Attractive | 100.0 | * | 86.1 |
| Women | Men | Interesting | 104.7 | ** | 89.4 |
| Men | Men | Interesting | 80.6 | ** | 97.1 |
| All | All | Warm | 105.0 | * | 97.4 |
| Men | All | Soft | 90.6 | | 85.7 |
| Women | All | Hard | 89.4 | ⇆ | 92.6 |
| All | Men | Emotional | 90.8 | | 86.7 |
| All | Women | Emotional | 105.5 | ⇆ | 110.4 |
| **Androstenone:** | | | | | |
| All | All | Attractive | 91.0 | * | 96.7 |
| Women | All | Intelligent | 87.4 | ** | 112.1 |
| Women | Women | Interesting | 81.1 | * | 100.5 |
| Women | Men | Soft | 75.7 | * | 88.8 |
| **Fatty acids:** | | | | | |
| All | All | Good | 95.2 | * | 108.3 |
| All | Men | Soft | 75.1 | ** | 85.6 |
| Men | All | Aggressive | 108.6 | * | 87.5 |
| Men | Men | Warm | 79.7 | ** | 97.1 |
| **Banana essence:** | | | | | |
| All | All | Intelligent | 105.3 | * | 108.1 |
| All | All | Interesting | 95.3 | * | 101.6 |
| All | All | Soft | 87.1 | * | 99.5 |
| All | All | Emotional | 98.9 | * | 108.6 |

* $p<0.05$
** $p<0.01$
⇆ interaction significant—no individual means significantly different.

**Table 3.** Odour ratings and description

| |
|---|
| Shalimar<br>Pleasant 59; Strength 20<br>Perfume 51, woman 26, sweet 26, musk 24, flowers 24, sex 19. |
| 1-Pyrroline<br>Pleasant 42; Strength 22<br>Musty 17, ammonia 17, animal 17, perspiration 13, underarm sweat 12, fish 12, man 10, body odour 10, musk 10, semen 9. |
| Androstenone<br>Pleasant 45; Strength 39<br>Musty 35, perspiration 32, animal 29, man 28, underarm sweat 25, body odour 24. |
| Fatty acids<br>Pleasant 39; Strength 34<br>Animal 38, musty 35, man 32, fish 29, stress 26, body odour 26. |
| Banana essence<br>Pleasant 58; Strength 39<br>Fish 24, animal 22, musty 22, perfume 21, perspiration 20, flowers 18. |

and the fatty acids had a negative effect on sexual or general factors of personal assessment, the effect being greater in female subjects for androstenone and greater in male subjects for the fatty acids. The banana essence also reduced ratings on general assessment of the photographed people.

## Discussion

The effects on mood and judgement appear to be related to the associations and strengths of the odorants. Shalimar and 1-pyrroline were perceived as weak in strength and had positive, sexual or unrecognized connotations.

Androstenone and the fatty acid mixture were strongly related to underarm sweat and perspiration, presumably with negative connotations. The experimenter was male and this may have cued responses to the odorants, e.g. the term man was given a significantly higher score than the term woman in the description of the fatty acid odour.

Shalimar and the banana essence were rated equally pleasant, yet caused very different effects. The less specific and sometimes negative effects of the latter possibly reflected incongruity of the association of this scent with people.

Shalimar exerted a marked and specific influence on subjects' attitude and mood, without their recognition of its presence. The main effects were to increase all subjects' sexy and soft ratings of the opposite sex, with subjects' mood becoming sexier. These positive changes of sexual factors appear to be related to the subjects' association of the odour with perfume and women.

The broad similarity in responses between men and women appears to be more consistent with the hypothesis that the responses have been learned than with the hypothesis of an innate process for which major differences in response might be expected to have had a selective value.

These studies indicate that hedonics are not sufficient to explain the effect of low levels of odorants on mood and judgement. Context, affective associations and cognitive set must also be considered (Kirk-Smith and Booth 1987; Baron 1988).

## References

Amoore, J.E. (1974). Four primary odour modalities of man: Experimental evidence and possible significance. In *Olfaction and taste*, V (ed. D.A. Denton and P. Coughlan), pp. 283–89. Academic Press, New York.

Amoore, J.E., Forrester, L. and Buttery, R.G. (1975). Specific anosmia to 1-pyrroline: The spermous primary odour. *J. Chem. Ecol.*, 299–310.

Baron, R.A. (1988). Perfume as a tactic of impression management in social and organisational settings. In *Perfumery: the psychology and biology of fragrance* (ed. S. Van Toller and G.H. Dodd), pp. 91–104. Chapman and Hall, London.

Bond, A. and Lader, M. (1974). The use of analogue scales in rating subjective feelings. *Br. Med. Psychol.*, 211–8.

Comfort, A. (1974). The likelihood of human pheromones. In *Pheromones* (ed. M.C. Birch). North Holland, Amsterdam.

Curtis, R.F., Ballantine, A., Keverne, E.B., Bonsall, R.W. and Michael R.P. (1971). Identification of primate sexual pheromones and the properties of synthetic attractants. *Nature*, 396–8.

Filsinger, E.E., Braun, J.J., Monte, W.C. and Linder, D.E. (1984). Human responses to the pig sex pheromone 5 alpha-androst-16-en-3-one. *J. Comp. Psychol.*, 98, 219–22.

Heimer, L. (1975). Olfactory projections to the diencephalon. In *Anatomical neuroendocrinology* (ed. W.E. Stumpf and L.D. Grant), pp.30–39. Karger, Basle.

Kirk-Smith, M.D. and Booth, D.A. (1980). Effect of androstenone on choice of location in others' presence. In *Olfaction and taste*, VII (ed. H. van der Starre), pp. 397–400. Information Retrieval Ltd, London.

Kirk-Smith, M.D. and Booth, D.A. (1987). Chemoreception in human behaviour: Experimental analysis of the social effects of fragrances. *Chem. Senses*, 12(1), 159–66.

Kirk-Smith, M.D., Booth, D.A., Carroll, D. and Davies, P. (1978). Human social attitudes affected by androstenol. *Res. Comm. Psychol. Psychiat. Behav.*, 3, 379–84.

Kirk-Smith, M.D., Van Toller, C. and Dodd, G.H. (1984). Unconscious odour conditioning in human subjects. *Biol. Psychol.*, 17, 221–31.

Norris, H. (1971). The action of sedatives on brain stem oculomotor systems in man. *Neuropharmacology*, 10, 181–91.

Osgood, C.E., Suci, G. and Tannenbaum, P.H. (1957). The measurement of meaning. University of Illinois Press, Urbana.

Paukner, E. (1965). Success and failure of odour classification applied to reactions to erogenous odours. *J. Soc. Cosmet. Chem.*, 16, 515–26.

Preti, G. and Huggins, G.R. (1975). Cyclical changes in volatile acidic metabolites of human vaginal secretions and their relation to ovulation. *J. Chem. Ecol.*, 1, 361–76.

Stevens, S.S. (1966). A metric for the social consensus. *Science*, 151, 530–41.

# I.5 16-Androstene Content of Apocrine Sweat and Microbiology of the Human Axilla

P.J. RENNIE, K.T. HOLLAND, A.I. MALLET, W.J. WATKINS AND D.B. GOWER

## Introduction

In recent years there has been increasing interest in the role of axillary odour in human olfactory communication. Body odour, which provides information as to gender and emotional state, is emitted from areas which are rich in apocrine sweat glands such as the axillary, ano-genital, sternal, and areolar region (for a review see Gower *et al.* 1988). These hirsute regions have been suggested as being ideal sites for odour production and dispersal (Montagna and Parakkal 1974).

In the quest for human pheromones an analogy has been postulated between the primer/releaser pheromones of the pig, 5α-androstenone and 3α-androstenol. The likelihood that these are indeed human pheromones has been discussed by Brooksbank *et al.* (1974) and Gower *et al.* (1986). 3α-Androstenol has been found in the urine and sweat of the human male (Brooksbank 1962; Brooksbank *et al.* 1974) and the presence of 5α-androstenone in male axillary secretions has been reported by Claus and Alsing (1976) and Bird and Gower (1981), although Labows *et al.* (1979) were unable to detect either of these 16-androstenes in pure apocrine secretion.

The characteristic 'axillary' odour appears to be released by the action of bacteria, present in the axilla, on the apocrine secretion. Pure apocrine secretion is thought to be odourless (Labows *et al.* 1979). However, when Shehadeh and Kligman (1963) cultured a variety of axillary bacteria in apocrine sweat the 'axillary' odour was produced only by resident corynebacteria and coagulase-negative micrococci and not by Gram-positive micro-organisms. This work was substantiated by Leyden *et al.* (1981) who showed a clear association between axillary microflora and intensity of axillary odour. Subjects with a coryneform-dominated microflora produced a higher level of axillary odour than those subjects with a *Micrococcaceae*-dominated axillary microflora.

We have attempted to establish, using gas chromatography-mass spectroscopy (gc-ms), the presence of various odorous 16-androstene steroids in pure apocrine secretions and in ether skin washings. We have assigned axillary odour intensity to axillary microflora and looked for correlation of odour intensity with the presence of odorous steroids in the apocrine extracts.

## Methods

Volunteers (9 male and 2 female) were recruited and asked to avoid the use of deodorants or antiperspirants for five days prior to the sweat collection. Approximately 12 h before treatment they were required to shave one axilla, wash both axillae with non-perfumed soap, and dry with clean paper towels. Clean clothing was then worn. At the time of collection, subjects were asked to lie supine with the axillae exposed. Underarm odour (UAO) was determined organoleptically and scored on a scale of 0–4. The axillary microflora was sampled using a modification of the method of Williamson and Kligman (1965). Immediately after collection the samples were serially diluted and plated on to selective media. The aerobic bacteria were then incubated for three days at 37°C and anaerobic bacteria for five days also at 37°C. Colonies were counted and the number of micro-organisms calculated as colony-forming units/$cm^2$ of skin. Representative colonies were picked off and subcultured for identification.

Apocrine sweating was induced by intra-dermal injection of 0.05 ml adrenaline, diluted 1:1000, into the vault of the shaved axilla. Droplets of apocrine sweat were collected into 2 $\mu$l Microcaps, the volume was determined, and the tubes were dropped into sterile bijou bottles containing internal standards (100 ng each of 5$\alpha$-androstanone and 3$\beta$-androstanol in 1 ml ethanol). Samples were stored at −20°C until analysed.

In order to determine the presence of 16-androstenes on the skin surface, the cup-scrubbing method (Williamson and Kligman 1965) was employed using 5 ml ether. The ether was recovered into bottles containing the internal standards and stored at −20°C to await analysis.

Samples were prepared for gc-ms and the spectra determined by the method of Nixon *et al.* (1988), with the exception of the 16-androstenols which were derivatized to the *tert*-butyldimethylsilyl ethers (Mallet *et al.* 1986) and monitored at $M^+ - 57$ (due to loss of the *tert*-butyl moiety) as follows: 5,16-androstadienol, 329 m/z; 3$\alpha$-androstenol, 331 m/z; 3$\beta$-androstenol, 331 m/z; and the internal standard, 3$\beta$-androstanol at 333 m/z. Data arising from the scans were processed using Teknivent (Vector/One System Options).

## Results

Table 1 shows the classification of subjects into underarm odour (UAO) types. The correlation between axillary odour score and mean bacterial count of organisms isolated from the axilla is shown in Fig. 1.

For low odour subjects, coryneform bacteria, *Micrococcacae*, and propionibacteria, counts were (mean of $\log_{10}$) 3.18, 4.61, and 4.29, respectively. Corresponding values for medium/high odour subjects were 5.50, 5.06, and 4.20, respectively. All low odour subjects had *Micrococcaceae*-dominated microflora whereas three out of five high odour subjects had coryneform-dominated flora.

**Table 1.** Classification of subjects into UAO types

| Description of UAO | none | slight | moderate | strong | very strong |
|---|---|---|---|---|---|
| Odour score | 0 | 1 | 2 | 3 | 4 |
| Odour type | low | low | medium | high | high |

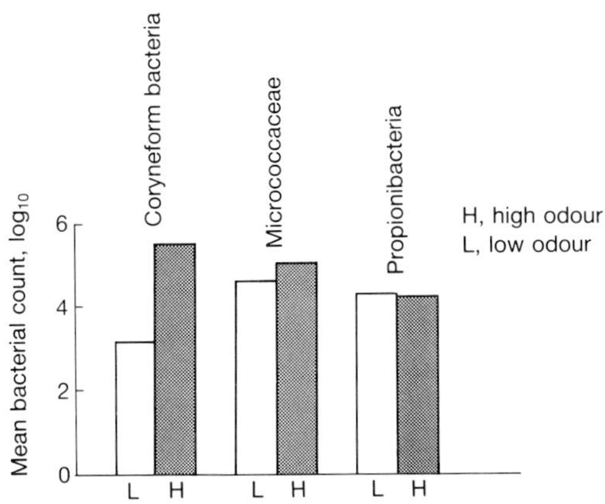


Fig. 1. Axillary microflora in human subjects.

The injection of adrenaline into the axillary vault resulted, in most individuals, in secretion of sweat from the apocrine glands. This could be clearly distinguished from eccrine secretion. Volumes collected ranged from 0.63 $\mu$l to 0.34 $\mu$l.

The results of selected ion monitoring of the 16-androstene derivatives by gc-ms in the apocrine secretions is shown in Table 2a. 5$\alpha$-Androstenone was found in the secretion of all subjects and the content ranged from 46 to 630 ng/$\mu$l of sweat; 4,16-androstadienone was found in four of the seven male apocrine collections and in both females; again there was a wide range of content. Of the 16-androstenols, 3$\beta$-androstenol and 5,16-androstadienol were found in the sweat of two males and in one of the female subjects; in the other female only 5,16-androstadienol was detected. 3$\alpha$-androstenol (18 ng/$\mu$l) was detected in the sweat of one female subject but not in the secretions of the other individuals. When ether skin washings were analysed for 16-androstene content (Table 2b), 4,16-androstadienone was present in greater amounts than the other 16-androstenes. The inter-subject variability of the individual 16-androstenes was generally less than that found in the apocrine secretions.

**Table 2(a)** Range of 16-androstenes (ng/$\mu$l) present in adrenaline-stimulated apocrine secretions from 9 subjects (7 male and 2 female)

| | |
|---|---|
| 5α-Androstenone | 46–603 |
| 4,16-Androstadienone | 62–764 (ND in 3 males) |
| 3α-Androstenol | 18 (only detected in one female) |
| 3β-Androstenol | 8–51 (ND in 5 males and 1 female) |
| 5,16-Androstadienol | 13–285 (ND in 5 males and the other female) |

ND = not detected

**Table 2(b)** Range of 16-androstenes (ng/cm$^2$) present in ether skin washings from 6 male subjects

| | |
|---|---|
| 5α-Androstenone | 0.5–1.2 (ND in 1 subject) |
| 4,16-Androstadienone | 2.8–7.2 |
| 3α-Androstenol | 0.4–7.0 |
| 3β-Androstenol | 0.1–1.9 |
| 5,16-Androstadienol | 0.2–1.1 |

No correlation between odour score and 16-androstene content of apocrine secretion was noted for the low and medium odour subjects; however one female with a high odour score did have higher concentrations of 5α-androstenone, 4,16-androstadienone, and 5,16-androstadienol than other subjects. One male with medium odour score had higher levels of the 16-androstenols than the other subjects in the ether skin washings and one subject who had been taking tetracycline 250 mg b.d. routinely had levels toward the lower end of the range.

## Discussion

The association of strong axillary odour with coryneform-dominated microflora has been noted previously (Leyden *et al.* 1981). In individuals with a coryneform-dominated axillary microflora, Jackman (1982) found the mean total aerobic bacterial count was five times higher than those with a micrococcal-dominated flora. A coccal-dominated axillary microflora was associated with shaving the axilla and the use of deodorants.

In the present study, where the use of deodorants was proscribed for five days before sampling, high-odour subjects were found to have a mean total aerobic bacterial count around 1.5 times higher than those with low UAO. The mean count of micrococcal organisms and propionibacteria was not significantly different for the two groups. It was unfortunate, for us, that none of the males and only one female had a high UAO score.

Using gc-ms (s.i.m.) and with authentic steroids for comparison, we found measurable quantities of 5α-androstenone in the sterile apocrine secretions of all our subjects (Table 2a). In some subjects, other odorous 16-androstenes such as 3β-androstenol, 5,16-androstadienol, and 4,16-androstadienone, were also found. These steroids have been quantified in human axillary sweat and hair, but to our knowledge this is the first indication that they are present in sterile apocrine secretions. In earlier work Labows *et al.* (1979) were unable to detect any of this group of steroids in freshly produced apocrine secretions. 4,16-Androstadienone was found in high concentrations in the apocrine secretion of some individuals, but this steroid was found in highest concentration in the ether skin washings of all the subjects. A similar situation occurs with axillary hair (Nixon *et al.* 1988). While 3α-androstenol was found only in the apocrine secretion of one female subject, it was present in higher concentrations than other androstenols in the axillary skin washings. It is significant that two axillary coryneform bacterial isolates converted 4,16-androstadienone into appreciable quantities of 3α-androstenol, whereas the former steroid was produced from 5α-androstenone, 3β-androstenol, and 5,16-androstadienol to a relatively minor extent (Rennie *et al.* 1989). These results could be consistent with 4,16-androstadienone being a precursor of other 16-androstenes, in particular 3α-androstenol, by axillary coryneform bacteria. 3α-Androstenol may be of special interest in the context of human social interaction, since it has been shown to have effects on judgement and change of mood (see Gower *et al.* 1988).

There did not appear to be any relationship between odour score and 16-androstene content of pure apocrine secretion. However, one subject with a medium score had higher concentrations of the three 16-androstenols, especially 3α-androstenol, in ether skin washings. It should be remembered, nevertheless, that there are numerous other odorous substances in axillary sweat, and the 16-androstenes are considered to make only some contribution to overall axillary odour.

## Acknowledgements

We are indebted to Prof. R.G. Spector for performing the adrenaline injections. The financial support (of W.J.W.) of the Dunhill Medical Trust is gratefully acknowledged.

## Abbreviations

| | |
|---|---|
| 5α-Androstenone | 5α-androst-16-en3-one |
| 4,16-Androstadienone | 4,16-androstadien-3-one |
| 3α-Androstenol | 5α-androst-16-en-3α-ol |
| 3β-Androstenol | 5α-androst-16-en-3β-ol |

| | |
|---|---|
| 5,16-Androstadienol | 5,16-androstadien-3$\beta$-ol |
| 5$\alpha$-Androstanone | 5$\alpha$-androstan-3-one |
| 3$\beta$-Androstanol | 5$\alpha$-androstan-3$\beta$-ol |

## References

Bird, S. and Gower, D.B. (1981). The validation and use of a radioimmunoassay for 5$\alpha$-androst-16-en-3-one in human axillary collections. *J. Steroid Biochem.*, 14, 213–19.

Brooksbank, B.W.L. (1962). Urinary excretion of androst-16-en-3$\alpha$-ol levels in normal subjects, and effects of treatment with trophic hormones. *J. Endocrinol.*, 24, 435–44.

Brooksbank, B.W.L., Brown, R. and Gustafsson, J.-A. (1974). The detection of 5$\alpha$-androst-16-en-3-ol in human axillary sweat. *Experientia*, 30, 864–5.

Claus, R. and Alsing, W. (1976). Occurrence of 5$\alpha$-androst-16-en-3-one, a boar pheromone in man and its relationship to testosterone. *J. Endocrinol.*, 68, 483–4.

Gower, D.B., Nixon, A., Jackman, P.J.H. and Mallet, A.I. (1986). Transformations of steroids by axillary coryneform bacteria. *Int. J. Cosmet. Sci.*, 8, 149–58.

Gower, D.B., Nixon, A. and Mallet, A.I. (1988). The significance of odorous steroids in axillary odour. In *Perfumery; the psychology and biology of fragrance* (ed. S. Van Toller and G.H. Dodd). Chapman and Hall, London and New York. pp. 47–76.

Labows, J.N., Petri, G., Hoelzle, E., Leyden, J. and Kligman, A. (1979). Steroid analysis of human apocrine secretion. *Steroids*, 34, 249–58.

Leyden, J.J., McGinley, K.J., Hoelzle, E., Labows, J.N. and Kligman, A.M. (1981). The microbiology of the human axilla and its relationship to axillary odor. *J. Invest. Dermatol.*, 77, 413–6.

Mallet, A.I., Barr, R.N. and Newton, J.A. (1986). Analysis of dihydroxyeicosatetranoic acids by gas chromatography-mass spectrometry. *J. Chromatogr.*, 378, 194–200.

Montagna, W. and Parakkal, P.F. (1974). In *The structure and function of skin.* Academic Press, New York, pp. 332–65.

Nixon, A., Mallet, A.I. and Gower, D.B. (1988). Simultaneous quantification of five odorous steroids (16-androstenes) in the axillary hair of men. *J. Steroid Biochem.*, 29, 505–10.

Rennie, P.J., Holland, K.T., Mallet, A.I., Watkins, W.J. and Gower, D.B. (1989). Interconversion of androst-16-ene steroids by human axillary aerobic coryneform bacteria. In *Chemical senses* (in press).

Shehadeh, N. and Kligman, A.M. (1963). The bacteria responsible for axillary odour. *J. Invest. Dermatol.*, 41, 3.

Williamson, P. and Kligman, A.M. (1965). A new method for the quantitative investigation of cutaneous bacteria. *J. Invest. Dermatol.*, 45, 498–503.

# I.6 Use of Selected Ion Monitoring in the Identification of Human Seminal 16-androstenes

T.K. KWAN AND D.B. GOWER

## Introduction

There is accumulating evidence to suggest that the odorous 16-androstenes may be human pheromones (for recent review see Gower *et al.* 1988). Some of the evidence has been criticized, however, on the basis that inadequate controls were used (see Labows 1988). Nevertheless, there are some recent data implicating the odour of axillary sweat in the alteration of menstrual cycle length (Cutler *et al.* 1986; Preti *et al.* 1986). The active principle involved is not known yet, but it is conceivable that one or more of the odorous 16-androstenes present (Nixon *et al.* 1988) may play a role in this intriguing process. We were thus prompted to analyse human semen for the possible presence of these steroids.

In this communication, mass spectrometric evidence is presented, using a modification of methods which have been published elsewhere (Nixon *et al.* 1988; Gower *et al.* 1989).

## Materials and methods

Human seminal fluids were collected from 11 individuals (whose sperm count was below $50 \times 10^6$/ml). These were pooled into one sample labelled A; samples B and C were from two different men whose sperm counts were greater than $60 \times 10^6$/ml. These semen samples were extracted three times, with ethyl acetate ($3 \times 5$ ml). The pooled ethyl acetate extracts were evaporated to dryness in a Buchler vortex-evaporator at 20 °C and then purified by the method of Nixon *et al.* (1986). Ethyl acetate (0.5 ml) was added to the dried residue; the resultant mixture was vortex-mixed and purified by running through a 2-part column. The upper part consisted of a Pasteur pipette containing anhydrous $Na_2SO_4$. The tip of this pipette fitted tightly on to a Florisil Sep-Pak cartridge. The eluate (6 ml) was evaporated to dryness, redissolved in methanol (30 $\mu$l) and vortex-mixed. This solution (20 $\mu$l) was injected into a high performance liquid chromatograph (HPLC Technology Model RR/035) equipped with Spherisorb 5 ODS column (25 cm × 4.6 mm, i.d.); methanol/dichloromethane (95:5, v/v; 0.9 ml/min) was used as mobile phase. The fractions between 3 and 11 ml were taken to ensure that all the

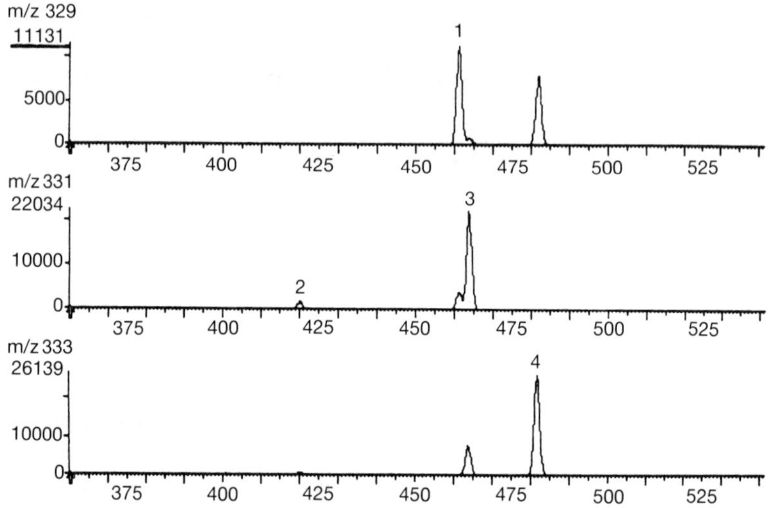


Fig. 1. Selected ion monitoring of 16-androstenols and 5α-androstan-3β-ol (TBDMS ethers). 1, androstadienol; 2, 3α-androstenol; 3, 3β-androstenol; 4, 5α-androstan-3β-ol. There is some cross-contamination of peak 4 in the m/z 329 channel and peak 3 in the m/z 333 channel.

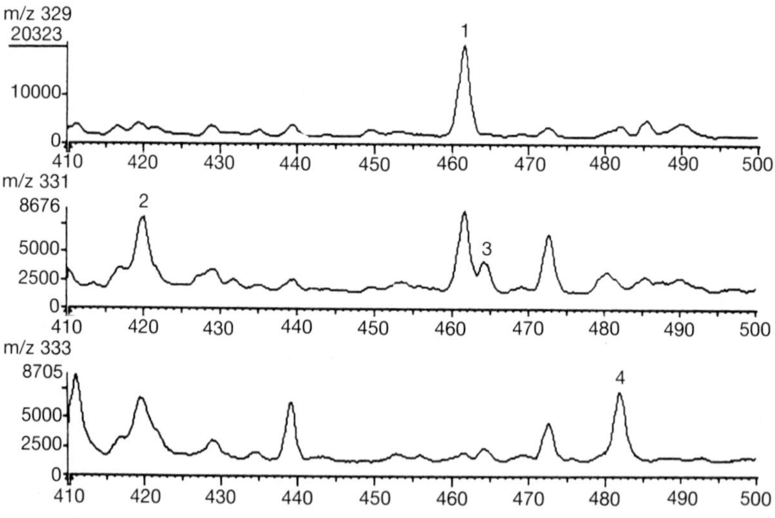


Fig. 2. Selected ion monitoring of semen extracts from healthy men. Identification of peaks as in Fig. 1. (Between ng and pg levels of steroids were analysed.)

16-androstenes were collected, since pilot studies had shown that 5α-[5α,6α-$^3$H] androstenone was eluted between 5 and 9 ml. The pooled eluate was evaporated to dryness, as before, and redissolved in 0.5 ml ethyl acetate. The solution was transferred to silanized vials and solvent removed using a Savant Speed Vac concentrator at 20 °C. The purified steroid solution was subjected to derivitization as *tert*-butyldimethylsilyl (TBDMS) ethers, as described previously (Gower *et al.* 1988).

The seminal steroids (2–5 μl) were analysed using a Packard Becker 419 gas chromatograph equipped with a flame-ionization detector and a fused silica OV-1 bonded column (25 m × 0.32 mm, i.d.). The column oven temperature was programmed between 175 and 295 °C. Two temperature ramps were used as follows: the initial temperature of 175 °C for 80 s was followed by a rise of 7.5 °C/min. for 320 s, after which a temperature rise of 30 °C/min. was used for 480 s. Tentative evidence for the presence of 5α-androst-16-en-3α-ol (3α-androstenol), 5,16-androstadien-3β-ol (androstadienol), 5α-androst-16-en-3β-ol (3β-androstenol), and 5α-androstan-3β-ol were noted in the pooled specimen A.

## Gas chromatography-mass spectrometry

Confirmation of results was sought using gas chromatography-mass spectrometry. A fused silica column (CP Sil 5; 25 m × 0.33 mm, i.d.), housed in a Varian 3400 gas chromatograph (F.I.D. detector), was directly interfaced into a VG MM-305 mass spectrometer equipped with a Technivent data system. A three-tier temperature ramp was used as follows: 175 °C for 2 mins followed by a 30 °C/min. rise to 240 °C and finally a 10 °C/min. rise up to 300 °C.

$[M - 57]^+$ ions were monitored at m/z 329, 331, and 333 for the TBDMS ethers of androstadienol, 3α(β)-androstenols, and 5α-androstan-3β-ol, respectively, where m/z 57 corresponds to loss of the *tert*-butyl group.

## Results

Fig. 1 shows the selected ion monitoring traces of authentic 16-androstenols and 5α-androstan-3β-ol (as TBDMS ethers). 3α-Androstenol, androstadienol, 3β-androstenol, and 5α-androstan-3β-ol (as TBDMS derivatives) of the pooled semen A eluted at 420, 462, 464, and 482 s, respectively (Fig. 2). The specimen analysis compared well with the retention times of the authentic standards. Similar elution patterns were obtained with samples B and C.

## Discussion

The finding in human semen of some 16-androstenes, opens up perspectives for further study and speculation. The 16-androstenes may be absorbed through the vaginal epithelium, by analogy with the absorption of substances through the buccal or nasal epithelium (Anand Kumar *et al.* 1977). Transport of the steroids to the hypothalamus could provide an effect on release of luteinizing hormone-releasing hormone and hence regulation of ovulation. Such a mode of absorption of 16-androstenes, as indicated above, would by-pass the portal circulation and prevent hepatic catabolism of the absorbed steroids.

Further studies are already in progress to investigate semen samples from more men in order to see if there are correlations between 16-androstene content and fertility or infertility.

## Acknowledgement

T.K.K. thanks the Wellcome Trust for a Fellowship.

## References

Anand Kumar, T.C., David, G.F. and Puri, V. (1977). Ovulation in rhesus monkeys suppressed by intranasal administration of progesterone and norethisterone. *Nature*, 270, 532–3.

Cutler, W.B., Preti, G., Krieger, A., Huggins, G.R., Garcia, C.R., and Lawley, H.J. (1986). Human axillary secretions influence women's menstrual cycles: the role of donor extract from men. *Horm. Behav.*, 20, 463–73.

Gower, D.B., Nixon, A. and Mallet, A.I. (1988). 'The significance of odorous steroids in axillary odour.' In *Perfumery: the psychology and biology of fragrance* (ed. S. van Toller and G.H. Dodd), pp. 47–76. Chapman and Hall, London.

Gower, D.B., Nixon, A., Watkins, W.J. and Mallet, A.I. (1989). Gas chromatography-mass spectometry (g.c.-m.s.) as a method for simultaneous measurement of odorous steroids in the axilla in health and disease. *Chem. Senses*, in press.

Labows, J.N. (1988). Odour detection, generation and etiology in the axilla. In *Antiperspirants and deodorants* (ed. C. Felgen and K. Laden), pp. 321–43. Marcel Dekker, New York.

Nixon, A., Mallet, A.I., Jackman, P.J.H. and Gower, D.B. (1986). Testosterone metabolism by isolated human axillary Corynebacterium spp: a gas-chromatographic mass-spectrometric study. *J. Steroid Biochem.*, 24, 887–93.

Nixon, A., Mallet, A.I. and Gower, D.B. (1988). Simultaneous quantification of five odorous steroids (16-androstenes) in the axillary hair of men. *J. Steroid Biochem.*, 29, 505–10.

Preti, G. Cutler, W.B., Garcia, C.R., Huggins, G.R. and Lawley, H.J. (1986). Human axillary secretions influence women's menstrual cycles: the role of donor extract of females. *Horm. Behav.*, 20, 474–82.

# I.7 Analysis of the Anal Sac Secretion of the Hyaenidae

ALAN J. BUGLASS, FRANCES M.C. DARLING AND JOHN S. WATERHOUSE

## Introduction

We report our analyses of the anal sac secretion of three different species of hyaena: brown (*Hyaena brunnea*), striped (*H. hyaena*) and spotted (*Crocuta crocuta*). All three species deposit a glandular secretion on grass stalks or, if confined to a cage, on a suitable object such as a water trough. This scent-marking behaviour, known as pasting, plays a crucial role in the social behaviour of these animals. Wheeler *et al.* (1975) reported the presence of an unusual sulphur-containing compound, 5-thiomethylpentane-2,3-dione (structure (1) in Fig. 1) in the anal sac secretion of the striped hyaena, together with a compound of molecular weight 224 identified as an unsaturated lactone of probable structure (2) (Fig. 1; Von Endt 1975; Lamb 1980).

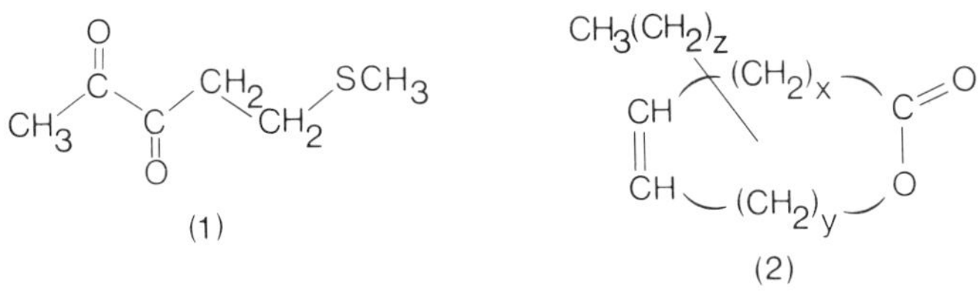

Fig. 1.

In their paper on the scent-marking behaviour of the brown hyaena, Mills *et al.* (1980) report finding two secretions from the anal sac: a white lipid-rich secretion produced by sebaceous glands situated centrally in the sac; and a black secretion produced by circular areas consisting almost entirely of epocrine, sudoriferous tissue. The function of these secretions is unknown although it has been suggested that the white mark, which has a longer lasting odour, may be used for territorial marking, whilst the black mark, which loses its odour relatively quickly, may be used to pass information between members of the same group (Mills 1978). Gas chromatographic analyses by Mills *et al.* of both these secretions from two different individuals revealed that there is a significant difference in the chemical components of the white and black secretions; although there is no difference in compounds between individuals, there is a marked difference in the relative amounts of these compounds for both white and black secretions.

Unpublished work of Lamb (1980) reported the presence of compound (*1*), as well as squalene, 2-tridecanone, octadecanoic acid, hexadecanoic acid, and possibly an unsaturated $\gamma$-undecalactone in the anal sac secretion of the brown hyaena (whether white or black is not specified). As in the other studies, analyses were carried out on dichloromethane solutions of the whole secretion (direct injection). The presence of (*1*) in both the brown and striped hyaena leads to the expectation that it might also be found in the spotted hyaena. One of our aims was to investigate this possibility. We also investigated three different methods of sampling for the chromatographic analysis: dissolving the secretion in dichloromethane (direct injection) as in the previous studies; collection of headspace volatiles; and collection of salted-out headspace volatiles. Our results allow us to compare these three methods, and they also provide further information on the nature of these semiochemically important secretions. Our aim was not to investigate differences between individual animals, because the methods of collection did not allow us to obtain information on their age, gender, etc. However, our own results and those of Mills *et al.* (1980) show that there should be no significant difference in the identity of compounds in different individuals of the same species.

## Methods

Pasted samples of the anal sac secretions of brown and striped hyaena were collected from San Diego Zoo, California, and from Port Lympne Zoo, Kent. Secretions of the brown hyaena were white upon collection but darkened before analysis. (N.B. All the analyses on the brown hyaena were carried out on the white and not the black secretion as this was unobtainable.) Spotted hyaena samples came from Dudley Zoo, West Midlands.

Samples for analyses were prepared as follows. Dichloromethane solutions (5 ml) of the whole secretion (1 g) were obtained and filtered prior to GC injection. Headspace volatiles were collected, using a Grob dynamic headspace apparatus (Grob and Zurcher 1976; Hachenberg and Schmidt 1986) on to an activated charcoal disc, from which the components were extracted into dichloromethane (35 $\mu$l). Headspace volatiles were also collected from a suspension of the secretion (500 mg) in saturated sodium sulphate solution (25 ml) for a period of 2 h, this being the salted-out procedure.

Chromatographic analyses were carried out on a Hewlett-Packard 5890 gas chromatograph equipped with a J & W DB-1 capillary column (length, 15 m; i.d. 0.25 mm) and connected to a 5970 Hewlett-Packard mass spectrometer detector. Sample volumes of 1 $\mu$l were injected under splitless conditions using a solenoid valve and a helium carrier gas flow rate of 2 ml per minute. The temperature programme used was 40° to 250° at 5°C per minute.

**Table 1.** Components of the anal sac secretion of brown, striped, and spotted hyaena

| Compound | Brown | Striped | Spotted |
|---|---|---|---|
| Hydrocarbons: | | | |
| Undecane | | | b |
| Dodecane | | | b |
| Tridecane | | | b |
| Pentadecane | | | b |
| 1,2-Dimethylbenzene | | b | |
| Isomers of ethylmethylbenzene | | b | |
| Alcohols: | | | |
| Hexan-1-01 | | | b |
| Nonan-4-01 | a b c | | |
| Aldehydes and Ketones: | | | |
| Nonanal | | b c | a b c |
| Decanal | | | b |
| 2,2-Dimethylpent-4-enal | a b c | c | |
| Benzaldehyde | b c | b c | |
| 2-Methyl-2,5-cyclohexadiene-1,4-dione | b c | | |
| Carboxylic acids, esters and lactones: | | | |
| Butanoic acid | b | b | |
| Pentanoic acid | b | b | |
| Hexanoic acid | a b c | b c | a c |
| Heptanoic acid | a b c | | c |
| Octanoic acid | b c | b c | a c |
| Nonanoic acid | | | c |
| Decanoic acid | | b c | a c |
| Undecanoic acid | | | c |
| Dodecanoic acid | a | b | a c |
| Tetradecanoic acid | a | c | c |
| Pentadecanoic acid | a | | |
| Hexadec-9-enoic acid | | | c |
| Hexadecanoic acid | a b c | a c | a c |
| Ethyl pentanoate | | c | |
| Methyl hexanoate | b c | | c |
| Methyl octanoate | | | b c |
| Methyl decanoate | | | b c |
| Methyl undecanoate | | | c |
| Methyl dodecanoate | | | a b c |
| Methyl tetradecanoate | | | a b c |
| Methyl hexadecanoate | | | b |
| Heptyl methanoate | | | b |
| Unsaturated lactone (*2*) | | a b c | |
| Possible isomer of (*2*) | a | b c | |

| Compound | Brown | Striped | Spotted |
|---|---|---|---|
| Sulphur compounds: | | | |
| 5-Thiomethylpentane-2,3-dione (*1*) | a b c | a b c | |
| Unknown sulphur compound (mol. wt.) 146) | a b | b c | |
| Dimethyltrisulphide | | b c | |

a = Direct injection, b = Headspace, c = Salted-out headspace

## Results and discussion

The results of our analyses on dichloromethane solutions of secretion (direct injection) (a), headspace volatiles (b), and salted-out headspace volatiles (c) are shown in Table 1. Compounds were identified mainly by comparison of their mass spectra with those of authentic compounds; therefore some of the identifications should be considered tentative and further work, such as co-injection on GC with authentic compounds, is required for confirmation. Significantly, all three methods show compound (*1*) to be absent in the spotted hyaena. Compound (*1*) is a major component of the secretion of the striped hyaena, but a relatively minor component of that of the brown hyaena. Its significance as a semiochemical has yet to be determined. Furthermore, the unsaturated lactone (*2*) is observed in the anal sac secretion of the striped hyaena only, where it is a major component. It is obvious that a large number of compounds are present in all three species, but surprisingly we find only five common to all three. Of the sixteen compounds that we have listed for the brown hyaena, only (*1*) and hexadecanoic acid were reported by Lamb (1980), and we were unable to detect his four other compounds in this study.

It is clear from our investigations that the analytical results depend considerably on the method. Of the three methods used, we found that the salted-out headspace technique produced the largest number of compounds on the GC trace. This is due to the increased concentration of volatile components in the headspace caused by increased ionic strength of the medium, and by concentration of these volatiles on the carbon disc. Results were far superior to those obtained by injection of a solution of the whole secretion (direct injection).

It is very difficult, without further work, to assess the biological significance of the analytical results; but assuming compound (*1*) is semiochemically significant, it seems reasonable to suggest that its absence in the spotted hyaena may be related in some way to the differences in social behaviour which exist between it and the other two species. These differences are manifest in the spotted hyaena by *more* co-operation in hunting and foraging, *less* frequent scent marking, *more* visual and vocal displays, and female dominance (Mills 1978).

## References


Grob, K. and Zurcher, F. (1976). Stripping of trace organic substances from water. Equipment and procedure. *J. Chromatog.*, 117, 285

Hachenberg, H. and Schmidt, A.P. (1986). *Gas chromatographic headspace analysis*. Wiley, Chichester.

Lamb, C.N. (1980). Brown hyaena anal sac secretion and the striped hyaena 224 compound. M.Sc. thesis, Howard University, Washington D.C.

Mills, M.G.L. (1978). The comparative socio-ecology of the hyaenidae. *Carnivore (Seattle)*, 1, 1.

Mills, M.G.L., Gorman, M.L. and Mills, M.E.J. (1980). The scent-marking behaviour of the Brown Hyaena *Hyaena brunnea*. *S. Afr. J. Zool.*, 15 (4), 240.

Von Endt, D.W. (1975). *Some aspects of chemical communications in insects and mammals*. Ph.D. thesis, Howard University, Washington D.C.

Wheeler, J.W., Von Endt, D.W. and Wemmer, C. (1975). 5-Thiomethylpentane-2,3-dione. A unique natural compound from the Striped Hyaena. *J. Amer. Chem. Soc.*, 97, 441.

# I.8 Host Odour Attractants for Tsetse Flies

DAVID R. HALL

## Introduction

Tsetse flies, *Glossina* spp. (Diptera: Glossinidae), are haematophagous vectors of trypanosomes which cause sleeping sickness in man and a similar disease, nagana, in cattle. Tsetse are found only in Africa, but it is estimated that over 11 million $km^2$ are rendered virtually uninhabitable because of their presence, and that over 45 million people are at risk from human trypanosomiasis (Hagan and Wilmshurst 1975).

Although drugs are available for treatment of trypanosomiases, these tend to be expensive and require an extensive infrastructure for effective administration. Eradication of the vector can provide a long-term solution to the problem: in the past, this was achieved by bush clearance and extermination of alternative, wild host animals, but these methods have been replaced by ground spraying of persistent insecticides and sequential aerosol spraying of non-persistent insecticides from the air (Allsopp 1984).

In recent years, attention has focused on use of trapping devices for the survey and control of tsetse. These devices utilize mechanisms by which the flies find the host animals on which they depend for blood meals. Riverine species of tsetse of the *palpalis* group are typically restricted to a linear, relatively localized habitat along river gallery forests, and it seems that they rely primarily on visual cues to locate their hosts. In many areas, these species can be effectively surveyed and controlled with unbaited traps or insecticide-coated screens of appropriate shape and colour which simulate a host animal (e.g. Politzar and Merot 1984).

Savannah species of tsetse are more wide-ranging: although simple, visual traps can be used to survey the incidence of these species, such devices cannot give effective control at economically viable densities (e.g. Politzar and Merot 1984). Over the years, there have been numerous isolated reports of attraction of tsetse flies to the odours of host animals and their residues, but it was the work of Vale (1974) in Zimbabwe that established the importance of odour in attracting savannah species of tsetse to their hosts and to stationary baits. This provided a basis for work on chemical identification of the attractive components, and the successful application of odour-baited trapping devices in survey and control of savannah species of tsetse.

## Identification of host odour attractants

Vale (1974) showed that the numbers of tsetse, *G. pallidipes* and *G. morsitans morsitans*, attracted to a simple model were increased by up to 50 times by odours from a single ox in an underground pit and that catches increased with the weight of the host animal used (Hargrove and Vale 1978). Starved animals were less attractive, but when such animals were returned to a fattening diet they rapidly became even more attractive than normal for a short period (Vale 1981).

Possible components of ox odour were screened as attractants for tsetse in the field (Vale 1980). Carbon dioxide increased catches of tsetse by between two and six times at release rates between 2.5 and 15 l $min^{-1}$. Acetone released at 0.3–30 g $h^{-1}$ increased catches by up to six times, and the effects of these two chemicals were additive. Of the other chemicals tested, butanone could replace acetone as an attractant, but acetophenone and some low molecular weight carboxylic acids were found to be repellent to tsetse.

Subsequent work concentrated on trapping, isolating, and identifying the components attractive to tsetse in natural host odour. For trapping the odours, solid adsorbents were preferred since it was found that the attractiveness of ox odour could be reduced to the levels shown by carbon dioxide and acetone by filtering it through activated charcoal (Vale and Hall 1985*a*).

For detecting active components in the mixtures collected, electroantennography (EAG) was used, because in a field experiment, Vale (1982) showed that removal of the antennae from tsetse essentially eliminated their ability to respond to acetone and carbon dioxide, indicating that the receptors for these and, presumably, other attractants are located on the antennae. In the EAG technique, originally developed for detecting components of the sex pheromones of Lepidoptera, microelectrodes inserted into the antenna record a depolarization that occurs when chemoreceptors on the antenna are stimulated by an active chemical (e.g. Roelofs 1977).

Volatiles from cattle were collected by drawing air over the animals and through Porapak resin. The materials trapped were desorbed with dichloromethane and analysed by linked gas chromatography–electroantennography (GC–EAG) in which the effluent from the gas chromatography column is split between the gas chromatography flame ionization detector and a tsetse EAG preparation, thereby allowing simultaneous recording of the normal gas chromatography trace and the EAG activity of the components separated.

One minor component of ox odour gave a strong EAG response from *G. m. morsitans*; it was identified by mass spectrometry and by comparison with synthetic material as 1-octen-3-ol (Hall *et al.* 1984). This compound stimulated upwind flight of tsetse in a laboratory wind-tunnel (Bursell 1984), and doubled the number of tsetse attracted by ox odour in the field (Vale and Hall 1985a). The enantiomeric composition of the 1-octen-3-ol isolated from ox odour was found to vary between 80:20 and 92:8 *R*:*S* in different collections (Hall *et al.* 1984): no difference was found between responses of tsetse to enantiomerically enriched samples in EAG, wind-tunnel or field tests.

Rates of production of acetone and 1-octen-3-ol by an ox as measured by air-sampling techniques were up to 24 mg $h^{-1}$, although typically 2–4 mg $h^{-1}$, for acetone, and up to 0.043 mg $h^{-1}$ for 1-octen-3-ol. A mixture of carbon dioxide, acetone, and 1-octen-3-ol released at naturally-occurring rates of 2 l $min^{-1}$, 5 mg $hr^{-1}$ and 0.05 mg $hr^{-1}$ respectively, was found to be nearly as attractive as the natural odour from one ox for *G. m. morsitans* but only half as attractive for *G. pallidipes* (Vale and Hall 1985*a*).

In the above field experiments, electrified nets were used to kill tsetse which had been attracted to the odour source, and thus the results are a measure of attraction from a distance only. Catching flies in a trap involves attraction from a distance and stimulation of the attracted flies to enter the trap. Experiments in which traps were baited with carbon dioxide, acetone, or 1-octen-3-ol showed that all three odours attract flies from a distance, while carbon dioxide also increased the trap-entering response. Baiting a trap with combinations of all three attractants increased catches of tsetse by up to 64 times (Vale and Hall 1985*b*).

More recent work has shown that cattle urine is a powerful attractant for savannah species of tsetse, particularly *G. pallidipes* (Owaga 1984, 1985), and that it further increases the catches of tsetse in traps already baited with acetone and 1-octen-3-ol (Vale *et al.* 1986; Dransfield *et al.* 1986). Solvent extracts of the urine were also found to be attractive: after chromatographic (Hassanali *et al.* 1986) or chemical (Bursell *et al.* 1988) fractionation, the attractiveness was found to be associated with the phenolic fraction. This contained up to eight simple phenols: phenol, 3- and 4-methylphenol, 3- and 4-ethylphenol, 3- and 4-propylphenol and 2-methoxyphenol (Hassanali *et al.* 1986; Bursell *et al.* 1988). All except 2-methoxyphenol were active by EAG, and several were active in stimulating take-off and/or upwind flight in a wind-tunnel (Bursell *et al.* 1988). Field testing showed that none of the phenols was particularly attractive when used alone, but combinations of 4-methylphenol and 3-propylphenol were as attractive as the total mixture (Vale *et al.* 1988*a*; Owaga *et al.* 1988). 2-Methoxyphenol markedly reduced trap catches of tsetse. The phenols are slightly less effective in increasing the numbers of tsetse caught at electrified nets when compared with their effect with traps, indicating that the phenols also increase trap-entering responses (Vale *et al.* 1988*a*).

The phenols seem to account for most of the attractiveness of the urine, but some minor components of the neutral fraction were found to cause strong EAG responses from *G. m. morsitans* and *G. pallidipes*. These components were identified as indole, 3-methylindole, and two carotenoid metabolites, isomers of 3,3,5-trimethyl-4-hydroxy-4-(3-oxobutyl)cyclohexanone (Gough *et al.* 1987). Despite some promising indications in the wind-tunnel bioassay, no attractiveness has yet been demonstrated for these compounds in the field under a variety of conditions.

Acetone and 1-octen-3-ol have been shown to increase catches of *G. m. submorsitans* in Burkina Faso (Politzar and Merot 1984). None of the above compounds has shown any attractiveness for the riverine species *G. palpalis palpalis*. However, the riverine species, *G. tachinoides*, is attracted by the odour of host animals, and passage of the odour through a charcoal filter removes most of the attractive components (Merot *et al.* 1986). Carbon dioxide (Galey *et al.* 1986) 1-octen-3-ol, and the mixture of phenols found in cattle urine are attractive to *G. tachinoides*, but acetone is not attractive. The most important phenol appears to be 3-methylphenol (Merot, personal communication), which is the major phenolic component of bushbuck urine (Gough, unpublished results), the favoured host of *G. tachinoides*.

## Use of tsetse attractants in survey and control

Of the known tsetse attractants, carbon dioxide, at doses of several litres per minute, is inconvenient to use on a large scale. However, baiting traps with acetone released at 100 mg $h^{-1}$, 1-octen-3-ol at 0.5 mg $hr^{-1}$, 4-methylphenol at 1.5 mg $hr^{-1}$.and 3-propylphenol at 0.2 mg $hr^{-1}$ can increase catches of *G. pallidipes* by 20–30 times, making the traps much more sensitive and convenient monitoring devices for this species than traditional ox rounds.

The 1-octen-3-ol and the two phenols are dispensed from sealed polythene sachets which allow the compounds to diffuse out at a controlled rate. The acetone is currently dispensed from a bottle which has an appropriately-sized hole in the lid, but sealed dispensers are being developed. Studies of the trap-orientated behaviour of tsetse and their responses to different colours have led to the design of very effective traps which have a high specificity for tsetse (Vale *et al.* 1985).

For control, simpler and cheaper cloth screens impregnated with insecticide are used, which are also baited with acetone, 1-octen-3-ol, 4-methylphenol, and 3-propylphenol. In an initial trial on a 5km$^2$ island in Lake Kariba, an introduced population of tsetse was eradicated with insecticide-impregnated screens baited with acetone and 1-octen-3-ol only, at a density of 4 per km$^2$ (Vale *et al.* 1986*b*). In a subsequent operation on an area of 600 km$^2$ in Zimbabwe, tsetse populations were reduced by over 99.99% with similar odour-baited screens (Vale *et al.* 1988*b*). Since the introduction of the phenolic attractants, the method has become even more effective, and large-scale control operations using screens impregnated with deltamethrin and baited with acetone, 1-octen-3-ol, 4-methylphenol, and 3-propylphenol are in progress in Zimbabwe and Zambia. In Kenya, local tribesmen are using traps baited with acetone and jars of cattle urine for tsetse control (Brightwell *et al.* 1987).

Odour-baited traps and screens now provide a valuable complement to existing methods of control of savannah species of tsetse based on ground spraying of persistent insecticide or aerial spraying of non-persistent insecticide. Ground spraying makes heavy demands on labour and requires organization, while aerial spraying requires sophisticated equipment and rigid scheduling. Traps and screens can be manufactured locally and deployed as resources permit and they operate continuously to monitor or reduce tsetse populations. The amounts of insecticide used on the screens are small, and environmental contamination is minimal. Furthermore, the synthetic odours and the designs of trap and screen are highly specific for tsetse.

The main problems and costs involved in using odour-baited traps and screens in tsetse control occur during the initial deployment and subsequent maintenance of these devices. These problems can be reduced if more attractive traps and screens are made available; such devices could then be placed at lower densities. Research into the identification of new attractants, and the improvement and simplification of the designs of traps and screens continues.

## Acknowledgements

Much of the work reviewed here is the result of a collaborative project between the Zimbabwe Department of Veterinary Services (Dr Glyn Vale, Dr John Hargrove, Stanley Flint), the Tsetse Research Laboratory, Bristol (Dr Tony Jordan, Prof. Einar Bursell and Andrew Gough) and ODNRI (Peter Beevor, Alan Cork, and Steven Torr).

## References

Allsopp, R. (1984). Control of tsetse flies (Diptera: Glossinidae) using insecticides: a review and future prospects. *Bull. ent. Res.*, 74, 1–23.

Brightwell, R., Dransfield, R.D., Kyorku, C., Golder, T.K., and Tarimo, S.A.R. (1987). A new trap for *Glossina pallidipes*. *Tropical Pest Management*, 33, 151–9.

Bursell, E. (1984). Effects of host odour on the behaviour of tsetse. *Insect Sci. Applic.*, 5, 345–9.

Bursell, E., Gough, A.J.E., Beevor, P.S., Cork, A., Hall, D.R. and Vale, G.A. (1988). Identification of components of cattle urine attractive to tsetse flies, *Glossina* spp. (Diptera: Glossinidae). *Bull. ent. Res.*, 78, 281–91.

Dransfield, R.D., Brightwell, R., Chaudhury, M.F., Golder, T.K. and Tarimo, S.A.R. (1986). The use of odour attractants for sampling *Glossina pallidipes* Austen (Diptera: Glossinidae) at Nguruman, Kenya. *Bull. ent. Res.*, 76, 607–19.

Galey, J.B., Merot, P., Mitteault, A., Filledier, J. and Politzar, H. (1986). Efficacité du dioxyde de carbone comme attractif pour *Glossina tachinoides* en savane humide d'Afrique de l'Ouest. *Rev. Elev. Med. vet. Pays trop.*, 39, 351–4.

Gough, A.J.E., Hall, D.R., Beevor, P.S., Cork, A., Bursell, E. and Vale, G.A. (1987). Attractants for tsetse from cattle urine. In *Proceedings of the 19th Meeting of the International Scientific Council for Trypanosomiasis Research and Control*. Lome, Togo.

Hagan, D.H. and Wilmshurst, E.C. (1975). The disease spread by tsetse. *World Crops*, 248–50.

Hall, D.R., Beevor, P.S., Cork, A., Nesbitt, B.F. and Vale, G.A. (1984). 1-Octen-3-ol: a potent olfactory stimulant and attractant for tsetse isolated from cattle odours. *Ins. Sci. Applic.*, 5, 335–9.

Hargrove, J.W. and Vale, G.A. (1978). The effect of host odour concentration on catches of tsetse flies (Glossinidae) and other Diptera in the field. *Bull. ent. Res.*, 68, 607–12.

Hassanali, A., McDowell, P.G., Owaga, M.L.A. and Saini, R.K. (1986). Identification of tsetse attractants from excretory products of a wild host animal, *Syncerus caffer*. *Ins. Sci. Applic.*, 7, 5–9.

Merot, P., Galey, J.B., Politzar, H., Filledier, J. and Mitteault, A. (1986). Pouvoir attractif de l'odeur des hotes nourriciers pour *Glossina tachinoides* en zone soudano-guineene (Burkina Faso). *Rev. Elev. Med. vet. Pays trop.*, 39, 345–50.

Owaga, M.L.A. (1984). Preliminary observations on the efficacy of olfactory attractants derived from wild hosts of tsetse. *Ins. Sci. Applic.*, 5, 87–90.

Owaga, M.L.A. (1985). Observations on the efficacy of buffalo urine as a potent olfactory attractant for *Glossina pallidipes* Austen. *Ins. Sci. Applic.*, 6, 561–6.

Owaga, M.L.A., Hassanali, A. and McDowell, P.G. (1988). The role of 4-cresol and 3-n-propylphenol in the attraction of tsetse flies to buffalo urine. *Ins. Sci. Applic.*, 9, 95–100.

Politzar, H. and Merot, P. (1984). Attraction of the tsetse fly *Glossina morsitans submorsitans* to acetone, 1-octen-3-ol and the combination of these compounds in West Africa. *Rev. Elev. Med. vet. Pays trop.*, 37, 468–73.

Roelofs, W.L. (1977). The scope and limitations of the electro-antennogram technique in identifying pheromone components. In *Crop Protection Agents: their Biological Evaluation* (ed. N.R. McFarlane). Academic Press, London, pp. 147–65.

Vale, G.A. (1974). The responses of tsetse flies (Diptera: Glossinidae) to mobile and stationary baits. *Bull. ent. Res.*, 64, 545–88.

Vale, G.A. (1980). Field studies of the responses of tsetse flies (Glossinidae) and other Diptera to carbon dioxide, acetone and other chemicals. *Bull. ent. Res.*, 70, 563–70.

Vale, G.A. (1981). An effect of host diet on the attraction of tsetse flies (Diptera: Glossinidae) to host odour. *Bull. ent. Res.*, 71, 259–65.

Vale, G.A. (1982). The role of the antennae in the availability of tsetse flies (Diptera: Glossinidae) to population sampling techniques. *Trans. Zimbabwe Sci. Assoc.*, 61, 33–40.

Vale, G.A. and Hall, D.R. (1985*a*). The role of 1-octen-3-ol, acetone and carbon dioxide in the attraction of tsetse flies *Glossina* spp. (Diptera: Glossinidae) to ox odour. *Bull. ent. Res.*, 75, 219–31.

Vale, G.A. and Hall, D.R. (1985*b*). The use of 1-octen-3-ol, acetone and carbon dioxide to improve baits for tsetse flies, *Glossina* spp. (Diptera: Glossinidae). *Bull. ent. Res.*, 75, 209–17.

Vale, G.A., Bursell, E. and Hargrove, J.W. (1985). Catching-out the tsetse fly. *Parasitology Today*, 1, 106–110.

Vale, G.A., Flint, S. and Hall, D.R. (1986). The field responses of tsetse flies, *Glossina* spp. (Diptera: Glossinidae), to odours of host residues. *Bull. ent. Res.*, 76, 685–93.

Vale, G.A., Hargrove, J.W., Cockbill, G.F. and Phelps, R.J. (1986). Field trials of baits to control populations of *Glossina morsitans* Westwood and *G. pallidipes* Austen (Diptera: Glossinidae). *Bull. ent. Res.*, 76, 179–93.

Vale, G.A., Hall, D.R. and Gough, A.J.E. (1988a). The olfactory responses of tsetse flies, *Glossina* spp. (Diptera: Glossinidae), to phenols and urine in the field. *Bull. ent. Res.*, 78, 293–300.

Vale, G.A., Lovemore, D.F., Flint, S. and Cockbill, G.F. (1988b). Odour-baited targets to control tsetse flies, *Glossina* spp. (Diptera: Glossinidae), in Zimbabwe. *Bull. ent. Res.*, 78, 31–49.

# I.9 Mammalian Semiochemistry and its Applications

ERIC S. ALBONE

## Introduction

A continuing characteristic of the study of mammalian chemical communication has been the lamentable under-representation of chemistry. Examples in this volume of the fruitful contribution of chemists, both in pure science and in its practical applications (see for example Abbott *et al.*, this volume on the development of a deer repellent), are rather more encouraging. The variety and complexity of mammalian chemical communication undoubtedly present a most formidable challenge to the semiochemist. The field is intrinsically interdisciplinary, and chemistry brings to the study of mammalian chemical communication powerful analytical techniques and important insights derived from flavour chemistry, insect pheromone research, and environmental chemistry.

The basic question is how chemists, whose interests predominantly lie in the unravelling of new chemical structures, can make their contribution to the study of mammalian chemical communication (Albone 1984). A knowledge of the chemical ambience of an animal can reasonably be expected to be a central piece of information; and yet, remarkably few chemists have risen to the challenge. Certainly natural product chemists have studied mammalian scents without reference to the biological context; one looks for example to the early work of Ruzicka (1926) on the macrocyclic musks. However, to achieve advances in the field of mammalian semiochemistry, the chemist requires the sensitive guidance of biologists.

## Chemical complexity

How is this to happen? I suggest a kind of 'chemical sandwich', which mirrors the interdisciplinary approach where chemical studies of the nature of the signal are sandwiched between studies of the signal source and studies of the biological significance of the signal for the receiving animal.

It is tempting to think that one has only to take some scent material and perform some simple chemical analyses to make progress. This view is mistaken on several levels. First, semiochemical substrate sources are very complex so that it is never possible to obtain a total chemical description and such descriptions as are obtained are bound to be selective. As Zlatkis *et al.* (1979) have written,

> A complete profile of all the constituents of biological fluid is at present impossible... The goal at which most workers aim is the development of a complete profile of a selected group of substances (such as steroids, aminoacids, etc.) or of compounds with similar physical properties (for example volatiles).

It is with this latter group of substances that semiochemists have tended to be most concerned.

Although the chemist needs be selective with regard to class of compound or scent source chosen for analysis, the animal is in reality confronted with a complex array of compounds from a multitude of scent sources, many of which may have been mixed together and may have reacted chemically. Scent glands commonly receive inputs from a variety of secretory tissues, as in the mink (Sokolov *et al.* 1980) where the anal sac receives the products of both sebaceous and apocrine tissues; these products could then be involved in further microbiological and chemical transformations both in the sac and after voiding to the environment. Also, the androgen-dependent, urinary, aversive pheromone of the mouse requires actual contact between the products of the coagulating glands and bladder urine (Jones and Nowell 1973), while the socially important chemical signal arising from the Harderian gland secretion of the mongolian gerbil is influenced by the saliva with which it is mixed during grooming (Thiessen *et al.* 1976).

The less volatile and involatile components may also play a role in mammalian chemical communication. Low volatility components may themselves have message value where physical contact with the semiochemical source occurs (Rivard and Klemm, this volume), and the vomeronasal organ (see this volume, Lepri *et al.*; Rasmussen and Hultgren; Schilling *et al.*) is known to be implicated in their detection. Volatile profiles may also be influenced directly by less volatile components modifying the release of volatiles from a source and possibly by acting as precursors for such volatiles (Albone 1984, pp. 13-17).

The volatile profile itself also requires careful assessment because the profile obtained experimentally depends heavily on the precise analytical procedure followed (Jennings and Filsoof 1977), while mammalian sensitivities to the different profile components can vary immensely, and trace components may dominate perceived odour, as for example the isobutylpyrazine in bell pepper oil (Buttery *et al.* 1969).

## Biological activity and the response-guided strategy

This chemical complexity is mirrored in the biological complexity of mammalian responses to the chemical signals they receive. Chemists would certainly prefer to embark on a semiochemical study using a simple, reproducable biological assay, which by repeated application can guide them through multiple fractionations of complex semiochemical substrates in search of biologically active components which may be present at very low levels. Let us call this a response-guided strategy. However, for mammals, this approach seems to be of only very limited value, since firm biological assays exist for only a minority of semiochemicals areas, such as with primer pheromones, where the effects are long term and physiological rather than short term and behavioural as for example the heat-labile, androgen-dependent, puberty accelerating pheromone present in male mouse urine, studied by Vandenbergh *et al.* (1975) by measurement of uterine weights.

Even where biological assays do exist, chemical fractionation (or even testing in an inappropriate context) can often lead to loss of activity. Although the activity sometimes seems to reside principally in simple pheromonal substances, as in the dimethyl disulphide male attractant in golden hamster vaginal secretion (O'Connell *et al.* 1979) and in the androstene steroids of boar submaxillary saliva (Booth 1980), more commonly, however, activity is associated with complex mixtures of unremarkable chemicals. Furthermore, a mammal responds not just to a chemical stimulus in isolation, but integrates the response with non-chemical cues and other factors, which is dependent on physiological state and past experience (see, for example, Reece Engel, this volume). Methods akin to the use of electoantennography which have proved so powerful in the study of insect semiochemistry, as was recently demonstrated in the identification by Hall and his colleagues (this volume) of 1-octen-3ol as a potent tsetse fly attractant in cattle odour, have not found much application in the study of mammalian semiochemistry, although Goodrich *et al* (1978) have used a radiotelemetric heart-rate monitor to select a series of C8 to C12 aldehydes for further study in the complex mixture that comprises rabbit anal gland headspace volatiles.

Animals commonly respond to a large range of novel stimuli whatever their natural biological significance. They rapidly learn to respond to test situations in certain ways, habituate, and show very considerable variations in response between one individual and another. This situation is likely to cause particular difficulties for those species where it is impractical to study a large number of individual animals. As Müller-Schwarze *et al.* (1986) have pointed out, the bioassay of a mammalian pheromone should ideally be undertaken with freely moving animals in a natural setting and intact social units.

## Scent source and chemical image strategy

A complementary approach to the response-guided strategy is for the chemist to obtain as much information as possible of the total *chemical image* that one mammal presents to another. Paying particular attention to features that vary with important biological parameters and which could be exploited semiochemically. This would accord with the view of Beauchamp *et al.* (1976) and many others that the concept of 'pheromone' as narrowly defined may be of little value in a mammalian context and that mammalian chemical signals are more likely to be encoded in the patterns of occurrence of unremarkable compounds rather than in the presence or absence of individual biologically active substances.

With regard to primates, Belcher *et al.* (1986) have described studies of scent marks from the highly developed circumgenital sebaceous and apocrine gland fields present in *Saguinus fuscicollis* tamarins. This work shows that the chemist can deduce *inter alia* the species, subspecies, and gender of the marking animal from the profiles of butanoate esters of simple C16 to C24 alcohols which the marks contain. Whether the animals themselves make similar use of these components to interpret the scent marks is less clear, although scent mark profiles certainly exhibit considerable stability with time (in the region of a year) and choice-test experiments show that the animals distinguish similarly between scent marks, between the total volatiles reassembled from scent marks on passing through a gas chromatograph, and between the butanoic acid ester fraction and the more volatile component fraction combined together, but not between these two fractions separately.

Headspace analysis of the more volatile components of scent materials often reveals the presence of a large number of rather ordinary components. Many of these can vary considerably in level from sample to sample, and interpretation of the information obtained is necessarily speculative. A recently published analysis of the headspace volatiles of female coyote urine obtained by Schultz *et al.* (1988) over a two-year period was typical in revealing a large number of components with great variation in the relative concentrations of components even over a very few days, although one component, 3-methylbut-3-enyl sulphide, commonly accounted for more than half the total volatiles observed; however, profile variations were not readily correlated with biological parameters. Such an approach can be occasionally more rewarding. Bailey *et al.* (1980) examining the volatile profiles of male red fox urine found that, although a number of sulphur compounds present accounted for the characteristic smell, just one sulphur compound peaked consistently during the breeding season, and was thus a suitable candidate for bioassay. In considering the nature of the chemical signal, the technique of dynamic solvent effect sampling, recently developed by Apps (this volume), opens up a very important new approach to the quantitative analysis of semiochemical volatiles.

The factors that modulate a signal source can be summarized as hormonal, bacterial, genetic, dietary, and environmental. The hormonal state of an animal influences not only secretion production, but also marking behaviour and responses of the recipient animal to any semiochemical. A gland whose activity is sensitive to the hormonal state of the animal and whose products are voided to the external environment, stands an excellent chance of being exploited semiochemically, as, for example, are the accessory sex glands. So chemists will naturally look particularly for those changes in the chemical image of an animal that correspond with hormonal changes.

Perhaps the most exciting developments in this direction have been made by

Novotny *et al.* (this volume). Their studies of hormonal dependence of the complex headspace profiles of male mouse urine revealed two unexpected testosterone-dependent trace components: one was a thiazole and the other a brevicomin derivative. These components were inactive separately, but together strongly elicited a range of distinct biological activity including inter-male aggression, female attraction, and synchronous ovulation in group suppressed females.

The potential contribution of bacteria to the generation of mammalian semiochemicals is increasingly being realized The mammalian body surface provides a variety of habitats within which micro-organisms flourish by utilizing and transforming the mammal's own secretions and excretions. Mammalian semiochemicals may in this manner be formed not by the mammal itself but by the microflora which its body supports. To be of potential use to the animal for communication, it is only necessary that the resulting volatile profile should be sufficiently stable with time and sufficiently responsive to a significant biological parameter. Important areas where evidence that the contribution of microbially derived odours could be semiochemically significant include axillary odours (Gower, this volume; Rennie *et al.*, this volume) vaginal odours, oral odours, skin odours, faecal odours, and the scents from anal sacs and related structures (Albone 1984, pp. 135–164). Consideration of the chemistry and microbiology of the anal sac of the red fox led us to suggest a fermentation hypothesis of chemical recognition (Albone and Perry 1976).

There is increasing interest in the genetic contribution to individual odour (Beauchamp *et al.*; Brown *et al.*, this volume). This possibility, which has long been realized by those concerned with the use of tracker dogs, has now been demonstrated using congenic rats and mice, differing only at the major histocompatibility locus: differences in odour have been shown that are of significance in mating preferences and in the induction of primer pheromone effects. Of particular fascination is the possibility that microorganisms may play a crucial role in generating these signals from precursors produced under genetic control.

## References

Albone, E.S. (1984). *Mammalian semiochemistry: the investigation of chemical signals between mammals.* Wiley-Interscience, Chichester and New York.

Albone, E.S. and Perry, G.C. (1976). Anal sac secretion of the red fox, *Vulpes vulpes.* Volatile fatty acids and diamines. Implications for a fermentative hypothesis of chemical recognition. *J. Chem. Ecol.*, 2, 101–11.

Bailey, S., Bunyan, P.J. and Page, J.M.J. (1980). Variation in the levels of some components of the volatile fraction of urine from captive red fox *(Vulpes vulpes)* and its relationship to the state of the animal. In *Chemical signals: vertebrates and aquatic invertebrates* (ed. D. Müller-Schwarze and R.M. Silverstein), pp. 391–403, Plenum, New York and London.

Beauchamp, G.K., Doty, R.L., Moulton, D.G. and Mugford, R.A. (1976). The pheromone concept in mammalian chemical communication: a brief critique. In *Mammalian olfaction, reproductive processes and behavior* (ed. R.L. Doty), pp. 143–160. Academic Press, New York.

Belcher, A.M., Smith, A.B., Jurs, P.C., Lavine, B. and Epple, G. (1986). Analysis of chemical signals in a primate species *(Saguinus fuscicollis)*: use of behavioral, chemical, and pattern recognition methods. *J. Chem. Ecol.*, 12, 513–31.

Booth, W.D. (1980). Endocrine and exocrine factors in the reproductive behaviour of the pig. *Symp. Zool. Soc. Lond.*, 45, 289–311.

Buttery, R.G., Seifert, R.M., Guadagni, D.G. and Ling, L.C. (1969). Characterization of some volatile constituents of bell peppers. *J. Agric. Food Chem.*, 17, 1322–7.

Goodrich, B.S., Hesterman, E.R., Murray, K.E., Mykytowycz, R., Stanley, G. and Sugowdz, G. (1978). Identification of behaviourally significant volatile compounds in the anal gland of the rabbit, *Oryctolagus cuniculus*. *J. Chem. Ecol.*, 7, 817–27.

Jennings, W.G. and Filsoof, M. (1977). A comparison of sample preparation techniques for gas chromatographic analysis. *J. Agric. Food Chem., 25*, 440–5.

Jones, R.B. and Nowell, N.W. (1973). The effect of urine on the investigatory behaviour of male albino mice. *Physiol. Behav.*, 11, 35–8.

Müller-Schwarze, D., Morehouse, L., Corradi, R., Cheng-hua Zhao and Silverstein, R.M. (1986). Odor images: responses of beaver to castoreum fractions. In *Chemical signals in vertebrates*, 4 (ed. D. Duvall, D. Müller-Schwarze and R.M. Silverstein), pp. 561–70. Plenum, New York and London.

O'Connell, R.J., Singer, A.G., Pfaffmann, C. and Agosta, W.C. (1979). Pheromones of hamster vaginal discharge. Attraction to femtogram amounts of dimethyl disulphide and to mixtures of volatile components. *J. Chem. Ecol.*, 5, 575–85.

Ruzicka, L. (1926). Zur Kenntnis des Kohlenstoffringes. I. Über die Konstitution des Zibetons. *Helv. Chim. Acta, 9*, 230–48.

Schultz, T.H., Flath, R.A., Stern, D.J., Mon, T.R., Teranishi, R., Kruse, S.M., Butler, B. and Howard, W.E. (1988). Coyote estrous urine. *J. Chem. Ecol.*, 14, 701–12.

Sokolov, V.E., Albone, E.S., Flood, P.F., Heap, P.F., Kagan, M. Z., Vasileva V.S., Roznov, V.V. and Zinkevich, E.P. (1980). Secretion and secretory tissues of the anal sac of the mink, *Mustela vison*: chemical and histological studies. *J. Chem. Ecol.*, 6, 805–25.

Thiessen, D.D., Clancy, A. and Goodwin, M. (1976). Harderian gland pheromone

in the mongolian gerbil, *Meriones unguiculatus*. *J. Chem. Ecol.*, 2, 231–8.
Vandenbergh, J.G., Whitsett, J.M. and Lombardi, J.R. (1975). Partial isolation of a pheromone accelerating puberty in female mice. *J. Reprod. Fertil.*, 43, 515–23.
Zlatkis, A., Lee, K.Y., Poole, C.F. and Holzer G. (1979). Capillary column gas chromatographic profile analysis of volatile compounds in sera of normal and virus-infected patients. *J. Chromatogr.*, 163, 125–33.

# SECTION II

## Neurophysiology and anatomy

# II.1 A Neural Mechanism for Olfactory Learning

E.B. KEVERNE AND H. KABA

## Introduction

In mammals, olfactory learning plays an integral part in the establishment of normal behavioural relationships, including those between a mother and her offspring (Fleming and Rosenblatt 1974), and those that occur in the context of consort interactions (Fillion and Blass 1986). One aspect of olfactory learning that has been investigated in some detail concerns the memory, established at mating, imprinted by the female mouse to the odours (pheromones) of the stud male (Keverne 1983). This olfactory memory is of considerable biological significance to the mouse, since it prevents any subsequent exposure to this male's pheromones from initiating neuroendocrine events that result in pregnancy loss (Keverne and Rosser 1986). Pheromones from strange males for which no memory has formed do, however, block the female's pregnancy (Bruce 1960).

Several features that characterize the neural basis of this olfactory memory have now been established. Pheromones that initiate the effect selectively activate the vomeronasal receptors rather than the main olfactory system (Lloyd-Thomas and Keverne 1982). This olfactory memory is established during a critical period of 4 h immediately after mating (Rosser and Keverne 1985) and endures the reproductive lifetime of the female (Kaba *et al.* 1988). The formation of this memory is dependent on intact noradrenergic projections from the locus coeruleus to the accessory olfactory bulb (Keverne and de la Riva 1982). The projections are active during the imprint phase, and are $\alpha$-adrenoceptor mediated (Kaba and Keverne 1988). These findings are consistent with the theory of a print now mechanism causing memory formation for sensory input coincident with high noradrenergic transmission at biologically significant times (Kety 1970).

## Olfactory recognition in the context of pregnancy block

Although a common neural and neuroendocrinal mechanism can account for all the pheromonal effects described in mice (Keverne 1983), the olfactory block to pregnancy is somewhat different in that only odours from a strange male, preferably of a different strain to the male that mated, are effective (Bruce 1968). The recently mated female mouse can be removed from her stud partner within a few hours of mating and be returned to him 24 h later, without disturbance to the pregnancy. This presupposes a recognition of the stud male, or at least an ability of the female to discriminate between the stud male made familiar by coitus, and the unfamiliar male to which she is secondarily exposed (Bruce 1968). Alternatively, the female could fail to respond to the stud male, not because he was the coital partner, but because he was genetically similar to the female, and therefore recognition of self could only be achieved with difficulty (Yamazaki *et al.* 1983).

In order to test this possibility females of the Balb C strain were all mated with Balb C males. These were removed immediately after mating (within 1 minute of ejaculation) and the female was then placed for 24 h in a cage containing recently soiled bedding from either Balb C or F1 (C57 × CBA) males. Each female was subsequently exposed to Balb C or F1 bedding. Although all of the females mated with the Balb C male, both Balb C and F1 pheromones were able to induce pregnancy block, but this depended upon the post-coital exposure of the female. If the female had been exposed to F1 bedding after mating, then Balb C male pheromones blocked pregnancy and conversely, if the female had been exposed to Balb C bedding, then F1 males blocked pregnancy. In other words the ineffectiveness of pregnancy block was not determined by the degree of relatedness to the stud male, but by those pheromones rendered familiar by coitus.

Recognition, particularly of something as complex as odours of differing strains of mice, may be thought to be represented by the cortical neuronal systems to which the main olfactory bulb projects. Experiments designed to address this issue failed to demonstrate that recognition, in the context of pregnancy block, had any relationship to the main olfactory system, and that this was not a necessary prerequisite to gate the pheromonal input from stud males. Although discrimination of urine containing male pheromones was severely impaired in comparison to castrate male urine by selective lesions of the main olfactory receptors, the ability of strange male urine to block pregnancy by contrast to stud male was unchanged. Selective destruction of the vomeronasal organ had the converse effect; it prevented pregnancy block but was ineffective on urine discriminatory ability (Lloyd-Thomas and Keverne 1982). Hence, the process of recognition which occurs following mating is a function of the vomeronasal system and its central projections.

The central projections of the vomeronasal system are distinct from those of the main olfactory system. Whereas the latter projects to piriform cortex, thalamus, entorhinal cortex, and orbito-frontal cortex, the vomeronasal system, having relayed in the accessory olfactory bulb has no cortical projection sites. The accessory bulb relays first in the amygdala and from here projects to the hypothalamus (BNST, pre-optic area, and ventromedial nucleus) before finally projecting to the arcuate nucleus where influence is brought to bear on neurons that regulate neuroendocrinal mechanisms. There are, however, parallel projections from the amygdala to the hippocampus, and hippocampal lesions are known to produce rapid forgetting of olfactory information (Staubli *et al.* 1984).

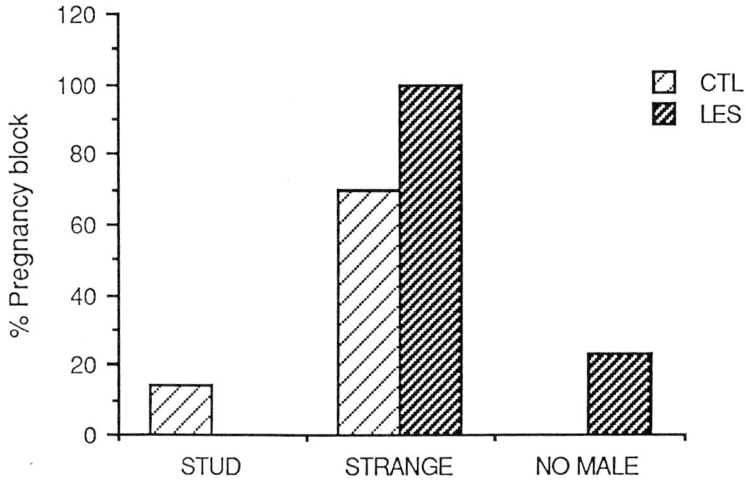


Fig. 1. Bilateral lesions to the hippocampus which impair olfactory learning to odour cues (butylacetate) are without effect on memory formation to the stud male. Only strange males block pregnancy in both lesioned and control groups. CTL, control; LES, hippocampal lesions.

Nevertheless, lesions to the hippocampus had no effect on either the ability of strange males to block pregnancy or the ability of the female to recognize the stud male and prevent pregnancy block (Fig. 1). This would support the view that the accessory olfactory system and its projections to the hypothalamus are sufficient, not only to produce the neuroendocrine changes necessary for pregnancy block, but also to sustain the process of recognition that occurs for the stud male.

The question therefore arises as to which of the projection sites of the accessory olfactory system are required for recognition of the stud males pheromones at mating, and because recall is brought into play some 48 h later, on the second exposure, then memory formation has also to be taken into account. One approach to addressing this question would be to lesion each relay site in the central neural projections of the vomeronasal system and investigate when recognition/memory fails to occur. However, such an approach would also prevent recall of the recognition/memory because pathways to the neuroendocrine effector system would be disrupted. An alternative and temporary disabling approach was therefore adopted whereby lignocaine anaesthetic was infused into each relay site of the centrally projecting accessory olfactory system. Such a procedure, when directed at the accessory olfactory bulb, prevented recognition/memory formation, as revealed by the stud male being able to block pregnancy. When lignocaine was infused into the medial amygdala immediately after mating during the period when memory formation occurs, there was no block to memory formation and only strange males were able to disrupt the females' pregnancy (Fig. 2) Taken together with all of the necessary control procedures for the non-specific effects of

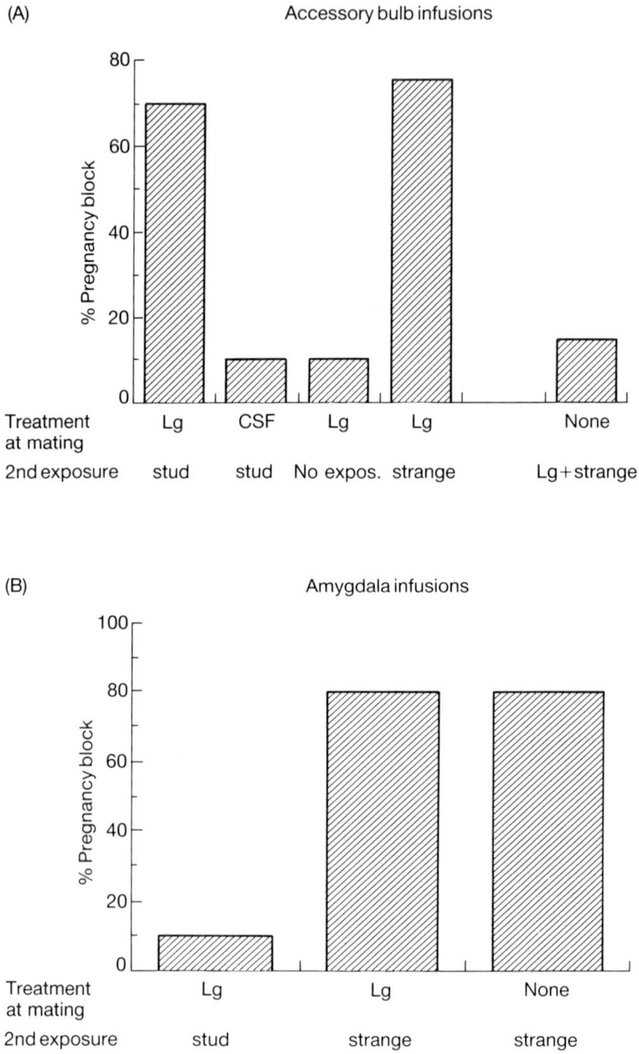


Fig. 2

A Animals receiving lignocaine (Lg) infusions during a 3 h period immediately after mating show high levels of pregnancy block when exposed to the strange or stud male some 24 h later. Lignocaine itself did not cause pregnancy block (Lg + no expos.) and neither did CSF infusions. However, lignocaine administered at the time of strange male exposure (Lg + strange) prevented pregnancy block revealing that this procedure blocked synaptic transmission.

B. Lignocaine infusions to the amygdala during sterile matings do not prevent memory formation to the stud male. A subsequent exposure to this male fails to block pregnancy, unlike strange males.

infusions and spontaneous pregnancy block, these experiments point to the accessory olfactory bulb itself as an important site, not only for the recognition process, but also for the memory trace which prevents the stud male from blocking his own pregnancy. Since lignocaine infusions to the amygdala disable transmission at this site, but do not prevent memory formation, then clearly this process must be occurring at some point prior to reaching the amygdala. Since the only relay prior to the amygdala for this pheromone mediated effect is the accessory olfactory bulb itself, our attention has focused on this structure.

The story that is emerging, therefore, focuses on the accessory olfactory bulb as the locus for events that are important for the recognition and the memory of the pheromones of the male at mating. This raises two important issues:

1. What is the mechanism whereby neural networks in the accessory olfactory bulb are informed that mating has occurred?
2. What is the structure of the accessory olfactory bulb and how might changes in its synaptic connection strengths account for the memory trace formed at mating?

## Centrifugal noradrenergic projections and olfactory memory

Both the main and accessory olfactory bulbs receive noradrenaline (NA), 5-hydroxytryptamine (5HT), dopamine (DA), and acetylcholine projections via the medial olfactory stria (Fallon and Moore 1978). In the mouse, NA terminals have been shown to be distributed to the external plexiform layer and granule cell layers (Rosser *et al.* 1986), and NA is a transmitter that has already been implicated in certain forms of sensory recognition. Thus modification of neural activity after exposure to certain types of visual stimuli occurs in early life in the rat (Kasamatsu and Pettigrew 1979), and the response of the female rat to tactile stimuli that provoke lordosis and pseudopregnancy are dependent upon noradrenergic mechanisms (Hansen *et al.* 1981). Indeed, the noradrenergic projections of the brain have been described as a subsystem which identifies situations of great survival value and provides the instructions for the storage of the relevant sensory information (Kety 1970). In the context of olfactory information, there are clearly many occasions when this is of survival advantage; for example, a mother's recognition of her own offspring; kin selection; mate choice; and of particular relevance to primer pheromones, recognition of the stud male's pheromones to avoid pregnancy block.

In the accessory olfactory bulb, the NA terminals synapse on inhibitory interneurons (Halász and Shepherd 1983; Rosser *et al.* 1986), and are important in modulating bulbar habituation to repeated presentation of a meaningful stimulus, whereas the serotoninergic input from the raphe nucleus is not (Gervais *et al.* 1984). A number of models have been developed to examine the significance of

NA terminals in the olfactory bulb on olfactory learning. In the rabbit, recording of the spatial EEG activity in a large area of the olfactory bulb has revealed a stable pattern of activity to a given odour, which is reproducible over several days (Freeman and Skarda 1985). However, when this same odour exposure was repeated together with either a positive or negative reinforcement, the spatial patterning of activity was modified, and restabilized after completion of the associative learning. The action of centrifugal NA innervation was of critical importance for this modification, since significant shifts in EEG activity were prevented when NA antagonists were infused during training sessions (Gray *et al.* 1986). Moreover, infusions of NA to the bulb during odour presentation without reinforcement also produced a modification in EEG shift similar to that seen with reinforcement alone. It would therefore seem likely that associative olfactory learning changes the pattern of bulbar neuronal firing, an event that is dependent on noradrenaline release.

In the rat pup, a very similar finding has been reported. Pairing a novel odour with a tactile stimulation that mimics parental licking produces a conditioned preference for that odour (Sullivan and Leon 1986). This learning preference is associated with a change in the responsiveness to the odour in selective olfactory glomeruli (Coopersmith and Leon 1984). Development of the preference and enhancement of the glomerular response is impaired by noradrenergic antagonists, but occurs in the absence of stroking when $\beta$-receptor agonists are given (Leon 1987; Wilson and Leon, 1988).

Our own studies have shown that formation of the olfactory memory to the stud male's pheromones is dependent upon centrifugal noradrenergic projections to the accessory olfactory bulb (Keverne and de la Riva, 1982). Selective lesions to this system in the female prior to mating result in the stud male blocking his own pregnancy. Moreover, mating induces a significant increase in NA turnover in the olfactory bulb, but not in the cortex. This increased actvity in noradrenergic terminals lasts for at least 3 h after mating, but is not recorded at 48 h when further exposure to pheromones produces pregnancy block (Rosser and Keverne 1985).

It is of interest that these increased rates of transmitter turnover in the noradrenergic terminals after mating correlate with the exposure time to pheromones that is required to form the memory. Moreover, at 48 h after mating when re-exposure to the stud male fails to block pregnancy, there is no enhanced NA turnover, and lesions studies made after mating revealed that NA is not required for memory recall (Rosser and Keverne 1985). Further studies have revealed that blockade of $\alpha$- but not $\beta$-receptors also prevents memory formation if the noradrenergic blocker is infused immediately after mating (Kaba and Keverne 1988).

## Structure of the accessory bulb and its relationship to olfactory memory

As already indicated, the main effect of NA in the accessory olfactory bulb is on the intrinsic granule cells and their terminal areas which have dendro-dendritic synapses with the mitral cells (Fig. 3). The granule cells release inhibitory gamma-aminobutyric acid (GABA) on to the mitral cell lateral dendrities at these synapses, whereas the mitral cell is excitatory to the granule cell (Shepherd 1972). The excitatory neurotransmitter is thought to be glutamate, aspartate (Halász and Shepherd 1983) or the dipeptide *N*-acetyl aspartylglutamate (Blakely *et al.* 1987). Intracellular recordings have clearly demonstrated that NA reduces the inhibition exerted by the granule cells on the mitral cells (Jahr and Nicholl 1982) which has been interpreted as enhancing the signal-to-noise ratio. Such effects would be anticipated following acute stimulation. However, chronic stimulation of the kind that is necessary for the formation of a memory trace to the stud male's pheromones, may act in a way that has much in common with long term potentiation (LTP) in the hippocampus. Behavioural experiments show that hippocampal LTP progressively develops over days during improvement of performance (Morris *et al.* 1986; McNaughton *et al.* 1986), while other studies have revealed the dependence of LTP on noradrenaline (Bliss *et al.* 1983; Stanton and Sarvey 1985).

The cellular mechanism supporting LTP has been ascribed to enhanced synaptic transmission using glutamate as an excitatory transmitter, and may involve variation in the number of *n*-methyl-*D*-aspartate receptors (NMDA). It is generally accepted that under normal conditions the NMDA receptors contribute very little to synaptic transmission, but can be brought into play under conditions whereby the membrance is depolarized (e.g. tetanic stimulation; contiguous NA stimulation). This not only permits monovalent ions to enter the NMDA channel, but also results in an influx of $Ca^2$ ions (MacDermott *et al.* 1986). The influx of $Ca^2$ is thought to activate biochemical changes involved in the long-term plastic changes of the post-synaptic neuron, which are brought about by secondary messenger systems. One second messenger system reported to be activated by LPT is the phosphorylated protein kinase C (Malenka *et al.* 1986). For this reason, and because of the resemblance of long-term pheromonal stimulation to LTP, we decided to examine the consequences of blocking protein kinase C on the formation of the olfactory memory.

Female mice with bilateral cannulae implanted to permit infusion of the accessory olfactory bulb were given polymyxin B (a protein kinase C inhibitor) immediately after mating and again 1.5 h later. The females were left in the presence of the stud males' pheromones for 6 h, which is normally sufficient for memory formation to occur, before removal to a clean cage. A control group received delayed infusions of polymyxin at 6 h and 7.5 h after mating, and both groups were reintroduced to the stud male for 48 h to test for pregnancy block.

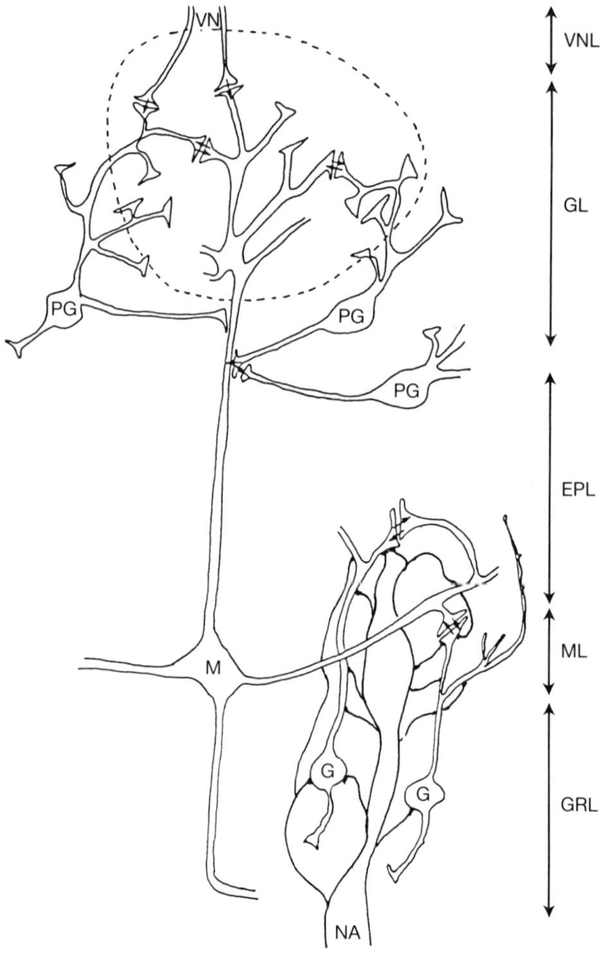


Fig. 3. Relationship between mitral cells (M) and intrinsic neurons (periglomerular, PG; granule, G) in the accessory bulb. Double arrows (↑↓) indicate dendro-dendritic synapses, single arrows show the direction of synaptic transmission. Centrifugal noradrenergic fibres terminate on granule cell perikarya and dendrites. VN represents incoming vomeronasal nerve. The layers are respresented on the right. VNL = vomeronasal nerve layer, GL = glomerular layer, EPL = external plexiform layer, ML = mitral cell layer, GRL = granule cell layer, M = mitral cell.

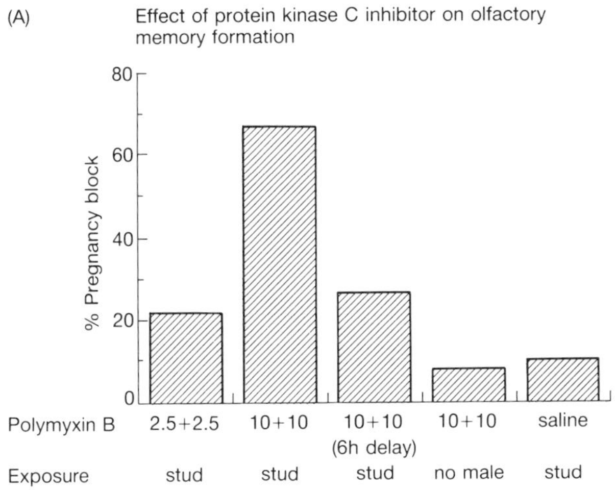


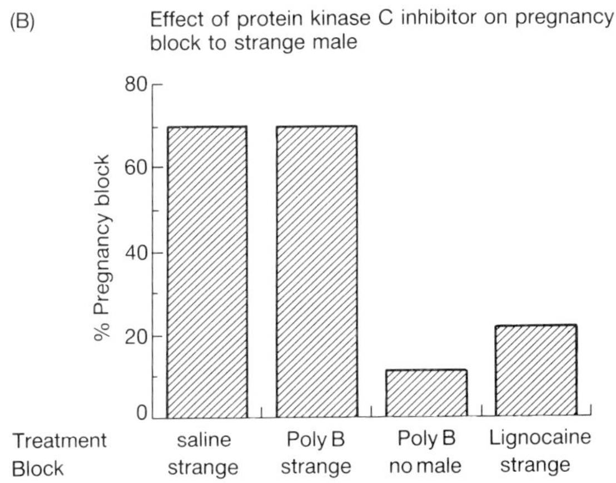


Fig. 4
A. Polymyxin B infusions to the accessory olfactory bulb contingent on mating prevented memory formation to the stud males unlike control infusions or polymyxin infusions and no male exposure. Polymyxin infusions given 6 h after mating did not prevent memory formation when compared with saline controls.
B. Control precedures to establish the effect of polymyxin infusions in neural transmission in the context of pregnancy block were without effect when compared with saline or lignocaine infusions, the latter having been established as blocking transmission. $\frac{2.5+2.5}{10+10}$ represent n. mol infusions of polymixin at 1.5 h intervals.

Further controls received saline and reintroduction to the stud, or polymyxin and no further male presentation (Fig. 4). The results show that protein kinase C inhibition during the critical period for memory formation, but not 6 h later, resulted in a failure of the females to recognize the males' pheromones and thus the stud male now blocked his own pregnancy.

Although the robust nature of olfactory memory formation in the context of pregnancy block has been demonstrated by the failure of numerous drug infusions to affect it (Kaba and Keverne 1988), it was necessary to demonstrate that polymyxin B in the accessory bulb did not produce its effects by interfering with neural transmission. This was achieved by infusions of the protein kinase C inhibitor at the time of exposure to a strange male after mating. Polymyxin B was shown to have no effect on direct transmission in the accessory bulb, since strange pheromones were able to block pregnancy, an effect not possible with infusions of drugs, e.g. lignocaine, known to interfere with transmission (Fig. 4).

Together with the noradrenergic effects, these studies implicate the mitral to granule cell dendro-dendritic synapse in memory formation. If noradrenaline reduces the feedback inhibition of granule to mitral cell, then sustained excitation of granule cells will occur at the dendro-dendritic synapses, producing a prolonged activation of this subset of granule cells over the 4 h period of NA release and pheromone activation. An alternative way to sustain excitation of granule cells could be achieved by blocking the feedback inhibitor of mitral cells using a GABA-ergic antagonist. Providing blockade is sustained for a prolonged period when the mitral cells are active, then it should be possible to create an olfactory memory without mating.

An experiment was designed to address this issue by infusing oestrous females with bicuculline bilaterally into the accessory olfactory bulb at 0 h and 1.5 h into a 6 h exposure to male bedding. At the next oestrus females were mated with a male of the same or a different strain to the bedding and were then re-exposed to pheromones of either a different or the same strain. The outcome of this experiment (Fig. 5) was the finding that pheromone exposure with blocked feedback inhibition to mitral cells does indeed create an olfactory memory, however this memory lacks the specificity of the memory formed at mating and appears to be generalized to the pheromones of at least one other strain. The lack of selectivity probably arises as a result of drug infusion blocking feedback inhibition to all granule cells, whereas mating-released noradrenaline finely tunes the subset of granule cells that are active at this time.

In conclusion, these findings together with the effect of noradrenergic lesions, implicate the granule to mitral cell dendro-dendritic synapse in memory formation. Since noradrenaline reduces the feedback inhibition of granule cells, there will be sustained excitation of a subset of granule cells over the several hours that mitral cells are active following mating and pheromonal stimulation. The consequences of enhancing synaptic connection strengths at the dendro-dendritic synapses would be a sustained inhibition of mitral cells. Mitral cell activity

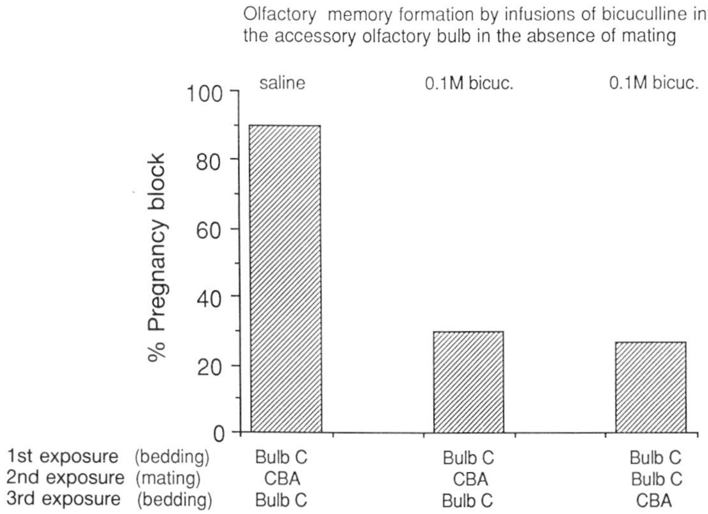


Fig. 5. It has proved possible to create olfactory memories without mating by blocking feedback inhibition from granule to mitral cell in the presence of an odour. Bicuculline infusions (gamma-aminobutyric acid receptor blocker) prevent pregnancy block on subsequent odour exposure (3rd exposure) some 48 h later. This memory is non-specific because other strains (CBA) to which there has been no previous exposure also block pregnancy.

matching that encountered following mating will, thus, be selectively gated (failure of the stud male to block pregnancy), while a level of mitral cell activity which is substantially different will not. Hence pheromones from strange males act on a population of mitral cells without sustained feedback inhibition, and promote neuroendocrinal mechanisms that are of significance in pregnancy failure. The mechanism for pheromone recognition is, therefore, a simple one whereby plastic changes in accessory olfactory bulb synaptic connectivity strengths produce a selective gating of input which serves a function in this long-term memory.

## References

Blakely, R.D., Ory-Lavollee, R., Grzanna, R., Koller, K.J. and Coyle, D.T. (1987). Selective immunocytochemical staining of mitral cells in rat olfactory bulb with affinity purified antibodies against N-acetyl-aspartate-glutamate. *Brain Res.*, 402, 373–8.

Bliss, T.V.P., Goddard, G.V. and Riives, M. (1983). Reduction of long-term potentiation in the dentate gyrus of the rat following selective depletion on monoamines. *J. Physiol. (Lond.)*, 334, 475–91.

Bruce, H. (1960). A block to pregnancy in the mouse caused by the proximity of strange males. *J. Reprod. Fertil.*, 1, 96–103.

Bruce, H.M. (1968). Absence of pregnancy-block in mice when stud males belong to an inbred strain. *J. Reprod. Fertil.*, 17, 407–8.

Coopersmith, R. and Leon, M. (1984). Enhanced neural response to familiar olfactory cues. *Science*, 225, 849–51.

Fallon, J.M. and Moore, R.Y. (1978). Catecholamine innervation of the basal forebrain. *J. Comp. Neurol.*, 180, 533–44.

Fillion, T.J. and Blass, E.M. (1986). Infantile experience with suckling odours determines adult sexual behaviour in male rats. *Science*, 231, 729–31.

Fleming, A.S. and Rosenblatt, J.S. (1974). Olfactory regulation of maternal behaviour in the rat. *J. Comp. Physiol. Psychol.*, 86, 233–46.

Freeman, W.J. and Skarda, C.A. (1985). Spatial EEG patterns, non-linear dynamics and perception. The neo-Sherringtonian view. *Brain Res. Rev.*, 10, 147–76.

Gervais, R., Araneda, S. and Pujol, J.F. (1984). Effect of local 5,6-dihydroxytryptamine on the rat olfactory bulb responsiveness during wakefulness and sleep. *Electroencephalogr. Clin. Neurophysiol.*, 57, 462–72.

Gray, C., Freeman, W.J. and Skinner, J.E. (1986). Chemical dependencies of learning in the rabbit olfactory bulb: acquisition of the transient spatial-pattern change depends on norepinephrine. *Behav. Neurosci.*, 100, 585–96.

Halász, N. and Shepherd, G.M. (1983). Neurochemistry of vertebrate olfactory bulb. *Neuroscience*, 10, 579–19.

Hansen, S., Stanfield, E.J. and Everitt, B.J. (1981). The effects of lesions of lateral tegmental noradrenergic neurons on components of sexual behaviour and pseudopregnancy in female rats. *Neuroscience*, 6, 1105–17.

Jahr, C.E. and Nicoll, R.A. (1982). Noradrenergic modulation of dendrodendritic inhibition of the olfactory bulb. *Nature*, 279, 227–8.

Kaba, H. and Keverne, E.B. (1988). The effect of microinfusions of drugs into the accessory olfactory bulb on the olfactory block to pregnancy. *Neuroscience*, 25, 1007–11.

Kaba, H., Rosser, A.E. and Keverne, E.B. (1988). Hormonal enhancement of neurogenesis and its relationship to the duration of olfactory memory. *Neuroscience*, 24, 93–8.

Kasamatsu, T. and Pettigrew, J.D. (1979). Restoration of visual cortical plasticity by local microperfusion of norepinephrine. *J. Comp. Neurol.*, 185, 163–82.

Kety, S.S. (1970). The biogenic amines in the central nervous system: their possible roles in arousal, emotion and learning. In *The Neurosciences*, 2nd Study Program (ed. F.O. Schmidt), pp. 324–36. Rockefeller University Press, New York.

Keverne, E.B. (1983). Pheromonal influences on the endocrine regulation of reproduction. *Trends Neurosci.*, 6, 381–4.

Keverne, E.B. and de la Riva, C. (1982). Pheromones: reciprocal interaction between the nose and brain. *Nature*, 296, 148–50.

Keverne, E.B. and Rosser, A.E. (1986). The evolutionary significance of the olfactory block to pregnancy. In *Chemical signals in vertebrates*, 4 (ed. D. Duvall, D. Müller-Schwarze, R.M. Silverstein), pp. 433–9. Plenum, New York.

Leon, M. (1987). Plasticity in olfactory output circuits related to early olfactory learning. *Trends Neurosci.*, 10, 434–8.

Lloyd-Thomas, A. and Keverne, E.B. (1982). Role of the brain and accessory olfactory system in the block to pregnancy in mice. *Neuroscience*, 7, 907–13.

MacDermott, A.B., Mayer, M.L., Westbrook, G.L., Smith, S.J. and Barker, J.L. (1986). NMDA-receptor activation increases cytoplasmic calcium concentration in cultured spinal cord neurones. *Nature*, 321, 519–22.

McNaughton, B.L., Barnes, G.A., Rao, G., Baldwin, J., and Rasmussen, M. (1986). Long-term enhancement of hippocampal synaptic transmission and the acquisition of spatial memory. J. Neurosci., 6, 563–71.

Malenka, R.C., Madison, D.V. and Nicoll, R.A. (1986). Potentiation of synaptic transmission in the hippocampus by phorbol esters. *Nature*, 321, 175–7.

Morris, R.G.M., Anderson, F., Lynch, G.S. and Baudry, M. (1986). Selective impairment of learning and blockade of long-term potentiation by an N-methyl-*D*-aspartate receptor antagonist, AP5. *Nature*, 319, 774–5.

Rosser, A.E. and Keverne, E.B. (1985). The importance of central noradrenergic neurones in the formation of an olfactory memory in the prevention of pregnancy block. *Neuroscience*, 15, 1141–7.

Rosser, A.E., Hokfelt, T. and Goldstein, M. (1986). LHRH, catecholamine neuronal systems in the olfactory bulb of the mouse. *J. Comp. Neurol.*, 250, 352–63.

Shepherd, G.M. (1972). Synaptic organisation of the mammalian olfactory bulb. *Physiol. Rev.*, 52, 864–917.

Stanton, P.C. and Sarvey, J.M. (1985). Depeletion of norepinephrine, but no serotonin, reduces long-term potentiation in the dentate gyrus of the rat hippocampal slices. *J. Neurosci.*, 5, 2169–76.

Staubli, U., Ivy. G. and Lynch, G. (1984). Hippocampal denervation causes rapid forgetting of olfactory information in rats. *Proc. Natl. Acad. Sci. USA*, 81, 5885–7.

Sullivan, R.M. and Leon, M. (1986). Early olfactory learning induces an enhanced olfactory bulb response in young rats. *Dev. Brain Res.*, 27, 278–82.

Wilson, D.A. and Leon, M. (1988). Noradrenergic modulation of olfactory bulb excitability in the post natal rat. *Dev. Brain Res.*, 42, 69–75.

Yamazaki, K., Beauchamp, G.K., Wysocki, C.J., Bard, J., Thomas, L. and Boyse, E.A. (1983). Recognition of H-2 types in relation to the blocking of pregnancy in mice. *Science*, 221, 186–8.

# II.2 Modulation of the Odour Detection Performance of Male Rats by Castration and the Administration of Dopaminergic Active Drugs: a Summary of Recent Findings

RICHARD L. DOTY

## Introduction

It is well established that complex relationships exist between endocrine factors and the olfactory function of mammals. Thus, odour preferences of many mammals are closely related to sexual status and endocrine state, as well as to prior social experience (Doty 1986). For example, oestrous—but not dioestrous—female lemmings prefer the odours of males which, in subsequent social tests, prove to be dominant over most of their male counterparts (Huck and Banks 1982). Likewise, castration attenuates or eliminates the sexually experienced male rat's preference for urine odour of oestrous vs. dioestrous females (Carr *et al.* 1965).

In contrast to odour preferences, few studies have examined the influences of endocrine factors on olfactory sensitivity, *per se*. Although castration has little short-term influence on the olfactory thresholds of laboratory rats (Carr *et al.* 1962), recent data indicate that this operation attenuates the long-term increase observed in odour detection performance of repeatedly tested sham-operated rats (Doty and Ferguson-Segall 1989). Furthermore, fluctuations occur in the odour detection performances of female rats across the phases of the oestrous cycle, as well as in women across the phases of the menstrual cycle (Pietras and Moulton 1974; Doty *et al.* 1981). Interestingly, the sensory changes in women, while correlated with perturbations in circulating oestrogen levels, appear not to be caused by the oestrogen fluctuations, *per se*, since they are also present in women taking oral contraceptives (Doty *et al.* 1981).

In view of these observations, we have recently initiated a research programme to explore the degree to which the olfactory sensitivity of both male and female rats can be altered by straightforward manipulations of the endocrine system and by pharmacological manipulations of neurotransmitter systems known to be closely associated with reproductive endocrine function. For example, we have shown that eliminating or greatly attenuating the olfactory bulb norepinephrine of male rats by intrabulbar injections of the neurotoxin 6-hydroxydopamine has no influence on their odour detection performance to the target odorant ethyl acetate (Doty *et al.* 1988*a*); however, systemic administration of drugs that influence D-2 dopamine receptors does affect such detection performance (Doty and Risser 1989). Such observations are of particular interest in the light of clinical studies which demonstrate an olfactory deficit in patients with Parkinson's disease (Doty *et al.* 1988*b*).

Three of the initial studies of this programme are reviewed in this paper. In the first, we demonstrate that castration influences the nature of the long-term odour-detection performance of male rats. In the second, we show that amphetamine, a drug that affects catecholamine release, degradation, and re-uptake at nerve terminals, significantly alters such detection performance. In the third, we demonstrate that the D-2 receptor agonist, quinpirole, produces a dose-related decrement in odour detection performance which can be blocked by the D-2 receptor antagonist spiperone.

## Olfactometer and test procedure

All three of these studies used a computer-controlled operant testing chamber connected to an air-dilution olfactometer described in detail elsewhere (Doty and Ferguson-Segall 1987). Five concentrations of the odorant ethyl acetate, spanning the $-5.50$ to $-3.50$ log concentration range, were generated by the olfactometer and delivered to two independent and simultaneously operable operant chambers housed within thermostatically controlled sound-attenuated boxes (Fig. 1). Each chamber consisted of a 19 cm long × 10.2 cm diameter tube fused to a glass funnel through which the stimuli were delivered (Fig. 2). A trial sequence was initiated when the rat interrupted a photobeam placed across the neck of the funnel. This resulted in a 1 s diversion of the airstream away from the chamber and the addition of odour (S+ trials) or air (S− trials) into an upstream mixing chamber. After this diversion, the odourized or non-odourized airstream was then delivered to the chamber. A bar touch under the S+ condition resulted in water reinforcement and termination of the trial. If no bar touch was made during a 5 s response window, the trial was automatically terminated and a 45 s no-response period was enforced before photobeam activation would initiate a new trial sequence

A daily test session consisted of a total of 260 trials per subject, the first 10 of which were warm-up trials not used in the performance calculations. The stimuli were presented in ascending and descending blocks (see Doty and Ferguson-Segall 1987). The primary dependent variable measured was the non-parametric

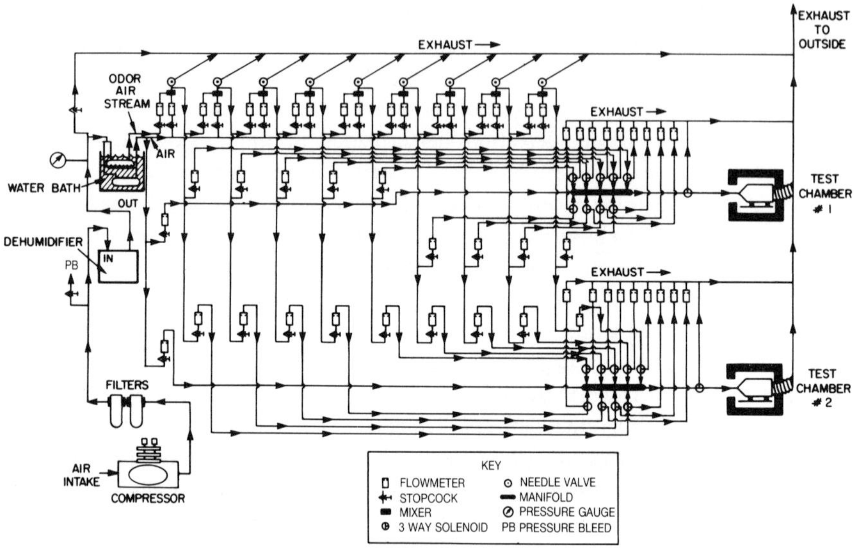


Fig. 1. Schematic representation of the air-dilution olfactometer and associated rat testing chambers. The three-way solenoid valves and both the stimulus and response contingencies are controlled by computer. From Doty and Ferguson-Segall (1987), with permission.

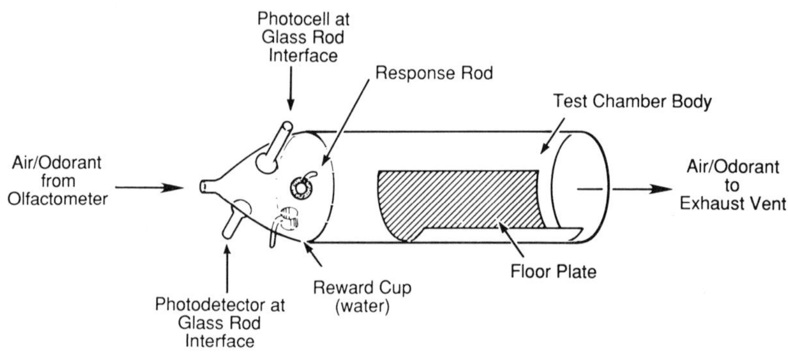


Fig. 2. Rat testing chamber which is housed within a temperature-controlled and sound-attenuated box, as indicated in Fig. 1. From Doty and Ferguson-Segall (1987), with permission.

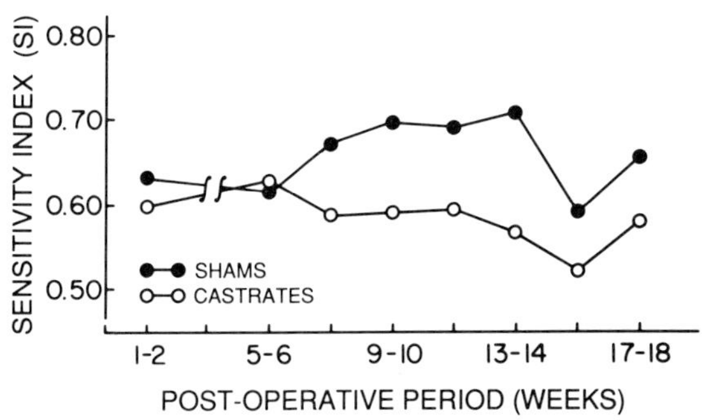


Fig. 3. Analysis of covariance-adjusted mean SI values as a function of post-operative period in sham and castrate male Long Evans rats. For details, see text. Reprinted from Doty and Risser (1989), with permission.

sensitivity index, SI, which is calculated from the proportion of HITS (i.e. bar contacts made under the S+ condition) and false alarms (FA; bar contacts made under the S− condition) using the formula of Frey and Colliver (1973):

$$SI = \frac{P(HIT) - P(FA)}{2[P(HIT) + P(FA)] - [P(HIT) + (FA)]^2}$$

## Influence of castration on long-term odour detection performance

In the first of our studies (Doty and Ferguson-Segall 1989), the odour detection performance of 15 adult male Long Evans rats was repeatedly assessed over an 18-week post-operative test period. Eight of the rats had been castrated, whereas the other seven had been sham-operated. As can be seen in Fig. 3, the sham-operated animals evidenced a significant tendency to improve their performance over time (with the exception of the unexplained drop in performance during weeks 15–16). The castrates, on the other hand, evidenced a non-significant tendency to decrease performance over the test period. These observations suggest that orchidectomy resulted in a mitigation of the tendency of the sham-operates to increase their detection performance across the test period.

## Influence of *d*-amphetamine on odour detection performance

In the second study (Doty and Ferguson-Segall 1987), the effects of *d*-amphetamine sulfate (0.2, 0.4, 0.8, and 1.6 mg/kg SC) on the odour detection performance of 16 adult male Long Evans rats was assessed. In this research, each animal served as his own control and the drug or saline was administered every 3rd day in a counterbalanced order derived from Latin squares, with the injection occurring 5 mins before each test session. As can be seen in Fig. 4, enhanced

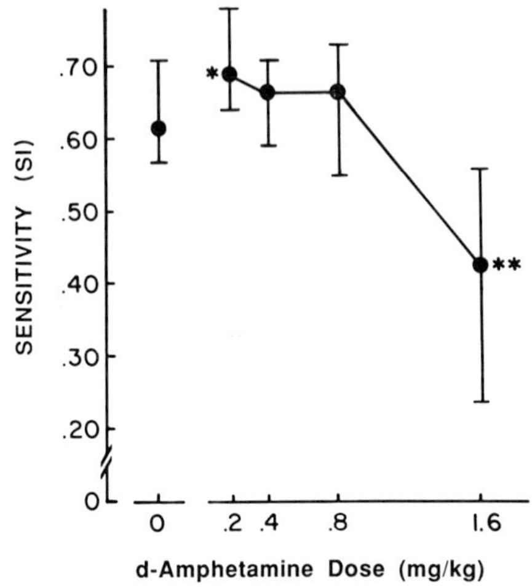


Fig. 4. Median SI (±) values and associated interquartile ranges for saline and four *d*-amphetamine dosage levels. * indicates difference from saline at 0.05 level. ** indicates difference at 0.001 level. From Doty and Ferguson-Segall (1987), with permission.

detection performance to ethyl acetate was observed following administration of 0.2 mg/kg of the drug, whereas decreased detection performance was observed following injection of 1.6 mg/kg of the drug.

## Influence of a D-2 receptor agonist and a D-2 receptor antagonist on detection performance

Since amphetamine is known to have a major influence on the release of dopamine, we followed up this finding in the next study by using a drug known to have relatively specific influences on the dopamine D-2 receptor. In this experiment, we tested the influence of five doses of the D-2 receptor antagonist quinpirole (0.025, 0.05, 0.75, 0.10, and 0.20 mg/kg IP) on the odour detection performance of 21 adult male Long Evans rats. Additionally, 10 rats were pre-treated with the D-2 receptor antagonist spiperone (0.062 mg/kg IP) and their performance monitored following quinpirole administration.

As in the case of amphetamine, treatments were administered every 3rd day in a counterbalanced order (with the quinpirole injections occurring 15 mins before, and the spiperone injections 35 mins before, the 260 trial test sessions). Quinpirole injection alone resulted in a dose-dependent decrease in odour detection performance, as measured by both SI and the more traditional performance measure of the per cent of trials for which correct performance occurred (Fig. 5). As can be seen in the two top left-hand sectors of this figure, prior treatment with spiperone eliminated these effects. Dose-related influences of quinpirole on (a) the

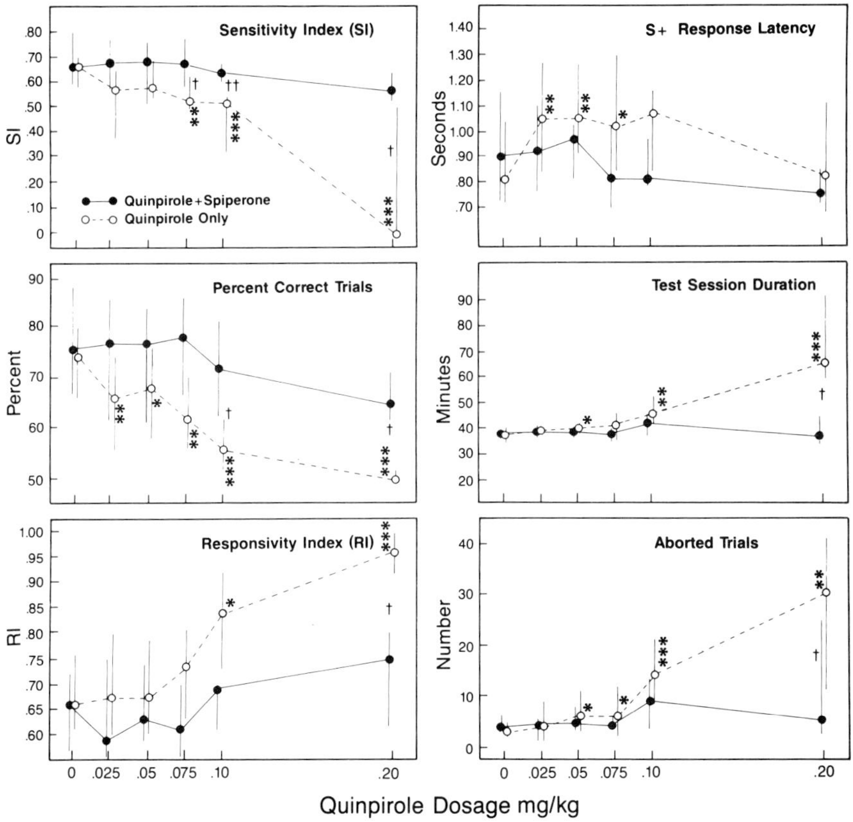


Fig. 5. Overall medians and interquartile ranges for six dependent variables as a function of quinpirole dosage. Asterisks denote doses at which scores differed significantly from saline (* $= p < 0.05$, ** $= p < 0.01$, *** $= p < 0.001$). Daggers indicate those doses at which scores of the 10 animals who served in both groups differed between treatments (†, ††, ††† = $p < 0.05$, 0.01, and 0.001, respectively). From Doty and Risser (1988), with permission.

average latency to initiate a detection response (i.e. the S+ response latency), (b) the total session duration, and (c) the number of aborted trials were also eliminated or greatly attenuated by prior spiperone injection. These results, along with those from other studies of our laboratory, suggest that dopaminergic-active drugs can modulate the odour performance of male rats.

## Discussion

The studies reviewed in this paper demonstrate that odour detection performance, determined quantitatively in a go/no go operant signal detection task, is influenced by (a) castration and (b) drugs known to influence dopamine D-2 receptor activity, including amphetamine. Although the basis of the castration-related effect is not known, it is apparent that it requires a relatively long post-castration period to be observed. Mechanisms potentially responsible for this decrement in performance include influences on olfactory sensitivity, general arousal, and learning (Doty and Ferguson-Segall 1989).

The aforementioned pharmacological studies do not definitely establish whether the influence of amphetamines on odour detection performance is mediated by dopaminergic or noradrenergic pathways. If this influence is mediated via arousal mechanisms, dopamine would appear to be the likely primary transmitter involved, since (a) alpha-or beta-adreno-receptor blocking agents do not alter amphetamine's central stimulant activities; (b) inhibition of the synthesis of norepinephrine, without altering dopamine synthesis, using inhibitors of the noradrenergic synthetic enzyme dopamine-beta-hydroxylase fails to block amphetamine's stimulant actions; (c) neuroleptic drugs which selectively block central dopaminergic receptors (e.g. chloropromazine, haloperidol, and pimozide) completely prevent the central stimulant effects of amphetamines; and (d) selective 6-hydroxydopamine (6-OHDA) lesions of dopaminergic pathways block amphetamine-induced stimulation, whereas noradrenergic lesions fail to alter the response (Iverson and Iverson 1981). Interestingly, high concentrations of dopamine and large numbers of dopaminergic receptor sites are found within key regions of the olfactory system (e.g. olfactory tubercle). Indeed, all olfactory tubercle laminae receive a dense dopaminergic innervation from fibres of the mesocortical dopamine system whose cell bodies lie within the medial substantia nigra and the ventral tegmental area (Fallon and Moore 1978; Fallon *et al.* 1978; Krieger *et al.* 1971; Lindvall and Bjorklund 1978). Current research at our centre is designed to establish whether D-1 and D-2 dopamine receptors participate, either alone or together, in the mediation of the amphetamine-induced alterations in detection performance.

## Acknowledgements

This work was supported by National Institute of Neurological and Communicative Disorders and Stroke grant NS 16365. I thank Dr Mark Ferguson-Segall and Ms Judy Risser for their collaboration on the primary studies discussed in this paper.

## References

Carr, W.J., Solberg, B. and Pfaffmann, C. (1962). The olfactory threshold for estrous female urine in normal and castrated male rats. *J. Comp. Physiol. Psychol.*, 55, 415–7.

Carr, W.J., Loeb, L.S. and Dissinger, M.L. (1965). Responses of rats to sex odors. *J. Comp. Physiol. Psychol.*, 59, 370–7.

Doty, R.L., Snyder, P., Huggins, G. and Lowry, R.D. (1981). Endocrine, cardiovascular, and psychological correlates of olfactory sensitivity changes during the human menstrual cycle. *J. Comp. Physiol. Psychol.*, 95, 45–60.

Doty, R.L. (1986). Odor-guided behavior in mammals. *Experientia*, 42, 257–71.

Doty, R.L. and Ferguson-Segall, M. (1987). Odor detection performance of rats following d-amphetamine treatment: A signal detection analysis. *Psychopharmacology*, 93, 87–93.

Doty, R.L., Ferguson-Segall, M., Lucki, I. and Kreider, M. (1988*a*). Effects of 6-hydroxydopamine intrabulbar injections on ethyl acetate odor detection in castrate and non-castrate male rats. *Brain Res.*, 444, 95–103.

Doty, R.L., Deems, D. and Stellar, S. (1988*b*). Olfactory dysfunction in Parkinson's disease: A general deficit unrelated to neurologic signs, disease stage, or disease duration. *Neurology*, 38, 1237–44.

Doty, R.L. and Ferguson-Segall, M. (1989). Influence of castration on the odor detection performance of male rats. *Behav. Neurosci.*, 103, 691–93.

Doty, R.L. and Risser, J. (1989). Influence of the D-2 dopamine receptor agonist quinpirole on rat odor detection performance before and after administration of spiperone. *Psychopharmacology*, 98, 310–15.

Fallon, J.H. and Moore, R.Y. (1978). Catecholamine innervation of the basal forebrain. III. Olfactory bulb, anterior olfactory nuclei, olfactory tubercle, and piriform cortex. *J. Comp. Neurol.*, 180, 533–44.

Fallon, J.H., Koziell, D.A. and Moore, R.Y. (1978). Catecholamine innervation of the basal forebrain. II. Amygdala, suprarhinal cortex, and entorhinal cortex. *J. Comp. Neurol.*, 180, 509–32.

Frey, P.W. and Colliver, J.A. (1973). Sensitivity and responsitivity measures for discrimination learning. *Learning Motiv.*, 4, 327–42.

Huck, U.W. and Banks, E.M. (1982). Differential attraction of females to dominant males: Olfactory discrimination and mating preference in the Brown Lemming (*Lemmus trimucronatus*). *Behav. Ecol. Sociobiol.*, 11, 217–22.

Iverson, S.D. and Iverson, L.L. (1981). *Behavioral endocrinology*, Oxford University Press, Oxford.

Krieger, N.R., Kauer, J.S., Shepherd, G.M. and Greengard, P. (1971). Dopamine-sensitive adenylate cyclase within laminae of the olfactory tubercle. *Brain Res.*, 131, 303–12.

Lindvall, O. and Bjorklund, A. (1978). The organization of the ascending catecholamine neurone system in the rat brain. *Acta Physiol. Scand., Suppl.*, 412, 1–48.

Pietras, R.J. and Moulton, D.G. (1974). Hormonal influences on odor detection of rats: changes associated with the estrous cycle, pseudopregnancy, ovariectomy, and administration of testosterone propionate. *Physiol. Behav.*, 121, 475–91.

# II.3 Posterolateral Gland Positions among Microtine Rodents

FREDERICK J. JANNETT, JR.

## Introduction

Many species of voles and lemmings (Rodentia: Muridae: Microtinae) possess scent glands on the posterolateral surface which are modified hypertrophied sebaceous glands. These scent glands have been considered as taxonomically important at the levels of genus, subgenus, and species (e.g. Miller 1896; Bailey 1897, 1900; Skurat 1969; Hall 1981). The taxonomic distributions, positions, and histological structure of many of these glands were reviewed by Quay (1968) who concluded that they are found on the tail (caudal gland of *Dicrostonyx*), rump (*Lemmus*), bilaterally on the hips (*Microtus*, subgenera *Lasiopodomys*, part of *Microtus*, part of *Pitymys*) and bilaterally on the flanks (*Neofiber*, *Clethrionomys*, *Aschizomys*, *Eothenomys*, *Alticola*, part of *Phenacomys*, *Lagurus*, *Arvicola*, and *Microtus* subgenera *Aulacomys*, *Stenocranius*, *Phaiomys*, *Neodon*, part of *Microtus*, *Pedomys* (?) and *Herpetomys* (?)). Von Claussen (1975) reported a rump gland on the wood lemming, *Myopus schisticolor*. Some species do not have any of these glands.

Quay (1968) relied heavily on published accounts and standard museum study skins. Unfortunately, the preparation of the latter distorts the position of bilateral hip and flank glands, and Quay (1968) left open the possibility of intermediate or overlapping positions of glandular areas in some species.

An intermediate position of glands between hips and flanks was reported by Jannett (1975) for *Microtus longicaudus*, a species that does not characteristically have posterolateral glands. This author induced glandular development by a pharmacological dose of testosterone propionate. Jannett (1975) also induced hip glands on *M. pennsylvanicus* which similarly does not characteristically have such glands.

This report reviews additional material, including more subjects given testosterone as mentioned above. More importantly, the precise locations of the glands for many species were localized by fixing fresh specimens in a uniform position, applying a radio-opaque paste to glands of representative specimens, and x-raying them. The hypothesis tested was that there is an essentially continuous zone of glandular responsiveness between flanks and rump.

**Table 1.** Fluid-preserved specimens examined.

| Genus (subgenus) species | Untreated specimens: Collections ♂♂ | Untreated specimens: Collections ♀♀ | Untreated specimens: Laboratory ♂♂ | Untreated specimens: Laboratory ♀♀ | Subjects given testosterone: Field-trapped ♂♂ | Subjects given testosterone: Field-trapped ♀♀ | Subjects given testosterone: Laboratory ♂♂ | Subjects given testosterone: Laboratory ♀♀ | Locality or source of breeding stock (*n*) |
|---|---|---|---|---|---|---|---|---|---|
| *Dicrostonyx rubricatus* | | | | 2 | | | 3 | 3 | USA, NE Alaska |
| *Synaptomys cooperi* | | | | | 3 | 3 | | | USA, Minnesota, Cook Co. |
| *Myopus schisticolor* | 3 | | | | | | | | Finland, Oulanka National Park (1); USSR, Ust'-Kamchatka District, Yelovka(2) |
| *Clethrionomys glareolus* | | | | | | | 3 | 3 | Great Britain |
| *C. smithii* | 3 | 2 | | | | | | | Japan, Honsu |
| *Lagurus curtatus* | | | 8 | | | | 8 | | USA, Wyoming, Albany Co. |
| *Arvicola terrestris* | 2 | 2 | | | | | 1 | 1 | Great Britain (2); USSR, Moscow District (1), Ulyanovsk District (1), Saratov District (1), Lithuanian SSR (1) |
| *Microtus (Aulacomys) richardsoni* | 18 | | | | 5 | 4 | | | USA, Wyoming, Teton Co. |
| *M. (Stenocranius) gregalis* | 4 | 5 | | | | | | | USSR, Krasnoyarsk area, Minusinsk District (3), Yamal Peninsula Morda River (6) |
| *M. (Sumeriomys) socialis* | 4 | 3 | 1 | 1 | | | | | USSR, Kazakh SSR, Turjev area, Emba (2), Kazakh SSR, Dzhambul District, Betpak-Dala (4), Uzbek, SSR, Samarkand (1), laboratory hybrids of two subspp. from Armenian SSR (2) |
| *M. (Chionomys) gud* | | | 2 | | | | 4 | 1 | USSR, RSFSR, Dagestan ASSR, Andiyskoye Koysu, near Khvarshe |
| *M. (Pedomys) ochrogaster* | | | | | | | 7 | 7 | USA, Missouri, Boone Co., S. of Columbia |
| *M.(Pitymys) pinetorum* | | | 11 | | | | 5 | 5 | USA, New York, Ulster Co., New Paltz |
| *M. (Blanfordimys) afghanus* | 4 | | | | | | | | USSR, Turkmenian SSR, Badhyz (3), Turkmenian SSR, Takhta-Bazar District, 10 km SW Kalai-Mor (1) |
| *M. (Microtus) agrestis* | | | | | | | 3 | 2 | Great Britain |
| *M. (M.) arvalis* | | | 2 | | | | 2 | 1 | USSR, Lithuanian SSR, near |

| | | | | | | | | | |
|---|---|---|---|---|---|---|---|---|---|
| | | | | | | | | | RSFSR, Krasnodar (2), RSFSR, Ladoga River area (2) |
| *M. (M.) californicus* | 9 | | 2 | | | | | | USA, California, Sonoma Co. Bodega Marine Laboratory |
| *M. (M.) chrotorrhinus* | 14 | 1 | | | | | | | USA, Minnesota, Cook Co. (13), New York, Essex Co. (2) |
| *M. (M.) evoronensis* | | | 1 | | | | 2 | | USSR, RSFSR, Primor'e, bank of Lake Evoron |
| *M. (M.) fortis* | 1 | | | | | | 2 | | USSR, RSFSR, Primorskiy Kray, Peter-the-Great Bay, Putyatin Island (1), Sakhando Reservation, Chita Oblast (2) |
| *M. (M.) hyperboreus* | | | | | | | 2 | | USSR, RSFSR, Yakut ASSR, 85 km from Adoucha at Agra River |
| *M. (M.) kirgisorum* | | | 1 | 1 | | | 2 | 2 | USSR, Uzbek SSR, near delta of Amu Dar'ya |
| *M. (M.) longicaudus* | 11 | 3 | 2 | 2 | 4 | 2 | 4 | 3 | USA, Wyoming, Teton Co. |
| *M. (M.) maximowiczii* | | | 1 | | | | 2 | | USSR, RSFSR, Primor'e, NW of Khabarovsk |
| *M. (M.) mongolicus* | 1 | | 4 | 1 | | | 4 | 3 | Mongolia, Khangai Ridge (1); USSR,RSFSR, Chita Oblast, Borzya (12) |
| *M. (M.) montanus* | 22 | | | | | | | | USA, Wyoming, Teton Co. |
| *M. (M.) oeconomus* | 5 | 2 | | | | | | | USA, Alaska, Alaska Peninsula, Izembek Lagoon, Round Island, 9 miles NNW Cold Bay |
| *M. (M.) pennsylvanicus* | 8 | | | | 8 | | | | USA, Wyoming, Teton Co. (12), New York, Tompkins Co. (4) |
| *M. (M.) rossiaemeridionalis* | | | 4 | 1 | | | 3 | 3 | USSR, RSFSR, Voronezh Oblast, Chernovoye (4), RSFSR, Leningrad area (7) |
| *M. (M.) xanthognathus* | 1 | 2 | | | | | | | USA, Alaska, near McGrath, Salmonberry Lake |
| *Prometheomys schaposchnikovi* | 2 | 2 | | | | | | | USSR, Georgian SSR, Adgariya Bakhmaro |
| *Ellobius fuscocapillus* | | | 9 | 3 | | | | | USSR, Turkmenian SSR, Kushka |
| *E. lutescens* | | | 9 | 3 | | | | | USSR, Armenian SSR, Volchabezd |
| *E. tancrei* | | | 3 | 3 | | | | | USSR, Pamiro-Altai Mtns, Vakhsh-Surkhob River valley |

## Materials and methods

Specimens studied were of four types, of which the first three are listed in Table 1.

Carcasses fixed in 10% formalin pinned flat on their backs and stretched gently between a pin in the tail and one in the nasal cartilage; the abdominal cavity was opened and the legs were not constrained.

2. Subjects that received a subcutaneous implant (75 mg pellet) of testosterone propionate in the nape region ('Oreton', Schering Corporation, Bloomfield, NJ). The light cycle for these subjects was usually 16L:8D, and the pellets were left in for 21 or 22 days. Carcasses were subsequently fixed as described above.
3. Select alcohol-preserved museum specimens with little contortion, some of which were suitable for radiography. When no posterolateral glands were present, body form was not relevant.
4. Standard museum study specimens and flat skins. These specimens in the Museum of Zoology of Moscow State University were used to supplement observations of the other specimens.

All other specimens reported here are in the collections of the Science Museum of Minnesota or of the author.

Posterolateral glands are generally most highly developed on adult males in reproductive condition (Quay 1968). For those included in Table 1, each species is represented by at least three adult males with scrotal testes unless fewer were available, but at least one of which had distinct glands. The exception was *Prometheomys schaposchnikovi*, represented by two adults of each sex, neither with glands. Females of at least some species develop posterolateral glands in response to the testosterone treatment. These individuals and females with naturally occurring glands were used to assess gland position. However, numbers were not sufficient to characterize frequency of occurrence for many species.

To localize the positions precisely, specimens were radiographed. A mid-dorsal cut was made in the skin to intersect a cut made on each side up from the venter and in front of the flanks. The skin was then peeled back on each side to expose the putative glands which, when present, from the inside appear strikingly bilateral and whitish. No attempt was made histologically to identify posterolateral glands that were not macroscopically obvious. On fresh specimens with posterolateral glands, there was always an oiliness on the hair above them. I followed the technique of Vrtis (1930) by applying a radio-opaque paste of barium sulphate to the inner area of the gland. The skin was then folded back into place and secured with an insect minutien pin. By blotting the inside of the skin and using a paste that did not run, there was no distortion of the localized spot. X-rays were recorded on mammography film. The procedure is described in greater detail by Tashjian and Jannett (ms.). Representative specimens (one or two per species) were radiographed with the following exceptions. Both of the two adult male *Arvicola terrestris* were radiographed; there was only one male *Microtus xanthognathus*, although the gland position was typical of that found on all other specimens (two parous adults, two subadult males, and three subadult females); the only alcohol-preserved adult male *Myopus schisticolor* was used; the least contorted *Microtus gregalis* alcohol-preserved specimen was used.

All species determinations of specimens from the Soviet Union were made by the Soviet museum and laboratory colleagues.

From the radiographs, the position of the posterolateral gland is categorized (Table 2) with respect to (i) the six lumbral vertebrae and hips, and (ii) the length of the femur. Information on *Clethrionomys rufocanus* in Tables 2 and 3 are from an unpublished radiograph, part of an extensive survey of numerous scent glands from all the Palearctic mainland species of *Clethrionomys* (to be presented elsewhere). Table 3 lists the absence or position of glands on all species examined.

## Results

Posterolateral glands occur from the anterior flank area to the tail. Representative specimens are presented in Figs. 1–5. There is essentially a continuous zone of hypertrophy seen when the specimens are aligned. Nevertheless, seven bilateral regions can be distinguished on the basis of the centre of the gland, for the purpose of comparison: anterior flank, mid-flank, posterior flank, knee, anterior hip, mid-hip, and posterior hip. The distinction between anterior flank glands and mid-flank glands is subjective but those in the first group appear far forward of the knee. All other regions appear distinct upon dissection. The posterior flank gland is just in front of the knee and may slightly overlap the knee. The knee gland is centred at the knee and overlaps the anterior margin of the hip and the last two or three lumbar vertebrae. The anterior hip gland is forward on the hip, barely reaches the sixth lumbar vertebra (L6), and does not extend to the acetabulum; with the femur in a forward position it appears along the mid-femur. The mid-hip gland appears along the middle of the hip and the proximate region of the femur, and reaches the acetabulum. The posterior hip gland is centred over the acetabulum.

Examination of alcohol-preserved *Myopus schisticolor* was supplemented by that of 74 standard study skins and 19 flat skins of both sexes and all age classes of this species from the eastern portion of its range. Six study specimens of adult males had bilateral patches of oil on what appeared to be the knee position, four of which were from Yakutia, Kolyina River, RSFSR, USSR. One flat skin of an adult male had bilateral patches averaging 15.5 mm long 7 mm wide in this position; the specimen was from Imandra Lake, Kola Peninsula, RSFSR, USSR.

No series was polymorphic for posterolateral gland location. The only species that appeared to be polymorphic was *Arvicola terrestris*. Although there were few

**Table 2.** Localization of posterolateral glands from radiographs.

| Genus (subgenus) species | Position of centre of gland | | Comments |
|---|---|---|---|
| | **With respect to axial skeleton** | **With respect to femur** | |
| *Clethrionomys rufocanus* | L2 + 3 | far forward | |
| *Arvicola terrestris* | L2 + 3 | " " | Great Britain; stretched by someone else |
| *Microtus (Chionomys) gud* | L2 + 3 | " " | |
| *Lagurus curtatus* | L3 + 4 | in front | |
| *Arvicola terrestris* | L4 + 5 | " " | USSR; not stretched |
| *Microtus (Aulacomys) richardsoni* | L3 + 4 | " " | |
| *Synaptomys cooperi* | L5 + 6 | in front, posterior margin barely touches knee | |
| *Microtus (M.) xanthognathus* | L5 + 6 | " " | not stretched |
| *Microtus (Stenocranius) gregalis* | L3–5 | " " | not stretched |
| *Myopus schisticolor* | L5 + 6 + anterior hip | over knee | |
| *Microtus (Pitymys) pinetorum* | L4–6 + anterior hip | " " | |
| *M. (M.) longicaudus* | L5 + 6 + anterior hip | " " | |
| *M. (M.) pennsylvanicus* | anterior hip | middle | reaches L6 but not acetabulum |
| *M. (M.) montanus* | mid-hip | proximate | reaches acetabulum; not stretched |
| *M. (M.) chrotorrhinus* | acetabulum | head | |
| *M. (M.) fortis* | " " | " " | |
| *M. (M.) oeconomus* | " " | " " | |

**Table 3.** Absence and positions of paired posterolateral glands among species.

| Genus (subspecies) species | A[a] | Location of posterolateral glands: Anterior flank | Mid-flank | Posterior flank | Knee | Anterior hip | Mid-hip | Posterior hip |
|---|---|---|---|---|---|---|---|---|
| *Dicrostonyx rubricatus* | X | | | | | | | |
| *Synaptomys cooperi* | | | | N, Fig.3b | | | | |
| *Myopus schisticolor* | | | | | N, Fig.3a | | | |
| *Clethrionomys glareolus* | X | | | | | | | |
| *C. smithii* | X | | | | | | | |
| *C. rufocanus* | | N[b] | | | | | | |
| *Lagurus curtatus* | | | N, Fig.3c | | | | | |
| *Arvicola terrestris* | | N, Great Britain, Fig.1a | N, USSR, Fig.3d | | | | | |
| *Microtus (Aulacomys) richardsoni* | | | N, Fig.1b | | | | | |
| *M. (Stenocranius) gregalis* | | | | N, Fig.4a | | | | |
| *M. (Sumeriomys) socialis* | X | | | | | | | |
| *M. (Chionomys) gud* | | I[c], Fig.2b | | | | | | |
| *M. (Pedomys) ochrogaster* | X | | | | | | | |
| *M. (Pitymys) pinetorum* | | | | | I, Fig.1c | | | |
| *M. (Blanfordimys) afghanus* | X | | | | | | | |
| *M. (Microtus) agrestis* | X | | | | | | | |
| *M. (M.) arvalis* | X | | | | | | | |
| *M. (M.) californicus* | | | | | | | N | |
| *M. (M.) chrotorrhinus* | | | | | | | | N, Fig.2e |
| *M. (M.) evoronensis* | | | | | | | | N |
| *M. (M.) fortis* | | | | | | | | N, Fig.1d |
| *M. (M.) hyperboreus* | | | | | | | | N |
| *M. (M.) kirgisorum* | X | | | | | | | |
| *M. (M.) longicaudus* | | | | | I, Fig.2c | | | |
| *M. (M.) maximowiczii* | | | | | | | | N |
| *M. (M.) mongolicus* | | | | | | | | N |
| *M. (M.) montanus* | | | | | | | N, Fig.2d | |
| *M. (M.) oeconomus* | | | | | | | | N, Fig.4b |
| *M. (M.) pennsylvanicus* | | | | | | I, Fig.5 | | |
| *M. (M.) rossiaemeridionalis* | X | | | | | | | |
| *M. (M.) xanthognathus* | | | | N, Fig.2a | | | | |
| *Prometheomys schaposchnikovi* | X | | | | | | | |
| *Ellobius fuscocapillus* | X | | | | | | | |
| *E. lutescens* | X | | | | | | | |
| *E. tancrei* | X | | | | | | | |

A[a] = absent; N[b] = natural glands; I[c] = induced glands

specimens and those from the USSR were not uniformly stretched, the glands of the specimens from Great Britain are nonetheless clearly further forward.

Administration of testosterone did not change the shape of already existing generally elliptical glands in *Synaptomys cooperi*, *Lagurus curtatus*, *Arvicola terrestris*, or *Microtus* (*Aulacomys*) *richardsoni*. The hormone treatment induced glandular hypertrophy on *Microtus* (*Pitymys*) *pinetorum*, *Microtus* (*Chionomys*) *gud*, and the original series of *M.* (*Microtus*) *pennsylvanicus* and *M.* (*M.*) *longicaudus*. It did not induce glands on *M.* (*Microtus*) *agrestis*.

## Discussion

### TAXONOMIC DISTRIBUTIONS

The taxonomic distributions of the posterolateral glands are not as clear-cut as summarized by Quay (1968). Voles of different species in the genus *Microtus* have posterolateral glands, natural or induced, along the entire region. Quay (1968) identified flank glands as having the widest taxonomic distribution: *Synaptomys*, *Neofiber*, *Clethrionomys*, *Aschizomys*, *Eothenomys*, *Alticola*, *Phenacomys* (in part), *Lagurus*, *Arvicola*, and *Microtus* (in part). This is surely true. However, within five of those genera reported here there are still considerable differences in location.

A few specific comments are pertinent. In addition to specimens of *M. chrotorrhinus* from Minnesota with posterior hip glands, a series of study skins of male *M. chrotorrhinus carolinensis* from West Virginia also have hip glands which appear to be in the posterior position (unpublished observations). The report of flank glands in this species perpetuated since Bailey (1900) would seem to be in error, so the position of the glands is not evidence for an affinity of this species with *M. xanthognathus*.

The North American pine vole, *pinetorum*, has here been referred to as *Microtus*, subgenus *Pitymys*, and not to the genus *Pitymys*, merely to follow the North American convention (e.g. Hall 1981). If fluid-preserved specimens of Palearctic *Pitymys* have knee glands, as a perusal of study skins indicates they do (unpublished observations), the shared character would support monophyly and the removal of *pinetorum* from *Microtus*, in as much as knee glands are rare among microtine rodents. At least, the shared character would fail to support the hypothesis that *Pitymys* is polyphyletic.

Specimens of *M.* (*Pedomys*) *ochrogaster* were taken from a laboratory colony removed from the wild for 12 years but which was occasionally outbred. Lack of a gland, even after testosterone administration, must be taken as a tentative conclusion in as much as the stock may have lost the trait or ability to respond. The pharmacological dose of testosterone did not induce glandular development on *Microtus agrestis*, but this species is known to possess posterolateral glands (Clarke and Frearson 1972). Similarly, lack of a posterolateral gland in some taxa is a tentative conclusion. However, mole-voles (*Ellobius*) without posterolateral glands are of particular interest, as Quay (1968) had no material to examine and some workers (e.g. Gromov and Polyakov 1977) have not included the genus among microtine rodents.

Lastly, the wood lemming (*Myopus schisticolor*) from the Scandinavian Peninsula reported by Von Claussen (1975) had a rump gland, whereas specimens reported here from Finland and the USSR had bilateral knee glands. The wood lemming in the USSR is not known to have a rump gland (J. Koslovski, personal communication). This is apparently the only species of microtine rodent that is clearly polymorphic for the position of posterolateral glands. This pattern may be correlated with geography in that the range of the species, in Norway and Sweden and then again from Finland across Soviet Asia, is disjunct. Similarly, the possibility of geographic variation in the position of *Arvicola terrestris* posterolateral glands should be explored.

## EVOLUTION OF POSTEROLATERAL GLANDS

The series of posterolateral glands reviewed here has been considered as taxonomically important for nearly a century but is perhaps more interesting in an evolutionary context. It should be borne in mind that microtine rodents are a relatively new assemblage in geological time, dating from the late Miocene (Carleton and Musser 1984). Many species show considerable karyotypic diversity (N.N. Vorontsov, personal communication). *Microtus*, which exhibits the most posterolateral gland diversity, is generally regarded to be the most highly evolved genus (Hinton 1926).

A preponderance of posterolateral glands in the general flank area and toward the rear of the hips is consonant with voles being able to employ these glands in active behavioural marking patterns which presumably effectively distribute their scent. At least three species of *Microtus* with well-developed hip glands actively mark with what is apparently the same pattern of rubbing the hips against tunnels as is performed in glass tubes (*M. montanus*, Jannett and Jannett 1974; *M. townsendi*, MacIsaac 1977; *M. californicus*, Lidicker 1980). Voles of numerous genera scratch the anterior flank glands, which transfers the scent to the feet and may stimulate secretion, and two species of *Microtus* also rub them against objects (Jannett and Jannett, ms. *b*). How glands in the intermediate knee position are employed is not known. They are perhaps poorly positioned for either active behavioural pattern.

The current study confirms only a very general correlation of gland site and taxonomy, and many species do not have posterolateral glands, even in the genus *Microtus*. These observations indicate that some other factors are significant in addition to possible phylogenetic constraints. Within species of the genus *Microtus*, Jannett (1986) found that preputial, anal, and posterolateral gland weights were correlated, and that among species with relatively large glands were several known or suspected to be polygynous, as would be expected by sexual selection theory. Other possible selection pressures were discussed by Jannett (1981).

The presence of glands seems to be plastic, as is strikingly demonstrated by the induction of glands on the four species (*M. pennsylvanicus*, *M. longicaudus*, *M. gud*, *M. pinetorum*) which do not normally have them or which perhaps have them in a very rudimentary state. More recently, naturally occurring hip glands have been identified in *M. pennsylvanicus* from Ontario (Boonstra and Youson 1982). Massachusetts (Tamarin 1981), and Wyoming (Jannett, unpublished). It was previously suggested (Jannett 1975) that *M. pennsylvanicus* and *M. longicaudus* had lost their glands. That interpretation has been accepted by others (e.g. Raikow *et al.* 1979). The induced glands have been termed atavistic (Coyne and Prout 1984). That interpretation would seem to have been premature. The current report indicates that, as a group, the microtine rodents exhibit a zone of especially androgen-responsive sebaceous glands along the posterolateral region, and not, as Quay (1968) concluded, distinct regions of scent organ development. An equally likely interpretation, not mutually exclusive, is that the zone is actively evolving. Induced glands may therefore be potential gland sites, not atavistic structures.

Previous work on posterolateral glands has implied or stressed that the glands are essentially alike and differ largely in their degree of development (Quay 1968; Skurat 1969, personal communication). However, there are several differences which may reflect different genetic control mechanisms. First, as shown by Quay (1968), the (mid-) flank glands of *M. richardsoni* have two kinds of sebaceous glands: tiny ones near the surface and large ones which form the bulk of the gland. In so far as has been reported, other well-developed posterolateral glands are not dimorphic (e.g Sokolov 1982).

Secondly, when given a large dose of testosterone, male *Microtus montanus* develop hip glands which change shape in that they grow dramatically to the base of the tail (Jannett 1978). Even more striking is the growth of the hip gland on some very old laboratory males: it extends from the tail to in front of the knee position, and thus encompasses posterior hip, mid-hip, anterior hip, and knee positions, and at least touches the posterior flank position. In contrast, the posterolateral glands of *M. richardsoni*, *Lagurus curtatus*, and *Synaptomys cooperi* do not change shape (Jannett and Jannett, ms. *a*; Jannett, personal observation).

Thirdly, as Quay (1968) pointed out, there is a gradient in sebaceous gland hypertrophy within many posterolateral glands such that anterior units are more developed than posterior units. This gradient is much less distinct in some species.

Lastly, the difference in hypertrophy is dramatic between species. Even after administration of a pharmacological dose of testosterone propionate, some glands remain hard to distinguish externally (e.g. *Clethrionomys glareolus*; Sokolov *et al.* 1989).

The above comparisons strengthen the conclusion that posterolateral gland development among microtine rodents as shown by induced glands is illustrative of stages in evolution (Jannett 1975). The present study indicates the following

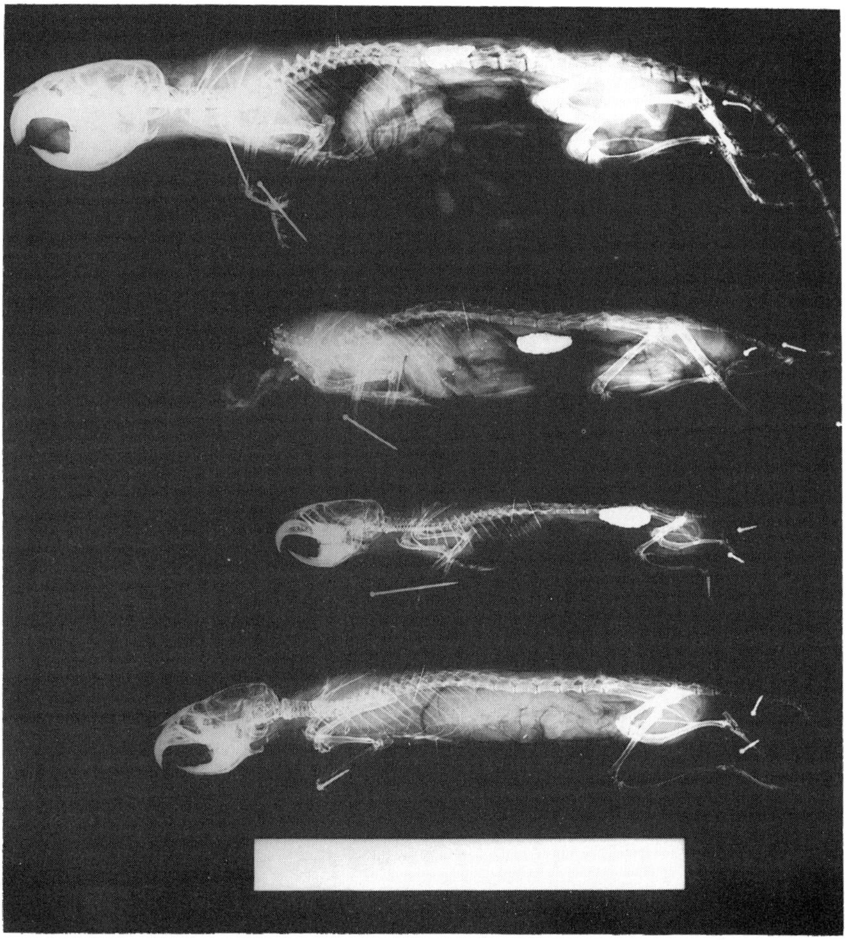

Fig. 1. Whole body radiographs of specimens of four species. a, (top) *Arvicola terrestris*; b, *Microtus richardsoni*; c, *M. pinetorum*; d, *M. fortis* (SMM catalog number Z84:22:40). All individuals received exogenous testosterone. Scale = 10 cm.

working hypothesis: there is a general zone of responsiveness of posterolateral gland development in the group. The mode of hypertrophy is regulated by genes determining which populations of sebaceous glands hypertrophy, how thick the gland becomes, how far it spreads, and how discrete it remains.

The above comparisons and hypothesized developmental sequence do not support Quay's (1968) conclusion that the flank glands are the primitive condition among microtine rodent posterolateral glands. Rather, there is a general zone of responsiveness where development occurs in the course of evolution. Posterolateral glands have probably evolved in several lineages and are still evolving.

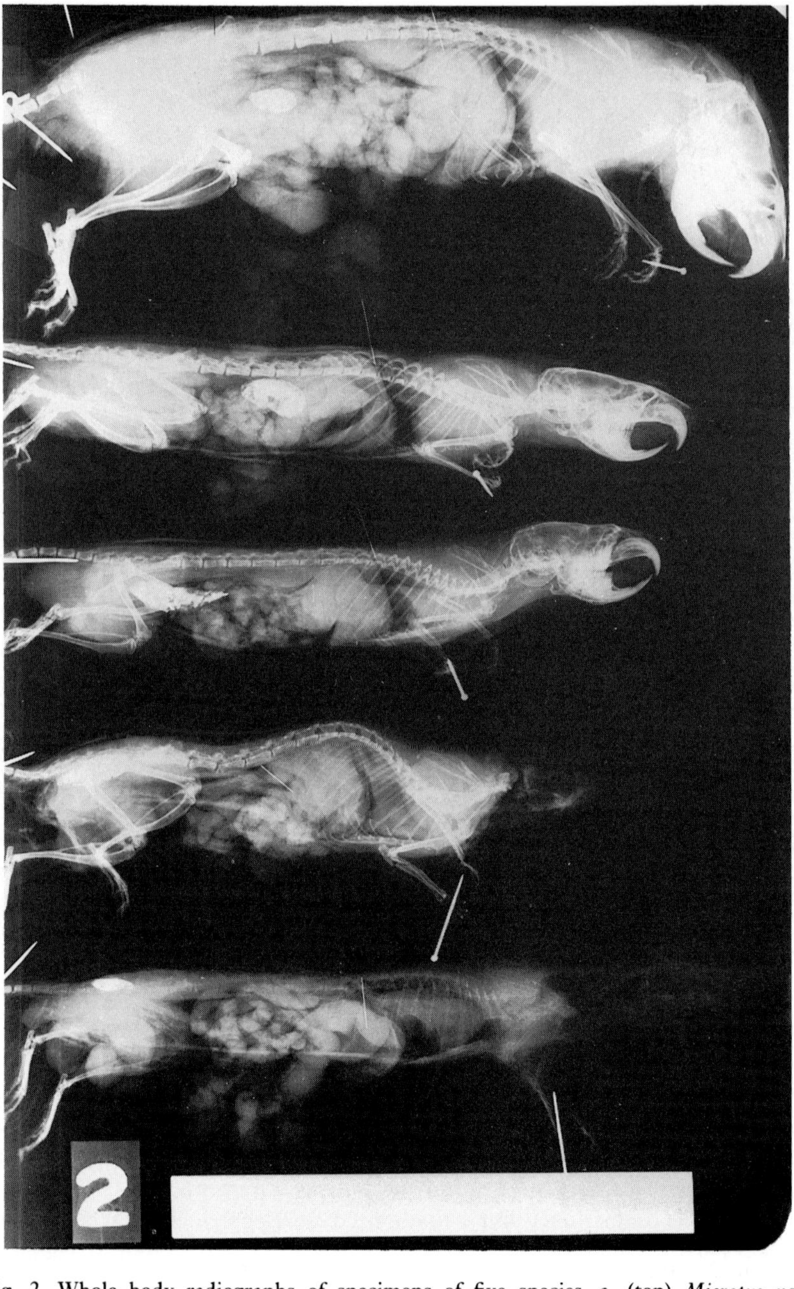

Fig. 2. Whole body radiographs of specimens of five species. a, (top) *Microtus xanthognathus* (Z86:8:1); b, *M. gud* (Z84:22:35) after testosterone treatment; c, *M. longicaudus* after testosterone treatment; d, *M. montanus*; e, *M. chrotorrhinus*. Scale = 10 cm.

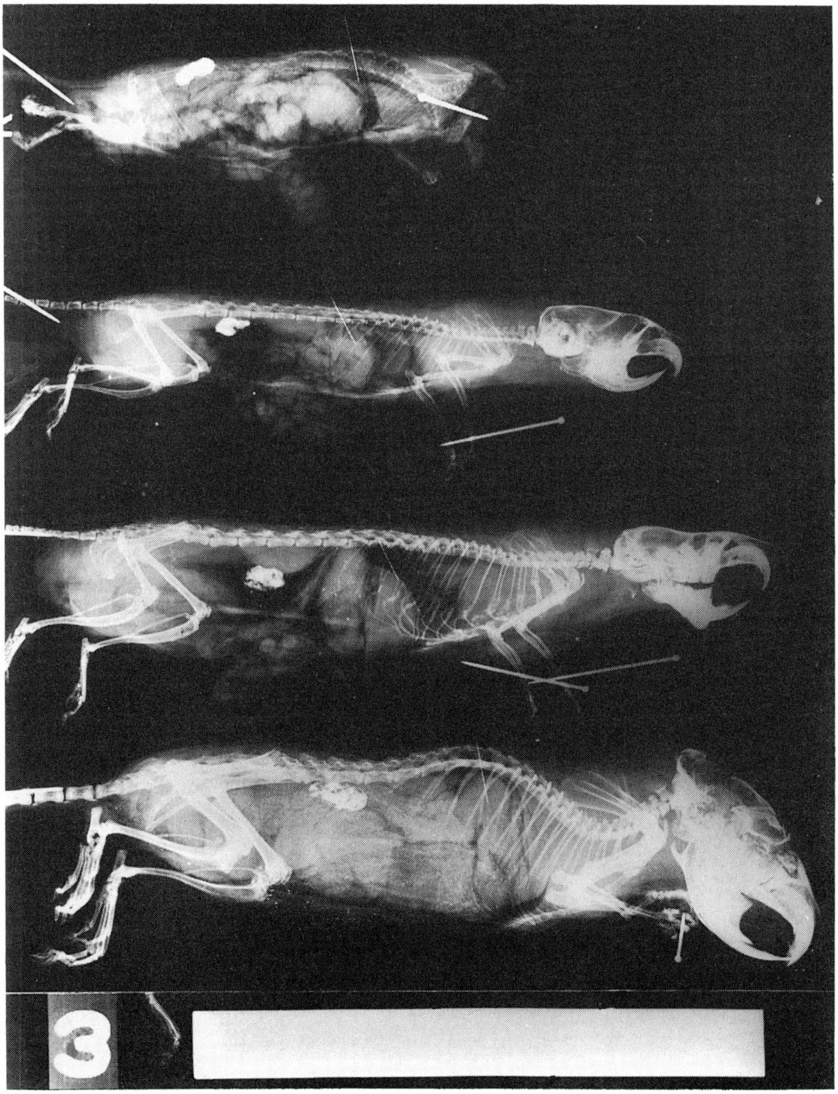

Fig. 3. Whole body radiographs of specimens of four species. a, (top) *Myopus schisticolor* (Z83:6:151); b, *Synaptomys cooperi* after testosterone treatment; c, *Lagurus curtatus* after testosterone treatment; d, *Arvicola terrestris* (Z84:21:1). Scale = 10 cm.

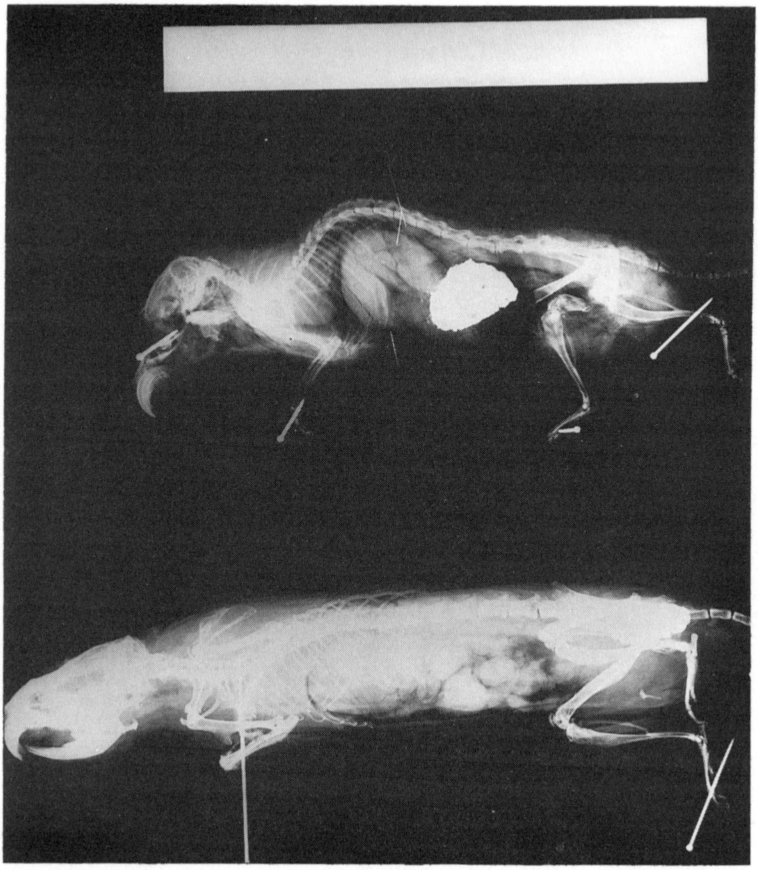

Fig. 4. Whole body radiographs of specimens of two species. a, (top) *Microtus gregalis* (Z84:21:2); b, *M. oeconomus* (Z87:1:1). Scale = 10 cm.

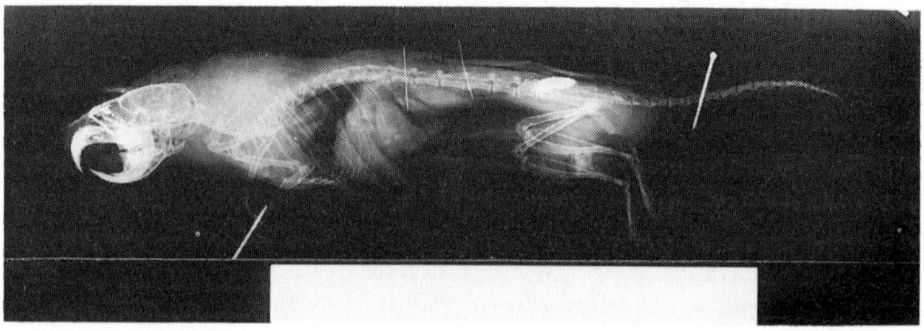

Fig. 5. Whole body radiograph of specimen of *Microtus pennsylvanicus* after testosterone treatment. Scale = 10 cm.

## Acknowledgements

This work was supported by the Ford Foundation Ecology of Pest Management Program at Cornell University, the New York Zoological Society, the New York Cooperative Wildlife Research Unit, the Society of the Sigmi Xi, the Science Museum of Minnesota (SMM), the Severtsov Institute, and the Academies of Science of the USA and USSR.

I particularly thank J. Tashjian and the Radiology Departments of the University of Minnesota and St Paul-Ramsey Hospital for radiographs. I also thank the National Park Service for permission to trap in Grand Teton National Park, and Susan Nowland for typing the manuscript.

I also thank F.N. Golenischev, R.S. Gourley, M.R. Hakhverdyan, G. Jarrell, A. Magoun, V. Malikov, V.M. Malygin, M.N. Meyer, R. Ostfeld, M.E. Richmond, M. Schadler, R.A. Stehn, N.N. Vorontsov for specimens, and N. Abramson, O. Rossolimo, P.P. Strelkov, and I.Ya. Pavlinov for access and assistance in collections. Lastly, special thanks to B. Blake and G. Jarrell for implanting testosterone in voles and lemmings.

## References

Bailey, V. (1897). Revision of the American voles of the genus *Evotomys*. *Proc. Biol. Soc. Washington*, 11, 113–38.

Bailey, V. (1900). Revision of American voles of the genus *Microtus*. *N. Amer. Fauna*, 17, 1–88.

Boonstra, R. and Youson, J.H. (1982). Hip glands in a field population of *Microtus pennsylvanicus*. *Can. J. Zool.*, 60, 2955–8.

Carleton, M.D. and Musser, G.G. (1984). *Muroid rodents*. In *Orders and families of recent mammals of the world* (ed. S. Anderson and J.K. Jones, Jr.), pp. 289–379. Wiley, New York.

Clarke, J.R. and S. Frearson (1972). Sebaceous glands on the hindquarters of the vole, *Microtus agrestis*. *J. Reprod. Fertil.*, 31, 477–81.

Coyne, J.A. and Prout, T. (1984). Restoration of mutationally suppressed characters in *Drosophila melanogaster*. *J. Hered.*, 75, 308–10.

Gromov, I.M. and Polyakov, I. Ya. (1977). *Polevki* [*Voles*] (*Microtinae*). Fauna SSSR Mlekopitayuschie [Mammals]. *Nauka, Moscow-Leningrad*, 3(8), 1–504.

Hall, E.R. (1981). The mammals of North America, Vol. II, 2nd edn. Wiley, New York.

Hinton, Martin A.C. (1926). *Monograph of the voles and lemmings (Microtinae) living and extinct*, Vol. 1. British Museum (Nat. Hist.), London.

Jannett, F.J., Jr. (1975). The 'hip glands' of *Microtus pennsylvanicus* and *M. longicaudus* (Rodentia: Muridae), voles 'without' hip glands. *Syst. Zool.*, 24, 171–5.

Jannett, F.J., Jr. (1978). Dosage response of the vesicular, preputial, anal, and hip glands of the male vole, *Microtus montanus* (Rodentia: Muridae), to testosterone propionate. *J. Mammal.*, 59, 772–9.

Jannett, F.J., Jr. (1981). Scent mediation of intraspecific, interspecific, and intergeneric agonistic behavior among sympatric species of voles (Microtinae). *Behav. Ecol. Sociobiol.*, 8, 293–6.

Jannett, F.J., Jr. (1986). Morphometric patterns among microtine rodents. I. Relative scent gland development in representative voles (*Microtus*). In *Chemical signals in vertebrates: ecology, evolution, and comparative biology* (ed. D. Duvall, D. Müller-Schwarze and R.M. Silverstein), pp. 541–50. Plenum, New York.

Jannett, F.J., Jr. and Jannett, J.A. (ms. *a*.) On scent marking patterns of small mammals.

Jannett, F.J., Jr. and Jannett, J.Z. (1974). Drum-marking by *Arvicola richardsoni* and its taxonomic significance. *Amer. Midl. Nat.*, 92, 230–4.

Jannett, J.A. and Jannett, F.J. Jr. (ms. *b*) Effects of gonadectomy and testosterone on scent glands and marking behaviors of the Richardson water vole (*Microtus richardsoni*).

Lidicker, W.Z., Jr. (1980). The social biology of the California vole. *Biologist*, 62, 46–55.

MacIsaac, G.L. (1977). Reproductive correlates of the hip gland in voles (*Microtus townsendii*). *Can. J. Zool.*, 55, 939–41.

Miller, G.S. (1896). Genera and subgenera of voles and lemmings. *N. Amer. Fauna*, 12, 1–85.

Quay, W.B. (1968). The specialized posterolateral sebaceous glandular regions in microtine rodents. *J. Mammal.*, 49, 427–45.

Raikow, R.J., Borecky, S.R. and Berman, S.L. (1979). The evolutionary re-establishment of a lost ancestral muscle in the bowerbird assemblage. *Condor*, 81, 203–6.

Skurat, L.N. (1969). The structure of sacral glands in some voles (Microtinae). *Zool. Zh.*, 48, 1221–6.

Sokolov, V.E. (1982). Mammal skin. University of California Press, Berkeley, California.

Sokolov, V.E., Skurat, L.N., Serbenyuk, M.A. and Jannett, F.J., Jr. (1989). Influence of testosterone on skin glands of the European bank vole, *Clethrionomys glareolus*. *Zool. Zh.*, (in press).

Tamarin, R.H. (1981). Hip glands in wild-caught *Microtus pennsylvanicus*. *J. Mammal.*, 62, 421.

Tashjian, J. and Jannett, F.J., Jr. (ms.) Mammographic film for the localization of structures in small animals.

Von Claussen, C.P. (1975). Der caudal Rückenfleck des Waldlemmings, *Myopus schisticolor* L. *Z. Säugetierk.*, 40, 368–71.

Vrtis, V. (1930). Glandular organ on the flanks of the water rat, their development and changes during breeding season. *Biologicke Spisy* (*Brno*), 9(4), 1–51.

# II.4 Alarm Chemosignals Suppress the Immune System

ROBERT COCKE AND DEL THIESSEN

## Introduction

There have been numerous studies linking stress with immune suppression in rodents. Immune suppression of lymphocyte proliferation is seen in rats exposed to electrical shock or loud noise (Monjan and Collector 1977). In addition, overcrowding, restraint, and acceleration stress reduce the response of splenic plaque forming cells to sheep red blood cells in rodents (Ader and Cohen 1985). Exposure to stressors activates the pituitary-adrenal axis and the adrenal secretions may contribute to immune suppression; for example, increased adrenal corticosterone levels suppress the proliferation of T lymphocytes in rats (Gillis *et al.* 1979). This suppression of T lymphocyte activity could subsequently result in the suppression of other components of the immune response.

Immunologists working with rodents have noted that some form of communication between stressed and non-stressed animals apparently suppresses the immune response. Stressed rats evoke adrenocortical activity in non-stressed rats housed with them, and the non-stressed rats subsequently display suppression of the immune response (Riley 1981). Recently it was reported that mice housed in cages near cages of stressed conspecifics display reduced numbers of natural killer (NK) cells (Irwin 1987). Odours or sounds from stressed mice may be responsible for the NK cell suppression. The possible mediation of this suppression may involve olfaction, as stress odour exposure in rodents results in adaptive physiological and behavioural responses (Cocke and Thiessen 1986; Mackay-Simm and Laing 1981). An increase in adrenal activity, which is immunosuppressive, is one of the physiological reactions observed following stress odour exposure (Ropartz 1967). Therefore, we speculated that the exposure to stress odours may result in immune suppression. The present experiments were conducted to demonstrate the effects of conspecific stress odours on the immune system and to specify the underlying mechanisms.

## Materials and methods

We selected the mitogen assay as the most appropriate measure of the immune response for this study. Mitogens are products derived from plants and bacteria which stimulate lymphocytes to undergo proliferation via a response similar to that of antigenic stimulation; however, unlike antigenic stimulation, mitogenic stimulation requires no prior exposure to antigen. The proliferation of mitogen stimulated lymphocytes is indicative of the immunocompetence of the cellular arm of the immune response. The mitogen assay is versatile, as different mitogens stimulate different subpopulations of lymphoctyes, and the media used in the assay can be supplemented with other factors (i.e. lymphokines) in order to determine possible mechanisms of action. The numbers of cells required for triplicate cultures in the mitogen assay are difficult to obtain from individual animals, thus isogenic mice were used in these studies since their cells can be pooled as culture samples. These experiments compare the proliferation response in the mitogen assay between groups of inbred mice exposed to stress and control odours.

### ANIMALS

BALB/c male mice, bred at the Animal Resources Center of the University of Texas at Austin were used in these studies. The mice were age-matched and housed together after weaning in groups of 16. For each experiment, one age-matched group was randomly divided into four equal groups and designated as follows: stress odour donor (SOD), control odour donor (COD), stress odour recipient (SOR), and control odour recipient (COR). The animals were housed under 12/12 h light dark cycle at 24° C ambient temperature; lab chow and tap water were available ad lib. Each group of four mice was housed in a cage of dimensions 17 × 13 × 27 cm. At the beginning of each experiment the animals were between 60 and 70 days old.

### OLFACTORY STIMULI PRESENTATION

The home cage of each group of four animals was placed in a clear sealed plastic chamber (21 × 26 × 50 cm). The SOD chamber was attached to a plastic tube (11 mm internal diameter) which directed filtered compressed air into the chamber. The exhaust from another opening at the opposite end of the SOD chamber was directed via plastic tubing into the SOR chamber located 3 feet from the SOD chamber. Air from the SOR chamber was drawn through a vacuum port and exhausted from the building. The COD and COR chambers were arranged in an identical fashion using independent air and vacuum sources. At the mid point of the light portion of the light/dark cycle, the four groups were placed in the apparatus and the odour exposure was initiated. In this apparatus visual and auditory communication was prevented between donors and recipients. At 6 h intervals for the duration of odour exposure one of the SOD animals was placed in a Laffayette Instruments foot shock apparatus and chamber and two 20 s shocks of 0.25 mA (600 v) each were administered, separated by a 1-min. interval. The tip of the animal's tail was then clipped with scissors and the animal returned to the cage. (Previous studies have shown that blood, urine, and whole body odours are involved in the transmission of alarm odours in rodents (Mackay-Simm and Laing 1981; Müller-Velten 1966)).

### IMMUNOLOGICAL METHODS

Upon termination of odour exposure the SOR and COR animals were sacrificed by cervical dislocation, and the spleen and thymus removed. The tissues were pooled for each group and transferred to a petri dish containing RPMI 1640 medium. The tissues were gently teased to form a single cell suspension and the spleen cells were separated by Ficoll-Hypaque sedimentation. The four samples were adjusted to a final concentration of $5 \times 10^6$ mononuclear cells/ml of RPMI 1640 (Irvine scientific, Santa Ana, CA.) supplemented with nonessential amino acids, streptomycin-penicillin, glutamine, sodium pyruvate, HEPES, and 2-mercaptoethanol. The T cell mitogens, phytohemagglutinin (PHA), and concanavalin A (ConA), and the B cell mitogens, *Salmonella typhimirum* (Salmonella Typhimirum Mitogen) and *Escherichia coli* (Lipopolysaccharide) were diluted in RPMI 1640 from stock aliquots to the appropriate concentrations. The mitogen solutions were placed in triplicate into wells of a microtiter plate for each concentration of the mitogens tested. A 0.20 ml aliquots of the cell suspensions in supplemented RPMI was added to each of the mitogen containing wells and controls wells containing mitogen free supplemented RPMI. The plates were placed in a humidified incubator at 37° C with 5% $CO_2$ in air. Forty-eight hours later, 0.5 $\mu$Ci of tritiated thymidine was added to each well. Eighteen hours later the cells were harvested with a MASH II cell harvester on to glass fibre filter paper. The incorporation of $^3$H-thymidine into newly synthesized DNA in the cell cultures was determined with a liquid scintillation counter (LBK RackBeta). Results are expressed in counts per minute, and given as means $\pm$ standard error (SEM).

## Results

### EXPERIMENT I

The duration of odour exposure in the first experiment was 48 h. Figure 1 shows the data on lymphocyte proliferation stimulation. The spleen lymphocytes were stimulated by B cell mitogens with maximum stimulation occurring at 5 $\mu$g STM/culture and 100 $\mu$gLPS/culture. There was suppression of the mean proliferation response of B lymphocytes in the SOR group compared to the COR group. There was no evidence of suppressed T cell proliferation in SOR animals at 48 h exposure.

Fig. 1.

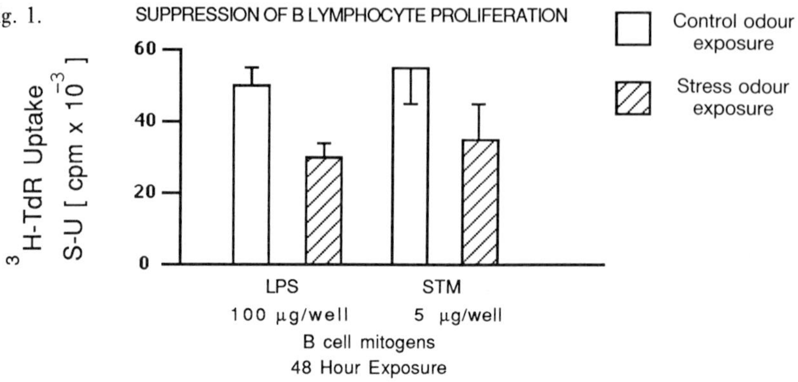


Fig. 2.

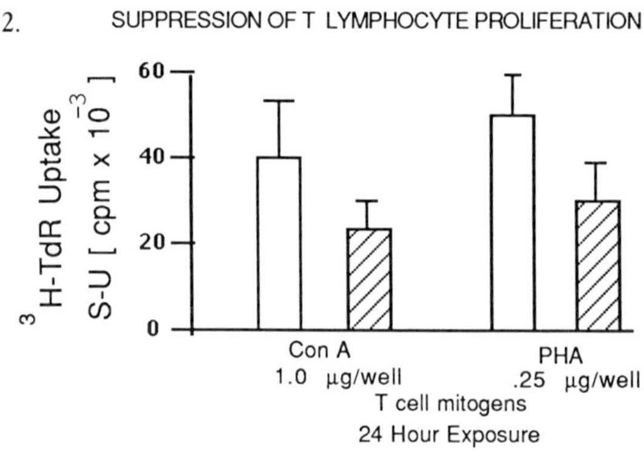


Fig. 3.

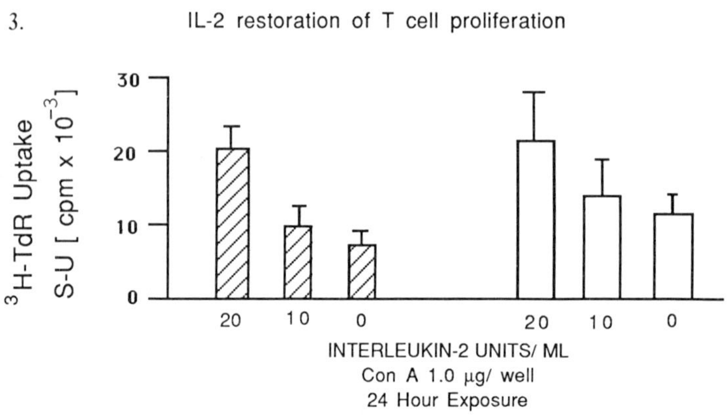


Figs. 1–3. Stimulation of lymphocytes by mitogens. Results, expressed as 3H-Tdr uptake (means ± std errors), are counts per minute (cpm) of triplicate measures of mitogen stimulated cultures (s), minus cpm of triplicate measures of unstimulated cultures (u). Time indicates length of odour exposure.

EXPERIMENT II

The suppression of B cells after 48 h of stress odour exposure suggests that there might be a previous effect on T cell proliferation. This would be expected, as T cell activity in the immune response is responsible in part for regulation of B cell proliferation (Hood *et al.* 1984). Therefore we conducted a second experiment to determine whether T cell mitogen stimulation would be suppressed in SOR animals at exposures of shorter duration than 48 h.

The procedures in Experiment II were identical to those in the first experiment, except that the odour recipients were removed and sacrificed after 24 h of exposure (Fig. 2). The mitogen stimulation of thymus lymphocytes resulted in increased proliferation of these cells. Peak stimulation of T cells occurred at PHA concentrations of 0.25 $\mu$g/culture, and Con A concentrations of 1.0 $\mu$g/culture. At both concentrations there was suppression of the SOR group compared to the COR group. These results indicate that T cell proliferation is suppressed following 24 h of stress odour exposure.

EXPERIMENT III

Experiments I and II indicate that exposure to the odour of stressed conspecifics results in the suppression of lymphocyte proliferation in BALB/c mice. The odour exposure seems to result first in suppression of T cell proliferation, followed by a suppression of B cell proliferation. There are several possible mechanisms that could account for lymphocyte suppression in the SOR group. Exposure to the odourant could result in release of adrenocorticotrophic hormone (ACTH) and eventually to increased levels of adrenal corticosteroids. Increased levels of adrenal glucocorticoids are known to inhibit the production of Interleukin-2 (IL-2), a lymphokine critical for the proliferation of T lymphocytes (Gillis *et al.* 1979). Reduced levels of IL-2 would account for the suppressed T cell response to mitogen stimulation. Furthermore, reduction in IL-2 activity has been implicated in the suppression of the immune response in stressed rodents. When lymphocytes from restraint stressed rodents were cultured in media supplemented with IL-2, the mitogen stimulated proliferation was restored (Sukhikh *et al.* 1984).

Therefore, in the third experiment the mitogen medium was supplemented with interleukin-2 to determine whether the addition of this factor could restore lymphocyte proliferation in the stress odour exposed group. The procedures and methods were the same as those described for Experiments I and II, except that the mitogen medium was supplemented with IL-2. Triplicate wells of stress or control lymphocytes were cultured with mitogen medium supplemented with 0, 10, and 20 units IL-2/ml (final concentration). These cultures were added to wells containing optimum concentrations of mitogens and to mitogen free control wells.

The addition of IL-2 at a concentration of 20 units/ml to the stress odour exposed samples appeared to restore the proliferation of lymphocytes to levels similar to those observed in non stressed control samples (Fig. 3). The restoration of proliferation suggests that the stress odour exposure may evoke physiological reactions suppressing production of IL-2, and thus attenuate the proliferation of T lymphocytes.

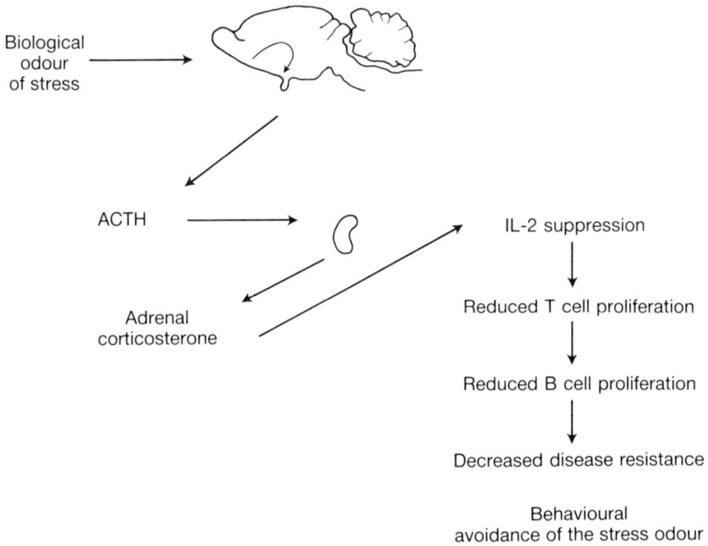


Fig. 4. Hypothetical mechanisms for olfactory suppression of the immune response.

## Discussion

The exposure to odours from foot-shocked mice results in the suppression of mitogen stimulated lymphocyte proliferation, relative to controls. The proliferation of T cells is suppressed following 24 h of stress odour exposure and B cell proliferation is suppressed at 48 h. Suppression of T cell proliferation can be restored by the addition of IL-2 (20 units/ml) to the culture media. We were able to replicate all the main effects discussed.

The results of these experiments suggest a possible sequence of physiological reactions following stress odour exposure resulting in suppression of the immune response (see Fig. 4). Exposure to the odours from stressed conspecifics may result in increased release of ACTH, as is seen with exposure to other stressors (Carlson *et al.* 1987). The resulting increase in adrenocorticord levels would then inhibit the production of IL-2 and suppress the proliferation of T lymphocytes. The reduced proliferation of T cells may then reduce the production of other lymphokines which are responsible for the subsequent proliferation of B lymphocytes. In future studies, adrenalectomy and adrenal steroid replacement in stress odour exposed animals will be employed to determine whether the adrenal response mediates immune suppression.

Whatever the mechanism of immune suppression, these experiments demonstrate that a biologically significant chemosignal, which directs adaptive behaviours and evokes adaptive physiological responses, is also capable of inducing significant alterations in the immune system. It is unclear from a functional standpoint why environmental stressors should produce immune suppression, or why such a relationship should be maintained in the evolution of the physiological systems; nevertheless environmental stressors do indeed compromise the immune system. We are currently conducting experiments to determine whether mice exposed to the odours from stressed conspecifics display a reduced resistance to disease.

## References


Ader, R. and Cohen, N. (1985). Cns-immune system interactions. *Behav. Brain Sci.*, 8, 379–95.

Carlson, S., Felten, D., Livnat, S. and Felten, S. (1987). Alteration of monoamines in specific nuclei following immunization in mice. *Brain Behav. Immunity*, 1, 52–63.

Cocke, R. and Thiessen, D. (1986). Chemocommunication among prey and predator species. *Anim. Learning Behav.*, 14, 90–2.

Gillis, S., Crabtree, G. and Smith, K. (1979). Glucocorticoid-induced inhibition of T cell growth factor production. *J. Immunol.*, 123, 624–31.

Hood, L., Weissman, I., Wood, W. and Wilson, J. (1984). *Immunology*. Benjamin and Cummings, Menlo Park.

Irwin, J. (1987). Worried sick. *Science News*, 132, 360.

Mackay-Simm, A. and Laing, D. (1981). Rat's responses to blood and body odor of stressed and non stressed conspecifics. *Physiol. Behav.*, 27, 503–10.

Monjan, A. and Collector, M. (1977). Stress induced modulation of the immune response. *Science*, 197, 307–8.

Müller-Velten, H.G. (1966) Uber den angstgeruch bie der hausmaus. *Zeitschrift fur Verleichende Physiologie*, 52, 401–29.

Riley, V. (1981). Psychoneuroendrocrine influences on immunocompetence and neoplasma. *Science*, 212, 1100–9.

Ropartz, M. (1967) L'effet de groupe chez les animaux. *Colloques International aux Centre National de Recherche Scientifique*, No. 173.

Sukhikh, G,. Malaitsev, V. and Bogdanova, I. (1984). Interleukin-2 and its possible role in the pathogenesis of stress changes in the immune system. *Proc. Acad. Med. Sci. USSR*, 278, 762–5 (translation).

# II.5 Behavioural and Physiological Responses to Alarm Pheromone by Ostariophysan Fishes and a Possible Modulating Role for Brain Benzodiazepine Receptors

BRADLEY G. REHNBERG AND R.J.F. SMITH

## Behavioural and physiological aspects of alarm

Alarm pheromone systems of ostariophysan fishes (e.g. minnows, catfishes, characins) include a variety of seemingly adaptive aspects of behaviour which share a common etiology. First, the alarm pheromone-releasing cells in the epidermis must be broken to permit a passive release of the pheromone. The water-soluble pheromone is then detected by olfaction which initiates alarm behaviour in the recipient. By responding behaviourally to the pheromone, the recipients provide a visual alarm signal which facilitates a rapid transmission of alarm to nearby conspecifics. Active alarm phases may involve activities such as rapid swimming, Mauthner-like C-turns (Pfeiffer *et al.* 1986), or jumping at the surface whereas inactive phases may include bottom-orientation or hiding among vegetation (Smith 1986).

Detection of alarm pheromone also has profound physiological effects on responding fish. Not surprisingly, the physiological effects reported in the literature have included elements of a stress response (Table 1). The primary response to stress in teleosts is a very rapid increase in circulating catecholamines and a slower elevation in plasma cortisol (Mazeaud and Mazeaud 1981). The time-course of cortisol elevation in response to alarm pheromone and the corresponding alteration of carbohydrate balance has been documented for the pearl dace (*Semotilus margarita*) (Rehnberg *et al.* 1987; Table 1). Although plasma catecholamines have not been measured in alarmed fish, secondary stress responses known to be induced by catecholamines have been demonstrated (e.g. bradycardia, altered concentrations of muscle metabolites and plasma cations; see Table 1). There is also evidence that sensory capacities of alarmed fish may be enhanced (Pfeiffer *et al.* 1985) which would make vigilance for predators more effective.

**Table 1.** Physiological responses by ostariophysan fishes after detection of alarm pheromone

| Species | Physiological effect | Reference |
|---|---|---|
| *Semotilus margarita* | ↑ plasma cortisol, glucose | Rehnberg *et al.* (1987) |
| *Phoxinus phoxinus* | bradycardia | Pfeiffer and Lamour (1976) |
| *Leuciscus idus* | ↑ muscle lactate, pyruvate, glucose, glucose-6-P | Gronow (1974) |
| *Abramis brama* | altered plasma cations | Malyukina *et al.* (1982) |
| *Gymnocorymbus ternetzi* | ↑optical alertness | Pfeiffer *et al.* (1985) |

## Neural control of alarm reactions

The neural mechanisms that initiate and control the behavioural and physiological components of alarm reactions are unknown. We have recently begun a neuropharmacological approach toward understanding the control of alarm in fathead minnows (*Pimephales promelas*) (Rehnberg *et al.* 1989). The benzodiazepines are a family of drugs that reduce anxiety in mammals and have been well characterized biochemically and clinically (Baldessarini 1985). The benzodiazepine receptor is part of a macromolecular complex which includes a receptor for gamma-aminobutyric acid (GABA) and a chloride channel (Braestrup *et al.* 1983). Benzodiazepines may function in mammals by enhancing chloride fluxes in response to GABA, possibly in serotonergic raphe nuclei in the rostral brain stem (Braestrup *et al.* 1983). Benzodiazepine treatment can reduce the stress-induced activation of the hypothalamic-pituitary-adrenal axis which suggests that the receptor complex may be involved in initiating stress responses (Trullas *et al.* 1987).

## Methods

Whole brains, including medulla and olfactory bulbs, were removed from fathead minnows and stored at $-70\,^{\circ}$C. A membrane fraction was prepared by homogenization and centrifugation using the procedure of Nielsen *et al.* (1978). A standard ligand binding assay, fully described by Rehnberg *et al.* (1989), was used to detect binding sites for [methyl-$^3$H]-diazepam. Specific binding was found by subtracting the non-specific binding of [$^3$H]diazepam (in the presence of $10^{-5}$ M non-radioactive diazepam) from total [$^3$H]diazepam binding.

Swimming activity of single fathead minnows was measured in a test tank ($45 \times 45 \times 20$ cm) in conjunction with a computer-automated activity tracking meter (Columbus Instruments) (Rehnberg *et al.* 1989). Photocells were located every 2.6 cm along two adjacent sides of the tank which created a grid of light beams. A computer collected data on light beam interruptions from photoreceptor cells positioned opposite the photocells. The resulting data were used to compute the total distance travelled (cm) by the fish during 2-min. intervals. After collecting baseline data for 8 mins, a drug control (water) or a dose of the benzodiazepine chlordiazepoxide hydrochloride (Hoffmann-La Roche, Calgary) was introduced into the tank. Beginning at 60 mins, data were collected in sequential 8-min. blocks after presentation of a stimulus control (water), an extract of alarm pheromone, and a chemical feeding stimulus. Activity during the 4 mins before and after presentation of stimuli was evaluated using Wilcoxon's matched-pairs signed-rank test.

Acclimation of fish and the presentation of chemical stimuli are described in Rehnberg *et al.* (1989). The activity data for each fish were normalized to its own baseline activity in order to help correct for the large range of activity across individuals (Rehnberg *et al.* 1989). This was done by determining the mean score for the 8 mins baseline for each animal. All data were then divided by this mean score to obtain the normalized score.

Alarm pheromone stimulus was prepared by removing a patch of skin ($1.0 \times 0.5$ cm) from both sides of the caudal peduncle of 12 fish. The pooled tissue was homogenized in 100 ml distilled water and diluted 1:10. The chemical feeding stimulus was made by stirring Tetramin Staplefood flakes in water at a concentration of 0.1 g flakes in 100 ml total volume. After 10 min, the extract was filtered through Whatman no. 2 qualitative filter paper. Both extracts were stored frozen; 1 ml of the thawed extracts was used for each behavioural test.

## Results

We found specific benzodiazepine binding sites in the brains of fathead minnows which had high affinity for [$^3$H]diazepam ($K_d = 10$ nM) and were apparently saturable ($B_{max}$ approximately 3.3 fmol/mg wet weight) (Fig. 1a). Chlordiazepoxide, a water-soluble benzodiazepine used in the behavioural tests, competed with diazepam for brain binding sites ($I.C._{50} - 1.6\ \mu M$; Fig. 1b).

In behavioural tests, introducing the drug control solution or 10 mg/l chlordiazepoxide resulted in elevated activity for reasons that are not known (Fig. 2). Presentation of a water stimulus caused a transient decrease in activity in controls but not in fish exposed to chlordiazepoxide (Fig. 2). After presentation of alarm pheromone, activity was significantly lower in fish exposed to drug control ($p < 0.01$) or to 10 mg/l chlordiazepoxide ($p < 0.05$), but not in fish exposed to 20 mg/l (a non-sedative dose) ($p = 0.79$). Presentation of a chemical feeding stimulus resulted in significantly greater activity for all groups of fish. Fatheads exposed to 20 mg/l became extremely active and explored all areas of the tank ($p = 0.004$) (Fig. 2).

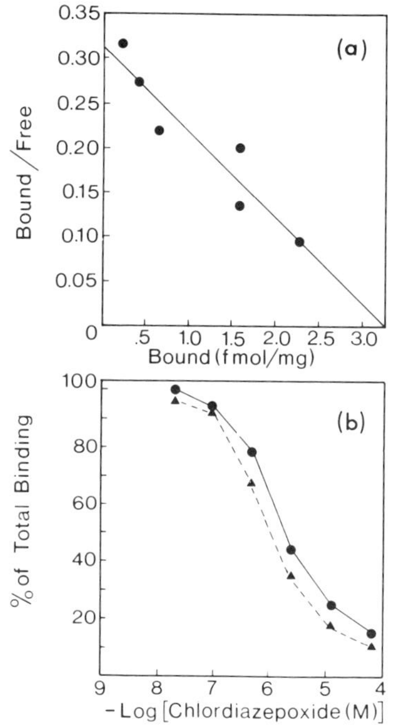


Fig. 1a. Scatchard plot of [$^3$H]diazepam binding to membranes prepared from the brains of fathead minnows. Data are means from two experiments run in quadruplicate. b. Displacement of [$^3$H]diazepam (1.5 nM) from its binding site by chlordiazepoxide at 6 concentrations. Results from two separate experiments are shown.

## Discussion

The behavioural data suggest that the anxiety-reducing property of chlordiazepoxide reduces or eliminates the behavioural response to alarm pheromone and thereby secondarily facilitates a vigorous foraging response. We propose a hypothetical neural model for the sensory inputs and the behavioural and physiological outputs of the alarm pheromone system (Fig. 3). At the core of the model is an anatomically undefined brain alarm centre which is benzodiazepine-sensitive. Limbic structures may be involved although there is evidence that emotion is not produced by distinct anatomical structures in the limbic system or in other discrete brain regions (Vanderwolf *et al.* 1988). Olfactory and visual pathways provide the predominant sensory inputs although other senses are probably also involved. The model proposes that the alarm message is conveyed to the alarm centre where it may be modulated by benzodiazepines or endogenous substances before reaching autonomic and motor nuclei. The prospect of an endogenous ligand received support from the discovery of diazepam binding

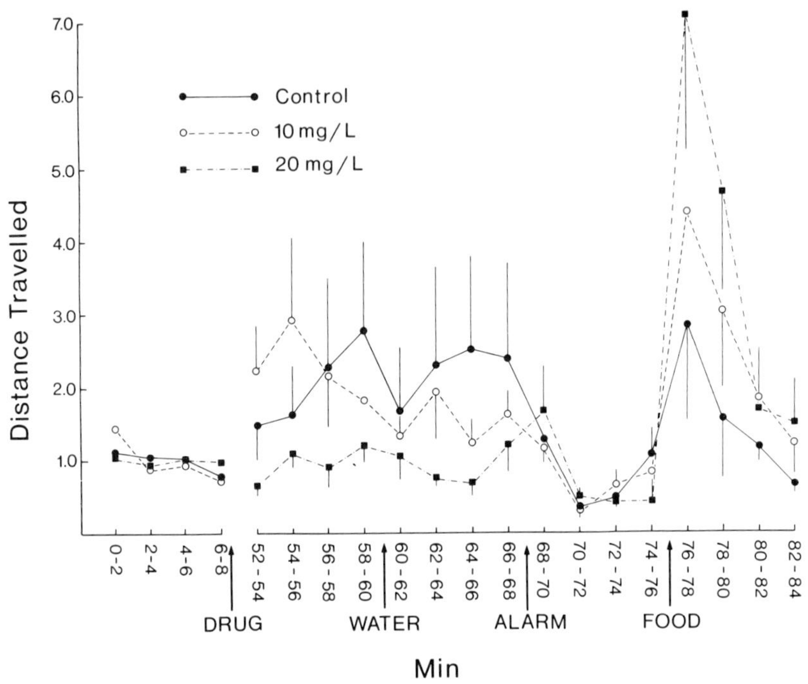


Fig. 2. Distance travelled by fathead minnows exposed to drug control (water) ($n = 27$) or to 10 mg/l ($n = 14$) or 20 mg/l ($n = 11$) chlordiazepoxide. Standard error bars that overlapped were omitted for clarity.

inhibitor (DBI) in the mammalian brain (Guidotti *et al.* 1983). DBI competes for binding sites with diazepam and apparently has anxiogenic properties.

Recent demonstrations of benzodiazepine binding sites in CNS preparations from elasmobranchs (Hebebrand *et al.* 1988) and insects (Robinson *et al.* 1986) indicate an early evolutionary origin for these receptors. Our results suggest that these phylogenetically ancient receptors can exert neural control over behaviour in anxiety-like states in fathead minnows.

## Acknowledgements

We thank S. Richardson for providing laboratory space and materials and E.H. Bates and D. Dyck for excellent technical support. The work was funded by an operating grant from NSERC and from the University of Saskatchewan President's Research Fund.

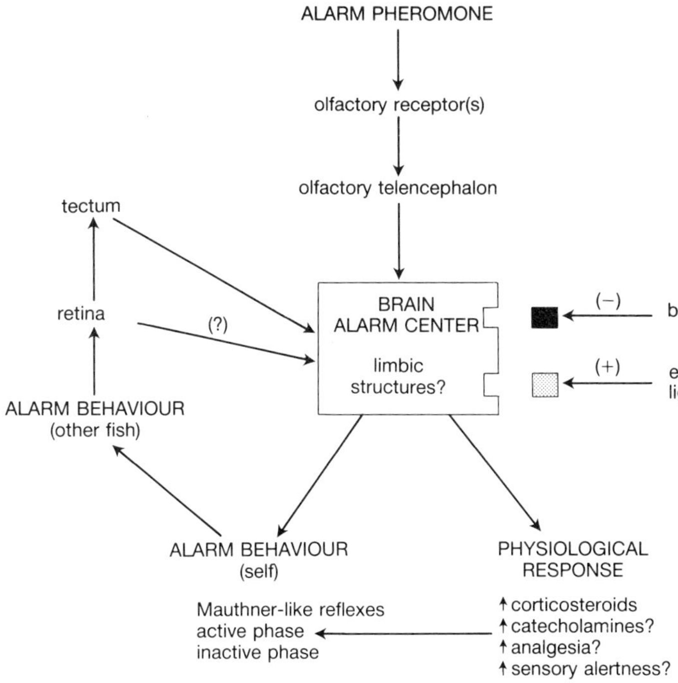


Fig. 3. Proposed model for the control of ostariophysan alarm responses by a hypothetical alarm centre in the brain.

## References

Baldessarini, R.J. (1985). Drugs and the treatment of psychiatic disorders. In *Goodman and Gilman's The pharmacological basis of therapeutics* (ed. A.G. Gilman, L.S. Goodman, T.W. Rall and F. Murad), pp. 387–445. MacMillan, New York.

Braestrup, C., Nielsen, M. and Honore, T. (1983). Benzodiazepine receptor ligands with positive and negative efficacy. In *CNS receptors—from molecular pharmacology to behavior* (ed. P. Mandel and F.V. DeFeudis), pp. 237–45. Raven Press, New York.

Gronow, G. (1974). Nukleinsäure- und Substratgehalte in der dorsalen Rumpfmuskulatur von Teleosteern wahrend eines 'biologischen Stress'. *Marine Biol. (Berlin)*, 24, 313–27.

Guidotti A., Forchetti, C.M., Corda, M.G., Konkel, D., Bennett, C.D. and Costa, E. (1983). Isolation, characterization, and purification to homogeneity of an endogenous polypeptide with agonistic action on benzodiazepine receptors. *Proc. Natl. Acad. Sci. USA*, 80, 3531–3.

Hebebrand, J., Friedl, W., Reichelt, R., Schmitz, E., Moller, P., and Propping P. (1988). The shark GABA-benzodiazepine receptor: further evidence for a not so late phylogenetic appearance of the benzodiazepine receptor. *Brain Res.*, 446, 251–61.

Malyukina, G.A., Martem'yanov, V.I. and Flerova, G.I. (1982). The alarm pheromone as a stress factor for fish. *J. Ichthyol.* (Engl. Transl.), 22, 147–50.

Mazeaud, M.M. and Mazeaud, F. (1981). The role of catecholamines in the stress response of fish. In *Stress and fish* (ed. A.D. Pickering), pp. 49–75. Academic Press, New York.

Nielsen, M., Braestrup, C. and Squires, R.F. (1978). Evidence for a late evolutionary appearance of brain-specific benzodiazepine receptors: an investigation of 18 vertebrate and 5 invertebrate species. *Brain Res.*, 141, 342–6.

Pfeiffer, W. and Lamour, D. (1976). Die Wirkung von Schreckstoff auf die Herzfrequenz von *Phoxinus phoxinus* (L.) (Cyprinidae, Ostariophysi, Pisces). *Rev. Suisse Zool.*, 83, 861–73.

Pfeiffer, W., Riegelbauer, G., Meier, G. and Scheibler, B. (1985). Effect of hypoxanthine-3(N)-oxide and hypoxanthine-1(N)-oxide on central nervous excitation of the black tetra *Gymnocorymbus ternetzi* (Characidae, Ostariophysi, Pisces) indicated by dorsal light response. *J. Chem. Ecol.*, 11, 507–23.

Pfeiffer, Von W., Denoix, M., Wehr, R., Gnass, D., Zachert, I. and Breisch, M. (1986). Slow motion analysis of the fright reaction of Ostariophysean fish and the significance of the 'Mauthner reflex'. *Zool. Jb. Physiol.* 90, 115–65.

Rehnberg, B.G., Smith, R.J.F. and Sloley, B.D. (1987). The reaction of pearl dace (Pisces, Cyprinidae) to alarm pheromone: time-course of behavior, brain amines, and stress physiology. *Can. J. Zool.*, 65, 2916–21.

Rehnberg, B.G., Bates, E.H., Smith, R.J.F., Sloley, B.D. and Richardson, J.S. (1989). Brain benzodiazepine receptors in fathead minnows and the behavioural response to alarm pheromone. *Pharm. Biochem. Behav.*, 33, in press.

Robinson, T., MacAllan, D., Lunt, G. and Battersby, M. (1986). $\gamma$-Aminobutyric acid receptor complex of insect CNS: characterization of a benzodiazepine binding site. *J. Neurochem.*, 47, 1955–62.

Smith, R.J.F. (1986). The evolution of chemical alarm signals in fishes. In *Chemical signals in vertebrates*, 4 (ed. D. Duvall, D. Müller-Schwarze and R.M. Silverstein), pp. 99–115. Plenum, New York.

Trullas, R., Havoundjian, H., Zamir, N., Paul, S. and Skolnick, P. (1987). Environmentally-induced modification of the benzodiazepine/GABA receptor coupled chloride ionophore. *Psychopharmacology*, 91, 384–90.

Vanderwolf, C.H., Kelly, M.E., Kraemer, P.. and Streather, A. (1988). Are emotion and motivation localized in the limbic system and nucleus accumbens? *Behav. Brain Res.*, 27, 45–58.

# II.6 The White Eye Secretion in Aplodontia

NANCY HACKMANN, C.S. ZAMORA AND E. STAUBER

## Introduction

Mountain beavers (*Aplodontia rufa*) occur endemically from northern California to southern British Columbia, primarily in moderate to dense vegetation of the Olympic, Cascade, and Sierra-Nevada mountain ranges. Monospecific, the mountain beaver has occurred from the upper Eocene and is considered to be the most primitive living rodent. At population densities of 1–2 per ha, beavers each maintain an extensive tunnel and burrow system. Males seek out females during the oestrous period (February–March). Adult mountain beavers are solitary at other times of the year.

The mountain beaver produces a distinctive white secretion from its eyes in some stressful situations, at death, and perhaps on other occasions. Active secretion is preceded by teeth-clicking and apparent agitation. When the secretion appears the beaver remains motionless. The secretion covers the entire eye and may remain for up to a minute. It is shaken on to the whiskers and may be washed from the face by the front paws. A similar secretion has been noted in pacaranas (Collins and Eisenberg 1980) and teneracs (Poduschka 1981), but its function is unknown.

The present study was carried out to assess the semiochemical and/or behavioural potential of the white eye secretion in wild caught mountain beaver.

## Materials and methods

Harderian and lacrimal glands were removed from 100 specimens and compared for differences in size, shape, and colour by age, sex, and season. Glands were manually expressed to determine whether the white secretion could be extruded and to locate the duct. Glands were removed from a further 20 specimens immediately post-mortem. One Harderian and lacrimal gland from each specimen was preserved in Telly's fixative, dehydrated in an ethanol-xylene series, embedded in paraffin, sectioned at 7 $\mu$m, and stained for light microscopy. The remaining glands were fixed in 2% gluteraldehyde followed by 1% osmium tetroxide, embedded in resin, and sectioned for transmission electron microscopy. Due to high lipid content, the Harderian glands needed to be stained *en bloc* with lead citrate.

Simple chemical tests were performed to establish the presence in both gland and secretion of lipid (anhydrous ether extraction), protein (Lowry determination), and carbohydrate (Jermyn modified anthrone method). Short-chain, volatile fatty acids (VFAs) were determined by GC in secretion samples diluted with distilled water and stabilized with 25% metaphosphoric acid (Anonymous 1975).

Behaviour trials involved placing eight mountain beavers separately in cages (3 × 1.2 × 1.5 m) where they established a nest and home territory for one week prior to the introduction of one of another eight beavers for an encounter lasting 30–60 mins. The roles of the beavers were reversed for a second set of trials. Encounters occurred between male and female, male and male, female and female, and juvenile and adult. Throughout 24 h, sampling periods of 10 mins of five beavers (total of 30 h observations) placed individually in the units were made to test for random occurrence of tear production.

## Results

### GROSS ANATOMY OF HARDERIAN GLAND

There were no noticeable gross morphological differences due to sex, age, or season. The Harderian gland is held by connective tissue to the orbital wall in several places and to the third eyelid. Lacking an orbital venous sinus and not being connected to the surrounding tissue, the gland lies tightly wedged within the orbit adjacent to the frontal bone and encloses the posterior wall of its eyeball and optic nerve (Fig.1).

A slight swelling can be detected on the inside surface of the gland, which when expressed from the bottom towards the third eyelid, extrudes the white secretion from a single excretory duct.

### LIGHT MICROSCOPY

The parenchyma of the Harderian gland, composed of tubulo-alveolar secretory units with wide lumina and showing few intra- and inter-lobular ducts, is divided into irregular lobules by a scanty interlobular connective tissue. A single layer of cuboidal–columnar cells forms each alveolus. Epon sections stained with toluidine blue reveal the lumina filled with secretory product. Alveoli joining interlobular ducts are readily seen, and thin myoepithelial cells are seen at the outer edges of the alveolus.

### ELECTRON MICROSCOPY

The parenchymal cells of the Harderain gland contain readily visible, round nuclei (N) and cytoplasm filled with numerous vacuoles (V), mitochondria (M), and

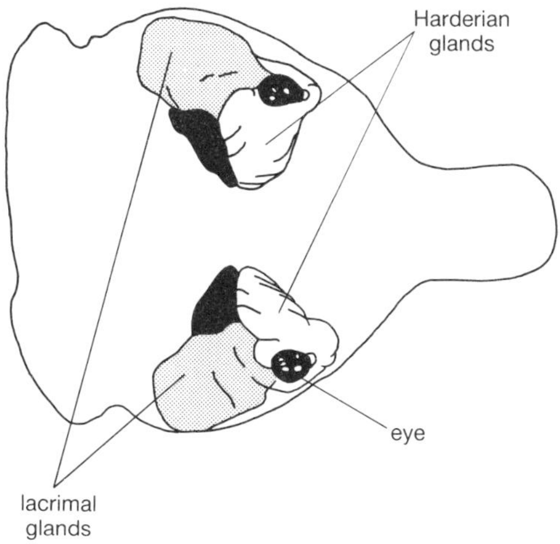


Location of the Harderian and lacrimal glands in the skull of the mountain beaver.

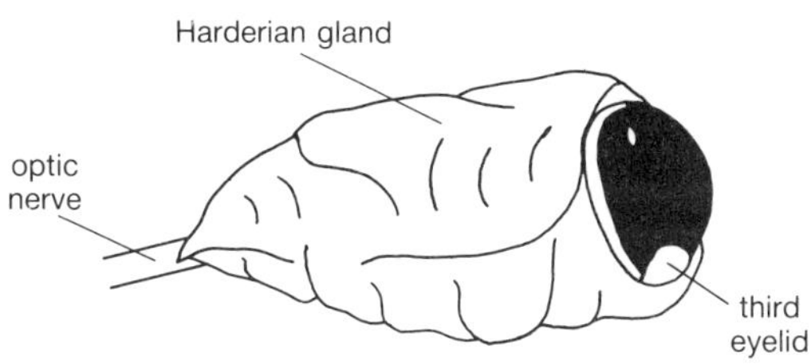


The Harderian gland is located around the posterior wall of the eye. Drawing A. (enlargement) shows the gland wrapped around the eyeball and the optic nerve. Photo B. shows the Harderian gland in place in the skull with and without the eyeball.

Fig. 1. Location of the Harderian and lacrimal glands in the skull of the mountain beaver.

well-developed tubular smooth endoplasmic reticulum (SER) (Fig. 2, × 5100; area A, × 12,000; area B, × 18,000). Adjacent cells appear in varying states of light to dark due to the density of the SER. Binucleated cells are occasionally seen and a few microvilli are found on the luminal membrane. Myoepithelial cell (MY) protrusions between adjacent parenchymal cells, and marked interdigitations between adjacent glandular cells, are common. The numerous dispersed mitochondria have lamellar cristae. There is no apparent difference in the number or shape of mitochondria in light or dark cells. All cells are vacuolated: some cells are almost devoid of vacuoles, while entire alveoli can be composed of cells densely

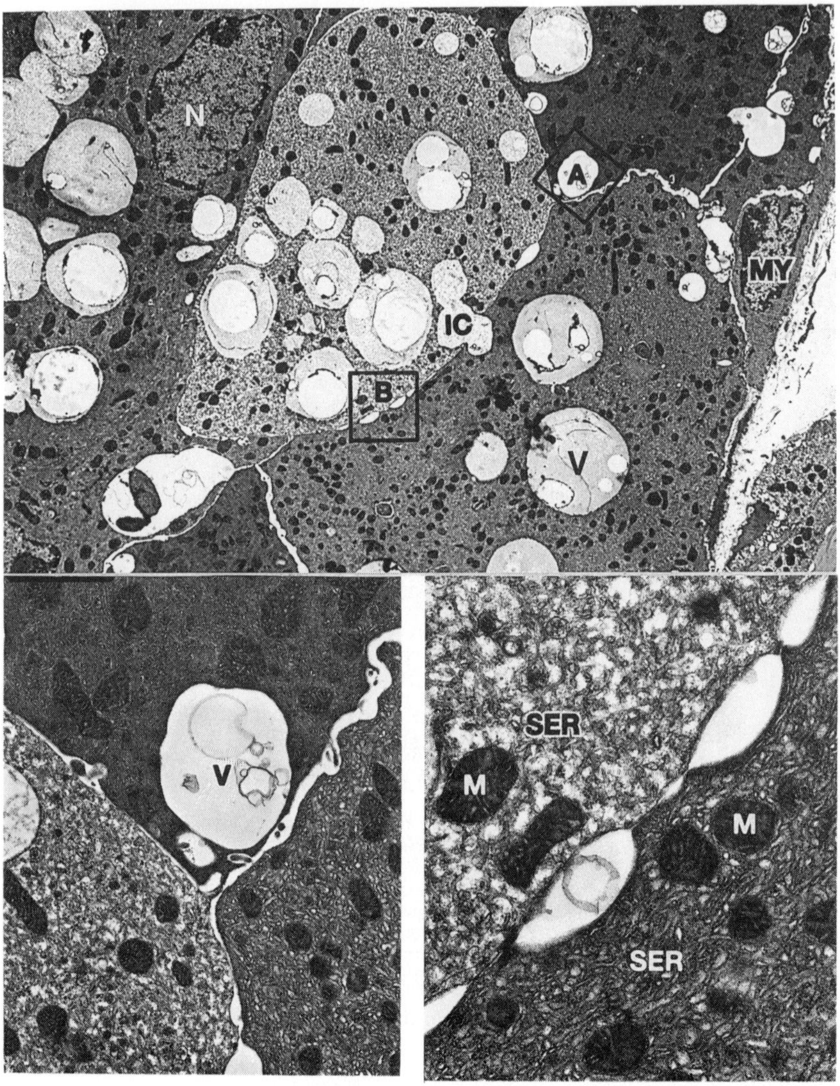


Fig. 2. Parenchymal cells of the Harderian gland are filled with vacuoles (V), mitochrondria (M) and tubular smooth endoplasmic reticulum (SER). Intracellular canaliculus (IC), myoepithelial cell (MY) and nuclei (N) are all visible (3825x).

vacuolated from the basal to apical membranes. However, this feature is not a means of distinguishing between light and dark cells. Intra-cellular canaliculi (IC) are seen connecting vacuoles within cells. Vacuoles and lumina are filled with electron-lucent to opaque substances. Cellular debris is sometimes seen in the

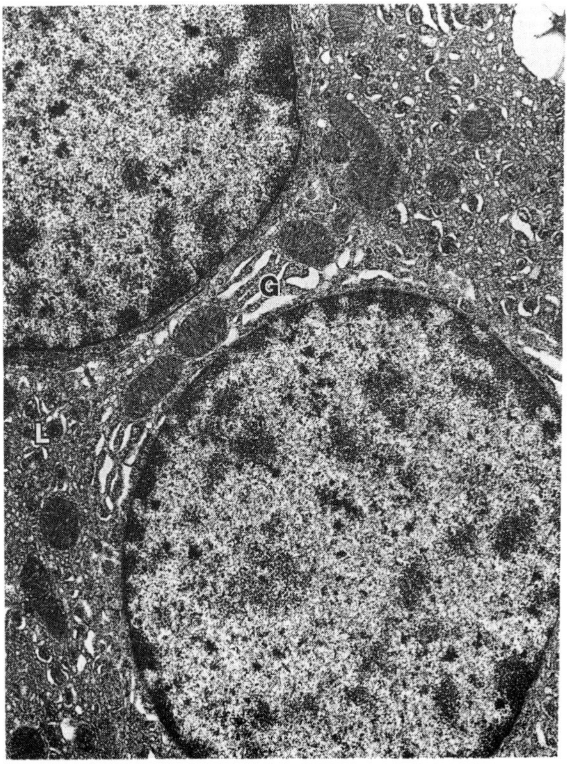


Fig. 3. Micrograph of a binucleated cell containing lysosome-like bodies (L) and a rare view of a Harderian gland Golgi apparatus (G) (35,000x).

irregularly shaped lumina. Interlobular ductal cells contain vacuoles that extend into the ductal lumen. Golgi complexes (G) are rare, appear loosely organized, and lie adjacent to the nuclei. Lysosome-like bodies (L) occur in the apical cytoplasm (Fig. 3). Rough endoplasmic reticulum (RER) was rarely observed, while binucleated cells were seen.

### CHEMISTRY OF SECRETION

The tear samples reacted with both acidic and basic paper chromatograpghy solvents and ninhydrin, but, unlike the standard free amino acids, the tear samples did not migrate up the paper. Lowry determination of total proteins in the tears showed 16.5 mg protein/ml of tear sample, while the anthrone method determined an average of 222.5 $\mu$g carbohydrates/ml. Ether extraction yielded an average of 55% crude fat in the Harderian gland. Sudan black-B-stained frozen sections revealed lipid droplets in the lumina and across the section. Short-chain VFAs were

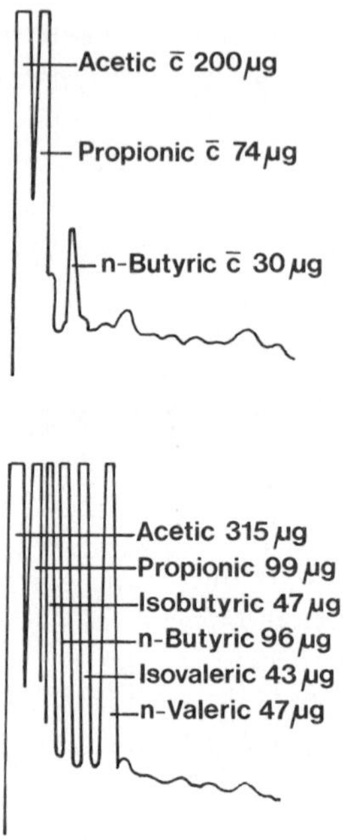


Fig. 4. Gas chromatograms of volatile fatty acids in a fresh sample of mountain beaver tear secretion (upper tracing) and in a commercial calibration standard (lower tracing).

identified in freshly collected secretion. Approximately 200 $\mu$g/ml acetic, 74 $\mu$g/ml propionic, and 30 $\mu$g/ml n-butyric and trace amounts of isobutyric, isovaleric, and n-valeric acids were present.

#### FUNCTION

Secretion occurred only during behavioural trials, when it was directed at an observer as he released a beaver into another's unit to initiate an encounter. Secretion was not seen during any of the 16 trials of encounters between beavers or during the 30 h of observation for its random occurrence. Secretion also was not observed in any beavers housed in artificial units where they could dig tunnels, place their nests in a location of their own choice and move that location at will, and select sites for eating and refuge.

## Discussion

The histology of the mountain beaver Harderian gland fits the Saki's (1981) general description for mammalian Harderian glands. However, the presence of an intralobular duct system is contrary to the findings for Harderian glands of other mammalian species, but is unusual for an exocrine gland. Discrete lobules, intracellular canaliculi, and a parenchyma containing only one type of secretory cell are unique to *Aplodontia*, while the Golgi apparati and RER commonly found in other mammalian Harderian glands are rare. The latter may indeed be of greater abundance than observed, but was obscured by the many vacuoles and abundant SER. Simple chemical analysis indicated the presence of protein, which may be bound as lipoprotein or proteinaceous lipid, thus eliminating the need for Golgi complexes. It would be expected that at least ribosomes, if not abundant RER, are present.

Ultrahistological investigation revealed that not all the cells extend to the lumen. Intercellular caniculi allow the secretory product of distal cells to be secreted into the lumen. Intracanalicular connections, connecting vacuoles within a cell, are more numerous near the lumen. The colour difference of light to dark cells appears to be due to the amount of SER present: the darker the cell the more abundant the SER. Thus the colour of the cells may not be an indication of different cell types, but rather of different maturation stages and cellular activity; the presence of lysosome bodies suggests enzymatic activity.

An abundance of lipid is present in the Harderian gland. Gas chromatographic analysis revealed the presence of acetic, propionic, and n-butyric acids with trace amounts of isobutyric, isovaleric, and n-valeric acids. Even trace amounts of VFAs have been indicated as an important component of scent marking secretion. The presence of short-chain VFAs and high levels of proteins and lipids suggests that the white eye secretion may have a semiochemical function. Since the tears stick to the vibrissae it may passively impart information to mountain beavers about tunnel usage.

Tonic immobility (TI) functions as an adaptive defence reaction to predator attack. Using cats and quail, Thompson *et al.* (1981) established that TI eliminates the movement stimuli that elicits an attack in the first place. Ewell *et al.* (1981) found that rabbits monitor the environment while in TI. Stimuli associated with the absence of a predator, nearness of nest, and availability of cover decreased the duration of TI, while the presence of a predator increased the duration. Observers of mountain beaver have noted a definite quieting of stressed behaviour after the eyes fill with the white eye secretion; TI could be a possible explanation.

Some authors (Wetterberg *et al.* 1970; Johnson *et al.* 1979; McMasters and Hoffman 1984) link a Harderian-pineal connection with season and sexual state. The Harderian gland may aid the pineal gland in determining a gonadal response to day length. Since our data do not reveal any seasonal differences in the size, shape, or colour of the Harderian gland, it is doubtful that a Harderian-pineal link exists in the mountain beaver.

The high lipid component of the Harderian gland secretion and the negative correlation of eye to gland size follows Sakai's (1981) theory that the funtion of the Harderian gland secretion is not lubrication of the eye. Secretion has been released only in response to the presence of humans in what appears to be a stressful situation for the beaver.

## Conclusion

The secretory glands of the mountain beaver eye have been morphologically and histologically identified as Harderian and lacrimal glands. The Harderian gland was found as the source of the white eye secretion, extruded from a single external duct which lies near the outer edge of the third eyelid.

The Harderian gland possesses abundant smooth endoplasmic reticulum, lipid-filled vacuoles, and mitochondria, in common with other mammalian Harderian glands. Unlike Harderian glands of other species, those of the mountain beaver have discrete lobules, an intralobular duct system, and alveolar cells containing intracellular canaliculi.

Chemical analysis of tissue sections, whole glands, and white eye secretion identified the presence of significant amounts of acetic, propionic, and n-butyric acids (VFAs), proteins, and lipids. At the present time the function of the Harderian white eye secretion cannot be determined. The presence of VFAs suggests a semiochemical function, while lubrication and extraretinal receptor function appear unlikely. Observations of mountain beavers in the wild are needed to determine the context(s) in which the white eye secretion occurs.

## References

Anonymous (1975). GC separation of VFA C2-C5. *Bellefonte, Pa. Supelco Bulletin* 749B.

Collins, L.R. and Eisenberg, J.F. (1980). Notes on the behavior and breeding of pacaranas in captivity. *Int. Zoo Yearbk.*, 12, 108–14.

Ewell, A.H., Cullen, J.M. and Woodruff, M.L. (1981). Tonic immobility as a predator defense in the rabbit (*Oryctolagus cuniculus*). *Behav. Neural Biol.*, 31, 483–9.

Johnson, D.D., Rudeen, P.K. and O'Steen, W.K. (1979). Photically induced experimental exophthalmos: role of the Harderian and pituitary glands. *Invest. Ophthalm. Visual Sci.*, 18, 1280–5.

McMasters, K.M. and Hoffman, R.A. (1984). Harderian gland: regulation of sexual 'type' by gonads and pineal gland. *Biol. Reprod.*, 31, 579–85.

Poduschka, V.W. (1974). Augendrusensekretionen bei den tenreciden Setifer setosus (Froreip 1806). Echinops telfairi (Martin 1838). Microgale dobsoni (Thomas 1918) und Microgale talazaci (Thomas 1918). *Z. Tierpsychol.*, 35, 303–19.

Sakai, T. (1981). The mammalian Harderian gland: morphology, biochemistry, function and phylogeny. *Arch. Histol. Jap.*, 44, 299–333.

Thompson, R.K.R., Foltin, R.W., Boylan, R.J., Sweet, A., Graves, C.A. and Lowitz, C.E. (1981). Tonic immobility in Japanese quail can reduce the probability of sustained attack by cats. *Anim. Learn. Behav.*, 9, 145–9.

Wetterburg, L., Geller, E. and Yuwiler, A. (1970). Harderian gland: an extra-retinal photoreceptor influencing the pineal gland in neonatal rats? *Science*, 167, 884–5.

# II.7 The Vomeronasal Organ of the Prairie Vole: Role in Chemosignal Activation of Female Reproduction and Observations of Fine Structure

JOHN J. LEPRI, CHARLES J. WYSOCKI, KAREN GERBER AND ROBERT D. LISK

## Introduction

Interest in the reproductive biology of voles has been stimulated by several factors. Prominent is the observation that over three-and four-year intervals, many species of *Microtus* exhibit dramatic fluctuations in population parameters (Frank 1957; Lidicker, 1988; Taitt and Krebs 1985). Another interesting observation concerns female reproduction: in most species of voles, females—when separated from males—lack spontaneous cycles of follicular maturation, sexual receptivity, and ovulation (Breed 1967; Carter *et al.* 1987; Richmond and Conaway 1969). Furthermore, the genus seems to have evolved nearly as much diversity in social organization and mating systems as the entire order Mammalia (see Keller 1985; Wolff 1985). Information on chemosignals and reproductive biology, such as that presented here, is of value in explaining the evolution and ecology of these interesting rodents.

Over the past 20 years, a variety of anatomical, physiological, and behavioural studies have convincingly demonstrated that many of the effects of chemosignals on rodent reproduction are mediated by the vomeronasal system (reviews by Halpern 1987; Wysocki and Meredith 1987). This chemosensory system is composed of chemoreceptors in the vomeronasal organ (VNO) in the nasal cavity, sensory afferent nerves from the VNO to the accessory olfactory bulb (AOB), and subsequent projections to brain nuclei that are involved in neuroendocrine regulation. Most studies have been conducted on golden hamsters, house mice, and Norway rats. Here we describe some of our recent anatomical and physiological studies of the vomeronasal system of prairie voles, *Microtus ochrogaster*.

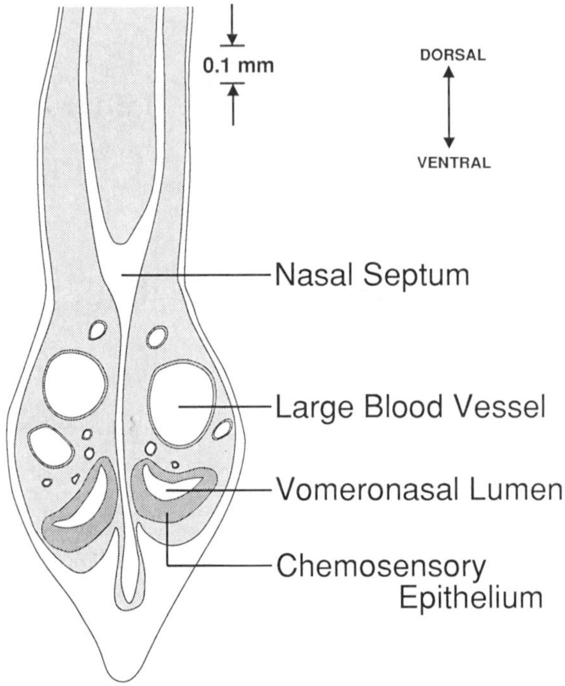


Fig. 1. Schematic representation of a coronal section through the vomeronasal organ of an adult male prairie vole, *Microtus ochrogaster*. Stimulus access is presumed to be via the crescent-shaped vomeronasal lumens, which are lined with ciliated, non-sensory epithelium along the lateral or dorso-lateral luminal surfaces, and with microvillar chemosensory epithelium along the medial or ventromedial luminal surfaces.

## Anatomy of the mammalian vomeronasal organ

The vomeronasal organ of rodents is found just above the palate, and consists of a bilateral pair of tubular structures along both sides of the nasal septum. Each 'tube' is composed of several large blood vessels, cavernous tissue, and a crescent-like structure that contains the chemosensory neurons of the system along the medial or ventro-medial surface of the lumen (Fig. 1). The lateral or dorso-lateral wall of the lumen is typically covered with ciliated non-sensory epithelium. Stimulus access to the chemosensory cells is thought to be at least partially regulated by the action of a 'vomeronasal pump' involving dilation and constriction of large blood vessels with corresponding alterations in VNO lumen volume; this results in the pumping in and out of nasal mucous and its dissolved odorants through the rostral opening of the vomeronasal organ (Meredith and O'Connell 1979).

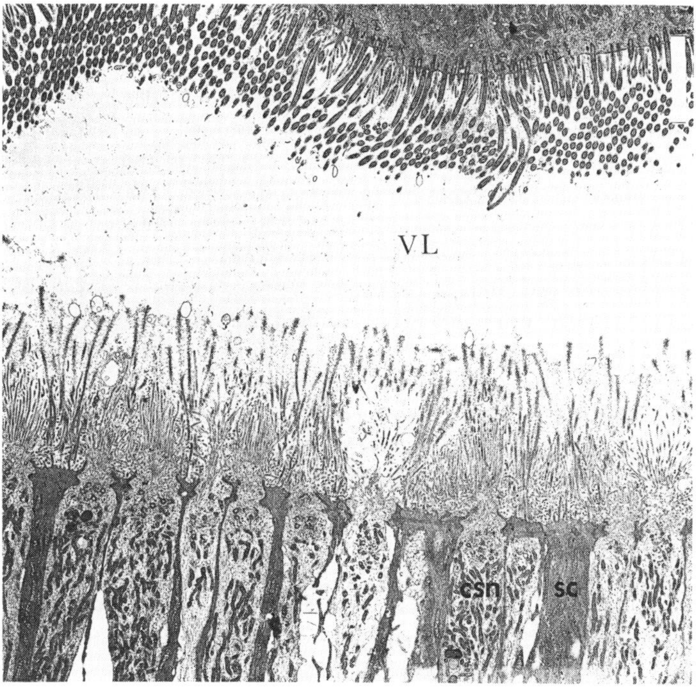


Fig. 2. Transmission electron micrograph of the non-sensory (top) and chemosensory (bottom) epithelia along the lumen (VL) of the vomeronasal organ of an adult male prairie vole. The non-sensory surface is predominantly covered with ciliated cells. The chemosensory epithelium is an alternating mixture of supporting cells (sc) and chemosensory neurons (csn). The chemosensory neurons are less electron dense (lighter grey) and have more mitochondria than the supporting cells (dark grey). Both types of cells project microvilli into the vomeronasal lumen, but those of the supporting cells are longer. Vertical bar in upper right corner represents approximately 2 $\mu$m.

From electron micrographs of vomeronasal tissue taken from three adult male prairie voles, we observed that microvillar chemosensory neurons alternate with supporting cells along the medial surface of the vomeronasal lumen, whereas the lateral surface is covered with ciliated non-sensory epithelium (Fig. 2). The chemosensory neurons contain a higher density of mitochondria than the supporting cells, and the terminal knob, from which microvilli project, is enlarged relative to the remainder of the dendrite. The alternation of microvillar chemosensory neurons and supporting cells observed in voles matches descriptions provided for other rodents (golden hamster, Taniguchi and Mochizuku 1982; house mouse, Nagura and Breipohl 1982; and Norway rat, Vaccarezza *et al.* 1981).

## Role of the vomeronasal organ in the activation of female reproduction

When isolated from males, female voles do not express a spontaneous cycle of oestrus and ovulation. Rather, physical contact with males or their urinary chemosignals is required for activation of the ovaries. The initial phases of reproductive activation, evidenced by follicular maturation and increased uterine weight, are observed after exposure to chemical signals from males; however, physical contact with males is normally essential for the expression of behavioural oestrus (Carter *et al.* 1987). We have recently shown (Lepri and Wysocki 1987) that removal of the vomeronasal organ (VNX) impairs the activation of female reproduction in prairie voles: in females that had been physically paired with stud males for 24 or 60 h, increases of ovarian and uterine weight were significantly lower in VNX females than in females with intact VNOs. The manifestations of impaired activation were not accompanied by any observable alterations in initial behavioural responses to males. Interestingly, 45% of VNX females paired with males for 60 h mated, as evidenced by the presence of sperm in the reproductive tract. This suggests that prolonged physical contact with males may partially ameliorate VNX-impaired activation.

The current study sought to determine whether initial physical contact with males, followed by prolonged exposure to their odours, can produce full activation of physiological and behavioural oestrus in VNX females. In a study by Carter *et al.* (1987), most female voles given physical contact with males followed by exposure to chemosignals responded with behavioural oestrus. A similar exposure-paradigm was used in the present study.

Adult female prairie voles aged 6–7 months were individually housed for this experiment. There were two treatment groups: 10 females whose vomeronasal organs were surgically removed (VNX); and 11 'control' females that underwent a sham surgical procedure (SHAM). Surgery was conducted under ketamine-acepromazine anaesthesia. The VNX procedure consisted of removing the vomeronasal organ from the nasal cavity through an incision in the palate, as previously described by Lepri and Wysocki (1987). The SHAM procedure consisted of most of the same manipulations except that the vomeronasal organ was left intact. The females were given at least three weeks to recover from surgery. Statistical analyses ($\alpha = 0.05$) were non-parametric U-tests for behavioural frequencies and t-tests for continuous variables such as duration and mass (Sokal and Rohlf, 1969).

Data collection began when individual VNX and SHAM females were paired with individually housed stud males, whose cages contained one week's accumulation of male odours. The stud males had been previously screened for sexual vigour using hormonally primed receptive females. Behavioural observations for the first 10 mins of contact revealed that VNX and SHAM females did not respond differently to the stud males, nor were they treated differently by the males. The duration of sniffing, whether nasal- or perianal-directed, and the frequencies of aggression, centre-line crossings (locomotor activity), and digging were also observed: VNX females did not differ significantly from SHAM females, and again the males did not behave differently toward them. Thus the effects of removing the vomeronasal organ were not apparent from observations of behaviour during initial contact.

After 12 h of physical contact, the males were removed from the cages and the females were left in the male-soiled cages for a further 18 h. The original stud males were then re-paired with their respective females for 2 h to determine whether the females would exhibit lordosis; only 1 of 11 SHAM and none of 10 VNX females did so. The males were then again removed from the dirty cages, leaving the females in contact with male-soiled bedding for an additional 22 h. When the males were again re-paired with their respective females for a final 2 h test, 6 of 11 SHAM, and 0 of 10 VNX females, exhibited lordosis.

The females were weighed, killed by pentobarbital overdose, perfused with fixative, and ovarian and uterine weights were obtained. There were no differences of body weight between treatments. All olfactory bulbs of VNX females will be cryostat-sectioned later for histological verification of complete VNX; hence, the present results must be considered preliminary.

**Table 1.** Reproductive organ weights of female prairie voles following removal of the vomeronasal organ (VNX) or SHAM surgery

| Treatment | | Ovarian weight* (mg tissue/40 g body weight) | Uterine weight* (mg tissue/40 g body weight) |
|---|---|---|---|
| VNX | ($n=10$) | 8.0 ($\pm$0.7) | 18.6 ($\pm$ 4.8) |
| SHAM | ($n=11$) | 12.8 ($\pm$1.0) | 98.7 ($\pm$14.4) |

* SHAM significantly greater than VNX; $p<0.005$, t-test with 19 d.f.

Physiological indices of reproductive activation revealed that SHAM females had higher ovarian and uterine weights than VNX females (Table 1). These observations suggest that the absence of vomeronasal chemoreception in VNX females inhibited their reproductive responsiveness to males, in support of earlier results (Lepri and Wysocki 1987). For female voles exposed to chemosignals from males, central projections from the intact vomeronasal organ presumably stimulate neural activity which regulates the secretion of luteinizing hormone-releasing hormone. This may occur at numerous sites, e.g. in the hypothalamus or perhaps in the accessory olfactory bulb (AOB) (Reger *et al.* 1987).

Together these anatomical and physiological results support the following hypothesis generated from tests of domestic species of rodents: vomeronasal chemoreception of social odours provides afferent information necessary to co-ordinate reproductive physiology and behaviour. Further research should be directed at exploring each step in the activation of reproduction in this induced ovulator, from initial events in vomeronasal chemoreception to synaptic processing in the AOB and more central targets. Research of chemosignal effects on reproduction in female voles may reveal the neurochemical mechanisms by which external cues modulate neuroendocrine activity. Such information could elucidate some of the general characteristics of the neuroendocrine control of fertility, and be useful in developing models of reproductive co-ordination and their implications for population biology.

## Acknowledgements

We are grateful to Elaine Lenk, electron microscopist, and to Jason Rudolph and Nicole DeLarato for assistance with the voles. The support of this project by the Department of Biology at Princeton University is especially appreciated.

## References

Breed, W.G. (1967). Ovulation in the genus *Microtus*. *Nature*, 214, 26.

Carter, C. S., Witt, D. M., Schnieder, J., Harris, Z.L. and Volkening, D. (1987). Male stimuli are necessary for female sexual behavior and uterine growth in prairie voles (*Microtus ochrogaster*). *Horm. Behav.*, 21, 74–82.

Frank, F. (1957). The causality of microtine cycles in Germany. *J. Wildl. Managm.*, 21, 113–21.

Halpern, M. (1987). The organization and function of the vomeronasal system. *Annu. Rev. Neurosci.*, 10, 325–62.

Keller, B.L. (1985). Reproductive patterns. In *Biology of New World microtus*, (ed. R.H. Tamarin), pp. 725–78. American Society of Mammalogists Special Publication, 8, Lawrence, Kan.

Lepri, J.J. and Wysocki, C.J. (1987). Removal of the vomeronasal organ disrupts the activation of reproduction in female voles. *Physiol. Behav.*, 40, 349–55.

Lidicker, W.Z., Jr. (1988). Solving the enigma of microtine 'cycles'. *J. Mamm.*, 69, 225–35.

Meredith, M. and O'Connell, R.J. (1979). Efferent control of stimulus access to the hamster vomeronasal organ. *J. Physiol.* 286, 301–16.

Nagura, T. and Breipohl, W. (1982). The vomeronasal epithelia of NMRI mouse. A scanning electron-microscopic study. *Cell Tissue Res.*, 227, 519–34.

Reger, R.L., Gerrall, A.A., Wysocki, C.J. and Carter, C.S. (1987). LHRH neuronal system in the accessory olfactory bulb of the prairie vole, *Microtus ochrogaster*. *Neurosci. Abst.*, 13, 993.

Richmond, M.E. and Conaway, C.H. (1969). Induced ovulation and estrus in

*Microtus ochrogaster*. *J. Reprod. Fert. Suppl.*, 6, 357–76.

Sokal, R.R. and Rohlf, F.J. (1969). *Biometry*. Freeman, San Francisco.

Taitt, M.J. and Krebs, C.J. (1985). Population dynamics and cycles. In *Biology of New World microtus* (ed. R.H. Tamarin), pp. 567–620. American Society of Mammalogists Special Publication, 8, Lawrence, Kan.

Taniguchi, K. and Mochizuku, K. (1982). Morphological studies on the vomeronasal organ in the golden hamster. *Jap. J. Vet. Sci.*, 44, 419–26.

Vaccarezza, O.L., Sepich, L.N. and Tramezzani, J.H. (1981). The vomeronasal organ of the rat. *J. Anat.*, 132, 167–85.

Wolff, J.O. (1985). Behavior. In *Biology of New World microtus* (ed. R.H. Tamarin), pp. 340–72. American Society of Mammalogists Special Publication, 8, Lawrence, Kan.

Wysocki, C.J. and Meredith, M. (1987). The vomeronasal system. In *Neurobiology of taste and smell* (ed. T. Finger and W. Silver), pp. 125–50. Wiley, New York.

# II.8 Gross and Microscopic Anatomy of the Vomeronasal Organ in the Asian Elephant (*Elephas maximus*)

L.E.L. RASMUSSEN AND B. HULTGREN

## Introduction

The vomeronasal organ (VNO) (Ruysch 1703; Jacobsen 1811) is present in the majority of amphibians, reptiles, and mammals. In present-day mammals, as described in rats (Vaccarezza *et al.* 1981), domestic cats (Eccles 1982), sheep (Kratzing 1971), cattle and horses (Taniguchi and Mikami 1985), and camels (Arnautovic *et al.* 1970), the organ is a bilateral, symmetrical structure encapsulated by cartilage lying on either side of the base of the nasal septum. Species differences in VNO anatomy include proportionate sizes, duct and organ interconnections, and relationships between the VNO and olfactory regions.

Recent studies (Estes 1972; Meredith and O'Connell 1979; Wysocki 1979) suggest that the VNO plays a role in sexual behaviour and reproduction rather than a supplementary olfaction (Read 1908) or taste reception function (Negus 1958) as formerly hypothesized. In the Asian elephant, *Elephas maximus*, flehmen responses are directed toward the paired incisive duct openings, which are connected to the VNO (Eisenberg *et al.* 1971; Rasmussen *et al.* 1982). The frequency of flehmen responses, during a variety of behavioural contexts (Rasmussen 1988), especially in the pre-mating sequence (Rasmussen *et al.* 1982), suggested that in the elephant the VNO may have roles similar to those described for other species.

Descriptions of the anatomy of the VNO of the Asian elephant are non-existent. This report describes the dissection procedure, gross anatomy, and histology of the VNO of three Asian elephants.

## Methods

The VNO of two female Asian elephants were dissected. By injection of fluorescein dye into the dual incisive duct openings, the connecting course of the bilateral ducts from the rostral openings in the oral cavity to the VNO and to the nasal area was traced (Figs. 1 and 2). As autolysis was probable, no microscopic

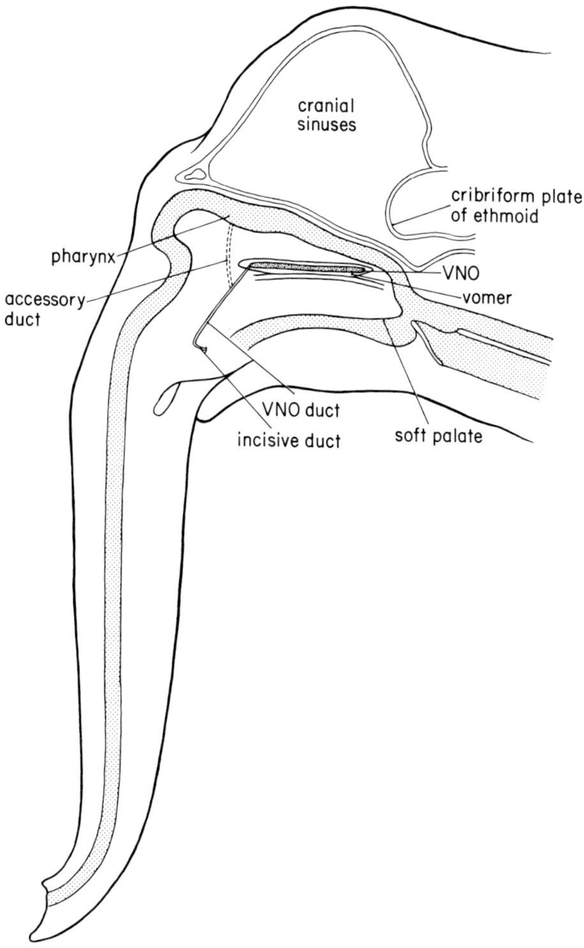


Fig. 1 Paramedial schematic view of the head of a female Asian elephant.

study was done on these specimens. Weight, length, and anatomical measurements were made, and relationships between the VNO and olfactory structures were studied (Figs. 1 and 2).

Duct orifices, ducts, the VNO (including associated respiratory and sensory epithelium), and the vomeronasal nerve were removed as rapidly as possible from a third Asian elephant and fixed in 10% buffered formalin. The tissues were embedded in paraffin; 6-mm-thick sections were stained with haematoxylin and eosin for light microscopy. One area of the VNO containing sensory epithelium was post-fixed in osmium tetroxide, dehydrated in alcohol, embedded in epoxy resin, and sectioned for electron microsopy.

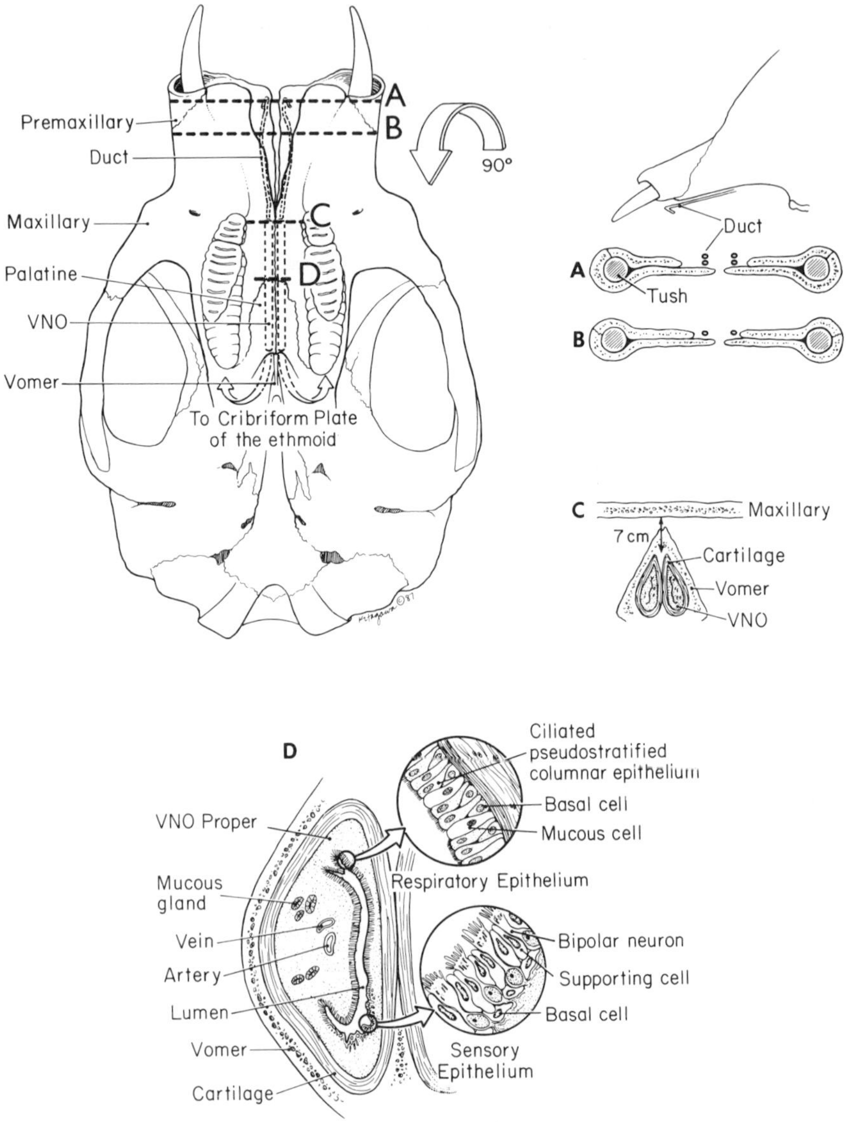


Fig. 2 Ventral view of the skull of an adult Asian elephant. The vomeronasal organ and its anatomical relationships are depicted. The arrow at the top right indicates that the teeth in the adult elephant have a 90° relationship to the anterior palate. A–D are cross-sectional views at the levels indicated; dorsal aspect is at bottom in A–C. A. Anterior and dorsal to incisive duct (ID) openings, at levels of tushes. B. Posterior to paired openings of ID. C. At start of VNO, at anterior teeth. D. At mid-region of VNO. The label VNO proper is essentially in the lamina propria of the VNO. Note three cell types in the epithelium and their locations.

# Results

### GROSS ANATOMY

The paired (1.5–2.0 cm apart) incisive duct (ID) openings were located anterio-rostrally on the roof of the mouth (Fig. 1) 1.5–4.0 cm posterior to the juncture of the mucosa of the mouth with the lip region. No flaps were associated with the ductal openings either in the 3 specimens examined or in 20 living Asian elephants examined of both sexes. The inner orifice diameter of the ID at the region of the oral cavity opening was 0.4–0.6 cm.

The paired ducts initially coursed anteriorly at a 45° angle from the oral cavity mucosa (Figs. 1 and 2A), then turned 160° and headed posteriorly, following an increasingly dorsal and somewhat more lateral course under the maxillary bone traversing the heavily vascularized fibrous connective tissue above the palate (Figs. 1 and 2B). The length of the ducts was 15–24 cm. In one elephant at about 6 cm posterior to the ID openings a side duct was seen extending up into the nasopharynx (Fig. 1).

The VNO was a paired tubular shaped structure (4 cm in diameter), surrounded by connective tissue and cartilage and partially ensconced in the vomer bone (Fig. 2C, D). The most anterior portion of the VNO was located about 15 cm posterior from the openings of the ID and about 5 cm anterior (rostral) to the most anterior projection of the dorsal upper teeth (Fig. 2C). The paired ID were about 1 cm in diameter where they joined the VNO proper. The VNO angled dorsally under the base of the upper teeth below the nasal septum and extended 12–17 cm posteriorly under the teeth. At its posterior end the VNO became composed of an increasing number of nerve fibres, eventually becoming the vomeronasal nerve about 20–25 cm posterior to the ID openings. Large arteries were associated with the nerves. The VNO nerves coursed continually dorsally and penetrated the cribriform plate as at least two discrete bundles on each side (Fig. 2, note arrows). The three female elephants, aged 23–42 years, weighed 3000–4000 kg; brain weights were 3.4–4.2 kgs. VNO weights of 0.025–0.030 kg and VNO lengths of 20–25 cm were recorded.

### HISTOLOGY

The lumina of the ducts were elliptical in cross section in the intact animal, and oblong and sickle-shaped in the fixed preparations. Squamous epithelium lined the openings of the ducts. The ducts were deeply embedded in fibrous connective tissue. At about 20 cm posterior to the ID openings and several centimetres from the VNO proper (just before the teeth), heavy concentrations of mucous cells were observed. Copious amounts of mucus oozed from the cut ducts during dissection. Numerous goblet cells and many mucus-producing glands were interspersed in the epithelium, especially at the junctional region of the ducts with the VNO.

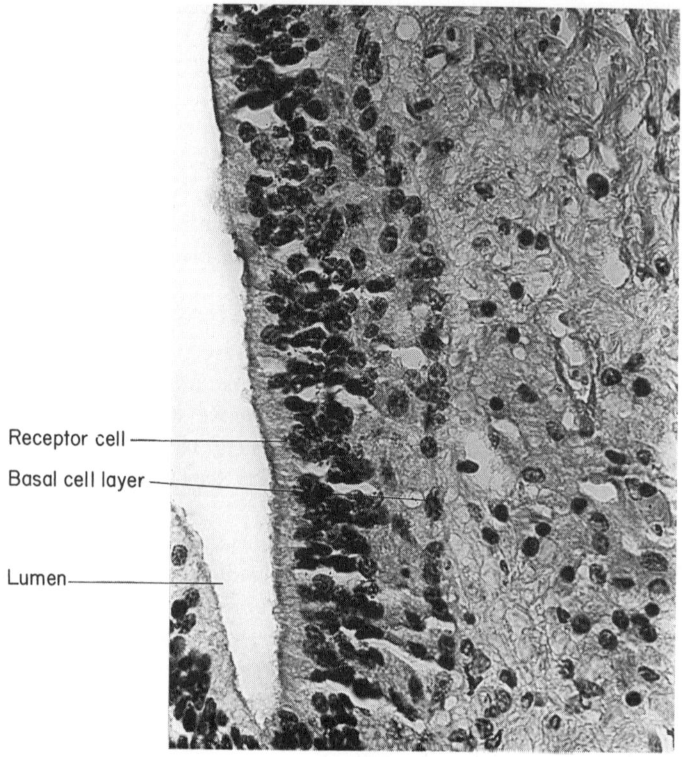


Fig. 3. Photomicrograph of the medial, ventral sensory epithelial region of the VNO. Note the layer of basal cells and receptor cells or bipolar neurons. The supporting cells are not labelled, but are interspersed among the sensory cells. Haematoxylin and eosin stain; original magnification, X400.

The VNO itself was completely encapsulated by fibrous connective tissue containing mucous-producing glands, blood vessels, and nerves. Each lumen was somewhat U-shaped with several folds and/or bifurcations present (Fig. 2D). As the ducts traversed posteriorly and became the VNO proper, two types of epithelium surrounded the paired central lumen, lateral respiratory-like epithelium and medial sensory epithelium. The pseudostratified, ciliated, columnar respiratory-like epithelium continually surrounding the paired lumina of the VNO (except in the ventral-medial region) had a row of basal cells and other cells that were supporting or mucous-secreting. Numerous nerves and blood vessels with pronounced walls and much glandular tissue were present in the surrounding fibrous connective tissue (Fig. 2D). The ventral, medial epithelium was characteristic of sensory epithelium (Fig. 3). A basal layer of cells with large nuclei was present (Figs. 2D and 3). Supporting cells and a layer of columnar neuronal cell bodies with terminals at the VNO lumen were seen (Fig. 3). This superficial

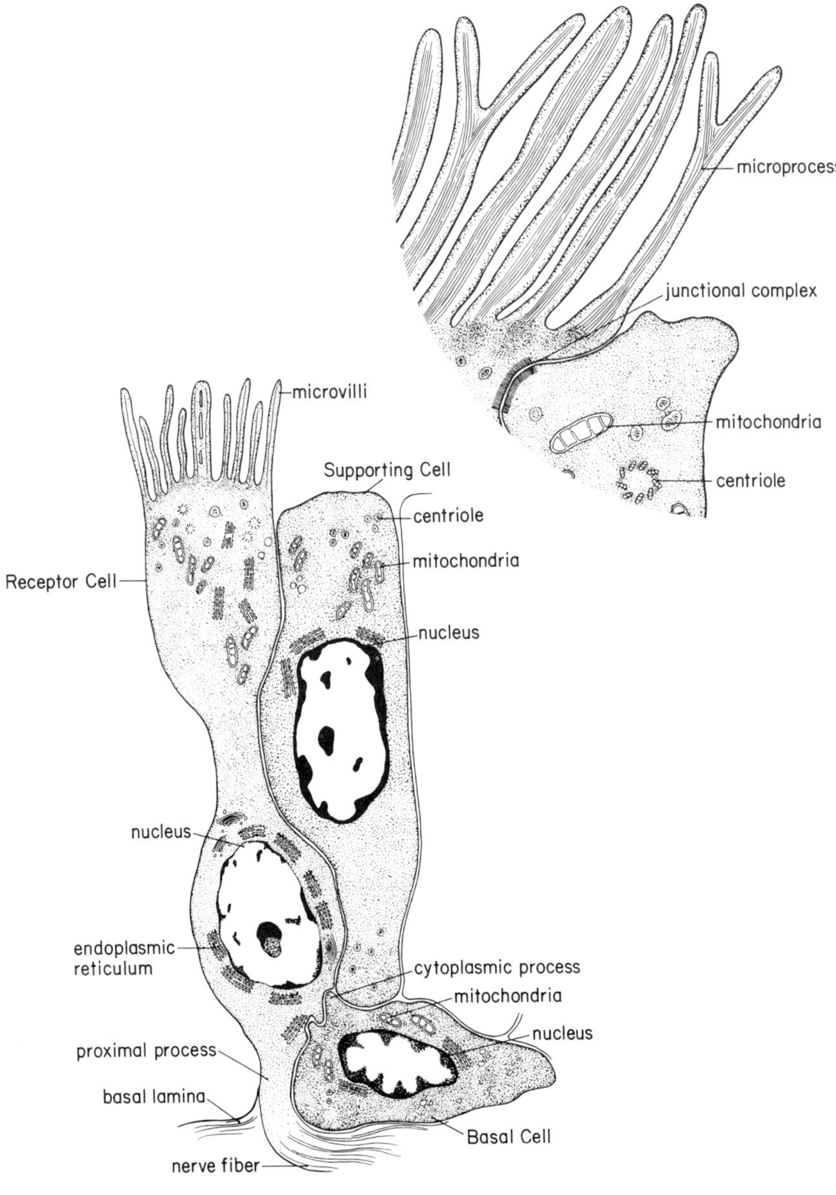


Fig. 4. Drawing of an electron micrograph of the cells of the sensory epithelium (medial, ventral) of the VNO. On the right is depicted a supporting cell. Basal cells are characterized by large nuclei. A bipolar neuronal receptor cell on the left has microvilli (at times branched) that project into the lumen of the VNO. Receptor cells also have many centrioles. Magnification X14,000. Insert on right is higher magnification of the sensory receptor cell. Note branched process of microvillus.

layer (lining the lumen) was composed of prolongations of both supporting and bipolar receptor cells (Fig. 2, view D; Fig. 4). Microvilli extended into the lumen from the receptor cells (Fig. 4). The paired VNO nerves, composed of numerous (up to 55) multi-sized myelinated fibre bundles, were large and of 0.55 cm in diameter.

## Discussion

Several distinguishing anatomical characteristics of the elephant VNO included its large size, the pathway of ID, and the large number of nerve bundles in the large vomeronasal nerve. The ID traversed forward from openings in the anterior oral cavity and after a 160° turn proceeded posteriorly, in a progressively dorsal direction and in a configuration somewhat similar to that seen in cattle (Jacobs *et al.* 1981). The protection of the VNO was extensive; it was completely encapsulated by cartilage (similar to the VNO of horses (Taniguchi and Mikami 1985), partially embedded in the vomer bone, and protected by the palate and teeth.

Distinguishing histological features include the presence of myelinated fibres in the vomeronasal nerve and the predominance of sensory epithelium in the dorsal medial aspect.

The abundance of mucus in the ID and the VNO was impressive. Occasionally, in the closely monitored elephant herd at Washington Park Zoo, mucus is spontaneously expressed from the ID openings. Future observations on the distribution frequency and relationships of mucous cells along the VNO lumen, and the possible presence of vomeronasal glands may help our understanding of how previously processed substances are cleared from the VNO.

## Acknowledgements

The support of the Friends of Washington Park Zoo (WPZ), dissection assistance by Jay Haight and Jill Mellen (WPZ), the donation of the deceased elephants by WPZ, and drawings and diagrams by Lynn Kitagawa are gratefully acknowledged.

## References

Arnautovic, I., Abdalla, O. and Fahnay, M.F.A. (1970). Anatomical study of the vomeronasal organ and nasopalatine duct of the one-humped camel. *Acta. Anat.*, 77, 144–54.

Eccles, R. (1982). Autonomic innervation of the vomeronasal organ of the cat. *Physiol. Behav.*, 28, 1011–5.

Eisenberg, J., McKay, G.M. and Jainudeen, M.R. (1971). Reproductive behavior of the Asiatic elephant, *Elephas maximus maximus*. *Behavior*, 38, 193–225.

Estes, R.D. (1972). The role of the vomeronasal organ in mammalian reproduction. *Mammalia*, 36, 315–41.

Jacobs, V.L., Sis, R.F., Chenoweth, P.J., Klemm, W.R. and Sherry, C.J. (1981). Structures of the bovine vomeronasal complex and its relationship to the palate: tongue manipulation. *Acta Anat.*, 110, 48–58.

Jacobsen, L. (1811). Description anatomique d'un organe observé dans les mammifères. *Ann. Mus. d'Histoire Nat. Paris* (reported by Cuvier), 18, 412–24.

Kratzing, J. (1971). The structure of the vomeronasal organ in the sheep. *J. Anat.*, 108, 247–60.

Meredith, M. and O'Connell, R.J. (1979). Efferent control of stimulus access to the hamster vomeronasal organ. *J. Physiol.*, 286, 301–16.

Negus, V. (1958). *The comparative anatomy and physiology of the nose and paranasal sinuses*. E. and S. Livingston, Ltd., London.

Rasmussen, L.E.L. (1988). Chemosensory responses in two species of elephants to constituents of temporal gland secretion and musth urine. *J. Chem. Ecol.*, 14, 1687–711.

Rasmussen, L.E., Schmidt, M.J., Henneous, R., Groves, D. and Daves, G.D. (1982). Asian bull elephants: flehmen-like responses to extractable components in female elephant estrous urine. *Science*, 217, 159–62.

Read, E.A. (1908). A contribution to the knowledge of the olfactory apparatus in dog, cat and man. *Amer. J. Anat.*, 8, 17–47.

Ruysch, F. (1703). *Thesaurus anatomicus III*. Amsterdam.

Taniguchi, K. and Mikami, S. (1985). Fine structure of the epithelia of the vomeronasal organ of horse and cattle. *Cell Tissue Res.*, 240, 41.

Vaccarezza, O.L., Sepich, L.N. and Tramezzani, H. (1981). The vomeronasal organ of the rat. *J. Anat.*, 132, 167–85.

Wysocki, C.J. (1979). Neurobehavioral evidence for the involvement of the vomeronasal system in mammalian reproduction. *Neurosci. Biohav. Rev.*, 3, 301–41.

# II.9 Physiological Determinism of Olfactory Attraction for Amniotic Fluids in Sheep

F. LEVY, E.B. KEVERNE, V. PIKETTY AND P. POINDRON

## Introduction

In sheep, strong interest for the newborn appears only at parturition and is stronger towards an uncleaned neonate than a washed one (Lévy and Poindron 1987). This variation in behaviour is the consequence of olfactory attraction for amniotic fluids (AF) which cover the coat of the neonate at birth. Indeed, AF are very attractive only for a few hours after parturition (Lévy *et al.* 1983) and at all other times (e.g. oestrous cycle, gestation) AF are strongly repulsive. Such behaviour appears to depend largely on olfaction, since anosmia suppresses both attraction and repulsion towards AF (Lévy *et al.* 1983).

The physiological factors that govern this sudden change of preference towards AF remain to be elucidated. However, recent studies have indicated that mechanical stimulation of the genital tract at parturition play an important role in the onset of maternal behaviour, especially regarding licking of the neonate (Keverne *et al.* 1983; Krehbiel *et al.* 1987). As this behaviour is associated with ingestion of AF, the possibility exists that both behaviours have common physiological determinants. The aim of the experiments presented here were to investigate: (i) the role of genital stimulation (GS) in regulation of attraction for AF; and (ii) the possible involvement of central oxytocin release in the manifestation of this behaviour.

## Effects of GS on attraction for amniotic fluids

Two series of experiments were carried out. In the first, the effects of GS deprivation were studied in peridural anaesthetized ewes. In the second, on the contrary, additional GS was provided to post-parturient females.

### EXPERIMENT I

Twelve ewes were studied. In the peridural group ($n = 7$), females received at the first signs of parturition an injection of 6 ml of Sylvocaïne in the peridural space between the sacrum and the first caudal vertebra. Previously, we have shown that this procedure blocks sensory input from the cervicovaginal region and inhibits maternal behaviour (Krehbiel *et al.* 1987). Lambs were extracted manually 15 to 60 min after anaesthesia. Ewes of the control group ($n=5$) received a saline injection instead of Sylvocaïne. To test all the ewes' attraction for AF, they were given a choice of feeding between one trough containing food moistened with water at 25°C and another trough containing food mixed with AF, also at 25°C (Lévy *et al.* 1983). The feeding time in each trough was recorded during the 4 mins of the test. Animals were tested 24 h before and just after parturition; their performances were compared using non-parametric statistics.

Results indicated that the *total* feeding time did not differ significantly between control and experimental groups, neither before (122 ± 71sec vs. 57 ± 38sec; $p=0.7$) nor after (66 ± 100sec vs. 54 ± 53sec; $p=0.2$) parturition, respectively. Before parturition, none of the females of either group ate in the trough containing AF. However, after parturition females of the peridural group spent significantly less time eating in the AF trough than females of control group (0 sec vs 46 ± 90 sec, $p=0.03$).

## EXPERIMENT 2

Our previous studies showed a decrease of attraction towards AF in less than 4 h after parturition in 50% of females (Lévy *et al.* 1983). We therefore sought to test the following hypothesis: if GS is a determining factor for this behaviour, additional stimulation should induce further attraction for AF. However, if such behaviour is dependent on the rapidly changing hormonal status of the post-parturient female, its effects might fade rapidly.

Using the two-choice test described for Experiment 1, multiparous ewes were selected that were not attracted to AF at different times after lambing (2 h, $n=10$; 4 h $n=11$; 8 h $n=9$). Immediately after, they received GS via a rubber bladder which was inserted into the uterus, then inflated and maintained in place for 5 mins before being expelled. A second two-choice test was then performed.

At 2 h and 4 h post-partum, the proportion of ewes showing a preference for the trough containing AF was significantly increased after GS (0/10 vs 7/10; 0/11 vs 4/11; $p<0.05$, respectively). No effect was observed 8 h after parturition.

These two experiments clearly indicate that GS plays a determinant role at parturition, namely to suppress aversion and induce attraction towards AF. Furthermore, the results of Experiment 2 also suggest that the effectiveness of GS decreases with time. This might be related to some of the rapid hormonal changes occurring after parturition; for example, oestradiol which falls to low concentrations within 4 h post-partum.

## Effects of intracerebroventricular injection of oxytocin on AF attraction

Recent studies have shown that oxytocin levels in cerebrospinal fluid and olfactory bulbs increase during parturition or after artificial GS (Kendrick *et al.* 1986, 1988*a*, *b*). Therefore we investigated the possible role of central oxytocin release in GS-induced attraction for AF. Primiparous parturient ewes implanted with bilateral cannulae in the lateral ventricles, received a peridural anaesthesia at the first signs of labour. Oxytocin (10μg dissolved in 0.5 ml sterile 0.9% NaCl, $n=7$) or saline (control group, $n=10$) was infused 60 mins later. A second intracerebroventricular injection was performed 90 mins after anaesthesia, just after the delivery of the lamb. One day before and just after parturition, AF attraction was assessed using the two-choice test described above.

Before and after lambing, *total* feeding time did not differ significantly between control and experimental group ($135 \pm 95$ sec vs. $114 \pm 73$ sec; $142 \pm 52$ sec vs. $78 \pm 53$ sec, respectively; $p > 0.1$). By contrast, after parturition the ewes which were injected with oxytocin spent significantly more time eating in the AF trough than did females of control group ($67 \pm 58$ sec vs. $20 \pm 51$ sec; $p=0.03$).

## Conclusion

These experiments clearly indicate that GS influences the olfactory attraction shown by parturient ewes for amniotic fluid. The results also suggest that this action is in part mediated via an intracerebral liberation of oxytocin, which is induced by the expulsion of the foetus. Together with studies showing an increase of oxytocin in the olfactory bulbs during parturition, this suggests that oxytocin could modulate olfactory processing at the level of the bulbs.

## References

Kendrick, K.M., Keverne, E.B., Baldwin, B.A. and Sharman, D.F. (1986). Cerebrospinal fluids levels of acetylcholinesterase, monoamines and oxytocin during labour, parturition, vaginocervical stimulation, lamb separation and suckling in sheep. *Neuroendocrinology*, 44, 149–56.

Kendrick, K.M., Keverne, E.B., Chapman, C. and Baldwin, B.A. (1988*a*). Intracranial dialysis measurement of oxytocin, monoamine and uric acid release from the olfactory bulb and substantia nigra of sheep during parturition, suckling, separation from lambs and eating. *Brain Res.*, 439, 1–10.

Kendrick, K.M., Keverne, E.B., Chapman, C. and Baldwin, B.A. (1988*b*). Microdialysis measurement of oxytocin, aspartate, γ-aminobutyric acid and glutamate release from the olfactory bulb of the sheep during vaginocervical stimulation. *Brain Res.*, 442, 171–4.

Keverne, E.B., Lévy, F., Poindron, P. and Lindsay, D.R. (1983). Vaginal stimulation: an important determinant of maternal bonding in sheep. *Science*, 219, 81–3.

Krehbiel, D., Poindron, P., Lévy, F. and Prud'homme, M.J. (1987). Effect of peridural anesthesia on maternal behavior in primiparous and multiparous parturient ewes. *Physiol. Behav.*, 40, 463–72.

Lévy, F. and Poindron, P. (1987). Importance of amniotic fluids for the establishment of maternal behaviour in relation with maternal experience in sheep. *Anim. Behav.*, 35, 1188–92.

Lévy, F., Poindron, P. and Le Neindre, P. (1983). Attraction and repulsion by amniotic fluids and their olfactory control in the ewe around parturition. *Physiol. Behav.*, 31, 687–92.

# SECTION III

## Reproduction and primers

# III.1 Pheromonal Contraception

D.H. ABBOTT, J. BARRETT AND C.G. FAULKES

## Introduction

The social environment can inhibit reproductive processes in many species of mammals, primate and non-primate (Abbott 1987; Abbott and Hearn 1978; Bowman *et al.* 1978; Dunbar 1980; Harcourt 1987; Jarvis 1981; McClintock 1983; Wasser and Barash 1983). There are two major ways in which socially induced infertility is brought about: through stress caused by behavioural intimidation (Dunbar 1989; Kaplan *et al.* 1986; Keverne 1982), and through the action of pheromones released by conspecifics (Marchlewska-Koj 1984; Massey and Vandenbergh 1980; Schilling *et al.* 1984). This paper is primarily concerned with the latter.

Pheromones in scent secretions, urine, or faeces, can have either signalling or priming effects on recipients. Signalling pheromones have a specific chemical composition and result in a defined species-specific behavioural response, such as the male hamster sexual attractant, dimethyl disulphide, found in the vaginal discharge of female hamsters, which induces investigation and sexual attraction in males (Singer *et al.* 1976). Primer pheromones vary quantitatively or qualitatively between individuals and produce behavioural or endocrine responses that are not necessarily species-specific (Keverne 1983; Müller-Schwarze 1977). The aim of this paper is to examine social situations where primer pheromones are involved in the suppression of reproduction in male and female mammals. Examples of clear, suppressive effects of primer pheromones on reproduction provide a natural experimental approach which can elucidate some of the physiological processes underlying mammalian reproduction.

## Males

Primer pheromone-induced reproductive suppression amongst male mammals involves puberty delay in young males or impaired testis function in subordinate adult males.

### DELAYED PUBERTY

One example of pheromonally delayed puberty in males can be seen in male deer mice, *Peromyscus maniculatus bairdii*, where sexual maturation of juvenile males was delayed after exposure to adult males or their soiled bedding (Lawton and Whitsett 1979). The growth of the testis and seminal vesicles was retarded in exposed juveniles. Urine from adult females was without such effect (Lawton and Whitsett 1979). The physiological mechanism initiating this pheromonal effect is not yet known but may involve the suppression of prolactin secretion found in puberty-delayed males (Kirkland and Bradley 1986). Certainly in female rats, reducing circulating prolactin concentrations will delay puberty in young females (Advis *et al.* 1981). However, other studies of puberty delay in juvenile male deer mice have not found such clear evidence for an inhibitory pheromonal effect (Terman 1987).

### IMPAIRED TESTIS FUNCTION

There are two examples of pheromonally induced suppression of testis function in adult males. In the marsupial sugar glider, *Petaurus breviceps papuana*, primer pheromones from dominant males cause reproductive inhibition in subordinate males (Schultz-Westrum 1969).

A more detailed account of reproductive suppression is given for subordinate male lesser mouse lemurs, *Microcebus murinus* (Schilling *et al.* 1984). Along with suppression of their sexual behaviour, subordinate males failed to show a rise of blood testosterone concentrations at the onset of the breeding season. This inhibition of testosterone secretion in subordinates was brought about experimentally by transferring the volatiles of dominant males' urine, or a solvent extract of the urine, into the subordinates' cages and thus identifying volatile urinary pheromones as the physiological suppressor. Only urine from dominant males produced this effect. Urine from subordinate males was not effective. The importance of dominance status rather than an individual specific odour was demonstrated when suppression of testosterone secretion in subordinates was achieved using the scent from the urine of dominant males which the subordinates had not met previously (Schilling and Perret 1987). Inhibition of gonadotrophin-releasing hormone (GnRH) from the hypothalamus of subordinate males is probably the cause of their testosterone suppression, and the inhibition of GnRH may be due to rapid and large increases in circulating prolactin concentrations at the onset of the breeding season (Schilling and Perret 1987; Perret and Schilling 1987).

## Females

Pheromonally induced suppression of reproduction in female mammals can be more striking because, unlike males, females may experience complete infertility (Vandenbergh 1986) and the pheromonally induced block to female reproduction can manifest itself through several different physiological mechanisms.

## DELAYED PUBERTY

Delayed puberty is perhaps the most common form of pheromonal contraception among female rodents. First described by van der Lee and Boot (1955) to explain the delay in sexual maturation seen in group-housed mice, the effect is due to a chemosignal present in the voided urine of group-housed female mice (Cowley and Wise 1972; Drickamer 1982). The puberty delaying pheromone in female mice is composed of several volatile compounds including ketones, esters, and a pyrazine (Novotny *et al.* 1986).

The efficacy of puberty delaying pheromones in the urine of wild mice depends on population density. Urine collected from adult females living in a high density population was able to suppress puberty in individually housed juvenile females in the laboratory, but urine collected when the density of the same population was low failed to delay puberty (Massey and Vandenbergh 1980; Coppola 1986). This density-dependent release of pheromone may therefore play an important role in regulating natural mouse populations by limiting the rate of female sexual maturation.

Pheromonal inhibition of female puberty is a common feature among high-density populations of microtine rodents, such as prairie deer mice (Lombardi and Whitsett 1980; Kirkland and Bradley 1986), bank voles, *Clethrionomys glareolus* (Kruczek and Marchlewska-Koj 1986), pine voles, *Microtus pinetorum* (Lepri 1986; Lepri and Vandenbergh 1986), Californian voles, *M. californicus*, and prairie voles, *M. ochrogaster* (Batzli *et al.* 1977). However, puberty delay is not normally seen in the meadow vole, *M. pennsylvanicus* (Batzli *et al.* 1977). Several authors have suggested that the occurrence of puberty delay among female rodents is related to habitat (Batzli *et al.* 1977; Vandenbergh 1986; Vandenbergh and Coppola 1986). The Californian vole and the prairie vole live in extensive grassland habitats where the best survival strategy would be for maturing offspring to leave the family before breeding. Puberty delay would enforce this. Conversely, the habitat of the meadow vole comprises small patches of moist meadow or marsh and new areas are continually invaded. In such an environment, breeding amongst siblings could be favoured because unrelated mates might not be readily available and reproductive suppression would therefore not be advantageous. Puberty delay would therefore not be necessary. A similar association between habitat type and the presence or absence of pheromonally induced puberty delay was found in the arctic dwelling brown lemming, *Lemmus sibiricus* (present), and the tundra vole, *M. oeconomus* (absent: Facemire and Batzli 1983).

Little is known of the physiological mechanisms that translate pheromonal chemoreception into puberty delay in females. Advis and colleagues (1981) have shown that hypoprolactinaemia will delay puberty in female rats but such a physiological mechanism has not been functionally linked with pheromonally mediated puberty delay.

## SUPPRESSION OF OESTRUS

Pheromones released by group-housed female mice and which suppress oestrus in adult females are dependent on the density and duration of grouping of the adult females (McClintock 1983; Coppola and Vandenbergh 1985). 'Suppression of oestrus' in mice refers to an increase in the time period between each bout of oestrous behaviour and does not usually mean the cessation of oestrous bouts or the suppression of ovulation (see below).

When adult female laboratory mice are kept singly, they undergo regular, 4-6 day oestrous cycles. Ovulation occurs spontaneously and, in the absence of mating, is followed by the development of non-functional corpora lutea. Like the rat, the female mouse depends on the stimulation of mating to activate neuroendocrine mechanisms that will support fully functional corpora lutea (McClintock 1983). However, when adult female mice are housed in groups, an increase in the length of their oestrous cycle occurs, usually called suppression of oestrus or the Lee-Boot effect (van der Lee and Boot 1955). Similar effects with grouped adult females have been found in deer mice (Lombardi and Terman 1980), Indian field mice, *Mus booduga* (Dominic and Pandey 1979) and wild hopping mice, *Notomys alexis* (Breed 1976).

McClintock (1983) indicated that the suppression of oestrus amongst group-housed female mice reflected either a lengthening of dioestrus, the luteal phase of the ovarian cycle, or spontaneous pseudopregnancy. Apparently the frequency of oestrus was reduced in grouped females by the formation of fully functional corpora lutea after ovulation (Ryan and Schwartz 1977). Whitten (1959) reported complete suppression of oestrous cycles and possibly suppression of ovulation in adult female mice living in groups of 30 animals. Ryan and Schwartz (1977), however, found pseudopregnancies in most female mice when housed in similar sized groups.

Primer pheromones in the urine of group-housed female mice act via the vomeronasal organ to suppress the frequency of oestrous cycles in these females. Bilateral lesions of the vomeronasal organ of group-housed female mice return the females to the normal 4–6 day oestrous cycle (Reynolds and Keverne 1979), while lesions of the main olfactory epithelium were without effect (Ingersoll, 1980, cited by Meredith 1983). The effects of the vomeronasal lesions could be replicated in intact group-housed female mice by administering bromocriptine, a dopamine agonist which inhibited prolactin secretion (Fig. 1). Conversely, giving haloperidol, a dopamine antagonist which elevated prolactin secretion, to group-housed females with vomeronasal organ lesions suppressed their oestrous cycles making them as infrequent as those of group-housed females without a vomeronasal organ lesion (Fig. 1). Thus, primer pheromones in the urine of group-housed female mice act via the vomeronasal organ to stimulate prolactin secretion. Since prolactin is a luteotrophic hormone in the mouse, the elevated prolactin levels supported fully functional corpora lutea which stimulated

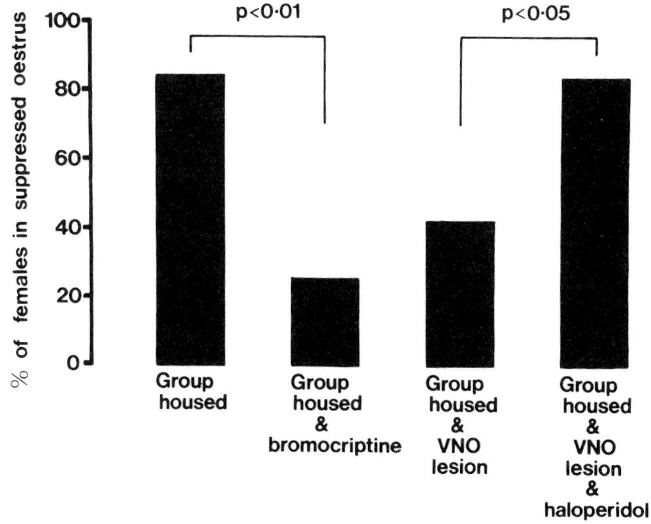


Fig. 1 The effect of lesions of the vomeronasal organ on the administration of bromocriptine or haloperidol on the percentage of grouped female mice in suppressed oestrus. Data were taken from Reynolds and Keverne 1979).

pseudopregnancy. The elevated prolactin levels were probably brought about by decreased dopamine release from the tubular-infundibular (TIDA) neurones in the hypothalamus and lesioning the vomeronasal organ apparently interrupted this (Reynolds and Keverne 1979). What neuroendocrine mechanism influences the TIDA neurones is not yet known. Nevertheless, as with puberty delay, pheromonally mediated suppression of oestrus could act as an effective population regulator because it also occurs in heterosexual groups of mice (Lamond 1959; Mody 1963; McClintock 1983). However, group housing does not affect all female mice equally since the length of the oestrous cycle of some females were less affected by increased population density than others (McClintock 1983). Whether or not the less affected females were socially dominant is not known.

SUPPRESSION OF OVULATION

Good examples of pheromonally induced suppression of ovulation in subordinate females by dominant females are found in the marmoset and tamarin monkeys. Social groups of marmoset and tamarin monkeys in the wild usually contain only 3–15 individuals (Stevenson and Rylands 1988) and each group possesses only one breeding female. Laboratory studies have shown that the socially dominant female in each group inhibits ovulation in all her subordinates (Abbott and Hearn 1978; Epple and Katz 1980; French *et al.* 1984). However, if subordinate females were

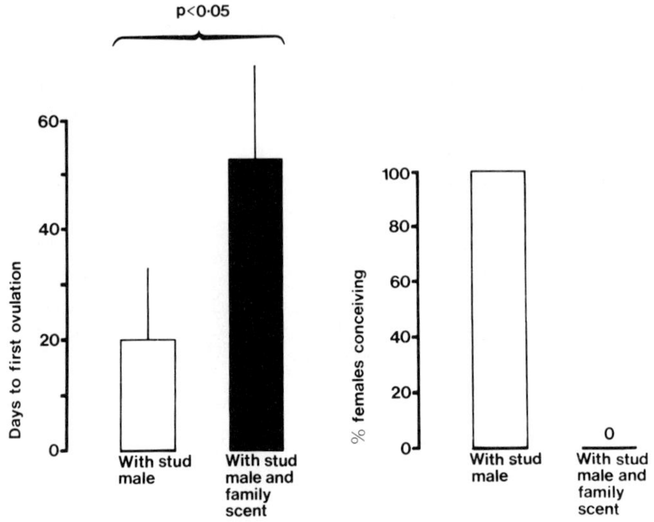


Fig. 2 The effects of scent transfer on (a) the time taken to the first ovulation and on (b) the number of females to conceive when acyclic, subordinate daughter female cotton top tamarins were removed from their family groups and paired with a male. The filled bars refer to the daily transfer of a scent-marked perch from the original family group and the open bars refer to the unmarked perch controls. Data were taken from Savage *et al.* (1988).

removed from their groups and housed on their own (Abbott *et al.* 1988*a*) or with a male mate (Abbott 1984; Epple and Katz 1980; French *et al.* 1984) they rapidly commenced ovulating and in the latter cases, became pregnant. Returning ovulating, non-pregnant females to subordinate status in a group quickly re-imposed suppression of ovulation (Abbott *et al.* 1988*a*).

Epple and Katz (1984) were the first to provide evidence for the true role of primer pheromones in mediating the social suppression of ovulation. Five subordinate daughters were removed from their captive family groups of saddle-back tamarins, *Saguinus fuscicollis*, and each was paired with a vasectomized male. In the ensuing 10-week period, all females showed an increase in urinary oestradiol. However, one of the five females had scent-marked plates from the home cage transferred daily to her cage, and this same female did not show the same increase in urinary oestradiol and cyclicity as her separated peers until the scent transfer from her family's cage was discontinued. Similarly, Savage and her colleagues (1988) found a delay in the onset of ovulation in subordinate daughter cotton top tamarins, *S. oedipus*, when they were housed with a male but maintained in scent contact with their families (Fig. 2). These females also failed to conceive during the scent transfer period (Fig. 2). However, once the females had conceived, further exposure to scent-marked perches from the original and an unfamiliar family did not disrupt the pregnancy.

Experiments with social groups of unrelated adult common marmoset monkeys, *Callithrix jacchus*, produced similar findings implicating primer pheromones in mediating the suppression of ovulation in subordinate females. In the control phase, subordinate females ($n=5$) were removed from their social groups and housed in clean cages. Thrice daily a cotton-bud moistened with saline was wiped over the external nares and nasopalatine canal opening of the subordinate female. This phase lasted for 6-8 weeks whereupon the subordinate female was returned to her original social group. In the scent transfer phase, subordinate females ($n=8$) were removed from their social groups and housed singly for 49 days, during which time scent communication was maintained with their group. Scent contact was maintained by alternating the group's cage with that of the isolated subordinate's thrice daily and on each occasion a saline-moistened cotton-bud was wiped over the vulval and anogenital region of the dominant female and then over the external nares and naeopalatine canal opening of the subordinate female. Two examples of subordinate females that underwent both phases are shown in Fig. 3. In the control phase both subordinate females ovulated within 2 weeks (388W, 9 days; 401W, 11 days) which was comparable to that seen in previous experiments when subordinate females ($n=5$) were isolated from their groups without further manipulations ($11 \pm 3.7$ days). During the period of scent transfer from their groups, neither subordinate female ovulated within the 2 weeks seen previously, 388W ovulated on day 45 and 401W ovulated on day 26. When the data collected from both the control and scent-transfer phase were examined, there was a significant ($p<0.05$) delay in the onset of ovulation when scent communication was maintained between the isolated subordinate female ($n=8$) and her group ($31.0 \pm 6.4$ days, mean $\pm$ S.E.M.) compared to controls ($10.4 \pm 0.8$, mean $\pm$ S.E.M., $n=5$).

However, primer pheromones may not be the only mechanism mediating social subordination into suppression of ovulation in marmoset monkeys. Six subordinate females remaining in their social groups, but each with bilateral ablations of the vomeronasal organ and the main olfactory epithelium, did not start ovulating in the following 49-day period (J. Barrett, D.H. Abbott and M. George, unpublished results). Visual and physical reinforcement of the females' subordinate status by their dominant female may therefore also play a part in maintaining the suppression of ovulation in subordinates, as previously suggested by Poole and Evans (1982) and Abbott (1984).

What physiological mechanisms mediated the pheromonally related suppression of ovulation in subordinate female marmoset monkeys? Suppressed pituitary luteinizing hormone (LH) secretion was indicated in both common marmosets and cotton top tamarins (Abbott 1986; Abbott *et al.* 1988*a*; Ziegler *et al.* 1987), but no information on levels of follicle stimulating hormone (FSH) is available. Inhibited GnRH release from the hypothalamus was apparently the cause of the LH insufficiency, because long-term GnRH replacement therapy given to subordinate female marmosets stimulated LH secretion (Ruiz de Elvira and Abbott 1986),

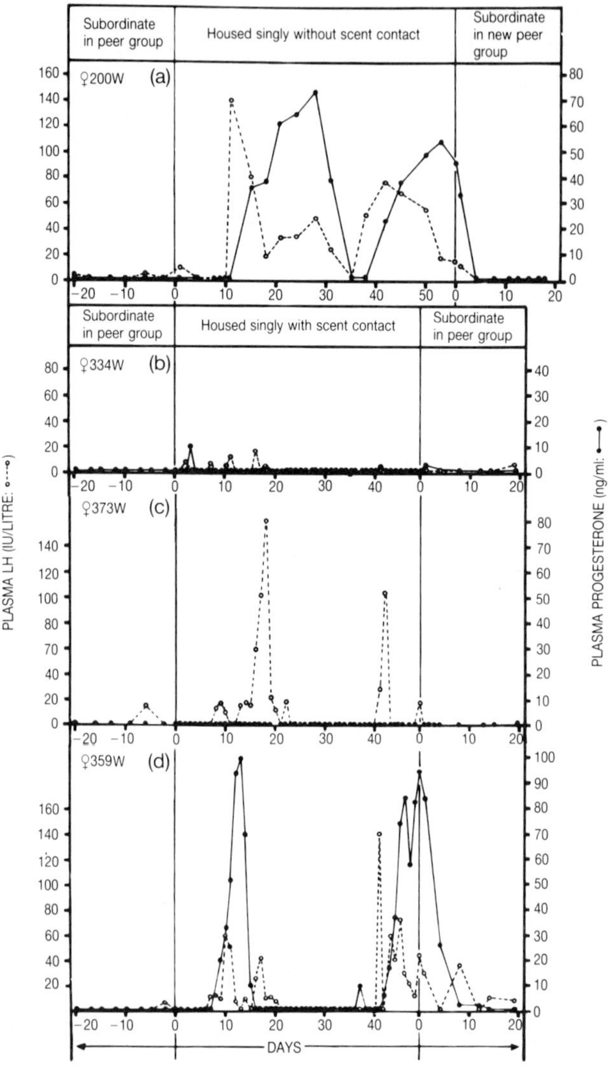


Fig. 3 Plasma LH and progesterone concentrations in acyclic, subordinate female marmoset monkeys removed from their social groups, housed on their own for 50 to 56 days and then returned to their original social group. In (a) female 200W is a typical example from a group of six subordinate females put through this procedure. In (b), (c) and (d), females 334W, 373W and 359W, respectively, the cages of the isolated females were alternated with those of the rest of the group thrice daily and the isolated females received a swab from the vulval-anogenital region of their dominant female on the nostrils, upper lip and nasal-palatine canal opening in the palate at the same times. The 'rest of the group' in each case was (b) dominant female, (c) dominant and one subordinate female and (d) dominant male and female.

ovarian follicular growth (Abbott, 1989), and ovulation (Abbott 1987). Neither plasma concentrations of prolactin and cortisol (Abbott *et al.* 1981) nor lowered body weight (Abbott 1986) were found in subordinate females and so could be ruled out as possible suppressors of hypothalamic GnRH secretion. Nevertheless, two possibly independent neuroendocrine mechanisms have been found. The more pronounced of the two is the development of an increased negative feedback sensitivity to oestradiol (Abbott 1988). The other is independent of the presence of ovarian hormones, such as oestradiol, and involves the suppression of GnRH secretion by endogenous opioid peptides (Abbott *et al.* 1988*b*; unpublished results).

### IMPLANTATION FAILURE

The male-induced block to implantation in mice (Bellringer *et al.* 1980) or the 'Bruce effect' (Bruce 1959) is probably the best known pheromonal contraceptive effect. In mice, implantation is blocked by non-volatile chemosignals emanating from the urine of unfamiliar males, but not from the urine of males with which the female has mated (Keverne 1983, 1989). Male-induced pregnancy disruption occurs in several rodent genera including *Mus*, *Peromyscus* and *Microtus* (Marchlewska-Koj 1983; Smale 1988). In mice, the pheromonal effect is mediated by the vomeronasal organ (Ingersoll 1980, cited by Meredith 1983; Lloyd-Thomas and Keverne 1982), and results in suppressed prolactin secretion in the exposed pregnant females. As a consequence, degeneration of the corpora lutea, and implantation failure of the embryos occurs (Dominic 1970; Keverne and de la Riva 1982; Bellringer *et al.* 1980).

A block to implantation induced by urinary chemosignals from older, probably dominant females, has been found in young female white-footed mice (Haigh *et al.* 1985, 1988). This is the first time that a female-induced, pheromonally related block to reproduction has been found to occur at a physiological stage following ovulation. All previous examples have been concentrated around delayed puberty or suppression of ovulation (Vandenbergh 1986). In natural populations of white-footed mice such a reproductive control could severely limit the effective number of breeding females (Haigh *et al.* 1988). The underlying physiological mechanism is not yet known. Pheromonally induced pregnancy blocks may also disrupt pregnancy after implantation has occurred, e.g. the prairie vole, *Microtus ochrogaster* (Stehn and Richmond 1975). Keverne (1989) deals further with the physiological and neural mechanisms operating the male-induced block to implantation in mice.

## Discussion

Primer pheromones thus mediate reproductive suppression in a variety of male and female mammals and in various social settings. The pheromonal contraceptives operate by delaying puberty in either sex, suppressing reproduction, and disrupting pregnancy. These studies have helped to elucidate some of the basic physiological, neuroendocrine, and neural events controlling reproduction including:

1. the role of prolactin secretion in rodents for (a) the timing of puberty, (b) spontaneously induced functional corpora lutea or pseudopregnancy, and (c) corpus luteum degeneration and implantation failure;
2. the neuroendocrine mechanisms controlling GnRH secretion and ovulation;
3. the learning process involved in recognising reproductively important olfactory signals (Keverne 1989).

The manipulation of primer pheromone action in mammals may, therefore, be of value in teasing apart the detailed mechanisms controlling reproduction, especially at the neural and molecular levels.

At present, there is no evidence for pheromonal contraception in the human, but solvent extracts of axillary secretions from women were found to synchronize the menstrual cycles of other women (Preti *et al.* 1986). Similar material obtained from men will also normalize the length of the menstrual cycle in women with aberrant cycle lengths (less than 26 days or more than 32 days; Cutler *et al.* 1986). Whether further research will reveal more far-reaching pheromonal influences on human reproductive physiology remains to be seen.

## Acknowledgements

The authors thank L.M. George for excellent technical assistance; M.J. Llovet, M. Bullock, J. Rozowski and J. Geddes for technical assistance and for care and maintenance of the marmoset monkeys; A.P.F. Flint and P.M. Summers for criticism of the manuscript; and M.R. Burr for typing the manuscript. This work was supported by grants from the Wellcome Trust, an MRC/AFRC Programme Grant and an SERC Studentship.

## References

Abbott, D.H. and Hearn, J.P. (1978). Physical, hormonal and behavioural aspects of sexual development in the marmoset monkey, (*Callithrix jacchus*). *J Reprod. Fertil.*, 53, 155–66.

Abbott, D.H., McNeilly, A.S., Lunn, S.F., Hulme, M.J. and Burden, F.J. (1981). Inhibition of ovarian function in subordinate female marmoset monkeys

(*Callithrix jacchus jacchus*). *J. Reprod. Fertil.*, 63, 335–45.

Abbott, D.H. (1984). Behavioural and physiological suppression of fertility in subordinate marmoset monkeys. *Am. J. Primatol.*, 6, 169–86.

Abbott, D.H. (1986). Social suppression of reproduction in subordinate marmoset monkeys (*Callithrix jacchus jacchus*). In *A Primatologia No Brasil*, 2 (ed. M.T. De Mello), pp. 1–16. Sociedade Brasileira de Primatologia, Brasilia.

Abbott, D.H. (1987). Behaviourally mediated suppression of reproduction in female primates. *J. Zool., Lond.*, 213, 455–70.

Abbott, D.H. (1988). Natural suppression of fertility. In *Reproduction and disease in captive and wild animals* (ed. J.P. Hearn and G.R. Smith), pp. 7–28. Symposia of the Zoological Society of London, Vol. 60. Oxford University Press, Oxford.

Abbott, D.H., Hodges, J.K. and George, L.M. (1988*a*). Social status controls LH secretion and ovulation in female marmoset monkey, (*Callithrix jacchus*). *J. Endocrinol.*, 117, 329–39.

Abbott, D.H., O'Byrne, K.T., Sheffield, J.W., Lunn, S.F. and George, L.M. (1988*b*). Neuroendocrine suppression of LH secretion in subordinate female marmoset monkeys (*Callithrix jacchus*). In *Comparative Reproduction in Mammals and Man* (ed. R.M. Eley), pp. 63–7. National Museums of Kenya, Nairobi.

Abbott, D.H. (1989). Social suppression of reproduction in primates. In *Comparative socioecology: the behavioural ecology of humans and other mammals* (ed. V. Standen and R.A. Foley), pp. 285–304. Blackwell Scientific, Oxford.

Advis, J.P., White, S.S. and Ojeda, S.R. (1981). Delayed puberty induced by chronic suppression of prolactin release in the female rat. *Endocrinology*, 109, 1321–30.

Barrett, J., Abbott, D.H. and George, L.M. (in press). The reproductive suppressive influence of pheromonal cues in subordinate female marmoset monkeys, *Callithrix jaccus*. *J. Reprod. Fertil.*

Batzli, G.O., Getz, L.L. and Hurley, S.S. (1977). Suppression of growth and reproduction of microtine rodents by social factors. *J. Mammal.*, 58, 583–91.

Bellringer, J.F., Pratt, H.P.M. and Keverne, E.B. (1980). Involvement of the vomeronasal organ and prolactin in pheromonal induction of delayed implantation in mice. *J. Reprod. Fertil.* 59, 223–8.

Bowman, L.A., Dilley, S.R. and Keverne, E.B. (1978). Suppression of oestrogen-induced LH by social subordination in talapoin monkeys. *Nature*, 275, 56–8.

Breed, W.G. (1976). Effect of environment on ovarian activity of wild hopping mice. *J. Reprod. Fertil.* 47, 395–7.

Bruce, H.M. (1959). An exteroceptive block to pregnancy in the mouse. *Nature*, 184, 105.

Coppola, D.M. (1986). The puberty delaying pheromone of the house mouse: field data and a new evolutionary perspective. In *Chemical signals in vertebrates, 4: Ecology, evolution, and comparative biology* (ed. D. Duvall, D. Müller-Schwarze and R.M. Silverstein), pp. 457–62. Plenum, New York.

Coppola, D.M. and Vandenbergh, J.G. (1985). Effect of density, duration of grouping and age of urine stimulus on the puberty delay pheromone in female mice. *J. Reprod. Fertil.*, 73, 517–22.

Cowley, J.J. and Wise, D.R. (1972). Some effects of mouse urine on neonatal growth and reproduction. *Anim. Behav.*, 20, 499–506.

Cutler, W.B., Preti, G., Kreiger, A., Huggins, G.R., Garcia, C.R. and Lawley, H.J. (1986). Human axillary secretions influence women's menstrual cycles: the role of donor extract from men. *Horm. Behav.*, 20, 463–73.

Dominic, C.J. (1970). Histological evidence for the failure of corpus luteum function in the olfactory block to pregnancy. *J. Anim. Morph. Physiol.*, 17, 126–30.

Dominic, C.J. and Pandey, S.D. (1979). Suppression of oestrus-inducing pheromone production in the Indian field mouse, *Mus booduga* Gray, by treatment with an antiandrogen. *Endokrinologie*, 74, 1-5.

Drickamer, L.C. (1982). Delay and acceleration of puberty in female mice by urinary cues from other females. *Dev. Psychobiol.*, 15, 433–42.

Dunbar, R.I.M. (1980). Determinants and evolutionary consequences of dominance among female Gelada baboons. *Behav. Ecol. Sociobiol.*, 7, 253–65.

Dunbar, R.I.M. (1989). Reproductive strategies of female gelada baboons. In *The sociobiology of sexual and reproductive strategies* (ed. A. Rasa, C. Vogel and E. Voland), pp. 74–92. Chapman and Hall, London.

Epple, G. and Katz, Y. (1980). Social influences on first reproductive success and related behaviours in the saddle back tamarin (*Saguinus fuscicollis*, Callitrichidae). *Int. J. Primatol.*, 1, 171–83.

Epple, G. and Katz, Y. (1984). Social influences on oestrogen excretion and ovarian cyclicity in Saddle Back tamarins (*Saguinus fuscicollis*). *Am. J. Primatol.*, 6, 215–27.

Facemire, C.F. and Batzli, G.O. (1983). Suppression of growth and reproduction by social factors in microtine rodents: tests of two hypotheses. *J. Mammal.*, 64, 152–6.

Fadem, B.H. (1986). Chemical communication in gray short-tailed opossums (*Monodelphis domestica*) with comparisons to other marsupials and with reference to montremes. In *Chemical signals in vertebrates, 4: Ecology, evolution and comparative biology* (ed. D. Duvall, D. Müller-Schwarze and R.M. Silverstein), pp. 587–607. Plenum, New York.

French, J.A., Abbott, D.H. and Snowdon, C.S. (1984). The effect of social environment on oestrogen excretion, scent marking and sociosexual behaviour in tamarins (*Saguinus Oedipus*). *Am. J. Primatol.*, 6, 155–67.

Haigh, G.R., Lounsbury, D.A. and Gordon, T.G. (1985). Pheromone induced reproductive inhibition in young female *Peromyscus leucopus*. *Biol. Reprod.*, 33, 271–6.

Haigh, G.R., Cushing, B.S. and Bronson, F.H. (1988). A novel postcopulatory block of reproduction in white-footed mice. *Biol. Reprod.*, 38, 623–6.

Harcourt, A.H. (1987). Dominance and fertility among female primates. *J. Zool.*, 213, 471–87.

Jarvis, J.U.M. (1981). Eusociality in a mammal: Co-operative breeding in naked mole-rat colonies. *Science*, 212, 571–3.

Kaplan, J.R., Adams, M.R., Koritnik, D.R., Rose, J.C. and Manuck, S.B. (1986). Adrenal responsiveness and social status in intact and ovariectomised *Macaca fascicularis*. *Amer. J. Primatol.*, 11, 181–93.

Keverne, E.B. (1982). The olfactory system and its role in pheromonally mediated changes in prolactin. In *Olfaction and endocrine regulation* (ed. W. Breipohl), pp. 127–40. IRL Press, London.

Keverne, E.B. (1983). Pheromonal influences on the endocrine regulation of reproduction. *TINS*, 6, 381–4.

Keverne, E.B. (1989)—this volume.

Keverne, E.B. and de la Riva, C. (1982). Pheromones in mice: reciprocal interaction between the nose and the brain. *Nature*, 296, 148–50.

Kirkland, L.E. and Bradley, E.L. (1986). Reproductive inhibition and serum prolactin levels in laboratory populations of the prairie deermouse. *Biol. Reprod.*, 35, 579–86.

Kruczek, M. and Marchlewska-Koj, A. (1986). Puberty delay of bank vole females in a high-density population. *Biol. Reprod.* 35, 537–41.

Lamond, D.R. (1959). Effect of stimulation derived from other animals of the same species on oestrous cycles in mice. *J. Endocrinol.*, 18, 343–9.

Lawton, A.D. and Whitsett, J.M. (1979). Inhibition of sexual maturation by a urinary pheromone in male prairie deer mice. *Horm. Behav.*, 13, 128–38.

Lepri, J.J., Wysocki, C.J. and Vandenbergh, J.G. (1984). Mouse vomeronasal organ: effects on chemosignal production and maternal behaviour. *Physiol. Behav.*, 35, 809–804.

Lepri, J.J. (1986). Familial chemosignals interfere with reproductive activation in female pine voles, *Microtus pinetorum*. In *Chemical Signals in Vertebrates 4: Ecology, Evolution and Comparative Biology* (ed. D. Duvall, D. Müller-Schwarze, and R.M. Silverstein), pp. 556–60. Plenum, New York.

Lepri, J.J. and Vandenbergh, J.G. (1986). Puberty in pine voles, *Microtus pinetorum*, and the influence of chemosignals on female reproduction. *Biol. Reprod.*, 34, 370–7.

Lloyd-Thomas, A. and Keverne, E.B. (1982). Role of the brain and accessory olfactory system in the block to pregnancy in mice. *Neuroscience*, 7, 907–13.

Lombardi, J.R. and Terman, C.R. (1980). The influence of the social environment on sexual maturation of female prairie deer mice (*Peromyscus maniculatus bairdii*). *Res. Popul. Ecol.*, 22, 93–100.

Lombardi, J.R. and Whitsett, J.M. (1980). Effects of urine from conspecifics on sexual maturation in female prairie deermice, *Peromyscus maniculatus bairdii*. *J. Mammal.*, 61, 766–8.

McClintock, M.K. (1983). Pheromonal regulation of the ovarian cycle: enhancement, suppression and synchrony. In: *Pheromones and reproduction in mammals* (ed. J.G. Vandenbergh), pp. 113–49. Academic Press, New York.

Marchlewska-Koj, A. (1983). Pregnancy blocking by pheromones. In *Pheromones and reproduction in mammals* (ed. J.G. Vandenbergh), pp. 151–74. Academic Press, New York.

Marchlewska-Koj, A. (1984) Pheromones and mammalian reproduction. In *Ox. Rev. Reprod. Biol.*, 6, 266–302.

Massey, A. and Vandenbergh, J.G. (1980). Puberty delay by a urinary cue from female house mice in feral populations. *Science*, 209, 821–2.

Meredith, M. (1983). Sensory physiology of pheromone communication. In *Pheromones and repoduction in mammals* (ed. J.G. Vandenbergh), pp. 199-252. Academic Press, New York.

Mody, J.K. (1963). Structural changes in the ovaries of I.F. mice due to age and various other states. *Anat. Rec.*, 145, 439–47.

Müller-Schwarze, D. (1977). Complex mammalian behaviour and pheromone bioassay in the field. In *Chemical signals in vertebrates* (ed. D. Müller-Schwarze and M.M. Mozell), pp. 413–33. Plenum, New York.

Novotny, M., Jemiolo, B., Harvey, S., Weisler, D. and Marchlewska-Koj, A. (1986). Adrenal-mediated endogenous metabolites inhibit puberty in female mice. *Science*, 231, 722–5.

Perret, M. and Schilling, A. (1987). Role of prolactin on the pheromone-like sexual inhibition in the male lesser mouse lemur. *J. Endocrinol.*, 114, 279–87.

Poole, T.B. and Evans, S.G. (1982). Reproduction, infant survival and productivity of a common marmoset colony (*Callithrix jacchus jacchus*) over a six year period. *Lab. Anim.*, 16, 88–97.

Preti, G., Cutler, W.B., Garcia, C.R., Huggins, G.R. and Lawley, H.J. (1986). Human axillary secretions influence women's menstrual cycles: the role of donor extract of females. *Horm. Behav.*, 20, 474–82.

Reynolds, J. and Keverne, E.B. (1979). The accessory olfactory system and its role in the pheromonally mediated suppression of oestrus in grouped mice. *J. Reprod. Fertil.*, 57, 31–5.

Ruiz de Elvira, M.-C. and Abbott, D.H. (1986). A backpack system for long-term osmotic minipump infusions into unrestrained marmoset monkeys. *Lab. Anim.*, 20, 329–34.

Ryan, K.D. and Schwartz, N.B. (1977). Grouped female mice: demonstration of pseudopregnancy. *Biol. Reprod.*, 17, 578–83.

Savage, A., Ziegler, T.E. and Snowdon, C.T. (1988). Sociosexual development, pair bond formation and mechanisms of fertility suppression in female cotton-top tamarins (*Saguinus oedipus oedipus*). *Am. J. Primatol.*, 14, 345–59.

Schilling, A., Perret, M. and Predine, J. (1984). Sexual inhibition in a Prosimian primate: a pheromone-like effect. *J. Endocrinol.* 102, 143–51.

Schilling, A. and Perret, M. (1987). Chemical signals and reproductive capacity in a male prosimian primate (*Microcebus murinus*). *Chem. Senses*, 12, 143–58.

Schultz-Westrum, T.G. (1969). Social communication by chemical signals in flying phalangers. In *Olfaction and taste* III, (ed. C. Pfaffmann), pp. 268–77. Rockefeller University Press, New York.

Singer, A.G., Agosta, W.C., O'Connell, R.J., Pfaffman, C., Bowen, D.V. and Field, F.H. (1976). Dimethyl disulfide; an attractant pheromone in hamster vaginal secretion. *Science*, 191, 948–50.

Smale, L. (1988). Influence of male gonadal hormones and familiarity on pregnancy interruption in prairie voles. *Biol. Reprod.*, 39, 28–31.

Stehn, R.A. and Richmond, M.E. (1975). Male-induced pregnancy termination in the prairie vole, *Microtus ochrogaster*. *Science*, 187, 1211–3.

Stevenson, M.F. and Rylands, A.B. (1988). The marmoset monkeys, genus *Callithrix*. In *Ecology and behaviour of neotropical primates*, Vol. 2 (ed. R.A. Mittermeier, A.B. Rylands, A. Coimbra-Filho and G.A.B. Fonseca), pp. 131–22. World Wildlife Fund, Washington, D.C.

Terman, C.R. (1987). Intrinsic behavioural and physiological differences among laboratory populations of prairie deermice. *Amer. Zool.*, 27, 853–66.

Vandenbergh, J.G. (1986). The suppression of ovarian function by chemosignals. In *Chemical signals in vertebrates, 4: Ecology, evolution and comparative biology* (ed. D. Duvall, D. Müller-Schwartze and R.M. Silverstein), pp. 423–32. Plenum, New York.

Vandenbergh, J.G., Whittsett, J.M. and Lombardi, J.R. (1975). Partial isolation of a pheromone accelerating puberty in female mice. *J. Reprod. Fertil.*, 43, 515–23.

Vandenbergh, J.G. and Coppola, D.M. (1986). The physiology and ecology of puberty modulation by primer pheromones. *Adv. Stud. Behav.*, 16, 71–07.

Van der Lee, S. and Boot, L.M. (1955). Spontaneous pseudopregnancy in mice. *Acta Physiol. Pharmac. N orl.*, 4, 442–3.

Wasser, S.K. and Barash, D.P. (1983). Reproductive suppression among female mammals: implications for biomedicine and sexual selection theory. *Q. Rev. Biol.*, 58, 513–38.

Whitten, W.K. (1959). Occurrence of anoestrus in mice caged in groups. *J. Endocrinol.*, 18, 102–107.

Ziegler, T.E., Savage, A., Scheffler, G. and Snowdon, C.T. (1987). The endocrinology of puberty and reproductive functioning in female cotton-top tamarins (*Saguinus oedipus*) under varying social conditions. *Biol. Reprod.*, 37, 618–27.

# III.2 Male Preferences for Females of Different Reproductive Conditions: a Critical Review

STEPHEN A. TAYLOR AND DONALD A. DEWSBURY

## Introduction

It is generally believed that male mammals discriminate oestrous (receptive) from dioestrous (non-receptive) females on the basis of olfactory cues (Johnston 1983; Brown 1979) and that these odours are differentially attractive to males (Johnston 1983). In this paper we critically review the literature on male preferences for females of different reproductive conditions. Male preferences are placed in a comparative perspective. Factors that may influence the preferences males display, including sexual experience, testing procedures, and cues available for discrimination in preference tests, are identified. Additionally, species differences in female reproductive physiology and social organization and mating system are implicated in the generation of species differences in preferences. We hope to identify questions that presently remain unanswered and indicate future directions for research in this area.

## How universal are preferences for rodent oestrous females?

Despite the widespread beliefs concerning the abilities of males to detect female reproductive condition on the basis of olfactory cues, relatively few species have been studied (see Table 1). Laboratory research concerned with these phenomena often takes the form of preference tests in which a male is presented with stimuli from an oestrous female and a dioestrous female. Behavioural responses (e.g. sniffing, proximity) to each set of stimuli from oestrous vs. dioestrous females are recorded. In some species preferences for oestrous over dioestrous females seem to be readily displayed, as in the case of laboratory rats (e.g. Carr *et al.* 1965; Lydell and Doty 1972; Pfaff and Pfaffmann 1969; Stern 1970; Landauer *et al.* 1977; Thody and Dijkstra 1978). Preferences for oestrous females have also been demonstrated in desert wood rats (Fleming *et al.* 1981), brown lemmings and collared lemmings (Huck and Banks 1984), and Mongolian gerbils (Block *et al.* 1981).

In other species, preferences for oestrous females are not so readily displayed. Male prairie voles (*Microtus ochrogaster*) display an oestrous female preference only under particular conditions, with the experiential history of males and testing conditions being important determinants (Taylor and Dewsbury 1988). Male golden hamsters seem equally attracted to oestrous and dioestrous females when the actual stimulus females are present (Landauer *et al.* 1978; Johnston 1980), but do exhibit differential responsiveness if soiled bedding is the odour source (Johnston 1980; 1983).

There are also species in which no preference for oestrous females has been demonstrated although some of them have received considerable study. Dewsbury *et al.* (1986) found no reliable preference for oestrous over dioestrous females in 11 experiments with deer mice (*Peromyscus maniculatus bairdi*) using a variety of methods and apparatus. Further work with deer mice has been conducted using intact, awake, behaving females as stimulus objects (Dewsbury, unpublished). Males could move to either end of an apparatus and interact with an oestrous or dioestrous female confined behind a hardware cloth barrier. No significant preferences were displayed in any of three experiments.

Randall (1985, 1986) has observed differences in the responsiveness to oestrous and dioestrous female urine in kangaroo rats. Male *Dipodomys spectabilis* prefer the urine of oestrous females, whereas male *D. merriami* are equally attracted to oestrous and dioestrous female urine.

Taylor (unpublished) studied preferences in sexually experienced male montane voles (*M. montanus*). Procedures and apparatus were the same as Experiment 5 of Taylor and Dewsbury (1988). Stimulus females were classified as oestrous or dioestrous based on vaginal smears (predominantly cornified cells = oestrus) and lordosis in response to brief ( $< 10$ mins) exposure to a stud male. Males were equally attracted to caged oestrous and dioestrous females in a 20 min. preference test (51% oestrous side). Difficulty was encountered with further work, however, as females that would have been previously classified as dioestrus (non-cornified smear and no lordosis with a stud male) were found to copulate with males after a longer period of exposure to a male (e.g. 45 mins). It may be that neither the experimenter nor male montane voles can quickly and easily determine the reproductive condition of female montane voles.

Guinea pigs can discriminate oestrous from dioestrous odours in a learning paradigm (Ruddy 1980), but apparently do not prefer oestrous to dioestrous odours (Beauchamp 1973; Beauchamp and Beruter 1973; Nyby 1983).

Thus, the general view that male rodents can determine reproductive condition and that a preference for oestrous females is readily manifested is perhaps unwarranted. In different species, the finding of a preference may be relatively straightforward, preferences may be dependent on factors such as testing conditions and/or the experiential history of subjects, or preferences for oestrous females have not been shown despite the use of methods found effective with other species.

**Table 1.** Male responses to females of different reproductive condition

| | Preference | Stimulus | Author |
|---|---|---|---|
| Rats (*Rattus norvegicus*) | | | |
| | O > D | ? | LeMagnen (1952), in Carr *et al.* (1965) |
| | O > D | Caged females, soiled bedding | Carr *et al.* (1965) |
| | O > D | Urine | Pfaff and Pfaffmann (1969) |
| | O > D | Anaesthetized females, caged females | Stern (1970) |
| | O > D | Urine | Lydell and Doty (1972) |
| | O > D | Anaesthetized females, soiled bedding | Landauer *et al.* (1977) |
| | O = D | Urine | Brown (1977) |
| | O = ProE > D | Preputial gland | Thody and Dijkstra (1978) |
| Hamsters (*Mesocricetus auratus*) | | | |
| | O = D | Body odours, anaesthetized females | Landauer *et al.* (1978) |
| | O > D | Soiled bedding (not body odours) | Johnston (1980) |
| | O > D | Primed caged females (see text) (not vaginal secretions) | Carmichael (1980) |
| Mice (*Mus musculus*) | O > D | Vaginal secretions (not urine preputial glands, uterus) | Hayashi and Kimura (1974) |
| | O > D | Urine | Rose and Drickamer (1975) |
| Desert wood rats (*Neotoma lepida lepida*) | O > D | Urine, vaginal secretions | Fleming *et al.* (1981) |

| | | | |
|---|---|---|---|
| Kangaroo rats | | | |
| (*Dipodomys merriami*) | O = D | Urine | Randall (1985) |
| (*D. spectabilis*) | O > D | Urine | Randall (1986) |
| Deer mice | | | |
| (*Peromyscus maniculatus*) | O = D | Soiled bedding, anaesthetized females, caged females | Dewsbury *et al.* (1986) |
| | O = D | Caged females (see text) | Dewsbury (unpublished) |
| Voles | | | |
| (*Microtus ochrogaster*) | O > D | Caged females, anaesthetized females, soiled bedding | Taylor and Dewsbury (1988) |
| (*M. montanus*) | O = D | Caged females (see text) | Taylor (unpublished) |
| Lemmings | | | |
| (*Lemmus sibiricus = trimucronatus*) | O > D | Body odours | Huck and Banks (1984) |
| (*Dicrostonyx groenlandicus*) | O > D | Body odours | Huck and Banks (1984) |
| Gerbils | | | |
| (*Meriones unguiculatus*) | O > D | Saliva | Block *et al.* (1981) |
| (*M. hurrianae*) | O – D discrimination | Soiled bedding | Kumari and Prakash (1984) |
| Columbian ground Squirrel (*Spermophilus columbianus*) | Pre-O > Post-O | Vaginal odours (not oral odours) | Harris and Murie (1984) |
| Guinea pigs | | | |
| (*Cavia porcellus*) | O – D discrimination | Anogenital odours | Ruddy (1980) (see text) |

O = oestrus; D = dioestrus

## Effects of experience

Experience can have profound effects on oestrous versus dioestrous female preferences; these may differ greatly between species. Although the effects of many different kinds of experience could be investigated (e.g. early rearing, social deprivation, etc.), most investigators have studied the role of sexual experience.

### IS SEXUAL EXPERIENCE NECESSARY?

There are data indicative of a beneficial role of sexual experience in generating preferences. Species in which sexually experienced, but not sexually naïve, males prefer oestrous over dioestrous females include laboratory rats (Carr *et al.* 1965; Lydell and Doty 1972; Stern 1970), house mice (Hayashi and Kimura 1974), brown lemmings and collared lemmings (Huck and Banks 1984), prairie voles (Taylor and Dewsbury 1988), hamsters (Johnston 1980), and dogs (Beach and Gilmore 1949; Doty and Dunbar 1974).

Other work indicates that sexual experience is not necessary for an oestrous preference however. This is evident in the literature on laboratory rats. LeMagnen (1952, in Carr *et al.* 1965) found that naïve males preferred the odours of oestrous females over those of dioestrous females. Landauer *et al.* (1977) showed that sexually naïve males that later proved to be copulators in mating tests preferred oestrous over dioestrous females, while naïve males that later proved to be non-copulators showed no reliable preference. Brown (1977) found no preferences for oestrous or dioestrous female urine in either experienced or naïve males. Fillion and Blass (1985) found differential responsiveness to oestrous and dioestrous stimuli in male laboratory rats (no preference actually tested) from late infancy to early adulthood independent of sexual experience. They contend that differential responsiveness is continuous from birth, even in males that have had no contact with females since weaning.

Most other rodent species have received much less study and generalizations must be made with caution. The effects of sexual experience may vary with the species or strain being considered. For example, Hayashi and Kimura (1974) found that only sexually experienced ICR-JCL mice preferred oestrous over dioestrous female odours. Rose and Drickamer (1975), however, found that both naïve and experienced male BDF1 mice preferred oestrous female urine.

### A CLOSER LOOK AT SEXUAL EXPERIENCE

Significant methodological differences make the interpretation of the effects of sexual experience even more difficult. Most investigators have used sexually experienced subjects obtained by one of three basic methods. Males may be exposed to only receptive oestrous females for one or more mating encounters (e.g. Lydell and Doty 1972; Stern 1970; Huck and Banks 1984; Rose and Drickamer

1975); thus, the male has only made contact with a receptive female(s). Alternatively, males may be exposed to both oestrous and dioestrous females in a series of encounters (e.g. Carr *et al.* 1965; Johnston 1980). Use of this procedure implies that an oestrous over dioestrous preference is acquired through encounters with both oestrous and dioestrous females. The third method involves housing a male together with a female long enough for mating to occur, the female being in oestrus for part of the housing period and non-receptive at other times (e.g. Hayashi and Kimura 1974; Dewsbury *et al.* 1986). The male is thus exposed to the same female in both receptive and non-receptive conditions. The details of these basic methods often differ considerably among investigations (e.g. number of sexual encounters, recency of sexual experience) and may confound comparison even when the same basic methods are used to obtain experienced subjects.

Taylor and Dewsbury (1988) found that the manner in which sexually experienced subjects are obtained influences oestrous-dioestrous preferences in prairie voles. Sexually naïve males were tested, as were sexually experienced males which were (a) experienced with both oestrous and dioestrous females, (b) housed monogamously with a female, (c) housed with two females, and (d) 'socially experienced' (see below). Naïve males, males housed monogamously with a female, and males housed with two females were all equally attracted to caged oestrous and dioestrous females, and the odours of bedding soiled by oestrous and dioestrous females. Males that were experienced with both oestrous and dioestrous females were found to prefer caged oestrous females, and anaesthetized oestrous females, but not the odours of bedding soiled by oestrous females. Only socially experienced males were found to display a preference for caged oestrous females and the odours of oestrous female soiled bedding.

Other investigators have used males with different experiential histories. Dewsbury *et al.* (1986) used male deer mice with different histories of sexual experience, but as noted previously, deer mice seem equally attracted to oestrous and dioestrous females. Johnston (1980) compared male hamsters experienced with oestrous females only and males experienced with both oestrous and dioestrous females. Neither group displayed a preference for oestrous over dioestrous females.

Brown (1977), in a well designed and careful study, did not find a preference for oestrous female urine in sexually experienced male laboratory rats. This finding contrasts with other published work on rats (see Table 1). He has suggested that his procedure for obtaining sexually experienced subjects may be responsible for the result (Brown 1977, 1979).

### SOCIAL EXPERIENCE

Males receiving sexual experience through one of the three basic methods described above are lacking some aspects of the experience an animal receives in its natural environment, including extensive experience with conspecifics. The latter may be beneficial in generating preferences. Taylor and Dewsbury (1988) defined 'social experience' as physical and interactive contact with both sexes of conspecifics who are not members of one's immediate family. In both deer mice (Dewsbury *et al.* 1986) and prairie voles (Taylor and Dewsbury 1988), socially experienced males (i.e. males that have encountered other males, dioestrous females, and oestrous females in a semi-natural enclosure) displayed preferences that sexually experienced males did not. Nyby and Whitney (1980) found that male house mice require experience with both males and females before they will show a difference in the frequency of ultrasonic 'courtship' calls in response to the urine odours of the two sexes. Restrictive laboratory conditions may limit the olfactory capabilities of subjects (Schultz and Tapp 1973).

## Cues available for discrimination

Important factors in both the outcome and interpretation of preference tests are the discriminatory cues available.

### CUE COMBINATIONS

The sexual behaviour of male mammals is thought to be directed by a variety of overlapping cues (Kwan and Johnston 1980; Izard 1983; Thor and Flannelly 1978; Carmichael 1980). Males may attend to female behavioural cues, such as vocalization, locomotion, and posturing, and bodily cues that are olfactory, gustatory, tactile, or visual (Landauer *et al.* 1977). These cues can be presented in different combinations, yielding information on the relative importance of different channels of information transmission.

Laboratory rats have been studied extensively, and the procedures used are illustrative of how different methods offer different combinations of cues. In some investigations the stimuli are olfactory, such as urine (Lydell and Doty 1972; Brown 1977) or the odours left on soiled bedding (Carr *et al.* 1965). Other investigators have used methods where subjects could contact anaesthetized stimulus females (Landauer *et al.* 1977). In this method tactile, visual, and gustatory cues are available in addition to olfactory cues, but behavioural responses of the female are prevented (Landauer *et al.* 1977; Fillion and Blass 1985). Still others have used methods where stimulus females were held in containers so that subjects could not see or touch them, but could hear and smell them (Stern 1970; Carr *et al.* 1965). Laboratory rats have been found to prefer oestrous females with all of the above complexes of stimuli.

In species that do not readily display preferences for oestrous females, one might suspect that preferences would more likely be found when a greater number of cues are available to the subject. Taylor and Dewsbury (1988) found that oestrous preferences were more often displayed in prairie voles when cues other than olfactory cues were available. The availability of cues seemed to interact with experiential history in the display of preferences. Males that were experienced with both oestrous and dioestrous females displayed oestrous female preferences when stimulus females were caged and awake, and when stimulus females were presented anaesthetized in an open field. These males did not prefer bedding soiled by oestrous females. Only males that were socially experienced (see above) preferred bedding soiled by oestrous females; a preference based solely on olfactory cues.

Differing availability of cues appears to be an important factor in hamster preferences. The most straightforward finding of an oestrous preference in hamsters resulted when the stimuli were odours from soiled bedding (Johnston 1980). In experiments where stimuli consisted of anaesthetized females behind screens (Johnston 1980), anaesthetized females in an open field where subjects could contact them (Landauer *et al.* 1978), or awake, caged females (subjects could see, hear, smell and, to a degree, touch the stimulus females) (Carmichael 1980), male hamsters displayed no such preference. Carmichael (1980) found that male hamsters reliably preferred caged oestrous over caged dioestrous females only when the females were 'primed' by brief contact with an alternate male immediately prior to testing. She speculated that 'primed' oestrous females increase the frequency and/or intensity of their ultrasonic calling which differentially attracts males to them.

Nyby (1983) has proposed an explanation for the different results obtained when different cues are available. He suggests female rodents may produce several chemosignals. One might be a long-range sex attractant that is ovarian-dependent and volatile. Another is thought to be a short-range promoter of sexual behaviour that is ovarian-independent and of low volatility. Nyby contends that in studies where the subject can contact the odour source, no preference is displayed because the low volatility cue masks the volatile, ovarian-dependent cue. Although there are results in the literature consistent with Nyby's hypothesis (see Nyby 1983), other results appear inconsistent. For example, male rats (Landauer *et al.* 1977) and prairie voles (Taylor and Dewsbury 1988) displayed an oestrous over dioestrous female preference when they could contact the anaesthetized female odour source.

Animals should be tested for preferences with different methods in which the cues available to the subject are varied. As illustrated by the work with prairie voles, deer mice, and hamsters described above, the clearest picture of the pattern of responsiveness of a given species may only emerge when a variety of different methods are used (Taylor and Dewsbury 1988).

## THE STIMULUS ANIMALS IN PREFERENCE TESTS

A major factor bearing upon the validity of preference tests is the nature of the stimuli presented to subjects; close consideration must be given to animals that provide stimuli. One common practice is the use of ovariectomized females as stimulus animals (e.g. Brown 1977; Lydell and Doty 1972; Rose and Drickamer 1975; Doty and Dunbar 1974). One group receives hormones and provides oestrous female stimuli. Another group receives control injections with no hormones, comprising the stimulus condition known variously as dioestrus, non-receptive, or anoestrus. The complex of stimuli produced by these 'oestrous' and 'dioestrous' females may not accurately represent the stimuli differences of intact oestrous and dioestrous females. For example, Beach *et al.* (1976) found that ovariectomized female hamsters injected with hormones exhibited less proceptive behaviour than females in natural oestrus. In a preference test Carmichael (1980) found that male hamsters preferred oestrous over ovariectomized (no hormones) caged females, but were equally attracted to caged oestrous and dioestrous females.

The odours of ovariectomized–hormone and ovariectomized–control females may not be representative of naturally receptive and non-receptive females. For example, ovariectomized non-receptive females have an internal hormonal milieu quite unlike an intact female in dioestrus (Feder 1981), and thus may produce very different odours. Potential differences in stimuli produced by ovariectomized and intact females may complicate naturalistic interpretation of studies using exclusively ovariectomized females.

Finally, to obtain a broader and more realistic picture of the function of sex odours in the field, investigators might study male responses to females that are pregnant, lactating (see Johnston 1980, 1983), in postpartum oestrus, reproductively immature, or other states. In most of the work cited in this paper the stimulus conditions are cycling oestrus versus dioestrus, yet it is not known if the cycling oestrus pattern occurs with any regularity in the field (Bronson 1979).

### OLFACTORY CUES: IDENTIFYING THE SOURCE

In some preference tests only olfactory cues are presented to subjects. There are a variety of potential sources of attractive odours; it may be desirable to identify the source of odours males use to discriminate oestrous from dioestrous females. Urine has been found to be a sufficient stimulus in several investigations (e.g. Lydell and Doty 1972; Pfaff and Pfaffmann 1969; Rose and Drickamer 1975). Results with other sources in rodents can be found in Table 1; these include vaginal secretions (e.g. Hayashi and Kimura 1974; Carmichael 1980), odours derived from the preputial glands (e.g. Thody and Dijkstra 1978; Hayashi and Kimura 1974), and the uterus (Hayashi and Kimura 1974). Dog preferences vary with the odour source (see Table 1); odour sources studied include urine, vaginal secretions, anal sac secretions, ear wax, and saliva (Dunbar 1977, 1978; Doty and Dunbar 1974).

Positive results with one stimulus source do not rule out the possibility of other sources. Saliva is a potentially important source of odours which has not received comparative investigation in rodents. Block *et al.* (1981) found that male

Mongolian gerbils preferred anaesthetized ovariectomized stimulus females whose lips were painted with oestrous female saliva to those whose lips were painted with dioestrous female saliva. This preference was only displayed when the saliva was presented on a female, saliva samples alone were not attractive.

## Species differences

Species differences in olfactory preferences for females of different reproductive conditions have not previously been given sufficient treatment. Two factors that may be responsible for species differences are the nature of female reproductive physiology in the species in question, and the social organization and mating system of that species.

### REPRODUCTIVE PHYSIOLOGY

Most of the species that have been studied with respect to oestrous-dioestrous preferences display spontaneous rat-like oestrous cycles in the laboratory. Preferences males display may be different in species that do not display spontaneous oestrous cycles. In voles (*Microtus*), females of some species exhibit long periods of leucocytic invasion (a diestrous smear) when housed in isolation (e.g. *M. ochrogaster*, *M. pinetorum*), and contact with a male is thought to induce vaginal cornification and behavioural receptivity (Richmond and Stehn 1976; Sawrey and Dewsbury 1985). Other species (e.g. *M. montanus*, *M. pennsylvanicus*) show long periods of vaginal cornification (an oestrous smear) when housed in isolation (see Sawrey and Dewsbury 1985). In the former one must question the general view that males are more attracted to oestrous than dioestrous females. Males might be attracted equally to each or even more attracted to dioestrous than oestrous females, as females that have attained oestrus may already be associated with another male. The variability in vaginal cytology may be indicative of variability in the dependence on male stimuli for receptivity among species (see Sawrey and Dewsbury 1985; Richmond and Stehn 1976). It is possible that male preferences for females of different reproductive conditions may vary with the dependence on male stimuli for receptivity in that species.

### SOCIAL ORGANIZATION AND MATING SYSTEM

Johnston (1980) proposed that the degree to which oestrous and dioestrous females differ in attractiveness to males should depend, at least in part, on the social organization and mating system of that species. This idea may prove fruitful with further work; at present a large comparative database is lacking (Johnston 1980; Dewsbury *et al.* 1986).

One possibility is that species in which the sexes are not normally in regular contact might be more dependent on olfactory cues than more social species

(Randall 1986). Randall (1985, 1986) has reported that male *Dipodomys merriami* do not prefer oestrous over dioestrous female urine, whereas male *D. spectabilis* do display this preference. She suggests that such differences may result from differences in social organization. Male and female *D. merriami* contact each other frequently and a female's reproductive condition may be communicated through olfactory cues arising from close contact. Female *D. spectabilis*, however, defend their territories and are not often in proximity to males, and olfactory cues in urine would be useful to advertise a female's reproductive condition.

Prairie voles appear to be a social species, as they exhibit evidence of monogamous pair-bonding under both laboratory and field conditions (Getz and Hofmann 1986). Taylor and Dewsbury (1988) found that male prairie voles housed monogamously with a female did not display preferences for oestrous over dioestrous females even under a variety of test conditions. Males experienced with females under different conditions did display preferences however. It was postulated that behavioural cues available during unrestricted contact in a monogamous housing condition might allow a male to discern his mate's reproductive condition, with olfactory cues then becoming less important.

Johnston (1983) has integrated much information in developing a scenario for sexual attraction in golden hamsters. He notes that female hamsters live solitarily, do not have a postpartum oestrus, and are very aggressive toward males when pregnant or lactating. Male hamsters do not display a preference for odours emanating from oestrous females, but they readily display a preference for oestrous females over pregnant or lactating females (Johnston 1980, and references therein). Thus, Johnston (1980) suggests that 'For hamsters the important sexual message may not be "oestrous" versus "dioestrous" but rather may be "will soon be oestrous" (i.e., cycling) versus "not soon oestrous" (i.e., pregnant, lactating, or acyclic).' He reasons that if a female becomes attractive before she is receptive (i.e. cycling), a means is provided for getting the male and female together for mating.

Thus, the importance of the olfactory transmission of information concerning reproductive state may vary depending on the normal relations of males and females in the field. Further testing of these ideas can only come about with further comparative work. Excellent opportunities exist within *Microtus*, as these species exhibit considerable variability with respect to social organization and mating system.

## Conclusions and future directions

The study of rodent male preference for females of different reproductive conditions is illustrative of the complexities involved in the study of mammalian chemical communication. While progress has been made, there are still many interesting questions. Some directions for future research which have emerged from this review are summarized below.

1. There are important species differences. A wider range of species should be studied in the future.
2. There are complex effects of experience on preferences for oestrous females. Further study of the role of experience is needed, particularly with relation to experience animals may have in the field.
3. The nature of the testing conditions has a profound effect on the appearance of preferences for oestrous females. Several issues concern the methodology used: (a) work is needed on the relative importance of cues from different modalities and their interaction in determining preferences; (b) stimulus conditions must be realistic with respect to stimuli produced by intact females; (c) the study of male preferences should be broadened beyond oestrus versus dioestrus to include females of other reproductive conditions; and (d) more work is needed in identifying the source of the attractive odour(s) in a wider range of species. Testing methods should be carefully developed, or perhaps even standardized, as this would more readily allow comparison across species.
4. The importance of the type of female reproductive physiology characteristic of a species needs emphasizing. More species that do not display a rat-like oestrous cycle should be studied. The reproductive condition of females to which males are primarily attracted may vary with differences in female reproductive physiology.
5. Finally, with a stronger comparative data-base with similar methods used and comparable experience given to animals of different species, ecological correlates of species differences may provide a more comprehensive understanding of the genesis of these preferences.

## Acknowledgements

We thank A. Salo and J. Pierce for helpful comments on the manuscript. This work was supported by Grant BNS-8520318 from the National Science Foundation.

## References

Beach, F.A., Stern, B., Carmichael, M. and Ranson, E. (1976). Comparisons of sexual receptivity and proceptivity in female hamsters. *Behav. Biol.*, 18, 473–87.

Beauchamp, G.K. (1973). Attraction of male guinea pigs to conspecific urine. *Physiol. Behav.*, 10, 589–94.

Beauchamp, G.K. and Beruter, J. (1973). Source and stability of attractive components in guinea pig (*Cavia porcellus*) urine. *Behav. Biol.*, 9, 43–7.

Block, M.L., Volpe, L.C. and Hayes, M.J. (1981). Saliva as a cue in the development of social behavior. *Science*, 211, 1062–4.

Bronson, F.H. (1979). The reproductive ecology of the house mouse. *Q. Rev. Biol.*, 54, 265–99.

Brown, R.E. (1977). Odor preference and urine-marking scales in male and female rats: Effects of gonadectomy and sexual experience on responses to conspecific odors. *J. Comp. Physiol. Psychol.*, 91, 1190–206.

Brown, R.E. (1979). Mammalian social odors: A critical review. *Adv. Stud. Behav.*, 10, 103–62.

Carmichael, M.S. (1980). Sexual discrimination by golden hamsters (*Mesocricetus auratus*). *Behav. Neural Biol.*, 29, 73–90.

Carr, W. J., Loeb, L.S. and Dissinger, M.L. (1965). Responses of rats to sex odors. *J. Comp. Physiol. Psychol.*, 59, 370–7.

Dewsbury, D.A., Ferguson, B., Hodges, A.W. and Taylor, S.A. (1986). Tests of preferences of deer mice (*Peromyscus maniculatus bairdi*) for individuals and their odors as a function of gender and estrous condition. *J. Comp. Psychol.*, 100, 117–27.

Doty, R.L., and Dunbar, I. (1974). Attraction of beagles to conspecific urine, vaginal and anal sac secretion odors. *Physiol. Behav.*, 12, 825–33.

Dunbar, I.F. (1977). Olfactory preferences in dogs: the response of male and female beagles to conspecific odors. *Behav. Biol.*, 20, 471–81.

Dunbar, I.F. (1978). Olfactory preferences in dogs: the response of male and female beagles to conspecific urine. *Behav. Biol.*, 3, 273–86.

Feder, H.H. (1981). Estrous cyclicity in mammals. In *Neuroendocrinology of reproduction, physiology and behavior* (ed. N.T. Adler), pp. 279–348. Plenum, New York.

Fillion, T.J. and Blass, E.M. (1985). Responsiveness to estrous chemostimuli in male rats (*Rattus norvegicus*) of different ages. *J. Comp. Psychol.*, 99, 328–35.

Fleming, A.S., Chee, P. and Vaccarino, F. (1981). Sexual behavior and its olfactory control in the desert woodrat (*Neotoma lepida lepida*). *Anim. Behav.*, 29, 727–45.

Getz, L.L. and Hofmann, J.E. (1986). Social organization in free-living prairie voles, *Microtus ochrogaster*. *Behav. Ecol. Sociobiol.*, 18, 275–82.

Harris, M.A. and Murie, J.O. (1984). Discrimination of estrus status by scent in Columbian ground squirrels. *Anim. Behav.*, 32, 939–40.

Hayashi, S. and Kimura, T. (1974). Sex-attractant emitted by female mice. *Physiol. Behav.*, 13, 563–67.

Huck, U.W. and Banks, E.M. (1984). Social olfaction in male brown lemmings (*Lemmus sibiricus = trimucronatus*) and collared lemmings (*Dicrostonyx groenlandicus*): I. Discrimination of species, sex, and estrous condition. *J. Comp. Psychol.*, 98, 54–9.

Izard, M.K. (1983). Pheromones and reproductions in domestic animals. In *Pheromones and reproduction in vertebrates* (ed. J.G. Vandenbergh), pp. 253–85. Academic Press, New York.

Johnston, R.E. (1980). Responses of male hamsters to odors of females in different reproductive states. *J. Comp. Physiol. Psychol.*, 94, 894–904.

Johnston, R.E. (1983). Chemical signals and reproductive behavior. In *Pheromones and reproduction in vertebrates* (ed. J.G. Vandenbergh), pp. 3–37. Academic Press, New York.

Kumari, S. and Prakash, I. (1984). Perception of oestrous odors by male *Meriones hurrianae*. *Anim. Behav.*, 32, 927–9.

Kwan, M. and Johnston, R.E. (1980). The role of vaginal secretion in hamster sexual behavior: Males' responses to normal and vaginectomized females and their odors. *J. Comp. Physiol. Psychol.*, 94, 905–13.

Landauer, M. R., Wiese, R.E. and Carr, W.J. (1977). Responses of sexually experienced and naive male rats to cues from receptive vs. nonreceptive females. *Anim. Learn. Behav.*, 5, 398–402.

Landauer, M.R., Banks, E.M. and Carter, C.S. (1978). Sexual and olfactory preferences of naive and experienced male hamsters. *Anim. Behav.*, 26, 611–21.

LeMagnen, J. (1952). Les phénomènes olfacto-sexuals chez le rat blanc. *Arch. Sci. Physiol.*, 6, 295–332.

Lydell, K. and Doty, R.L. (1972). Male rat odor preferences for female urine as a function of sexual experience, urine age, and urine source. *Horm. Behav.*, 3, 205–12.

Nyby, J. (1983). Volatile and nonvolatile chemosignals of female rodents: differences in hormonal regulation. In *Chemical signals in vertebrates 3* (ed. D. Müller-Schwarze and R.M. Silverstein), pp. 179–93. Plenum, New York.

Nyby, J. and Whitney, G. (1980). Experience affects behavioral responses to sex odors. In *Chemical signals: vertebrates and aquatic invertebrates* (ed. D. Müller-Schwarze and R.M. Silverstein), pp. 173–92. Plenum, New York.

Pfaff, D. and Pfaffmann, C. (1969). Behavioral and electrophysiological responses of male rats to female urine odors. In *Olfaction and taste* (ed. C. Pfaffmann), pp. 258–67. Rockefeller University Press, New York.

Randall, J. (1985). Role of urine in coordinating reproductions in a desert rodent (*Dipodomys merriami*). *Physiol. Behav.*, 34, 199–203.

Randall, J. (1986). Preference for estrous female urine by male kangaroo rats (*Dipodomys spectabilis*). *J. Mamm.*, 67, 736–39.

Richmond, M. and Stehn, R. (1976). Olfaction and reproductive behavior in microtine rodents. In *Mammalian olfaction, reproductive processes, and behavior* (ed. R.L. Doty), pp. 197–217. Academic Press, New York.

Rose, E. and Drickamer, L.C. (1975). Castration, sexual experience, and female urine odor preferences in adult BDF1 male mice. *Bull. Psychon. Soc.*, 5, 84–6.

Ruddy, L.L. (1980). Discrimination among colony mates' anogenital odors by guinea pigs (*Cavia porcellus*). *J. Comp. Physiol. Psychol.*, 94, 767–74.

Sawrey, D.K. and Dewsbury, D.A. (1985). Control of ovulation, vaginal estrus, and behavioral receptivity in voles (*Microtus*). *Neurosci. Biobehav. Rev.*, 9, 563–71.

Schultz, E.F. and Tapp, J.T. (1973). Olfactory control of behavior in rodents. *Psychol. Bull.*, 79, 21–44.

Stern, J.J. (1970). Responses of male rats to sex odors. *Physiol. Behav.*, 5, 519–24.

Taylor, S.A. and Dewsbury, D.A. (1988). Effects of experience and available cues on estrous versus diestrous preferences in male prairie voles, (*Microtus ochrogaster*). *Physiol. Behav.*, 42, 379–88.

Thody, A.J. and Dijkstra, H. (1978). Effect of ovarian steroids on the preputial gland odours in the female rat. *J. Endocrin.*, 77, 397–403.

Thor, D.H. and Flannelly K.J. (1978). Sex-eliciting behavior of the female rat: Discrimination of receptivity by anosmic and intact males. *Behav. Biol.*, 23,

# III.3 Pregnancy Disruption by Unfamiliar Males in Meadow Voles: A Comparison of Chemical and Behavioural Cues

ANNE E. STOREY

## Introduction

Pregnancy blocks are mediated in mice by androgen-dependent pheromones (Bruce 1965; Dominic 1965; Hoppe 1975), and the assumption has been that chemical cues are also critical for male-induced pregnancy disruption in other rodents. In most studies with voles (genus *Microtus*), unfamiliar males have had physical contact with females (e.g. Clulow and Langford 1971; Stehn and Richmond 1975; Stehn and Jannett 1981). Three published studies have compared the relative effectiveness of chemical cues and physical presence in *Microtus* species. Schadler (1981) showed that chemical cues were as effective as male presence for pine voles (*M. pinetorum*), whereas Milligan (1976) for field voles (*M. agrestis*) and Storey (1986*a*) for meadow voles (*M. pennsylvanicus*) showed that chemical cues were much less effective than direct exposure to the unfamiliar male.

The sensory mechanisms mediating pregnancy disruptions in mice may differ from those of microtine voles. Voles are induced ovulators, in contrast to the spontaneous ovulation of mice, and thus female voles may be more sensitive to physical cues from unfamiliar males during, as well as before, pregnancy. Pregnancy disruptions in mice do not occur after embryo implantation, whereas in voles pregnancy disruptions can occur well after implantation (Stehn and Richmond 1975; Schadler 1981; Stehn and Jannett 1981, Storey 1986*a*). Sensory cues mediating pregnancy disruptions may also change with stage of pregnancy, such that more intense stimulation from unfamiliar males may be necessary as pregnancy progresses.

The role of chemical and physical cues were compared for pre- and post-implantation disruptions in meadow voles (*M. pennsylvanicus*). In these studies, pre-implantation exposure means that an unfamiliar male was introduced on days 2–4 of pregnancy, and post-implantation exposure means that an unfamiliar male was introduced on days 10–12 of the 21-day pregnancy. The experiments also attempted to determine whether specific behaviour patterns such as mating are important antecedents of pregnancy disruption.

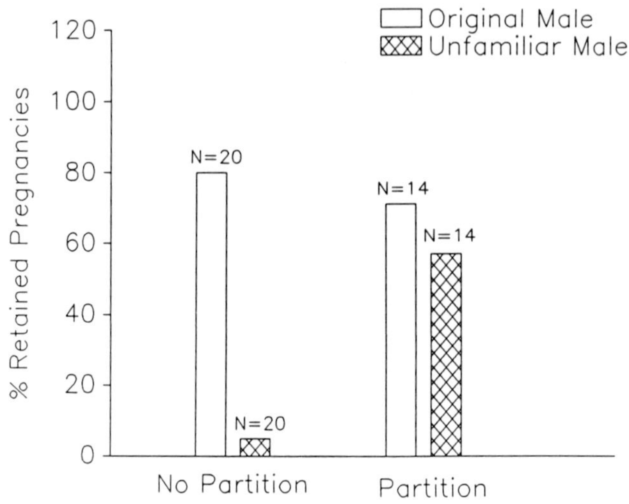


Fig. 1. Effects of male caged either directly with the female with no separating partition or when separated by a partition. The physical presence of the unfamiliar male results in fewer retained pregnancies than both the continued presence of the original male, and the presence of the unfamiliar male on the other side of the partition (from Storey 1986(b)).

## Role of chemical cues in pregnancy disruptions of meadow voles

In the studies to be described, females were placed on one side of a wire mesh partition, and sexually experienced males were placed on the other side. This initial exposure period brought the females into oestrus, so that mating and parturition were synchronized among females (Richmond and Conaway 1969), and it was then possible to determine when disruptions occurred.

The partitions separating the pairs were removed after four days, and this was called day 1 of the experiment. Females were placed in clean cages on day 4, and unfamiliar males were either placed with them or on the other side of the partition. Unfamiliar males remained with females until day 18 of the 21-day pregnancy.

Figure 1 shows the proportion of litters retained for females exposed directly to unfamiliar males, and for females separated from the males by a wire partition. The unfamiliar males' physical presence more effectively disrupted pregnancy than did exposure to non-physical cues. The only situation that produced any reduction in the proportion of retained pregnancies was when the female was in a central compartment with an unfamiliar male in each of the other two compartments. Even in this situation, 50% of the females remained pregnant, compared to 6% when females were placed with a single unfamiliar male (Storey 1986*a*).

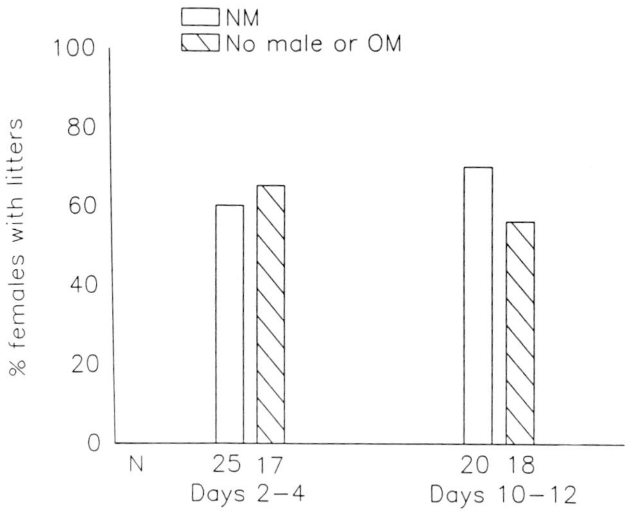


Fig. 2. Comparison of exposure to chemical cues from unfamiliar males versus the original or no male on days 2–4 or days 10–12. Exposure to chemical cues during either exposure period did not significantly reduce the incidence of pregnancy disruption relative to controls.

Because there are several reasons why side-by-side separation from unfamiliar males in the experiment described above might not disrupt pregnancy, the role of chemical cues was retested, with several procedural changes. These changes included: (a) examining the effect of chemical cues both before and after implantation; (b) starting the pre-implantation exposure on day 2 for consistency with previous findings of the time of greatest susceptibility in mice; (c) placing the female in a cage with bedding previously soiled by the unfamiliar male; and (d) placing the unfamiliar male in a cage over the female's cage so that his urine and faeces would drop into the female's cage. Both (c) and (d) should have increased the females' direct contact with chemical cues from the new male. The final change involved the genetic background of the unfamiliar males: in the first experiment unfamiliar males were from the same stock as original studs, and it is likely that genetic, and presumably chemical, differences among individuals were not great. In this experiment, the unfamiliar males were wild caught or first generation males from a separate population of voles. These unfamiliar males had been housed in a different room so that their odours would be novel to the females.

For both the pre- and post-implantation exposures, original studs were removed on day 2. Unfamiliar males were placed in cages above the female during either days 2–4 (pre-implantation) or days 10–12 (post-implantation). For the controls in the pre-implantation groups, a cage either empty or containing the original male was placed above the female. Empty cages were also placed above the cages of all females in the post-implantation control group.

Females that did not become pregnant in the experiment were subsequently tested for fertility, and so the proportion of disrupted pregnancies is based on the total number of fertile females. Figure 2 shows the proportion of females in each group that gave birth. For either days 2–4 ($\chi^2 = 0.1$, d.f. $= 1$, $p > 0.05$) or days 10–12 ($\chi^2 = 0.8$, d.f. $= 1$, $p > 0.05$), incidence of pregnancy was not significantly affected by the presence of an unfamiliar male in the cage above (Fig. 2).

Litter size was also examined because females may not implant as many embryos or may undergo a partial reabsorption in response to chemical stimuli from unfamiliar males. For both pre- ($t = 0.8$, d.f. $= 24$, $p > 0.05$) and post-implantation ($t = 0.0$, d.f. $= 21$, $p > 0.05$) exposure, there were again no significant differences between experimental and control groups (Table 1). Even with changing the procedure to increase female exposure to chemical signals, it appears that meadow voles do not disrupt pregnancies in response only to chemical cues from unfamiliar males.

**Table 1. Mean Litter Size (±SE) for females exposed to chemical cues from unfamiliar males**

| Exposure condition | Exposure Days: Days 2–4 | *N* | Days 10–12 | *N* |
|---|---|---|---|---|
| Unfamiliar male[a] | 3.9 ± 0.6 | 15 | 5.0 ± 0.5 | 13 |
| No male or original male | 4.2 ± 0.5 | 11 | 5.0 ± 0.8 | 10 |

[a] experimental-control group differences are not significant.

## Effects of presence of the unfamiliar male

The question of why not all females reabsorb embryos in response to unfamiliar males was then investigated. I hypothesized that behavioural interactions among pairs may indicate to females when it would be adaptive to reabsorb the pregnancy. In these experiments the tested groups had previously been found to have about a 50% disruption rate, so that there would be roughly equal number of pairs with retained and disrupted pregnancies. In the day 4 group, each female was placed with an unfamiliar male, and the original stud was placed on the other side of a partition. All pairs were observed with the original male for 1 h immediately after the dividers were removed on day 1; for 1 h on the day the unfamiliar male was introduced; and for 1 h on each of the two subsequent days.

Whether or not the original pregnancy was retained, there was more mating on day 1 than on day 4. However, no mating was observed on day 4 for females that retained the pregnancy, whereas for females with disrupted pregnancies, 16% of pairs were observed mating within the first hour after the introduction of the unfamiliar male (Storey, submitted).

For the day 12 groups, female-unfamiliar male pairs were observed. Pairs with disrupted pregnancies fought more, mated more, and had more social contact than pairs where females retained pregnancies (Storey and Snow, submitted). Prior to introduction to the female, unfamiliar males were placed with an unrelated pup for 10 mins and pup care and aggression by the unfamiliar male were recorded. Pup vocalization was taken as a sign of aggression from the male, and if this occurred the pup was removed immediately. Males that were aggressive to unrelated pups were also more aggressive to females, indicating that the level of male aggressiveness could signal to females about when the unfamiliar male would be likely to harm pups if the pregnancies were retained. Unlike the day 4 groups, the relative aggressiveness of the two males did not appear to be a factor in the disruption.

Mating frequency was the most striking difference among groups for the day 12 pairs. Mating within an hour of introduction of the unfamiliar male was observed in 44% of the pairs where the female disrupted the pregnancy and was reimpregnated by the unfamiliar male; 29% of pairs where a disruption without reimpregnation occurred; but only in 6% of pairs where the female retained the pregnancy (Storey and Snow, submitted). It is difficult, however, to show from these data that mating is causally related to disruption, because it is correlated with the other behaviour patterns, aggression and social contact, that also occur more frequently when the pregnancy is disrupted.

## The role of mating in pregnancy disruption

Isolation of the mating component of the behavioural interaction was attempted in order to clarify whether mating is a trigger, rather than a consequence, of pregnancy disruption. Giving the males low doses of a phenothiazine tranquillizer was attempted, but the drug did not reliably suppress mating; also, since non-mating pairs were not showing any aggressive interactions, I was not able to tell whether decreased mating or decreased aggression prevented a disruption. Experiments on larger rodents have used techniques for covering the male's penis to prevent intromission. Our attempts at this included constructing a small jacket to cover the males' genital area. We succeeded in designing a jacket that would stay on, but the voles did not behave normally while wearing these jackets. The best approach finally seemed to be to observe the animals, and for some pairs, to separate the male and female as soon as mating was initiated, i.e. they were not allowed to go beyond the first mount.

There is always the problem with pregnancy disruption experiments in determining which females are actually pregnant prior to the manipulation. The only information on pregnancy available by day 4 is whether females mated with the original male, and this was the criterion used for the day 4 group. For the day 12 group, information on day 1 mating was supplemented by weight gain and blood test data (Mihok 1987).

Pairs were initiated as before, and on the morning of day 12 each female was anaesthetized with ether, a blood sample was taken, and the female was weighed and placed in a clean cage. Three hours later, the unfamiliar male was placed in the cage and a continuous 3 h session of behavioural observations followed. The frequencies of fights, mating, vocalization, and social contact were recorded. At the first mount, pairs were assigned to one of two conditions: an experimental group, in which a partition was put into the cage to separate the pair; and a control group, that were allowed to continue mating. At the end of the 3 h, partitions were inserted into all cages, and on the following two days the partitions were removed and the pairs were observed for an additional 6 h. If pairs in the experimental groups started to mate during any of the observation periods, the dividers were again reinserted. The same observations were also recorded for pairs in which the unfamiliar male was introduced on day 4.

**Table 2.** Frequency of females that mated on day 1 and retained the original litter when no mating was initiated with the unfamiliar male, or mating with the unfamiliar male was terminated after the first mount, or mating continued to satiety

| | No mating | Mating halted after 1st mount | Mated to satiety |
|---|---|---|---|
| Day 4 | 25/31 (81%) | 6/11 (54%) | 3/12 (25%) |
| Day 12 | 25/34 (74%) | 6/10 (60%) | 1/7 (14%) |
| Overall | 50/65 (77%) | 12/21 (57%) | 4/19 (21%)* |

* $\chi^2 = 4.0$, $p < 0.05$.

Separating pairs after the initial mount resulted in more pregnancies being retained than if pairs were allowed to continue mating (Table 2). Of the 105 females that were observed to mate on day 1, 40 (38%) started to mate with the unfamiliar male on either day 4 ($n = 23$ females) or day 12 ($n = 17$ females). There was a significant difference in the proportion of females giving birth in the groups in which no mating was initiated, mating was terminated, and where mating to satiation was allowed ($\chi = 19.8$, d.f. $= 2$, $p < 0.001$). A higher proportion of females gave birth when mating was stopped than when it was allowed to continue ($\chi = 4.0$, d.f. $= 1$, $p < 0.05$; Table 2). There was no significant difference in the proportion of parturient females when no mating was initiated, and when mating was terminated ($\chi = 1.2$, d.f. $= 1$, $p > 0.05$).

Females that mated with the unfamiliar male and still retained the pregnancy sired by the original male received significantly fewer intromissions ($\bar{X} = 6.75$, S.E. $= 3.9$) than females that did not retain the original pregnancy ($\bar{X} = 18.6$, S.E. $= 3.95$, $t = 2.2$, d.f. $= 18$, $p < 0.05$). These data also suggest that females are more likely to reabsorb pregnancies with increasing amounts of copulatory stimulation.

While mating appears to be involved in these pregnancy disruptions, it is not clear why males are successful in copulation attempts with some but not all of the females. There appeared to be no group differences in the degree to which females solicited unfamiliar males for copulation. In fact, frequent approaches to the male by the female were usually made by females that retained pregnancies (70% of frequently approaching females), rather than those that did not give birth (30%). Thus, these data suggest that prevention of mating facilitates pregnancy retention, and that without mating no pregnancy disruptions occur.

Mating during pregnancy has not typically been observed in rodents. In the only other previous study, Jeppsson (1986) reported that some pregnant water voles (*Arvicola terrestris*) emigrating to new territories mate with the resident males, and retain the original pregnancy. We recently performed experiments in C57 mice, a strain in which females are relatively unsusceptible to disruption by chemical cues (Marsden and Bronson 1965). Female mice were placed with males of a different strain, and observations showed that sexual activity was correlated with pre-implantation disruption, even in the spontaneously ovulating mouse (de Catanzaro and Storey, in press).

## Adaptive significance of pregnancy disruptions

Because male-induced pregnancy disruptions occur so readily, and so far into pregnancy in voles, it is reasonable to assume that they play a role in the overall reproductive strategies of individuals, particularly females. I have previously suggested that the function of pregnancy disruptions may change with the timing of the introduction of the unfamiliar male (Storey 1986*b*). When the unfamiliar male was introduced on day 4, and remained with the female for the rest of pregnancy and pup rearing, there were no differences in reproductive success for females that retained or reabsorbed the pregnancy. Females that retained the pregnancy (unfamiliar male is not sire of litter) had as high litter survival and males were with pups as much as females that disrupted and mated with the unfamiliar male (unfamiliar male is sire of litter). Prevention of infanticide has been hypothesized as the function of pre-implantation pregnancy disruption in mice (Huck 1984; Labov 1981), but my data suggest that when the unfamiliar male remains with the female for most of the pregnancy, he does not kill pups, and even if he has not sired them, will show nest attendance. It appears, as suggested by Schwagmeyer (1979), that these early pregnancy disruptions may benefit females, not directly in terms of pup survival, but rather as a form of post-fertilization mate-selection, so that the most aggressive males are likely to sire the litters (Storey, submitted).

In contrast to the day 4 results, I found that when the unfamiliar male was introduced on day 12, pup survival was higher, and males spent more time with pups if the female reabsorbed the original litter and was reimpregnated by the new male. The results indicate that reabsorption benefits female reproductive success when the unfamiliar male is introduced later in pregnancy (Storey and Snow 1987; Storey and Snow, submitted). On the other hand, when pregnancies were not disrupted, relatively aggressive females paired with relatively non-aggressive males had surviving litters, indicating that retaining the pregnancy can be a successful response to the presence of the unfamiliar male (Storey and Snow, submitted). The results of these studies suggest that the positive interactions between males and pups, as well as negative ones, should be considered in assessing the benefits to females of male-induced pregnancy disruptions.

There is accumulating evidence of monogamy in some vole species, with males contributing to pup care, and the timing of parturition coincident with the presence or absence of particular males may therefore be particularly important. Within voles, species differences in susceptibility to disruption correlate well with the amount of pup attendance by males. Pine voles and prairie voles (*M. ochrogaster*) have high rates of pregnancy disruption (Stehn and Richmond 1975, Schadler 1981). These species show extensive paternal care in the laboratory (Oliveras and Novak 1986), and there is also evidence of monogamous relations in the field (Getz *et al.* 1981; FitzGerald and Madison 1983). In contrast, meadow voles are less susceptible to pregnancy disruption (Kenney, *et al.* 1977; Storey 1986; cf. Stehn and Richmond 1975). They are also less paternal in both laboratory (Oliveras and Novak 1986) and field (Madison 1980), although males do nest share with females and pups in the fall and winter (Madison *et al.* 1984), a time when their presence would make the greatest thermoregulatory contribution to the developing pups. Susceptibility to disruption throughout most of pregnancy may have evolved so that females could respond to changes in which males were likely to be present when the pups were born, given that males are making post-insemination investment in offspring either seasonally or throughout the year.

The results of this work indicate that in meadow voles chemical cues are not sufficient to produce pregnancy disruptions either before or after implantation of the embryos. The pre-implantation results are consistent with my earlier work (Storey 1986*a*), and with the work of Milligan (1976) for field voles. The lack of a post-implantation disruption via chemical cues contrasts with the finding of Schadler (1981) that chemical and physical cues were equally potent for pine voles.

The sensory mechanisms underlying pregnancy disruption in voles do not appear to change as pregnancy advances, in spite of changes in possible advantages for females. Having pregnancy initiation and disruption triggered by mating probably allows the female more flexibility in the timing of reproduction. The high level of susceptibility to pregnancy disruption in voles has been suggested as a minor disadvantage associated with being an induced ovulator. I suggest that both induced ovulation and pregnancy disruption, which appear to be mediated by the same male cues, help female voles produce offspring at optimum times.

## Acknowledgements

I thank Kathleen and Donald McKay for teaching us the techniques for obtaining and analysing the blood smears, and Tammy Joyce for assistance with data collection. This research was funded by operating grant A7475 from the Natural Sciences and Engineering Research Council of Canada and a Challenge Canada grant.

## References

Bruce, H.M. (1965). Effect of castration on the reproductive pheromones of male mice. *J. Reprod. Fertil.*, 10, 141–3.

Clulow, F.V. and Langford, P.E. (1971). Pregnancy-block in the meadow vole, *Microtus pennsylvanicus*. *J. Reprod. Fertil.*, 24, 275–7.

De Catanzaro, D. and Storey, A. (1989). Partial mediation of strange male-induced pregnancy blocks in mice by sexual activity. *J. Comp. Psychol.* (in press).

Dominic, C.J. (1965). The origin of the pheromones causing pregnancy blocks in mice. *J. Reprod. Fertil.*, 10, 469–72.

FitzGerald, R.W. and Madison, D.M. (1983). Social organization of a free-ranging population of pine voles, *Microtus pinetorum*. *Behav. Ecol. Sociobiol.*, 13, 183–7.

Getz, L.L., Carter, C.S. and Gavish, L. (1981). The mating system of the prairie vole, *Microtus ochrogaster*: field and laboratory evidence for pair-bonding. *Behav. Ecol. Sociobiol.*, 8, 189–94.

Hoppe, P.C. (1975). Genetic and endocrine studies of the pregnancy-blocking pheromone of mice. *J. Reprod. Fertil.*, 45, 109–15.

Huck, U.W. (1984). Infanticide and the evolution of pregnancy block in rodents. In *Infanticide* (ed. G. Hausfater and S.B. Hrdy), pp. 323–65. Hawthorne, New York.

Jeppsson, B. (1986). Mating by pregnant water voles (*Arvicola terrestris*): a strategy to counter infanticide by males? *Behav. Ecol. Sociobiol.*, 19, 293–6.

Kenney, A.M., Evans, R.L. and Dewsbury, D.A. (1977). Postimplantation pregnancy disruption in *M. ochrogaster*, *M. pennsylvanicus* and *Peromyscus maniculatus*. *J. Reprod.Fertil.*, 49, 365–7.

Labov, J.B. (1981). Pregnancy blocking in rodents: adaptive advantages for females. *Amer. Nat.*, 118, 361–71.

Madison, D.M. (1980). Space use and social structure in meadow voles *Microtus pennsylvanicus*. *Behav. Ecol. Sociobiol.*, 7, 65–71.

Madison, D.M., FitzGerald, R.W. and McShea, W.J. (1984). Dynamics of social nesting in overwintering meadow voles (*Microtus pennsylvanicus*): possible consequences for population cycling. *Behav. Ecol. Sociobiol.*, 15, 9–17.

Marsden, H.M. and Bronson, F.H. (1965). Strange male block to pregnancy: its absence in inbred mouse strains. *Nature*, 207, 878.

Mihok, S. (1987). Pregnancy test for meadow voles (*Microtus pennsylvanicus*) based on blood azurocytes counts. *Can. J. Zool.*, 65, 2830–2.

Milligan, S.R. (1976). Pregnancy blocking in the vole, *Microtus agrestis* I. Effect of the social environment. *J. Reprod. Fertil.*, 46, 91–5.

Oliveras, D. and Novak, M. (1986). A comparison of the paternal behaviour in the meadow vole *Microtus pennsylvanicus*, the pine vole *M. pinetorum*, and the prairie vole *M. ochrogaster*. *Anim. Behav.*, 34, 519–26.

Richmond, M. and Conaway, C.H. (1969). Induced ovulation and oestrus in *Microtus ochrogaster*. *J. Reprod. Fertil. Suppl.*, 6, 357–76.

Schadler, M.H. (1981). Postimplantation abortion in pine voles (*Microtus pinetorum*) induced by strange males and pheromones of strange males. *Biol. Reprod.*, 25, 295–7.

Schwagmeyer, P.L. (1979). The Bruce effect: an evaluation of male/female advantages. *Amer. Nat.*, 114, 932–8.

Stehn, R.A. and Jannett, F.J. (1981). Male-induced abortion in various microtine rodents. *J. Mamm.*, 62, 369–72.

Stehn, R. and Richmond, M. (1975). Male-induced pregnancy termination in the prairie vole, *Microtus ochrogaster*. *Science*, 75, 1211–3.

Storey, A.E. (1986*a*). Influence of sires on male-induced pregnancy disruptions in meadow voles (*Microtus pennsylvanicus*) differs with stage of pregnancy. *J. Comp. Psychol.*, 100, 15–20.

Storey, A.E. (1986*b*). Adaptive significance of male-induced pregnancy disruptions for female rodents. In *Reproduction: a behavioral and neuroendocrine perspective* (ed. B.R. Komisaruk, H.I. Siegel, M.F. Cheng and H.H. Feder), pp. 135–40. New York Academy of Science.

Storey, A.E. and Snow, D.T. (1987). Male identity and enclosure size affect paternal attendance of meadow voles *Microtus pennsylvanicus*. *Anim. Behav.*, 35, 411–9.

Storey, A.E. Adaptive significance of preimplantation pregnancy disruptions for female meadow voles: Post fertilization mate selection? (Submitted).

Storey, A.E. and Snow, D.T. Post-implantation pregnancy disruptions in meadow voles: Relationship to variation in male sexual and aggressive behaviour. (Submitted).

# III.4 Chemical Interaction Between Adult Female Mice: the Role of Ovarian and Adrenal Hormones

ANNA MARCHLEWSKA-KOJ

## Introduction

Grouped female mice release urinary chemosignals that inhibit ovulation and evoke pseudopregnancy in other females (van der Lee and Boot 1955; Ryan and Schwartz 1977). The experiments presented in this paper were designed to investigate the role of ovarian and adrenal hormones in the production of, and the reaction by female mice, to such chemicals.

## The production of a chemosignal

### MATERIALS AND METHODS

The number of oestrous phases in outbred females was determined following exposure to the bedding of eight adult females during 20 days and was used as an indicator of pheromonal activity. The influence on oestrous cycling of exposure to the chemosignal was determined by housing two adult females per cage and transferring them every morning for 20 days either to a clean cage, or to a cage previously occupied for 24 h by eight non-operated females or ovariectomized (OVX) females. The group of ovariectomized females were injected daily with oil, 50 ng oestrogen, 0.5 mg progesterone or a mixture of 50 ng oestrogen and 0.5 mg progesterone. Oestrus was indicated when vaginal smears contained only cornified epithelial cells.

### RESULTS AND DISCUSSION

The data summarized in Table 1 show that bedding soiled by grouped non-operated and OVX females suppressed oestrus in tested females. Injection-induced increases of oestrogen and progesterone levels inhibited release of the oestrous-suppressing factor.

From these results it is possible to hypothesize that the release of the oestrous-suppressing factor, as with the maturation-delaying factor (Drickamer

**Table 1.** Effect of bedding from non-operated and ovariectomized (OVX) females injected with estrogen (E) or progesterone (P) on number of oestrus phases in female mice. Values are mean ±SEM; $n = 14$

| Treatment/housing conditions | | No. of oestrus phases per female |
|---|---|---|
| Non-operated | + oil | $2.7 \pm 0.19^a$ |
| OVX | + oil | $2.6 \pm 0.23^a$ |
| Control (clean cage) | | $3.7 \pm 0.19^b$ |
| Non-operated | + E | $3.6 \pm 0.22^b$ |
| OVX | + | $3.6 \pm 0.20^b$ |
| Non-operated | + P | $4.0 \pm 0.18^b$ |
| OVX | + P | $3.6 \pm 0.22^b$ |
| Non-operated | + E + P | $3.2 \pm 0.32^b$ |
| OVX | + E + P | $4.1 \pm 0.18^b$ |

$^{a\ b}$ Those means marked by different letters are significantly different: $F_{(8,116)} = 5.69$; $p < 0.01$. (ANOVA and Duncan's new multiple range test.)

1985), could be stimulated by adrenal glands and inhibited by a high level of ovarian hormones. It is possible that both such effects are evoked by the same pheromone released by grouped female mice.

## The reaction to a chemosignal

### MATERIALS AND METHODS

Multiparous outbred female mice were reared one per cage after weaning. Oestrous cycles were determined by vaginal smears according to the procedure described for the first experiment. Females with two consecutive, four-day oestrous cycles were used for further investigation.

For the investigation of pseudopregnancy, single-reared oestrous or dioestrous females were housed with either eight adult females, or with just the bedding of a second group of eight adult females for the next eight days. The single-reared females were mated with vasectomized males or were stressed by electric foot-shock (1.2 mA, 30 V) once over 20 mins on the day of the oestrous or dioestrous phase.

Females were anaesthetized with ether four days later and their right uterine horn was traumatized; three days after the surgical procedure both uterine horns were dissected and weighed. The difference in weight between traumatized (right) and intact (left) horns was taken as the decidual cell reaction (DCR) indicator of pseudopregnancy.

The influence of chemosignals on corticosterone level in oestrous or dioestrous females was also investigated. Single-reared females were transferred for 20 mins to (i) a clean cage, (ii) a cage with eight adult females, (iii) a cage with the bedding of eight adult females, or (iv) were stressed by electric foot-shock. Females were then returned to their own cage and left undisturbed for the next 30 mins. After that time blood samples were taken from anaesthetized females by retro-orbital puncture through heparinized tubes. Plasma corticosterone level was measured by radioimmunoasay using anti-corticosterone antibody, corticosterone-3-MCO-BSA (a-B/R 556), obtained from the Institute of Sport, Warsaw.

## RESULTS AND DISCUSSION

Females in oestrus are sensitive to the pheromone released by grouped females and the stressor, electric foot-shock. Both factors evoked a similar effect, manifested as establishment of pseudopregnancy (Table 2).

**Table 2.** Pseudopregnancy estimated by increase of the weight of traumatized uterine horn in oestrus female mice after copulation with vasectomized males, exposure to 8 females, exposure to bedding of 8 females, or after foot-shock. Values are means ± SEM

| | Uterine horn weight[a] (mg) | | |
|---|---|---|---|
| | **Intact** | **Traumatized** | |
| Copulation (7/8)[b] | 42.3 ± 3.7 | 109.0 ± 20.9 | $p<0.001$ |
| 8 females (7/8) | 49.8 ± 4.6 | 86.0 ± 11.2 | $p<0.001$ |
| Bedding of 8 females (4/12) | 47.7 ± 7.7 | 86.3 ± 8.3 | $p<0.001$ |
| Footshock (8/9) | 64.7 ± 11.3 | 110.6 ± 15.1 | $p<0.001$ |

[a] The differences between uterine horn weight were compared by Wilcoxon's signed test (Sokal and Rohlf 1969).
[b] The values in prenthesis are the no. of pseudopregnant/no. of tested females.

There were differences in the reaction of adrenal glands to these factors. Presence of females or their bedding caused an acute decrease of plasma corticosterone, whereas foot-shock increased the release of this hormone (Table 3). In contrast, Nichols and Cheyvins (1981) reported that male urine accelerated the oestrus and acutely increased the plasma corticosterone level in female mice. It is possible that the function of ovaries and adrenal glands is inhibited by a female chemosignal or stimulated by a male chemosignal. Both effects can be mediated by the same neurohormonal pathway in female mice.

**Table 3.** Plasma corticosterone level (ng/ml) in female mice transferred to a clean cage (control), a cage with 8 females, or a cage with bedding of 8 females or after footshock. Values are means ± SEM

| Treatment | Females in oestrous | (*n*) | dioestrous | (*n*) |
|---|---|---|---|---|
| Clean cage | $169 \pm 34^{ab}$ | (5) | $165 \pm 11^{f}$ | (6) |
| 8 females | $29 \pm 2^{ace}$ | (6) | $72 \pm 7^{fgh}$ | (6) |
| Bedding of 8 females | $73 \pm 22^{bde}$ | (4) | $168 \pm 26^{g}$ | (5) |
| Footshock | $249 \pm 21^{cd}$ | (6) | $161 \pm 23^{h}$ | (6) |

$^{a-h}$ Those means marked by the same letter are significantly different, *a,c,d,f,g,h,* $p<0.001$ *b,e,* $p<0.05$. The data were compared by Kruskal–Wallis analysis of variance (Sokal and Rohlf 1969).

## Conclusion

The results presented above indicate that females reared at high population density release chemosignals which inhibit ovulation in other females. Release of this chemosignal is not under control of the ovarian hormones. Moreover, the effect of the pheromone is not mediated by stimulation of adrenals since the plasma corticosterone level was found to be decreased in the recipients.

## Acknowledgement

This work was supported by the Polish Academy of Sciences, grant no. CPBP 05.06.

## References

Drickamer, L.C. (1985). Puberty-influencing chemosignals in house mice: ecological and evolutionary consideration. In *Chemical signals in vertebrates*, 4 (ed. D. Duvall, D. Müller-Schwarze and R.M. Silverstein), pp. 441–57. Plenum, New York.

Lee, S. van der and Boot, L.M. (1955). Spontaneous pseudopregnancy in mice. I. *Acta Physiol.Neerl.*, 4, 442–6.

Nichols, D.J. and Cheyvins, P.F.D. (1981) Adrenocortical responses and changes during the oestrous cycle in mice: effect of male presence, male urine and housing conditions. *J. Endocrinol.*, 91, 263–9.

Ryan, K.D. and Schwartz, N.B. (1977). Grouped female mice demonstration of pseudopregnancy. *Biol. Reprod.*, 17, 578–83.

Sokal, R.R. and Rohlf, F.J. (1969) *Biometry*. Freeman, New York.

# III.5 Female Urinary Chemosignals Stimulate Scent Marking Behaviour in Male Mongolian Gerbils

B. PROBST

## Introduction

Mongolian gerbils (*Meriones unguiculatus*) scent mark objects in their environment with the secretions from a specialized scent gland. The frequency of this scent-marking behaviour was quantified during a standardized open-field test (open-field: $60 \times 60$ cm, 9 marking pegs). In the absence of females, males scent marked at a rate that was more or less constant for any one individual but differed between individuals. When males and females were present inside the same room but without direct contact, the males' scent-marking frequencies approximately doubled. After removal of the females the marking frequencies decreased to initial levels.

This increase in scent marking is mediated by chemical signals contained in the urine of females (Probst 1985). Application of 40 $\mu$l female urine/day to the external nares of males increases marking frequencies to the same extend as the presence of females. The increased level of marking activity is achieved after two days of urine application (Probst 1985). Using this procedure as a relatively fast and very sensitive bioassay, the chemical constitution of the signal was further characterized by different analytical techniques.

Head-space distillation of female urine was performed to examine the influence of volatile and non-volatile cues on scent marking. The volatile fraction of female urine, however, was ineffective in stimulating scent marking, even when applied in higher dosages. Dialysing female urine (mol. weight cut off at 8000 Dalton) retained the active signal in the non-dialysable fraction (Probst and Lorenz 1987).

Lipids from female urine were separated by liquid-liquid extraction using Extrelut columns and ether as solvent. Neither the acidic lipophilic extract nor the basic lipophilic extract contained scent marking stimulatory activity. However, the dialysed aqueous residue, reconstituted to the original volume with physiological saline, significantly increased scent-marking activity (Wilcoxon's matched-pairs test: $p < 0.01$; Probst and Lorenz 1987).

The polypeptides contained in female urine were reversibly precipitated by

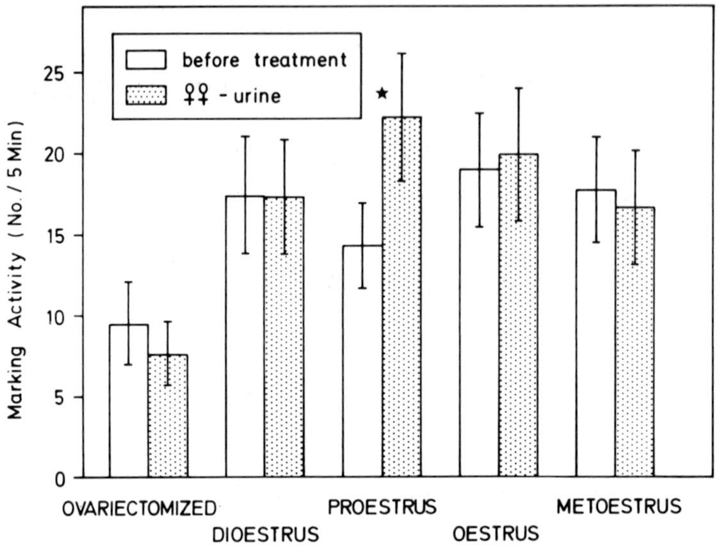


Fig. 1. Influence of urine from ovariectomized females and urine from females of different estrous states on scent marking activity of male gerbils (mean $\pm$ SEM, $n=9$, $\times : p<0.01$).

addition of ammonium sulphate at a final concentration of 80% at 4 °C. The redissolved and dialysed precipitate significantly stimulated scent marking while the supernatant significantly reduced the scent marking activity (Wilcoxon's matched-pairs test: $p<0.001$; Probst and Lorenz 1987).

In total these experiments have indicated that the active fraction of the female scent-marking stimulatory chemosignal is hydrophilic and non-volatile, and is contained in the non-dialysable, high molecular weight polypeptide fraction of female gerbil urine (Probst and Lorenz 1987).

## Relationship between oestrous cycle and scent-marking stimulation

Since urine of ovariectomized female gerbils did not stimulate males scent marking, the efficacy of urine collected from females with known oestrous states was determined. A female permitting a sexually active male to copulate was rated to be in oestrus. While cycles of 3 to 6 days clearly dominated (more than 75% of individuals), 7 of 28 females displayed anoestrous periods of 7 to 23 days, which were attributed by Marston and Chang (1965) to pseudopregnancy. Urine of pro-oestrous females increased male scent-marking activity most markedly, while urine from oestrous females resulted in minor activity (Fig. 1). Dioestrous and metoestrous female urine did not increase male-marking activity. Concentrations of free oestradiol and progesterone in individual urine samples collected over

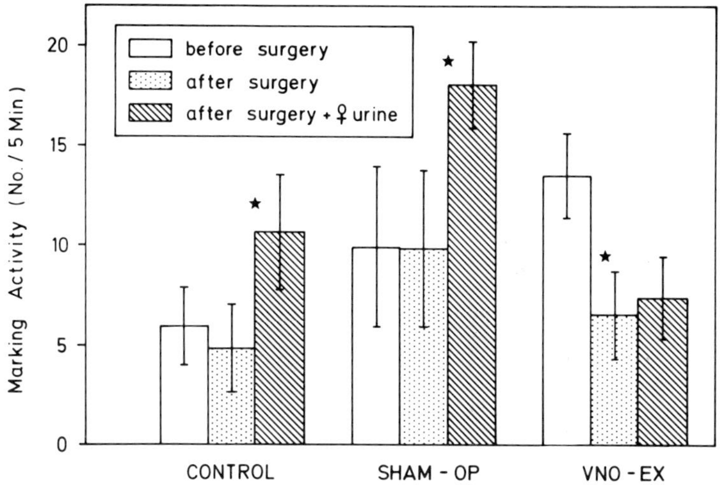


Fig. 2. Effect of vomeronasalectomy upon scent marking activity of male gerbils during the control period before surgery, after surgery, and after surgery and treatment with female urine (mean $\pm$ SEM, control, $n=6$; sham, $n=4$; VNX, $n=9$).

several oestrous cycles showed no correlation with the oestrous states of the females (Fenske and Probst, unpublished data).

## Perception of the signal

The perception of high molecular weight polypeptides is probably mediated by the vomeronasal organ (Wysocki *et al.* 1985). After surgical removal of the vomeronasal organ no increase in scent-marking frequencies occurred during or following application of female urine (Fig. 2). Other open-field behaviours were not affected by the surgical procedure.

To examine the influence of the primary olfactory epithelium upon male scent marking, the nasal mucosa was irrigated by intranasal influsion with $ZnSO_4$. However, this treatment caused a drastic reduction in scent-marking frequencies and other open-field behaviour patterns such as locomotor activity. In addition, the $ZnSO_4$-treated males lost about 20% body weight. During application of female urine the scent marking activities of these males further decreased (Fig. 3). Therefore, the $ZnSO_4$ treatment apparently induced a more or less general injury, which prevented the action of the scent-marking stimulatory chemosignal.

## Discussion

The biological significance of scent marking in Mongolian gerbils is unknown. However, the increase in scent-marking frequencies of male gerbils by chemical signals of females and oestrous-related excretion of the chemical signal point to a

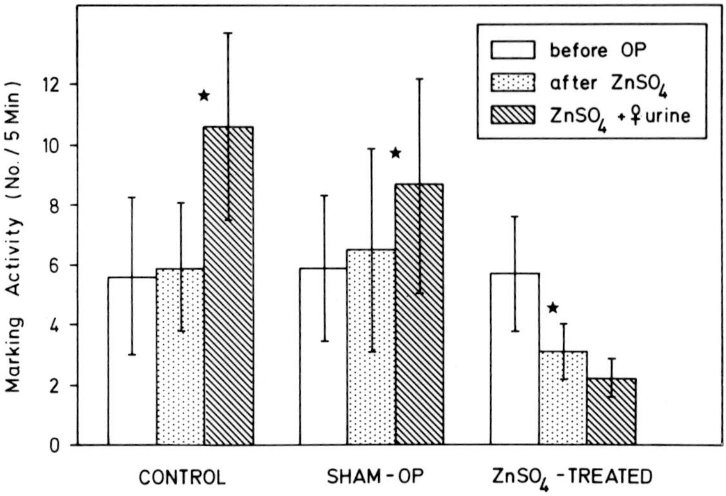


Fig. 3. Effect of intranasal infusion with $ZnSO_4$ upon scent marking activity of male gerbils during the control period before infusion, after infusion, and after treatment with female urine (mean $\pm$ SEM, control, $n=6$; NaCl, $n=4$; $ZnSO_4$, $n=9$).

major role of scent marking in a sexual (Brown *et al.* 1988) rather than territorial (Thiessen 1968) context. As the action of the signal is reversible and a fast and very sensitive biotest is available, the gerbil may be a valuable model for the study of action of chemical signals.

## References

Brown, R.E., Hausschild, M., Holman, S.D. and Hutchison, J.B. (1988). Mate recognition by urine odors in the Mongolian gerbil (*Meriones unguiculatus*). *Behav. Neural Biol.*, 49, 174–83.

Marston, J.H. and Chang, M.C. (1965). The breeding, management and reproductive physiology of the Mongolian gerbil. *Lab. Anim. Care*, 15, 34–48.

Probst, B. (1985). Female urinary chemosignals influence scent marking behavior in male Mongolian gerbils. *Z. Naturforsch, C*, 40, 936–8.

Probst, B. and Lorenz, M. (1987). Increased scent marking in male Mongolian gerbils by urinary polypeptides from female conspecifics. *J. Chem Ecol.*, 13, 851–61.

Thiessen, D.D. (1968). The roots of territorial marking in the Mongolian gerbil: A problem of species-common topography. *Behav. Res. Methods Instru.*, 1, 70–6.

Wysocki, C.J., Beauchamp, G.K., Reidinger, R.R. and Wellington, J.L. (1985). Access of large and non-volatile molecules to the vomeronasal organ of mammals during social and feeding behaviors. *J. Chem. Ecol.*, 11, 1147–59.

# III.6 Short Term Memory for Vaginal Odour in the Syrian Hamster

ELIZABETH STEEL

## Introduction

The vaginal secretion of female hamsters is strongly attractive to males (Johnston 1974). Males sniff and lick this secretion, resulting in an elevation of plasma testosterone (Macrides *et al.* 1974) and an increased likelihood of mounting and intromission (Murphy 1973). The attractive function of vaginal odour has been less well documented. However, Lisk *et al.* (1983) have shown that mating takes place mostly in or near the nest of the female; they suggested that deposited vaginal secretion forms a 'trail' which the male can follow to the nest where he is only tolerated shortly before the female becomes receptive. Although all females mark in the presence of a male, pro-oestrous females also mark in his absence (Steel 1985).

There is little information concerning the natural ecology and behaviour of the Syrian hamster in the wild (Murphy 1973; Johnston 1977); the inference from the behaviour of captive animals (chiefly their extreme aggressiveness) is that they may be mainly solitary animals living at low densities. The peak of female vaginal marking 24 h before oestrus (Johnston 1977) suggests that, where males are widely dispersed, mate attraction is too important to be left until the day of female receptivity. Deposited vaginal odour persists in the captive environment and is capable, unlike visual and auditory signals, of conveying information long after the depositing female has left (Johnston and Schmidt 1979). The question addressed in this paper is: does repeated contact with female vaginal odour (in the absence of the marking female) have an effect on the male's behaviour towards females that are subsequently encountered? Sexual arousal would be inappropriate towards dioestrous, scent-marking females, but behaviour enabling males to find and stay with such pro-oestrous females would seem to be adaptive.

## Olfactory recognition by male hamsters

### EFFECT OF VAGINAL SECRETION ON MALE BEHAVIOUR

A male hamster, after exposure to female vaginal secretion, shows less subsequent interest in an unfamiliar female than in a familiar female (i.e. the donor of the secretion)(Steel 1984). Furthermore, males can even distinguish between individuals of the same hormonal state. However, the hormonal status of females also matters: familiar and unfamiliar females are better discriminated after exposure of males to the secretion from pro-oestrous females than to secretion from females on the day after oestrus (Steel 1985).

Males approach, sniff and leave, or stay with the female. The probabilities of any one of these behaviours being shown are independent of one another and are thought to be partly under separate hormonal control (Steel 1982; Steel and Hutchison 1988). In tests we compared the interest shown in a novel female under two conditions: (a) no prior exposure to vaginal odour (*control*); and (b) pre-exposure to vaginal odour such that the novel female was a mismatch to the recently encountered odour (*unfamiliar*). The response to an *unfamiliar* female is manifest as a decreased tendency to leave the female (visits get longer) and an increased tendency to sniff her (increase in sniffing time as a percentage of time spent near the female). Significant changes in one or both of these parameters form evidence for the existence of 'recognition' as referred to below. The time courses of the two responses are not identical but have not been studied in detail.

### FORMATION OF A SHORT-TERM MEMORY FOR VAGINAL ODOUR.

Pre-exposure to the odour resulted in subsequent discrimination between the *control* and the other, *unfamiliar* females. In summary, investigations showed that several exposures to vaginal odour were needed, the inter-exposure interval was critical, and the behavioural effect (change in interest in the female, as above) lasted only a few hours after the last previous encounter with the odour.

The following results were obtained:

1. With five exposures to odour at 15 min. intervals, recognition of the *unfamiliar* female was present at between 1 to 2 h after the last odour exposure. It began to decline between 3 and 4 h after the last exposure and was absent 24 h later.
2. With presentations at 15 min. intervals and testing time at 1 to 2 h, it was found that two presentations had no effect, five had a maximal effect, whilst three presentations were intermediate in their effectiveness for recognition.
3. With the number of presentations held at five and with testing time at 1 to 2 h, a spacing of 2 mins between odour presentations was ineffective whilst intervals of 5 and 15 mins had a maximal effect on recognition.
4. A single forced exposure to the secretion (vaginal swab wiped across the nostrils and oro-nasal groove) resulted in marked discrimination. Sniffing was reduced after 3 mins and amount of time spent with the female was reduced 6 mins after the pre-exposure; intervals of 12 and 24 mins were no more effective than 6 mins.
5. Males who were exposed only to the volatile components of the secretion did not discriminate. An opportunity to lick the secretion was essential.
6. Males with ablated vomeronasal organs did not discriminate (Fig. 1). When

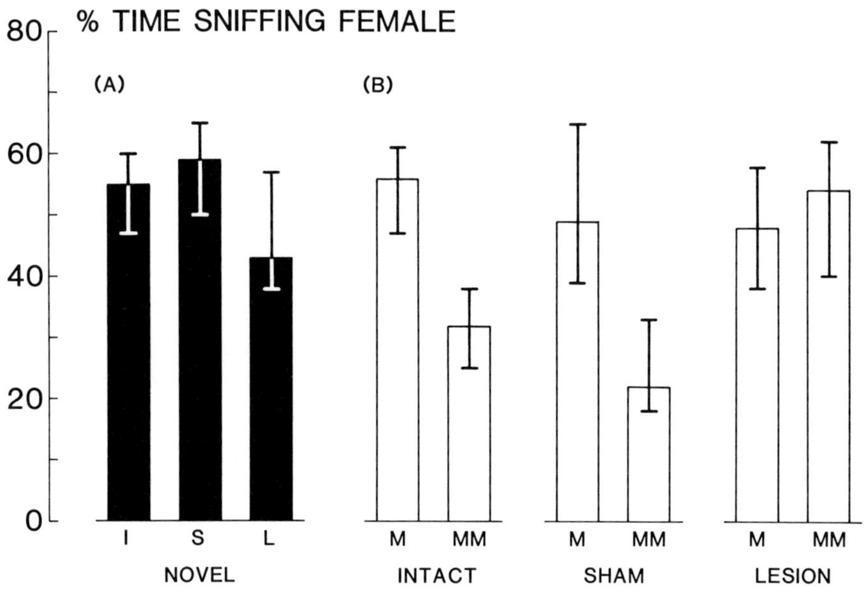


Fig. 1. The amount of sniffing directed at females by intact and vomeronasal-lesioned male hamsters. Group medians and inter-quartile ranges are shown.

A. Response of three treatment groups to novel females; no pre-exposure to odour. I = intact males; S = sham-operated males; L = lesioned males. Kruskal-Wallis one-way analysis of variance not significant.

B. Response of the three treatment groups (after pre-exposure to odour) to matching (M) and mismatching (MM) females. Wilcoxon matched pairs test, for intact and sham, $P<0.001$; lesion, n.s.

there was no pre-exposure to odour, percentage of time spent sniffing a novel female was the same for three experimental groups (Fig. 1A) (Steel and Keverne 1985).

After pre-exposure to odour, intact males decreased their amount of sniffing towards a novel female who did not match the odour (MM) relative to a novel female who matched (M) the odour; the response of sham-operated males was similar to those left intact. However, lesioned males directed similar amounts of sniffing towards both M and MM females (Fig. 1B).

The finding that a massive forced exposure to the olfactory stimulus can alter the males' subsequent behaviour contrasts with the effect of encountering the same stimulus deposited in the environment; in the latter case, one exposure had no marked effect whilst several exposures resulted in a change of behaviour. This raises the question of how males sample the stimulus and how they pace their own

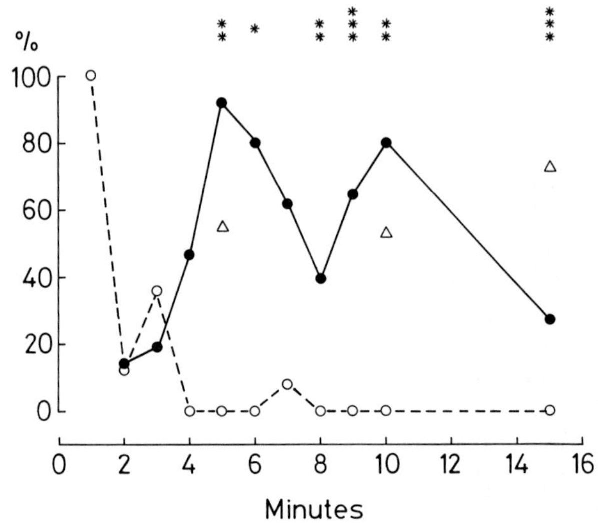


Fig. 2. Time spent investigating a vaginal smear in minutes 2 to 15 as a percentage of time spent investigating the smear in minute one (100% represents a median duration of 19 s; IQR 11–28).

Open circles, same smear present for 15 mins (habituation);
closed circles, fresh smear from the same female present for 1 min.;
open triangle, fresh smear from novel female present for 1 min.

All data are the median scores for independent groups of hamsters (n = 10). Mann-Whitney U-test between habituated smear (open circles) and a fresh smear from the same female (closed circles) in any one minute, *$P<0.05$; **$P<0.01$; ***$P<0.001$.

exposure to vaginal scent marks. There seem to be at least two methods available to them.

Firstly behavioural sampling, which was studied by looking at the time spent sniffing and licking a standard presentation of vaginal secretion. Males, in a small bare arena, given fresh secretion spread over a 76 × 38 mm glass plate, spent on average 19 seconds in the first minute of exposure licking the secretion; thereafter they rarely returned for further investigation although much of the stimulus remained (Fig. 2). A new slide with fresh secretion from the same female induced a sharp increase in interest during the 5th min. (but not the 2nd or 3rd). This suggests that enough information is gathered in 19 s to lay down the initial part of the memory trace or to reinforce a trace formed 4 mins or more previously. It also accords with the finding that exposures to odour do not summate if they are less than about 4 mins apart. Until a male has habituated to the first odour (for about 4 mins) a second, fresh odour from the same female is not perceived as a separate event.

Secondly, males may also pace their exposure to odour by means of the autonomically driven vomeronasal pump (Meredith 1983), which by altering the volume within the lumen could allow more or less fluid to be admitted to the receptor neurones. Signals from the main olfactory system, on the initial detection of an odour, may be followed selectively by vomeronasal sampling. Thus it may not follow that, because a male is licking the secretion, he is automatically allowing access to the vomeronasal organ. In Fig. 2 it can be seen that, after habituation to the initial odour, neither novel nor familiar odour at 5 and 10 differ in the amount of licking that they elicit. It would be interesting to know whether admission of vaginal odours from an unfamiliar female to the vomeronasal cavity was reduced after a memory of the familiar odour had been formed.

In a functional sense, it would be reasonable for a male who had already encountered the odour marks of a pro-oestrous female to ignore other females who were not actually in oestrus. Also, he may already have invested time and energy in following one female, either because of random encounters with her scent marks or because of specific mate choice; distraction by fortuitous encounters with dioestrous females would be disadvantageous.

The physiological mechanisms underlying this rapid alteration of behaviour in response to vomeronasal stimulation are at present unknown; however, the action of ascending noradrenergic fibres (projecting to the main and accessory olfactory bulbs) are important in modulating bulbar habituation to repeated presentation of a biologically meaningful odour (Gervais *et al.* 1988). Increased noradrenalin release in the olfactory bulbs induced by arousing stimuli (e.g. mating, parturition, or maternal licking) in the presence of the odour has been shown to be associated with memory formation for selective olfactory cues. Further studies are needed to determine whether a similar mechanism operates in the hamster and to identify the relevant arousing stimuli.

## References

Gervais, R., Holley, A. and Keverne, B. (1988). The importance of central noradrenergic influences on the olfactory bulb in the processing of learned olfactory cues. *Chem. Senses*, 13, 3–12.

Johnston, R.E. (1974). Sexual attraction function of golden hamster vaginal secretion. *Behav. Biol.*, 12, 111–7.

Johnston, R.E. (1977). The causation of two scent marking behaviours in female hamsters. *Anim. Behav.*, 25, 317–27.

Johnston, R.E. and Schmidt, T. (1979). Responses of hamsters to scent marks of different ages. *Behav. Neural. Biol.*, 26, 64–75.

Lisk, R.D., Ciaccio, L.A. and Catanzaro, C. (1983). Mating behaviour of the golden hamster under seminatural conditions. *Anim. Behav.*, 31, 659–66.

Macrides, F., Bartke, A., Fernandez, F. and D'Angelo, W. (1974). Effects of the exposure to vaginal odour and receptive females on plasma testosterone in the male hamster. *Neuroendocrinology*, 15, 355–60.

Meredith, M. (1983). Sensory physiology of pheromone communication. In *Pheromones and Reproduction in Mammals*, (ed. J.Vandenbergh), pp. 199–252. Academic Press, New York.

Murphy, M.R. (1971). Effects of female hamster discharge on the behavior of the male hamster. *Behav. Biol.*, 9, 367–75.

Steel, E. (1982). Testosterone-dependent non-copulatory behaviour in male hamsters (*Mesocricetus auratus*). *J .Endocrinol.*, 95, 387–96.

Steel, E. (1984). Effect of the odour of vaginal secretion on non-copulatory behaviour of male hamsters (*Mesocricetus auratus*). *Anim. Behav.*, 32, 597–608.

Steel, E. (1985). Effect of male proximity on vaginal marking by the female Syrian hamster (*Mesocricetus auratus*). *Z. Tierpsychol.*, 67, 69–77.

Steel, E. and Keverne, E.B. (1985). Effect of female odour on male hamsters mediated by the vomeronasal organ. *Physiol. Behav., 35*, 195–200.

Steel, E. and Hutchison, J.B. (1988). Pre-copulatory behaviour in the male hamster: effect of the 5α-reductase inhibitor 4-MA (17B-N, N-diethylcarbamoyl-4-aza-5α-androstan-3-one). *J. Endocrinol.*, 119, 483–91.

# III.7 Effect of Male Presence and Urine on the Sexual Development of Female Field Voles (*Microtus agrestis*)

ISABELLE WERNER AND J.R. CLARKE

## Introduction

Microtine rodents pose important problems for population ecologists (Tait and Krebs 1985). Understanding their population dynamics, in particular, requires a detailed knowledge of factors determining fertility. The principal environmental determinant of sexual development is photoperiod, but social factors appear to modulate its effects (Spears and Clarke 1988). The present experiments have therefore been designed to examine the influence of (i) the presence, and (ii) the urine, of sexually experienced males on the sexual development of females.

## Materials and methods

All animals were obtained from the main breeding colony of field voles (*Microtus agrestis*) kept in the Department of Zoology, Oxford University. The animals were fed whole oats and SDS Diet rat/mouse No 3 *ad libitum*; water was constantly available. The experiments were of a randomized block design, each block being composed of weanling females ($16 \pm 1$ days old), which were matched for age, body weight and absence of sexual development. Animals of any one block were allocated at random to the three treatments of each experiment. All treatments continued for three weeks under a photoperiod of 16 h light/8 h dark. Females were then killed, and the uterus, ovaries, adrenals, and pituitary gland were removed and weighed.

### EXPERIMENT I: EFFECT OF MALE PRESENCE

Treatments were: (i) Presence of a sexually experienced male; (ii) presence of a sexually mature but naïve male; and (iii) no male present.

Females were housed in plastic cages ($44 \times 28 \times 21$ cm) each of which contained a small wire mesh cage ($16 \times 16 \times 8$ cm). The inner cage contained a sexually mature male, or was empty, and allowed only slight tactile contact between the male and the female.

### EXPERIMENT 2: EFFECT OF MALE URINE

Urine was obtained from males placed in a metabolic cage (Jencons), with samples being collected every 3 h. The urine was stored in glass vials (2.5 ml) in a deep freeze and was used within 1 week of sampling. A separate vial was opened each day for each treatment.

The treatments consisted of painting once daily the dorsal surface of the nose of female field voles with: (i) urine of sexually experienced males; (ii) urine of naïve mature males; and (iii) water.

For all experiments, body growth rates were calculated from the difference between final and initial weight measurements divided by the time in days for which the measurements were made. The results were analysed using factorial analysis of variance (Bailey 1981) and Duncan's multiple range test (Ridgman 1975).

## Results

### EXPERIMENT 1: EFFECT OF MALE PRESENCE

The treatments had no significant effect on body growth rate or ovarian weights (Table 1). There was a significant effect of the presence of males on uterine weights ($p<0.01$) and pituitary weights ($p<0.005$); the treatment effect on adrenal weights was not significant ($p<0.1$).

Females reared with sexually experienced males had significantly heavier uteri, adrenals, and pituitary than females reared in the absence of males. Those reared in the presence of naïve males had organ weights that were intermediate between females reared with sexually experienced males and those reared alone; uterus and adrenal weights, in particular, were more variable than for either of the other treatments. The pituitary glands of females reared with sexually naïve males were significantly smaller than those of females reared with sexually experienced males.

### EXPERIMENT 2: EFFECT OF MALE URINE

Urinary treatments had no significant effect on either body growth rate or ovarian, adrenal, and uterine weights (Table 2). The effect of the treatments on pituitary gland weights was also not significant ($p<0.1$). Females treated with urine of sexually naïve males had significantly heavier pituitary glands than those of females receiving urine from sexually experienced males.

**Table 1.** Body growth rate (g/day), and ovarian, adrenal, uterine, and pituitary gland weights (mg) of females after 3 weeks of Experiment 1

| | Treatment[a] | | | |
|---|---|---|---|---|
| | +M.SE<br>$n=5$ | +M.N<br>$n=$ | -M<br>$n=6$ | Male effect |
| Growth rate | $00.47^c \pm 0.04$ | $00.41^c \pm 0.06$ | $00.39^c \pm 0.06$ | n.s.[b] |
| Ovarian weight | $02.86^c \pm 0.38$ | $03.20^c \pm 0.58$ | $02.40^c \pm 0.37$ | n.s. |
| Adrenal weight | $10.76^c \pm 0.79$ | $09.23^{c+} \pm 1.09$ | $07.95^+ \pm 0.28$ | n.s. |
| Uterine weight | $65.42^c \pm 2.99$ | $44.98^{c+} \pm 11.31$ | $21.13^+ \pm 6.46$ | $p<0.01$ |
| Pituitary weight | $01.66^c \pm 0.18$ | $01.10^+ \pm 0.17$ | $00.67^+ \pm 0.16$ | $p<0.005$ |

[a] Wealing females were reared in the presence of sexually experienced males (+M.SE) or naïve mature males (+M.N), or were without males (-M).
[b] n.s., not significant.
[c+] Values are means ±SE; means in rows with different symbols in brackets are significantly different.

**Table 2.** Body growth rate (g/day), and ovarian, adrenal, uterine, and pituitary gland weights (mg) of females after 3 weeks of Experiment 2

| | Treatment[a] | | | |
|---|---|---|---|---|
| | +U.SE<br>$n=3$ | +U.N<br>$n=4$ | -W<br>$n=4$ | Male effect |
| Growth rate | $00.38 \pm 10.04$ | $00.58 \pm 0.09$ | $00.42 \pm 0.06$ | n.s.[b] |
| Ovarian weight | $02.90 \pm 0.26$ | $03.60 \pm 0.47$ | $03.38 \pm 0.40$ | n.s. |
| Adrenal weight | $07.47 \pm 0.35$ | $09.10 \pm 1.61$ | $08.50 \pm 0.62$ | n.s. |
| Uterine weight | $41.13 \pm 5.82$ | $52.05 \pm 6.09$ | $44.85 \pm 14.08$ | n.s. |
| Pituitary weight | $00.87^* \pm 0.17$ | $01.53^+ \pm 0.20$ | $01.45^{*+} \pm 0.15$ | n.s. |

[a] Wealing females were exposed to urine of experienced male (+U.SE), urine of naïve mature males (+U.N), or water (-W). Values are means ±SE.
[b] n.s., not significant.
[*+] Pituitary weight means with different symbols in brackets are significantly different.

## Discussion

The results of Experiment 1 show a 'whole' male effect upon the sexual maturation of young female voles. The presence of a sexually experienced male stimulates greater uterine, adrenal, and pituitary gland development than that found in females reared alone. Furthermore, pituitary gland development is significantly greater in the presence of sexually experienced males than it is in the presence of sexually naïve males.

These results extend those obtained by Spears and Clarke (1986): using the same species of vole, they found a significant male effect on uterine and pituitary gland weights but no effect on ovarian weights, as in the present study. Females reared with a naïve male did not show such a clear effect: effects were intermediate and showed a greater variability of uterine and adrenal weights. Body growth rates of the females were also unaffected by the presence of a mature male.

Environmental stimuli (visual, tactile, olfactory, and auditory) presumably induce the hypothalamus to liberate releasing factors (follicle-stimulating hormone-/luteinizing hormone-release factor) which control the production of pituitary hormones. The production of oestrogen by the ovaries is under the control of these glycoproteins and causes an enlargement of the uterus (Hogarth 1978). An increased uterine weight suggests higher oestrogen release from the ovaries, and the weight of the uterus was, in fact, found to be a reliable index for measuring the onset of puberty in female field voles (Akhtar 1987).

The acceleration of females' maturation in the presence of a mature male has been reported for several other rodent species (for review, see Marchlewska-Koj 1984). In most cases exposure to male urine has been shown to be important; for example, urine applied to the nose is known to affect the sexual development of weanling female mice (Drickamer 1987). Akhtar (1987) showed that young female field voles had heavier uterine weights when painted with male urine. The results of our Experiment 2 do not show such an effect. Several factors may have contributed to the disparity between our results and those obtained by Akhtar (1987):

1. The urine used for the treatments was collected every 3 h from males housed in a metabolic cage. It is possible that some of the postulated pheromones would have decomposed in this time.
2. Application of urine to the nose stresses the animals, which may tend to obscure any urine effects. Introducing the urine tube of the metabolic cage directly into the cage where the experimental animal is housed, is one way of overcoming both these problems (M. Kruczek, personal communication).
3. The variability, even among siblings, of the initial state of sexual development is great: both the smallest uterus (8.40 mg) and the largest one (77 mg) were found for the water treatment. Closer matching of females in a randomized block design may make the test more sensitive. If this then demonstrated a urinary effect, the implication would be that the presence of a male has a stronger effect than the presence of male urine alone.
4. Akhtar's results (1987) may have been confounded by non-urinary olfactory signals: although caged separately, all her experimental and control animals were housed in the same experimental room. The results presented here need to be supported by further work to determine whether or not the presence of male urine is involved in the sexual maturation of female voles.

In these experiments, the presence of a sexually experienced male was associated with greater pituitary development, compared to that found in females kept with naïve males, yet the opposite was found for the urinary treatments.

Drickamer (1981) has shown that prior sexual experience affects a male's ability to accelerate puberty in female mice, both by his presence and by excretion of urinary pheromone. Females caged with sexually experienced males were younger at first oestrus than the females caged with a naïve male or the females painted with urine from experienced males. In Experiment 2, body growth rates and all organ weights were less for females painted with the urine of experienced males than for females painted with the urine of naïve males. Although this trend is not significant, except for pituitary gland weights, it is opposite to that found in other studies (Drickamer 1981), and is currently the subject of further research.

Perhaps the most likely explanation for these results is that olfactory signals are involved in sexual development. Chemosignals can produce physiological effects by influencing endocrine co-ordination (Marchlewska-Koj 1984). If they are the major cause of the effects reported here, then, given that they can stimulate or inhibit gonadal activity, they could play an important role in population regulation.

## References

Akhtar, N. (1987). Effects of odour on the sexual development of young female field voles (*Microtus agrestis*). Ph.D. thesis, University of London.

Bailey, N.T.J. (1981). *Statistical methods in biology*. Hodder and Stoughton, London.

Drickamer, L.C. (1981). Pheromones, social influences and population regulation in rodents. In *Environmental factors in mammal reproduction* (ed. D. Gilmore and B. Cook). Macmillans, London and Basingstoke.

Drickamer, L.C. (1987). Seasonal variations in the effectiveness of urinary chemosignals influencing puberty in female house mice. *J. Reprod. Fert.*, 80, 295–300.

Hogarth, P.J. (1978). *Biology of reproduction*. Blackie, Glasgow and London.

Marchlewska-Koj, A. (1984). Pheromones and mammalian reproduction. In *Oxford review of reproductive biology*, Vol. 6 (ed. J.R. Clarke), pp. 266–302. Clarendon Press, Oxford.

Ridgman, W.J. (1975). *Experimentation in biology: an introduction to design and analysis*. Blackie, Glasgow and London.

Spears, N. and Clarke, J.R. (1986). Effect of male presence and of photoperiod on the sexual maturation of the field vole (*Microtus agrestis*). *J. Reprod. Fert.*, 78, 231–8.

Spears, N. and Clarke, J.R. (1988). Selection in field voles (*Microtus agrestis*) for gonadal growth under short photoperiods. *J. Anim. Ecol.*, 57, 61–70.

Tait, M.J. and Krebs, C. (1985). Population dynamics and cycles. In *Biology of New World Microtus* (ed. R.H. Tamarin), pp. 567–620. Special Publication No. 8, American Society of Mammalogists.

# III.8 The MHC and Individual Odours in Rats

RICHARD E. BROWN, BRUCE ROSER AND PRIM B. SINGH

## Individual recognition and reproductive behaviour

Each animal has an individual body odour or chemical fingerprint (Brown 1979; Halpin 1986) and individual recognition plays an important, but often ignored, role in the reproductive behaviour of mammals. Olfactory signals have both a releasing and priming function in mammalian reproduction (Wilson and Bossert 1963; Macdonald and Brown 1985) and recognition of individuality is influential in each of these functions.

With respect to releaser effects, individual recognition is important for mate selection (Yamazaki *et al.* 1976; Shapiro *et al.* 1986), reduction of agonistic responses in potential mates (Coopersmith and Banks 1983), avoidance of incestuous breeding (Dewsbury 1982), mate recognition (Brown *et al.* 1988) and sexual rearousal in the Coolidge effect (Dewsbury 1981). Likewise, with respect to the primer effects, individual recognition influences the induction of puberty in juvenile females (Lendrem 1985), the pregnancy block effect (Yamazaki *et al.* 1983*a*; for review see Brown 1985, pp. 299–308), and may also be important for induced oestrus synchrony in adult females (McClintock 1983).

Finally, recent interest in the theory of kin selection has stimulated research on the salience of olfactory signals for kin recognition (Blaustein *et al.* 1987; Hepper 1986; Sherman and Holmes 1985; Waldman 1987). Four possible mechanisms have been proposed to account for kin recognition: (i) spatial contiguity; (ii) familiarity or prior association, including association with a common third party such as a mate or mother; (iii) phenotype matching; and (iv) recognition alleles (Blaustein 1983). Except for spatial contiguity, each of these mechanisms depends on individuals and their kin possessing a uniquely identifiable attribute and, in mammals, this attribute is usually an individual body odour. The individual odour of an animal is a phenotypic trait which has been linked to genotypic individuality of the major histocompatibility complex (Beauchamp *et al.* 1986; Boyse *et al.* 1987).

## The Major Histocompatibility Complex (MHC)

The MHC is a linked cluster of genes which code for structural glycoproteins in the membranes of virtually every cell of the body. These membrane molecules

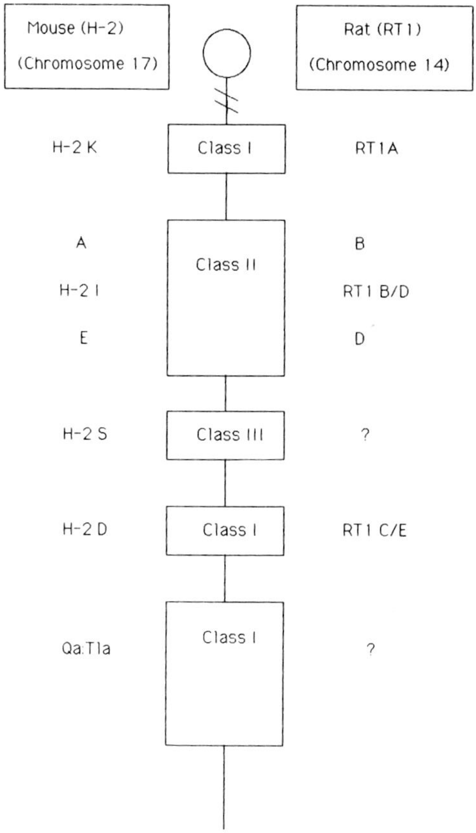


Fig. 1. Comparative maps of the mouse (H-2) and rat (RT1) MHC, showing the strong overall similarity of organization (from Butcher and Howard 1986; Hood *et al.* 1983; Smith *et al.* 1986).

carry the major histocompatibility antigens, which are responsible for the rapid rejection of tissues transplanted from one individual to another in nearly every species so far examined. The MHC genes are of two major types: the class I genes encode a 45 Kd heavy chain which associates non-covalently on all cell surfaces with a 13 Kd light chain ($\beta 2$ microglobulin); the class II genes encode a dimeric molecule of two roughly equal-sized peptide chains expressed predominantly on the surface of certain cells of the immune system (Klein 1986). These antigens constitute the most polymorphic known family of protein molecules, there being up to 100 allelic forms of each of the six class I genes in each diploid cell. In an outbreeding population this polymorphism can result in the expression of more than $3.6 \times 10^9$ different class I phenotypes, thus acting essentially as markers of genetic individuality (Singh *et al.* 1987).

## The MHC and odours of individuality in mice

The mouse MHC, located on chromosome 17, has three different class I loci that have been shown to confer an individual specific odour: H-2K, H-2D, and the Qa:Tla region (Fig. 1). Discrimination between individuals of two congenic inbred strains of mice, which differ by only one MHC locus, indicates the contribution of that region to the individual body odour of the mouse (Table 1).

**Table 1.** Results of mouse odour discrimination experiments using urine from congenic strains

| | | MHC regions | | | | |
|---|---|---|---|---|---|---|
| Mouse strain | H-2 type | Class Region | I K | II I(A/E) | I D | I Qa:Tla |
| 1. Two strains which differ at all H-2 loci[1] | | | | | | |
| C57BL/6J | (H-$2^b$) | | b | b | b | |
| versus | | | | | | |
| B6-H-2k | (H-$2^k$) | | k | k | k | k |
| 2. Two strains which differ only at class IK region[2] | | | | | | |
| B10.A | (H-$2^a$) | | k | k | d | d |
| versus | | | | | | |
| B10.A (18R) | (H-$2^{18r}$) | | b | k | d | d |
| C57BL/6J | (H-$2^b$) | | b | b | b | b |
| versus | | | | | | |
| B6-H-$2^{bm1}$ | (H-$2^{bm1}$) | | *bm1* | b | b | b |
| 3. Two strains which differ only in the Qa:Tla region[3] | | | | | | |
| C57BL/6J | (H-$2^b$) | | b | b | b | b |
| versus | | | | | | |
| B6-*Tla*$^a$ | (H-$2^a$) | | b | b | b | a |

[1] Yamazaki *et al.*, 1979; Yamaguchi *et al.*, 1981
[2] Yamazaki *et al.*, 1983*a,b*
[3] Yamazaki *et al.*, 1982

Mice which differ at all H-2 loci [C57BL/6J versus B6-H-$2^k$)], only in the classic class I H-2K region [B10.A versus B10.A (18R) and C57BL/6J versus B6-H-$2^{bm1}$] and only in the oligomorphic class I Qa:Tla region [C57BL/6J versus B6.*Tla*$^a$] can be discriminated by their urine odours (Yamaguchi *et al.* 1981; Yamazaki *et al.* 1979, 1982, 1983*a*, *b*). Yamazaki *et al.* (this volume) have shown that the class II region of the mouse MHC also influences the production of unique urinary odours. The class III (H-2S) region of the mouse MHC is homogenous across all mice (Klein 1986), so is not likely to be involved in individual odour production.

The X and Y chromosomes also influence the individual odour of mice (Yamazaki *et al.* 1986), but not all regions of the mouse genotype are important for the production of individually unique urine odours. Genes controlling the major urinary proteins (MUPs), for example, do not appear to influence odours (Yamazaki *et al.* 1986; but see Beauchamp *et al.* this volume).

## The MHC and individual recognition in rats

The rat MHC, termed RT1 and located on chromosome 14 (Oikawa *et al.* 1984), has three regions: A, B/D, and C/E (see Fig. 1) which correspond to the mouse H-2 regions K, I (A/E), and D/Qa:Tla (Butcher and Howard 1986). In previous papers (Brown *et al.* 1987*a*; Singh *et al.* 1987), we have shown that differences can be discriminated in urine odours of two congenic strains of rat which differ only in the class IA region (PVG versus PVG.R1).

In this paper, we report our experiments on the contribution of different loci of the rat MHC to the determination of distinct urinary odours. Our original experiments were conducted at Babraham, England, using PVG-$RT1^{u}$ and Lister hooded rats as subjects. This study replicates and extends these results using Wistar albino and Long-Evans hooded rats tested at Dalhousie University.

## Experimental design and procedure

### SUBJECTS

We used 24 male Wistar albino rats and 42 male Long-Evans hooded rats as subjects. The Wistars were born at Dalhousie University to parents purchased from Canadian Hybrid Farms. After weaning at 22 days of age, these rats were housed in littermate groups of four in hanging metal cages measuring 43.5 × 24.5 × 19 cm. They were kept on a reversed 12 h light/12 h dark cycle with lights off at 8:00 a.m. and received Purina lab chow and water *ad libitum*. They were tested for odour preferences starting at 90 days of age. The Long-Evans rats were purchased from Charles River, Canada when 70 days of age and housed in groups of four in hanging cages under the same conditions as the Wistars. They were tested when 90–120 days old.

### URINE DONORS

Urine was collected from 4 adult male PVG, 4 adult male PVG.R1 and 4 adult male PVG-$RT1^{av1}$ rats which had been purchased from OLAC, Bicester,

**Table 2.** Differences in congenic rat strains at each region of the MHC (RT1)

| Congenic strains | RT1 | MHC regions Class I A | Class II B/D | Class I C/E |
|---|---|---|---|---|
| 1. Two strains which differ at all MHC regions | | | | |
| PVG | (P) | *c* | *c* | *c* |
| versus | | | | |
| PVG-$RT1^{av1}$ | (A) | *av1* | *av1* | *av1* |
| 2. Two strains which differ only in the class IA region | | | | |
| PVG | (P) | *c* | *c* | *c* |
| versus | | | | |
| PVG.R1 | (R) | *av1* | *c* | *c* |
| 3. Two strains which differ in all MHC regions except class IA | | | | |
| PVG-$RT1^{av1}$ | (A) | *av1* | *av1* | *av1* |
| versus | | | | |
| PVG.R1 | (R) | *av1* | *c* | *c* |

England, and were housed in groups of 2 to 4 in the Pathology Department, Dalhousie University. Urine was collected from individual males in metabolism cages and frozen at −20 °C until used in the tests.

### APPARATUS AND PROCEDURE

The habituation-dishabituation tests were conducted in arenas similar to those described previously (Brown *et al.* 1987*a*). These were 38.5 × 32 × 17.5 cm opaque plastic animal housing cages covered by a wire mesh top which was 11 cm tall, making the cage 28.5 cm high. During the experiments, a 7-cm circle of filter paper (Whatman no. 1) was taped to the back of the cage top and 0.1 ml liquid was squirted on to the filter paper from a syringe.

In each test, the subjects received nine 2-min. odour presentations: the first three with water on the filter paper; the next three with urine sample 1; and the final three with urine sample 2. The time that the subject spent rearing and sniffing at the urine on the filter paper was recorded with a stopwatch. After each 2-min odour presentation, the top of the cage was removed and replaced with another top containing a new piece of odourized filter paper. The used top was then washed and allowed to dry before being used again.

Subjects participated in three different tests. These examined their ability to discriminate between urine from two individual male rats which differed in:

1. all three RT1 regions (PVG versus PVG-$RT1^{av1}$ donors;
2. only the class IA region (PVG versus PVG.R1);
3. in all MHC regions except in the class IA region (PVG-$RT1^{av1}$ versus PVG.R1).

These comparisons are summarized in Table 2. Odour presentation was counterbalanced within each experiment.

## The habituation-dishabituation test

The habituation-dishabituation test is a simple and powerful technique for determining whether the subject can detect a difference between two odours (Gregg and Thiessen 1981; Sundberg *et al.* 1982). We present water on the filter paper for the first three 2-min trials in order to give the rat time to explore the test arena and become familiar with it. The first urine odour (trial 4) is then a novel stimulus in an environment to which the subject has been exposed for 6 min. We expect each subject to respond to the urine presentation of trial 4 by orienting to the odour, rearing on its hind legs, and sniffing it. If there is no response, the session is terminated and that subject is not used for the experiment. Very few subjects (less than 5%) fail to respond on trial 4. After the rat has had three trials (6 mins) with the first urine, it no longer rears to investigate it, but when a novel urine sample is presented on trial 7, there is a rapid dishabituation of the rearing response. The rat lifts its head, sniffs the air, and then moves toward the new odour and rears to investigate it. If the second odour is not different from the first, the rat does not show this dishabituation. Thus, the time spent investigating the odour on trial 7 is a critical measure of whether or not the two urine odours can be discriminated. A randomized block ANOVA was conducted over the nine odour presentations for each experiment for each strain. Newman-Keuls *post hoc* tests with a criterion of $p = 0.01$ were then performed to investigate differences among the means after a significant F-test.

## Discrimination between individuals of two different strains

When the urine was from individuals of two different strains, whether PVG and PVG-$RT1^{av1}$ (Fig. 2a), PVG and PVG.R1 (Fig. 2b), or PVG-$RT1^{av1}$ and PVG.R1 (Fig. 2c), both Wistar ($n = 8$) and Long-Evans ($n = 14$) subjects showed a significantly different amount of time investigating the odours over all nine presentations (Wistars: all F's $(8, 56) \geqslant 9.14$ $p < 0.001$; Long-Evans: all F's $(8, 104) \geqslant 4.53$ $p < 0.001$). In each of the three tests, both Wistar and Long-Evans subjects spent more time investigating the odours on trials 4 and 7 than on any other presentation ($p < 0.01$), indicating that the first urine presented was discriminated from water (trial 4 versus trial 3) and the second urine was distinguished from the first (trial 7 versus trial 6).

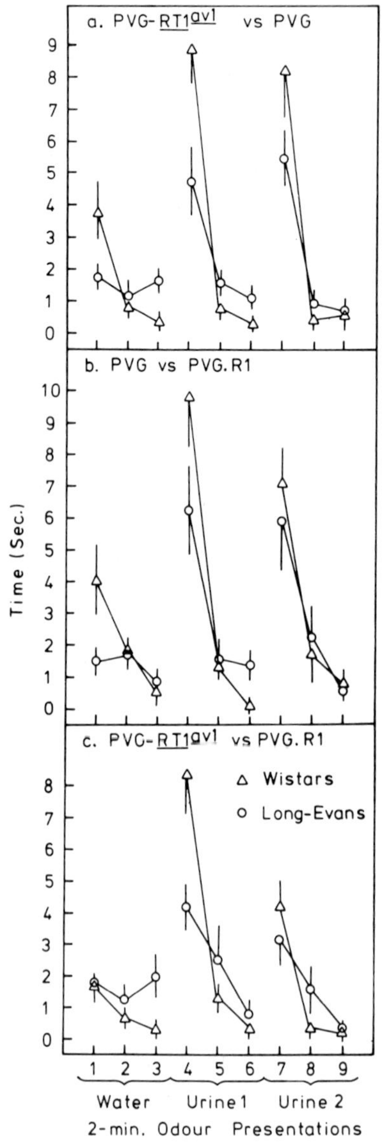


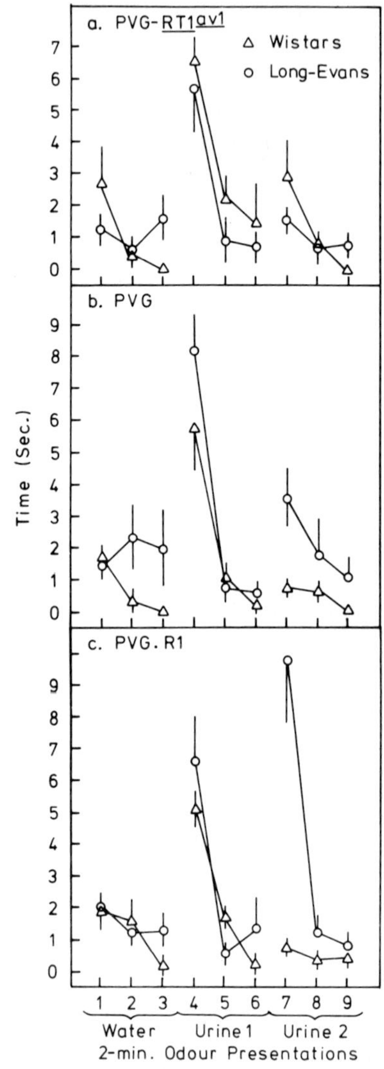


Fig. 2. Mean time (± SEM) spent by male Wistar (△) and Long-Evans (○) subjects rearing and sniffing at urine odours from males of two different strains in each 2-min. odour presentation.

Fig. 3. Mean time (± SEM) spent by male Wistar (△) and Long-Evants (○) subjects rearing and sniffing at urine odours from two different males of the same strain in each 2-min. odour presentation.

Thus, the urine odours from each of the three pairs of congenic strains of rats were discriminable. Those strains which differ in all three MHC regions (PVG-$RT1^{av1}$ and PVG) and in only the classical class I RT1A region (PVG and PVG.R1) are readily discriminable (Figs. 2a and b), since the difference between the investigation time on trials 6 and 7 is as large as that between the first urine presentation and water (trials 3 and 4). Although still significant, the difference between the two strains which are the same at the classical class I RT1A region but differ in the class II B/D and class I C/E regions (PVG-$RT1^{av1}$ and PVG.R1) is much less (Fig. 2c).

## Discrimination between individuals within a strain

Because the urine donors belong to congenic inbred strains produced by extensive backcrossing and selection, each individual is genetically identical to all the others of the same strain as determined by skin grafting and other methods (for review, see Butcher and Howard 1986). Since the individuals are genetically identical, like human identical twins, they should have the same odour (Kalmus 1955). Thus, we would not expect individual males within a strain to have odours that are discriminable. In the second series of experiments, we presented our test subjects with urine from two individual males from each donor strain.

When urine odours were from two different individuals within the PVG-$RT1^{av1}$ (Fig. 3a) or PVG (Fig. 3b) strains, there were significant differences over the nine trials in the time spent investigating odours for both strains of subjects[Wistars($n=4$): all F's$(8,24) \geqslant 4.55, p<0.005$; Long-Evans($n=7$): all F's$(8, 48) \geqslant 4.97$, $p<0.001$]. Both groups of subjects spent significantly more time investigating the odours of trial 4 than for any other presentation in these two tests ($p<0.01$). Trial 7 did not differ from trials 5 and 6, indicating that the urine odours from two individuals within these strains were not discriminable.

When the urine donors were two individual PVG.R1 males (Fig. 3c), there was a significant difference over the nine presentations in the time spent investigating urine odours [Wistars ($n=4$): F $(8, 24)=9.01$, $p<0.001$; Long-Evans ($n=7$): F $(8, 48)=10.82$, $p<0.001$]. For the Wistar males, only the investigation time of trial 4 differed from that of the other odour presentations ($p<0.01$). The Long-Evans subjects showed significant increases in sniffing on both trials 4 and 7, indicating that they could discriminate between the two individual PVG.R1 males by their urine odours (Fig. 3c).

We have examined odours from individuals of four different congenic strains of rats in a number of experiments. For three of these strains, we have never found that two individuals of the same strain can be discriminated by their odours. These results are from four experiments with PVG donors, three experiments with PVG-$RT1^{av1}$ donors, and two experiments with PVG.R19 donors (Brown *et al.* 1987*a*, *b*, 1989; Singh *et al.* 1987, 1988*a*, *b*). With the PVG.R1 males, however, we found a discrimination between individuals in two of four experiments. Originally we concluded that this discrimination could be due to 'individual differences in metabolic products of gonadal steroids or other chemicals in the urine, or to contamination of the urine with faeces, food droppings, hair, etc., during collection in the metabolism cage' (Brown *et al.*1987*a*, p. 68). We no longer think that this explanation is tenable, and would like to suggest that either the PVG.R1 strain may not be as genetically homogenous as the other strains or that a threshold effect is operating. The PVG.R1 strain has less class I antigen than the PVG strain and is more variable (Howard *et al.* 1978). Whatever the explanation, this interesting exception to the general finding is amenable to experimentation.

## Discrimination between two urine samples from the same individual donor

As a final control for the habituation-dishabituation method, we used two separately collected urine samples from the same individual donor for each strain. If the response to the odour habituates completely, then the greatest response should be on trial 4 and subsequent responses should decline over the six urine presentations.

When the two urine samples were from the same individual donors, whether PVG-$RT1^{av1}$ (Fig. 4a), PVG (Fig. 4b), or PVG.R1 (Fig. 4c), there was a significant difference in the time spent investigating the odour over the nine trials for both Wistar ($n=4$; all F's $(8, 24) \geqslant 15.15$, $p<0.001$) and Long-Evans subjects ($n=7$; all F's $(8, 48) \geqslant 3.36$, $p<0.005$) and in each test, for both groups of subjects, only trial 4 differed from the others ($p<0.01$).

Thus, when the six odour presentations (trials 4 to 9) involve urine odours from the same individual, habituation of the rearing-investigation response occurs after three presentations and there is no dishabituation with the second urine sample (Fig. 4). When the urine odours come from two different individuals within the same strain, however, there is often a slight, but non-significant, increase in the investigation time on trial 7 (Fig. 3), indicating that there are some differences in odour between individuals within an inbred strain which do not occur when the odours are from the same individual.

## Each region of the MHC contributes to the individuality of the urine odour

Male rats of both Wistar and Long-Evans strains are thus able to discriminate between urine odours from congenic strains of rats which differ in all three RT1 regions of the MHC, only the class IA region, and all MHC regions except the class IA region. These results replicate those of our previous studies (Brown *et al.* 1987*a*; Singh *et al.* 1987, 1988*a*, *b*) and indicate the robustness of the phenomenon. We have now demonstrated this effect with both inbred and outbred strains of

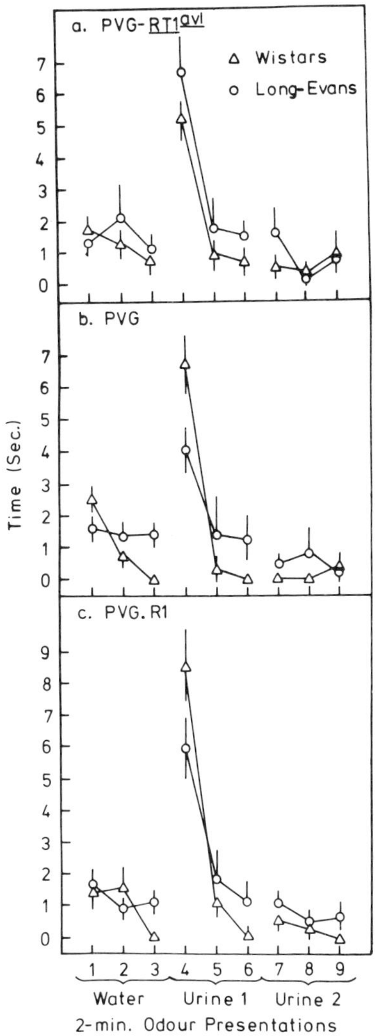


Fig. 4. Mean time (± SEM) spent by male Wistar (△) and Long-Evans (○) subjects rearing and sniffing at two urine samples from the same male of each strain in each 2 min. odour presentation.

subjects in two different laboratories, and with three different experimenters conducting the research.

In addition to these findings, recent research at Babraham with PVG-$RT1^{u}$ subjects (Brown *et al.* 1987*b*, 1989) has demonstrated that congenic rat strains, which differ only in the class I C/E region of the MHC (PVG.R19 versus PVG-$RT1^{av1}$, and strains which differ only in the class II B/D region (PVG.R1 versus PVG.R19) can also be discriminated by their urine odours (Table 3).

**Table 3.** Congenic rat strains which differ in the class I C/E and class II B/D regions*

| Congenic strains | RT1 | MHC region Class I A | Class II B/D | Class I C/E |
|---|---|---|---|---|
| 1. Two strains which differ only in the class I C/E region | | | | |
| PVG.R19 | (R19) | *av1* | *av1* | *c* |
| versus | | | | |
| PVG-$RT1^{av1}$ | (A) | *av1* | *av1* | *av1* |
| 2. Two strains which differ only in the class II B/D region | | | | |
| PVG.R1 | (R) | *av1* | *c* | *c* |
| versus | | | | |
| PVG.R19 | (R19) | *av1* | *av1* | *c* |

* Brown, Singh and Roser, 1987*b*

## What is the source of these urinary odours of individuality?

There are still a number of questions to ask concerning the distinctive urinary odours that allow discrimination to be made between these strains. We would, for example, like to know what chemicals in the urine are discriminable. We originally felt that these might be the class I molecules themselves, but this is not so since the urine from PVG and PVG.R1 males with the class I molecules removed by affinity chromatography can still be discriminated (Brown *et al.* 1987*a*). We now suspect that the odour may be due to a small volatile fragment of the class I molecule, or a volatile secondary metabolite which is associated with a fragment of the class I molecule (Singh *et al.* 1987, 1988*a*, *b*).

We know that the critical chemical is volatile because we removed all of the volatile components from the urine of PVG and PVG.R1 males by bubbling inert nitrogen gas through the urine for 48 h. Test subjects could not discriminate between urines which had the volatile components removed in this fashion (Singh *et al.* 1988*b*). We suspect that the volatile component in the urine has, or is associated with, a fragment of molecular weight between 10 and 43 Kd and we are conducting a series of dialyses to determine the molecular weight more exactly.

Individual differences in dominance status or gonadal hormone levels may provide some degree of odour differences, even though individuals can be discriminated by their odours independently of hormone levels (Brown 1988). It is also possible that each rat has a unique bacterial flora in the gut and its body odour may be influenced by this bacterial population (Howard 1977). We have recently found that congenic strains of rats reared in a germ-free environment cannot be discriminated by their urine odours (Singh *at al.* 1989), indicating an interaction between the MHC antigens and commensal bacteria in the production of the odour of individuality.

## Future directions

How general are these results? Our experiments used urine from congenic inbred and recombinant strains of rats, which are hardly a natural phenomenon. Our results, however, reproduce and extend those reported for mice by Yamazaki and co-workers. The MHC of the rat and mouse are comparable (Fig. 1) and differences in homologous regions of the rat and mouse MHC result in discriminable urine odours in both species. Since all animals have an MHC or its analogue, we postulate that this is a mechanism of early evolutionary significance for identification of conspecifics, self, and kin.

We know that outbred Long-Evans hooded rats can discriminate between individual littermates by their odour (Hopp *et al.* 1985), and urine odours from individual male and female Long-Evans rats can be discriminated using our habituation-dishabituation procedure (Brown 1988). Outbred animals have a very polymorphic MHC, with over $10^9$ different MHC types possible in the rat population as a whole.

Given that recognition in protochordates may be mediated by an MHC-like gene system (Scofield *et al.* 1982) and that a similar mechanism can be used for kin recognition in colonial invertebrates (Grosberg and Quin 1986), it is likely that some MHC-mediated signal is the genetic mechanism controlling individual and kin recognition by odour in most animals. By using radiation chimaeras, congenic recombinant strains, and other gametic manipulations, we can directly test whether excretion of the protein products of the highly polymorphic MHC genes in the urine does indeed provide the basis for a unique individuality signal. The interaction between the MHC antigens and the volatile products of commensural bacteria in the production of the individuality signal must also be examined. Such a signal will be of interest in the study of reproduction and other aspects of social behaviour, including aggression and parental behaviour.

## Acknowledgements

This research was supported by the Natural Sciences and Engineering Research Council of Canada grant A7441 (R.E.B.), an AFRC New Initiatives Grant (P.B.S.), the Medical Research Council of Great Britain (B.R.), an NSERC–Royal Society Collaborative Research Fellowship, and a NATO Collaborative Research Grant. We would like to thank Donna Sutton, Don Scott, Lynda Wisker, and Joe Cox for their assistance. We also thank Dr Stan Stepkowski and Dr Bill Duncan of the Department of Surgery at Dalhousie University for allowing us to collect urine from their PVG congenic rats.

## References

Beauchamp, G., Gilbert, A., Yamazaki, K. and Boyse, E.A. (1986). Genetic basis for individual discriminations: The major histocompatibility complex of the mouse. In *Chemical signals in vertebrates*, 4 (ed. D. Duvall, D. Müller-Schwarze and R.M. Silverstein), pp. 414–22. Plenum, New York.

Blaustein, A.R. (1983). Kin recognition mechanisms: Phenotype matching or recognition alleles? *Amer. Naturalist*, 121, 749–54.

Blaustein, A.R., Bekoff, M. and Daniels, T.J. (1987). Kin recognition in vertebrates (excluding primates): Mechanisms, functions and future research. In *Kin recognition in animals* (ed. D.J.C. Fletcher and C.D. Michener), pp. 333–57. Wiley, New York.

Boyse, E.A., Beauchamp, G.K. and Yamazaki, K. (1987). The genetics of body scent. *Trends Genet.*, 3, 97–102.

Brown, R.E. (1979). Mammalian social odors: A critical review. *Adv. Study Behav.*, 10, 103–62.

Brown, R.E. (1985). The rodents. I: Effects of odours on reproductive physiology (primer effects). In *Social odours in mammals*, Vol. 1, (ed. R.E. Brown and D.W. Macdonald), pp. 254–344. Clarendon Press, Oxford.

Brown, R.E. (1988). Individual odors of rats are discriminable independently of changes in gonadal hormone levels. *Physiol. Behav.*, 43, 359–63.

Brown, R.E., Singh, P.B. and Roser, B. (1987*a*). The major histocompatibility complex and the chemosensory recognition of individuality in rats. *Physiol. Behav.*, 40, 65–73.

Brown, R.E., Singh, P.B. and Roser, B. (1987*b*). Both class I and class II regions of the Major Histocompatibility Complex influence the distinctive urinary odours of congenic rat strains. *Amer. Zoologist*, 27(4), 47A.

Brown, R.E., Hauschild, M., Holman, S.D. and Hutchison, J.B. (1988). Mate recognition by urine odors in the Mongolian gerbil (*Meriones unguiculatus*). *Behav. Neural Biol.*, 49, 174–83.

Brown, R.E., Roser, B. and Singh, P.B. (1989). Class I and class II regions of the major histocompatibility complex both contribute to individual odours in congenic inbred strains of rats. *Behav. Genet.*, 19,

Butcher, G.W. and Howard, J.C. (1986). The MHC of the laboratory rat, *Rattus norvegicus*. In *Handbook of experimental immunology*, Vol. 3, Genetics and molecular immunology (ed. D.M. Weir), pp. 101.1–101.18. Blackwell, Oxford.

Coopersmith, C.B. and Banks, E.M. (1983). Effects of olfactory cues on sexual behavior in the brown lemming, *Lemmus trimuconatus*. *J. Comp. Psychol.*, 97, 120–6.

Dewsbury, D.A. (1981). Effects of novelty on copulatory behaviour: The Coolidge effect and related phenomena. *Psychol. Bull.*, 89, 464–82.

Dewsbury, D.A. (1982). Avoidance of incestuous breeding between siblings in two species of *Peromyscus* mice. *Biol. Behav.*, 7, 157–69.

Gregg, B. and Thiessen, D.D. (1981). A simple method of olfactory discrimination of urines for the Mongolian gerbil, *Meriones unguiculatus*. *Physiol. Behav.*, 26, 1133–6.

Grosberg, R.K. and Quin, J.F. (1986). The genetic control and consequences of kin recognition by the larvae of a colonial marine invertebrate. *Nature*, 322, 456–8.

Halpin, Z.T. (1986). Individual odors among mammals: Origins and functions. *Adv. Study Behav.*, 16, 39–70.

Hepper, P.G. (1986). Kin recognition: Functions and mechanisms. A review. *Biol. Rev.*, 61, 63–93.

Hood, L., Steinmetz, M. and Malissen, B. (1983). Genes of the major histocompatibility complex of the mouse. *Ann. Rev. Immunol.*, 1, 529–68.

Hopp, S.L., Owren, M.J. and Marion, J.R. (1985). Olfactory discrimination of individual littermates in rats (*Rattus norvegicus*). *J. Comp. Psychol.*, 99, 248–51.

Howard, J.C. (1977). H-2 and mating preferences. *Nature*, 266, 406–8.

Howard, J.C., Butcher, G.W., Galfre, G. and Milstein, C. (1978). Monoclonal anti-rat MHC (H-1) alloantibodies. *Curr. Topics Microbiol. Immunol.*, 81, 54–60.

Kalmus, H. (1955). The discrimination by the nose of the dog of individual human odours and in particular the odours of twins. *Br. J. Anim. Behav.*, 3, 25–31.

Klein, J. (1986). *The natural history of the Major Histocompatibility Complex*. Wiley, New York.

Lendrem, D.W. (1985). Kinship affects puberty acceleration in mice (*Mus musculus*). *Behav. Ecol. Sociobiol.*, 17, 397–9.

Macdonald, D.W. and Brown, R.E. (1985). Introduction: The pheromone concept in mammalian chemical communication. In *Social odours in mammals*, Vol. 1, (eds. R.E. Brown and D.W. Macdonald). pp. 1–18. Clarendon Press, Oxford.

McClintock, M.K. (1983). Pheromonal regulation of the ovarian cycle: Enhancement, suppression, and synchrony. In *Pheromones and reproduction in mammals*, (ed. J.G. Vandenbergh), pp. 113–49. Academic Press, New York.

Oikawa, T., Yoshida, M.C., Satoh, H., Yamashina, K., Sasaki, M. and Kobayashi,

H. (1984). Assignment of the rat major histocompatibility complex to chromosome 14. *Cytogenetics and Cell Genetics*, 7, 558ff.

Scofield, V.L., Schumpberger, J.M., West, L.A. and Weissman, I.L. (1982). Protochordate allorecognition is controlled by a MHC-like gene system. *Nature*, 295, 499–502.

Shapiro, L.E., Austin, D., Ward, S.E. and Dewsbury, D.A. (1986). Familiarity and female mate choice in two species of voles (*Microtus ochrogaster* and *Microtus montanus*). *Anim. Behav.*, 34, 90–7.

Sherman, P.W. and Holmes, W.G. (1985). Kin recognition: Issues and evidence. In *Experimental behavioral ecology and sociobiology* (ed. B. Holldobler and M. Lindauer), pp. 437–60. Sinauer, Sunderland, Mass.

Singh, P.B., Brown, R.E. and Roser, B. (1987). MHC antigens in urine as olfactory recognition cues. *Nature*, 327, 161–4.

Singh, P.B., Brown, R.E. and Roser, B. (1988*a*). Class I transplantation antigens in solution in body fluids and in the urine: Individuality signals to the environment. *J. Exp. Med.*, 168, 195–211.

Singh, P.B., Brown, R.E. and Roser, B. (1988*b*). Soluble class I MHC antigens in solution in the body fluids. In *MHC × X: Complex formation and antibody induction* (ed. P. Ivanyi), pp. 226–40. Springer–Verlag, New York.

Singh, P.B., Herbert, J., Roser, B., Arnott, L., Tucker, D.K. and Brown, R.E. (1989). Rearing rats in a germ-free environment eliminates their odours of individuality. Submitted for publication.

Smith, L., Steinmetz, M. and Hood, L. (1986). The major histocompatibility complex of the mouse. In *Handbook of experimental immunology*, Vol. 3, *Genetics and molecular immunology* (ed. D.M. Weir), pp. 87.1–87.10. Blackwell, Oxford.

Sundberg, H., Doving, K., Novikov, S. and Ursin, H. (1982). A method for studying responses in habituation to odors in rats. *Behav. Neural Biol.*, 34, 113–9.

Waldman, B. (1987). Mechanisms of kin recognition. *J. Theor. Biol.*, 128, 159–85.

Wilson, E.O. and Bossert, W.H. (1963). Chemical communication among animals. *Recent Prog. Horm. Res.*, 19, 673–710.

Yamaguchi, M., Yamazaki, K., Beauchamp, G.K., Bard, J., Thomas, L. and Boyse, E.A. (1981). Distinctive urinary odors governed by the major histocompatibility locus of the mouse. *Proc. Natl. Acad. Sci. USA*, 78, 5817–20.

Yamazaki, K., Boyse, E.A., Mike, V., Thaler, H.T., Mathieson, B.J., Abbott, J., Boyse, J., Zayas, Z.A. and Thomas, L. (1976). Control of mating preferences in mice by genes in the major histocompatibility complex. *J. Exp. Med.*, 144, 1324–34.

Yamazaki, K., Yamaguchi, M., Baranoski, L., Bard, J., Boyse, E.A. and Thomas, L. (1979). Recognition among mice: Evidence from the use of a Y-maze differentially scented by congenic mice of different major histocompatibility types. *J. Exp. Med.*, 79, 755–60.

Yamazaki, K. Beauchamp, G.K., Bard, J., Thomas, L. and Boyse, E.A. (1982). Chemosensory recognition of phenotypes determined by the *Tla* and *H-2K* regions of chromosome 17 of the mouse. *Proc. Natl. Acad. Sci. USA*, 79, 7828–31.

Yamazaki, K., Beauchamp, G.K., Wysocki, C.J., Bard, J., Thomas, L. and Boyse, E.A. (1983*a*). Recognition of H-2 types in relation to the blocking of pregnancy in mice. *Science*, 221, 186–8.

Yamazaki, K., Beauchamp, G.K., Egorov, I.K., Bard, J., Thomas, L. and Boyse, E.A. (1983*b*). Sensory distinction between $H\text{-}2^{b}$ and $H\text{-}2^{bm1}$ mutant mice. *Proc. Natl. Acad. Sci. USA*, 80, 5685–8.

Yamazaki, K., Beauchamp, G.K., Matsuzaki, O., Bard, J., Thomas, L. and Boyse, E.A. (1986). Participation of the murine X and Y chromosomes in genetically determined chemosensory identity. *Proc. Natl. Acad. Sci. USA*, 83, 4438–40.

# III.9 Genetic Determination of Individual Mouse Odour

GARY K. BEAUCHAMP, KUNIO YAMAZAKI, HEATHER DUNCAN, JUDITH BARD AND EDWARD A. BOYSE

## Introduction

The Major Histocompatibility Complex of genes (MHC, labelled H-2 in the mouse and HLA in humans), can potentially impart to each individual a unique chemical signature. This set of genes is probably found in all vertebrates, attesting to its conservation during evolution. The MHC of the mouse occupies a segment of chromosome 17 and comprises up to 50 linked genes, but accounts for less than 0.5% of the total mouse genome (Klein 1986).

The genes of the MHC also provide animals with an olfactory signal characteristic of the animal's genotype (Yamazaki *et al.* 1979, 1982, 1984; Yamaguchi *et al.* 1981; Beauchamp *et al.* 1985*a*, 1986, 1988; Boyse *et al.* 1987; Singh *et al.* 1987; Brown *et al.* 1987). As the most variable gene complex known, the MHC is subject to high mutation or gene conversion rates (Pease *et al.* 1983; Weiss *et al.* 1983; Egrov and Egrov 1988). A mutation of a single class I MHC gene (the K gene) renders mice discriminable by odour from non-mutant mice demonstrating that at least one known class I gene contributes to the individuality of odour. If one were to imagine a genetic system that would permit an organism to gauge individuality and genetic relatedness, one with the characteristics of the MHC would be most likely to be selected (Boyse *et al.* 1983, 1989; Beauchamp *et al.* 1986). In a natural population of mice, it is likely that every individual has a unique MHC genotype.

While the MHC is involved in the specification of genetically controlled body odour, other loci also contribute. Strains of mice which are identical throughout the MHC but which differ at many other loci of the genome can be distinguished on the basis of the odour of their urine (unpublished data). More specifically, genes on the sex chromosomes, X and Y, have been shown to influence urine odour (Yamazaki *et al.* 1986). Thus, the MHC in combination with other genes provides the basis for genetically determined individuality of odour.

The specification of individual mouse odour by MHC genes has been demonstrated within three contexts: (i) natural mating preferences; (ii) interruption of pregnancy by novel male stimuli; and, (iii) most extensively, training in a Y maze. In all cases, mouse odours have been detected by other mice and the genetic constitution of the donor has been critical. It is likely that the genotype of the recipient of the olfactory signals is not important, since we have not found any differences in the ability of mice of various genotypes to learn to respond to the odours in the Y maze (unpublished data). More broadly, preliminary studies with rats (Beauchamp *et al.* 1985*b*) and humans (Gilbert *et al.* 1986) have demonstrated that at least some MHC-controlled olfactory distinctions can be made by individuals of different species.

To explore further the ability of another species to discriminate MHC-determined mouse odour and, following this, to begin an evaluation of the relative importance of MHC-determined odour compared with odour determined by other genes throughout the genome, new studies have been conducted on trained rats' responses to mouse odours. First, it is shown that trained rats are capable not only of distinguishing mice differing throughout H-2 (as has been previously reported; Beauchamp *et al.* 1985*b*), but also of discriminating between strains of H-2 congenic mice differing only at the K gene. Second, studies on the relative potency of different genetic regions demonstrate that both H-2 and the remainder of the genome (background) contribute to differences in mouse odour characteristic of inbred strains: the two regions (H-2 and background) each independently make important contributions to odour characteristics.

## General methods

### SUBJECTS

In Experiment 1, two adult female Wistar rats were used as subjects. For Experiments 2 and 3, two male rats (R11 and R31) were of the inbred Wistar-Furth strain and two male rats (R12 and R32) were Harlan Sprague-Dawley strain. All subjects were approximately two months old and weighed 300–350 g at the beginning of the experiment. Rats were housed individually in large tub cages and had access to Purina Rat Lab Chow at all times. During the experiment, rats were maintained on 12–15 ml water per day. After a training session, in which 2–3 ml of water was earned, rats received supplemental water in the home cage.

### DONORS

Inbred and inbred congenic strains of 3- to 12-month-old male mice were used as urine odour sources for Experiment 2 (Table 1). The inbred strains used were C57BL/6 (denoted B6) and AKR, which had been derived by serial brother-to-sister matings over more than 20 generations to the point that each strain is completely homozygous for all alleles. The B6 and AKR strains carry different alleles at many hundreds of loci throughout the genome (Klein 1986). At the MHC, the B6 strain is H-2 type b, or H-$2^b$, while the AKR strain is H-2 type k, or

**Table 1.** Strains of mice used

| Stimuli | Background type | H-2 type |
|---|---|---|
| *Training stimuli* | | |
| C57BL/6 | B6 | b |
| AKR | AKR | k |
| *Test stimuli* | | |
| C57BL/6-H-$2^k$ | B6 | k |
| AKR-H-$2^b$ | AKR | b |

H-$2^k$. Inbred congenic strains are produced by crossing two inbred strains of different H-2 types (e.g. B6 × AKR) and then backcrossing many times to one of these strains, serologically selecting for the donor H-2 type in each generation (as explained in Beauchamp *et al.* 1985*a*). The final congenic strain (e.g. B6-H-$2^k$) is genetically identical to the base strain (e.g. B6) except for both copies of the segment of chromosome 17 bearing the H-2 haplotype introduced from the donor strain (AKR in this example). Any difference between the base strain and its congenic partner strain, if proven to be genetic, must necessarily be due to genes in the H-2 region. Finally, mutant strains (e.g. C57BL/$6^{bm1}$) are strains that differ from the parent strain by a known genetic difference restricted, in the case of the bm1 mutant, to a single gene. The bm1 mutant differs from its parental strain (B6) by only three amino acids of the protein encoded by the K gene. In all other ways, the mice of the two strains are genetically identical.

### APPARATUS

Rats were trained to detect and discriminate odours in a test chamber adapted from Slotnick and Katz (1974), a prototype of which is described in Beauchamp *et al.* (1985*b*). Essentially, compressed filtered air is passed over urine samples and is briefly delivered to a tunnel-shaped test chamber housing the test animal. The test chamber is provisioned with a response bar connected to a touch circuit and a water reservoir for delivery on reinforcement. Details of the apparatus will be published elsewhere (Duncan *et al.*, in preparation).

### ODORANT COLLECTION

Urine was collected from male mice in group metabolism cages. All animals from a cage (3–6) were placed together in a metabolism cage overnight with food and water available *ad libitum*. The following morning the mice were returned to their home cage and aliquots of their urine were pipetted into test tubes for storage. All urine was identified by date and donor cage. Urine that was clearly contaminated by faeces or food particles was discarded. Urine that was not to be used on the day of collection was frozen at −4 °C.

In a discrimination training session, urine from one of two strains of interest was assigned to be the S+ stimulus and bar touches in the presence of this stimulus were reinforced. For each rat, the other strain was designated as S−; bar touches in the presence of this odour were never reinforced. S+ and S− trials alternated in a random manner with the restriction that no more than three of either type would occur consecutively. S+ and S− trials were each drawn from one of four stimulus jars. The individual samples used came from different urine donors each day.

A generalization test was basically a modification of a training session. That is, during a training session in which accuracy to S+ and S− stimuli were being monitored, occasional probe trials were given. Bar touch responses were recorded, but there were no programmed consequences. Ten to twenty trials from each of two test stimuli could be given in one test session.

## Experiment 1

The purpose of this experiment was to determine whether rats could discriminate odours of mice differing only in a single gene within the MHC. To accomplish this, two rats were first trained to discriminate panels of two inbred congenic strains, B6 and B6-H-$2^k$ (see Table 1). Following demonstration of successful generalization to $F_2$ panels, these rats were then trained to discriminate panels of B6 vs. bm1 mutant mice.

### RESULTS AND DISCUSSION

The results of this experiment are straightforward. Both rats learned to discriminate urines of mutant mice from non-mutant mice and this discrimination was evident during critical generalization trials. Thus, trained rats can discriminate between very small genetic differences in two mouse strains, and in this regard do not differ from trained mice. Consequently, for the following experiments the relative potency was assessed of H-2 vs. background genetic differences in urine odour.

## Experiments 2 and 3

One way of assessing the salience or importance of an odour difference is to examine the ease with which the differences can be used as discriminative cues in a learning task. The number of training trials to reach criterion performance is one such measure. We have found that mice trained in a Y maze to distinguish between odours of mice differing only at H-2, learn more rapidly than they learn discrimination based on some other specific genetic differences, such as those on the X and Y chromosomes. H-2 differences are therefore presumed to be more salient than differences derived from any of the other genetic variations tested. In unpublished studies, it was found that the number of trials required to reach criterion peformance was about equal, when the odour source mice differed at H-2, to the number required when the odour source mice were identical at H-2 but differed at many other loci throughout the genome. Without a direct comparison of the relative potency of the H-2 and the rest of the genome (the background) it is premature, however, to conclude that these two regions are equally potent. To study this issue, a new method was developed to compare the salience of genetically controlled odorants (Duncan *et al.*, in preparation). In this method, rats were trained to discriminate odours of panels of mice differing both at H-2 and at many other loci throughout the genome. The trained animals were then given generalization or probe trials with test stimuli, selected to compare the relative importance of genetic variation at H-2 with variation in the sum total of all other genetic regions differentiating the inbred strains.

## Experiment 2

### METHODS

Rats received preliminary training and discrimination training. Rats R11 and R12 were given samples of urine from B6(H-$2^b$) mice as S+ stimuli, and samples of AKR(H-$2^k$) mouse urine as S− stimuli. Rats R31 and R32 were trained with the opposite configuration of S+ and S− stimuli. Test stimuli for all four animals came from the two congenic strains, B6-H-$2^k$ and AKR-H-$2^b$. Consequently, urine donor mice either (a) carried the same background as the S+ strain, but differed at H-2 (i.e. were congenic to the S+ strain); or (b) carried the same H-2 of the test strain but with different background (i.e. test stimulus strains were congenic to the S− strain; see Table 1). If generalization was more difficult with urines from mice of identical H-2 types but different genetic backgrounds than with urines from mice whose backgrounds were identical but differed at H-2, it would suggest that the urinary components controlled by H-2 were more potent than those controlled by the multitude of other varying genes, and vice versa.

### RESULTS AND DISCUSSION

All subjects learned to discriminate between the AKR and B6 strains. As in Experiment 1, strain discrimination was proven by the success of generalization trials. The critical test in Experiment 2 was of H-2 versus background in controlling response to urine odours. Figure 1 presents the per cent response to the test stimuli for each animal. As a comparison, responses to the training stimuli in these test sessions is also included. It is evident that with training performance intact, some generalization from S+ to test stimuli occurs, but only for two animals.

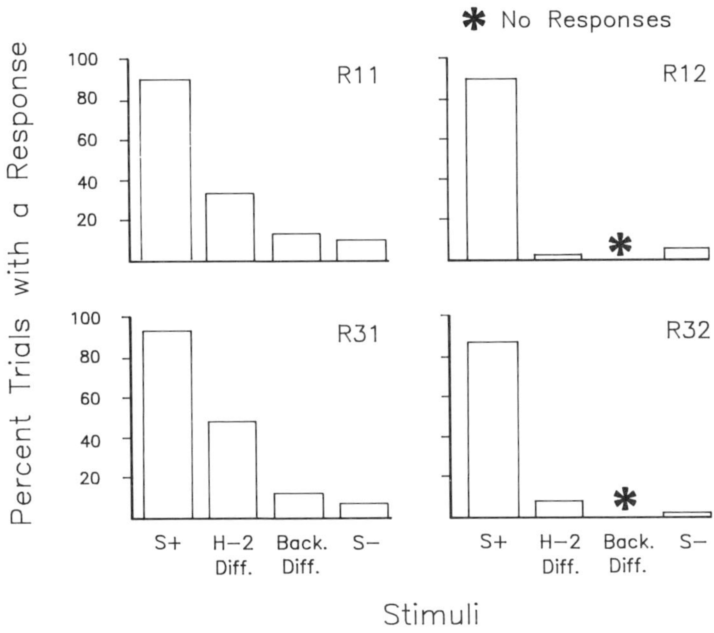


Fig. 1. Per cent olfactometer response for four rats (R11, R12, R31, and R32) to S + urine (far left bar), S − urine (far right bar), and two generalization urine odours, one of varying H-2 while keeping background constant (second from left), the other of varying background while keeping H-2 constant (third from left).

For these two rats, R11 and R31, similarity of genomic background between training and test stimuli elicited more responding than similarity of H-2 alone. These data suggest that for these two animals, the odour contribution of the H-2 complex is less significant to the overall odour profile than background genetic difference. Data from rats R12 and R32, on the other hand, suggest an equal contribution of H-2 and background to the odour profile. When stimulus mice differed from S + mice by only the H-2 region, the resulting difference in urine odour cues was enough to almost completely suppress responses.

Clearly, both H-2 and background contribute to the stimulus animal's urine odour. It is not possible, however, to determine which of the two areas was most salient since there was no overall significant difference in their potency to inhibit responding. What is needed is a method of confronting the trained animals with graded variations of H-2 vs. background in order to evaluate further the relative contribution of the two odour sources. Experiment 3 was directed at testing stimuli that differed from S + and S − stimuli in less dramatic ways than urines from the congenic strains.

## Experiment 3

### METHODS

Experiment 3 was conducted with the same trained rats and urine donor mice used in Experiment 2. All apparatus and training and testing procedures were the same. Urine was collected from panels of mice from each of the four strains listed in Table 1, until approximately 10 ml from each congenic strain and 50 ml from each of the two base strains (S+ and S−) had been obtained. Graded stimuli were then created by mixing the S+ stimulus urine with that of one of the congenic strains in the ratio 1:3, 1:1, or 3:1. For each of the two S+ stimuli (AKR or B6) two series were created; the adulterating urine either came from mice with the same genetic background but a different H-2 (Table 2, i–iii), or the mice carried the same H-2, but differed elsewhere in the genome (Table 2, iv–vi).

**Table 2.** Mixtures of test urine used in experiment 3

| | S+urine[a] | H-2 difference only[a] | Background difference only[a] |
|---|---|---|---|
| i | 75% | 25% | — |
| ii | 50% | 50% | — |
| iii | 25% | 75% | — |
| iv | 75% | — | 25% |
| v | 50% | — | 50% |
| vi | 25% | — | 75% |

[a] For two trained rats S+ was B6 (H-$2^b$), the urine donors (H-2 difference) were B6-H-$2^k$ and the urine donors (background difference) were AKR-H-$2^b$; for the other trained rats S+ was AKR (H-$2^k$), and urine donors were AKR-H-$2^b$ and B6-H-$2^k$ respectively.

### RESULTS AND DISCUSSION

Results for the four test animals are shown in Fig. 2. Adulteration of the S+ stimuli both with urine from mice with differences at H-2 only or with urine from mice differing in background only caused a decrease in response. For two of the rats (R11 and R12), the change in response was generally greater following changes in background compared with comparable changes in H-2. The reverse was the case for one rat (R32) while the fourth rat (R31) demonstrated essentially equal responses to adulteration. The values of 100% dilution in Fig. 2 are taken from Experiment 2.

These data demonstrate that alteration of the S+ stimuli either by adulteration with urine from strains differing only in background or only at H-2 alters the odour characteristics of the mixtures, as evidenced by suppression of bar pressing

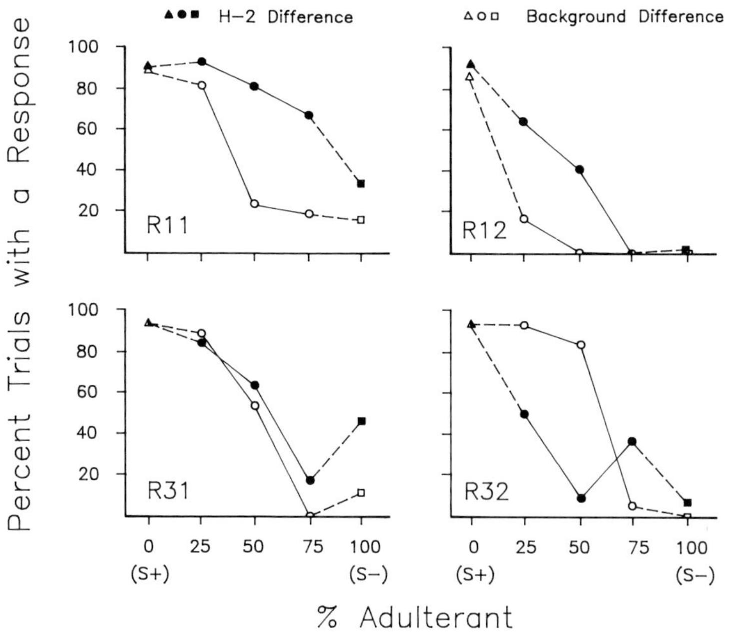


Fig. 2. Per cent response for same four rats as in Fig. 1. For each rat, the urine was diluted with either H-2 signals only (filled circles) or background signals only (open circles). The responses for 0% are taken from S− training (non-generalization) sessions while the responses for 100% come from the previous experiment, S+.

during test trials. The data do not permit one to conclude that either H-2 or background is more salient in the specification of strain-specific urine odour, at least as monitored by trained rats. We conclude that both H-2 differences and other differences in non-H-2 genes are implicated in constitution of individual odour phenotypes in inbred mouse strains, with neither being decisively more potent than the other when rats are used as the odour sensors.

## Conclusions

Our previous studies, as well as those of others, have shown that genes within the MHC are involved in the control of individual odour. We have also demonstrated that genes located on X and Y chromosomes independently control some aspects of olfactory individuality. The current studies attempted to test the relative potency of the genetic background with that of differences at H-2, of which the latter constitutes less than 0.5% of the genome. The results indicate that the sum total of non-H-2 genetic differences are at least as potent as differences at H-2. There are two major implications of these results.

First, the relatively small H-2 region is remarkably potent in its effect on characteristic urine odour. Such potency is likely to reflect the extreme importance of this region in genetically determined chemosensory individuality. This is in keeping with our hypotheses concerning the importance of maintenance of the great individual variation at this gene complex (see Beauchamp *et al.* 1985*a*; Boyse *et al.* 1987). Second, H-2 is not the single control of genetically determined odour. It is not known, however, whether there are one or a few scattered potent loci which involve themselves in individuality of odour or whether many marginally effective loci are responsible. There are in fact many genetic differences between the strains. An informal survey of 118 loci coding for various factors selected from throughout the mouse genome shows that the B6 and AKR strains carry same alleles at about 54% of the loci (Duncan *et. al.*, in preparation; Roderick 1982).

A genetic basis for individual and kin recognition provides scope for altruistic behaviour and assortive mating (see Trivers 1972; Blaustein 1983; Waldman 1987). Our studies have demonstrated that the latter is based in part on H-2 controlled genetic differences (Yamazaki *et al.* 1976), with male mate choice being determined by familial imprinting (Yamazaki *et al.* 1988). Here we have shown that non-H-2 genes contribute to odour but whether these genes are also involved in control of mating preferences remains to be investigated.

## Acknowledgements

This work was supported by NIH grants GMCA 32096 and NS 22623.

## References

Beauchamp, G.K., Yamazaki, K. and Boyse, E.A. (1985*a*). The chemosensory recognition of genetic individuality. *Sci. Amer.*, 253, 86–92.

Beauchamp, G.K., Yamazaki, K., Wysocki, C.J., Slotnick, B.M., Thomas, L. and Boyse, E.A. (1985*b*). Chemosensory recognition of mouse major histocompatibility types by another species. *Proc. Natl. Acad. Sci. USA*, 82, 4186–88.

Beauchamp, G.K., Gilbert, A.N., Yamazaki, K. and Boyse, E.A. (1986). Genetic basis for individual discriminations: The major histocompatibility complex of the mouse. In *Chemical signals in vertebrates*, 4 (ed. D. Duvall, D. Müller-Schwarze and R.M. Silverstein), pp. 413–22. Plenum, New York.

Beauchamp, G.K., Yamazaki, K., Bard, J. and Boyse, E.A. (1988). Preweaning experience in the control of mating preferences by genes in the major histocompatibility complex of the mouse. *Behav. Genet.*, 18, 537–47.

Blaustein, A.R. (1983). Kin recognition mechanisms: Phenotypic matching or recognition in alleles. *Amer. Nat.*, 121, 749–54.

Boyse, E.A., Beauchamp, G.K. and Yamazaki, K. (1983). The sensory perception of genotypic polymorphism of the major histocompatibility complex and other genes. Some physiological and phylogenetic implications. *Hum. Immunol.*, 6, 177–83.

Boyse, E.A., Beauchamp, G.K. and Yamazaki, K. (1987). The genetics of body scent. *Trends Genet.*, 3, 97–102.

Boyse, E.A., Beauchamp, G.K., Yamazaki, K. and Bard, J. (1989). Genetic components of kin recognition. In *Kin recognition* (ed. P.G. Hepper), Cambridge U. Press (in press).

Brown, R.E., Singh, P.B. and Roser, B. (1987). The major histocompatibility complex and the chemosensory recognition of individuality in rats. *Physiol. Behav.*, 40, 65–73.

Egrov, I.K. and Egrov, O.S. (1988). Detection of new MHC mutations in mice by skin grafting, tumor transplantation and monoclonal antibodies: A comparison. *Genetics*, 118, 287–98.

Gilbert, A.N., Yamazaki, K., Beauchamp, G.K. and Thomas, L. (1986). Human olfactory discrimination of the mouse (*Mus musculus*) strains and major histocompatibility types. *J. Comp. Psychol.*, 100, 262–5.

Klein, J. (1986). *Natural history of the major histocompatibility complex.* Wiley, New York.

Pease, R., Schulze, D.H., Pfaffenbach, G.M. and Nathenson, S.G. (1983). Spontaneous H-2 mutants provide evidence that a copy mechanism analogous to gene conversion generates polymorphism in the major histocompatibility complex. *Proc. Natl. Acad. Sci. USA*, 80, 242–6.

Roderick, T.H. (1982). *Handbook on genetically standardized mice*, pp. 2.21–2.48. The Jackson Laboratory, ME.

Singh, P.B., Brown, R.E. and Roser, B. (1987). MHC antigens in urine as olfactory recognition cues. *Nature*, 327, 161–4.

Slotnick, B.M. and Katz, B.J. (1974). Olfactory stimulus control evaluated in a small animal olfactometer. *Percept. Mot. Skills*, 39, 583–97.

Trivers, R.L. (1972). Parental investment and sexual selection. In *Sexual selection and the descent of man* (ed. B. Campbell), pp. 136–79. Chicago University Press, Chicago.

Waldman, B. (1987). Mechanism of kin recognition. *J. Theor. Biol.*, 128, 159–85.

Weiss, E.H., Mellor, A., Golden, L., Fahrner, K., Simpson, E., Hurst, J. and Flavell, R.A. (1983). The structure of a mutant H-2 gene suggests conversion-like events. *Nature*, 301, 671–4.

Yamaguchi, M., Yamazaki, K., Beauchamp, G.K., Bard, J., Thomas, L. and Boyse, E.A. (1981). Distinctive urinary odors governed by major histocompatibility locus of the mouse. *Proc. Natl. Acad. Sci. USA*, 78, 5817–20.

Yamazaki, K., Boyse, E.A., Mike, V., Thaler, H.T., Mathieson, B.J., Abbott, J., Boyse, J., Zayas, Z.A. and Thomas, L. (1976). Control of mating preferences in mice by genes in the major histocompatibility complex. *J. Exp. Med.*, 144, 1324–35.

Yamazaki, K., Yamaguchi, M., Baranoski, L., Bard, L., Boyse, E.A. and Thomas,

L. (1979). Recognition among mice: Evidence from the use of a Y maze differentially scented by congenic mice of different major histocompatability types. *J. Exp. Med.*, 150, 755–60.

Yamazaki, K., Beauchamp, G.K., Bard, J., Thomas, L. and Boyse, E.A. (1982). Chemosensory recognition of phenotypes determined by the Tla and H-2K regions of chromosome 17 of the mouse. *Proc. Natl. Acad. Sci. USA*, 79, 7828–31.

Yamazaki, K., Beauchamp, G.K., Thomas, L. and Boyse, E.A. (1984). Chemosensory identity of H-2 heterozygotes. *J. Mol. Cell Immunol.*, 1, 79–82.

Yamazaki, K., Beauchamp, G.K., Matsuzaki, O., Bard, J., Thomas, L. and Boyse, E.A. (1986). Participation of the murine X and Y chromosomes in genetically determined chemosensory identity. *Proc. Natl. Acad. Sci. USA*, 83, 4437–40.

Yamazaki, K., Beauchamp, G.K., Kupniewski, D., Bard, J., Thomas, L. and Boyse, E.A. (1988). Familial imprinting determines H-2 selective mating preferences. *Science*, 240, 1331–2.

# III.10 Single MHC Gene Mutations Alter Urine Odour Constitution in Mice

KUNIO YAMAZAKI, GARY K. BEAUCHAMP, JUDITH BARD AND EDWARD A. BOYSE

## Introduction

The Major Histocompatibility Complex of genes (MHC) encompasses a region of chromosome 17 in the mouse containing a large number of closely linked genes, many of which control graft rejection and other immune responses (Klein 1986). We previously found that differences in MHC genes affect the mating choices of mice (Yamazaki *et al.* 1976, 1978; Andrews and Boyse 1978; Yamaguchi *et al.* 1978), and that mice can be trained to distinguish between arms of a Y-maze scented by odours from mice differing only in this genetic region (Yamazaki *et al.* 1979, 1982; Yamaguchi *et al.* 1981). We also showed a raised incidence of blockage of pregnancy or pseudopregnancy in mated females exposed to the scent of alien males whose MHC type differed from that of the mate (Yamazaki *et al.* 1983). As no other known gene complex exhibits such vast diversity as the MHC, its potential as a major source for genetically based signals of individual identity is great.

It was decided to evaluate whether very small differences in the genetic make-up of mouse strains are responsible for individual variation in chemosensory identity, and to evaluate the role of specific genes. We have therefore studied the ability of trained mice to distinguish the odour of non-mutant strains from that of congenic mutant strains differing only by mutations of Class I or Class II MHC genes.

As in previous studies of the sensory discrimination of MHC phenotypes, mice were trained by reward in a Y maze to distinguish the odours of urine samples from donors with various mutations (see Table 1). Successful training was confirmed by tests of transfer of training (generalization) using coded samples of urine from genetically equivalent donor mice which the trained mice had not previously encountered (for a description and rationale of this technique, see Yamaguchi *et al.* 1981).

## Participation of class I genes in individual odour constitution

H-2K belongs to one group of MHC genes (Class I), which play a vital part in immunological recognition and response. H-2K is one of the most polymorphic genes known. More than 50 H-2K alleles or mutants have been identified, and for some of these the complete amino acid sequence of the protein component of the H-2K cell surface glycoprotein (the 45 kDa class I heavy-chain coded by H-2K) is known. Distinctions of B6 [C57BL/6 (H-$2^b$)] from the H-2K mutants, bm1, bm5, bm7, and bm8 (Table 1) were studied in the first series of tests. Fourteen mice were used in this series; seven were trained to choose the odour of the inbred strain (B6) over the mutant mice and the other seven were trained to make the reverse choice.

Cosegregation of odour phenotype with H-$2^b$ and H-$2^{bm1}$ was demonstrated by successful generalization to typed H-$2^b$ and H-$2^{bm1}$ homozygous segregants of $F_2$ generations of appropriate crosses. Distinction of the other H-2K mutant strains from the inbred B6 strain was also demonstrated by training and generalization tests. Thus mice could be trained to discriminate B6 vs. bm1, B6 vs. bm5, B6 vs. bm7 and B6 vs. bm8 (Fig. 1). Fig. 2 shows that animals trained to discriminate B6 from bm5 generalized this training successfully to B6 vs. bm8 and B6 vs. bm1. However, as also shown in Fig. 2, there was no evidence of discrimination between two mutants (Fig. 2, bm5 vs. bm8), even though it is known that these mutants differ at several DNA base pairs.

Based on these limited data, non-mutants appear to have an odour distinguishing them from K mutant mice; however, K mutants may not be mutually discriminable.

### SENSORY DISTINCTION OF CLASS II GENES IN ODOUR CONSTITUTION

The Class II genes occupy the I region between H-2K and H-2D and code for the protein moiety of the two glycoprotein chains (alpha and beta) that compose the cell surface molecule. The I-A mutant congenic mouse strain B6.C-H-$2^{bm12}$ (bm12) has been used to probe several structure-function correlates of immune recognition of Ia molecules.

Successful olfactory discrimination of B6 vs. bm12 by trained mice has been accomplished. Four mice were used: two trained to respond to B6 over bm12 and two trained to make the opposite discrimination. Discrimination was confirmed by generalization tests (Fig. 1, far right).

In sum, the data demonstrate that both Class I and Class II MHC genes are involved in individual odour constitution. The odour phenotype is thus a composite determined by several, perhaps many, genes within the MHC and, as demonstrated elsewhere, is also dependent upon non-MHC genes (see Beauchamp *et al.*, this volume). Genetic determination of olfactory individuality provides a basis for kin recognition and assortative mating, of which the latter is known to be influenced by genes within H-2 (Yamazaki *et al.* 1976, 1988).

## Acknowledgement

This work was supported by NIH grants GMCA 32096 and NS 22623.

**Table 1.** H-2 mutants[a] used in Y-maze testing

| Strain | Mutated H-2 locus | Amino acid interchange[b] |
|---|---|---|
| B6.C-H-$2^{bm1}$ | K | 152 (Glu→Ala), 155 (Arg→Tyr), 156 (Leu →Tyr) |
| C57BL/6-H-$2^{bm5}$ | K | 116 (Tyr→Phe) |
| B6.C-H-$2^{bm7}$ | K | 116 (Tyr→Phe), 121 (Cys→Arg) |
| C57BL/6-H-$2^{bm8}$ | K | 22 (Tyr→Phe), 23 (Met→Ile), 24 (Glu→Ser) |
| B6.C-H-$2^{bm12}$ | $IA_{\beta}$ | |

[a] Data from Nathenson *et al.* 1986.
[b] The numbers represent the amino acid coded for by DNA of the K gene (bm1–bm8). For example, the bm1 mutant differs from its B6 parent strain by only 3 amino acids (152, 155, 156) out of almost 350 which make up the protein coded for by K. B6 and bm1 strains are genetically identical at all other loci throughout the genome.

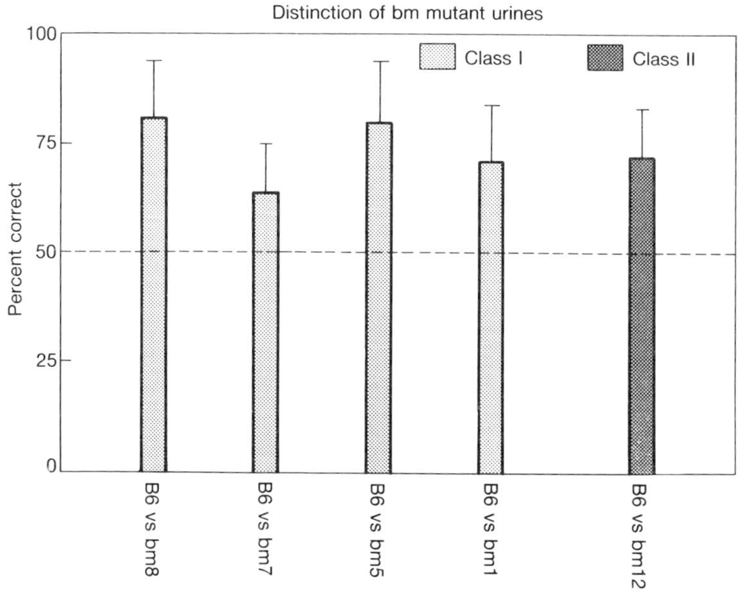


Fig. 1. Results of generalization trials. Mice were trained to discriminate B6 from each mutant (bm1, bm5, bm7, bm8, and bm12) and given generalization trials with novel urine donors of identical MHC types. The bars represent 95% confidence intervals.

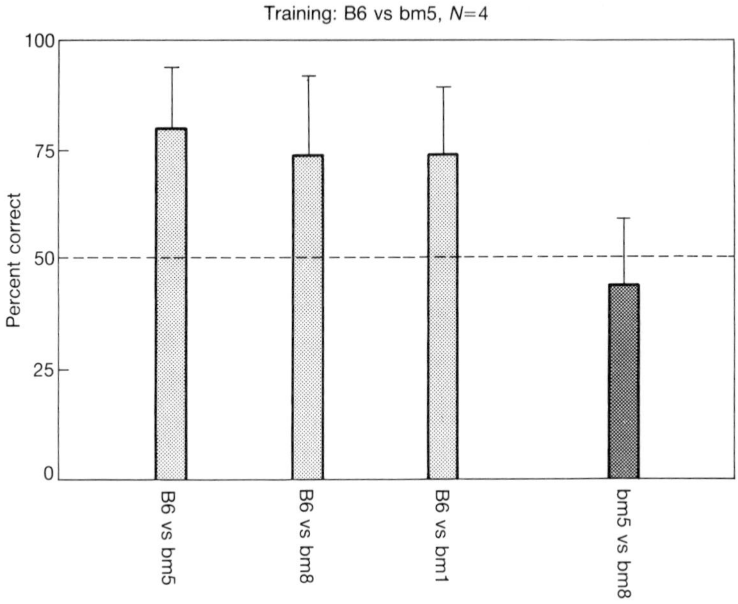


Fig. 2. Same as Fig. 1 except mice were trained to discriminate B6 vs. bm5. While there was successful generalization to B6 vs bm8 and B6 vs. bm1, no discrimination was evident in the bm5 vs bm8 generalization tests (far right).

## References

Andrews, P.W. and Boyse, E.A. (1978). Mapping of an H-2 linked gene that influences mating preference in mice. *Immunogenetics*, 6, 265–8.

Klein, J. (1986). *Natural history of the major histocompatibility complex* John Wiley, New York.

Nathenson, S.G., Geliebter, J., Pfaffenbach, G.M. and Zeff, R.A. (1986). Murine major histocompatibility complex class I mutants: molecular analysis and structure-function implications. *Annu. Rev. Immunol.*, 4, 471–502.

Yamaguchi, M., Yamazaki, K. and Boyse, E.A. (1978). Mating preference tests with the recombinant congenic strain BALB.HTG. *Immunogenetics*, 6, 261–4.

Yamaguchi, M., Yamazaki, K., Beauchamp, G.K., Bard, J., Thomas, L. and Boyse, E.A. (1981). Distinctive urinary odors governed by major histocompatibility locus of the mouse. *Proc. Natl. Acad. Sci. USA*, 78, 5817–20.

Yamazaki, K., Boyse, E.A., Mike, V., Thaler, H.T., Mathieson, B.J., Abbott, J., Boyse, J., Zayas, Z.A. and Thomas, L. (1976). Control of mating preferences in mice by genes in the major histocompatibility complex. *J. Exp. Med.*, 144, 1324–35.

Yamazaki, K., Yamaguchi, M., Andrews, P.W., Peake, B. and Boyse, E.A. (1978). Mating preferences of $F_2$ segregants of crosses between MHC-congenic mouse strains. *Immunogenetics*, 6, 253–9.

Yamazaki, K., Yamaguchi, M., Baranoski, L., Bard, L., Boyse, E.A. and Thomas, L. (1979). Recognition among mice: Evidence from the use of a Y maze differentially scented by congenic mice of different major histocompatibility types. *J. Exp. Med.*, 150, 755–60.

Yamazaki, K., Beauchamp, G.K., Bard, J., Thomas, L. and Boyse, E.A. (1982). Chemosensory recognition of phenotypes determined by the Tla and H-2K regions of chromosome 17 of the mouse. *Proc. Natl. Acad. Sci. USA*, 79, 7828–31.

Yamazaki, K., Beauchamp, G.K., Wysocki, C.J., Bard, J., Thomas, L. and Boyse, E.A. (1983). Recognition of types in relation to the blocking of pregnancy in mice. *Science*, 221, 186–8.

Yamazaki, K., Beauchamp, G.K., Kupniewski, D., Bard, J., Thomas, L. and Boyse, E.A. (1988). Familial imprinting determines H-2 selective mating preferences. *Science*, 240, 1331–2.

# III.11 Exposure Early in Life to Urine from Mice with Translocation Chromosomes and Isolated Mice has Long Term Effects on Development and Behaviour

J.J. COWLEY, R.K. PEWTRESS, AND S. McDONALD

## Introduction

The urine of adult male and female mice has pheromonal properties which change the timing of onset of puberty (Vandenbergh 1967, 1969) and the movements of mice in an open-field test and frequency of defaecation (Fullerton and Cowley 1971; see also review, Cowley 1980). There is a growing realization that the donor's reproductive state and the size and social composition of the group may also be instrumental in accelerating or delaying the age of puberty and subsequently the behaviour of the mice. Thus, the exposure of infant and juvenile female mice to the urine of adult female conspecifics delays the onset of puberty (Cowley and Wise 1972; McIntosh and Drickamer 1977), while the urine of lactating mice and those in oestrus has an accelerating effect (Drickamer and Hoover 1979; Drickamer 1986). Furthermore, the length of time that the donor females have been housed together and the number caged together have been implicated in timing of the onset of puberty in recipients exposed to their urine (Massey and Vandenbergh 1980; Drickamer 1983; Coppola and Vandenbergh 1985); the age of the donors does not seem to be a factor involved (Drickamer 1974). Adrenalectomy (Drickamer and McIntosh 1980) abolishes the action of the urine from female donors in delaying the onset of first oestrus; ovariectomy (Drickamer *et al.* 1978) does not. Several metabolites have been identified which, when added to the urine of adrenalectomized mice, are able to restore the delaying effect on puberty (Novotny *et al.* 1986).

Elsewhere we have suggested that the behavioural characteristics of donor mice, which presumably reflect their hormonal and metabolic state, may modify, through the action of their urine, the behaviour of recipient mice in a direction in keeping with that of the donors' behaviour. Infant female recipient mice exposed, for example, to the urine of lactating donors with low activity scores, as tested in an arena, likewise gave low scores in a similar test situation when adult. In contrast, infant female mice exposed to urine voided by highly active donors were themselves highly active when tested. Exposure to the urine from mice in the mid-activity range was subsequently associated with recipients occupying a position between the other two treatments (Cowley and Smale 1986).

This paper describes two experiments. In Experiment 1 we report on the effects of exposing infant female recipient mice to the urine of male donors with known chromosonal anomalies (see also Cowley *et al.* 1988). In the second experiment (Experiment 2) we describe for the first time the effects of exposing infant female mice to the urine of female donors that have been housed in isolation and hence would be expected to have altered adrenal secretions. In both experiments all the recipient mice are of the same TO (Theiler Original) strain; the effects of exposure to the urine are examined in relation to the onset of puberty and their frequency of movements and defaecation in an open arena.

## Experiment 1

The interaction between genetic and pheromonal factors in modifying the onset of first oestrus and behaviour has received little attention, although Drickamer (1981) has selected mice, over successive generations, for early and late onset of first oestrus. After six generations the lines selected failed to respond further to an acceleration or delay of first oestrus when exposed to the urine of male and female conspecifics. The donors used, in this experiment, were Robertsonian stock [Rb9(9,16)Rma and Rb(11,16)2H] mice carrying spontaneously occurring fusion chromosomes (translocation chromosomes) which can be transferred, through mating, to laboratory strains. The stock used are of clinical interest in that, when bred, the hybrids can be mated with normal mice to produce trisomic conceptuses that have many of the morphological and biochemical characteristics found in Down's syndrome (Epstein 1985; Epstein *et al.* 1985; Searle 1985).

The question we have asked, is whether exposing normal female infant TO strain mice to the urine of adult males from the Robertsonian stock would modify the age of onset of puberty and change their movements and frequency of defaecation. The method used is illustrated in Fig. 1, which also provides information about the Robertsonian stock used in the experiment and details of sample size. The urine from male donors was diluted (1:5 with water) and applied gently to the dorsal region of the nose of the infant recipients from 1 to 21 days of age using a fine artist's paintbrush. The mice, after being separated from their mothers, were monitored daily, and from the time of vaginal opening vaginal smears were taken over 10 successive days. Movement in an open arena was selected as a behavioural index on the basis of our experience with the test in a number of different pheromonal experiments (Cowley and Smale 1986; Cowley and Pewtress 1986).

The automated test has high reliability; although arena activity may have a strong genetic component similar to that reported for the open-field test, this does

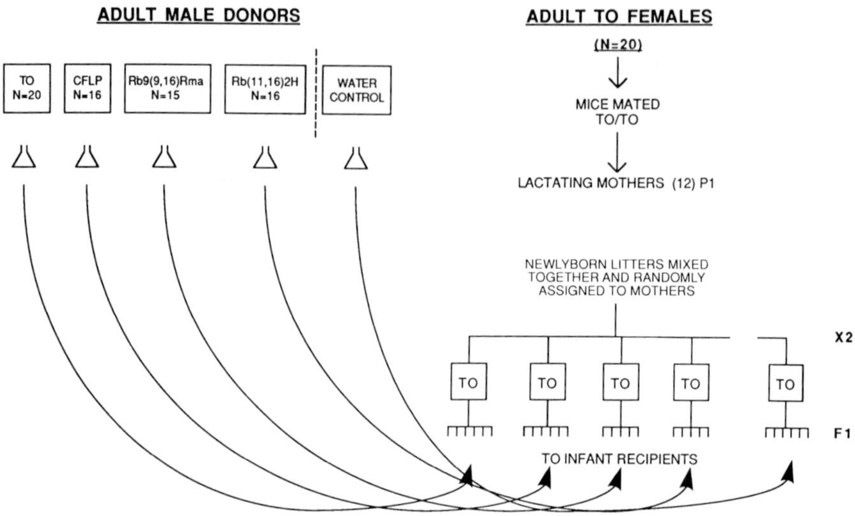


Fig. 1. History and composition of sample. Exposure of recipient mice to urine from laboratory strains and mice with translocated chromosomes.

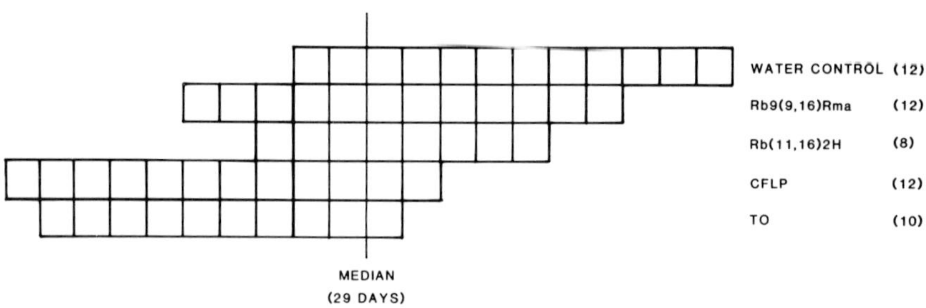


Fig. 2. The number of mice attaining first oestrus have been distributed about the median age of onset for the whole sample (29 days). Each square represents one mouse; the age at which the individual mice reach puberty is not shown.

not preclude the test from having sufficient sensitivity to record changes associated with the modification of independent variables, such as exposure to pheromones. The mice were tested in the open arena for four successive days (2 mins/day) when 63–73 days of age. In recording movements about the open arena two levels of sensitivity were used ('4' and '7'). The former describes more forceful movements including hopping and startled responses, while the latter describes more general exploratory movements on a horizontal plane similar to those that

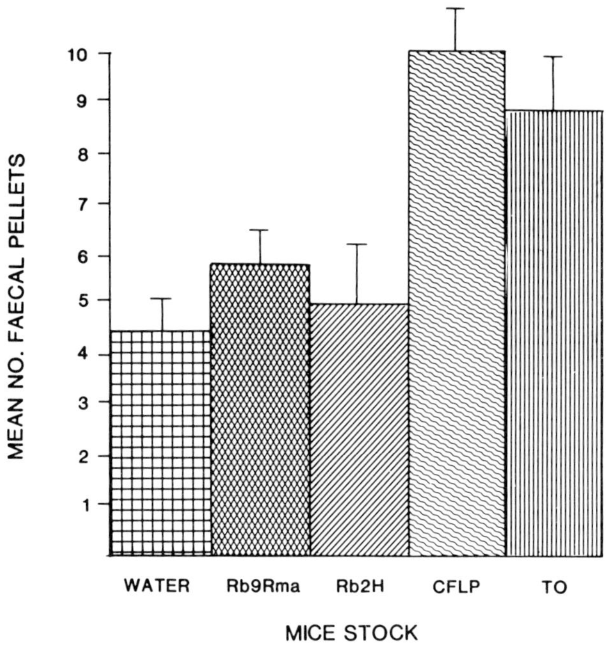


Fig. 3. Mean $\pm$SD of total number of faecal pellets passed by recipients exposed to urine voided by laboratory strains and mice with translocated chromosomes. The mice were tested over 4 successive days in an open arena.

would be recorded in a open-field test. The number of faecal pellets deposited for each test trial was also noted. For statistical analysis one-way analysis of variance and Duncan's new multiple range test were carried out with the use of SPSS (Nie *et al.* 1975) and a DEC 1091 computer.

Mice exposed to the urine from donors with translocated chromosomes, and those exposed to the water control treatment, were significantly lighter at the age when treatment was discontinued ($p < 0.003$); subsequently there were no differences between treatments. These mice, unlike those exposed to the urine of the laboratory strains, also showed no evidence of accelerated puberty. In Fig. 2 each square represents one mouse and the number of mice attaining first oestrus is distributed about the median age of onset for the whole sample (29 days).

These findings suggest that exposure to urine voided by the mice with the chromosonal translocations affects the age of onset of puberty and, within the limited 10-day period over which the mice were monitored, fails to establish vaginal cyclicity. There are associated changes in the cell content of the vaginal smears, suggesting that the pattern of hormonal release in the Robertsonian- and laboratory-exposed mice differs and that urine voided by one is able to modify responses of the other. The exposure to urine of the chromosomically translocated mice is associated with a high instance of dioestrus, but even between the two Robertsonian groups there were differences; Rb(11,16)2H showed fewer smears categorized as metoestrus than the Rb9(9,16)Rma stock. Mice exposed to the Robertsonian urine also defaecate least, suggesting that in addition to producing changes in the balance of sex hormones there is an associated change in autonomic activity. The number of faecal pellets passed did not differ between control treatments, but mice exposed to the urine of the combined Robertsonian stock passed fewer pellets that the other treatments (Fig. 3; $p < 0.01$). Mice exposed to the Robertsonian urine were also less active in both measures of activity but the only sensitivity measure to approach statistical significance ($p < 0.06$) was '4'. The results show that genetic factors, associated with the translocation chromosomes, are responsible for the absence of the puberty accelerating effect and the accompanying behavioural differences. There is, however, a need for caution; in the second experiment we show that exposure of infant TO mice to the urine of donors stressed by living in isolation may similarly affect some, but not all, of the same parameters.

## Experiment 2

A similar design was used (Fig. 4) but the donor and recipient mice were of the same strain (TO). The mice were kept under the same conditions of housing, artificial lighting (12:12 h night/day cycle with lights on at 6.00 a.m.) and temperature (approximately 21 °C) as in the previous study. The infant recipient mice were from mothers of the same age and were randomly assigned to treatment groups. Mice were daily exposed from 1 to 21 days of age to urine that had been collected at intervals from donors isolated at 80 days (long term) and a second treatment group where donors were isolated for 59–65 days (short term). Mice housed in groups of five per cage were kept together as groups for a time interval corresponding to that of mice of the isolated treatment. The urine from isolated mice was pooled as was that for the grouped mice. The timing of the isolation (and the corresponding grouping) was so arranged that the mice were of equivalent age at the end of the isolation period. Recipient mice were separated from their mothers at 21 days of age; thereafter the mice were examined daily and vaginal smears taken for 10 days from vaginal introitus.

The mice were tested in the open arena for four consecutive days at 60 days of age and again, for a single trial, at 100 days of age. The number of faecal pellets passed during the test trials was noted.

There was a significant difference in body weight between the treatments at the time of vaginal opening [$F(3,62) = 3.9$; $p < 0.01$]. Mice exposed to the urine from grouped donors were somewhat heavier than those exposed to the urine of the long-term isolates or distilled water ($p < 0.05$ in both instances). These differences were transient and by 150 days of age when the mice were killed, their body weights did not differ significantly between treatments. There were no differences

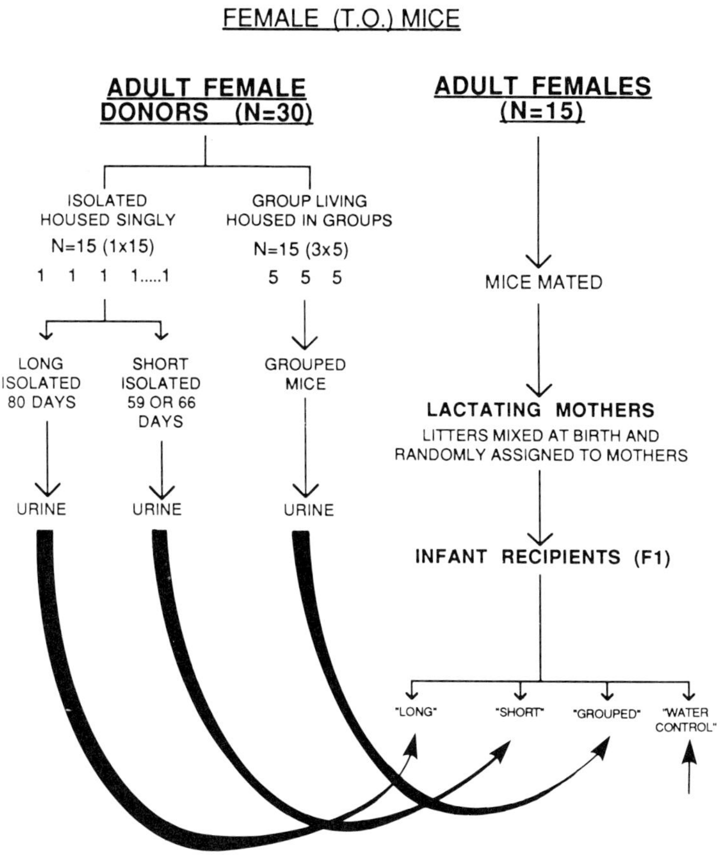


Fig. 4. History and composition of the sample. Exposure to mice living in group conditions and with a history of isolation.

in the number of faecal pellets between the treatment groups. Furthermore, neither the age of vaginal opening nor that of first oestrus differed significantly between treatments.

There were marked differences for the test measure of sensitivity '4' over the 4 days of testing [$F(3,36) = 5.4$; $p < 0.004$]. The major difference was between the high mean scores of mice exposed to the urine from short-term isolated mice and the other treatments ($p < 0.05$ in all cases). In the first trial (overall $p < 0.02$), all treatments showed lower activity than the short-term isolated mice ($p < 0.05$ in all cases). The differences between treatments are less marked (and not significant) for the second and third trials but for the fourth trial the differences between treatments re-emerge (overall $p < 0.01$); the mean differences between the short-term isolate and all other treatments again is statistically significant ($p < 0.05$ in all instances). Finally, when tested at the age of 100 days, for a single test trial the short-term treatment sample was significantly more active than the long-term ($p < 0.05$). Mean scores ($\pm$SE) of the various treatments for test sensitivity '4', as a function of days, are shown in Fig. 5.

The pattern of results for '7' was similar to that for sensitivity '4' but there were fewer significant differences. The long- and short-term isolation groups differed when all trials were combined and a similar difference was apparent on the first and second days of the trial ($p < 0.05$ in all instances). No other differences were significant. When the mice were re-tested in a single trial at 100 days of age, the differences between treatments were marked [$F(3,36) = 4.7$; $p < 0.01$]. Mice exposed to the urine of the long-term isolated group again showed less activity than the other treatments ($p < 0.05$in all cases). The mean scores ($\pm$SE) of the different treatments for test sensitivity level '7', as a function of days tested, are shown in Fig. 5.

## Discussion

The results shows that female mice exposed early in life to the urine of female mice with a history of isolation, subsequently behave differently, in an automated activity testing device, from comparable control mice. Mice exposed to the urine of the short-term isolates had enhanced activity while those exposed to the urine of mice isolated for a longer period of time showed lower levels of activity for the duration of the testing period.

The results are not dissimilar to those in which mice exposed to the urine of highly active lactating donors enhanced activity and those exposed to the urine of low activity donors reduced the activity of recipients (Cowley and Smale 1986). The donors used in the isolation experiment had been tested in an open arena using much the same procedure to that used with the recipients; there was some evidence that the isolation lowered their activity on sensitivity '4' level over the 4 days ($p < 0.06$). The differences largely arose from the results of the first trial, where the group-housed mice were more than twice as active as the isolated ($p < 0.02$).

The literature on the effects of isolation on rodent activity shows inconsistencies between different studies (Archer 1969, and reviewed by Archer 1970). In keeping with the present findings are those of Einon *et al.* (1981) who found that mice housed alone were less active than those living in groups. What needs to be emphasized is that where we have taken a behavioural criterion such as level of activity (or isolation which affects activity), then the urine from these mice is without effect on the age of onset of puberty and there is no increase in defaecation. However, the corresponding activity changes in the recipients are marked. Selecting donors with a genetic difference based on a chromosome

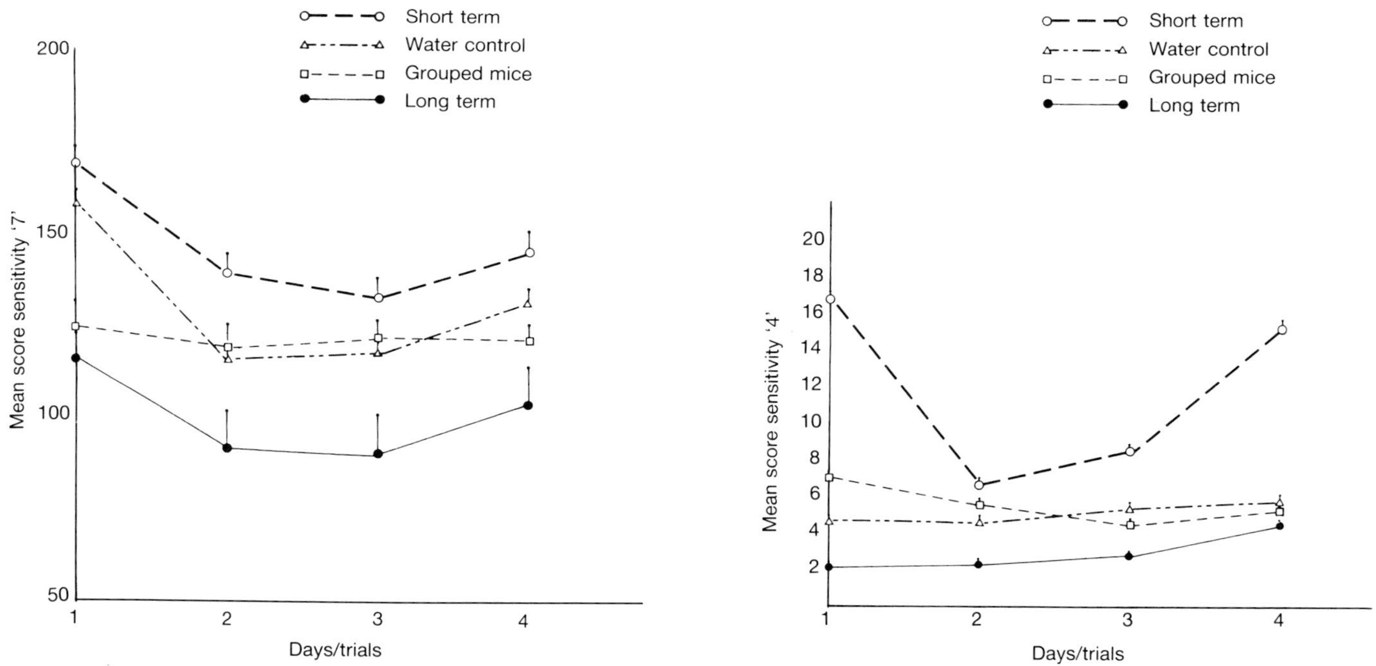


Fig. 5. Mean activity score ($\pm$SEM) of the recipient mice as a function of days tested. The mice were exposed to the urine from long-term and short-term isolated and grouped mice from day 1 to 21 days of age. The activity scores of a comparable control group of mice exposed to distilled water are also shown. The activity units were automatically monitored in an open arena, with the recording equipment set at a high (sensitivity '4') and low (sensitivity '7') level.

translocation, was not without effect on the behaviour of the recipients (particularly in terms of frequency of defaecation and to some extent also in terms of activity); the most striking feature was the incapacity of male mice, compared to the laboratory strains, to modify the onset of puberty. In mice, exposure to male urine is a potent factor in accelerating the onset of puberty whereas the female urine, as used in the isolation experiment, does not necessarily lead to a delay in puberty onset (see also Bronson and Rissman 1986, for reviews and discussion of social regulation of puberty).

Insofar as all recipients were of the same strain, we cannot preclude the possibility that the domesticated laboratory strains lack appropriate sensory receptors and or receptor cells that respond to the urine of the feral mice or indeed that the feral mice lack the accelerating pheromone. The block to pregnancy is more effective when the male is from a different strain from the stud male (Bruce 1961). If a similar mechanism is involved in the acceleration of puberty, then the pheromonal action of these genetic anomalies, and the hybrids that can be used for production of trisomic mice, may have a contribution to make to our understanding of the pathology of development. Overall the studies emphasize that the history of the parent, its phenotype, and genotype contribute to the spectrum of secretions that have been implicated in modifying the growth, development, and behaviour of mice.

## References

Archer, J. (1969). Contrasting effects of group housing and isolation on subsequent open field exploration in laboratory rats. *Psychon. Sci.*, 14, 234–5.

Archer, J. (1970). Effects of population density on behaviour of rodents. In *Social behaviour in birds and mammals. Essays on the social ethology of animals and man* (ed. J.H. Crook), pp. 169–210. Academic Press, London.

Bronson, F.H. and Rissman, E.F. (1986). The biology of puberty. *Biol. Rev.*, 61, 157–95.

Bruce, H.M. (1961). An olfactory block to pregnancy in mice. Part 1: Characteristics of the block. *Proc. 4th Int. Congr. Anim. Reprod.*, pp. 159–62. The Hague, Netherlands.

Coppola, D.M. and Vandenbergh, J.G. (1985). Effect of density, duration of grouping and age of urine stimulus on the puberty delay pheromone in female mice. *J. Reprod. Fert.*, 73, 517–22.

Cowley, J.J. and Wise, D.R. (1972). Some effects of mouse urine on neonatal growth and reproduction. *Anim. Behav.*, 20, 499–506.

Cowley, J.J. (1980). Growth and maturation in mice (*Mus musculus*). *Symp. Zool. Soc. Lond.*, 45, 213–50.

Cowley, J.J. and Pewtress, R.K. (1986). Post-parturition and late lactation urine: long-term effects of exposure on mouse activity and sniffing behaviour. *Reprod. Nutr. Develop.*, 26, 113–26.

Cowley, J.J. and Smale, S. (1986). Long term effects on activity in female mice following exposure to urine from lactating mothers. *Physiol. Behav.*, 36, 115–21.

Cowley, J.J., Pewtress, R., and McDonald, S. (1988). Effects of stock differences in the acceleration of puberty in female mice. *J. Reprod. Fert.*, 83, 407–12.

Drickamer, L.C. (1974). Sexual maturation of female house mice: social inhibition. *Devl. Psychobiol.*, 7, 257–65.

Drickamer, L.C., McIntosh, T.K., and Rose, E.A. (1978). Effects of ovariectomy on the presence of a maturation-delaying pheromone in the urine of female mice. *Horm. Behav.*, 11, 131–7.

Drickamer, L.C. and Hoover, J.E. (1979). Effects of urine from pregnant and lactating female house mice on sexual maturation of juvenile females. *Dev. Psychobiol.*, 12, 545–51.

Drickamer, L.C. and McIntosh, T.K. (1980). Effects of adrenalectomy on the presence of a maturation-delaying pheromone in the urine of female mice. *Horm. Behav.*, 14, 146–52.

Drickamer, L.C. (1981). Acceleration and delay of sexual maturation in female house mice previously selected for early and late first vaginal oestrus. *J. Reprod. Fert.*, 63, 325–9.

Drickamer, L.C. (1983). Effect of period of grouping of donors and duration of stimulus exposure on delay of puberty in female mice by a urinary chemosignal from grouped females. *J. Reprod. Fert.*, 69, 723–7.

Drickamer, L.C. (1986). Effects of urine from females in oestrus on puberty in female mice. *J. Reprod. Fert.*, 77, 613–22.

Einon, D.F., Humphreys, A.P., Chivers, S.M., Field, S., and Naylor, V. (1981). Isolation has permanent effects upon the behaviour of the rat, but not the mouse, gerbil or guinea pig. *Dev. Psychobiol.*, 14, 343–55.

Epstein, C.J. (1985). Mouse monosomies and trisomies as experimental systems for studying mammalian aneuploidy. *Trends Genet.*, 1, 129–34.

Epstein, C.J., Cox, D., and Epstein, L.B. (1985). Mouse trisomy 16: an animal model of human trisomy 21 (Down syndrome). *Ann. N.Y. Acad. Sci.*, 450, 157–68.

Fullerton, C. and Cowley, J.J. (1971). The differential effect of the presence of adult male and female mice on the growth and development of the young. *J. Genet. Psychol.*, 119, 89–98.

Massey, A. and Vandenbergh, J.G. (1980). Puberty delay by a urinary cue from female house mice in feral populations. *Science.*, 209, 821–2.

McIntosh, T.K. and Drickamer, L.C. (1977). Excreted urine, bladder urine, and the delay of sexual maturation in female house mice. *Anim. Behav.*, 25, 999–1004.

Nie, N.H., Hull, C.H., Jenkins, J.G., Steinbrenner, K. and Bent, D.H. (eds) (1975). *Statistical package for the Social Sciences (SPSS)*. 2nd edn. McGraw-Hill, New York.

Novotny, M., Jemiolo, B., Harvey, S., Wiesler, D., and Marchlewska-Koj, A.

(1986). Adrenal-mediated endogenous metabolites inhibit puberty in female mice. *Science*, 231, 723–5.

Searle, A.G. (1985). Chromosonal variants. In *Issues and reviews in teratology* (ed. H. Kalter), pp. 324–357. Plenum, New York.

Vandenbergh, J.G. (1967). Effect of the presence of a male on the sexual maturation of female mice. *Endocrinology*, 81, 345–9.

Vandenbergh, J.G. (1969). Male odor accelerates female maturation in mice. *Endocrinology*, 84, 658–60.

# III.12 Dominance Status and its Influence on the Bruce Effect in House Mice

A.A. HAFER

## Introduction

The Bruce Effect is a phenomenon whereby exposure to the odour of an unfamiliar male may lead to pregnancy blockage in recently copulated female rodents (Bruce 1959, 1960). It appears that the inducing of ovulation when females are not pregnant, and the blocking of pregnancy when they are, are encompassed by the same system and are subject to the same influences (Bronson, 1979). Both phenomena result from the release of gonadotrophins subsequent to exposure to a strange male (Hoppe and Whitten 1972).

It has been found that the urinary cuing which induces ovulation is done solely by dominant males (Bronson 1979). The urinary pheromones by which males produce these effects are androgen-dependent (Bronson and Whitten 1968), and cannot be found in castrated mice (Bruce 1965). Social subordination has an inhibitory effect on gonadal function in male laboratory mice (Bronson 1973), suppresses scent marking (Desjardins *et al.* 1973), and nullifies the effect of a male's urine in accelerating the commencement of puberty in females (Lombardi and Vandenbergh 1977).

Hence, if there is a link between male dominance status and these other pheromonal effects which are physiologically and behaviourally closely related to it, it seems likely that dominance will also affect a male's ability to induce pregnancy blockage. In this study, the relation between dominance and a male's ability to produce pre-implantation pregnancy blockage was the phenomenon under investigation.

## Materials and methods

### ANIMALS

The mice used were Swiss-Webster, a randomly mated inbred strain of albino house mice, which had been bred at Swarthmore College. Male and female subjects were housed individually in plastic cages measuring 29 × 18 × 13 cm, with wire-mesh tops and sawdust bedding. These cages had water and lab chow available at all times.

Male mice were ranked by exposing two mice to one another in a small aquarium for 10 mins, with chase, attack, and fight scores being recorded. Only males whose dominance relationship remained constant, and which had been reaffirmed on the day prior to experimentation were used.

### EXPERIMENTAL PROCEDURE

Females were divided into three treatment groups—two experimental and one control—with vaginal smears being taken daily in all cases. When vaginal smears revealed that a female had come into oestrus she was mated with a stud male.

In the control group, the stud male remained with the female for another 2–3 days. In the two experimental groups, the stud male was removed, and was replaced by a strange male, which remained with the female for another 2–3 days. In Experimental Group A, the stud male was subordinate and the strange male was dominant to that stud. In Experimental Group B, the stud male was the dominant and the strange male the subordinate.

Vaginal smears were taken daily throughout the experiment, until the female either came into oestrus, aborted, or gave birth. The number of days between the introduction of the last male and the occurrence of oestrus, abortion, or birth was recorded.

As a practical frame of reference, the data from the different treatment groups could be divided into those females which came into oestrus within $7\frac{1}{2}$ days from the introduction of the last male, and those which came into oestrus later, or gave birth. The females in the former category were assumed to have been subject to pre-implantation blockage, since 7–8 days represents the time necessary for two oestrous cycles; one to be interrupted, which may or may not make itself evident with an oestrous cell smear, and one normal one, which would become evident to an investigator taking daily vaginal smears. Animals coming into oestrus after that time may not have been subject to pre-implantation blockage, having begun cycling again for extraneous reasons.

## Results

Among the three females in the control group, one became oestrous after 13 days, and two gave birth, with gestation times of 20 days each. Among the six females in Experimental Group A, most came into oestrus again fairly quickly. The number of days required were 2 days, 4 days, 5 days, 7 days (in two cases), and 11 days. Experimental Group B, showed a distinctly different pattern, with females resuming cycling again much later, or in one case, giving birth. The numbers of days between last exposure to a male and oestrus in this case were as follows: 8 days (in two cases), 10 days (in two cases), 13 days, and one birth (20 days). These results are illustrated in Fig. 1, and summarized in Table 1.

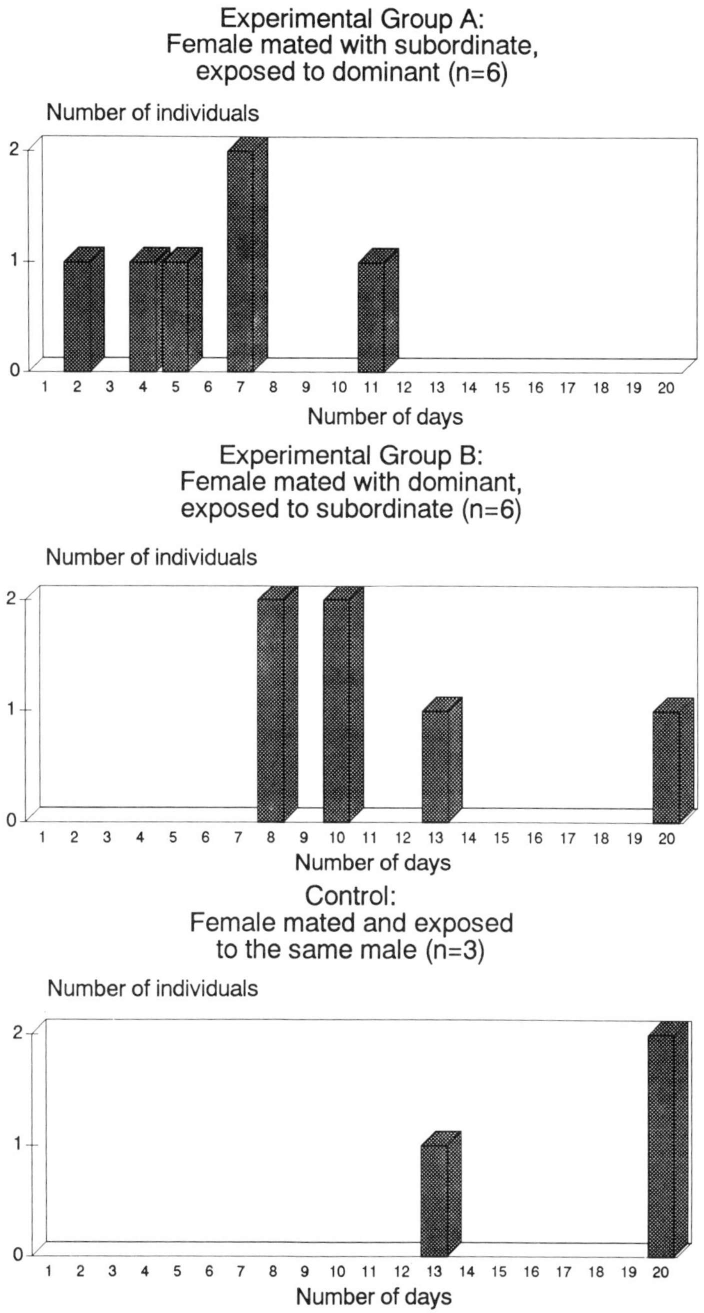


Fig. 1. Numbers of females in each group going into oestrus on any given day.

**Table 1.** Summary of results

| | Number of females | Females returning to oestrus in $<7\frac{1}{2}$ days | Females returning to oestrus after $7\frac{1}{2}$ days | Average number of days to oestrus |
|---|---|---|---|---|
| Experimental Group A | 6 | 5 | 1 | 4.17 |
| Experimental Group B | 6 | 0 | 6 | 11.50 |
| Control | 3 | 0 | 3 | 17.70 |

The various individual treatments, as recorded in Fig. 1, were compared with one another using a Mann-Whitney U test. My hypothesis, that a dominant male should be more likely to induce pre-implantation blockage than a subordinate one, predicts that there should be a significant difference between the scores found in Experimental Group A and the control group. On the other hand, the scores for Experimental Group B should not differ significantly from those of the control group. Finally, the scores for the two experimental groups should differ significantly from one another.

The first comparison, between Experimental Group A and the control group yielded a $p$ value of 0.012, using a one-tailed test (significant). The second comparison, between Experimental Group B and the control group yielded a $p$ value of 0.066, using a one-tailed test (not significant). Finally, the comparison between the two experimental groups yielded a $p$ value of 0.013, using a one-tailed test (significant).

## Discussion

Although the sample sizes in these tests were small, these results strongly suggest that dominance status has a noticeable effect on a male's ability to block implantation of ova fertilized by another male. This would increase his fitness in the next generation both by eliminating competitors' offspring, and by making the female more readily available to him.

Mice in nature live in colonies with one dominant male, which monopolizes mating, together with some subordinate males and many females (Lacy 1978, Bronson 1979). If a female copulates with a male, and then comes into proximity to a male that is dominant to her mate, it is likely that the old dominant has been deposed; it is also likely that the new dominant male will kill any pups sired by the old stud (Labov 1980; Huck *et al.* 1982).

Thus, it is energetically to the female's advantage to absorb the litter rather than carrying the pregnancy to term. She will also come into oestrus sooner, thereby increasing her fitness in the next generation by more quickly producing offspring that have a high chance of survival. Thus, natural selection would favour a female who can perceive differences in male dominance status, and who can then use this information to her best advantage.

## Acknowledgements

I would like to thank Dr U.W.E. Huck for his initial assistance with this project, and for his help in preparing the original manuscript of this paper. I would also like to thank Dr Mark Ridley for his later suggestions.

## References

Bronson, F.H. and Whitten, W.K. (1968). Oestrus-accelerating pheromone of mice: assay, androgen-dependency, and presence in bladder urine. *J. Reprod. Fertil.*, 15, 131–54.

Bronson, F.H. (1973). Establishment of social rank among grouped male mice: relative effects on circulating FSH, LH, and corticosterone. *Physiol. Behav.*, 10, 947–51.

Bronson, F.H. (1979). The reproductive ecology of the house mouse. *Q. Rev. Biol.*, 54, 265–99.

Bruce, H.M. (1959). An exteroreceptive block to pregnancy in the mouse. *Nature*, 54, 105.

Bruce, H.M. (1960). A block to pregnancy in the mouse by proximity of strange males. *J. Reprod. Fertil.*, 1, 96–103.

Bruce, H.M. (1965). Effect of castration on the reproductive pheromones of male mice. *J. Reprod. Fertil.*, 10, 141–3.

Desjardins, C., Maruniak, J.A. and Bronson, F.H. (1973). Social rank in house mice: differentiation revealed by ultraviolet visualization of urinary marking patterns. *Science*, 182, 939–41.

Hoppe, P.C. and Whitten, W.K. (1972). Pregnancy block; imitation by administered gonadotropin. *Biol. Reprod.*, 7, 254–9.

Huck, U.W., Soltice, R.L. and Coopersmith, C.B. (1982). Infanticide in male laboratory mice: effects of social status, prior sexual experience, and basis for discrimination between related and unrelated young. *Anim. Behav.*, 30, 1158–65.

Labov, J.B. (1980). Factors influencing infanticidal behaviour in wild male housemice (*Mus musculus*). *Behav. Ecol. Sociobiol.*, 6, 297–303.

Lacy, R.C. (1978). Dynamics of t-alleles in *Mus musculus* populations: review and speculation. *Biologist*, 60, 41–67.

Lombardi, J.R. and Vandenbergh, J.G. (1977). Pheromonally induced sexual maturation in females: regulation by the social status of the male. *Science*, 196, 545–6.

# III.13 Amniotic Fluid and Dodecyl Propionate Pheromone: Influence on the Onset and Regulation of Maternal Anogenital Licking in the Rat

I. BROUETTE-LAHLOU AND E. VERNET-MAURY

## Introduction

It is now well established that mammalian maternal behaviour is under multisensory and hormonal control. The onset of maternal behaviour depends on oestrogens, particularly during the period around parturition, and in the postpartum period, on relayed stimuli from the pups. Rosenblatt (1967) demonstrated that all rats have a basic level of maternal responsiveness independent of hormonal stimulation: stimuli emanating from pups are able to induce maternal behaviour. Various factors may facilitate the induction of maternal behaviour; in sheep, for instance, Keverne *et al.* (1983) have emphasized vaginal stimulation as an important determinant of maternal bonding. Different sensory processes involved in mice were analysed by Chantrey and Jenkins (1982). Among them, placenta and amniotic fluid were suspected to be active. The influence of amniotic fluids on the onset of maternal behaviour in parturient ewes was well demonstrated by Levy and Poindron (1984). In rats the situation is less obvious but many facts lead to the same conclusion. Firstly, dams seem to be strongly attracted to foetal fluids at parturition, since they usually consume at once the placenta and membrane and lick amniotic fluid from the skin of the pups (Rosenblatt and Lehrman 1963). Secondly, Kristal and Graber (1976) demonstrated that a large percentage (45%) of virgin female rats avidly and immediately ingest donor placenta the first time it is presented to them. In contrast, maternal behaviour appears readily towards cleaned foster pups in caesarean delivered rats (Moltz *et al.* 1966).

In order to understand our previous results concerning preputial pheromone (Brouette-Lahlou *et al.* 1989), it was of interest to know exactly how amniotic fluids induce maternal behaviour in the rat. Indeed, a chemical agent from the preputial gland (PG) of the rat pup was found to be an important link in the mother–young bond, as it regulates anogenital licking (AGL)—a pattern of maternal behaviour crucial to pup survival. This result was obtained by removing preputial glands and observing effects on pups at birth before any AGL on the part of the dam. Enhanced pup mortality and increased AGL time obtained after PG ablation may be due to adventitious suppression of amniotic fluid. In order to elucidate this, the following study was designed to determine the respective influence of amniotic fluid or PG ablation *per se* on maternal AGL.

## Methods

The experimental animals were 24 SPF (specific pathogen free) primiparous female Wistar rats, weighing approximately 250 g obtained from the IFFA-CREDO breeding stock. They were housed in groups of five until mating at three months. Dams were divided into four groups according to the pups' treatment. Before being licked by the dam all operated pups were submitted to PG ablation at birth. During surgery, pups of the first group (A) were cleaned and washed so as to remove amniotic fluid; in contrast, pups of the second group (B) were carefully treated to prevent any removal of amniotic fluid; amniotic fluid from the first group was added to the surgical incision. PG ablation was performed as described elsewhere (Brouette-Lahlou *et al.* 1989). Groups C and D served as control: skin was excised under ether anaesthesia but the PG was not removed. The control pups were divided into two groups: one with washed pups (C) and the other one with amniotic fluid left on (D).

The pups were kept in a warm place and returned to their dam after recovery from anaesthesia. Observations began 20 h after parturition. The parents were left with their litters. All AGL was recorded on stop-watches from 8 a.m. to 1 p.m., on days 1, 3, and 5 after birth. In order to test for maternal behaviour, the following aspects of behaviour were observed: nest building, retrieving, crouching, nursing. Pups were regularly weighed.

## Results

The different aspects of maternal behaviour were present in the four groups, A, B, C and D: nest building occurred, all pups were retrieved, and crouching and nursing were also observed. However, their intensity varied between groups A and B on the one hand (PG ablation) and groups C and D on the other (controls). Some preliminary observations lead us to assume that the importance of each item is dependent on litter size, which differed as a result of PG ablation.

In contrast, AGL was disturbed in groups A and B whose pups were submitted to PG ablation at birth. The median duration of each AGL behaviour was calculated for days 1, 3 and 5 after birth.

Using Friedman rank analysis of variance, no significant difference emerged between the three days' observations for groups A, B, C and D. ($\chi^2$ rA $= 4.66, p =$

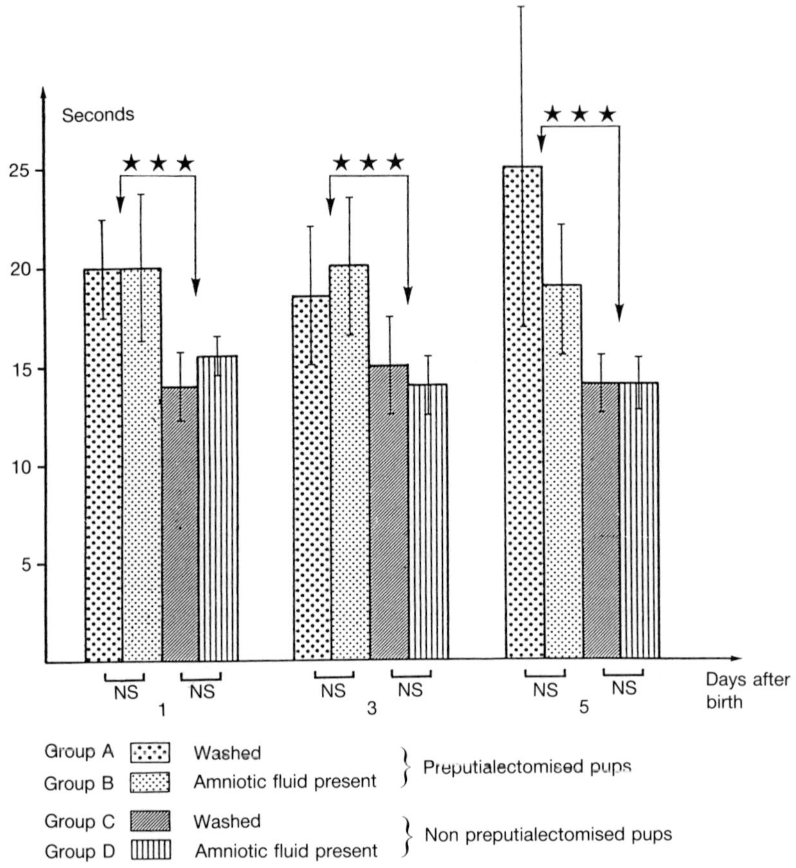


Fig. 1. Median anogenital licking time.

0.194 ; $\chi^2$ rB = 4.5, $0.361 < p < 0.194$ ; $\chi^2$ rC = 4.666, $p = 0.194$; $\chi^2$ rD = 2.166, $0.361 < p < 0.528$.) These groups were homogeneous, the median and (semi-interquartile ranges) of licking time being respectively : A = 20 (7.5); B = 20 (7); C = 14 (3); D = 14 (2.5). Mann-Whitney U tests between group A (PG ablation pups) and control pups (C) indicated a significant effect on AGL duration ($Z = -5.30$ $p = 0.00003$). Similarly, groups B and D were significantly different ($Z = 5.84$, $p = 0.00003$) (see Fig. 1). Thus, PG ablation *per se* enhanced anogenital licking time.

The groups also showed differences in per cent mortality (see Fig. 2). $\chi$-$^2$ tests revealed a significant difference between groups A (36.9%) and C (3.1%) ($\chi^2$ = 31.95, $p$ =0.01) and groups B (35.7%) and D (0%) ($\chi^2$ = 40, $p$ = 0.01). PG ablation increased the pups' mortality rate.

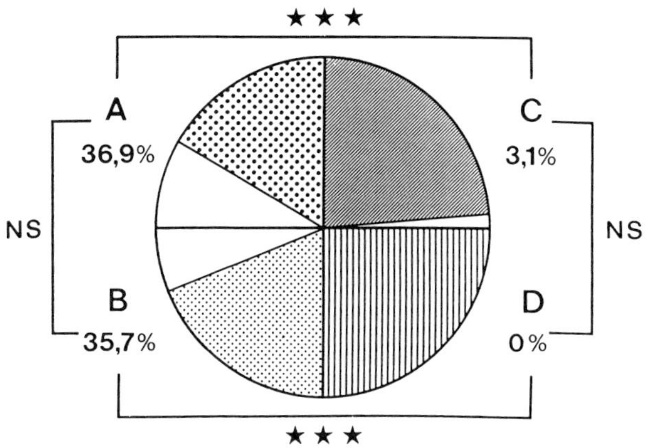


Fig. 2. Mortality percentage.

The influence of amniotic fluid could be evaluated by matching groups with (B, D) and without this fluid (A, C), as well as PG ablation pups (A versus B) and no PG ablation ones (C versus D). Using the Mann-Whitney U test no significant difference emerged between groups C and D ($Z = 0.44, p = 0.33$). Amniotic fluid did not influence AGL duration. The same result was obtained for pups without PG from groups A and B ($Z = 0.52, p = 0.30$). Removal of amniotic fluid played no role on pups' mortality ($\chi$-$^2$ test : A versus B, $\chi^2 = 0.022$; C versus D, $\chi^2 = 3.5$).

## Discussion

The enhanced anogenital licking time observed is in accordance with previous work concerning results from PG ablation (Brouette-Lahlou, *et al.* 1989). The pheromonal-like compound from the preputial gland has been isolated and identified as dodecyl propionate (DP, Vernet-Maury *et al.* 1989). An ontogenetic analysis (Vernet-Maury and Brouette-Lahlou 1987) confirmed DP as the only active compound present in the preputial gland at birth. From these results we suggested that PG secretion regulates licking time and allows the mother to know which pup needs AGL: without the pheromone some pups were never licked and died with big, full intestines. This hypothesis would explain the increased mortality observed after PG ablation. Previous results from this work confirmed that the per cent mortality differed from one litter to another (Brouette-Lahlou 1984). Furthermore, the presence of the pheromone does not lead to any increase in AGL by the dam.

No differences emerged between the different aspects of maternal behaviour towards pups with or without amniotic fluid. This result allows us to conclude that amniotic fluid is not important in the onset of maternal behaviour in rats and is in accordance with previous work (La Briola 1953; Moltz *et al.* 1966; Bridges 1977; Mayer and Rosenblatt 1980). In fact, results leading to the conclusion that amniotic fluid plays a role in the acceptance of newborn rats were obtained in the virgin and not in the lactating female. So, amniotic fluid shortens the maternal sensitization latency in virgins but seems to be a negligible stimulus in lactating females. Thus, in the rat the dam is guided by active stimuli from the pups and amniotic fluid is not efficient in this role. Conversely, amniotic fluid is involved in the initial identification of the nipple by pups (Teicher and Blass 1977). One possibility is that the pup saliva, maternal saliva and amniotic fluid all contain dimethyl disulfide (Leon 1983) and that the pups respond to a single cue throughout lactation. Concerning the onset of the dam's maternal behaviour, the present results indicate that amniotic fluid is inefficient, while dodecyl propionate, the pheromone from the preputial gland is confirmed as a fundamental regulating factor in terms of an enhanced percent mortality rate due to misdirected AGL after PG ablation.

## References

Bridges, R.S. (1977). Parturition: its role in the long-term retention of maternal behavior in the rat. *Physiol. Behav.*, 18, 487–90.

Brouette-Lahlou, I. (1984). Thesis, University of Lyon.

Brouette-Lahlou, I., Vernet-Maury, E. and Chanel, J. (1989). Relation entre petits chez le rat: odeur préputiale et léchement anogénital. (Submitted.)

Chantrey, D.F. and Jenkins, B.A.B. (1982). Sensory processes in the discrimination of pups by female mice. *Anim. Behav.*, 30, 881–5.

Keverne, E.B., Levy, F., Poindron, P. and Lindsay, D.R. (1983). Vaginal stimulation: an important determinant of maternal bonding in sheep. *Science*, 219, 81–3.

Kristal, M.B. and Graber, G.C. (1976). Placentophagia in nonpregnant rats: influence of estrous cycles and birth place. *Physiol. Behav.*, 17, 599–605.

La Briola, J. (1953). Effects of caesarean delivery upon maternal behavior in rats. *Proc. Soc. Exp. Biol. Med.*, 83, 556–7.

Levy, F. and Poindron, P. (1984). Influence du liquide amniotique sur la manifestation du comportement maternel chez la brebis parturiente. *Biol. Behav.*, 9, 271–8.

Leon, M. (1983). Mother-young interactions. In *Pheromones and reproduction in mammals* (ed. J.G. Vandenbergh). Academic Press, New York, pp. 39–77.

Mayer, A.D. and Rosenblatt, J.S. (1980) . Hormonal interaction with stimulus and situational factors in the initiation of maternal behavior in non pregnant rats. *J. Comp. Physiol. Psychol.*, 94, 1040–59.

Moltz, H., Robbins, D. and Parks, M. (1966). Caesarean delivery and maternal behavior of primiparous and multiparous rats. *J. Comp. Physiol. Psychol.*, *61*, 455–60.

Rosenblatt, J.S. (1967). Non hormonal basis of maternal behavior. *Science*, 156, 1512–4.

Rosenblatt, J.S. and Lehrman, D.S. (1963). Maternal behavior of the laboratory rat. In *Maternal behavior in mammals* (ed. H.L. Rheingold). Wiley, New York, pp. 8–57.

Teicher, M.H. and Blass, E.M. (1977). First suckling response of the newborn albino rat: the roles of olfaction and amniotic fluid. *Science*, 198, 635–6.

Vernet-Maury, E. and Brouette-Lahlou, I. (1987). Ontogenic analysis of the rat pup pheromone implicated in perigenital licking. *Chem. Senses*, 12, 186.

Vernet-Maury, E., Brouette-Lahlou, I., Chastrette, F., Amouroux, R., Stofelsma, J. and Cosnier, J. (1989). Dodecyl propionate, a pheromone emitted by the rat pup. Characterization and identification. (Submitted.)

# III.14 Foetal Olfaction

PETER G. HEPPER

Odour stimuli form an important part of the newborn vertebrate's environment (Doty 1986; Coopersmith and Leon 1988; Walser 1985), and olfaction is often the best developed of the newborn's senses (Brunjes and Frazier 1986). It is not surprising, therefore, to find that responsiveness to chemosensory stimulation commences pre-natally (e.g. see Hepper 1987). The present paper discusses functional aspects of foetal olfaction; readers are referred to Bradley and Mistretta (1975), Brunjes and Frazier (1986) and Pedersen *et al.* (1986) for reviews of the physiological development of the foetal olfactory system.

Early views (e.g. Preyer 1937) that there could be no olfactory sensation whilst the nasal cavities were bathed in fluid are now known to be false, as olfactory receptors can be stimulated by odours dissolved in fluid (Bradley and Mistretta 1975). However, it is difficult to separate the effects of gustatory and olfactory stimulation in the foetus, since both types of receptor will be bathed in fluid containing the stimulating chemicals. Thus, even though olfactory stimuli are used post-natally to investigate the influence of pre-natal chemosensory experience, the resulting response will, in part, be determined by gustatory experience *in utero* and this should be borne in mind during the following discussion.

Most demonstrations of foetal chemosensory responsiveness have involved mammals, in particular the rat. Rat pups, which as 20-day-old foetuses have been aversively conditioned to an odour by pairing it *in utero* with lithium chloride, e.g. apple juice + lithium chloride, will decrease their running speed in a runway when the odour of apple is introduced (Smotherman 1982), and will not attach to nipples that have been coated with apple (Stickrod *et al*, 1982). Besides being able to learn about odours pre-natally rat foetuses have also been shown to be capable of discriminating different chemical substances *in utero*, e.g. milk, lemon, saline (Smotherman and Robinson 1987).

Pre-natal chemosensory responsiveness is not restricted to mammals. Although much of the work examining perceptual capabilities in embryonic birds has concentrated upon audition (e.g. see Shindler 1984) there is evidence that, prior to hatching, birds are responsive to chemical signals. Tolhurst and Vince (1976) found the domestic fowl, one day before hatching, exhibited an increase in heart rate, beak clapping and head shaking in response to odour.

The human foetus can detect changes in the chemical composition of the amniotic fluid, e.g. increasing the sweetness of the fluid results in increased swallowing by the foetus (De Snoo 1937), whereas injection of the noxious substance Lipidiol (iodinated poppy seed) results in a reduction in swallowing (Liley 1972). Whether this response involves olfactory as well as gustatory receptors is not known. Premature infants have been observed to respond to the olfactory stimulation as early as 35 weeks (Hepper, unpublished observation) indicating that olfactory receptors are mature and responsive to stimulation at this age.

This present review shows that the foetus is responsive to chemosensory stimulation. For most of the studies chemicals were artificially introduced into the uterine environment; while this demonstrates that the olfactory system is functioning pre-natally, questions remain about the extent of olfactory stimulation to which the foetus will be naturally exposed. Whether or not the organism is likely to experience odour stimulation pre-natally is a question that should therefore be considered.

There are a number of possible routes by which the foetus may be exposed to odour stimulation (Fig. 1). Stimulation may come directly from the environment,

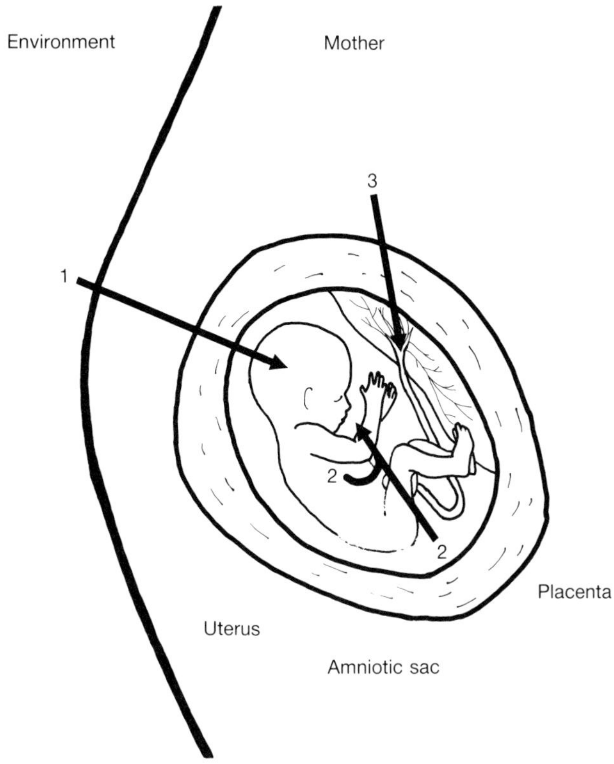


Fig. 1 Schema of possible routes by which vertebrate foetus may experience its olfactory environment.

through the womb or shell (Route 1). In this case the limiting factor that determines the amount of stimulation received is the permeability of the membranes surrounding the developing organism. The fluid surrounding the developing individual (Route 2) may be a source of chemical stimulation. Its composition will be affected by both the mother and the organism via glandular and other excretions and secretions. Mammalian amniotic fluid is a rich source of chemicals containing for example glucose, fructose, lactic, pyruvic and citric acids, fatty acids, phospholipids, creatinine, urea, uric acid, amino acids, polypeptides, proteins, and salts. Of these, lactic and citric acid, fatty acids, uric acid and the amino acids may be the most likely to stimulate olfactory receptors. The consistency of the amniotic fluid is continually changing, exhibiting both gradual changes during the course of pregnancy and more short-term changes, e.g. as a result of foetal micturation. The human foetus swallows amniotic fluid during pregnancy, and this stimulates vomeronasal and gustatory receptors. It is not known whether there is movement or change of the fluid within the nasal cavities, but breathing movements exhibited by the foetus from the 3rd month of gestation may induce changes in or flow through the nasal cavities.

Finally, the olfactory receptors may be stimulated by volatile compounds carried in the blood (Route 3). These blood-borne odours may stimulate olfactory receptors by diffusing out of the blood stream in the nasal capillary bed (e.g. see Maruniak *et al.* 1983). Thus volatile compounds present in the maternal blood supply may cross the placental barrier and stimulate the foetus (Hepper 1988). Such compounds may be excreted by the foetus into the amniotic fluid and thereby have an extended effect by stimulating the olfactory receptors from this medium (Route 2). Thus the developing organism is naturally exposed to a diverse and changing olfactory environment.

Since the individual is exposed to odours pre-natally these could be expected to influence post-natal preferences. Rat pups will avoid odours to which they have been aversively conditioned pre-natally (e.g. see Smotherman 1982; Stickrod *et al.* 1982); however, the effects of simple exposure have been far more difficult to demonstrate. For example, Pedersen and Blass (1982) found that rat pups attached to nipples coated with citral if they had experienced this odour pre-natally *and* post-natally but not if they had experienced it only pre-natally or only post-natally. The foetus may require a longer period of association with the odour for preferences to be affected by pre-natal exposure alone. Rat foetuses were exposed to orange by injection into the amniotic fluid on either day 17, day 19, or day 21 of gestation. At 12 days of age pups were given a two-choice simultaneous olfactory discrimination test for their preference between orange (to which they had been exposed pre-natally) or water. Testing took place in a rectangular arena 30 × 15 × 15 cm with a mesh floor. Orange and water were placed under the floor and the time spent by the pups over the orange or water was recorded in a 2 min. trial. The results (see Fig. 2) were clear. Pups exposed to orange on day 19 of gestation showed a preference for orange, whereas pups exposed to orange on day

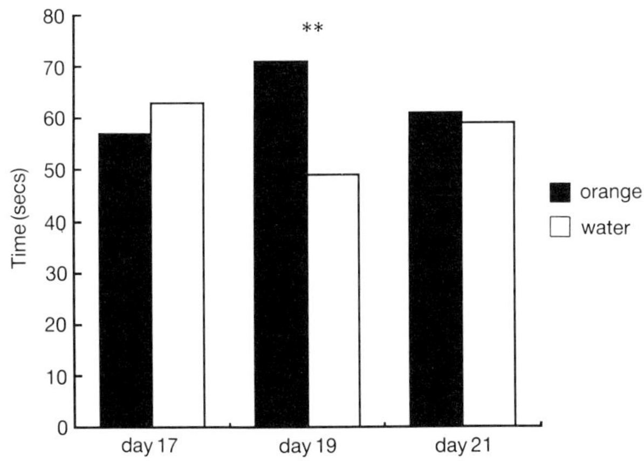


Fig. 2 Time (in seconds) spent investigating the odour of orange or water for 12-day-old pups after being exposed to orange on either day 17, 19 or 21 of gestation. Pups investigated orange significantly longer only after exposure on day 19 of gestation (** $p < 0.001$ matched pairs $t$-test).

17 or day 21 did not. Control pups, either those injected with saline on day 17, 19 or 21, or those left completely alone during gestation, exhibited no preference for the odour of orange. Furthermore, the preference was specific to the odour of pre-natal exposure: pups exposed to orange pre-natally and then given a choice between strawberry or water, and lemon or water, exhibit no preference for either odour. Similar effects have been obtained using pre-natal injections of strawberry, lemon, and banana. It would appear that simple exposure of an odour pre-natally is sufficient to establish a preference as long as the pup is exposed to the odour for an adequate period. The reasons for no preference in 17- and 21-day-old pups may be different. On day 21 the amniotic fluid is more viscous than on day 19 (personal observation) and any injected odour may not diffuse through the fluid as easily, thus stimulating the olfactory receptors less. Seventeen-day-old foetuses may show no preference due to the rapid turnover of the amniotic fluid, which restricts the time the odour is present in its environment. Indeed, when a greater volume of odour is introduced into the amniotic fluid at day 17 a preference for this odour is observed after birth.

Since foetal chemosensory experience can affect post-natal behaviour the role of 'normal' pre-natal olfactory experience should be considered. By 'normal' I refer to conditions were the foetus will be exposed to odours naturally and not by experimental manipulation. Three possible functions are considered below.

## Dietary preference

One possibility recently examined in this laboratory was that the diet consumed by the mother during her pregnancy may influence the post-natal preferences of her offspring. Garlic contains volatile compounds that are carried in the blood and can be perceived from their presence in the blood (Maruniak *et al.* 1983). Thus the foetus may experience garlic consumed by the mother via molecules crossing the placental barrier, entering the foetal blood system and stimulating the olfactory receptors by diffusion from the nasal capillary bed.

Pregnant rats were fed garlic during gestation and their offspring, who had never experienced garlic post-natally, were examined for their preference for garlic or onion at 12 days of age. The results (Hepper 1988) indicate that the offspring of mothers who had eaten garlic during pregnancy showed a preference for garlic, whereas the offspring of mothers who had not been exposed to garlic did not. Thus the mother's diet during pregnancy can affect the post-natal preferences of her offspring.

## Successful suckling

The importance of learning from the amniotic fluid for the initiation of suckling is well established (e.g. see Pedersen and Blass 1982). However, pre-natal learning may also be important. Mothers, during gestation, were either fed garlic along with their normal diet or just fed their normal rat pellets. All pups were removed from their mothers at birth and placed with an unfamiliar female: either one that had been fed the same diet as their mother, or one that had been fed the alternative diet. Pups showed no difference in latency to attach to the nipple of the new female: however pups placed with females who had been fed a different diet during pregnancy gained less weight in the first 24 h than pups placed with females fed the same diet; there were no differences in time spent suckling by the females (unpublished observation). Since the mother's milk contains information regarding her diet (Galef and Sherry 1973), pups placed with females fed different diets will have had no experience of this diet pre-natally and thus initially do not consume as much milk. However, pups placed with mothers fed the same diet have encountered this diet pre-natally and show no hesitation in consuming the milk.

## Socialization

The chemical environment experienced pre-natally can profoundly influence conspecific preferences after birth. For example, rat pups recognize their amniotic fluid at birth (Hepper 1987) and this is important in mediating their reactions to kin and non-kin (Hepper 1986). If the chemical composition of the amniotic fluid is altered, for example by injection of orange extract, individuals no longer exhibit a preference for normally scented conspecifics at birth but prefer the conspecifics scented with orange. These studies indicate that individuals encounter the 'smell' of their conspecifics pre-natally, probably from excretions or secretions contained in the surrounding fluid (Hepper 1987) and this ability is highly advantageous in learning about their conspecifics (Hepper 1986).

Behavioural evidence of a functioning olfactory system before birth may have implications for our understanding of olfactory system function. I shall briefly touch on two points. First, it is interesting to note that while the stimuli experienced before birth are waterborne, the preference exhibited post-natally is for an airborne stimulus. It would thus appear that the quality of the stimulus is unaffected by this change in mode of perceptual experience. Furthermore, it seems that different olfactory structures are stimulated by pre- and post-natal olfactory experience (Pedersen et al. 1986). The accessory olfactory bulb is activated by chemosensory experience pre-natally, whereas post-natally it does not respond to stimulation. Post-natal olfactory experience appears to stimulate the modified glomerular complex and main olfactory bulb but not the accessory olfactory bulb. Since odours experienced pre-natally are responded to post-natally, this would suggest that such sites (accessory olfactory bulb, modified glomerular complex, and main olfactory bulb) are processing stations and the memory of the odour is stored elsewhere, presumably at a site of common input from these systems.

In summary, the vertebrate foetus does not live in an environment devoid of chemosensory stimulation but one of constantly changing stimulation. Furthermore, chemosensory experience before birth may influence the post-natal behaviour of the organism.

## Acknowledgements

I thank Prof. Ken Brown for providing facilities for this research and Fiona Hepper for comments on the manuscript.

## References

Bradley, R.M. and Mistretta, C.M. (1975). Fetal sensory receptors. *Physiol. Rev.*, 55, 352–82.

Brunjes, P.C. and Frazier, L.L. (1986). Maturation and plasticity in the olfactory system of vertebrates. *Brain Res Rev.*, 11, 1–45.

Coopersmith, R. and Leon, M. (1988). The neurobiology of early olfactory learning. In *Handbook of behavioral neurobiology* (ed. E.M. Blass), Vol. 9, pp. 283–308. Plenum, New York.

De Snoo, K. (1937). Das trinkende Kind im Uterus. *Monatsschr. Geburtsh. Gynaekol.*, 105, 88–97.

Doty, R.L. (1986). Odor-guided behavior in mammals. *Experientia*, 42, 257–71.

Galef, B.G. and Sherry, D.F. (1973). Mother's milk: a medium for transmission of cues reflecting the flavour of mothers diet. *J. Comp. Physiol. Psychol.*, 83, 374–8.

Hepper, P.G. (1986). Kin recognition: Functions and mechanisms. A review. *Biol. Rev.*, 61, 63–93.

Hepper, P.G. (1987). The amniotic fluid: an important priming role in kin recognition. *Anim. Behav.*, 35, 1343–6.

Hepper, P.G. (1988). Adaptive fetal learning: pre-natal exposure to garlic affects post-natal preferences. *Anim. Behav.*, 36, 935–6.

Liley, A.W. (1972). Disorders of amniotic fluid. In *Pathophysiology of gestation. Fetal-placental disorders* (ed. N.S. Assali), Vol. 2, pp. 157–206. Academic Press, New York.

Maruniak, J.A., Silver, W.L. and Moulton, D.G. (1983). Olfactory receptors respond to blood-borne odorants. *Brain Res.*, 265, 312–6.

Pedersen, P.E. and Blass, E.M. (1982). Pre-natal and post-natal determinants of the first suckling episode in albino rats. *Dev. Psychobiol.*, 15, 349–55.

Pedersen, P.E., Greer, C.A. and Shepherd, G.M. (1986). Early development of olfactory function. In *Handbook of behavioral Neurobiology* (ed. E.M. Blass), Vol. 8, pp. 163–203. Plenum, New York.

Preyer, W. (1937). Embryonic motility and sensitivity. *Monogr. Soc. Res. Child. Dev.*, 2, 1–115.

Shindler, K.M. (1984). A three year study of fetal auditory imprinting. *J. Wash. Acad. Sci.*, 74, 121–4.

Smotherman, W.P. (1982). Odor aversion learning by the rat fetus. *Physiol. Behav.*, 29, 769–71.

Smotherman, W.P. and Robinson, S.R. (1987). Pre-natal expression of species-typical action patterns in the rat. *J. comp. Psychol.*, 101, 190–6.

Stickrod, G., Kimble, D.P. and Smotherman, W.P. (1982). In utero taste/odor aversion conditioning in the rat. *Physiol. Behav.*, 28, 5–7.

Tolhurst, B.E. and Vince, M.A. (1976). Sensitivity to odours in the embryo of the domestic fowl. *Anim. Behav.*, 24, 772–9.

Walser, E.S. (1985). Neonatal sensory development and exploratory behaviour. In *Ethology of farm animals* (ed. A.F. Fraser), pp. 127–32. Elsevier, Amsterdam.

# III.15 Responses of Human Neonates to Odours of Lactating Females

RICHARD H. PORTER , JENNIFER W. MAKIN AND BENOIST SCHAAL

## Introduction

The role of maternal olfactory cues in mediating the growth and behaviour of mammalian neonates is best documented for several species of rodents. Newborn rat pups rely on odours of maternal saliva applied to the mother's ventrum for nipple localization and attachment (Blass and Teicher 1980). Lactating house mice, rats, and spiny mice emit chemical signals ('maternal pheromones') which are attractive to neonates and function to keep them in the vicinity of the nest or home area (Porter 1983). Human newborns may be less dependent than rodent pups on maternal odours; nonetheless, as elaborated below, lactating females of our own species produce olfactory signals to which neonates are selectively responsive.

## Neonatal responsiveness to breast odours

Breasts of lactating females provide a rich source of secretions that could potentially function as semiochemicals. Aside from releasing colostrum and milk, the nipple and areola contain dense concentrations of eccrine and apocrine sweat glands, as well as sebaceous glands. Diffusion of odorous substances could be enhanced by the relatively high surface temperature of the areola which results from the pronounced vascularization of this region (Schaal 1988).

### INDIVIDUAL DISCRIMINATION

The possibility that infants may orient towards their (nursing) mother's breast odour has long been a subject of speculation (e.g. Darwin 1877); however, empirical verification of this hypothesis was not published until 1975 (Macfarlane, 1975; see also Ripin *et al.* 1930 for an earlier, but unsuccessful experiment). In Macfarlane's (1975) pioneering study, breast-feeding infants were tested with two gauze odour-pads suspended simultaneously alongside each cheek. When presented with a soiled breast pad from their own mother and a clean pad, 17 of 20 neonates (age range: 2–7 days) spent a longer period of time with their head turned towards the maternal pad.

A subsequent experiment tested infants' abilitity to discriminate between the odours of their own mother and an unfamiliar lactating female. At both 6 and 8–10 days of age significantly more infants oriented preferentially to the odour of their own mother than to that of an unfamiliar female.

Reports from two additional laboratories provide corroborative evidence of individually discernible odours (olfactory signatures) emanating from the breasts of lactating females. Russell (1976) tested infants successively with their own mother's breast pad, a pad worn by an unfamiliar mother, and a clean control pad (in a random sequence). At 6 weeks of age, 7 of 10 infants displayed sucking and orienting movements when the maternal pad was held under their nostrils, whereas only one infant responded to the odour of the unfamiliar female. Differential responsiveness to the olfactory stimuli was not observed in the same infants when 2 days and 2 weeks old. Using a simultaneous two-choice testing procedure similar to that developed by Macfarlane, Schaal, *et al.* (1980) observed reduced head and arm movements when an active infant's nose came into contact with its mother's breast pad but not with a clean pad or one soiled by an unfamiliar nursing female. Such discriminative reduction in ongoing motor activity was evident in neonates as early as 3 days postpartum.

### GENERAL ATTRACTANT

In addition to serving as individually distinctive chemical signatures by which infants can discriminate their own mother, breast odours of lactating females also appear to function as a general attractant for neonates—at least for those within certain age ranges. The breast-fed infants tested by Russell (1976) responded in a comparable manner to breast odours from their own and an unfamiliar nursing mother (presented one at a time) when two weeks old. Only at six weeks did these infants react differentially to breast odours from these two sources. Likewise, two-week-old girls who had been bottle-fed since birth oriented preferentially towards a breast pad worn by an unfamiliar lactating female when paired with either that same female's axillary pad or a breast pad from an unfamiliar non-parturient female (Makin 1987).

Direct contact with a nursing female's breasts is, therefore, not necessary for infants to develop a tendency to orient towards lactation-associated breast odours. To the extent that the odours of breast milk resemble that of the infants' familiar formula, stimulus generalization could account for the selective responsiveness to the odour of the lactating female. Alternatively, this early olfactory preference could reflect an inborn response bias developing independently of specific postnatal experience.

## Other maternal odours

### INDIVIDUAL DISCRIMINATION

Odours other than those of the nipple and areola may also provide a basis for maternal discrimination by neonates. A study with five infant subjects found significant reductions in head and arm movements when the nose was turned towards a pad impregnated with the mother's neck odour compared to similar odours from a strange mother (Schaal 1986). It is not clear whether the above infants were responding to their mothers' natural body odours,however, since the author noted that most of the stimulus pads had a noticeable 'perfume scent' (Schaal 1988).

A related series of experiments investigated whether two-week-old infants are capable of discriminating their own parents by axillary odours alone (Cernoch and Porter 1985). In each of five experiments, a different group of infants was tested with two odourized gauze pads—one that had been taped in the axillae of the baby's mother (or father) and a second axillary pad worn by an appropriate control person. All individuals who served as sources of test odours were instructed to refrain from using deodorant while wearing the stimulus pads.

Breast-feeding infants responded preferentially (head-turning response) to their mother's axillary odour over comparable chemical cues from unfamiliar lactating or non-parturient females. In contrast, bottle-feeding infants displayed no behavioural indication that they recognized their own mothers' axillary odours when these were presented along with the axillary pad from a non-parturient female or an unfamiliar non-breast-feeding female (also at two-weeks post-partum). Furthermore, breast-fed infants did not respond differentially to the odour of their own father versus that of an adult male stranger.

### BASIS FOR DIFFERENCES BETWEEN BREAST- AND BOTTLE-FED INFANTS

Even though the underlying basis for the observed differences between bottle- and breast-fed infants in their responsiveness to maternal axillary odours has not been well elucidated, several logically plausible hypotheses can be proposed:

1. Chemical signals, emanating from skin glands distributed over various regions of the body, may differ (either qualitatively or quantitatively) between lactating and non-lactating females. Accordingly, axillary odours from lactating females could be particularly salient to neonates and therefore more readily discriminated than such cues from non-lactating women. Excretion rates of sebaceous glands, for example, are elevated during pregnancy and remain high following parturition in breast-feeding females (Burton *et al.* 1973).

   At the same time, sebum excretion rates of females who do not breast feed their infants decline to the same level as in non-pregnant control women. In the above study, sebum samples were collected from the forehead, which has a very high density of sebaceous glands. Sebaceous secretions are also produced in the axillae, where they are believed to function as 'odour fixatives' by prolonging the effectiveness of semiochemicals (Mykytowycz 1985).

2. Response differences of breast-fed compared to bottle-fed infants might be a function of differing perceptual or motor capabilities in these two sub-populations. This seems an unlikely possibility, however, especially in view of the above-mentioned preferential orientation to breast odours of lactating females by bottle-fed female infants. These results indicate that bottle-fed neonates, like breast feeders, are capable of perceiving olfactory cues, and head orientation is an appropriate measure of their responsiveness to odours (at least for two-week-old girls).
3. Two-week-old breast-fed infants may simply be more familiar with their mother's characteristic body odour than are bottle-fed infants of the same age. While suckling at the breast, neonates are exposed directly to their mother's bare flesh and associated odours for relatively long periods of time. Infants who are bottle fed are not likely to be exposed routinely to maternal odours in this manner. Because of their increased contact with the mother's olfactory signature, breast-fed infants may be capable of discriminating that odour cue at an earlier age than bottle-feeders.

The hypothesized relationship between odour familiarity and the development of neonatal olfactory preferences was assessed further by Balogh and Porter (1986). Within the first two days postpartum, infants were exposed to an artificial odorant for approximately 23 h. During the exposure period, a gauze pad treated with either cherry or ginger was taped to the inside of the bassinet, behind the infant's head. At the end of the treatment period, each infant was tested by simultaneously presenting a cherry- and ginger-scented pad, one on each side of the baby's face. For one-half of the subjects, ginger was the familiar odour and cherry the unfamiliar scent, while the reverse was true for the remaining 12 neonates.

Overall, the 24 infants oriented significantly longer towards the odour to which they had been previously exposed, regardless of whether they were breast- or bottle-feeding. There was a marked sex difference in preferential orientation to the familiar odour, however. Whereas the 12 girls showed a strong directional bias in favour of the familiar odour, boys did not respond differentially to the familiar and novel olfactory stimuli. This sex difference is consistent with the superior performance by females which is often reported for odour discrimination and detection tests with older subjects (ranging in age from 5 to over 90 years) (e.g. Doty 1981; Doty *et al.* 1984).

For two-day-old girls, at least, familiarization through mere exposure results in a preference for specific artificial odorants. This same process could also be involved in the development of a discriminative preference for the mother's unique odour signature.

## Conclusions

In humans, as for other species of mammals, maternal odours may play an important role in early mother–infant interactions. Lactating females of our own species produce two functionally distinct categories of chemical signals for newborn infants:

1. General attractant—odours emanating from the nipple and areola region, towards which neonates (regardless of their feeding history) display preferential orientation.
2. Odour signature—individually unique odours ('olfactory fingerprints'), produced in the breast and axillary regions (and possibly elsewhere), by which breast-fed infants can discriminate their own mother.

## Acknowledgements

Preparation of this paper was supported in part by NATO collaborative research grant 0366/87, and grant HD-15051 from the U.S. National Institute of Child Health and Human Development.

## References

Balogh, R.D and Porter, R.H. (1986). Olfactory preferences resulting from mere exposure in human neonates. *Infant Behav. Develop.*, 9, 395–401.

Blass, E.M. and Teicher, M.H. (1980). Suckling. *Science*, 210, 15–22.

Burton, J.L., Shuster, S., Cartlidge, M., Libman, L.J. and Martell, U. (1973). Lactation, sebum excretion and melanocyte-stimulating hormone. *Nature*, 243, 349–50.

Cernoch, J.M. and Porter, R.H. (1985). Recognition of maternal axillary odors by infants. *Child Develop.*, 56, 1593–8.

Darwin, C. (1877). A biographical sketch of an infant. *Mind*, 7, 285–94.

Doty, R.L. (1981). Olfactory communication in humans. *Chem. Senses*, 6, 351–76.

Doty, R.L., Shaman, P., Applebaum, S.L., Giberson, R., Siksorski, L. and Rosenberg, L. (1984). Smell identification ability: changes with age. *Science*, 226, 1441–3.

Macfarlane, A. (1975). Olfaction in the development of social preferences in the human neonate. In *Parent–infant interaction*, Ciba Foundation Symposium 33, pp. 103–113. Elsevier, New York.

Makin, J.W. (1987). Bottle-feeding infants' responsiveness to lactating maternal breast and axillary odors. Ph.D. thesis, Vanderbilt University.

Mykytowycz, R. (1985). Olfaction—a link with the past. *J. Human Evolution*, 14, 75–90.

Porter, R. H. (1983). Communication in rodents: Adults to infants. In *Parental behaviour of rodents* (ed. R.W. Elwood), pp. 95–125. Wiley, Chichester.

Ripin, R., Hetzer, H. and Bühler, C. (1930). Frühestes Lernen des Säuglings in der Ernährungssituation. *Zeitschr. Psychol.*, 118, 82–127.

Russell, M.J. (1976). Human olfactory communication. *Nature*, 260, 520–2.

Schaal, B. (1986). Presumed olfactory exchanges between mother and neonate in humans. In *Ethology and psychololgy* (ed. J. Le Camus and J. Cosnier), pp. 101–10. Privat, Toulouse.

Schaal, B. (1988). Olfaction in infants and children: developmental and functional perspectives. *Chem. Senses*, 13, 145–90.

Schaal, B., Montagner, H., Hertling E., Bolzoni, D., Moyse, A., and Quichon, R. (1980). Les stimulations olfactives dans les relations entre l'enfant et la mère. *Reprod. Nutr. Dévelop.*, 20, 843–58.

# III.16 Behavioural Components of White-Tailed Deer Scent Marking: Social and Seasonal Effects

R. LARRY MARCHINTON, KAREN L. JOHANSEN AND KARL V. MILLER

## Introduction

White-tailed deer (*Odocoileus virginianus*) have a complex system of scent communication that is still not well understood (Marchinton and Hirth 1984). Various aspects of marking behaviour performed by males during the breeding season have been examined (Kile and Marchinton 1977; Moore and Marchinton 1974; Miller *et al.* 1987*a*, *b*; Ozoga and Verme 1985; Woods 1988), but no investigation of scent communication in white-tailed deer throughout both breeding and non-breeding seasons has been conducted. The primary objectives of our study were to describe types and frequencies of scent marking, as well as relationships among aggression, social status, and the various behavioural components of marking throughout the year.

## Methods

Twenty captive white-tailed deer were studied at the University of Georgia's Whitehall Deer Research Facility. The deer were maintained in a 1.0-ha enclosure containing about equal amounts of meadow and mixed hardwood forest. The enclosure was divided into three paddocks each containing a group of 3–4 adult bucks, 2–3 adult does, and 3–4 fawns. Screening was erected to prevent visual interaction between paddocks. In addition to natural browse, deer had access to water and a balanced dry ration (16% crude protein) at all times.

All-occurrences behavioural sampling (Altmann 1974) began after antler shedding in early April 1986, and continued until antlers were shed again in late March 1987. Rutting activity in this area peaks during October and November and parturition occurs in May or June. Sampling consisted of eighteen 60-min. sessions each month. Pens were observed in random sequence within 90 mins after sunrise or 90 mins before sunset. The number of observation periods was equally distributed among the groups each month.

Wins and losses in agonistic encounters (see Forand *et al.* 1985) were summarized in sociometric matrices (Altmann 1974) and dominance relationships, determined by a binomial test for significant asymmetry, were used to construct hierarchies. Five aspects of behaviour related to scent marking were recorded: (1) antler rubbing, (2) pawing, (3) rub-urination, (4) overhead branch marking, and (5) the full scrape sequence (for descriptions see Moore and Marchinton 1974; Johansen 1987). Two-way analyses of variance using months as the within-subjects variable and hierarchical ranks as the between-subjects variable were conducted for average hourly rates of each type of marking behaviour recorded. Because of unequal sample sizes, Fisher's Protected Test for least significant difference was used for pairwise comparison of frequency means between dominant and subordinate deer (Ott 1984).

## Results and discussion

### DOMINANCE HIERARCHIES

The alpha buck in each group tended to be older (4, 7, and 9 years old), larger, and more socially experienced. Aside from three dominance fights, hierarchies were relatively stable and maintained by threats or sparring.

Aggression by does usually was directed at other does, fawns, or low-ranking bucks. Although does engaged in agonistic interactions throughout the year, most aggression by does was observed immediately prior to parturition. Ozoga *et al.* (1982) reported that free-ranging does selected an isolated fawning site within their home range and aggressively excluded other deer.

### ANTLER RUBBING

Most rubbing of antlers and forehead on the trunk of small trees (91%) occurred when bucks were in hard antler. Rubbing frequencies of subordinates were lower ($p < 0.05$, 70 df) than those of alpha bucks during September, October, December, and February. This is compatible with previous reports that older bucks rub more frequently (Ozoga and Verme 1985; Miller *et al.* 1987*a*) and that dominant males tend to have high forehead gland activity during the rut (Atkeson and Marchinton 1982). Dominant bucks rubbed most in September prior to breeding, while rubbing by subordinates peaked in October and November (Fig. 1). Ozoga and Verme (1985) also found that yearling bucks marked later.

Temporal differences in rubbing between ranks suggest functional differences. Rubbing by alpha bucks may help establish their dominance and mark the areas where they are dominant prior to breeding (Moore and Marchinton 1974; Kile and Marchinton 1977). Rubbing also may physiologically prepare does for the rut (Miller *et al.* in press). The early rubs by alpha bucks probably allow does to

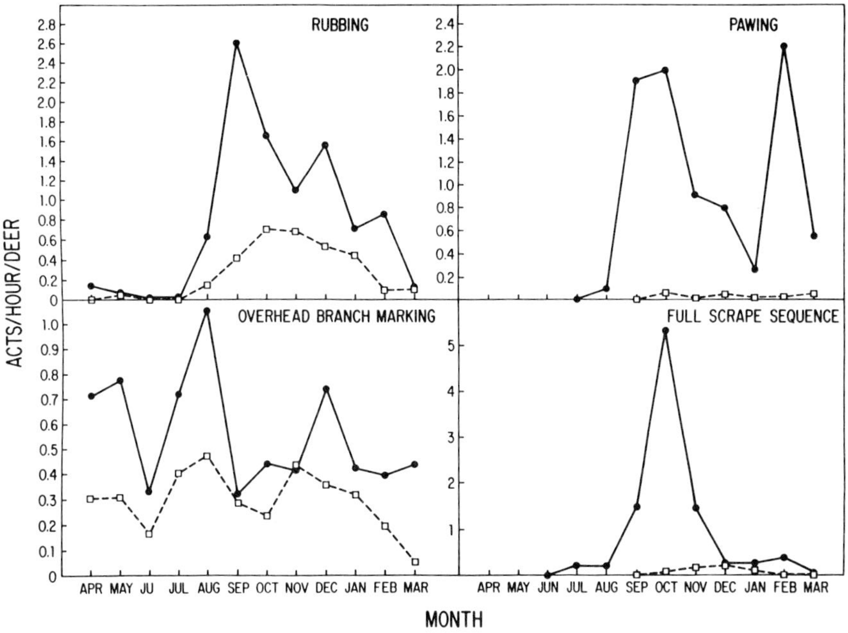


Fig. 1. Mean monthly frequencies of rubbing, pawing, overhead branch marking, and the full scrape sequence by dominant (●) and subordinate (□) white-tailed bucks. Pawing and overhead branch marking occurring as a part of full scrape sequences are reported under that category only.

recognize the scent of dominant males prior to actual breeding. Later rubbing by subordinates may serve as practice for assuming dominant status in the future (Forand *et al.* 1985).

## PAWING

Pawing of soil not associated with the full scrape sequence occurred frequently before or after agonistic interactions, but also occurred when bucks were approached by dogs or humans. Pawing by alpha bucks peaked in September and October, and was much more frequent ($p < 0.05$, 46 df) than pawing by subordinates (Fig. 1).

An association between pawing and testosterone levels is suggested because (i) it was seen only in males; (ii) it usually was accompanied by agonistic behaviour; (iii) it was rare in spring and summer; and (iv) bucks that pawed at a high frequency late in the season also were the last to cast antlers, indicating maintenance of high testosterone levels (Lincoln 1984). Also, high testosterone levels are correlated with scraping (Miller *et al.* 1987*b*), and pawing is the component of scraping that seems to be performed only by dominants.

The communicative significance of pawing is unclear, but we believe that pawing, as a separate aspect of behaviour, is an unlikely source of olfactory cues and is primarily a visual threat or dominance display (see Walther 1984).

### RUB-URINATION

In rub-urination, deer urinate on their tarsal or hock glands. This mixes urine with tarsal gland secretions (Müller-Schwarze 1971) but also may be a visual display (Marchinton and Hirth 1984). Adult bucks rub-urinated most often when in hard antler, especially during threats and full scrape sequences. However, deer of all social ranks rub-urinated in the absence of any of these aspects of behaviour. The relationship between rub-urination and other aspects of marking behaviour was complicated by bucks occasionally urinating in a normal stance while scraping, pawing, or overhead branch marking. Neither form of urination was seen in conjunction with rubbing.

On three occasions a doe rub-urinated after being chased out of a fawning territory, suggesting an agonistic or displacement context for this form of marking.

### OVERHEAD BRANCH MARKING

In overhead branch marking, the buck mouthed a branch about 1.5 m in height and rubbed it with forehead, preorbital area, nose, and chin, while pausing frequently to sniff and lick the branch. Overhead branch marking was performed both as part of the full scrape sequence and separately. Unlike rubbing and pawing, it occurred throughout the year (Fig. 1). Bucks of all social ranks commonly marked branches and inspected those that had been marked by other bucks. Branch marking was performed at higher frequencies by alpha bucks than subordinates during May and August ($p < 0.05$, 73 df).

Although a licking branch is described in a popular article (McGuire 1984), overhead branch marking occurring separately from scraping has not been previously reported for white-tailed deer in the scientific literature (see also Johansen *et al.* 1988). Forehead rubbing throughout the year was described for black-tailed deer (*O. hemionus columbianus*) by Müller-Schwarze (1972), who determined that rubbing and sniffing twigs occurred in both sexes and in all age classes. Müller-Schwarze's study did not distinguish between antler rubbing of tree boles and overhead branch marking. Our results indicate that these appear to be separate aspects of behaviour in white-tailed deer and may convey different information.

The temporal distribution of overhead branch marking and the communal use of marking sites suggest that it communicates a buck's identity. In the fall, scrapes were usually associated with branches that had been marked during the summer, but not all marked branches became scrape sites.

### FULL SCRAPE SEQUENCE

The full scrape sequence is a combination of overhead branch marking, pawing, and urination. It apparently involves several sources of scent and may have multiple functions and a number of types of chemosignals. Scraping was observed from July through March (Fig. 1). Occasional scraping outside of the rut also has been noted by Ozoga (1989). Scraping frequency was higher ($p < 0.05$, 32 df) for alpha bucks versus subordinates during July through November and again in February. Scraping by alpha males peaked sharply in October, just prior to maximum breeding activity, while subordinates scraped later in the rut and then only at very low levels. Miller *et al.* (1987*b*) also found that only mature, dominant males produced a significant number of scrapes. They reported higher testosterone levels in scrapers and concluded that testosterone, high dominance rank, and age are all factors contributing to high levels of scraping.

The low level of scraping by subordinates may be due in part to behavioural or pheromonal suppression by alpha bucks (Marchinton *et al.* 1988). On several occasions, after making eye contact with an alpha buck, a subordinate turned from a path leading to a scrape. Also, as the peak of breeding approached, an alpha buck was seen chasing subordinate bucks that came near his scrape. Occasional defence of scrapes by free-ranging bucks has been interpreted as a form of facultative territoriality (Moore and Marchinton 1974; Marchinton and Atkeson 1985).

Full scrape sequences by subordinates were very rare. While they often mouthed the branches over scrapes, they usually did not paw the ground or urinate in conjunction with branch marking; hence their use of scrapes was recorded only as overhead branch marking. Perhaps the pawing and urination components of the full scrape sequence conveys a dominant's superior status, while branch marking relates to identity or home range in all classes of bucks. The occurrence of peak scraping by dominants early in the rutting season, coupled with reports of doe inspection of scrapes (Woods 1988; Sawyer *et al.* 1989), also is consistent with the hypotheses that scrapes assist in mate selection by does (Moore and Marchinton 1974) or help to synchronize oestrus cycles (Miller *et al.* in press).

### DOE AND FAWN MARKING

Marking by fawns consisted of thirteen observations of overhead branch marking by males. Their infrequent marking probably is due to physiological and behavioural immaturity. Other than rub-urination, only one instance of possible marking by a doe was recorded. In early September, an adult doe was seen mouthing an overhead branch previously marked by bucks. Additionally, on eight occasions does were observed sniffing branches marked by bucks and twice sniffing the pawed depression in a scrape.

Doe responses to rubs and scrapes have been reported for free-ranging deer (Woods 1988; Sawyer *et al.* 1989). Our infrequent observation of does responding to signposts are probably due to the effects of confinement.

## Conclusions

Relatively little scent marking by does and fawns was observed. Most rubbing, pawing, and scraping was done by dominant males during the breeding season. Subordinate males displayed considerably less rubbing than dominants and very little scraping or pawing. Overhead branch marking was performed all year round by bucks of all ranks. Scraping is a combination of overhead branch marking, pawing, and urination indicating multiple functions. Behavioural observations suggest several scent sources including urine and saliva, as well as tarsal, preorbital, forehead, interdigital, and nasal glands.

Scent-marked signposts made by this species apparently express identity and social status throughout the year and aid in mate selection during the breeding season. They, also, may be a source of male pheromones that prime the oestrus cycle in females and suppress testosterone levels in subordinate males. It is anticipated that further investigation into the behaviour, semiochemicals, and associated hormones will clarify these and other functions of marking in this species.

## Acknowledgements

We acknowledge funding of the Agricultural Experiment Station, McIntire-Stennis Project No. GEO-0030-MS-H, and express special appreciation to Bob McGuire for his ideas and financial support.

## References

Altmann, J. (1974). Observational study of behavior: sampling methods. *Behavior*, 49, 227–67.

Atkeson, T.D. and Marchinton, R.L. (1982). Forehead glands in white-tailed deer. *J. Mammal.*, 63, 613–7.

Forand, K.J., Marchinton, R.L. and Miller, K.V. (1985). Influence of dominance rank on the antler cycle of white-tailed deer. *J. Mammal.*, 66, 58–62.

Johansen, K.L. (1987). *Seasonal variation in marking behavior of white-tailed deer.* Thesis, University of Georgia, Athens.

Johansen, K.L., Marchinton, R.L., Miller, K.V. and McGuire, R.C. (1988). Seasonal variation in marking behavior of white-tailed deer. *Southeast Deer Study Group*, 11, 15.

Kile, T.L. and Marchinton, R.L. (1977). White-tailed deer rubs and scrapes: spatial, temporal, and physical characteristics and social role. *Amer. Midl. Nat.*, 97, 257–66.

Lincoln, G.A. (1984). Antlers and their regeneration—a study using hummels, hinds and haviers. *Proc. Roy. Soc. Edinburgh*, 82B, 243–59.

Marchinton, R.L. and Atkeson, T.D. (1985). Plasticity of socio-spatial behavior of white-tailed deer and the concept of facultative territoriality. In *Biology of deer production* (ed. P.F. Fennessy and K.R. Drew). *Roy. Soc. New Zealand Bull.*, 22, 375–7.

Marchinton, R.L. and Hirth, D.H. (1984). Behavior. In *White-tailed deer: ecology and management* (ed. L.K. Halls), pp. 129–68. Stackpole Books, Harrisburg, Pa.

Marchinton, R.L., Miller, K.V., Hamilton, R.J. and Guynn, D.C. (1988). (Abstract.) Sociobiological implications of quality management—evidence from the Mt. Holly experiment. *Southeast Deer Study Group*, 11, 7–8.

McGuire, B. (1984) The licking branch. *Michigan Archery*, 2(4), 9–12.

Miller, K.V., Kammermeyer, K.E., Marchinton, R.L. and Moser, E.B. (1987*a*). Population and habitat influences on antler rubbing by white-tailed deer. *J. Wildl. Managm.*, 51, 62–6.

Miller, K.V., Marchinton, R.L., Forand, K.J. and Johansen, K.L. (1987*b*). Dominance, testosterone levels, and scraping activity in a captive herd of white-tailed deer. *J. Mammal.*, 68, 812–7.

Miller, K.V., Marchinton, R.L. and Knox, W.M. (in press). White-tailed deer signposts and their role as a source of priming pheromones: a hypothesis. *Cong. Int. Union Game Biol.*, 18.

Moore, W.G. and Marchinton, R.L. (1974). Marking behavior and its social function in white-tailed deer. In *The behaviour of ungulates and its relation to management* (ed. V. Geist and F.R. Walther). *IUCN New Ser. Publ.*, 24, 447–56, IUCN, Morges, Switzerland.

Müller-Schwarze, D. (1971). Pheromones in blacktailed deer. *Anim. Behav.*, 19, 141–52.

Müller-Schwarze, D. (1972). Social significance of forehead rubbing in blacktailed deer (*Odocoileus hemionus columbianus*). *Anim. Behav.*, 20, 788–97.

Ott, L. (1984). *An introduction to statistical methods and data analyses*, 2nd edn. Duxbury Press, Boston.

Ozoga, J.J. and Verme, L.J. (1985). Comparative breeding behavior and performance of yearling vs. prime-age white-tailed bucks. *J. Wildl. Managm.*, 49, 364–72.

Ozoga, J.J., Verme, L.J. and Bienz, C.S. (1982). Parturition behavior and territoriality in white-tailed deer: impact on neonatal mortality. *J. Wildl. Managm.*, 46, 1–11.

Ozoga, J.J. (1989). Temporal pattern of scraping behavior in white-tailed deer . *J. Mammal.*, 70, 633–6.

Sawyer, T.G., Marchinton, R.L. and Miller, K.V. (1989). Response of female white-tailed deer to scrapes and antler rubs. *J. Mammal.*, 70, 431–3.

Walther, F.R. (1984). *Communication and expression in hoofed mammals.* Indiana University Press, Bloomington, Ind.

Woods, G.R. (1988). *Scrape behavior and physical characteristics of scrapes and rubs of white-tailed deer.* Thesis, S.W. Missouri St. University, Springfield.

# III.17 Identified Hormonal Pheromones in the Goldfish: The Basis for a Model of Sex Pheromone Function in Teleost Fish

PETER W. SORENSEN AND NORMAN E. STACEY

## Introduction

Pheromones are defined as 'substances that are excreted to the outside by an individual and received by a second individual of the same species, in which they release a specific reaction, for example a definite behaviour or developmental process' (Karlson and Lüscher 1959). Living in an aquatic environment filled with chemicals and often devoid of light, teleost fish have evolved highly developed chemosensory systems and pheromonal signalling systems. These pheromonal systems have a wide variety of functions (Liley 1982; Colombo *et al.* 1982), often including the promotion of behavioural and physiological synchrony of reproduction (Stacey *et al.* 1986, 1987). Recently, we discovered that goldfish use several common sex hormones and their metabolites as sex pheromones. Here we use the goldfish as a model to develop the hypothesis that sex hormones commonly function as pheromones in fish; we then address the various questions of definition, function, evolution, and functional diversity this hypothesis raises.

## The goldfish model

In temperate climates, female goldfish ovulate in the spring in response to rising water temperatures and emerging aquatic vegetation (Stacey *et al.* 1979). Females ovulate in the early morning when light levels are low and, like other oviparous teleosts, become sexually receptive at the time of ovulation (Stacey 1987). Because females spawn (release) their eggs within a few hours, male–female reproductive physiology and behaviour must be synchronized to ensure reproductive success. This synchrony appears to be mediated by two pheromones released in a sequential manner by ovulatory females. Because both pheromones function initially as hormones, they serve as precise signals of the female's reproductive state.

About 10 h prior to ovulation and spawning, female goldfish synthesize increased quantities of the steroidal hormone 17α,20β-dihydroxy-4-pregnen-3-one (17,20βP) to induce final oocyte maturation (nuclear migration and breakdown; Goetz 1983). Subsequently 17,20βP is cleared to the water (Dulka *et al.* 1987) where it functions as a potent olfactory stimulant (threshold $10^{-12}$–$10^{-13}$M; Sorensen *et al.* 1987) with pheromonal actions. Males detecting water-borne 17,20βP experience rapid increases in their gonadotropin (GtH; Dulka *et al.* 1987) and this in turn stimulates increased milt (sperm and seminal fluid) by the time of spawning (Stacey and Sorensen 1986; Dulka *et al.* 1987). Because the rate of 17,20βP release returns to basal levels by the time of ovulation (Dulka *et al.* 1987; Stacey *et al.* 1989), and water-borne 17,20βP has little effect on male behaviour (Sorensen *et al.* 1989*a*), this pheromone may be considered a pre-ovulatory 'primer'. The principal function of 17,20βP may be to increase male fertility, providing an advantage in the intense intermale competition of goldfish spawning.

At the time of ovulation, females become sexually receptive and release another pheromone thought to be composed of prostaglandin $F_{2\alpha}$ and its metabolites (PGFs). This pheromone also functions initially as a hormone; prostaglandin $F_{2\alpha}$ ($PGF_{2\alpha}$) modulates follicular rupture and stimulates female spawning behaviour through direct actions on the brain (see Stacey and Goetz 1982). Electrophysiological and behavioural studies have suggested that circulating PGFs are rapidly metabolized and excreted, and that male goldfish possess exceedingly sensitive olfactory receptor mechanisms for $PGF_{2\alpha}$ and one of its metabolites (probably 15-keto-prostaglandin $F_{2\alpha}$; Sorensen *et al.* 1988). Mature males exposed to water-borne PGFs become sexually aroused, but their GtH does not increase unless they are allowed to sexually interact (Sorensen *et al.* 1989*a*). Accordingly, this pheromone may be considered a post-ovulatory 'releaser', possibly allowing males to detect ovulated females and locate them within spawning groups.

We are only beginning to understand pheromonal function in goldfish. For example, the actions of 17,20βP may be more complex than those of a simple male stimulant. Females also detect 17,20βP (Sorensen *et al.* 1987) and when exposed to it they often ovulate (Sorensen and Stacey 1987); and males exposed to water-borne 17,20βP increase their blood 17,20βP (Dulka *et al.* 1987). Therefore, 17,20βP could be a 'bisexual' pheromone which functions as a group synchronizer. In addition, we recently found that water-borne androgens also function as olfactory stimulants and that when mature males are exposed to these compounds their milt production is inhibited (unpublished results). However, neither the source(s) nor the physiological mechanism of this potential inhibitory pheromone is known. As yet nothing is known about the gender-specific pheromone described by Yamazaki and Watanabe (1979).

## Theoretical implications of the goldfish model

### HORMONAL PHEROMONES: QUESTIONS OF DEFINITION

The issue of whether hormonal metabolites can be considered as 'true' pheromones has been raised (Liley 1982). These compounds do not appear to fulfil the criteria for 'true' communication which are that the signal used be 'specialized' and emitted in a controlled manner directly benefiting the signaller (Burghardt 1970; Liley 1982). At present, we do not have any data to suggest that either 17,20$\beta$P or PGFs are specialized byproducts, or that their release is directly controlled. However, Karlson and Lüscher's (1959) definition merely requires that a pheromone 'release a specific action, for example a definite behaviour or developmental process'. Also, their analogy of hormonal and pheromonal function suggests that they did not interpret 'communication' in the strict sense it has recently developed. Accordingly, as goldfish hormonal pheromones represent potent chemical signals released in very specific manners and conveying specific, potentially beneficial information to conspecifics, we consider them to be 'true' pheromones. However, what is of interest to us is not whether the goldfish pheromonal system represents 'true' communication, but that it almost certainly represents a necessary intermediate step in the evolution of chemical communication.

### HORMONAL PHEROMONES: EVOLUTIONARY QUESTIONS

Our findings that goldfish use at least two types of hormones as pheromones strongly reinforce theoretical arguments that fish may have evolved to use hormones and their metabolites as pheromones (Døving 1976; Colombo *et al.* 1982). Briefly, these theories identify three reasons why hormones and their metabolites are likely to develop pheromonal functions: they represent pre-existing signals produced in synchrony with discrete reproductive events; they are readily excreted to the water; and the evolution of olfactory receptor mechanisms for pheromones might be simplified by the prior existence of internal endocrine receptors. For pheromonal function to arise, mutations are initially required only in the recipient; evolution *de novo* of an entire specialized pheromonal system is not required. Later, if the reproductive success of the recipient increases the success of the donor, the evolution of a specialized signalling system might occur.

### HORMONAL PHEROMONES: QUESTIONS OF FUNCTION

As the diversity of teleost fish exceeds the diversity of their hormone systems, not all hormonally derived pheromones can be species-specific. Several examples of non-species-specific fish pheromones are known (Hunter and Hasler 1965; Rossi 1969; Chien 1973; Chen and Martinich 1975; McKinnon and Liley 1987). For

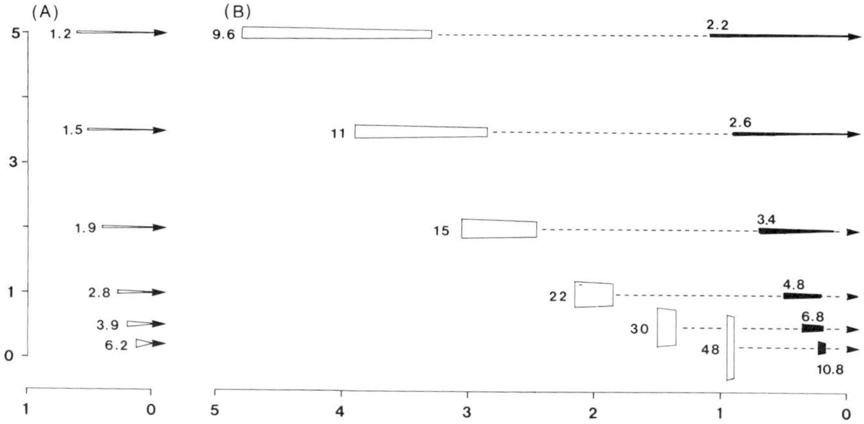


Fig. 1. Effects of swimming speed and mechanism of pheromone release on the detectability of hypothetical 17,20$\beta$P plumes released by female goldfish (arrows). The length of the detectable plume and the time (s) elapsed before the plume reaches undetectable concentrations ($10^{-12}$M) is shown under two release modes: (A) continuous release; (B) pulsatile release (one 3-s pulse every 3 mins; see text) by ovulatory (open arrows) and non-ovulatory females (closed arrows). The Y-axis is swimming speed (body lengths $s^{-1}$).

those species whose reproductive activities are sympatric, there is the potential problem that a common sex pheromone could trigger maladaptive interspecific responses. At least in the case of goldfish, however, the dynamics of pheromone release appear to minimize this problem. For example, a 20 g ovulatory goldfish releases approximately 650 ng of 17,20$\beta$P in the 9-h period prior to ovulating whereas a non-ovulatory female releases only 30 ng (Stacey *et al.* 1989). If we assume that 17,20$\beta$P is released at a constant rate (20 pg $s^{-1}$ and 1 pg $s^{-1}$ respectively), the odour plume's diameter increases 1 cm $s^{-1}$, and males detect 17,20$\beta$P at $10^{-12}$M, then over the normal range of swimming speeds (0.2–5.0 body lengths $s^{-1}$), the dimensions of the ovulatory odour plume will range from 20 cm to 1.2 cm in diameter and become undetectable 1.2 to 6.0 s after the fish has passed (Fig. 1a). Such a small and transient active space is only likely to be encountered by an interested conspecific.

The ephemeral nature of the goldfish's odour plume appears to contradict predictions (Wilson 1975; Liley 1982) that aquatic pheromones (like aerial pheromones) should affect large areas and have slow transmission and fade-out times. Because fish pheromonal signals may encorporate temporal coding, it becomes important to understand the dynamics of pheromone release and the minimum exposure time required to elicit a response. We do not know how goldfish release pheromones but if, for example, goldfish release 17,20$\beta$P within pulses of urine (a realistic possibility; Colombo *et al.* 1982) a very different pattern of detectability emerges. Should the 20 g female void her urine as 3-s pulses every 3 mins the resulting odour cloud would be detectable for up to 4 m and non-ovulatory females would be detectable (Fig. 1b)—but considerable distances (36 m at a speed of 2 body lengths $s^{-1}$) would separate the pulses. Such a signal would be inappropriate for identifying individuals, but would function well as a group stimulant/synchronizer.

Although controlled release may frequently eliminate the need for species-specific pheromones, closely related sympatric species might still benefit from species-specific signals. For example, McKinnon and Liley (1987) found that while female belontiids with a history of sympatric spawning release species-specific pheromones, those with an allopatric history do not. Species-specificity might be achieved either by using pheromonal mixtures which include non-hormonal compounds, or by modifying hormone metabolic pathways. Should the extreme specificity of the goldfish olfactory system to 17,20βP (Sorensen *et al.* 1989→b) be representative of other species, then very small changes to a metabolite's structure could readily impart species-specific information.

## Potential diversity of fish sex pheromone systems

Fish pheromone systems are likely to be characterized by three types of variation: variation in the types of hormones and metabolites (e.g. 17,20βP-like steroids) utilized as pheromones, variation in the structure of the most potent metabolites, and variation in the nature of responses to pheromone types. Structural variation is likely to arise primarily from natural variation in hormone synthesis pathways and, as discussed, from selective pressure for reproductive isolating mechanisms. Whether a particular hormone type comes to function as a pheromone and what its biological effects might be are likely to be determined by the relationship between hormone synthesis pathways, reproductive strategies, and the potential benefit and cost (including the likelihood of an inappropriate response) of using the available metabolic signals. Although very little is known about these relationships, it presently appears that oviparous teleosts, like most vertebrates, use two basic types of sex pheromones: 'primers', the principal effects of which are physiological; and 'releasers', the principal effects of which are behavioural (Wilson and Bossert 1963).

### PRIMERS

It is probable that 17,20βP developed a primer function in goldfish because it is synthesized at precisely the right time to signal impending ovulation and because male and female goldfish are closely associated during the time of oocyte final maturation. Because many fish use a similar reproductive strategy, and all fish studied to date use 17,20βP-based hormones for final oocyte maturation (Scott and Canario 1987), we expect that 17,20βP-like compounds are commonly used as pre-ovulatory pheromones. Peak circulating levels of 17,20βP during oocyte maturation vary enormously among teleosts (0.2–1000 ng $ml^{-1}$; Scott and Canario 1987) with the highest levels being found in salmonids, which also maintain these levels for several days after ovulation. Although this extended period of elevated 17,20βP synthesis may reflect an unknown hormonal function, a pheromonal function should be considered. For example, in studies of wild rainbow trout (*Salmo gairdneri*) it has been found that 17,20β production increases in sexually active females and that males interacting with these females have increased milt production (Liley *et al.* 1987). If, however, 17,20βP levels are indicative of pheromonal function, it is likely the steroid is metabolized prior to its release because 17,20βP itself does not appear to be an olfactory stimulant for salmonids (Hara and Sveinsson, personal communication).

Male fish may also release hormonal primers. The odour of mature males stimulates ovarian growth and ovulation in both the zebrafish (*Brachydanio rerio*) and a cichlid (*Pterophyllum scalare*) (Chen and Martinich 1975; Chien 1973). More recent work indicates that the zebrafish pheromone is a conjugated C19 steroid (van den Hurk *et al.* 1987). Similarly, as previously discussed, 17,20βP may also be released by male goldfish to function as a 'bisexual' primer. Similar bisexual primers may account for the tightly synchronized spawning activity of many fish.

### RELEASERS

Females of many teleost fish which have recently ovulated have been found to release potent pheromones that stimulate male reproductive behaviour (Liley 1982; Stacey *et al.* 1986, 1987). We believe that most, if not all, of these post-ovulatory releasers are prostaglandins (PGs). The reason for this is simple—PGs play fundamental roles in mediating the two principal prerequisites for spawning: they modulate follicular rupture (Goetz 1983) and stimulate female spawning behaviour in a wide variety of fish (Stacey 1987). Additionally, PGs are rapidly metabolized and released by fish (Sorensen *et al.* 1988), and studies in mammals indicate that PG metabolic pathways may be flexible enough to permit species-specificity (Granstrom and Kindahl 1982). PG pheromones could have a variety of functions: identifying ovulated individuals in competitive group spawning situations (e.g. the goldfish); facilitating male–female bonds during prolonged pair spawning (e.g. salmonids); and allowing males to distinguish sneaker males (e.g. salmonids) or female-mimic males (e.g. centrachids) from ovulated conspecifics. Although work in this field is just beginning, there is evidence that a number of species other than goldfish employ PG pheromones. Courtship responses of male fathead minnows (*Pimephales promelas*) are stimulated by the odour of PG-injected females (Cole and Smith 1987); male milkfish (*Chanos chanos*) become sexually aroused when exposed to water-borne $PGF_{2\alpha}$ (Kelley, Tamaru and Lee, personal communication); the olfactory sense of male arctic charr, (*Salvelinus alpinus*) detects water-borne PGFs (Sveinsson and Hara, personal communication).

Males of many species are also known to produce releasers (see: Liley 1982; Stacey *et al.* 1986). Where studied, these pheromones appear to be hormonal in origin. Perhaps the best documented example is the black goby, *Gobius jozo*. Males of this species are territorial and maintain nest sites whereas the females are free-ranging and actively seek out males when ovulated. Male *G. jozo* possess a testis which is specialized for the production of 5$\beta$-reduced androgens, in particular etiocholanolone glucuronide which is attractive to ovulated females (Colombo *et al.* 1982). More recently, it has been found that male African catfish, *Clarius gariepinus*, also have specialized testicular structures (seminal vessicles ) which produce 5$\beta$-pregnane-3$\alpha$,17$\alpha$-diol-20-one glucuronide, which functions as a potent olfactory stimulant and an attractant for ovulated females (Resink *et al.* 1987).

Male releasing pheromones can also serve less specialized functions. For example, milt released by the Pacific herring, *Clupea harengus*, triggers the release of both milt and eggs from proximate conspecifics (Stacey and Hourston 1982). Interestingly, water-borne $PGF_{1\alpha}$ has recently been shown partially to mimic this effect (Kyle and Sherwood, personal communication). Although bisexual releaser pheromones have not yet been described in fish, they are clearly a possibility for pelagic species with very simple spawning strategies.

#### OTHER POSSIBILITIES

Fish hormonal pheromones, like fish reproductive strategies, appear to have great inherent flexibility. It is likely that hormonal pheromones will be found to serve a wide variety of functions which eclipse those of simple primers and releasers. The chemical signals that mediate complex social interactions between mature males (see: Todd 1971; Liley 1982) could be based on hormones that mediate behavioural function. Similarly, the social attractant released by female zebrafish appears to be comprised of conjugated sex steroids (van den Hurk and Lambert 1983). It also seems reasonable to speculate that 'crowding' pheromones (Chen and Martinich 1975) may be metabolites of 'stress' hormones. Finally, there is no reason why some pheromones could not have both behavioural and physiological functions, or why pheromones that function as primers in one species could not be used as releasers by another. We anticipate that future studies of fish pheromone systems will reveal the dynamic relationships that can link the physiology, behaviour, and ecology of aquatic organisms.

## Acknowledgements

Much of the goldfish research was supported by the Alberta Heritage Foundation for Medical Research, the Department of Fisheries and Oceans Canada, and the Natural Sciences and Engineering Council of Canada (Grant A2903 to N.E.S.). The senior author (P.W.S.) was supported by the Minnesota Agricultural

Experimental Station at the time this manuscript was written. Contribution No. 16,178 of the Minnesota Agricultural Experiment Station based on research supported by the Station.

## References

Burghardt, G.M. (1970). Defining 'communication'. In *Advances in chemoreception*, Vol. 1 (ed. J.W. Johnson, D.G. Moulton and A. Turk), pp.5–18. Appleton-Century-Crofts, New York.

Chen, L.C. and Martinich, R.L. (1975). Pheromonal stimulation and metabolite inhibition of ovulation in the zebrafish, *Brachydanio rerio*. *Fish. Bull.*, 73, 889–94.

Chien, A.K. (1973). Reproductive behavior of the angelfish *Pterophyllum scalare* (Pisces: Cichlidae). II. Influence of male stimuli upon the spawning rate of females. *Animal Behav.*, 21, 457–63.

Cole, K.S. and Smith, R.J.F. (1987). Release of chemicals by prostaglandin treated female fathead minnows, *Pimephales promelas*, that stimulates male courtship. *Horm. Behav.*, 21, 440–56.

Colombo, L., Belvedere, P.C., Marconato, A. and Bentivegna, F. (1982). Pheromones in teleost fish. In *Proceedings of the second international symposium on the reproductive physiology of fish*. (ed. C.J.J. Richter and T.J.Th. Goos), pp. 84–94. Pudoc, The Netherlands.

Døving, K.B. (1976). Evolutionary trends in olfaction. In *The Structure-activity relationships in chemoreception*. (ed. G. Benz), pp. 149–59. IRL Press, London.

Dulka, J.G., Stacey, N.E., Sorensen, P.W. and Van Der Kraak, G.J. (1987). Sex steroid pheromone synchronizes male–female spawning readiness in the goldfish. *Nature* (London), 325, 251–3.

Goetz, F.Wm. (1983). Hormonal control of oocyte final maturation and ovulation in fishes. In *Fish physiology* Vol. IXB (ed. W.S. Hoar, D.J. Randall and E.M. Donaldson), pp. 117–70. Academic Press, New York.

Granstrom, E. and Kindahl, H. (1982). Species differences in circulating prostaglandin metabolites: relevance for the assay of prostaglandin release. *Biochim. Biophys. Acta*, 713, 555–69.

Hunter, J.R. and Hasler, A.D. (1965). Spawning association of the redfin shiner, *Notropis umbratilis*, and the green sunfish, *Lepomis cyanelleus*. *Copeia*, 265–81.

Karlson, P. and Lüscher, M. (1959). 'Pheromones': a new term for a class of biologically active substances. *Nature* (London), *183*, 55–6.

Liley, N.R. (1982). Chemical communication in fish. *Can. J. Fish. Aquat. Sci.*, 39, 22–35.

Liley, N.R., Caldwell, J. R. and Rouger, Y. (1987). Current status of hormones and sexual behavior in fish. In *Proceedings of the third International symposium on the reproductive physiology of fish*. (ed. D.R. Idler, L.W. Crim and J.M. Walsh), pp. 142–9, Memorial University Press, St. John's, Newfoundland.

McKinnon, J.S. and Liley, N.R. (1987). Assymetric species specificity in response to female pheromone by males of two species of trichogaster (Pisces: Blontidae). *Can. J. Zool.*, 65, 1129–34.

Resink, J.W., van den Hurk, R., Peters, R.C. and van Oordt, P.G.W.J. (1987). Steroid glucuronides as sex attracting pheromones in the African catfish, *Clarius gariepinus*. In *Proceedings of the third International symposium on the reproductive physiology of fish*. (ed. D.R. Idler, L.W. Crim and J.M. Walsh), p. 163. Memorial University Press, St. John's, Newfoundland.

Rossi, A.C. (1969). Chemical signals and nest building in two species of *Colisa* (Pisces: Anabantidae). *Monit. Zool. Ital.*, 3, 225–37.

Scott, A.P. and Canario, A.V.M. (1987). Status of oocyte maturation-inducing steroids in teleosts. In *Proceedings of the third International symposium on the reproductive physiology of fish*. (ed. D.R. Idler, L.W. Crim and J.M. Walsh), pp. 224–34. Memorial University Press, St. John's, Newfoundland.

Sorensen, P.W. and Stacey, N.E. (1987). 17$\alpha$,20$\beta$-dihydroxy-4-pregnen-3-one functions as a bisexual priming pheromone in goldfish. *Amer. Zool.*, 27, 412.

Sorensen, P.W., Hara, T.J. and Stacey, N.E. (1987). Extreme olfactory sensitivity of mature and gonadally-regressed goldfish to a potent steroidal pheromone, 17$\alpha$,20$\beta$-dihydroxy-4-pregnen-3-one. *J. Comp. Physiol. A*, 160, 305–13.

Sorensen, P.W., Hara, T.J., Stacey, N.E. and Goetz, F.Wm. (1988). F prostaglandins function as potent olfactory stimulants that comprise the postovulatory female sex pheromone in goldfish. *Biol. Reprod.*, 39, 1039–105.

Sorensen, P.W., Stacey, N.E. and Chamberlain, K.J. (1989*a*). Differing behavioral and endocrinological effects of two female sex pheromones on male goldfish. *Horm. Behav.*. (in press).

Sorensen, P.W., Hara, T.J., Stacey, N.E. and Dulka, J.G. (1989*b*). Extreme olfactory specificity of male goldfish to the preovulatory pheromone. 17$\alpha$,20$\beta$-dihydroxy-4-pregnen-3-one. *J. Comp. Physiol.*

Stacey, N.E. (1987). Roles of hormones and pheromones in fish reproductive behavior. In *Psychobiology of reproductive behavior*. (ed. D. Crews), pp. 28–69. Prentice-Hall, New York.

Stacey, N.E. and Goetz, F.Wm. (1982). Role of prostaglandins in fish reproduction. *Can. J. Fish. Aquat. Sci.*, 39, 492–8.

Stacey, N.E. and Hourston, A.S. (1982). Spawning and feeding behavior of captive Pacific herring, *Clupea harengus pallasi*. *Can. J. Fish. Aquat. Sci.*, 39, 489–98.

Stacey, N.E. and Sorensen, P.W. (1986). 17$\alpha$,20$\beta$-dihydroxy-4-pregnen-3-one: a potent steroidal primer pheromone which increases milt volume in the goldfish *Carassius auratus*. *Can. J. Zool.*, 64, 2412–7.

Stacey, N.E., Cook, A.F. and Peter, R.E. (1979). Spontaneous and gonadotropin-induced ovulation in the goldfish, *Carassius auratus L.*: effects of external factors. *J. Fish Biol.*, 15, 349–61.

Stacey, N.E., Kyle, A.L. and Liley, N.R. (1986). Fish reproductive pheromones.

In *Chemical signals in vertebrates*, Vol. IV, *Ecology, evolution and comparative biology*. (ed. D. Duvall, D. Müller-Schwarze and R.M. Silverstein), pp. 119–33. Plenum Press, New York.

Stacey, N.E., Sorensen, P.W., Van Der Kraak, G.J. and Dulka, J.G. (1989). Direct evidence that 17α,20β-dihydroxy-4-pregnen-3-one functions as a goldfish primer pheromone: Preovulatory release is closely associated with male endocrine responses. *Gen. Comp. Endocrinol.*, 75, 62–70.

Stacey, N.E., Sorensen, P.W., Dulka, J.G., Van Der Kraak, G.J. and Hara, T.J. (1987). Teleost sex pheromones: recent studies on identity and function. In *Proceedings of the third international symposium on the reproductive physiology of fish*, (ed. D.R. Idler, L.W. Crim and J.M. Walsh), pp. 150–4. Memorial University Press, St. Johns, Newfoundland.

Todd, J.H. (1971). The chemical language of fishes. *Sci. Amer., 224*, 98–108.

van den Hurk, R. and Lambert, J.G.D. (1983). Ovarian steroid glucuronides function as sex pheromones for male zebrafish, *Brachydanio rerio*. *Can. J. Zool.*, 61, 2381–7.

van den Hurk, R., Schoonen, W.G.E.J., van Zoelen, G.A. and Lambert, J.G.D. (1987). Biosynthesis of steroid glucuronides in the testis of the zebrafish *Brachydanio rerio*, and their pheromonal function as ovulation inducers. *Gen. Comp. Endocrinol.*, 68, 179–88.

Wilson, E.O. (1975). *Sociobiology: The new synthesis*. Harvard University Press, Cambridge, MA.

Wilson, E.O. and Bossert W.H. (1963). Chemical communication among animals. In *Recent progress in hormone research*, Vol. 19 (ed. G. Pincus), pp. 673–716. Academic Press, New York.

Yamazaki, F. and Watanabe, K. (1979). The role of steroid hormones in sex recognition during spawning behavior of the goldfish, *Carassius auratus* L. *Proc. Indian Natl. Sci. Acad.* Part B., 45, 505–11.

# SECTION IV

## Social behaviour

# IV.1 Scent Marking by Resource Holders: Alternative Mechanisms for Advertising the Costs of Competition

L.M. GOSLING

## Introduction

Although there is great variation in the location of specialized scent glands, scent marking occurs in a remarkably uniform fashion amongst male mammals. Deposits or smears of glandular secretion, and often faeces and urine, are placed at conspicuous places in ranges or territories, often in lines along paths or boundaries. Marking is most commonly carried out by those males that have established dominance over their peers in contests and is almost universally linked to agonistic behaviour, forming a central part of many ritualized contests between males, or competing groups: the aptly named 'stink fights' which occur between neighbouring groups of lemurs, *Lemur catta* (Jolly 1966), are a well known example. Despite this essential uniformity there has been an enormous proliferation of ideas about the function of scent marks. The reason for this apparent discrepancy may be that some early and influential ideas were formulated in such a restricted way that their predictions were easily falsified. These early proposals were restricted to scent marking in territories (Uexküll and Krisz t 1934; Hediger 1949) and suggested that marks should repel intruders. When it subsequently became clear that many non-territorial animals marked (Ralls 1971) and that many intruders were not deterred by scent marks (Walther 1979; Franklin 1980) alternative explanations were needed.

These problems are avoided if the central idea in the early treatments of scent marking as territory defence is reformulated in terms of the selective advantage of status advertisement in male intrasexual competition. The costs to a breeding male of defending its status include time, energy, and the risk of injury and death. These costs can be reduced by prior assessment of the likely outcome of any interaction; this assessment will include the competitive ability (used here as a synonym of Resource Holding Power; Parker 1974) of potential opponents (Maynard Smith and Price 1973; Parker 1974; Maynard Smith and Parker 1976; Zahavi 1975, 1977). The principal function of scent marking in male intrasexual competition may thus be for status advertisement, or, more generally, for resource holding males to reduce the costs of agonistic encounters by providing a means of assessment to potential competitors. This proposition has been discussed elsewhere in relation to scent marking in territories (Gosling 1982, 1985; Erlinge *et al.* 1982; Gorman 1984). In this paper scent marking in territories is treated as a special case, although one that is relatively easy to understand because of the simple link between dominance and a scent marked area.

It is proposed that there are only three basic mechanisms involving scent marks by which resource holding males could transmit information about their status to potential competitors. These are (i) that the information is an intrinsic property of the mark; (ii) that receivers learn to associate the odour of a mark with that of a known resource holder; and (iii) that receivers compare the memorized odour of marks with potential opponents and thus identify resource holders. The argument used to discuss the deployment of these possible mechanisms assumes that the mechanisms employed should depend on the costs and benefits to the signaller; costs would arise principally from agonistic encounters and benefits either directly or indirectly from enhanced reproductive success.

A curious property of signalling by scent marks is that it may depend partly on the receiver seeking out the marks. In other signalling modes, for example where a dominant male calls, subordinates have little choice but to receive the information that is transmitted. This paper considers the significance of this difference, both in terms of the value to the receiver in assessing opponents and the constraints that this imposes on the accuracy of the information that a resource holder transmits.

There are two principal actors in the arguments presented. One is the 'resource holder' (following Parker 1974), usually a male which has breeding status because it has established dominance over competing males. Such animals live either alone or in groups containing subordinate males and in either situation there may be varying degrees of area defence; in its most simple form territories with clearly defined boundaries may be defended. These males are often referred to as 'signallers' in this paper since it is these animals that send information in scent marks. The other actor is a male that does not have breeding status; these are the so-called 'receivers' because they are the animals that detect and use the information in scent marks. They are 'subordinate' males in dominance mating systems and non-territorial males where breeding males occupy territories; non-territorial males are referred to as 'intruders' when they are within a territory.

## Scent marking, competitor assessment, and the advertisement of costs

Explanations of scent marking, particularly scent marking in territories abound in the literature. Commonly cited interpretations are:

1. Marking deters potential intruders (Uexküll and Kriszat 1934; Hediger 1949).
2. Marking intimidates intruders (Geist 1965; Mykytowycz 1965).

3. Marking enhances the confidence of owners (Mykytowycz 1973, 1975; von Richter 1974).
4. Intrinsic properties of marks provide intruders with information about the status of the owner (Mykytowycz 1970).
5. Marks orientate the resident within its territory (Hediger 1949; Walther 1978).
6. Marks attract or stimulate mates (Noble 1939).
7. Marks assist in pair-bond formation (Rothman and Mech 1979).
8. Marks assist in population regulation (Wynne-Edwards 1962; Aleksiuk 1968).
9. Marks assist owners to forage optimally (Henry 1977).

These various suggestions have been reviewed elsewhere (Gosling 1982, 1985). Here it is emphasized that these diverse suggestions refer to what appears to be an essentially unitary phenomenon. While undoubtedly there are complexities, and possibly multiple functions in some cases, in general it is not useful to suggest different explanations for every minute variation of a phenomenon, but rather to seek explanations that have broad explanatory value. As suggested in the introduction, the excessive proliferation of ideas may have occurred because early hypotheses linking marking with intrasexual competition were too narrowly formulated and, as a result, were easily falsified. When systematic observations of wild animals became common, it was quickly seen that animals without territories often scent marked and that intruders readily entered scent marked territories when the territory owner was elsewhere.

In fact, the observation that males, for example non-territorial male antelopes, routinely walk through and feed in scent marked areas is not surprising. When resources are dispersed and the owner is not close by, the likely costs of intrusion are low and would often be outweighed by the benefits of utilizing the resource. Cost/benefit considerations of this kind may often predict when an animal should use or ignore the information present in a scent mark. Where costs are potentially very high in relation to likely benefits animals might be deterred by scent marks. For example, costs could potentially be high when a limiting resource is concentrated or the chance of injury in a contest is high. In this sort of situation it may pay an intruder to withdraw when it detects any fresh marks rather than to risk meeting a potential opponent. For example, the chance of injury when moles meet underground may be high and this may be why territorial moles, *Talpa europaea*, appear to avoid both territory neighbours and their scent marks (Stone and Gorman, this volume). However, empirical data suggest that such cases are rare compared to those where intrusion is common.

Despite the fact that many intruders in territories are not deterred by scent marks, a link between scent marking and territory defence seems likely. Territory owners often mark during agonistic encounters with intruders and neighbours (Jolly 1966; Estes 1969), many mark the boundaries of their territories, and they leave more marks where intruder pressure is greatest (Gilbert 1973; Walther 1978; Gosling 1981, 1985). An alternative to the idea that marks deter intruders is that marks provide a means of reducing the costs of defence by helping to establish an asymmetry between the owner and an intruder. Marks could function in this way if, in game theory terms, they allowed the intruder to assess the owner, or, in this case, to identify it as an owner. Since owners generally have a better knowledge of the resources of their territory than an intruder, they have more to lose than the intruder has to gain and will be prepared to escalate an encounter further than the intruder. Since owners have been able to take over and hold on to the territory, they must also be of at least reasonably high competitive ability. It would thus pay most intruders to cut their losses and withdraw early when they identify a potential opponent as a territory owner. The costs of territory defence would thus be reduced for the owner (Gosling 1986) and so it would pay to advertise its identity in this way.

This hypothesis can readily be extrapolated to dominance and harem defence mating systems because it refers essentially to status advertisement. The advertisement of territory ownership is a special case of the general phenomenon of status advertisement by a resource holder. In its general form the hypothesis does require that resource holders are sufficiently sedentary to be present in the same general area as the scent marks they have left. However, this is not an obstacle to the generalization stated because the mark, whatever the function, would be of little use if the animal that made the mark had left the area. Subordinate males could thus assess potential opponents from information in scent marks that a dominant male had placed either in conspicuous places throughout its home range or near to a defended group of females. Thus the idea that scent marks are used for competitor assessment could apply simply to such animals as bison and white-tailed deer which have dominance mating systems and red deer which defend loosely defined areas around harems.

The term 'status advertisement' is often used rather loosely; it is important to define it carefully if the spatial characteristic of scent marks is to be understood. When an animal advertises its 'status' it is regarded here as displaying its competitive ability (formally, 'resource holding power'; Parker 1974), or, more precisely, the likely costs to the receiver if an encounter with the signaller is escalated. The unique property of scent marking as a type of animal signalling is that this information is projected not wherever the signaller happens to be, but in predetermined places where the scent-marks are placed. Scent marks are thus conditional signals that operate only at particular locations.

What are the characteristics of these locations? If marks do advertise status, as suggested and, if, as discussed elsewhere (Gosling 1981) the number of marks available is limited by economic considerations, then marks should only be placed where they are most important, that is, where there are limiting resources. These resources may be simply food or water but more often they are directly or indirectly concerned with reproduction. Marks are usually placed in a territory that females need to enter, perhaps for feeding purposes (resource defence territoriality; Emlen and Oring 1977), in a territory that contains one or more resident females or in the area around a group of females.

Fig. 1. A male gerenuk, *Litocranius walleri*, marks a female with the secretion of the antorbital gland during a mating interaction. Males may thus establish an owner advantage over any male that attempts to compete for the female, a context for scent matching which could be particularly important in dominance mating systems or in other situations where mate guarding is common. (From Gosling 1985; reprinted with permission from Oxford University Press.)

The link between scent marking and reproductive success is shown most explicitly in cases where males directly mark females. Following the above argument, the resource holder may deposit these marks to advertise the potential costs to other males of any attempt to compete for the female. Badgers, *Meles meles*, mark females with glandular secretion (Kruuk *et al.* 1984) and the males of a number of hystricomorph rodent species spray females with urine or mark them with glandular secretion (Kleiman 1974) during courtship. Male gerenuk, *Litocranius walleri*, mark females with antorbital gland secretion during mating interactions (Fig. 1) and male Gray's waterbuck, *Kobus megaceros*, mark females by rubbing their urine soaked manes on to the female's body (Walther 1979). This could also be the reason that males spray urine on their underparts in a number of other species, including goats *Capra* spp. (Shank 1972; Coblenz 1976), since the urine would be transferred to the female during copulation. This behaviour could be particularly important for mate guarding in dominance systems. Marking could help to prevent other males from mating with a female, which they have themselves mated, for long enough to ensure that the female has been fertilized (see Jewell *et al.* 1986).

The conditional spatial significance of scent marks becomes particularly clear when trying to understand why scent marking occurs at all when two competitors interact frequently. Territorial neighbours have agonistic encounters at least daily at common boundaries as do males that grow up within a social group. Intuitively, it seems likely that such males learn each others' competitive ability in some detail, and that they know each others' odour and that of their scent marks. If these opponents are so familiar with each other, why do they mark? Clearly no further assessment of competitive ability is needed. The probable answer is that these males are signalling the costs to their opponent of competition for the resource at the point where the mark is deposited. There is no advantage in unconditional assessment between neighbouring territorial males because the likely costs of escalation depend on where the male is in relation to its defended resource. This variation probably occurs not only in relation to whether or not a male is in its territory but also in relation to variation in resource value within the territory. At a territory boundary, where males and social groups often mark at high intensity during agonistic encounters, marks signal to the neighbour the costs of attempting to expand the territory. Marks may thus place a spatial limit on the owner advantage of males in neighbouring territories (Gosling 1987). Within social groups, marks may signal to subordinate group members the costs of any attempt to escalate a contest and challenge for breeding status.

## How do males transmit information about status?

Assuming that the function of scent marks in a male intrasexual context is for status advertisement, how could a resource holder use scent marks to transmit this information to competitors? Three basic mechanisms are possible.

### INTRINSIC MEANING

The information about status could be an intrinsic part of the odour in the scent mark. This is what was envisaged in the original concept of a pheromone and what seems to be a common feature of invertebrate olfactory signals and of mammalian priming pheromones. In the present context, the scent mark could contain the information that the mark was made by a male of high competitive ability, for example if it indicated, or contained, high androgen levels. The receiver might withdraw after detecting such marks to avoid meeting a potentially dangerous opponent.

As mentioned above, the main evidence against the idea of intrinsic meaning is that field observations show almost unanimously that intruders into territories do not usually avoid scent marked areas. Similar results have sometimes been obtained in captivity (e.g. gerbils, *Meriones unguiculatus*; Fullenkamp *et al.* 1985) although such negative results are probably not usually reported and may be under-represented in the literature.

In spite of this there is experimental evidence that a number of mammals can discriminate between the odour of different categories of animals, including males versus females (Bowers and Alexander 1967), castrates versus intact males (Landauer *et al.* 1977; Sawyer 1981), and dominant versus subordinate males (Jones and Nowell 1974). Most of this work demonstrates sensory ability rather than function but in a few cases there is evidence of an adaptive response. For example, male laboratory mice tend to avoid the smell of dominant males but to ignore that of subordinates (Jones and Nowell 1974). In general, it seems likely that males can detect broad properties of a signaller from its scent marks so that some information about status could be transmitted directly.

### LEARNT ASSOCIATION

Males could learn to associate the smell of a resource holding male with its marks. When interacting with the resource holder (and learning its odour) the subject would also learn its competitive ability. If the male was potentially dangerous, a male of relatively low competitive ability could then reduce the chance of costly encounters by avoiding the area it had marked (Brown 1979, p. 138).

The main evidence that has been presented for the existence of this mechanism is the link between aggressive behaviour and scent marking (Ralls 1971) and the demonstration that animals can learn to distinguish between the odour of individuals. Rasa (1973) and Kruuk *et al.* (1984) showed, respectively, that mongooses, *Helogale undulata*, and badgers can distinguish between individuals using only the odour of glandular secretion. However, these demonstrations of sensory capacity do not necessarily mean that wild animals form, and use, learnt associations between a resource holder and its marks.

It is also assumed that animals interacting regularly with each other would learn each others' identity. This seems likely, particularly in the cases of neighbouring territorial males, or groups, and in the case of social groups whose members live with each other for long periods.

### SCENT MATCHING

The low status male could memorize the smell of the marks as it enters a new area and compare this odour with that of any potential opponent that it met. When the scents matched, the male would know that the opponent was a resource holder and, depending on its own competitive ability, it could withdraw from the encounter at an early stage (Gosling 1982). Resource holders would thus benefit from reduced costs of intrasexual competition.

Direct demonstrations of scent matching are rare although the hypothesis is generally supported by observations of scent marking behaviour in the literature (Gosling 1982, 1985; Kruuk *et al.* 1984). Most important, it explains the widespread observation that territory owners and other resource holders anoint their own bodies with the substance used for scent marking and make themselves available for scent matching at the start of many encounters (Fig. 2); many males evert their scent glands as they approach opponents (e.g. hyaenas, *Crocuta crocuta*; Kruuk 1972). The existence of scent matching has been shown experimentally in another functional context: male hamsters are more likely to select females as mates if they match the odour of scent marks, made by the females, which the males had previously encountered (Steel 1984). This suggests not only that

Fig. 2. An encounter between a territorial male hartebeest, *Alcelaphus buselapus*, (right) and an intruder. The territorial male stands immobile as the intruder smells an areas of the pelage where owners frequently rub their own antorbital gland secretion. This secretion is also used to mark objects in their territories. The territorial male may be giving the intruder an opportunity to match its odour with that of marks in the vicinity which the intruder has memorized. (From Gosling 1982; reprinted with permission of Verlag Paul Parey, Berlin and Hamburg.)

hamsters use scent matching but also that they can remember marks. This latter ability, a prerequisite of the scent matching hypothesis, has also been shown experimentally by Roeder (1983) using genets (*Genetta genetta*). When two rabbits, *Oryctolagus cuniculus*, are placed in an arena with the scent marks of only one of the animals (chin and anal gland secretion) the animal that matches the marks proves to be dominant in the contests that follow (Mykytowycz 1973, 1975); these results are consistent with competitor assessment by scent matching although they were originally interpreted in terms of 'confidence enhancement'.

## Functional considerations: when would these mechanisms be favoured by selection?

Responses to the information in marks can occur (i) when the receiver detects a scent mark, in the absence of the signaller (the resource holder), and (ii) when it meets the signaller, having previously detected some marks. The receiver may also detect additional marks during its encounter with the signaller because marking is very common during agonistic encounters.

### RESPONSES TO MARKS IN THE ABSENCE OF THE SIGNALLER

Considering first the situation where the receiver only detects marks, and assuming that the function of scent marking is to convey the signaller's status to potential opponents, how should the receiver behave if it used each of the three mechanisms outlined above? Two responses are possible, one to leave the scent marked resource and the other to continue to use it. If an animal leaves, it must have used either intrinsic information in the mark or information from a learnt association between the signaller and its marks. It withdraws to avoid an encounter with the signaller because either intrinsic information (such as high androgen levels) or learnt information (from being defeated in previous encounters) suggests that the encounter would be costly. However, as discussed earlier, the response of the receiver depends on benefits as well as costs: if the value of the scent marked resource outweighs the risk of meeting the signaller then the receiver should stay. Thus, even when intrinsic or learnt information is used there is no simple prediction about the receiver's behaviour. A further complication is that the cost/benefit ratio is also affected by variation in the receiver's status: a male that is seeking to take over a resource, or gain breeding status, will be prepared to incur high costs (in an escalated contest) because it may gain high benefits.

In spite of these complications, if either intrinsic or learnt information were used frequently on their own, we would expect to see animals withdraw from scent marks at a certain frequency in all populations where these mechanisms were used. In fact such behaviour is extremely rare, suggesting that these mechanisms are correspondingly uncommon. In the small number of cases where withdrawal from marks occurs it would be expected that costs should outweigh benefits to an unusual extent. One example may be that of moles where receivers might respond to intrinsic information in marks by withdrawal because the costs of contests in underground burrows are often very high. Mice in experimental arenas may avoid scent marked areas (e.g. Jones and Nowell 1974) because the marks signal the possibility of some costs (if the subject were to meet the animal which left the marks) while there are no resources in the marked area and thus no benefits at all.

If scent matching is the primary mechanism for obtaining information about the signaller's status then animals that detected the scent marks would not be expected to withdraw. The fact that such responses are so rarely seen is consistent with this expectation.

A further theoretical problem concerning the use of intrinsic information in marks is that it may usually be of too general a character to give precise information to a male about the likely costs of a contest: for example, androgen levels could indicate that an opponent was in a generally dangerous category but probably could not indicate the small differences in competitive ability which determine the outcome of contests. Thus any potential for using intrinsic information might only arise where the differences in status between the signaller and receiver are large and thus easily perceived and assessed. Conversely, intrinsic information at such poor resolution could often lead to a low status male avoiding areas that can be utilized safely. If intruders show any tendency to remain in an area after detecting a mark they would eventually become habituated to the signals under these circumstances. Perhaps such generalized information should only be used as an interim stage in assessment. One possibility is that if males recognize that a particular mark is, say, of an adult male, they will then spend more time investigating it in order to memorize the odour. The information gained could then be processed using another mechanism (see below) when the male meets a potential opponent.

### RESPONSES AFTER MEETING THE SIGNALLER

The principal danger for an animal encountering a resource holder is that owners are generally prepared to escalate contests to risky levels in defence of their resource. One explanation for this is that because of its experience in utilizing the resource the owner has more to lose than the competitor has to gain. Thus the most important fact that the receiver should establish on meeting a potential opponent is whether or not it is a resource holder. This can be established in all situations, regardless of whether or not the male has met its potential opponent before, by matching the scent of the opponent with the memorized odour of nearby scent marks. For this reason alone, scent matching would be expected to be the primary mechanism but there are two further reasons why scent matching would be selected for owner identification. First, it does not require the costs of the agonistic encounters required to establish a learned association between its odour and that of its marks. These costs would be particularly high when a large proportion of the male sector of the population is not breeding at any one time; this state of affairs is characteristic of the polygynous mating systems that predominate in mammals. Second, for reasons discussed at more length below, scent matching is more cheat-proof than signalling using marks with intrinsic meaning. It would be relatively easy for animals to fabricate intrinsic components of signals to deceive receivers and examples of such deception abound in the invertebrates. This sort of cheating could be avoided by selection for very costly signals but if this occurred selection might again favour the economic mechanism of scent matching.

There seems little doubt that a male could obtain information about a potential opponent from both intrinsic properties of its odour and from experience of past encounters with it. However, a role for scent marks does not fit easily into this picture. It could be argued that there is intrinsic information in marks which receivers do not use until they meet the territory owner. However, since the intruder then has the owner itself to assess there is little information that could not be gained directly. Similarly it is difficult to see any great advantage for a receiver in detecting the scent marks of a known individual before meeting it, as opposed to simply recognizing it by its odour when they meet. If males can get information about competitive ability directly from potential opponents, this suggests that intrinsic or learnt information could only be useful through responses to marks in the absence of the signaller (see previous section).

## Honesty in signalling: a pay-off for the receiver

The functional interpretation that males advertise their status in a way that involves scent marks, depends to some extent on whether or not low status males seek out marks. This is an unusual property of a social signal but Müller-Schwarze (1974) has observed that black-tailed deer, *Odocoileus hemionus*, do actively seek out scent marks as they enter a new area. When a signaller displays visually or by calling, the receiver has no option but to accept the information in the call. Scent marks are placed where intruders are most likely to find them (for example, on trails and around territory boundaries: Gosling 1981; Robinson 1988) but there is still a need for the low status male to seek out these marks and to investigate them; this could be particularly important if an essential part of the mechanism involved is to memorize complex properties of the odour. Thus the pay-off to the resource holder of reduced costs of intrasexual competition depends on an advantage for the low status male in inspecting scent marks.

One advantage for the low status male is clearly to avoid costly fights. However, such information would only be of value to the low status male if it was also reliable. There may thus be an advantage for resource holding males in providing honest information. If this is the case, it would be consistent with selection for the mechanism of scent matching. Other signals by resource holders, such as the deep croaks of toads, *Bufo bufo* (Davies and Halliday 1978) may have been selected for because they are difficult to mimic. In contrast, advertisement by scent matching is impossible to mimic because, to a unique extent amongst animal signals, scent marks record what an animal has actually done. A male that advertises resource holding status in this way must have achieved this status in competition with other males and then successfully defended it for long enough to have marked the area comprehensively.

## References

Aleksiuk, M. (1968). Scent-mound communication, territoriality, and population regulation in Beaver (*Castor canadensis* Kuhl). *J. Mammal.*, 49, 759–62.

Bowers, J.M. and Alexander, B.K. (1967). Mice: individual recognition by olfactory cues. *Science*, 158, 1208–10.

Brown, R.E. (1979). Mammalian social odours: a critical review. *Adv. Stud. Behav.*, 10, 103–62.

Coblenz, B.E. (1976). Functions of scent-urination in ungulates with special reference to feral goats (*Capra hircus* L.). *Amer. Nat.*, 110, 549–57.

Davies, N.B. and Halliday T.R. (1978). Deep croaks and fighting assessment in toads, *Bufo bufo*. *Nature*, 391, 56–8.

Emlen, S.T. and Oring, L.W. (1977). Ecology, sexual selection, and the evolution of mating systems. *Science*, 197, 215–23.

Erlinge, S., Sandell, M. and Brinck, C. (1982). Scent-marking and its territorial significance in stoats, *Mustela erminea*. *Anim. Behav.*, 30, 811–8.

Estes, R.D. (1969). Territorial behavior of the wildebeest (*Connochaetes taurinus* Burchell, 1823). *Z. Tierpsychol.*, 26, 284–370.

Franklin, W.L. (1980). Territorial marking behavior by the South American vicuna. In *Chemical signals: vertebrates and aquatic invertebrates* (ed. D. Müller-Schwarze and R.M. Silverstein), pp. 53–66). Plenum, New York.

Fullenkamp, A., Fischer, R.B., Vance, R.A. and Duffey, K.A. (1985). The failure to demonstrate avoidance of ventral gland odors in male gerbils (*Meriones unguiculatus*). *Physiol. Behav.*, 35, 763–5.

Geist, V. (1965). On the rutting behaviour of the mountain goat. *J. Mammal.*, 45, 551–68.

Gilbert, B.K. (1973). Scent marking and territoriality in pronghorn (*Antilocapra americana*) in Yellowstone National Park. *Mammalia*, 37, 25–33.

Gorman, M.L. (1984). Scent marking and territoriality. *Acta Zool. Fennica*, 171, 49–53.

Gosling, L.M. (1981). Demarkation in a gerenuk territory: an economic approach. *Z. Tierpsychol.*, 56, 305–22.

Gosling, L.M. (1982). A reassessment of the function of scent marking in territories. *Z. Tierpsychol.*, 60, 89–118.

Gosling, L.M. (1985). The even-toed ungulates: order Artiodactyla. Sources, behavioural context and function of chemical signals. In *Social odours in mammals* (ed. R.E. Brown and D.W. Macdonald), pp. 550–618. Oxford University Press.

Gosling, L.M. (1986). Economic consequences of scent marking in mammalian territoriality. In *Chemical signals in vertebrates*, 4 (ed. D. Duvall, D. Müller-Schwarze and R.M. Silverstein), pp. 385–95. Plenum, New York.

Gosling, L.M. (1987). Scent marking in an antelope lek territory. *Anim. Behav.*, 35, 620–2.

Hediger, H. (1949). Säugetier-Territorien und ihre Markierung. *Bijdr. Dierkd.*, 28, 172–84.

Henry, J.D. (1977). The use of urine marking in the scavenging behaviour of the red fox (*Vulpes vulpes*). *Behaviour*, 61, 82–105.

Jewell, P.A., Hall, S.J.G. and Rosenberg, M.M. (1968). Multiple mating and siring success during natural oestrus in the ewe. *J. Reprod. Fert.*, 77, 81–9.

Jolly, A. (1966). Lemur behaviour: A Madagascar field study. Chicago University Press.

Jones, R.B. and Nowell N.W. (1974). A comparison of the aversive and female attractant properties of urine from dominant and subordinate male mice. *Anim. Learn. Behav.*, 2, 141–4.

Kleiman, D.G. (1974). Patterns of behaviour in hystricomorph rodents. *Symp. Zool. Soc. Lond.*, 34, 171–209.

Kruuk, H. (1972). *The Spotted Hyaena. A study of predation and social behaviour.* Chicago University Press.

Kruuk, H., Gorman, M. and Leitch, A. (1984). Scent-marking with the subcaudal gland by the European badger, *Meles meles* L. *Anim. Behav.*, 32, 899–907.

Landauer, M.R., Banks, E.M. and Carter, C.S. (1977). Sexual preferences of male hamsters (*Mesocricetus auratus*) for conspecifics in different endocrine conditions. *Horm. Behav.*, 9, 193–202.

Maynard Smith, J. and Price, G.R. (1973). The logic of animal conflict. *Nature*, 246, 15–18.

Maynard Smith, J. and Parker, G.A. (1976). The logic of asymmetric contests. *Anim. Behav.*, 24, 159–75.

Müller-Schwarze, D. (1974). Social functions of various scent glands in certain ungulates and the problems encountered in experimental studies of scent communication. In *The behaviour of ungulates and its relation to management* (ed. V. Geist and F.R. Walther), pp. 107–13. IUCN, Morges, Switzerland.

Mykytowycz, R. (1965). Further observations on the territorial function and histology of the submandibular subcutaneous (chin) glands in the rabbit, *Oryctolagus cuniculus* (L.). *Anim. Behav.*, 13, 400–12.

Mykytowycz, R. (1970). The role of skin glands in mammalian communication. In *Advances in chemoreception*. I. *Communication by chemical senses* (ed. J.W. Johnston, D.C. Moulton and A. Turk), pp. 327–60. Appleton-Century-Crofts, New York.

Mykytowycz, R. (1973). Reproduction of mammals in relation to environmental odours. *J. Reprod. Fert.*, 19, 433–46.

Mykytowycz, R. (1975). Activation of territorial behaviour in the rabbit, *Oryctolagus cuniculus*, by stimulation with its own chin gland secretion. In *Olfaction and taste*, V (ed. D.A. Denton and J.P. Coghlan), pp. 425–32. Academic Press, New York.

Noble, G.K. (1939). The role of dominance in the life of birds. *Auk*, 263–73.

Parker, G.A. (1974). Assessment strategy and the evolution of fighting behaviour. *J. Theor. Biol.*, 47, 223–43.

Ralls, K. (1971). Mammalian scent marking. *Science*, 171, 443–9.

Rasa, O.A.E. (1973). Marking behaviour and its social significance in the African dwarf mongoose, *Helogale undulata rufula*. *Z. Tierpsychol.*, 32, 293–318.

Richter, W. von (1974). Territorial behaviour of the black wildebeest. *Zool. Afr.*, 7, 207–31.

Robinson, I.H. (1988). The distribution of faeces by the Spanish lynx, *Felis pardina*. *J. Zool., Lond.*, 216, 577–82.

Roeder, J.-J. (1983). Memorisation des marques olfactives chez la genette (*Genetta genetta* L.): durée de reconnaissance par les femelles de marques olfactives de males. *Z. Tierpsychol.*, 61, 311–4.

Rothman, R.J. and Mech, D. (1979). Scent-marking in lone wolves and newly formed pairs. *Anim. Behav.*, 27, 750–60.

Sawyer, T.F. (1981). Learned aversion to the odors of male mice: effects on agonistic behavior. *Physiol. Behav.*, 27, 19–26.

Shank, C.C. (1972). Some aspects of social behaviour in a population of feral goats (*Capra hircus* L.). *Z. Tierpsychol.*, 30, 488–528.

Steel, E. (1984). Effect of the odour of vaginal secretion on non-copulatory behaviour of male hamsters (*Mesocricetus auratus*). *Anim. Behav.*, 32, 597–608.

Stone, R.D. and Gorman, M.L. (1989). Mutual avoidance by European moles *Talpa europaea*. This volume.

Uexküll, J.V. and Kriszat, G. (1934). Streifzüge durch die Umwelten von Tieren und Menschen. Berlin. Quoted in Hediger (1944).

Walther, F.R. (1978). Mapping the structure and the marking system of a territory of the Thomson's gazelle. *E. Afr. Wildl. J.*, 16, 167–76.

Walther, F.R. (1979). Das Verhalten der Hornträger (Bovidae). *Handb. Zool. VIII*, 10(30), 1–184.

Wynne-Edwards, V.C. (1962). *Animal dispersion in relation to social behaviour.* Oliver and Boyd, London.

Zahavi, A. (1975). Mate selection—a selection for a handicap. *J. Theor. Biol.*, 53, 205–14.

Zahavi, A. (1977). Reliability in communication systems and the evolution of altruism. In *Evolutionary ecology* (ed. B. Stonehouse and C. Perrins), pp. 253–9. Macmillan, London.

# IV.2 Scent Marking, Residency, and Female Choice in the European Rabbit (*Oryctolagus cuniculus*)

CINDY REECE ENGEL

## Introduction

The lifetime reproductive success of a female is more strongly influenced by the 'quality' rather than the quantity of mates with which they copulate. In a species such as the rabbit, in which the male makes little contribution to parental care, the only apparent benefit the female derives from mating is male genes. Females might therefore be expected to select males on the basis of certain cues indicative of genetic quality, e.g. older males (which might reflect heritable differences in general efficiency), or high-ranking males (reflecting competitive ability). Female rabbits do show a degree of active choice in the selection of mating partners. Copulation cannot occur without lordosis and females may reject one male whilst actively soliciting another (Hafez 1970; Adams 1972).

In a two-choice situation, females spend longer investigating dominant (high-frequency chin-marking) males than subordinate (low-frequency chin-marking) males, and are able to make the discrimination on the basis of urinary odour cues (Bell 1977, 1981). Social status and frequency of chin-marking are correlated with aspects of gonadal physiology (testicular weight and plasma testosterone level) which, in turn, reflect the properties of the ejaculate (number of sperm/ejaculate and gelatinous plugs in the semen important for sperm transport). Bell (1983) therefore proposed that androgen-dependent scents may allow females to assess the relative fertility of males (Hafez 1970; Bell 1977, 1979; Reece 1985). In the field, those males that enjoy a priority of access to females are, indeed, dominant with higher frequencies of scent marking, larger testes and scent glands, and higher plasma testosterone levels (Mykytowycz 1965; Myers *et al.*, 1971; Reece 1985).

In the field, however, a male may possess additional properties by which he is assessed. When, for example, a dominant male (with large testes and high plasma testosterone level) is removed from an established colony of enclosed wild rabbits, the dominant niche is rapidly filled by the second ranking male (Mykytowycz 1958; Lockley 1961; Reece 1985). When reintroduced, these (previously dominant) males are rejected (i.e. attacked) by their former female partners, despite being heavier and having larger testes and higher plasma testosterone levels than the new dominant male (Reece 1985). In such a situation, in order to assess the males, females must use criteria other than those described previously.

One possibility is that, in the field, territory ownership is assessed by females. Dominant males are responsible for the majority of territorial marking (Mykytowycz 1968; Bell 1977, 1980; Reece 1985), and appropriation of dominant status involves intense scent marking which appears to precede any association with females (Reece 1985). I therefore tested the hypothesis that females assess the residency of males based on the predominance of his odour in the environment. The number of animals used was kept to a minimum in accordance with the *Guidelines for the use of animals in research* (1986).

## Materials and methods

Female response to suitably matched males was investigated in a two-choice situation. Each male was alternately the depositor of odour cues in the immediate environment of the test doe (i.e. the resident) and an intruder in that environment (i.e. non-resident). All subjects were adult, sexually experienced, saddle-backed domestic rabbits.

### ESTABLISHMENT OF RESIDENCY

Two males, matched for age (20 months) and previous experience (i.e. housing and sexual) were selected from a pool of 13 animals. The two males (MA and MB) were selected on the basis of behavioural and morphological characteristics known to be correlated with high social status, i.e. high chin-marking frequencies (Mykytowycz 1965; Black-Cleworth and Verberne 1975; Bell 1977), large testes (Myers *et al.* 1971; Bell 1977; Reece 1985) and low body weight (Bell 1977; Bray 1987). The two males also had comparable plasma testosterone levels (Table 1). Neither male had previously encountered the test females.

**Table 1.** Characteristics of the two experimental males (MA and MB) showing their respective ranking (in parentheses) within a pool of 13 males

| Male | Body weight (g) | Testicular Weight: in grams | Testicular Weight: as % of body weight | Plasma testosterone in nmol/l | Chin -marking frequency: mean no. of acts/5 min |
|---|---|---|---|---|---|
| MA | 1900 (9) | 5.31 (3) | 0.28 (2) | 2.3 (10.5) | 70 (3) |
| MB | 1700 (7) | 5.62 (2) | 0.33 (1) | 0.9 (13) | 52 (4) |

The test environment consisted of two, neighbouring, indoor gravel-based enclosures (A and B), each 1 × 1 × 2 m separated by a 1 × 2 m hardboard divide. These purpose-built enclosures had not previously held any other rabbits. Residency was established by housing the males in the test enclosures for seven days prior to experimentation: MA was housed in enclosure A and MB in enclosure B. Proprietary pelleted food and water were provided *ad libitum*. The scent-marking behaviour of each male was recorded over 2 h (14.00–16.00 h) on days 1, 3 and 5 for the seven days of enclosure housing.

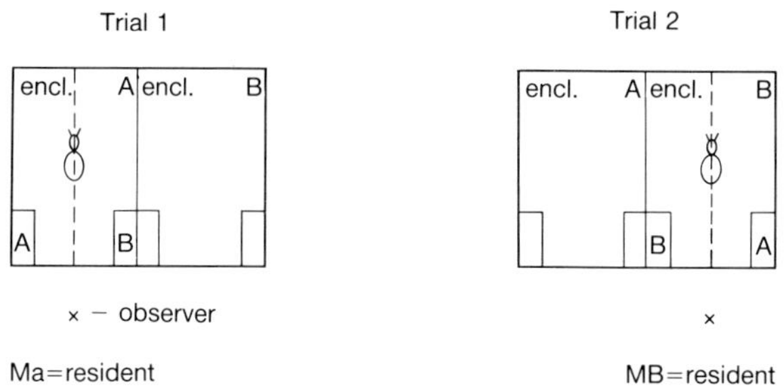


Fig. 1. Experimental enclosures showing positions of animals during trials 1 and 2 for each female.

ASSESSMENT OF FEMALE CHOICE

The six female subjects were sexually receptive (determined from vulva appearance; Chudnovskii 1957). Each female received two 15-min. trials, one in enclosure A and one in enclosure B; the males were held in individual wire mesh 'restraining' cages (65 × 18 × 18 cm) on either side of the test enclosure (Fig. 1). Thus, each male acted as both resident and non-resident for each female. Side preference was eliminated by alternating the sides on which resident and non-resident males were presented between trials. All food, water, and refuges were removed before introducing the female into the test enclosure. Females were introduced at a standard, central location facing the rear of their enclosures (Fig. 1) and their behaviour (Table 2) recorded for the following 15 mins. Observations were made 1 m from a central position (marked by the dotted line in Fig. 1) between the two test males. Three of the females received their first trial in enclosure A and three in enclosure B. Males were given a minimum of 1 h between trials in which to re-mark their enclosures. Wilcoxon matched-pairs signed-ranks tests (2-tailed) were applied to the data.

**Table 2.** Recorded behaviour of female rabbits (towards both resident and non-resident males)

1 Approaching a male = moving towards male whilst in his half of the enclosure (see dotted line in Fig. 1).
2 Bowing = lowering of head, with outstretched neck, towards restraining cage of male.
3 Sniffing male = nose visibly twitching towards body or restraining cage of male.
4 Nose to nose = naso-naso investigation within 10 cm of male.
5 'Near' male = any behaviour performed within one rabbit length of male.
6 Scent marking near male = any act of scent-marking (chin marking, urine spraying, body rolling, and pawscraping; Bell 1980) performed within one rabbit length of male.
7 Urination near male = urinating within one rabbit length of male.
8 Total investigation = total time spent in behaviours 1–7.
9 'Other' = any behaviours not directed towards males, i.e. not 1–8.

## Results

### ESTABLISHMENT OF ODOUR ENVIRONMENT/RESIDENCY

Both MA and MB showed intense scent-marking activity during the initial seven-day period of 'establishment of residency' in the test enclosure (Table 3).

**Table 3.** Number of scent marking acts (chin marking, urine squirting, pawscraping, and body rolling) performed per hour, by males MA and MB during the 7-day 'establishment of residency' period

| Male | Days | | | Total |
|---|---|---|---|---|
| | 1 | 3 | 5 | |
| MA | 48 | 89 | 18 | 155 |
| MB | 56 | 98 | 22 | 176 |

### FEMALE CHOICE

The females spent significantly longer approaching, being near (i.e. within one rabbit length of), scent-marking near, and investigating the resident male than the non-resident male ($p < 0.05$; Table 4).

**Table 4.** Mean time ± SE (in seconds) spent by females in performance of various behaviours towards resident and non-resident males

| Behaviour | Resident | Non-resident | $p$* |
|---|---|---|---|
| Approaching | 30.2±11.2 | 21.2±13.0 | <0.05 |
| Bowing | 7.7±10.7 | 2.4± 4.4 | n.s. |
| Sniffing | 41.5±24.4 | 37.1±17.3 | n.s. |
| Nose/nose | 8.9± 9.2 | 10.2± 7.3 | n.s. |
| Near | 31.1±18.3 | 11.6± 8.6 | <0.05 |
| Scent marking | 22.9± 9.7 | 9.6± 4.4 | <0.05 |
| Urination | 0.6± 1.7 | 0.0± 0.0 | n.s. |
| Investigating | 154.0±43.5 | 101.8±26.8 | <0.05 |

* Wilcoxon matched-pairs signed-ranks tests.
n.s. = not significant.

## Discussion

Adult, sexually experienced females spent longer investigating a male which had deposited its odour in their immediate environment (resident) than a male which was an intruder in that environment (non-resident). This was despite the fact that the same individual males acted as both residents and non-residents during the experiment. The results indicate that females:

1. are able to discriminate between a resident male (which is surrounded by its own odour) and a non-resident male (which is surrounded by the odour of another male); and
2 spend longer associating with the male whose odour predominates in that environment.

In the light of these results one may now offer an explanation for the phenomenon described earlier, i.e. that reintroduced (previously dominant) males are rejected by their former female partners despite having larger testes and higher plasma testosterone levels than the new dominant male. The reintroduced males may be rejected because they are not the depositors of the odour permeating the female's environment, and as such are no longer the territory owners.

Such discrimination may play an important role in mate choice since, in the field, the frequency of male scent marking is correlated with both social and physical status (Mykytowycz 1965; Myers *et al.* 1971; Bell 1977; Cowan 1983; Reece 1985). Therefore, choosing a male whose scent predominates in the female's immediate environment will automatically select a male that is both fertile and a successful intrasexual competitor.

It is known that the behaviour of rabbits is affected by their olfactory familiarity with the environment (Adams 1972; Mykytowycz 1975). Therefore one could suggest that, despite the restraining cages of the present study, the females were reacting to subtle differences in behaviour between resident and non-resident males. The working assumption here is that the males have formed an olfactory assessment of their environment, and behaved accordingly. I suggest that the underlying mechanism of residency assessment is likely to be the same in both sexes, i.e. to be olfactory in nature. By simply comparing the odour of an individual with the predominant odour in the environment, a female may reliably assess a male's residency: a match identifies the resident. Thus, the female is provided with an unbluffable indicator of male quality. Scent matching, as a method of identifying territory owners, was first proposed by Gosling (1982), and further investigation is warranted to assess fully its role in mate choice.

## Acknowledgements

I wish to thank Dr D.J. Bell and Dr D.W. Macdonald for critical discussion, and the Science and Engineering Research Council for financial support.

## References

Adams, C.E. (1972). The Rabbit. In *UFAW Handbook of care and management of laboratory animals* (4th edn). Churchill Livingstone, Edinburgh.

Bell, D.J. (1977). Aspects of the social behaviour of wild and domesticated rabbits *Oryctolagus cuniculus* L. Ph.D. thesis, University of Cardiff, Wales.

Bell, D.J. (1979). Chemical communication and social status—some intercorrelations. *XVI Int. Ethol. Conf.* University of British Columbia, Vancouver, Canada.

Bell, D.J. (1980). Social olfaction in lagomorphs. *Symp. Zool. Soc. Lond.*, 45, pp. 141–64.

Bell, D.J. (1981). Chemical communication in the European rabbit: urine and social status. In *Proceedings of the world lagomorph conference, 1979* (ed. K. Myers and C.A. MacInnes), pp. 271–9. Guelph.

Bell, D.J. (1983). Mate choice in the European rabbit. In *Mate choice* (ed. P. Bateson). Cambridge University Press, Cambridge.

Black-Cleworth, P. and Verberne, G. (1975). Scent-marking, dominance and flehmen behaviour in domestic rabbits in an artificial laboratory territory. *Chem. Senses Flavour 1*, pp. 465–95.

Bray, G. (1987). Physiological correlates of social status in the domestic rabbit (*Oryctolagus cuniculus*). Ph.D. thesis, University of East Anglia, Norwich.

Chudnovskii, L.A. (1957). Methods of determination of rabbits' sexual activities. *Karakulevodstro i zverovodstro*, 10, pp. 89–118.

Cowan, D. P. (1983). Aspects of the behavioural ecology of a free-living population of the European wild rabbit *O. cuniculus* L. in southern England. Ph.D. thesis, Royal Holloway College.

Gosling, L.M. (1982) A reassessment of the function of scent-marking in territories. *Z. Tierpsychol.*, 60, pp. 89–118.

Guidelines for the use of animals in research (1986). In *Anim. Behav.*, 34, pp. 315–8.

Hafez, E.S.E. (1970). Rabbits. In *Reproduction and breeding techniques for laboratory animals* (ed. Hafez). Lea and Feibiger, Philadelphia, Pennsylvania.

Lockley, R.M. (1961). Social structure and stress in the rabbit warren. *J. Anim. Ecol.*, 30, pp. 385–423.

Myers, K., Hale, C.S., Mykytowycz, R. and Hughes, R.L. (1971). The effects of varying density and space on sociality and health in animals. In *Behaviour and environment* (ed. A.H. Esser), pp. 148–87. Plenum, New York.

Mykytowycz, R. (1958). Social behaviour of an experimental colony of wild rabbits (*Oryctolagus cuniculus*). I. Establishment of the colony. *C.S.I.R.O. Wildl. Res.*, 3, pp. 7–25.

Mykytowycz, R. (1965). Further observations on the territorial function and histology of the submandibular cutaneous (chin) gland in the rabbit, *Oryctolagus cuniculus*. *Anim. Behav.*, 13, pp. 400–12.

Mykytowycz, R. (1968). Territorial marking by rabbits. *Sci. Am.*, 218, pp. 116–216.

Mykytowycz, R. (1975). Activation of territorial behaviour in the rabbit (*Oryctolagus cuniculus*), by stimulation with its own chin gland secretion. In *Olfaction and taste, V* (ed. D.A. Denton and J.P. Coghlan), pp. 421–32. Academic Press, New York.

Reece, C. (1985). Aspects of reproduction in the European rabbit (*Oryctolagus cuniculus*). Ph.D. thesis, University of East Anglia, Norwich.

# IV.3 Group Discrimination on the Basis of Urine in a Farm Cat Colony

WARNER C. PASSANISI AND DAVID W. MACDONALD

## Introduction

Many recent studies of mammalian olfactory communication in natural populations have attempted to determine the properties and functions of social odours (e.g. Eisenberg and Kleiman 1972; Mech and Peters 1977; Brown 1979; Brown and Macdonald 1985). There is mounting evidence for individual differences in odours, which may also be important for individual discrimination (e.g. Müller-Schwarze 1974; Gorman 1976; Halpin 1980, 1986), especially within social species (Wilson 1970; Rasa 1973; Gorman 1980; Brown and Macdonald 1985).

The Carnivora exhibit a high degree of flexibility in social and spatial organization, a feature commonly explained by the contrasting ecological circumstances of different populations (Gittleman 1983; Macdonald 1983; and various chapters in Gittleman 1989). Reflecting this variability, the properties and functions of scent marking (and other forms of communication) in carnivores appear to vary inter- and intra-specifically (reviewed by Macdonald 1985). The social and spatial organization of the domestic cat (*Felis catus* L.), varies markedly between populations (reviewed by Kerby and Macdonald 1988; Liberg and Sandell 1988), with group size ranging from solitary individuals (e.g. Jones 1977; Corbett 1979), to large groups at high densities (e.g. Dards 1979; Izawa *et al.* 1982; Natoli 1985*b*; Kerby 1987). The patterns of deployment and the possible functions of cats' scent marks are also variable between different habitats and populations (Dards 1979; Corbett 1979; Natoli 1985*a*; Macdonald *et al.* 1987; reviewed in Kerby and Macdonald 1988).

The domestic cat appears to use olfactory cues to acquire information about its environment (De Boer and Verberne 1981), for intra-specific communication (e.g. Corbett 1979; Macdonald *et al.* 1987), and perhaps for the detection of prey during hunting (Turner and Meister 1988). Both free-ranging and laboratory cats are able to discriminate between conspecifics using their urine odours. Urine from oestrous females elicits longer sniffing durations than from anoestrous females (Verberne and De Boer 1976), and cats sniff males' 'fresh' spray marks for longer than they do 'old' marks (De Boer 1977). Natoli (in press) has also demonstrated that the reproductive state of females affects their investigation of sprayed urine from strange males.

Natoli (1985*a*) demonstrated that, for a group of feral cats, adult males sniffed experimentally presented urine for significantly longer than did adult females, and that the sprayed urine of an unknown tom elicited longer sniffing durations in both sexes, than did excretory or squat urine. Here we seek to examine further the factors involved in Natoli's findings, and to investigate whether, in the wild, cats are able to discriminate between urine marks on the basis of the familiarity, sex of the marker, and behavioural context of urination.

## Methods

### STUDY AREA AND COLONY STRUCTURE

The study was carried out on Barley Park Farm, a mixed stock and arable farm in Oxfordshire, UK, between June and August 1988. The farm supported a population of up to 70 individually identifiable, semi-dependent, free-ranging farm cats. At any given time during the study, adults were classed as older than 12 months, juveniles aged from 6 to 12 months, and kittens were less than 6 months old.

The farmyard was divided into three mutually exclusive *communal feeding areas* (an area encompassing numerous denning sites, shelter, latrines, and a feeding site, which was provisioned daily). Within each communal feeding area, cats were associated into a relatively stable *feeding group* (Izawa *et al.* 1982). Although both males and females occasionally moved between the communal feeding areas, they predominantly interacted with, and spent considerably more time in close proximity to, members of their own feeding group. The feeding groups described here consisted predominantly of inter-related matrilines with associated males, structured around highly predictable resources. Immigration and emigration from the population as a whole was rare, and the majority of cats stayed within their natal feeding groups. The social organization and patterns of scent marking in this colony are described by Passanisi and Macdonald (in preparation).

### URINE DONORS

The urine samples used in this experiment were collected from six sexually mature adults under intensive direct observation; a male and female from each of three locations:

1. *unknown*, unrelated individuals that were residents of an area approximately 15 miles away;
2. *unfamiliar* individuals from a different feeding group to the test animals, centred in another communal feeding area approximately 100 m away;
3 *familiar* individuals from the same feeding group as the test animals.

We classified the donors as unknown, unfamiliar, and familiar as an indication of the increasing frequency of an encounter (from zero to high) between them and the test animals. Therefore, we speak of increasing familiarity between donor and test animal along this social continuum. Donors voided urine in different behavioural contexts; males donated both squat and spray urine, whereas females were donors only of squat urine.

### URINE COLLECTION

Urine was collected within 30 s of urination, using large cotton buds, which were handled with forceps. A median of 4 (range: 1–6) cotton buds were used per urine sample, and each bud was stored separately in a small, stoppered plastic tube. Frequently males were found to spray on material such as empty fertilizer bags, or plastic sheets, which facilitated urine collection.

### TEST ANIMALS

Twenty-one sexually mature, adult cats (14 females and 7 males) were chosen as test animals (or urine recipients). The 14 females were known to have previously produced litters, and the 7 males (at least two years old in June 1988) had previously been observed mating at least once. All 21 test animals were members of one of the three feeding groups.

For both sexes, a crude measure of social status was assigned to each test animal. Each cat was termed as (i) high ranking, (ii) middle ranking or (iii) low ranking according to the net flow of social interactions between colony members, observed in detail from August 1987 to August 1988. Thus, using this simplified classification of intra-sexual social relationships, a high-ranking female would tend to supplant other females from a certain location, and would win those agonistic encounters where both females did not retreat. A middle-ranking female would tend to dominate some females and be supplanted by others, whereas a low-ranking female would tend to be supplanted from resource areas by other, more dominant females.

### URINE PRESENTATION TO TEST ANIMALS

Each presentation involved the positioning of a urine-soaked cotton bud on a bare patch of ground in the communal feeding area, as close as possible to the test animal. The observer retreated to at least 5 m away and recorded, with a stop-watch, subsequent sniffing duration (defined as the cumulative time the cat's nose was within 10 cm of the cotton bud, generally actively sniffing, during an investigatory period punctuated by intervals of less than 10 s in duration where the cat was more than 10 cm away from the cotton bud). The cotton bud was retrieved after the test animal had voluntarily ceased to sniff and had walked away from the bud. A brief jet of water from an accurate high-pressure water-sprayer, aimed in front of an advancing cat, was used to restrict any interference with the test animal by other cats whilst it was investigating the cotton bud. The water jet did not alter the behaviour of the test animal.

Between June and August 1988, all 21 test animals were presented with each of the nine different types of fresh conspecific urine, classified according to mutually exclusive combinations of familiarity and behavioural context. None of the donors was a test animal. Each of the nine different types of urine was presented in random order to an individual chosen at random from the population of test animals. All 21 test animals were tested with each urine type before re-testing. Each urine type, taken from separate urination events, was presented to each test animal on three different days; a total of 567 successful urine presentations evoking a measurable behavioural response from the test animal. All urine samples tested in this experiment were fresh, in the sense that each was presented to the test animal within 90 mins of urination, and the maximum exposure time during testing was 10 mins (although generally under 3 mins). Only one test animal was tested per sample presentation.

The data were analysed primarily using analyses of variance and t-tests (see the results for details). Means are calculated from untransformed data, and quoted together with standard deviations and sample size. All t-tests are two-tailed and statistical significance is taken as the 5% level.

## Results

Sniffing durations recorded during this experiment were highly variable, ranging from 1.0 to 57.8 s [mean: 13.6 s $\pm$ 6.3 ($n = 567$)]. Preliminary analysis of the data revealed that the variance increased as the means increased. The most appropriate data transformation was found to be a square-root transformation, which effectively made the variances independent of the means. The transformed sniffing durations were then subjected to two analyses of variance: (i) a two-factor (test animal sex by social status) ANOVA; and (ii) a four-factor ANOVA design incorporating three repeated measures (familiarity (or donor location), behavioural context (or urine type), and replicate), with the non-repeated, between factor of test animal sex (see below for details).

In the two-factor analysis, there was a highly significant difference between male and female test animals in sniffing duration over all urine presentations [$F_{(1,15)} = 14.29, p = 0.002$]. Males [mean: 16.3 s $\pm$ 8.2 ($n = 63$)] sniffed urine samples for significantly longer than did females [mean: 12.2 s $\pm$ 6.0 ($n = 126$)], considering all experimental presentations together, averaged between replicates [$t_{(187)} = 3.943$, $p < 0.001$]. This sex difference was apparent for all of the six familiarity/behavioural context combinations of urine taken from individuals outside the feeding group; there was no difference between the males' and females' sniffing durations of the

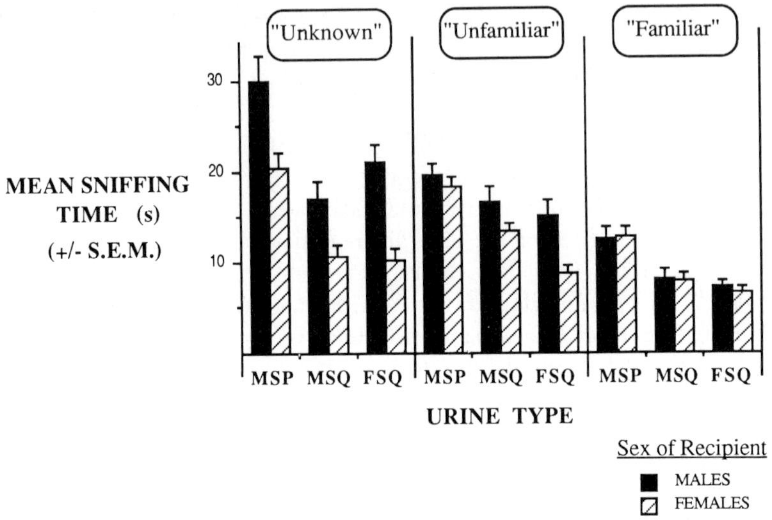


Fig. 1. Variation in the sniffing response of cats presented with various types of conspecific urine. Urine donors were: (i) unknown, from a strange area; (ii) unfamiliar, from a different feeding group; or (iii) familiar, from the same feeding group as the test animal. The urine samples were from one of three mutually exclusive behavioural contexts, either sprayed urine from males (MSP), or squat urine from males (MSQ) or females (FSQ).

three combinations within the same feeding group (see Fig. 1). Neither the social status of the test animal [$F_{(2,15)}=0.38$, $p>0.05$] nor the test animal sex × social status interaction [$F_{(2,15)}=1.47$, $p>0.05$] had significant effects on urine investigation, and thus the social status factor was eliminated from the subsequent analysis.

In the four-factor analysis, sniffing duration differed significantly between the three classes of donor familiarity [$F_{(2,38)}=61.85$, $p<0.001$]. Urine taken from unknown cats elicited the longest sniffing durations [mean: 16.8 s ± 8.6 ($n=63$)]; this was not significantly different from that voided by unfamiliar individuals [mean: 14.7 s ± 5.6 ($n=63$); $t_{(124)}=1.638$, $p>0.05$], but was significantly longer than that originating from familiar cats in the same feeding group [mean: 9.3 s ± 3.8 ($n=63$); $t_{(124)}=6.364$, $p<0.001$]. Urine originating from unfamiliar donors was also sniffed for significantly longer than that from familiar donors [$t_{(124)}=6.360$, $p<0.001$]. There was a significant interaction between the sex of the test animal and the donor familiarity [$F_{(2,38)}=10.94$, $p<0.002$]. The pattern described above was consistent for both sexes, except that males distinguished urine voided by unknown donors (mean: 22.7 s ± 8.6; $n=21$) from that voided by unfamiliar cats [mean: 17.1 s ± 5.0 ($n=21$); $t_{(40)}=2.572$, $p<0.02$]. No significant difference was found between the three replicates [$F_{(2,38)}=2.23$, $p>0.05$], nor was there any significant interaction involving replicates.

The behavioural context of urination also had a significant influence on sniffing duration [$F_{(2,38)} = 60.28$, $p < 0.001$]. Considering both sexes of test animal together, males' sprayed urine elicited the longest sniffing durations [mean: 18.3 s $\pm$ 7.4 ($n = 63$)]; this was significantly longer than both males' squat [mean: 11.8 s $\pm$ 5.3 ($n = 63$); $t_{(124)} = 5.703$, $p < 0.001$] and females' squat urine [mean: 10.6 s $\pm$ 5.8 ($n = 63$); $t_{(124)} = 6.543$, $p < 0.001$]. There was no significant difference in sniffing duration between the squat urine from males and females ($t_{(124)} = 1.186$, $p > 0.05$]. There was a significant interaction between sex and behavioural context [$F_{(2,38)} = 4.20$, $p > 0.05$]. The response to behavioural context from both males and females was the same as described above, except that females spent more time sniffing males' squat urine [mean: 10.7 s $\pm$ 4.1 ($n = 42$)] than females' squat urine [mean: 8.7 s $\pm$ 4.1 ($n = 42$); $t_{(82)} = 2.284$, $p < 0.05$].

## Discussion

The evidence presented here establishes that cats respond differentially to conspecific urine from different sources. For both males and females, the behavioural context and familiarity of urine donor were significant factors affecting sniffing durations. Our results confirm three of Natoli's (1985*a*) findings. Firstly, adult male cats investigated virtually all types of conspecific urine for longer than adult females. Secondly, both sexes investigated sprayed urine more than squat urine of either sex, even from totally unknown individuals. And, thirdly, sprayed urine from unknown conspecifics elicited significantly longer sniffing durations from both sexes than did that from more familiar individuals from their own population.

Furthermore, we demonstrate a differential response to urine within the same population, depending upon the feeding group of its provenance. Both sexes investigated urine from unfamiliar individuals of a different feeding group for significantly longer than that from familiar individuals of their own feeding group. Males further investigated the urine from unknown individuals for significantly longer than that from unfamiliar animals, whereas females apparently did not distinguish between unknown and unfamiliar individuals on the basis of urine. However, only females appeared to distinguish between the sexes of the squat urine donors. It is further evident that the difference between male and female sniffing durations decreases along the unknown to familiar continuum.

It is apparent that, as far as unrestrained cats are concerned, sprayed and squat urine smell quite different to each other (see also Natoli 1985*a*). It has been suggested that this difference may, at least in part, be due to the presence of anal gland secretions in sprayed urine (Fox 1975), and that these secretions may be essential for an odour identification system. This would also account for the fact that squat urine seems to be less important in olfactory communication, and is probably primarily excretory in function (De Boer 1977).

Although this paper only deals with the interaction between the recipient and the odour, any interpretation of the distribution and probable functions of scent marks must also take into account the fact that scent marks are effectively signals placed for the benefit of the signaller, the intention of which is to manipulate the recipient, often to the recipient's disadvantage (Dawkins and Krebs 1978; Krebs and Dawkins 1984; Gosling 1982, 1985, 1990). All animals were tested within their own communal feeding areas; thus the differential response to the odours presented is due primarily to intrinsic properties of the odours. The ability to match an individual odour with that of a resource holder, and to respond appropriately (Gosling 1982, 1985, this volume), would account for the differential response of cats to fellow members of the same feeding group and to those from outside that group, and might also be a factor in the maintenance of amicable relationships within feeding groups. In the light of Gosling's (1982) scent matching hypothesis, it is probable that behaviours such as allo-marking (by rubbing and licking) and the rubbing of fresh scent marks with the head and flanks (Macdonald *et al.* 1987) may further enable members of a feeding group to be identified as owners of their communal feeding areas. Such a mechanism would also explain the heightened behavioural responses of both males and females to strange individuals, as observed in many studies of free-ranging cats (e.g. Corbett 1979; Liberg 1980; Kerby 1987).

Mate choice could be facilitated by the use of these same olfactory mechanisms, and it seems probable that oestrous females use spray marks to advertise their reproductive condition (Verberne and De Boer 1976; Liberg and Sandell 1988; Passanisi and Macdonald, in preparation). The ability of females to distinguish the sprayed urine of unfamiliar individuals could further be of adaptive significance when considering observations of male infanticide (Macdonald *et al.* 1987).

In summary, the evidence presented here establishes that free-ranging farm cats respond differentially to conspecific urine from different sources, depending upon the familiarity of the urine donor and the behavioural context of urination. Our results confirm Natoli's (1985*a*) findings, in that

1. Adult male farm cats investigated virtually all types of conspecific urine for longer than did adult females.
2. Both sexes investigated sprayed urine more than squat urine, even from unknown individuals.
3. Urine from unknown conspecifics elicited longer sniffing durations than that from familiar individuals.

Furthermore, we demonstrate a differential response to urine depending upon the feeding group of its provenance. Both sexes investigated urine from unfamiliar individuals for significantly longer than that from familiar individuals; further, males investigated urine from unknown individuals for significantly longer than that from unfamiliar animals. We discuss the possible consequences of the group significance of odour identification.

## Acknowledgements

This work was partly funded by the Waltham Centre for Pet Nutrition. Mr H.F. Parker kindly allowed us the freedom of his farm. Drs P.J. Apps, R.E. Brown, L.M. Gosling, G. Kerby, O. Liberg, H. McKay, and M.E.L. Berdoy, E. Natoli, and an anonymous referee greatly improved an earlier draft. Dr S. Mehta provided statistical advice.

## References

Brown, R.E. (1979). Mammalian social odors: a critical review. *Adv. Study Behav.*, 10, 103–62.

Brown, R.E. and Macdonald, D.W. (1985). *Social odours in mammals*, Vols. I and II. Oxford University Press, Oxford.

Corbett, L.K. (1979). Feeding ecology and social organization of wildcats (*Felis silvestris*) and domestic cats (*Felis catus*) in Scotland. Ph.D. thesis, University of Aberdeen.

Dards, J.L. (1979). The population ecology of feral cats (*Felis catus* L.) in Portsmouth Dockyard. Ph.D. thesis, University of Bradford.

Dawkins, R. and Krebs, J.R. (1978). Animal signals: information or manipulation. In *Behavioural ecology: an evolutionary approach*, 1st edn. (ed. J.R. Krebs and N.B. Davies), pp. 282–309. Blackwell, Oxford.

De Boer, J.N. (1977). The age of olfactory cues functioning in chemo-communication among male domestic cats. *Behav. Proc.*, 2, 209–25.

De Boer, J.N. and Verberne, G. (1981). Slow and fast wave activity in the olfactory system in cats during perception of pheromones. *Physiol. Behav.*, 27, 521–35.

Eisenberg, J.F. and Kleiman, D.G. (1972). Olfactory communication in mammals. *Annu. Rev. Ecol. Syst.*, 3, 1–32.

Fox, M.W. (1975). The behaviour of cats. In *The behaviour of domestic animals*, 3rd edn. (ed. E.S.E. Hafez), pp. 410–36. Williams and Williams, Baltimore.

Gittleman, J.L. (1973). The behavioural ecology of carnivores. Ph.D. thesis, University of Sussex.

Gittleman, J.L., ed. (1989). *Carnivore behavior, ecology and evolution.* Cornell University Press.

Gorman, M.L. (1976). A mechanism for individual recognition by odour in *Herpestes auropunctatus*. *Anim. Behav.*, 24, 141–5.

Gorman, M.L. (1980). Sweaty mongooses and other smelly carnivores. *Symp. Zool. Soc. Lond.*, 45, 87–105.

Gosling, L.M. (1982). A reassessment of the function of scent marking in territories. *Z. Tierpsychol.*, 60, 89–118.

Gosling, L.M. (1985). The even-toed ungulates: order Artiodactyla. Sources, behavioural context and function of chemical signals. In *Social odours in mammals* (ed. R.E. Brown and D.W. Macdonald), pp. 550–618. Oxford University Press, Oxford.

Gosling, L.M. (1990). Scent-marking by resource holders: alternative mechanisms for advertising the costs of competition. In *Chemical communication in vertebrates*, V (ed. D.W. Macdonald, D. Müller-Schwarze and S. Natynczuk). Oxford University Press, Oxford.

Halpin, Z.T. (1980). Individual odors and individual recognition: review and commentary. *Biol. Behav.*, 5, 233–48.

Halpin, Z.T. (1986). Individual odors among mammals: origins and functions. *Adv. Study Behav.*, 16, 39–70.

Izawa, M., Doi, T. and Ono, Y. (1982). Grouping patterns of feral cats (*Felis catus*) living on a small island in Japan. *Jap. J. Ecol.*, 32, 373–82.

Jones, E. (1977). Ecology of the feral cat, *Felis catus* (L), (Carnivora: Felidae) on Macquarie Island. *Aust. Wildl. Res.*, 4, 249–62.

Kerby G. (1987). The social organization of farm cats (*Felis catus*, L.). D.Phil. thesis, University of Oxford.

Kerby, G. and Macdonald, D.W. (1988). Cat society and the consequences of colony size. In *The domestic cat: the biology of its behaviour* (ed. D.C. Turner and P. Bateson), pp. 67–81. Cambridge University Press, Cambridge.

Krebs, J.R. and Dawkins, R. (1984). Animal signals: mind reading and manipulation. In *Behavioural ecology: an evolutionary approach*, 2nd edn. (ed. J.R. Krebs and N.B. Davies), pp. 380–402. Blackwell, Oxford.

Liberg, O. (1980). Spacing patterns in a population of rural, free roaming, domestic cats. *Oikos*, 35, 336–49.

Liberg, O. and Sandell, M. (1988). Spatial organisation and reproductive tactics in the domestic cat and other felids. In *The domestic cat: the biology of its behaviour* (ed. D.C. Turner and P. Bateson), pp. 83–98. Cambridge University Press, Cambridge.

Macdonald, D.W. (1983). The ecology of carnivore social behaviour. *Nature*, 301, 379–89.

Macdonald, D.W. (1985). The carnivores: Order Carnivora. In *Social odours in mammals* (ed. R.E. Brown and D.W. Macdonald), pp. 619–722. Oxford University Press, Oxford.

Macdonald, D.W., Apps, P.J., Carr, G.M. and Kerby, G. (1987). Social dynamics, nursing coalitions and infanticide among farm cats, *Felis catus*. *Adv. Ethol.*. 28. 1–64.

Mech, L.D. and Peters, R.P. (1977). The study of chemical communication in free-ranging mammals. In *Chemical signals in vertebrates* (ed. D. Müller-Schwarze and M.M. Mozell), pp. 321–31. Plenum, New York.

Müller-Schwarze, D. (1974). Olfactory recognition of species, groups, individuals, and physiological states among animals. In *Pheromones* (ed. M. Birch), pp. 316–26. Elsevier, New York.

Natoli, E. (1985*a*). Behavioural responses of urban feral cats to different types of urine mark. *Behaviour*, 94, 234–43.

Natoli, E. (1985*b*). Spacing patterns in a colony of urban stray cats (*Felis catus*, L.) in the historic centre of Rome. *Appl. Anim. Ethol.*, 14, 289–304.

Natoli E. (in press). Reactions of female domestic cats (*Felis catus* L.) to sprayed urine of a strange male. 1. Olfactory behaviour. *Ethol. Ecol. Evol.*

Passanisi, W.C. and Macdonald, D.W. (in preparation). Scent marking by farm cats: spatial patterns and social behaviour.

Rasa, O.A.E. (1973). Marking behaviour and its social significance in the African dwarf mongoose *Helogale undulata rufula*. *Z. Tierpsychol.*, 32, 293–318.

Turner, D.C. and Meister, O. (1988). Hunting behaviour of the domestic cat. In *The domestic cat: the biology of its behaviour* (ed. D.C. Turner and P. Bateson), pp. 111–21. Cambridge University Press, Cambridge.

Verberne, G. and De Boer, J.N. (1976). Chemocommunication among domestic cats, mediated by the olfactory and vomeronasal senses. 1. Chemocommunication. *Z. Tierpsychol.*, 42, 86–109.

Wilson, E.O. (1970). Chemical communication within animal species. In *Chemical ecology* (ed. E. Sondheimer and J.B. Simeone), pp. 133–55. Academic Press, New York.

# IV.4 Urine Marking in Female Wolves (*Canis lupus*): an Indicator of Dominance Status and Reproductive State

JENNY RYON AND RICHARD E. BROWN

## Introduction

Wolf packs appear to adjust their reproduction rate to the carrying capacity of the environment (Mech 1970). Not all females in a wolf pack produce pups (Harrington *et al.* 1982). The dominant female is the most reproductively successful. Reproduction among female wolves may be regulated by aggression between females, direct interference during copulation attempts, or through deferred reproduction by younger females (Packard *et al.* 1983).

Urine-marking may be involved in the establishment and maintenance of pair-bonds in wolves (Rothman and Mech 1979) and may indicate dominance status (Macdonald 1985). Since the chemical constituents of the urine of both male and female wolves change seasonally (Raymer *et al.* 1983), urine marks may also provide information on reproductive status. We examined urine marking of wolves by sex, dominance status and season, and the responses of known wolves to these marks.

## Methods

We observed a pack of wolves (six females and nine males) aged from 20 months to 10 years in a 3.8 ha semi-natural enclosure. One-hour observation periods were made 2–4 times per week for 65 months with up to 621 hours of observations on individual wolves. The dominance status, year of birth, and total number of hours for which each wolf was observed, are shown in Table 1.

Data from three dominant and three subdominant females and two dominant and seven subdominant males were analysed. The dominance status of each wolf was determined by body postures during spontaneous pairwise interactions. The urine marks of each wolf and the responses of other wolves (sniff, lick, and over-mark) to these marks were recorded. As each wolf was observed for a different number of observation periods, the data were normalized to 100 hours of observation per wolf. The data were analysed in a 2 (sex) by 2 (dominance status) factorial ANOVA.

**Table 1.** Wolves observed. The names of each wolf observed, their date of birth and the number of hours they were observed. As each wolf was observed for a different length of time, the data were normalized for 100 hours of observation per wolf

| Dominant females | Subdominant females |
|---|---|
| Kluane (b. 1974) 397 h | Etain (b. 1977) 175 h |
| Quill (b. 1982) 207 h | Raven (b. 1977) 621 h |
| Pawnee (b. 1984) 17 h | Ursula (b. 1980) 621 h |
| **Dominant males** | **Subdominant males** |
| Milford (b. 1974) 442 h | Thor (b. 1974) 165 h |
| Fingal (b. 1980) 179 h | Benjy (b. 1976) 355 h |
| | Cuchulain (b. 1977) 111 h |
| | Draemon (b. 1979) 621 h |
| | Ashley (b. 1980) 521 h |
| | Yukon (b. 1982) 481 h |
| | Voochko (b. 1984) 243 h |

## Results

Dominant males and females marked more than subdominants ($F(1, 11) = 36.58$, $p<0.01$) and dominant females marked more than dominant males ($F(1, 11) = 10.44$, $p<0.01$) (Fig. 1). A total of 1221 female urine marks and 671 male urine marks are recorded in this figure. The score for each wolf was calculated as: (Total number of marks/Hours of observation) × 100.

There were significant changes in the frequency with which female wolves urined marked over the year (Fig. 2). Marking was most frequent from November to February, during which courtship and mating begin (repeated measures ANOVA, $F(5, 20) = 2.95$, $p<0.05$). The scores for each cell in the analysis were calculated as: (Number of marks/Number of females observed/Hours of observation × 100.

The responses of all other wolves (male and female) to the urine marks of females over the seasons are shown in Fig. 3. The frequency of investigating (sniffing) female urine marks changed only slightly over the seasons, but the frequency of marking over the female's marks showed a significant change ($F(5,20) = 4.29$, $p<0.01$), with the most overmarks occurring from November to February. Likewise, the frequency with which female's marks were licked changed significantly over the seasons ($F(5, 20) = 11.95$, $p<0.01$), with the most responses occurring in the period from January to mid March, and decreasing at the end of oestrus. Data for each wolf were calculated as: (Responses/Number of wolves/Number of initial marks) × 100.

The responses of all wolves to urine marks of dominant and subdominant males and females are shown in Fig. 4. Marks of females were investigated

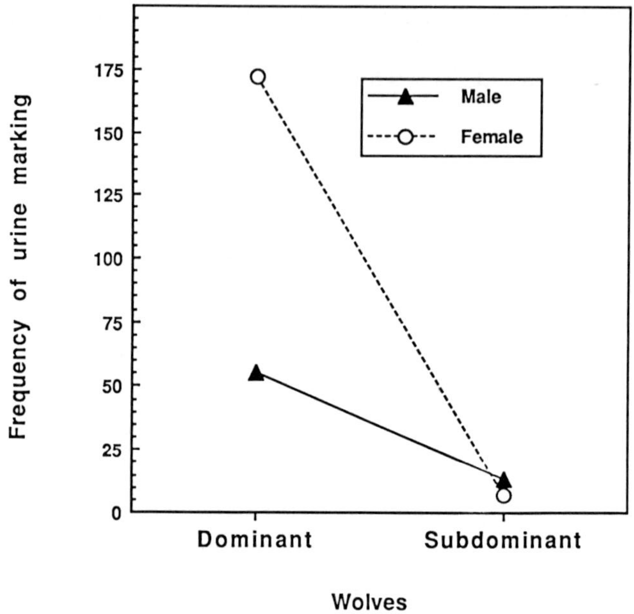


Fig. 1. Mean frequency of urine marking by dominant and subdominant male and female wolves.

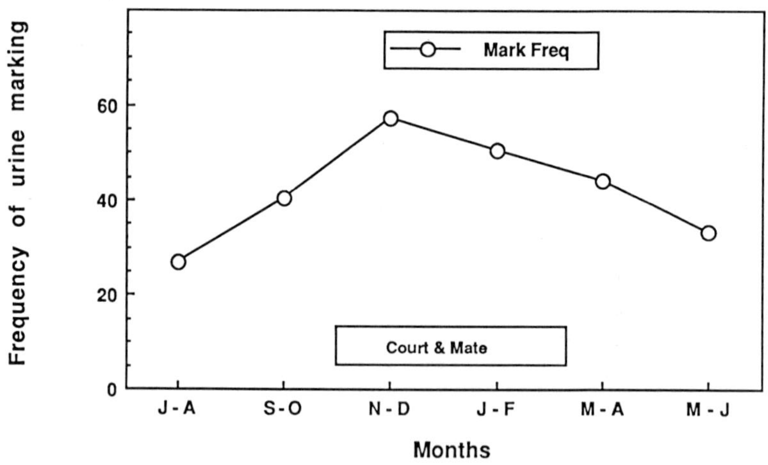


Fig. 2. Mean frequency of urine marking by female wolves over the year, shown in two month blocks, starting in July–August.

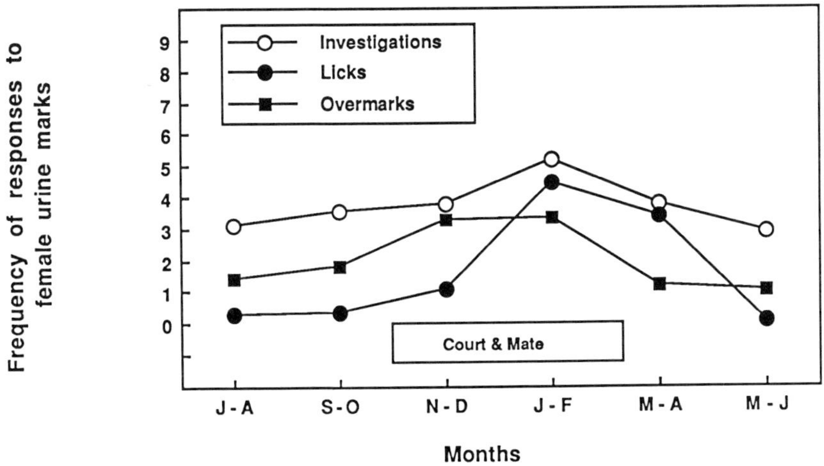


Fig. 3. Mean frequency of sniff investigations, licks, and overmarks directed by all wolves to the urine marks of females, shown in two month blocks, starting in July–August.

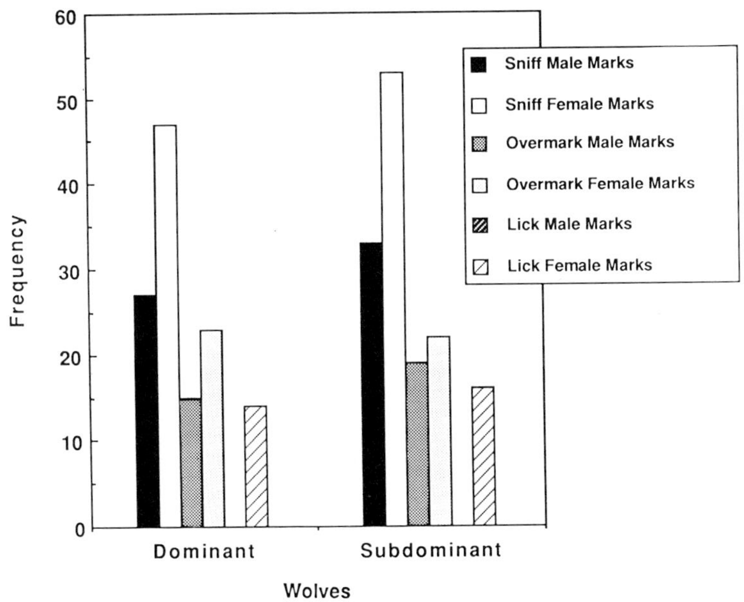


Fig. 4. Mean frequency of sniff investigations, licks, and overmarks directed by all wolves to the urine marks of dominant and subdominant males and females.

[(Sniffs/Marks) × 100] more than those of males ($F(1, 11) = 5.55, p<0.05$) but the dominance status of the marker had little effect on the responses to the mark by other wolves. Neither the sex nor dominance status of the urine marker influenced the frequency of over-marks [(Over-marks/Mark) × 100] by other wolves. Urine marks of dominant and subdominant males and females were all marked equally often. Urine marks of females were licked [(Licks/Mark) × 100] more than those of males ($F(1, 11) = 7.50, p<0.02$). We rarely observed any wolves licking the urine marks of males.

The responses of dominant and subdominant male and female wolves to the urine marks of females are shown in Fig. 5. Dominant wolves sniffed [(Sniffs/Mark) × 100] female urine marks more than subdominant wolves ($F(1, 11) = 6.46, p<0.05$), but there was no sex difference in the frequency of sniffing. Likewise, dominant wolves marked over female urine marks [(Over-marks/Mark) × 100] more than subdominants ($F(1, 11) = 17.62, p<0.01$). Although males marked over female marks more frequently than did females, this sex difference was not significant. Males licked female urine marks [(Licks/Mark) × 100] more than females ($F(1, 11) = 16.28, p<0.01$), but there was no effect of dominance on urine-licking.

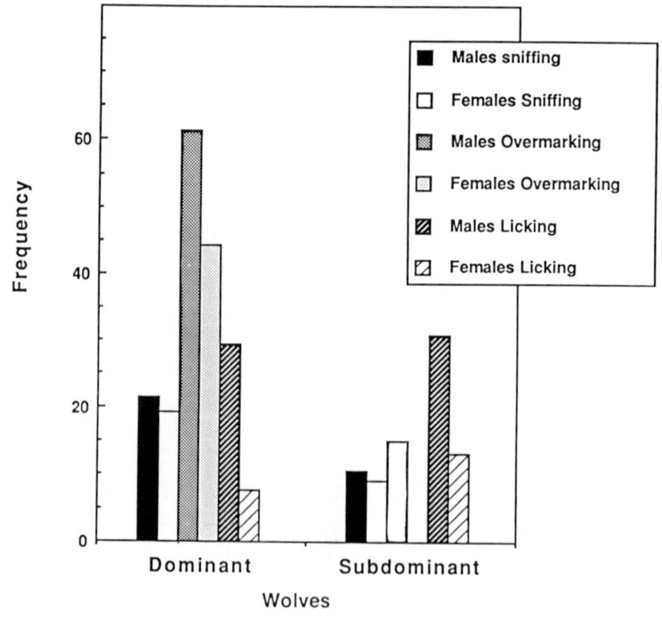


Fig. 5. Mean frequency of sniff investigations, licks, and overmarks performed by dominant and subdominant male and female wolves toward the urine marks of other wolves.

## Conclusions

Dominant wolves mark more than subdominants, and female wolves urine-mark more than males. Urine marks of female wolves are responded to more frequently than those of males by other wolves in the pack, thus it appears that these marks provide information on the sex of the marker.

Dominant wolves sniff and over-mark female urine marks more than subdominants, but there does not appear to be any evidence that the urine marks of dominant wolves of either sex are treated differently than the marks of subdominants.

The urine-marking frequency of female wolves increases during the courtship and breeding season, and the responses of other wolves to the marks of the females also increases during these periods. This suggests that urine marking is used to convey information about reproductive status. The high level of marking by dominant females may indicate their social rank. The function of urine marks in wolves, therefore, appears to indicate dominance status, sex and reproductive state. Licking and overmarking show seasonal changes in frequency. Most overmarking was done by the alpha male and female.

## Acknowledgements

This research was supported by an NSERC if Canada Infrastructure grant to J.C. Fentress and NSERC of Canada operating grant No. A7441 to R.E.B.

## References

Harrington, F.H., Paquet, P.C., Ryon, C.J. and Fentress, J.C. (1982). Monogamy in wolves: a review of the evidence. In *Wolves of the world: perspectives on behavior* (ed. F.H. Harrington and P.C. Paquet), pp. 209–22. Noyes Publications, Park Ridge, N.J.

Macdonald, D.W. (1985). The carnivores: Order Carnivora. In *Social odours in mammals* (ed. R.E. Brown and D.W. Macdonald), pp. 619–722. Clarendon Press, Oxford.

Mech, L.D. (1970). *The wolf: ecology and behavior of an endangered species.* Doubleday, New York.

Packard, J.M., Mech, L.D. and Seal, U.S. (1983). Social influences on reproduction in wolves. In *Wolves of Canada and Alaska: their status, biology and management* (ed. L.N. Carbyn). Canadian Wildlife Service Report Series, No. 45.

Raymer, J.D., Weisler, D., Novotny, M., Asa, C., Seal, U.S. and Mech, L.D. (1983). Volatile constituents of wolf (*Canis lupus*) urine as related to gender and season. *Experientia*, 40, 707–9.

Rothman, R.V. and Mech, L.D. (1979). Scent-marking in lone wolves and newly formed pairs. *Anim. Behav.*, 27, 750–60.

# IV.5 Effect of a Conspecific's Urine on Reproduction in Arctic Foxes

L. BARTOŠ AND P. RÖDL

## Introduction

The present study was performed at the request of breeders of the Arctic fox (*Alopex lagopus*). On a fur farm at Dublovice, Czechoslovakia, problems occurred with reproduction: by comparison with bitches, breeding males tended to be delayed in the start of their reproductive activity. As the Arctic fox belongs to the group of monoestrous animals (Jöchle 1973), bitches with early heat failed to be mated and hence failed to have progeny during that season.

We tried to apply olfactory stimulation in order to synchronize the seasonal reproductive activity of both sexes and we have extended this manipulation to a larger study of olfaction in this species (Bartoš *et al.* 1989; Rödl *et al.* 1989). Urine was tested as a possible source of olfactory stimulation.

The pheromonal properties of urine have been described for many species including Canids (Mykytowycz 1977; Brown 1979; Adams 1980; Stoddart 1980). It is generally assumed that an inspection of urine marking allows males to recognize approaching oestrus in the female (Estes 1972; Brown 1979; Stoddart 1980). In the red fox, the odour of the urine of both sexes increases abruptly at the onset of breeding (Henry 1980). A communicative role for urine in Arctic foxes has been suggested already by Heimburger (1959). In our previous study on the species, it appeared that the urine of sexually active bitches was attractive compared to a variety of other urine samples (e.g. the urine of sexually active males, of intact animals of both sexes out of reproductive season, and of castrates of both sexes). Both males and females showed a significant olfactory preference for this urine (Rödl *et al.* 1989). Synchrony of the reproductive status of males and females might be important in those species that reproduce throughout the year, but could be of critical significance in species that mate seasonally (Drickamer 1980), as in the case of the Arctic fox.

In Canids, with some limitations, urine marking is thought to be associated with the acquisition and maintenance of individual or group territories (Wells and Bekoff 1981). Territorial boundaries of wild Arctic foxes appear to be maintained by both vocalization and scent marking (Eberhardt *et al.* 1982). We have previously suggested that urine odours of reproductively active bitches may affect reproduction of other females and/or attract males, while in the male they may play a more important role in other situations such as territorial marking (Bartoš *et al.* 1989; Rödl *et al.* 1989). We have already demonstrated pheromonal activity of the sexually active female's urine, which stimulated puberty of immature Arctic foxes of both sexes. On the other hand, the pheromonal activity of the sexually active males was much less apparent (Bartoš *et al.* 1989).

In the present study we tested the hypothesis that (i) urine of sexually active Arctic foxes would synchronize seasonal reproductive activity of the two sexes and/or stimulate breeding performance of conspecifics of both sexes (in terms of increased probability to conceive and higher number of resulting progeny); and (ii) urine of sexually active females would be more potent in this respect than that of sexually active males.

## Materials and methods

Twelve castrated adult female and twelve male Arctic foxes were individually housed. Their sexual behaviour was induced by external hormones as described elsewhere (Bartoš *et al.* 1988), in order to obtain a behaviourally defined source of urine. Animals exhibiting complete sexual behaviour in daily tests after the hormonal treatment were considered as sexually active. Urine samples were collected from beneath the wire-mesh floor of the animal's cage (11.00 × 11.50 × 10.00 cm), using a plexiglass collector. The upper portion of the collector slightly exceeded the area of the cage's floor, oblique walls lead into a plastic bottle. The bottle was changed every morning and its content frozen at −24°C. If an animal proved to be displaying sexual behaviour, its sample was collected the morning following the day of observation.

For technical reasons, all experimental and/or control groups of foxes had to be placed at the same farm. However, within the farm they were located at least 100 m apart. Both sexes were tested and data were statistically analysed independently in different seasons.

The breeding females were divided into three experimental groups: group F ($n = 47$)was exposed to the urine of sexually active female conspecifics; group M ($n = 46$) was exposed to the urine of sexually active male conspecifics; and a control group C ($n = 80$) was exposed to water. In the following season, the breeding males were divided into three experimental groups: group F ($n = 22$) was exposed to the urine of sexually active female conspecifics; group M ($n = 20$) was exposed to the urine of sexually active male conspecifics; and a control group C ($n = 41$) was exposed to water.

Pooled samples of urine were applied beginning on 1 February (about one month before the beginning of the mating season). Twenty millilitres of the fluid was deposited in a plastic bottle attached to the outer wall of the cage, where the bottles were left until the termination of mating (end of April). The bottles were protected against biting by wire mesh.

Various confounding factors that might affect reproduction of the bitches tested were eliminated by least squares (LS) multifactorial analysis after Harvey (1966) according to the equation

$$Y_{ijklmnp} = \mu + A_i + B_j + C_k + \& D_e + E_m + F_p + e_{ijklmnp}$$

where $\mu$ = mean; $A_i$ = an influence of individual keepers; $B_j$ = number of previous litters of the bitch or number of previous seasons when the male was sexually active; $C_k$ = age of the sire; $D_e$ = age of the bitch; $E_m$ = number of copulations achieved (the keepers on the farm released a bitch into a male's cage for about 1 h when she first showed external signs of oestrus. This was repeated during the following few days even if the bitch copulated. Thus this variable is equal to the number of days when the bitch copulated); $F_p$ = experimental group; $e_{ijklmpn}$ = remainder. Detailed analysis was performed by a modified version of Duncan's test (Kramer 1957).

The bitch's breeding performance was estimated by *date of delivery* and *number of pups/litter*. Male reproductive success was measured by: *number of copulated bitches* (number of individual females with which the male copulated within the season); *per cent of yeld bitches* (bitches which failed to have any progeny after being copulated); *number of offspring/male* (a mean number of progeny of all the bitches copulated by the male).

## Results

### FEMALES' BREEDING PERFORMANCE

LS ANOVA analysis showed that there was no significant influence ($p > 0.05$) of individual keepers, number of previous litters, and age of the sire. Figure 1 shows LS means (± S.E.) of the reproductive characteristics of females in relation to different factors (with the influence of other factors eliminated mathematically).

Bitches exposed to urine of sexually active females gave birth later in the season than bitches exposed to urine of sexually active males ($p < 0.05$), and they had larger litters than control bitches ($p < 0.05$) or bitches exposed to the male urine ($p < 0.05$). The age of a breeding bitch influenced the date of delivery so that the older the bitch, the earlier the date of delivery (ages 1 vs. 2, $p < 0.01$; 1 vs. 3 $p < 0.01$; 1 vs. 4, $p < 0.01$; 2 vs. 3, $p < 0.01$; 2 vs. 4 $p < 0.01$). Bitches one and two years old produced similar numbers of young per litter, both having larger litters than those aged three ($p < 0.01$) and four ($p < 0.01$) years. The number of copulations influenced both characteristics of reproduction. Bitches that copulated twice gave birth later in the season, and had larger litters. Date of delivery was correlated with number of pups/litter (residual coefficient of correlation, $r = 0.69$, $p < 0.001$).

### MALES' BREEDING PERFORMANCE

Male reproductive success was not influenced by individual keepers, age of male,

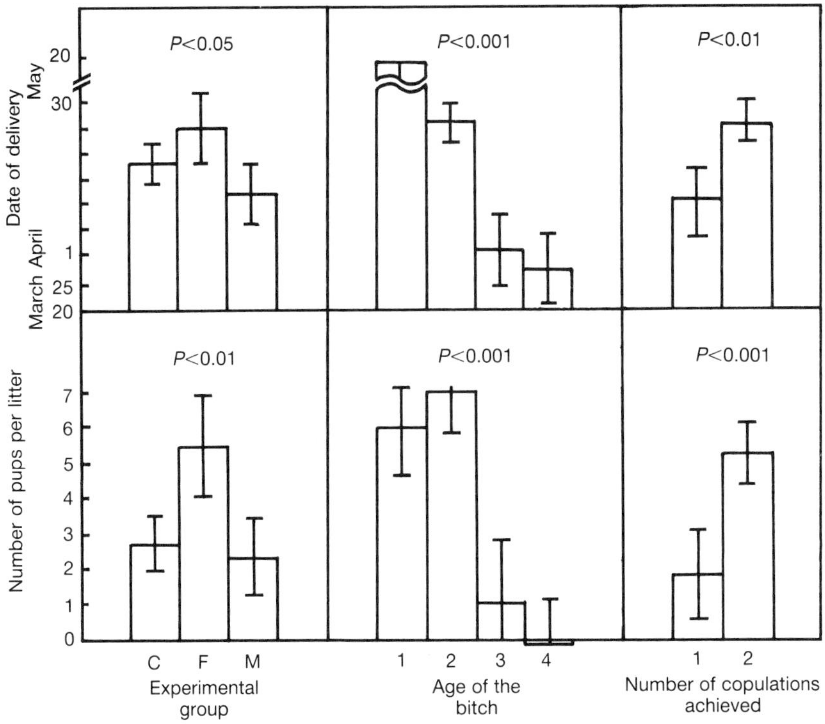


Fig. 1. Males breeding performance. Least-squares means ($\pm$SE) of dates of deliveries and number of pups per litter according to experimental group, age of the bitch, and number of copulations achieved.

or the number of previous seasons for which the male was sexually active ($p > 0.05$). The influence of the experimental group was not statistically significant ($p > 0.05$), but an influence was indicated in the number of copulated bitches, in that the males exposed to the urine of sexually active females tended to copulate with the highest number of bitches (Fig. 2). The number of times the male copulated with a bitch was the only other significant factor (Fig. 2).

In previous breeding seasons less than 20% of yearling males were able to copulate with receptive bitches. Therefore we compared proportions of males able to copulate according to their age (Table 1). While older males (two and three years old) did not respond significantly to exposure to any urine sample, yearling males exposed to the urine of sexually active females performed better than those exposed to the urine of sexually active males ($\chi$}2 $= 3.88$, $p < 0.05$) or the water control ($\chi$}2 $= 5.98$, $p < 0.01$). Performance of yearling males was not affected by exposure to male urine compared to water controls ($\chi$}2 $= 0.03$, $p > 0.05$).

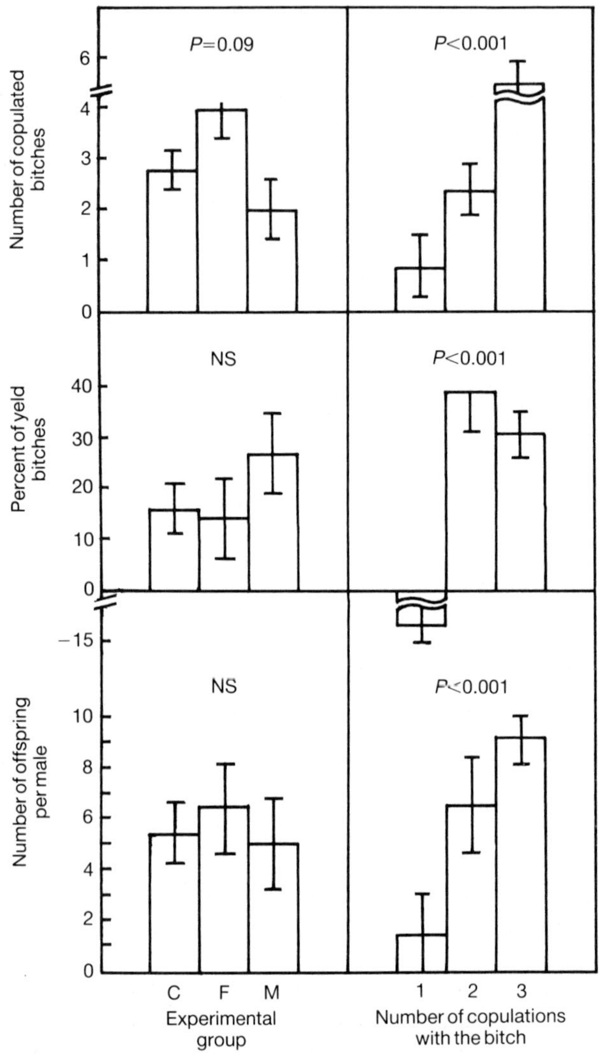


Fig. 2. Females' breeding preformance. Least-squares means ($\pm$SE) of the characteristics of reproduction according to experimental group and the number of copulations with the bitch.

**Table 1.** Proportion of males of the three experimental groups which were able to copulate with bitches, according to age

| Age (years) | C $n$ | C % | F $n$ | F % | M $n$ | M % | $\chi^2$ | $p$ |
|---|---|---|---|---|---|---|---|---|
| 1 | 19 | 15.79 | 10 | 60.00 | 11 | 18.18 | 7.08 | $p<0.03$ |
| 2 | 4 | 75.00 | 5 | 60.00 | 5 | 40.00 | 1.14 | $p>0.05$ |
| 3 | 18 | 88.89 | 7 | 71.43 | 4 | 100.00 | 2.03 | $p>0.05$ |

## Discussion

In agreement with our hypothesis, the results suggest an effect of the urine of sexually active female Arctic fox on a conspecific's reproduction in both sexes, the effect being less pronounced in males. The seasonal reproductive activity of the two sexes seemed to be synchronized. The bitches subjected to female urine presentation tended to give birth later, when the males were able to copulate with them, thus solving the major practical problem encountered by the fur farm. Exposure to the urine of sexually active males had no effect. No inhibitory effect was detected of exposure to any type of urine.

The demonstration that pheromones produced by one sex may affect both sexes is of rather rare occurrence. Vandenbergh and Drickamer (1974) applied hormones to two ovariectomized rhesus macaque females during the part of the season when the animals were in reproductive quiescence. Within eight to ten weeks, it was recorded that the males of the group started to behave sexually earlier in the season than is normal and levels of their androgens increased. Intact females of the group gave birth the following season earlier than the control group. On the other hand, when intact female rhesus macaques were paired outside the regular reproductive season with males in which sexual activity was artificially induced, no changes in ovarian cyclicity were detected (Vandenbergh and Post 1976). The authors concluded that the interactions between females significantly affected such changes.

Contrary to other reports under similar conditions (Kukla 1985), the higher number of copulations achieved with a bitch did not have a positive influence on pregnancy rates. In contrast, our results indicate a lower percentage of yeld bitches after one copulation compared with further copulations (Fig. 2), indicating that the bitch's reproductive condition is an important determinant of breeding performance. On the other hand, the number of copulations did affect the number of offspring per male, which is in strong agreement with the results of Kukla (1985).

Our results seem to be further evidence for the suggestion expressed earlier (Bartoš *et al.* 1989) that in Arctic foxes, urine odours of sexually active bitches may affect reproductive functions among bitches and/or attract males. It is also suggested that male urine odours may play a major role in territorial marking, although further field testing is needed for verification. For fur farmers, the use of oestrous female urine may provide a potent means to improve breeding success on the farm.

## Acknowledgements

We are grateful Dr G. Horalek for her kind help with preparation of the English version of the manuscript and D. Řehák for computing the statistics. Technical assistance of M. Bartůněk and the staff of the fur farm at Dublovice is gratefully acknowledged.

## References

Adams, M.G. (1980). Odour-producing organs of mammals. *Symp. Zool. Soc. Lond.*, 45, 57–86.

Bartoš, L., Rödl, P. and Madlafousek, J. (1988). Induction of copulatory behavior in female and male Arctic foxes. *Horm. Behav.*, 22, 467–73.

Bartoš, L., Rödl, P. and Fiedler, J. (1989). Influence of urine odours on puberty in male and female Arctic foxes. *Proc. XVIIIth Congress I.U.G.B., Krakow*, in press.

Brown, B.E. (1979). Mammalian social odors: a critical review. In *Advances in the study of behavior* (ed. J.S. Rosenblatt, R.A. Hinde, C. Beer and M.C. Busnel), Vol. 10, pp. 103–60. Academic Press, New York.

Drickamer, L.C. (1980). Social cues and reproduction—rodents and primates. In *Biosocial mechanisms of population regulation* (ed. M.N. Cohen, R.S. Malpass and H.G. Klein), pp. 37–53. Yale University Press, New Haven.

Eberhardt, L.E., Hanson, W.C., Eberhardt, L.E., Hanson, W.C., Bengtson, J.L., Garrott, R.A. and Hanson, E.E. (1982). Arctic fox home range characteristics in an oil-development area. *J. Wildlife Managm.*, 46, 183–90.

Estes, R.D. (1972). The role of the vomero-nasal organ in mammalian reproduction. *Mammalia*, 36, 315–41.

Harvey, W.R. (1966). Least squares analysis of data with unequal subclass numbers. *Agric. Res. Serv., USDA.*

Heimburger, N. (1959). Das Markierungsverhalten einiger Caniden. *Z. Tierpsychol.*, 16, 104–13.

Henry, J.D. (1980). The urine marking behavior and movement patterns of red foxes (*Vulpes vulpes*) during a breeding and post-breeding period. In *Chemical signals in vertebrates* (ed. D. Müller-Schwarze and R.M. Silverstein), pp. 11–27. Plenum, New York, London.

Jöchle, W. (1973). Coitus-induced ovulation. *Contraception*, 7, 523–564.

Kramer, C.Y. (1957). Extension of multiple range tests to group correlated adjusted means. *Biometrics*, 13, 13–18.

Kukla, F. (1985). Některé údaje o potenci pesců. In *Sborník produkcia a zdravie v chove kožušinových zvierat*, pp. 69–73. Dôm techniky ČSVTS, Košice.

Mykytowycz, R. (1977). Olfaction in relation to reproduction in domestic animals. In *Chemical signals in vertebrates* (ed. D. Müller-Schwarze and M.M. Mozell), pp. 207–24. Plenum, New York and London.

Rödl, P., Bartoš, L. and Fiedler, J. (1989). Olfactory preference of the male and female urine in Arctic foxes. *Proc. XVIIIth Congress I.U.G.B., Krakow*, in press.

Stoddart, D.M. (1980). *The ecology of vertebrate olfaction*. Chapman and Hall, London and New York.

Vandenbergh, J.H. and Drickamer, L.D. (1974). Reproductive coordination among free-ranging rhesus monkeys. *Physiol. Behav.*, 13, 373–6.

Vandenbergh, J.H. and Post, W. (1976). Endocrine coordination in rhesus monkeys: female responses to the male. *Physiol. Behav.*, 17, 979–84.

Wells, M.C. and Bekoff, M. (1981). An observational study of scent marking in coyotes, *Canis latrans*. *Anim. Behav*. 29, 332–350.

# IV.6 Discrimination of Conspecific Chemosignals by Female Ringtailed Lemurs, *Lemur catta L.*

S.J. DUGMORE AND C.S. EVANS

## Introduction

*Lemur catta* is an opportunistic lemur living in gallery and deciduous forest and marginal areas of brush and scrub in south-eastern Madagascar. Troops of *L. catta* defend home ranges, of which there is variation in size and exclusiveness (Jolly 1972). These lemurs are diurnal; they spend much of the day on the ground which allows the use of some visual signals, although their communication is predominantly olfactory, as is that of most prosimians (Epple 1984). In troops the females are dominant, displacing any male at feeding and resting sites, and are the principal home range defenders (Mertl-Millhollen *et al.* 1979).

*L. catta* troop females form an apparently linear and stable core group which is maintained with little overt aggression. Here we report on a study of the behavioural responses of females presented with various conspecific scents in order to examine the functions of olfactory cues in the regulation of conflict or other interactions within and between troops.

## Scent glands and scent marking

For scent marking, female *Lemur catta* use the median surface of the genital labia, which is an area well supplied with apocrine and sebaceous glands. The entire labial region is applied to a vertical surface and a gentle rubbing transfers the secretion to the object. Male *L. catta* have three scent sources: the bare skin of the scrotum, the brachial gland, and the ante-brachial gland spur complex. As with the female labial region, the male scrotum has abundant sebaceous glands and secretion is applied to objects by rubbing. Brachial gland secretion is applied only in conjunction with ante-brachial secretion with which it is mixed by rubbing the ante-brachial gland over the brachial gland. The secretion is then applied by placing the ante-brachial complex distal to a branch; rotation of the forearm and shoulder leaves a crescent-shaped scar impregnated with 'mixed' secretion (Evans and Goy 1968; Schilling 1974).

## Experimental methods

Four adult female *L. catta* (Cheeky, CH; Lucy, LU; Venus, VS; Rossi, RO) were tested in a free-choice, paired presentation of conspecific scents. Each female served as both subject and scent donor, with five adult males providing male scent. The scents were presented in one of two ways, either naturally deposited marks on wooden dowels (no swab present), or on a sterile cotton swab presented in association with a wooden dowel. Scent marks were obtained by placing clean dowels in the donors' cage and watching until the required number of marks were deposited: six for male arm-marks and one for female genital marks, both based on the number of marks left in a normal marking session. Both males' and females' genital secretions were collected by briefly restraining the animal in a squeeze cage. The swabs obtained were then frozen at $-15$ °C until required. Tests were carried out in small observation cages (Bailey 1979), to which the animals were accustomed, in a partially sound-proofed test room. A separate ventilation system was used for the animals' home rooms and test room, and for the room used to house 'strange' lemurs. The latter animals were separated socially from test animals for a minimum of three months.

Each test was of 5 min. duration. The responses measured were the duration of total investigation (sniffing plus licking) and the frequencies of licking, flehmen, and all forms of scent-marking behaviour. Data analysis was by the Binomial Test or the more powerful Wilcoxon Matched-Pairs Signed-Ranks Test where $N$ values were sufficiently large.

## Results

### EXPERIMENT A: FEMALE RESPONSES TO SIMULTANEOUS PRESENTATION OF THEIR OWN AND A STRANGE FEMALE CONSPECIFIC'S LABIAL MARKS

The overall result of these trials shows a strong discrimination between the presented scents by all subjects (Fig. 1), with strange female scents eliciting higher levels of investigation (for lumped data, $p < 0.001$). Lumped data for licking, marking, and flehmen all showed statistically significant differences, each at a higher level to the strange female scent (licking, $p < 0.001$; marking, $p < 0.05$; flehmen, $p < 0.02$). All subjects showed higher frequencies of licking and three subjects (VS, LU, RO) also showed higher frequencies of flehmen.

### EXPERIMENT B: FEMALE RESPONSES TO SIMULTANEOUS PRESENTATION OF THEIR OWN AND A STRANGE MALE'S SCROTAL SCENT

There was no observed discrimination between these two scents for any of the measured responses for lumped or individual data. Three subjects showed slightly

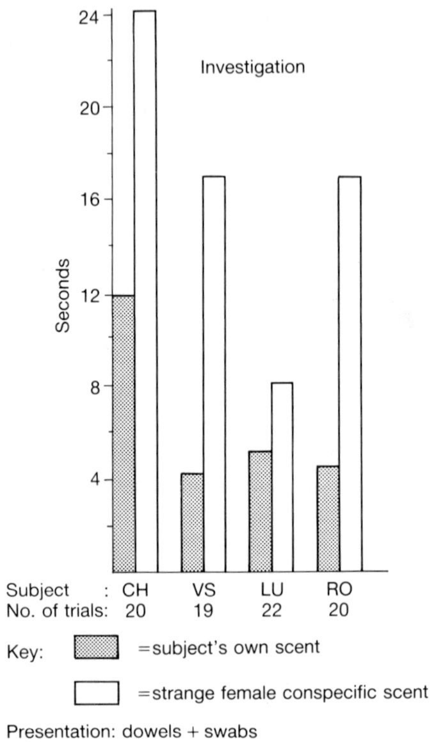


Fig. 1. Chemo-investigations and marking to own vs. other labial scent marks.

higher frequencies of investigation of male scrotal scent (Fig. 2), and two (CH, VS) showed slightly higher frequencies of licking. Marking and flehmen frequencies were low and showed no trends in direction of response.

## EXPERIMENT C: FEMALE RESPONSES TO SIMULTANEOUS PRESENTATION OF A STRANGE FEMALE'S LABIAL SCENT PAIRED WITH A STRANGE MALE'S SCROTAL SCENT

Overall, strange female labial scent was preferred ($p < 0.003$). Two subjects showed a significant preference for investigation of strange female scent (VS, $p < 0.01$; RO, $p < 0.01$). One other subject showed the same direction of preference; however the fourth exhibited a very slight preference for male scrotal scent (Fig. 3). Only subject VS showed a significant difference in licking frequency ($p < 0.02$). Marking levels were very low with no significant differences. Flehmen levels were also low with two subjects (CH, VS) showing slightly higher levels towards the female scent.

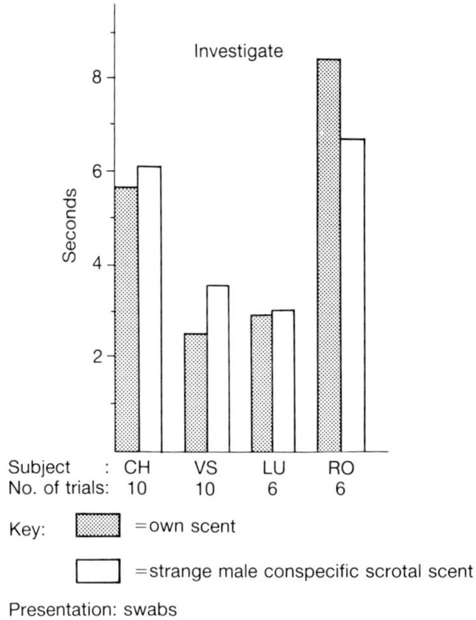


Fig. 2. Chemo-investigation and marking to genital scents: own vs. strange male.

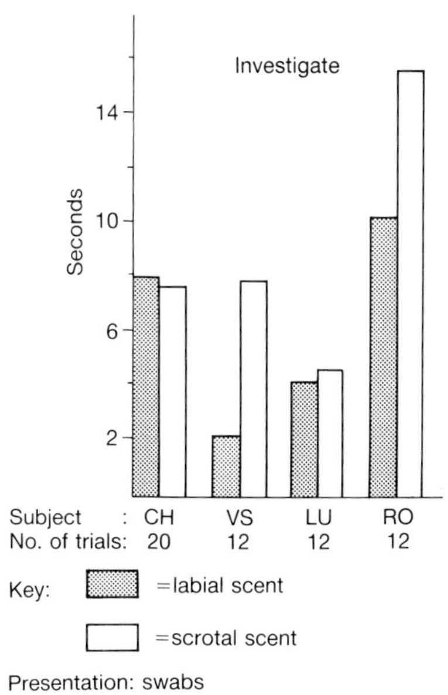


Fig. 3. Chemo-investigation and marking to strange genital scents: Male vs. female.

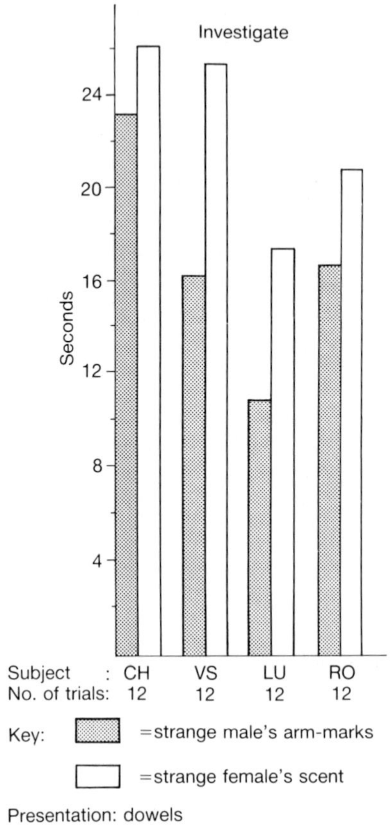


Fig. 4. Chemo-investigation and marking to labial vs. ante-brachial/brachial marks.

### EXPERIMENT D: FEMALE RESPONSES TO SIMULTANEOUS PRESENTATION OF A STRANGE FEMALE'S LABIAL SCENT PAIRED WITH A STRANGE MALE'S ANTE-BRACHIAL MARK

Overall the subjects showed a preference for the strange female labial scent (Fig. 4); $p < 0.01$), with two subjects showing significant discrimination (LU, $p$ 0.01; RO, $p < 0.05$). Licking showed the same directionality as did investigation, with lumped data significant at $p < 0.004$. All four subjects showed the flehmen response during these trials with three exhibiting significantly higher levels to the female scent source (CH, $p < 0.05$; VS, $p < 0.05$; RO, $p < 0.02$).

## Discussion

The female *L. catta* used in this study showed discrimination between their own and 'strange' females' labial scent, with longer investigation and higher frequencies of licking, genital marking, and flehmen, all of which suggest a strong preference for this stimulus. In contrast, no discrimination was shown between male scrotal secretions and the subjects' own scent marks on any behavioural measure. The latter result suggests that neither scent stimulus was attractive to the subjects, in the context in which they were presented. This is supported by the results of Experiment C where an overall preference for 'strange' female scent was shown when compared with 'strange' male scrotal scent, demonstrating the ability to distinguish between male and female genital scents.

It was known that males show a preference for female scent over male arm-marks (Dugmore *et al.* 1984). Experiment D indicates that this is also true for females, with all subjects showing longer investigation times to 'strange' female scent. Marking levels were low during all experimental trials, with only the pairing of labial scents (own versus strange) eliciting a strong difference in marking levels. This suggests that female marking is selectively elicited by any strange conspecific labial scent. The flehmen response showed an interesting pattern of occurrence, following that seen in males (Bailey 1979), with strange female scent eliciting higher levels than either male scent source.

The pattern of high levels of interest shown by adult males and females to conspecific female scent marks, both in this study and previously (Dugmore *et al.* 1984) may be due, at least in part, to the pattern of female dominance observed in this species (Jolly 1966). Olfactory signals have been suggested as containing information on dominance or rank position in many male mammals (Brown 1979). It is likely that among females exhibiting group ranking systems chemosignals will show a parallel use in the transmission of such information. The establishment of matrilines in a captive troop of *L. catta* (Taylor and Sussman 1985) supports this contention.

## References

Bailey, K. (1979). Differential responses of ring-tailed lemurs (*Lemur catta*) to olfactory stimuli from conspecifics. Ph.D. thesis, Dept. of Biological Sciences, Glasgow College, Scotland.

Brown, R.E. (1979). Mammalian social odours. *Adv. Study Behav.*, 10, 103–62.

Dugmore, S.J., Bailey, K. and Evans, C.S. (1984). Discrimination by ring-tailed lemurs (*Lemur catta*) between the scent marks of male and those of female conspecifics. *Int. J. Primatol.*, 5, 535–45.

Epple, G. (1984). Communication by chemical signals. In *Comparative primate biology*, 2A, (ed. G. Mitchell), pp. 531–80. Liss, New York.

Evans, C.S. and Goy, R.W. (1968). Social behaviour and reproductive cycles in captive ring-tailed lemurs. *J. Zool.*, 156, 181–97.
Jolly, A. (1966). *Lemur behaviour*. Chicago University Press.
Jolly, A. (1972). Troop continuity and troop spacing in *P. verreauxi* and *L. catta* at Berenty. *Folia Primatol.*, 17, 335–62.
Mertl-Millhollen, A.S. *et al.* (1979). Population and territorial stability of *Lemur catta* at Berenty. *Folia Primatol.*, 31, 106–22.
Schilling, A. (1974). A study of marking behaviour in *Lemur catta*. In *Prosimian biology* (ed. R.J. Martin *et al.*), pp. 347–64. Duckworth, London.
Taylor, L. and Sussman, R.W. (1985). A preliminary study of kinship and social organisation in a semifree ranging group of *Lemur catta*. *Int. J. Primatol.*, 6, 601–15.

# IV.7 Mutual Avoidance by European Moles *Talpa europaea*.

MARTYN L. GORMAN AND R. DAVID STONE

## Introduction

European moles, *Talpa europaea*, despite their subterranean mode of life, are very much aware of the presence and movements of their neighbours. Here, we present evidence for this statement and describe experiments designed to determine how moles might obtain information about the movements and behaviour of their neighbours. First, some background information is provided on the social organization of moles and their activity patterns (Stone and Gorman 1985; Stone 1986).

Moles were trapped in Friesian traps and radio-tracked at the Haddo Estate on the Buchan plain in the north-east of Scotland. The radios, weighing *c*. 2.0 g and transmitting at 173.2 MHz, were attached to the dorsal surface of the tail and made location possible to an accuracy of $\pm$ 0.25 m. This technique is described in Stone (1986). Individual moles were tracked for several days, locating them every 10–15 mins. The radio fixes were then used to describe home ranges using the harmonic mean transformation (Dixon and Chapman 1980).

For most of the year moles are solitary and sedentary creatures that occupy a mosaic of home ranges containing complex networks of subterranean tunnels (Fig. 1). The ranges of individuals are largely exclusive but there is a degree of overlap between some neighbours: this involves mainly the interdigitation of tunnels in the soil column but also some sharing of tunnels. Typically, the ranges of males do not overlap, but each male's range may overlap with those of several females, and females may overlap with other females (Fig. 2). This pattern breaks down during the breeding season when males dig extensive outward tunnels from their home ranges to break into the tunnel systems of receptive females (Godfrey and Crowcroft 1960).

Radio-tracking has also provided information on patterns of activity. Usually, each mole has a single underground nest where it sleeps. By placing recorders over the nests of radio-tagged moles it has been possible to monitor, over long periods of time, when they are asleep and when they are active in the tunnels. For most of the year moles show a regular periodicity with three bouts of activity per 24 h, each separated from the next by a period of sleep (Fig. 3).

Two lines of evidence support our contention that moles are aware of their neighbours: (i) removal experiments and (ii) temporal patterns of exploitation.

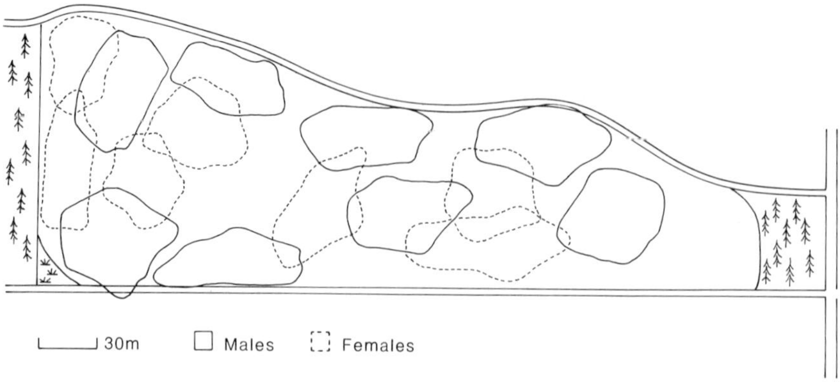


Fig. 1. The home ranges of a population of moles living in a single field in north-east Scotland. The ranges were computed using the harmonic mean method and are depicted as the isopleth containing 95% of the radio fixes. Data from Stone (1986).

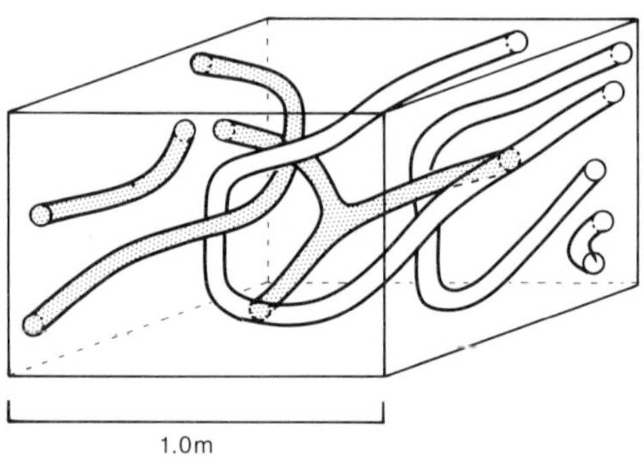


Fig. 2. A cross-section through the ground in the area of overlap between two neighbouring moles. The tunnels of one individual are shown stippled, the other without stippling. Note that in addition to the interdigitation of tunnels, the tunnels of the two animals are connected at two points.

During the non-breeding season, moles normally remain strictly within the confines of their own tunnel system. However, data exist to show that if a mole dies, or if it is removed from its range, then neighbouring moles will move, often rapidly, into the vacated area (Stone 1986). This is exemplified in Fig. 4 (a and b) which shows the ranges of two sets of four neighbouring moles. The upper part of the figures show the spatial organization of the moles' ranges over a period of several weeks. When the moles, whose ranges are shown stippled, were trapped and removed one or more neighbours moved into, and occupied parts of, the vacated areas. In both cases, the detection of the trapped animal's disappearance was rapid, with the first neighbours encroaching on the empty area within 15 h.

If moles meet during the non-breeding season, serious fighting usually ensues.

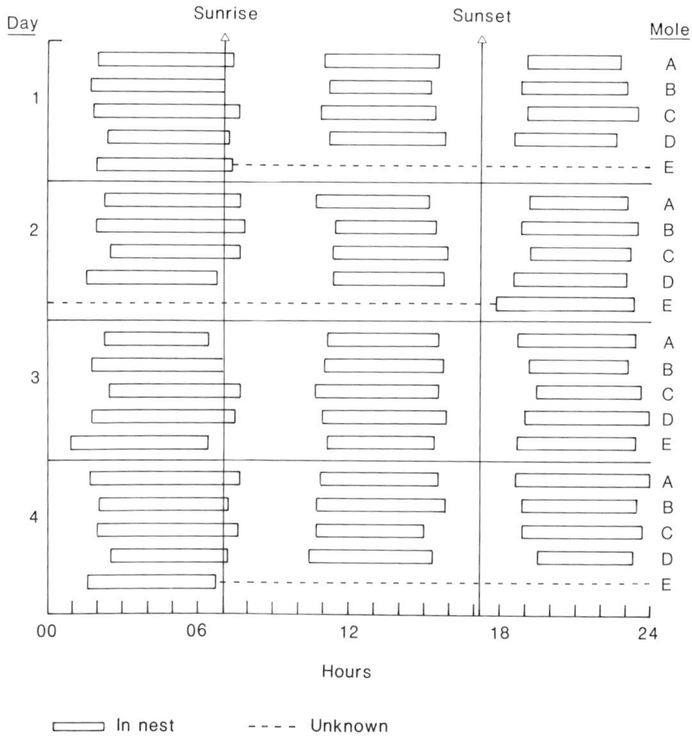


Fig. 3. The activity patterns of five moles (A–E) over 4 consecutive days. The closed bars represent the time periods when the moles were in their nests. The record for mole E is incomplete due to equipment failure and the missing data are shown as a broken line.

However, this is a rare occurrence, despite the overlap between ranges, because neighbouring moles show temporal avoidance, with only one individual using an area held in common at any one time (Stone and Gorman 1985). This is clearly shown by five radio-tagged moles whose ranges and patterns of activity are shown in Figs. 3 and 5. Their activity was highly synchronized, all five leaving and returning to their nests at about the same time. The area of overlap between the ranges of any two neighbours was exploited by only one individual during any given period of activity (Fig. 6).

Clearly, moles were not only aware of the territorial occupancy of their neighbours, but they also had sufficient appreciation of their movements to avoid meeting in areas of overlap. That they quite quickly detected the disappearance of a neighbour implies that the information about a mole's occupany is short-lived. One likely route of signalling is by scent marking. Male and female moles possess paired preputial glands which contain apocrine and sebaceous elements, and which drain via ducts into the urethra (Fig. 7). The glands add to the urine a

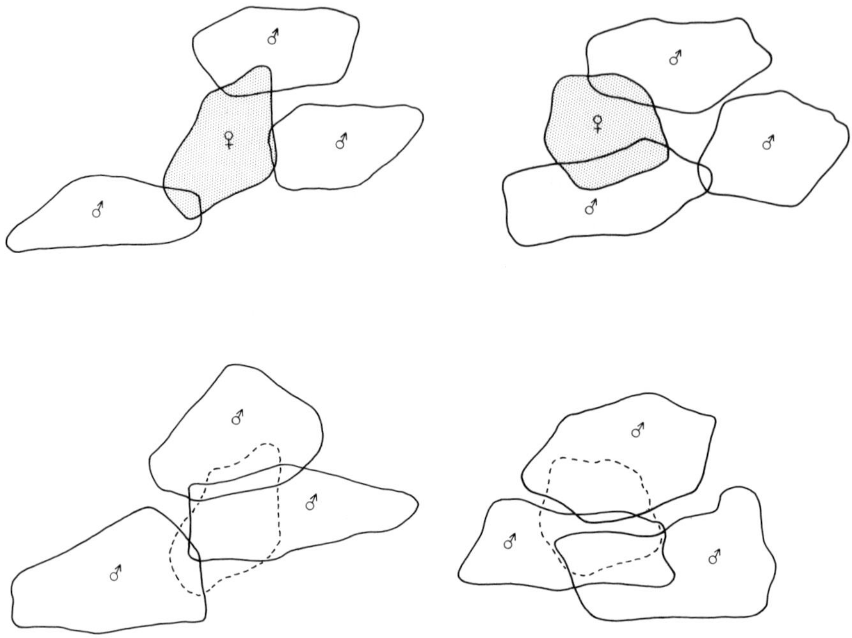


Fig. 4 (a & b). The home ranges of two sets of four neighbouring moles computed and drawn as in Fig. 1. The upper part of each figure shows the situation prior to the removal of one of the females, the lower part the situation following their demise.

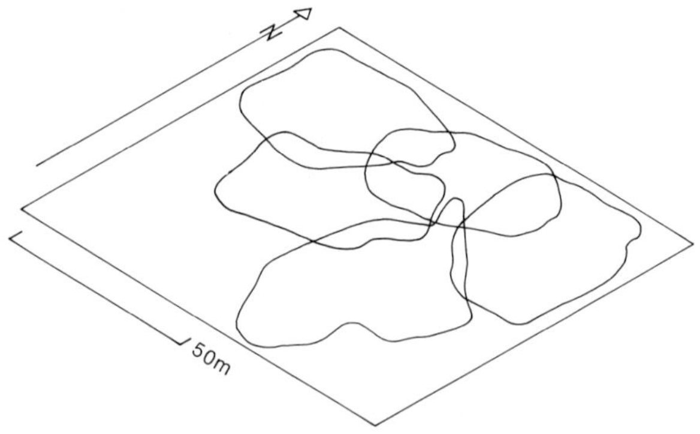


Fig. 5. The home ranges, shown as the harmonic mean isopleths containing 95% of radio fixes, of five neighbouring moles. The map is shown as if seen from the south-west at an elevation of 20° above the horizontal.

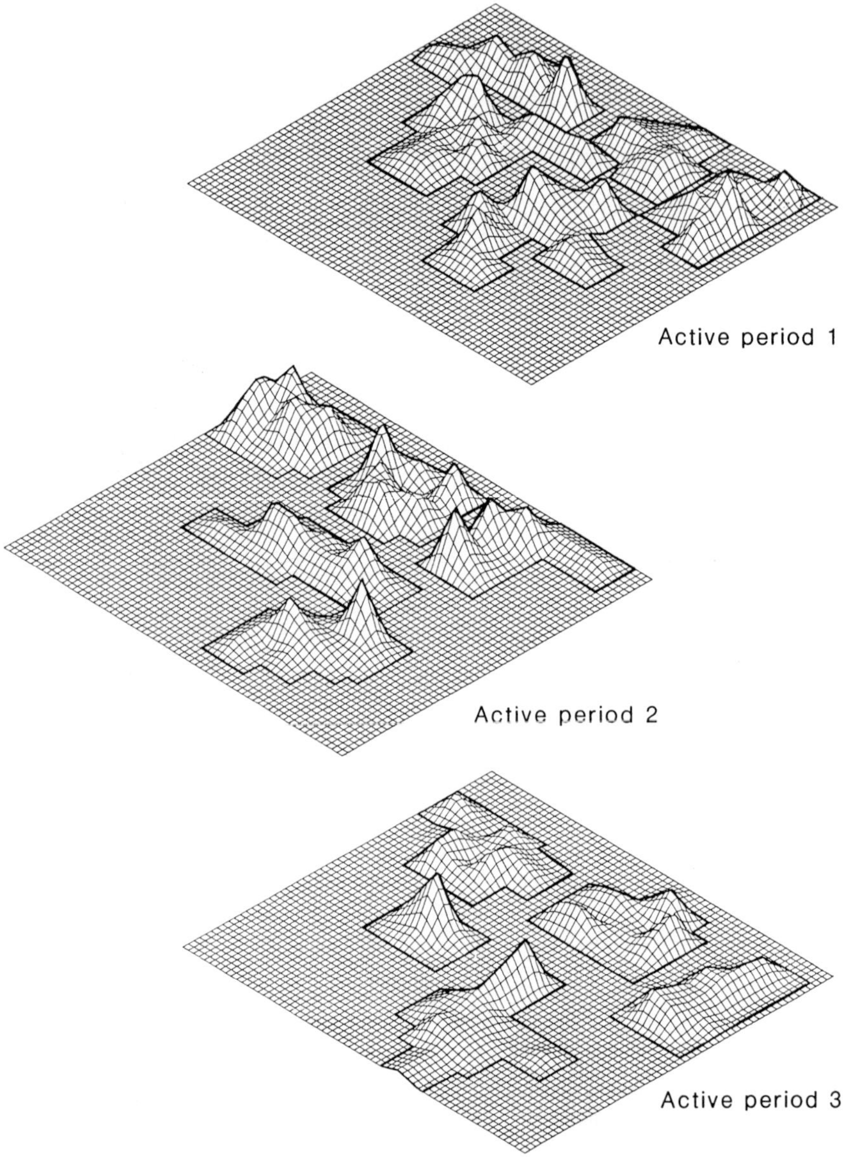


Fig. 6. The pattern of movements and range utilization, of the five males whose ranges are shown in Fig. 5 over three consecutive activity periods. The plots were prepared by dividing the map into a matrix of 5 by 5 metre squares and counting the number of times moles were located in each element. The surfaces were computed with the SURFACE II suite of routines (Sampson 1978) and are drawn as if seen from the south-east with the observer placed 20° above the horizontal and 10000 map units away from the centre of the matrix.

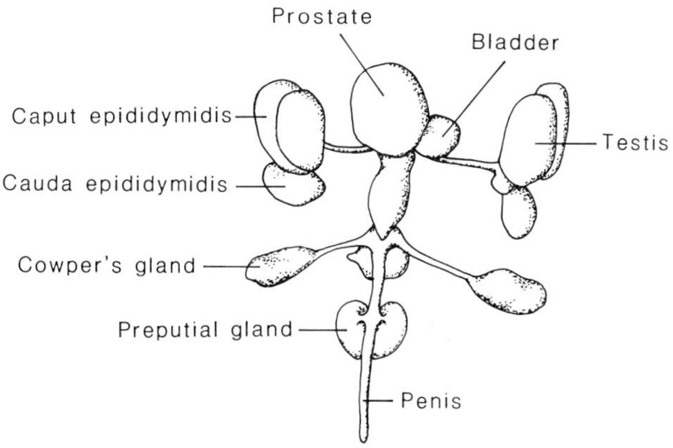


Fig. 7. The preputial glands showing their bodily position (a), their relationship to other organs (b), and their apocrine (c) and sebaceous (d) elements. (b) drawn from a photograph in Racey (1978).

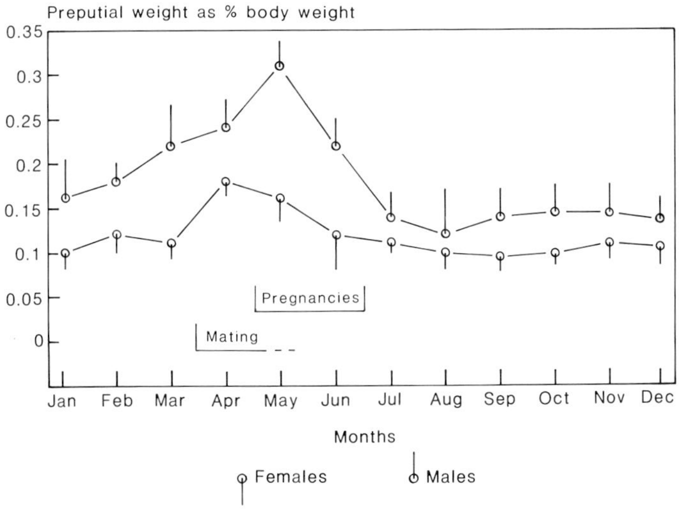


Fig. 8. Seasonal changes in the mean ($\pm$SEM) weight of male and female preputial glands. Sample sizes were 204 males and 152 females. Data from Stone (1986).

secretion with an odour reminiscent of the general smell of a mole. The glands of males are larger than those of females, even allowing for differences in body weight, but they increase in size in both sexes during the breeding season (Fig. 8) (Stone 1986). The secretion contains a complex mixture of volatile chemicals with marked individual and sexual differences in composition (Gorman and Stone 1989). There is the potential, therefore, for moles to use urine and preputial

secretions to signal that an area is occupied or that a resident has passed by in the recent past. Mole ranges contain several hundreds of metres of tunnels and thus quite large quantities of urine would be required if the range is to be marked regularly. The average mole consumes around 60 g of earthworms each day, containing some 50 g of water (Gorman and Stone 1990). Given the high humidity of the tunnels, most of this will be voided as urine.

We have performed field experiments to determine whether moles do disperse urine through their tunnels and if so, in what pattern. We have also carried out field trials to determine the responses of moles to preputial gland secretions.

## Materials and methods

### DISPERSION OF URINE

To investigate the spatial pattern of urination, a male mole's home range was determined by radio-tracking. The animal was then recaptured and injected with $^{32}P$ at a dosage of 1.0 Ci$\mu$ g$^{-1}$ body weight. The mole was then released into its tunnels for 48 h, after which time a section of the range was excavated and the tunnels in the area checked for radio-activity using a hand-held monitor.

### REACTIONS TO PREPUTIAL SECRETIONS

The behavioural experiments were designed to determine whether free-living moles actively avoid mole odours. The first approach was to compare the catching efficiency of clean Friesian traps with that of traps that had recently contained a mole. All traps were first cleaned by washing in 10% Decon 75, rinsing in water, and then being left to dry. To produce soiled traps a captive male mole was placed in a clean trap for 3–4 h just prior to use.

Overall, an equal trapping effort was made with each trap type, and equal numbers of clean and soiled traps were set in any given month. Only one trap was set in any one mole territory and each was left in place for up to 72 h, with daily resetting if blocked off with soil by a mole.

The second approach was to place cotton buds, clean or smeared with preputial secretion, into the tunnels of radio-tagged moles and to determine their response on encountering the stimuli. Three males and two females were used in the trials and each was tracked to obtain a clear idea of their tunnel systems and normal patterns of movement. Particular sites which the moles visited on a regular basis, and which contained shallow surface tunnels, were then identified. Stimuli were then placed in these tunnels, via a small hole in the roof, some 10–15 mins before a mole was likely to visit the spot. The reaction of a mole to a stimulus was scored as: (i) a positive retreat back down the tunnel; (ii) a negative response with the mole passing the stimulus and continuing along the tunnel; and (iii) an unclear response. The person doing the radio-tracking was unaware of the nature of the

stimulus. Only one presentation was made to any one mole on a given day; the stimulus was then removed and the tunnel repaired. Subsequent presentations were made in different parts of the moles' ranges. The secretions came from trapped animals whose glands had been removed shortly after death and stored at −25 °C. The glands were then crushed and the secretion applied to the cotton buds. All of the donors were strangers to the test moles with the exception of one male who had been a neighbour of male number 3 (Table 1).

**Table 1.** The Trapping Efficiency of Friesian Traps

| Number of captures in | |
|---|---|
| Clean traps | Soiled traps |
| 28 | 11 |

The experiments described above can only provide information on the gross response of moles to social odours. A field experiment was therefore carried out to gain insight into whether preputial secretions are involved in the day-to-day maintenance of territorial integrity; this involved the removal of a resident animal and the subsequent marking of some of its tunnels with its own preputial secretions. The trial involved two males and a female, and was carried out in October—the non-breeding season. The ranges of the three moles were initially determined by radio-tracking over a period of 10 days. The tunnels in the areas of range overlap, which were used by two animals, were identified and marked with wooden pegs. One animal was then trapped and its preputial glands removed. A number of cotton buds were then smeared with the secretion and distributed in the tunnels used by both the trapped mole and by one of its neighbours. An equal number of clean cotton buds were spread in the tunnels in the area of overlap with the second neighbour. The remaining animals were then tracked for several more days.

## Results and discussion

### THE DISPERSION PATTERN OF URINE MARKS

That part of the the mole's range excavated is shown in Fig. 9, as are all the locations where the level of radiation was at least 100% above background and where there was no evidence of faecal material. It is clear that urine was deposited at distinct sites throughout the tunnel system; particularly noteworthy is the fact that all tunnel junctions were marked, thus maximizing the chance of scent marks being discovered by an intruding animal, regardless of its direction of movement.

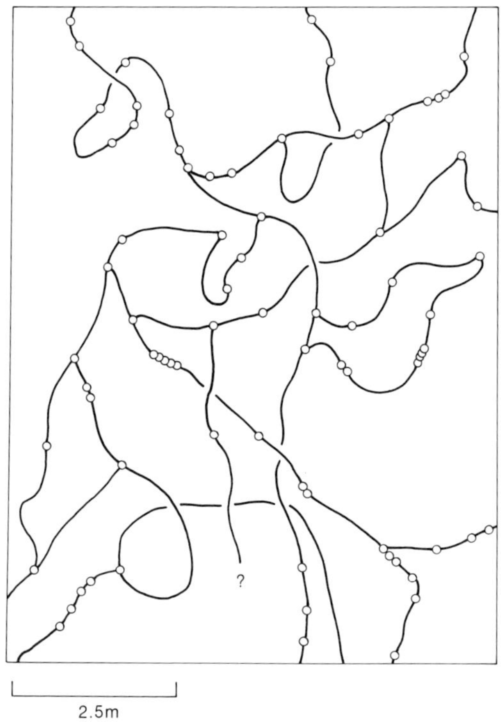


Fig. 9. Map showing the positions of excavated tunnels in part of the range of a male mole. The open circles indicate the positions of radioactive deposits.

### RESPONSES TO PREPUTIAL SECRETIONS

Trapping efficiency in clean and soiled traps is shown in Table 1. Clearly the traps that had recently contained a mole were markedly less efficient at catching than were clean traps. We conclude that, outside the breeding season at least, moles detect, and avoid, the odours left by other moles.

The reactions of the radio-tagged moles to the single stimuli placed in their tunnels are shown in Table 2. The responses were marked and clear-cut, with moles almost always retreating when they encountered preputial secretions, regardless of the sex of the donor, or whether it was a stranger or familiar neighbour.

The result of the single experiment on territorial maintenance is shown in Fig. 10. The upper part of the figure shows the ranges of the three animals prior to the removal of female 28 at 07.00 h on 10 October. Following her demise, cotton buds smeared with her preputial secretions were placed in the area of overlap with male 41. The ranges used by the two males in the week following the removal of the female are shown at the bottom part of the figure. At 03.00 h on 11 October male

**Table 2.** The response of free-living moles to either clean cotton buds or to cotton buds smeared with secretions of preputial glands.

| Mole number | Type of stimulus | Response: Positive | Negative | Unclear |
|---|---|---|---|---|
| Male 1 | Strange male | 15 | 2 | 3 |
| | Control | 2 | 16 | 2 |
| Male 2 | Strange female | 15 | 1 | 1 |
| | Control | 0 | 15 | 1 |
| Male 3 | Familiar male | 10 | 0 | 0 |
| | Control | 1 | 10 | 0 |
| Female 1 | Strange female | 11 | 2 | 3 |
| | Control | 0 | 16 | 0 |
| Female 2 | Strange male | 10 | 1 | 4 |
| | Control | 1 | 7 | 2 |

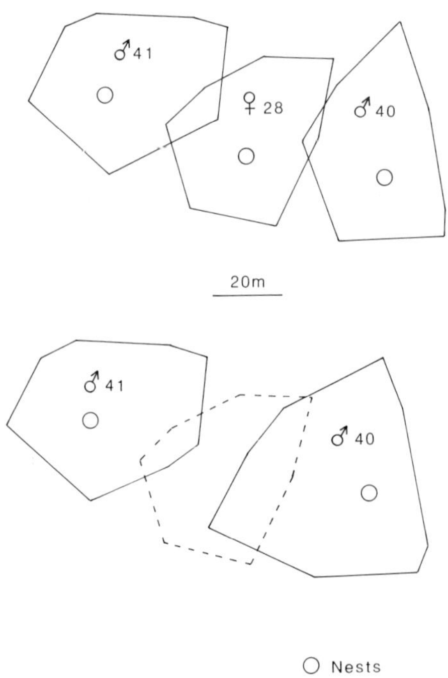


Fig. 10. The home ranges, shown as convex polygons, of three neighbouring moles, before and after the removal of female 28. See text for further details.

40 entered, for the first time, the now unoccupied range and remained there for 2.5 h before retiring to his own nest. At the beginning of the next activity period male 40 again entered the deceased female's range and remained there for the entire bout of activity. Male 40 continued to use parts of the female's range until at least 21 October when the radio failed. Throughout this period there was no indication that male 41 entered the empty range.

Clearly our experimental data are very limited, but it does seem that the preputial glands are likely to play a central role in advertising the presence of movements of moles in free-living populations.

## References

Dixon, K.R. and Chapman, J. A. (1980). Harmonic mean measures of animal activity areas. *Ecology*, 61, 1040–4.

Godfrey, G.K. and Crowcroft, P. (1960). The life of the mole (*T. europaea* Linnaeus). Museum Press, London.

Gorman, M.L. and Stone, R.D. (1989). Repelling moles. In *Mammals as pests*, (ed. R. Putman). Symposium, Zoological Society of London, pp. 81–97.

Gorman, M.L. and Stone, R.D. (1990). *The natural history of moles*. Croom Helm, London.

Racey, P.A. (1978). Seasonal changes in testosterone levels and androgen dependent organs in male moles (*Talpa europaea*). *J. Reprod. Fert.*, 52, 195–200.

Stone, R.D. (1986). The social ecology of the European mole (*Talpa europaea* L) and the Pyrenean desman (*Galemys pyrnaicus* G): a comparative study. Unpublished Ph.D., University of Aberdeen.

Stone, R.D. and Gorman, M.L. (1985). Social organisation of the European mole (*Talpa europaea*) and the Pyrenean desman (*Galemys pyrenaicus*). *Mammal Rev.*, 15, 35–42.

# IV.8 Scent Marking and Territoriality in the Aardwolf

P.R.K. RICHARDSON

## Introduction

The aardwolf *Proteles cristatus* is a highly specialized carnivore, feeding almost exclusively on termites and primarily on a few species of the genus *Trinervitermes* (Kruuk and Sands 1972; Cooper and Skinner 1979). In the northern Cape of South Africa, *T. trinervoides* provides over 90% of the aardwolf's diet (Richardson 1987*a*). I have argued elsewhere (Richardson 1985) that these termites constitute a limited resource, that aardwolves protect by being territorial.

*Trinervitermes trinervoides* forage in dense concentrations on the soil surface while collecting dry grass, making them highly susceptible to predation by insectivorous predators (Richardson 1987*a*). Soldiers possess a highly effective chemical defence system, of squirting fine threads of noxious and distasteful terpenes at any predator that attacks them (Prestwich 1983). The aardwolf is possibly unique in its ability to tolerate these terpene secretions (Richardson 1987*b*) which persist in its faeces and retain their characteristic smell. Kruuk and Sands (1972) have suggested that aardwolves defaecate at middens in order to concentrate faeces by burying them at known sites, and this reduces the chances of aardwolves confusing faeces with foraging termites. Burying the faeces may also reduce their smell (Kruuk and Sands 1972; Nel and Bothma 1983).

For the major form of scent marking, the aardwolf squats over a grass stalk and by drawing it through the anal pouch a thin smear of secretion is deposited on the stalk (Kruuk and Sands 1972; Nel and Bothma 1983). Gorman and Mills (1984) have described this behaviour in the Hyaenidae as 'pasting'. Before the start of this study (Richardson 1985), the only evidence of territoriality in the aardwolf was from incidental observations on scent-marking behaviour in the Serengeti made by Kruuk and Sands (1972). They noticed that individuals pasted with increased frequency along the border of their range and, on one occasion, two individuals avoided each other along this border.

The primary aim of this paper is to describe the dispersion pattern of scent marks in five aardwolf territories and to examine them in the context of what one may expect from a territorial species which uses scent marks to advertise its presence (Gosling 1982; Erlinge *et al.* 1982; Gorman and Mills 1984). Secondly, the pattern of scent marking at, and the use of, middens, is considered in terms of the olfactory confusion hypothesis (Kruuk and Sands 1972).

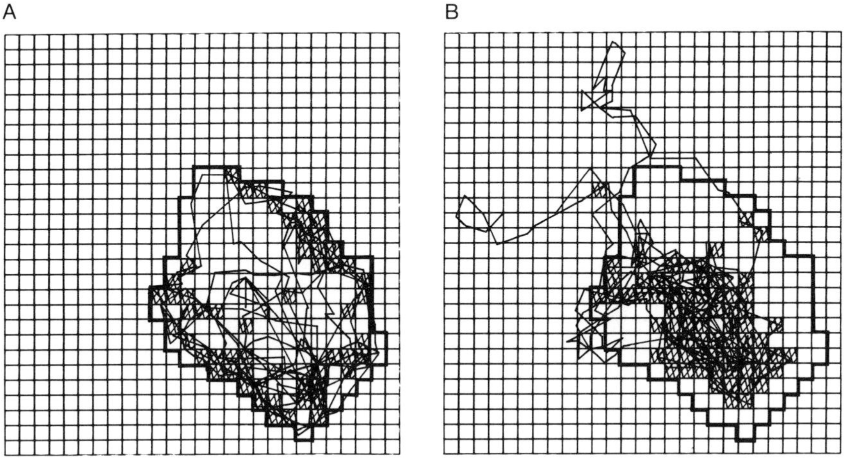


Fig. 1 Maps illustrating the paths followed and patterns of scent marking by the adult residents of Territory A during the non-mating season of 1982–1983. The territory boundary is shown in bold. In map A the hatched areas represent cells in which the male scent marked ⩾ 5 marks/ha. Map B illustrates the paths followed by the female and include several feeding excursions outside the territory; the hatched areas represent ⩾ 1 mark/ha.

## Materials and Methods

The study was carried out on Benfontein (28 50′S, 24 50′E), an 11.300 ha game farm near Kimberley, South Africa. The main study area, of 2015 ha, follows a gradual incline from a pan basin in the north-west to an area of Kalahari thornveld in the south-east. The slopes are dominated by various perennial grasses and short karoo bushes. *Trinervitermes* mounds are present throughout the study area, except in the pan and its surrounding basin.

The resident aardwolves occupying five territories were observed between July 1981 and September 1984. They were followed in a four-wheel-drive vehicle at a distance of 15–30 m. Since the aardwolves were undisturbed by our observations, at night visual contact was maintained by using the headlights of the vehicle and a spotlamp; all observations were recorded on a pocket dictaphone. The movements of aardwolves were plotted with reference to 420 colour-coded stakes placed throughout the territories and analysed in terms of distance travelled and number of scent marks per hectare. All scent marks were recorded on a counter and totalled for each observation. Observations were recorded whenever an aardwolf passed near a stake, changed direction, or showed any noticeable change of behaviour. Further details on the identification of individuals, nomination of focal animals, and the mapping of their movements are given in Richardson (1987*a*).

The data presented in this paper are derived largely from observations of aardwolves pasting on grass stalks and their use of dens and middens. Middens are defined as sites where there were at least two separate defaecations and which

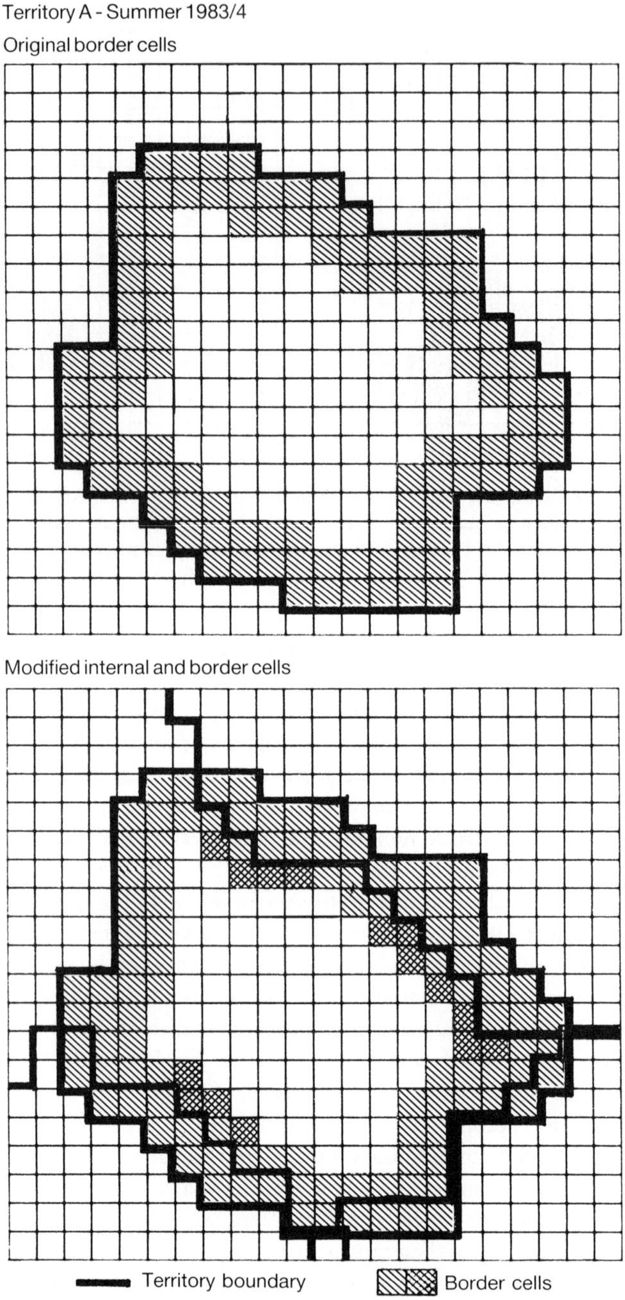


Fig. 2 Maps illustrating how the internal and border cells of a territory are defined.

showed signs of sand having been scratched over the faeces. When an aardwolf visited a den or midden and pasted at it, pastings constituted a separate event from the last previously recorded observation.

Territory boundaries were defined by analysis of the data for the distribution and density of scent marks by each adult. For each day a map was drawn of the path followed by the focal animal and the number of scent marks per hectare. The territories of the male and female of each pair overlapped entirely, so these data were combined when determining the territory boundaries (Fig. 1). For each pair, all the cells that had one or more pastings/cell were plotted on the same map. In most cases these maps clearly defined the territory of each pair; where the boundary line was not marked it was determined by joining up the cells that formed the shortest distance between the cells defined on either side. This was verified by referring to the actual paths followed by the aardwolves. Thus cells that had one scent mark but were only just entered were excluded from the territory. Territory boundaries were defined as being fixed for a year, and were redefined after the end of each mating season. Because of the changes that occurred during the mating season (Richardson 1987*c*), the data presented here are taken only from the non-mating (summer) seasons (August–May).

Mills *et al.* (1980) placed a matrix of 2.5 × 2.5 km cells over the territories of the brown hyaena, *Hyaena brunnea*, and designated those cells on the periphery of the territory as being border cells and the others as internal cells. The same method has been used here but using 1 ha cells, which were considered to be the smallest sized cell to still give a high degree of accuracy.

Because adjacent territories often had a number of cells common to both, and to allow for slight changes in boundaries, the border zone was defined as being 200 m wide. This was possible when the border ran north or east, but where it ran diagonally across a cell, the border zone was defined as being at least the diagonal width of a cell (141 m). Internal cells were defined as never being in contact with the border cells of an adjacent territory. Where this occurred because of overlapping territories, the internal cell concerned was reclassified as a border cell (Fig. 2).

## Results

### SCENT MARKS

Twelve adult aardwolves were followed for 375 km, and a total of over 5700 scent marks were recorded during the summer seasons of 1982–3 and 1983–4 (see Fig. 3). Adults scent marked very little outside their territories—on average by less than 5%. When one takes into account the fact that a territory cannot be perfectly defined, the results suggest that the incidence of aardwolves scent marking outside their established territories may be much less than 5%. Some aardwolves did make specific excursions outside their territories, but when they did so, they stopped

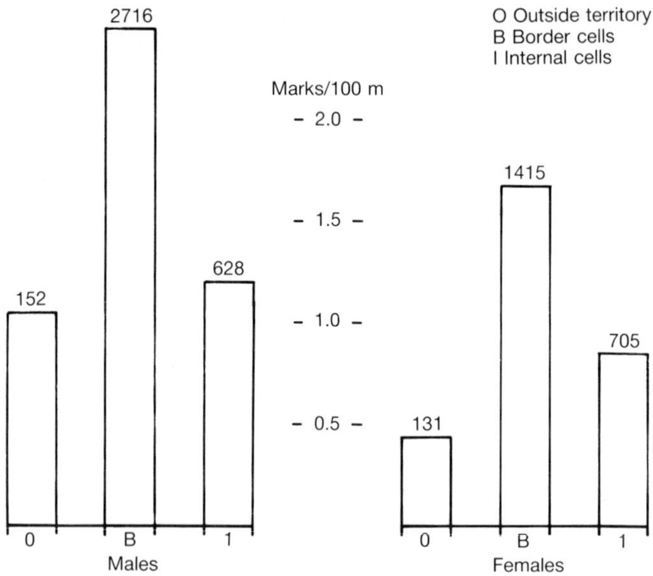


Fig. 3 Rates of scent marking (scent marks/100 m) by male and female aardwolves in different zones of their territories during the two summer seasons of 1982–1984. The total numbers of marks are indicated for each zone.

scent marking. This was a particular feature of the female in territory A, who spent 22% of her time mostly feeding outside the territory (Fig. 1).

A second feature manifest from Fig. 3 is that the aardwolves scent marked at a significantly higher rate in the border cells than in the internal cells of their territories. Males visited the border cells significantly more frequently ($\chi^2 = 43.93$, $p < 0.001$) and scent marked there at a significantly higher rate (Kruskal-Wallis H test, chi-squared approximation, [KWH test], $\chi^2 = 133.2$, $p < 0.001$) than in the internal cells of their territories. The result of this pattern was that they deposited a significantly higher density of scent marks in the border zone ($\chi^2 = 345$, $p < 0.001$).

By contrast, females visited the internal cells of their territories at a significantly higher frequency than the border cells ($\chi^2 = 67.5$, $p < 0.001$). Nevertheless, because females still scent marked at a significantly higher rate in the border cells (KWH test, $\chi^2 = 54.3$, $p < 0.001$) the net result was that they deposited a significantly higher density of scent marks in the border zones of their territories ($\chi^2 = 17.49$, $p < 0.001$) and complemented the male pattern of scent marking.

### DENS AND MIDDENS

During the study period, 143 dens and 118 active middens were located. The positions of 188 aardwolf defaecations were recorded during the two summer seasons; of these, 140 were deposited at middens and 110 were buried with sand. All large defaecations (the first of the evening) were deposited and buried at middens. All large defaecations were also urinated over, although aardwolves also frequently urinated by simply squatting while foraging. Urination was never immediately preceded or followed by scent marking and appeared to occur at totally random positions in the territory, thus having no role in scent marking during the non-mating season. Urination did, however, appear to play some role during the mating season when males occasionally left small deposits of urine at well-used dens (Richardson 1985).

The positioning (internal vs. border) of all the dens and middens in the most intensively studied territory is illustrated in Table 1. Although there tended to be more dens and middens in the border cells, none of the differences was significant ($p > 0.1$).

**Table 1.** Distribution of dens and middens and the relative areas of internal (I) and border (B) zones of territory A

| | Year | Number | | Zone area (ha) | | Observed vs. expected | |
|---|---|---|---|---|---|---|---|
| | | I | B | I | B | $\chi^2$ | $p$ |
| Middens | 1982 | 9 | 22 | 70 | 108 | 2.6 | 0.11 |
| | 1983 | 13 | 15 | 78 | 119 | 0.95 | 0.33 |
| | Total | 22 | 37 | 148 | 227 | 0.06 | 0.81 |
| Dens | 1982 | 13 | 13 | as above | | 0.61 | 0.44 |
| | 1983 | 13 | 15 | | | 0.29 | 0.59 |
| | Total | 26 | 28 | | | 0.83 | 0.36 |

The data for the frequency of scent marking at dens and middens (and the use of middens) are presented in Table 2. Aardwolves scent marked at dens and middens in the border cells significantly more ($p < 0.001$) than would be expected if they were scent marking at a frequency in proportion to the areas of the two zones (Table 2). This pattern was also significant ($p < 0.05$) for the frequency at which different middens were used (defaecating and urinating). Furthermore, the relative frequency (border vs. internal) at which the border middens were scent marked was significantly greater ($\chi^2 = 22.32, p < 0.001$) than the frequency at which they were used for sanitation purposes. This suggests that not only did aardwolves use border middens in preference to internal middens, but they also visited them frequently simply to scent mark.

The scent marking activities of both males and females tend to be concentrated

**Table 2.** Frequency of scent marking (pasting) and use of dens and middens in the internal (I) and border (B) zones of all five territories during the non-mating seasons of 1982 to 1984. (The total areas of the two zones are: internal cells = 753, border cells = 1158)

| | Middens | | | | Dens | |
|---|---|---|---|---|---|---|
| | Pasting | | Usage | | Pasting | |
| | I | B | I | B | I | B |
| Males | 17 | 169 | 12 | 42 | 36 | 171 |
| Females | 11 | 79 | 24 | 47 | 23 | 78 |
| All Adults | 28 | 248 | 36 | 89 | 59 | 249 |
| | Chi-squared test | | | | | |
| | $\chi^2$ | p | $\chi^2$ | p | $\chi^2$ | p |
| Males vs. females | 0.63 | 0.43 | 2.01 | 0.16 | 1.27 | 0.26 |
| Adults: observed vs. expected | 89.9 | 0.001 | 5.56 | 0.018 | 46.9 | 0.001 |

in the border cells. The results for their activities at dens and middens confirm this pattern, suggesting that these sites may also serve some territorial function. As aardwolves defaecate on average 3–6 times on a summer night, middens act as focal points for aardwolf activity, and therefore could easily assume territorial importance. The same argument may be applied to dens. An example of this is that on the two summer nights when the male of territory A was followed for the whole night, he visited 5 and 4 dens, and 7 and 6 middens, respectively, although he defaecated only 3 times per night.

Where middens were located on the territory boundary, aardwolves from both territories would use them and paste over each others' scent marks. This resulted in grass stalks around these middens being heavily covered with scent marks (Fig. 4).

## Discussion

### SCENT MARKS

The chemical properties (Apps *et al.* 1989) and distribution of aardwolf scent marks are compatible with the theory that scent marks are used as an unambiguous cue advertising territory ownership, thereby establishing an asymmetry of contest between residents and intruders (Gosling 1982). The behaviour of intruders confirmed this asymmetry of contest as they refrained from marking outside their territories and usually retreated immediately after the resident was seen, without showing any inclination to fight (Richardson 1985, 1987*c*). If residents saw

Fig. 4 Aardwolf midden with a number of defaecations in the background, and a heavily marked grass stalk in the foreground.

intruders in their territory they invariably chased the intruder, usually right to the territory border (Richardson 1985).

In terms of chemical properties, marked differences were found both in the occurrence and relative areas of the chromatographic peaks of different individuals' scent marks (Apps *et al.* 1989), suggesting that they are most probably identifiable to their producer. The odour of scent marks may also persist for up to six months (Apps *et al.* 1989). These are desirable properties of a signal to be used as an unambiguous cue of territory ownership.

The pattern of scent marking by pasting at a higher frequency and density along the border zone has been previously described by Gorman and Mills (1984). It is the most effective strategy to be employed by an animal living in a relatively small territory where border cells can be regularly visited.

## DENS AND MIDDENS

For a number of carnivore species, defaecation sites (middens or latrines) serve an important territorial function as scent-marking sites (Kruuk 1972; Peters and Mech 1975; Macdonald 1980, Gorman and Mills 1984). The function of middens is not clear in striped hyaenas *Hyaena hyaena* (Kruuk 1976; Macdonald 1978). However, in the spotted *Crocuta crocuta* and brown hyaenas, faeces are generally white and very conspicuous, and together with pastings appear to have a territorial signalling function (Kruuk 1972; Gorman and Mills 1984; Mills and Gorman 1987). The preference of aardwolves for defaecating and scent marking at dens and middens in the border cells of their territories also suggests a territorial function. That these sites were located throughout the territory may be explained by the fact that they are reused over many years (Richardson 1985). As territorial boundaries shift around, so their positions relative to the boundaries will also move.

If one accepts that middens do have some territorial function, the question arises as to why aardwolves bury their faeces—unlike the hyaenas (Kruuk 1972; Gorman and Mills 1984)—since this presumably reduces their signal value. The most conspicuous feature of an aardwolf's faeces is its smell of terpenes. The use of smell in detecting terpene secretions produced at low concentrations by foraging termites is possibly the aardwolf's primary method of locating its food (Richardson 1985, 1987*a*). Therefore, if faeces were distributed randomly throughout the territory the aardwolf may well confuse them with foraging termites and waste time and energy investigating them. Kruuk and Sands' (1972) suggestion that aardwolves concentrate their faeces at middens in order to reduce olfactory confusion may therefore be valid. These authors have also suggested that middens may actually facilitate the decomposition of aardwolf faeces. They found that the termite *Macrotermes vitrialatus* was very common at middens in the Serengeti and presumed it was feeding on vegetable and insect residues in the faeces. No such termites were present at middens in the study area, but fly larvae were quite frequently found in the faeces.

By concentrating faeces at a few known localities (middens), such sites automatically become focal points of aardwolf activity and could therefore assume territorial importance, which need not necessarily contradict the olfactory confusion hypothesis. Alternatively, middens may originally (as they do now in the hyaenas) have served mainly a territorial function, and the burying of faeces may have evolved with the aardwolf's increasing preference for *Trinervitermes*.

## Acknowledgements

I would like to thank De Beer's Consolidated Mines for access to their game farm Benfontein, the use of computer facilities, and technical help from various members of staff. This research was conducted while working as an employee and latterly Research Associate of the McGregor Museum, Kimberley. The project received financial support from the following organizations or funds: British Council; Cape Department of Nature and Environmental Conservation; Endangered Wildlife Trust; Johannesburg Consolidated Investment Company; Kango Wolf Power Tools; Wolf Garden Machinery Manufacturers; Oliver Ashe Overseas Scholarship Fund; and Overseas Research Student Awards.

## References

Apps, P., Viljoen, H.W., Richardson, P.R.K. and Pretorius, V. (1989). Volatile components of the anal gland secretion of the aardwolf (*Proteles cristatus*). *J. Chem. Ecol.*, 15, 1681–1688.

Cooper, R.L. and Skinner, J.D. (1979). Importance of termites in the diet of the aardwolf Proteles cristatus *in South Africa. S. Afr. J. Zool.*, 14, 5–8.

Erlinge, S., Sandell, M. and Brinck, C. (1982). Scent-marking and its territorial significance in stoats, *Mustela erminea. Anim. Behav.*, 30, 811–8.

Gorman, M.L. and Mills, M.G.L. (1984). Scent marking strategies in hyaenas (Mammalia). *J. Zool. Lond.*, 202, 535–47.

Gosling L.M. (1982). A reassessment of the function of scent marking territories. *Z. Tierpsychol.*, 60, 89–118.

Kruuk, H. (1972). *The spotted hyena. A study of predation and social behavior.* University of Chicago Press, Chicago.

Kruuk, H. (1976). Feeding and social behaviour of the striped hyaena (*Hyaena vulagris* Desmarest). *East Afr. Wildl. J.*, 14, 91–111.

Kruuk, H. and Sands, W.A. (1972). The aardwolf (*Proteles cristatus* Sparrman) 1783 as a predator termites. *E. Afr. Wildl. J.*, 10, 211–27.

Macdonald, D.W. (1978). Observations on the behaviour and ecology of the striped hyaena, *Hyaena hyaena*, in Israel. *Israel J. Zool.*, 27, 189–98.

Macdonald, D.W.(1980). Patterns of scent marking with urine and faeces amongst carnivore communities. *Symp. zool. Soc. Lond.*, 45, 106–21.

Mills, M.G.L. and Gorman, M.L. (1987). The scent-marking behaviour of the spotted hyaena *Crocuta crocuta* in the southern Kalahari. *J. Zool. Lond.*, 212, 483–97.

Mills, M.G.L., Gorman, M.L. and Mills, M.E.J. (1980). The scent marking behaviour of the brown hyaena *Hyaena brunnea. S. Afr. J. Zool.*, 15, 240–8.

Nel, J.A.J. and Bothma, J. du P. (1983). Scent marking and midden use by aardwolves (*Proteles cristatus*) in the Namib Desert. *Afr. J. Ecol.*, 21, 25–39.

Peters, R.P. and Mech, L.D. (1975). Scent-marking in wolves. *Amer. Sci.*, 63, 628–37.

Prestwich, G.D. (1983). The chemical defenses of termites. *Sci. Amer.*, 249(2), 68–75.

Richardson, P.R.K. (1985). The social behaviour and ecology of the aardwolf, *Proteles cristatus* (Sparrman, 1783) in relation to its food resources. D.Phil. thesis, University of Oxford.

Richardson, P.R.K. (1987*a*). Food consumption and seasonal variation in the diet of the aardwolf *Proteles cristatus* in southern Africa. *Z. Sägetierk.*, 52, 307–25.

Richardson, P.R.K. (1987*b*). Aardwolf: The most specialized myrmecophagous mammal? *S. Afr. J. Sci.*, 83, 643–6.

Richardson, P.R.K. (1987*c*). Aardwolf mating system: Overt cuckoldry in an apparently monogamous mammal. *S. Afr. J. Sci.*, 83, 405–10.

# IV.9 Spatial Utilization and Scent Mark Location in a Territorial Lagomorph

CARRON A. MEANEY

## Introduction

North American pikas (*Ochotona princeps*) are small lagomorphs that inhabit montane regions in western North America. They are diurnal and habitat-specific to talus slopes where both males and females maintain individual territories. Pikas feed on vegetation adjacent to their territories on the talus. The breeding season is April, May, and June. In mid-July pikas start to collect vegetation which they bring back to their territories and store in hay-piles (Kawamichi 1976). These hay-piles serve as a potential winter food source (Conner 1983).

Pikas have a scent gland located at the angle of the jaw, which they rub against rocks in their environment. This behaviour is termed cheek marking. Both males and females cheek mark, although males generally at a higher rate and particularly during the breeding season (Meaney 1983).

## Methods

A 5 × 5 m grid of pin flags was erected on the 1 ha study site, located on Niwot Ridge in Colorado at 3600 m (11,800 ft). All pikas were captured and ear-tagged with coloured plastic discs. Marked pikas were followed on foot and focal animal sampling was used (Altmann 1974). An animal's location was noted every 5 mins during 15-min. observation periods. Cheek mark locations were noted whenever cheek marking behaviour occurred. Observations were made from 7.00 to 11.00 a.m. and from 3.30 to 8.30 p.m., these being times of peak animal activity. The data were collected during June, July, August, and September over a three-year period. In all, 643 15-min. observation periods were completed during 605 hours at the site on 97 days.

A computer program was used to generate convex polygons that encompassed 50%, 80%, and 95% of the location points. These three polygons were generated for each animal in each month. The 50% polygon rejected the 50% location points most distant from the centre and encompassed the 50% most centrally located points. The same technique was used to generate the 80% and 95% polygons.

Each polygon described the limits within which the animal was seen for the prescribed percentage of observations. In certain analyses only the 80% polygons were used; this eliminated the extensive overlap seen in the 95% polygons but revealed patterns of overlap not seen in the 50% polygons. The 80% polygons are assumed to be comparable in size to pika territories. All analyses involving mark densities used transformed data ($\log^{10}$ of marks per square meter × 100) to meet assumptions of normality (Sokal and Rohlf 1981).

## Results

### SEASONAL CHANGES IN THE 80% POLYGON

The size of the 80% polygon decreased significantly from 681 $m^2$ in June to 308 $m^2$ in September (Table 1; Kruskal-Wallis test $H = 7.875$, $df = 3$, $p < 0.05$). The area of overlap and the per cent overlap also decreased significantly in September compared to June ($H = 29.164$, $df = 3$, $p < 0.0001$; and $H = 27.861$, $df = 3$, $p < 0.0001$ for area of overlap and per cent overlap, respectively). In June, the breeding season is underway, and large overlap areas facilitate contact between breeding pairs. During August, territories contract as pikas gather their hay-piles and defend them.

**Table 1.** Size of 80% polygon ($m^2$), total overlap area ($m^2$), and per cent overlap for June through September (±SE) for all pikas

| | June | July | August | September |
|---|---|---|---|---|
| Number of pikas | 10 | 18 | 18 | 12 |
| Mean area of 80% polygons | 681 ± 106.9 | 421 ± 63.2 | 347 ± 39.4 | 308 ± 32.7* |
| Mean area of overlap | 437 ± 89.6 | 219 ± 42.5 | 100 ± 18.5 | 20 ± 6.5† |
| Mean per cent overlap | 68% ± 8 | 52% ± 8 | 27% ± 4 | 6% ± 2† |

* $p < 0.05$
† $p < 0.0001$

### CHEEK-MARK LOCATIONS

*From the resident's point of view*

This approach evaluated the frequencies of cheek marks inside the 50% polygon, between the 50% and 80% polygons, between the 80% and 95% polygons, and outside the 95% polygon. This distribution of marks was compared with an expected distribution generated from location points for each region. The 50% polygon contained, by definition, 50% of the location points; the area between the 50% and 80% polygons contained 30% of the location points; and the area between the 80% and 95% polygons contained 15% of the location points. Table 2 shows the frequencies of cheek mark points in the various polygon regions compared to an expected frequency based on the proportion of location points. A Chi-squared test for goodness-of-fit was used to compare the observed and expected distributions. There were significantly fewer cheek mark points inside the 50% polygon, about the same between the 50% and 80% polygons, and significantly more between the 80% and 95% polygons and outside the 95% polygon than expected ($\chi = 198.21$, df$=3$, $p < 0.001$). This suggests that, as a group, pikas are concentrating their marks in the perimeter regions of their territories.

**Table 2.** Frequencies of cheek mark points in various regions of the polygons compared to an expected frequency based on the proportion of location points. Data were pooled over 3 years. $\chi^2 = 198.21$, df$=3$, $p < 0.001$

| | Expected proportion | Observed frequency | Expected frequency |
|---|---|---|---|
| Within 50% polygon | 0.50 | 454 | 645 |
| Between 50–80% | 0.30 | 423 | 387 |
| Between 80–95% | 0.15 | 263 | 193.5 |
| Outside 95% polygon | 0.05 | 150 | 64.5 |
| **Total** | | **1290** | |

*From the intruder's point of view*

An intruder probably assesses absolute mark density, rather than perceiving proportions. Table 3 presents the mean number of cheek marks per square meter in the different polygon regions for June through September. Within each month there is a consistent decrease in density from the 50% polygon out towards the exterior perimeter with an increase again in the overlap region. Within each region, there is a slight increase in density from June to September. A two-way analysis of variance, which did not include the overlap region, revealed a significant effect due to region ($F = 7.129$, df$=2$, $p < 0.05$) but not for month ($F = 1.605$, df$=3$, $p > 0.05$).

A synthesis of the information on mark location from the points of view of resident and intruder suggests that pikas mark at a slightly greater rate in the perimeter regions of their polygons. The animals often went on 'marking excursions' (Gosling 1982) during which the rate of marking was notably high. While more marks were deposited at the periphery than was predicted by the

**Table 3.** Mean number of cheek marks per square metre in different polygon regions for all pikas from June through September (±SE)

| Month | Region: Within 50% polygon | Between 50–80% polygon | Between 80–95% polygon | In overlap region |
|---|---|---|---|---|
| June ($n=10$) | $0.037 \pm 0.01$ | $0.018 \pm 0.005$ | $0.014 \pm 0.01$ | $0.024 \pm 0.005$ ($n=10$) |
| July ($n=23$) | $0.034 \pm 0.01$ | $0.034 \pm 0.01$ | $0.017 \pm 0.005$ | $0.033 \pm 0.01$ ($n=22$) |
| August ($n=23$) | $0.043 \pm 0.01$ | $0.027 \pm 0.005$ | $0.015 \pm 0.003$ | $0.028 \pm 0.01$ ($n=19$) |
| September ($n=12$) | $0.061 \pm 0.03$ | $0.031 \pm 0.01$ | $0.017 \pm 0.01$ | $0.010 \pm 0.01$ ($n=7$) |

proportion of location points, the rate of marking was not elevated sufficiently to affect absolute mark density, which is a very robust measure. Thus actual mark density decreases with increasing distance from the core (50% polygon) area. The pattern of mark location is to a large degree a consequence of where the animals spend their time.

*Marks in the overlap region*

The overlap region is defined as the region of overlap of the 80% polygon with other individuals' 80% polygons. The notable high density of cheek marks in the overlap region of the 80% polygon led to a comparison of mark density in overlap versus non-overlap regions. Almost all 80% polygons had some amount of overlap. Non-overlap is that portion of the 80% polygon that does not overlap with any other 80% polygons. Although not statistically significant (two-way ANOVA, $F=0.927$, $df=4$, $p>0.05$), the mark density in the region of overlap decreases through the season while the mark density in the region of non-overlap increases consistently through the season (Table 4).

## Discussion

The interrelationship of spatial utilization and scent mark location reveals a pattern of scent marking that shifts in focus through the summer season. During the breeding season in June, territories are large and individual pikas roam extensively seeking mates. They tend to situate their territories such that individuals are surrounded by members of the opposite sex (Smith and Ivins 1983). A large amount of territorial overlap is tolerated and utilized for scent marking purposes which acts to facilitate breeding. Previous work has shown that

**Table 4.** Mean number of cheek marks per square metre in the overlap and non-overlap regions of the 80% polygons for all pikas from June through September (±SE)

| Month | Region | |
|---|---|---|
| | Overlap | Non-overlap |
| June ($n=10$) | 0.024±0.005 | 0.013±0.01 |
| July ($n=22$) | 0.033±0.01 | 0.032±0.02 |
| August ($n=19$) | 0.028±0.01 | 0.034±0.01 |
| September ($n=7$) | 0.010±0.01 | 0.044±0.02 |

females are more likely to mate with males whose cheek gland odours are familiar (Meaney 1986). As the season progresses, territories become smaller, pikas start hay-piling in late July, and areas of overlap decrease.

Temporal and energy resources for scent marking and defending a large territory are limited (Gosling 1982, 1986). The end of the summer presents three new demands, or factors, relating to pika resources. The first is haying behaviour, which takes up 55% of their time (Conner 1983). The second is the occurrence of the young of the year who are setting up their own territories, often between existing territories. The third factor is the increased resource value of a territory with a hay-pile on it. These factors combine to reduce the area of pika territories to a size that is realistically defendable. At this time there is a decreased focus on marking in the areas of overlap, with greater concentration of activities and scent marking in the core area and areas of non-overlap, where hay-piles are located.

The pattern of concentric circles of decreasing mark density is similar to that of territory owners exhibiting aggressive or dominant behaviour in concentric circles of decreasing aggression outwards from the centre of their territories (Desjardins *et al.* 1973). Female Richardson's ground squirrels (*Spermophilus richardsonii*) show an increase in the ratio of aggressive to submissive interactions as they get closer to their core area, or 50% polygon (Michener 1979). These ground squirrels are most likely to be dominant to conspecifics within the core area, where pikas concentrate their scent marks. A correlation of these two patterns within one species would be of considerable interest. Testosterone levels are known to play an important mechanistic role in both aggressive and scent marking behaviours (Thiessen *et al.* 1971; Yahr 1981).

In conclusion, much can be learned by a detailed analysis of scent-mark locations in relation to convex polygons describing spatial utilization. Resident pikas spend more of their time marking in the outer perimeters of their territories, and intruders encounter marks at an increasing rate towards the centre of a territory. The mark locations thus play an important role in understanding the functions of scent marking.

## References

Altmann, J. (1974). Observational study of behavior: sampling methods. *Behaviour*, 49 (3,4), 227–65.

Conner, D.A. (1983). Seasonal changes in activity patterns and the adaptive value of haying in pikas (*Ochotona princeps*). *Can. J. Zool.*, 61, 411–16.

Desjardins, D., Maruniak, J.A. and Bronson, F.H. (1973). Social rank in house mice: differentiation by ultraviolet visualization of urinary marking patterns. *Science*, 182, 939–41.

Gosling, L.M. (1982). A reassessment of the function of scent marking in territories. *Z. Tierpsychol.*, 60, 89–118.

Gosling, L.M. (1986). Economic consequences of scent marking in mammalian territoriality. In *Chemical signals in vertebrates*, 4 (ed. D. Duvall, D. Müller-Schwarze and R.M. Silverstein), pp. 385–95. Plenum, New York.

Kawamichi, T. (1976). Hay territory and dominance rank of pikas. *J. Mamm.*, 57, 133–48.

Meaney, C.A. (1983). Rate and location of scent marking by pikas during the breeding season. In *Chemical signals in vertebrates*, 3 (ed. D. Müller-Schwarze and R.M. Silverstein), pp. 329–32. Plenum, New York.

Meaney, C.A. (1986). Scent-marking in pikas (*Ochotona princeps*): Test of a breeding-facilitation hypothesis. In *Chemical signals in vertebrates*, 4 (ed. D. Duvall, D. Müller-Schwarze and R.M. Silverstein), pp. 571–7, Plenum,, New York.

Michener, G.R. (1979). Spatial relationships and social organization of adult Richardson's ground squirrels. *Can. J. Zool.*, 57, 125–39.

Smith, A.T. and Ivins, B.L. (1983). Colonization in a pika population: dispersal vs. philopatry. *Behav. Ecol. Sociobiol.*, 13, 37-47.

Sokal, R.R. and Rohlf, F.J. (1981). *Biometry: the principles and practice of statistics in biological research*. Freeman, San Francisco.

Thiessen, D.D., Owen, K. and Lindzey, G. (1971). Mechanisms of territorial marking in the male and female Mongolian gerbil (*Meriones unguiculatus*). *J. Comp. Physiol. Psych.*, 77, 38–47.

Yahr, P. (1981). Scent marking, sexual behaviour and aggression in male gerbils: comparative analysis of endocrine control. *Amer. Zool.*, 21, 143–51.

# IV.10 Scent Marking and Territoriality in Wolves of Riding Mountain National Park

PAUL C. PAQUET AND WILLIAM A. FULLER

## Introduction

Scent marking as a means of defending an area should be most frequent at the periphery of a territory (Johnson 1973). Intruders could then recognize the edge of a territory by the rate at which they encounter scent marks or by a changing odour gradient (Bowen and McTaggart Cowan 1980). Peters and Mech (1975) reported an increased rate of marking by wolves (*Canis lupus*) near territory boundaries, and concluded that the distribution of scent marks was related to territory advertisement and enforcement. Conversely, Carbyn (1980) found no difference between peripheral and interior marking rates for a wolf pack in Riding Mountain National Park, Manitoba, Canada.

Bowen and McTaggart Cowan (1980) found the density of coyote (*C. latrans*) scent marks to be highest at the edge of the territory and lowest at the centre. Neighbouring conspecifics, however, failed to respect territorial boundaries. Camenzind (1978) recorded 68% of all scent marks on territorial boundaries.

No assessment of the spatial distribution of scent marks within individual wolf territories has been carried out. In this paper, we tested the hypothesis that the frequency of scent marking by wolves increases from the centre to the periphery of the territory, with the expectation that change in deposition frequency is related to boundary demarcation.

## Study area

The study was conducted in Riding Mountain National Park (RMNP) between 1983 and 1986. The park is located 177 km north of the international boundary between Manitoba, Canada and North Dakota, USA. Dominant vegetation is aspen (*Populus tremuloides*) and white spruce (*Picea glauca*), interspersed with black spruce (*Picea mariana*) bogs, and jack pine (*Pinus banksiana*) forest. Other large predators in the area include coyotes, lynx (*Lynx lynx*) and black bears (*Ursus americanus*). Elk (*Cervus elaphus*), deer (*Odocoileus virginianus*), moose (*Alces alces*), and beaver (*Castor canadensis*) are the primary prey species for wolves (Carbyn 1980; Meleshko 1986; Paquet 1988).

## Methods

Urine and scratch marks were recorded in four wolf territories from December through March. Tracking was done by snowmobile, cross-country skis, snowshoes, or on foot. Distances were calculated by pedometer when travelling on foot and by odometer when travelling by snow machine. All distances were rounded to the nearest kilometre. Trails were selected haphazardly (Lehner 1979) within designated territories and followed for a minimum distance of 5 km. In 1983–4, data were collected from trails used exclusively by wolves to avoid confusion with coyote urine deposits and scratches. During the winter 1984–5 marking information was collected from trails used by wolves only, and by both wolves and coyotes.

As there is disagreement as to what constitutes a scent mark, we did not attempt to discriminate between the behavioural and physiological rates of urination and defaecation (Kleiman 1966; Peters and Mech 1975; Barrette and Messier 1980; Bowen and McTaggart Cowan 1980; Bekoff and Wells 1986). Urine deposits were classified as Directed (DU) and Non-Directed (NDU). DUs were defined as marks produced by urine that had been directed toward a target (Kleiman 1966), either a physical object or previous urine deposit. Deposits without obvious orientation were classified as NDUs. Scratches (SCR) were defined as extended parallel tracks produced by the claws of wolves (Kleiman 1966). Data collected within 25 m of an ungulate killed by a wolf or coyote were discarded.

Territories were delineated by extensive and repeated ground tracking of selected wolf packs (Paquet 1988). Monitoring began when a permanent winter snow cover was established. In all cases consistent track patterns were evident by mid-December, allowing for easy recognition of individual packs. Adjacent packs were also monitored, which helped identify questionable tracks, for example, when a pack temporarily divided into subunits. Visual observations of wolves from the air and on the ground provided additional information as to location of packs and their range of movement.

Maps of territories were subdivided into 1 km wide concentric zones (shells) for analysis. Zone 1 was designated the outermost shell. The contour of each zone followed the outline of the territorial boundary. Hence, the outside edge of each shell was a constant distant from the boundary.

The frequency distribution of urine and scratches within each territory was evaluated using the G-test for goodness of fit (Sokal and Rohlf 1981, p. 692). Frequency values were adjusted by means of Williams' correction for continuity. We assumed that wolves mark at a constant rate, regardless of physiography or speed of travel. Multiple comparisons of the same data were carried out on: (1) deposition frequencies among individual shells; and (2) pooled data from the outer two shells (edge) and pooled data from the remaining inner shells (centre) (Peters and Mech 1975). Critical $p$-values were adjusted by Sidak's multiplicative inequality (Sokal and Rohlf 1981, p. 728). The null hypotheses stated that the frequency of urine and scratch marks is constant regardless of the location within a territory. Expected frequencies were calculated by dividing the distance travelled within each shell by the total distance travelled within the territory: the product of the resulting quotient and the total number of marks recorded in the territory provided the expected number of marks for each shell.

## Results

This analysis is based on 922 urine and scratch marks recorded in 492 km of tracking (Table 1). The Audy Lake pack (1984–5) had the highest deposition rate (2.4 marks/km) and the Clear Lake pack (1983–4) the lowest (1.2 marks/km). Comparison of the overall marking frequencies between packs showed significant differences ($G=63.968$, $p<0.001$, d.f. $=3$). Pairwise unplanned tests (Sokal and Rohlf 1981, p. 728) indicated that most of the variability was attributable to differences in deposition frequencies between Clear Lake (1984–1985) and the other 3 packs. The remaining subsets were homogeneous.

There was no significant difference in the spatial distribution of marks recorded along wolf/coyote trails and wolf only trails for either the Audy L. ($G=2.153, p>0.500$, d.f. $=4$) or Clear L. (1984–1985) ($G=1.738, p>0.500$, d.f. $=4$) territories. Therefore, the data were pooled for further analyses.

### ANALYSIS OF CONCENTRIC 1 KM SHELLS

In three of the four territories, the highest frequency of marking (all marks combined) occurred in intermediate shells, i.e. shells that were neither on the periphery nor at the centre ($p<0.001$). Note that in 1984–5 the Clear Lake pack did not use the central zone, although they marked with highest frequency in the innermost zone that they used (Table 1). The null hypothesis was rejected for all packs for DUs ($p<0.05$) and NDUs ($p<0.001$). The distribution of SCRs was random in the Audy L. territory ($p<0.05$) and differential in the two Clear L. territories (1983–4, $p<0.001$; 1984–1985, $p<0.05$). Insufficient data precluded analysis for the Baldy L. territory.

### ANALYSIS OF 2 KM EDGE AND CENTRE

For three of the four territories there was a highly significant difference in overall marking frequency between peripheral and central areas. Clear L. (1983–4) and Audy L. wolves marked most often near territorial boundaries ($p<0.001$ and $p<0.005$ respectively); the Baldy L. pack left more marks in the central part of their territory ($p<0.001$); the Clear L. (1984–5) pack marked trails throughout the two designated areas of the territory at the same frequency, regardless of location ($p>0.05$). In all cases, differences were attributable to uneven distribution of DUs ($p<0.001$) and NDUs ($p<0.001$). The frequency of SCRs never departed significantly

**Table 1.** Summary of the distribution of urine and scratch marks for five wolf packs in RMNP. Shells are numbered in ascending order from the outside to the inside, so that shell number 1 is outermost

| Pack (year) | Shell number | Number of scent marks | | | | Total Distance | Marks/ km |
|---|---|---|---|---|---|---|---|
| | | DU | NDU | SCR | Total | | |
| Clear Lake | 1 | 18 | 31 | 9 | 58 | 39.0 | 1.5 |
| 2–3 wolves | 2 | 51 | 16 | 9 | 76 | 31.0 | 2.5 |
| (1983–1984) | 3 | 19 | 17 | 5 | 41 | 56.0 | 0.7 |
| | 4 | 21 | 0 | 10 | 31 | 21.0 | 1.5 |
| | 5 | 0 | 0 | 0 | 3 | 27.0 | 0.1 |
| | | *111* | *65* | *33* | *209* | *174.0* | *1.2* |
| Clear Lake | 1 | 39 | 45 | 19 | 103 | 41.0 | 2.5 |
| 2–3 wolves | 2 | 42 | 12 | 7 | 61 | 43.0 | 1.4 |
| (1984–1985) | 3 | 12 | 7 | 2 | 21 | 27.0 | 0.8 |
| | 4 | 15 | 14 | 5 | 34 | 10.0 | 3.4 |
| | 5 | 0 | 0 | 0 | 0 | 0.0 | 0.0 |
| | | *108* | *78* | *33* | *219* | *121.0* | *1.8* |
| Audy Lake | 1 | 22 | 19 | 8 | 49 | 17.0 | 2.8 |
| 5 wolves | 2 | 51 | 23 | 9 | 83 | 23.0 | 3.7 |
| (1984– 1985) | 3 | 9 | 0 | 5 | 14 | 7.0 | 2.0 |
| | 4 | 29 | 13 | 10 | 52 | 29.0 | 1.8 |
| | 5 | 0 | 2 | 0 | 2 | 6.0 | 0.3 |
| | | *111* | *57* | *32* | *200* | *82.0* | *2.4* |
| Baldy Lake | 1 | 0 | 1 | 0 | 1 | 16.0 | 0.6 |
| 10 wolves | 2 | 6 | 5 | 2 | 13 | 8.0 | 1.6 |
| (1983– 1984) | 3 | 0 | 0 | 0 | 0 | 2.0 | 0.0 |
| | 4 | 40 | 2 | 7 | 49 | 13.0 | 3.8 |
| | 5 | 56 | 37 | 3 | 96 | 31.0 | 3.1 |
| | 6 | 34 | 12 | 2 | 48 | 24.0 | 2.0 |
| | 7 | 33 | 5 | 6 | 44 | 21.0 | 2.1 |
| | | *169* | *62* | *20* | *251* | *115.0* | *2.2* |

DU = directed urine deposits; NDU = non-directed urine deposits; SCR = scratch marks.

from expectation ($p < 0.05$). Portions of each wolf territory examined included sections of park boundary, along which there were no neighbouring packs.

It has been postulated that the accumulation of scent marks on the periphery of a territory is due to the combined contributions of neighbouring packs and an increased stimulus for marking (Peters and Mech 1975; Bowen and McTaggart Cowan 1980; Barrette and Messier 1980). A reduced marking rate would therefore be expected in sections of territory boundary without neighbouring packs. We accordingly re-evaluated the data for winter territories from 1983–4 and 1984–5. After eliminating trail sections adjacent to the park boundary, there was no change in the relative marking frequencies throughout the territories, analysing by either the shell ($G = 1.738, p > 0.500$, d.f. = 3) or centre/edge ($G = 0.788, p > 0.500$, d.f. = 3) method. In addition, there were no differences in marking frequencies along park boundaries and boundaries adjacent to neighbouring wolf packs ($G = 1.341$, $p > 0.500$, d.f. = 1).

## Discussion and conclusions

Peters and Mech (1975) concluded that wolf-marking sites are more numerous along territorial peripheries, and proposed as a model an 'olfactory bowl' of scent, in which the number of marks decreases from the edge to the middle of a territory. Bowen and McTaggart Cowan (1980) described a similar distribution of scent marks in coyote territories. Barrette and Messier (1980) questioned whether these models were appropriate, because both were based on marks/km rather than marks/km$^2$.

In this study, analysis of marking frequency in concentric 1 km shells gave no support to the olfactory bowl model. Comparisons of marking frequency between a 2 km-wide peripheral band and the remaining central portion of each territory gave conflicting results. For two of the packs, the frequency was greater at the periphery than in the centre; in another case, the reverse was true; and for the fourth pack, the observed frequency showed an equal distribution.

Peters and Mech (1975) pooled data from three wolf packs. When we pooled our data we found greater than expected numbers in band 3, fewer than expected in zones 2 and 4, and about the expected numbers in zones 1 and in the centre. Overall, the greatest frequency was in zone 4 ($G = 82.6, p < 0.001$, d.f. = 4). When we compared the periphery with the centre in our pooled data, we found no difference in the frequency of marks ($G = 4.87, p < 0.050$, d.f. = 1).

The high density of coyotes in RMNP might disrupt or alter the 'normal' marking behaviour of wolves. Although the data used in the analysis were carefully collected from trails with visible wolf tracks only, residual or undetected coyote scent may have been present beneath the snow. However, the distribution of marks on mixed-use trails was similar to wolf only trails, suggesting that coyotes do not affect the placement of marks within wolf territories. In addition, concurrent studies of wolf/coyote interactions showed little response by wolves to coyote marks (Paquet 1988).

Peters and Mech (1975) noted that physiography influences the frequency of scent marking. Marking studies concurrent with this study (Paquet 1988) confirm that observation. Established roads and trails exhibit the highest number of marks and frozen waterways the lowest. Therefore, territories encompassing large bodies of water and/or an unusual configuration of roads and trails would probably show an uneven distribution of scent marks.

Finally, we agree with Barrette and Messier (1980) that the relative amount of time spent in each part of a territory must influence marking frequency. Assuming that the frequency of marking is positively associated with the time spent in an area, the non-random distribution of marks observed might reflect foraging behaviour by wolves.

At Riding Mountain the predominant prey of wolves is elk (Carbyn 1980; Meleshko 1986, Paquet 1988). Winter distribution of elk is clumped and uneven (Rounds 1980, Meleshko 1986), and this is partially related to cover, food resources (Trottier *et al.*1983), and snow thickness (Paquet *et al.*, in preparation). Consequently, marking frequency could be elevated in areas of high elk density simply as a function of the time wolves spend hunting there. The shell method of analysis used in this study would not detect these areas of concentration because several concentric zones could overlap portions of the same elk range.

However, we reject the suggestion of Barrette and Messier (1980) that marking data should be expressed as marks/unit area rather than marks/unit length. Travel routes of wolves are predictable because wolves consistently use the same trails. If we assume that frequent marks (or the resulting high olfactory stimulus) are sufficient to discourage intruders, then wolves ought to increase the marking rate on existing trails. To increase the overall density of marks would require a much more extensive network of trails, which would be costly in both time and energy. No such system of trails exists in RMNP, nor has one been reported in the literature.

Our results show clearly that wolves mark parts of their territory differentially. When data for each pack are analysed separately, we conclude that the olfactory bowl model does not describe the marking behaviour of wolves in RMNP. Considering the influence of variable topography and the hypothesized importance of patterns of prey distribution, we think it unlikely that the model is universally applicable.

## References

Barrette, C. and Messier, F. (1980). Scent-marking in free-ranging coyotes, *Canis latrans*. *Anim. Behav.*, 28, 814–9.

Bekoff, M. and Wells, M.C. (1986). Social ecology and behavior of coyotes. In *Advances in the study of behavior*, Vol. 16.

Bowen, W.D. and McTaggart Cowan, I. (1980). Scent marking in coyotes. *Can. J. Zool.*, 58, 473–80.

Camenzind, F.J. (1978). Behavioral ecology of coyotes on the national Elk Refuge, Jackson, Wyoming. In *Coyotes: biology, behavior, and management* (ed. M. Bekoff), pp. 267–94. Academic Press, N.Y.

Carbyn, L.N. (1980). Ecology and management of wolves in Riding Mountain National Park, Manitoba. Final Rep. Large Mamm. Sys. Stud., *10*. Canadian Wildlife Service, Edmonton, Alta.

Johnson, R.P. (1973). Scent marking in mammals. *Anim. Behav.*, 21, 521–35.

Kleiman, D.G. (1966). Scent marking in the Canidae. *Symp. Zool. Soc. Lond.*, 18, 167–77.

Lehner, P.N. (1979). Handbook of ethological methods. Garland STPM Press, N.Y. and London.

Meleshko, D.W. (1986). Feeding habits of sympatric canids in an area of moderate ungulate density. M.S. thesis, University of Alberta, Edmonton. 89 pp.

Paquet, P.C. (1988). Behavioral ecology of sympatric wolves and coyotes in Riding Mountain National Park, Manitoba. Ph.D. thesis, University of Alberta, Edmonton, Alberta.

Peters, R. and Mech, L.D. (1975). Scent-marking in wolves. *Amer. Sci.*, 63, 628–37.

Rounds, R. (1980). Aggregation behavior of wapiti (*Cervus elaphus*) in Riding Mountain National Park, Manitoba. *Can. Field-Nat.*, 94, 148–53.

Sokal, R.R. and Rohlf, F.J. (1981). Biometry, the principles and practice of statistics in biological research, 2nd edn. W.H. Freeman and Company, N.Y.

Trottier, G.C., Rollans, S.R. and Hutchinson, R.C. (1983). Range, habitat and foraging relationships of ungulates in Riding Mountain National Park. *Large Mamm. Sys. Stud.*, Report 14. Can. Wild. Serv., Edmonton, AB.

# IV.11 The Network of Olfactory Communication Operating in Populations of Wild House Mice

JANE L. HURST

## Introduction

The urine-marking behaviour of house mice, and their behavioural and physiological responses to urine and other olfactory stimuli, have been researched extensively under stringently controlled laboratory conditions. Most behavioural studies have been conducted in neutral test arenas using isolated or paired laboratory mice (reviewed by Brown 1985*a*). Under these conditions dominant adult male mice deposit numerous small spots of urine on the substrate; they deposit many more marks than subordinate males (Desjardins *et al.* 1973) or adult females (Maruniak, *et al.* 1975→b). The high frequency of marking is largely a response to novel social or non-social odours, such as the presence or urine of another mouse, or the smell of a clean arena (Maruniak *et al.* 1974; Maruniak, *et al.* 1975→a). After repeated exposure to the same type of odour, the marking response can habituate dramatically (Maruniak *et al.* 1974; though see Powell and Wolff 1982).

Mice can gain considerable social information from urine marks. An impressive series of studies has revealed that each mouse has an effectively unique individual odour that is genetically determined and largely imparted by the highly polymorphic MHC genes; the odour is carried in the urine and can be easily distinguished by other mice of both sexes (see review by Boyse *et al.* 1987; also Beauchamp *et al.*, Brown *et al.*, this volume). Mice differentially investigate urine depending on the social and reproductive status of both urine donor and receiver (see Brown 1985*a*), and on the familiarity and relatedness between the mice (e.g. Barnard and Fitzsimons 1988).

Many laboratory tests demonstrate, however, that marking and responses to marks are highly dependent on the precise experimental conditions and on the previous experiences of subjects and urine donors. A mouse's natural social and physical environment is considerably different from clean laboratory test conditions. This paper outlines some recent investigations of urine marking within free-living and enclosed populations of house mice. Results indicate that laboratory arena tests may give a rather misleading picture of urine marking behaviour and the functions of those marks within socially organized mouse populations (e.g. Bronson 1976; Smith 1981).

## Urine marking within free-living populations

The extent and importance of substrate marking within low-density feral mouse populations is not yet known, but in commensal habitats urine marking is extensive. In stable habitats where surfaces are protected from weathering, any surfaces used by mice become completely smeared with their urine, and small posts of solidified urine mixed with dust build up in certain sites (Welch 1953; Hurst 1987*a*). Urine posts look like small stalagmites and can build up to 2 or 3 cm high over a few weeks or months, though there is considerable variation between populations and sites (personal observation). Posts are formed in sites used frequently by a number of mice, often on topographical edges or in strings along narrow ledges, and usually near feeding and nesting sites and on well-used pathways. Welch (1953) reported that posts form on faeces stuck to the substrate, but observations by the present author both in the field and in population enclosures indicate that faeces are only rarely a component.

Any clean surfaces introduced into an infested habitat are rapidly marked by the mice with thin streaks, smears and spots of urine (Hurst 1987*a*). Experimental translocation of these marks between defended territories showed that this marking is largely a response to a clean patch of substrate: any marks already on the substrate inhibit further marking, whether deposited previously by resident mice or by potential intruders. However, the greater the density of marks on adjacent substrate, the more marks are deposited on clean surfaces. Hurst (1987*a*) suggested that urine smears and posts could constitute a dual system of marking behaviour, as posts of urine could only accumulate if mice were stimulated to counter-mark the same spot consistently over a long period of time. Low-density substrate smears could provide cues for orientation, while posts could act as loci for the exchange of social information.

## Urine marking within population enclosures

It was decided to examine individual marking behaviour and responses to marks more closely, and analyse the potential functions of urine marking in socially organized populations. To do this house mice were established in enclosed populations where individuals could be tested while living within a more natural social and olfactory environment than that provided by laboratory arena tests. Mice were established and tested within 8 isolated (Hurst 1989) and 8 territorial family groups (Hurst in press *a–c*). A grid of interchangeable tiles ($0.15 \times 0.15$ m) on the floor of enclosures ($1.2 \times 1.2$ m) each housing a single family group allowed experimental manipulation of the substrate once it had been marked by the resident mice. A variety of test odours could also be incorporated temporarily into the family-marked substrate by replacing a home area tile at random with a marked test tile. In control tests the locations of two home area tiles were interchanged to allow for disruption in the mark pattern already deposited on the

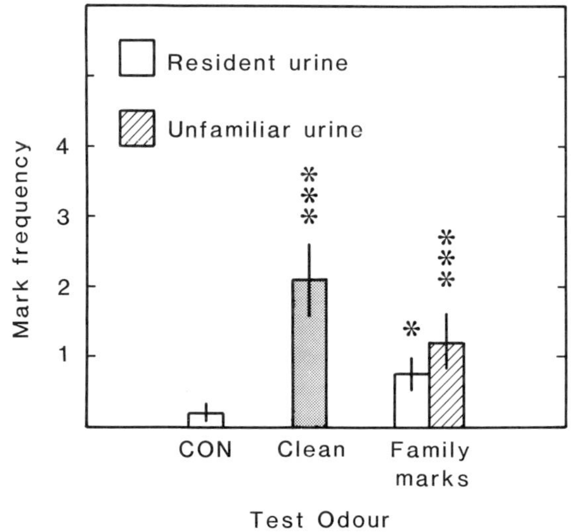


Fig. 1. Frequencies of marks deposited on test odours by individual mice within 8 family groups during 3-min. trials (mean ± S.E., $n = 73$). Family marks = clean floor tile exposed to the resident or an unfamiliar family for previous 24 h; Clean = clean tile marked with water; Con = control home area tile. Significantly more marks were deposited on introduced odours than on a control tile (Wilcoxon one-tailed tests, *** $p < 0.001$, ** $p < 0.01$, * $p < 0.05$), with no significant difference in counter-marking on unfamiliar vs. resident family marks ($p = 0.240$, Wilcoxon two-tailed test). Mice were in contact with test odours for an average of 7.23 ± 0.20 s per trial. Data derived from Hurst (1989).

substrate by mice, which were resident in their enclosures for 10–18 days (isolated groups) or 36–50 days (territorial groups) before tests were conducted. Individual responses to each test odour were measured in random order, and without interference from other family members, by isolating a test area (0.3 × 1.2 m) within each enclosure immediately before each test. Trials were kept much shorter (3 mins) than standard arena tests and behaviour was video recorded in the dark under infra-red light. Full details are given in Hurst (1989).

The general responses of all mice living in isolated family groups showed that mice stopped to investigate substrate marks of an unfamiliar family encountered within their home area, even though their marking responses were similar to those in free-living populations: a patch of clean substrate stimulated the greatest mark response, with no significant difference between counter-marking on substrate marks of either a resident or an unfamiliar family (Fig. 1). Males and females of all ages responded in the same way, but dominant adult males and breeding adult females deposited significantly more marks than their juvenile and subadult offspring (Hurst 1989). Fig. 2 shows however that resident mice, which rarely stopped to investigate their familiar (family-marked) home substrate, all stopped

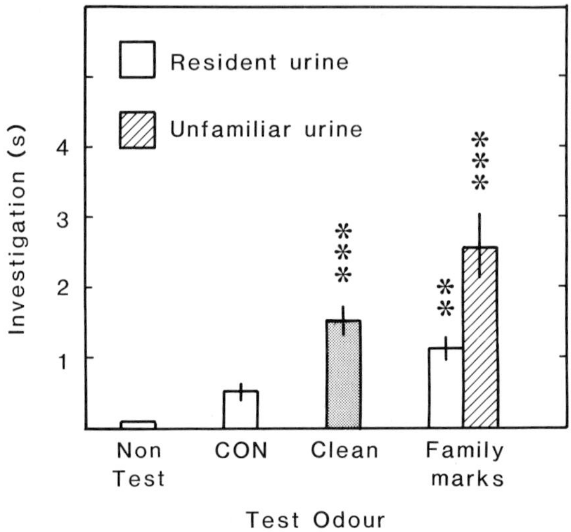


Fig. 2. Time spent investigating test odours and surrounding home substrate by individual mice in 8 family groups during 3-min. trials (mean $\pm$ S.E., $n=73$). Mice spent longer investigating test odours than a control tile (Wilcoxon one-tailed tests, *** $p<0.001$, ** $p<0.01$, * $p<0.05$), and longer investigating unfamiliar family marks than resident family marks ($P=0.0001$) or a clean floor tile ($p=0.026$, Wilcoxon two-tailed test). Key as Fig.1; non-test shows investigation of undisturbed surrounding substrate tiles. Data derived from Hurst (1989).

to investigate the marks of an unfamiliar family, and spent longer investigating such marks than a clean patch of substrate.

To test responses to the marks of specific individuals, test tiles were marked artificially with urine collected from individual family members, using 15 $\mu$l per test tile to provide a strong mark signal that imitated the largest streaks deposited naturally by resident mice (Hurst 1989). When these individually marked test tiles were introduced into the home substrate, particular classes of mice responded more strongly to some types of urine than to others; in general, mice responded as much to a concentrated patch of familiar urine as to any other local change in the family-marked substrate. For example, Fig. 3 shows that mice within isolated family groups stopped to investigate most types of unfamiliar urine significantly more than a control disruption of the resident mark pattern (though not unfamiliar urine from sub-adult males); similarly, a patch of the same types of urine from resident mice stimulated significant investigation despite the familiarity of these odours. Stationary investigation of test odours was brief, occurring on the first one or two visits only, and mice usually stopped very suddenly when in contact with the odour to investigate. Mice were not usually attracted to a test odour from a distance and generally spent the same amount of time on test tiles compared to surrounding home area tiles. Attraction or avoidance was found only in a few

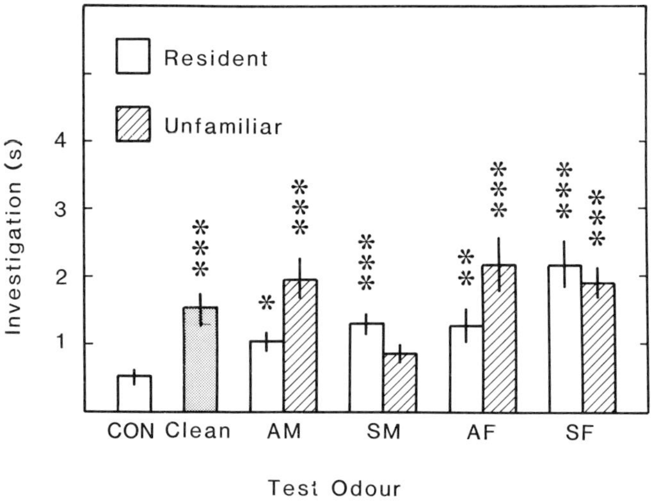


Fig. 3. Time spent investigating a patch of urine from different individuals by mice in 8 family groups (mean $\pm$ S.E., $n = 73$). Asterisks indicate investigation significantly greater than control tiles, as in Fig.2. AM = dominant adult male urine; SM = subadult male urine; AF = breeding adult female urine; SF = subadult female urine; Clean = clean tile; Con = control home area tile. Mice spent longer investigating unfamiliar versus resident AM ($p = 0.029$), resident versus unfamiliar SM ($p = 0.039$), unfamiliar versus resident AF ($p = 0.044$), with no significant difference between unfamiliar versus resident SF ($p = 0.766$, Wilcoxon two-tailed tests). Data derived from Hurst (1989).

specific cases (see below). However, all types of introduced odours stimulated a significant local increase in marking by resident mice (Fig. 4). In contrast to family marks deposited in response to a clean substrate (Fig. 2), a concentrated patch of most types of individual urine generated as much counter-marking as a clean patch of substrate (Hurst 1989, in press *a–c*). Significant differences in counter-marking between test odours and between classes of mice were due to the frequency rather than size of marks deposited (Hurst 1989). Generally there was little correlation between the amount of time spent investigating an odour and the counter-marking response: mice did not have to stop to investigate a test odour closely for the odour to stimulate an increase in mark rate (Hurst 1989, in press *a–c*).

## A system of marking for orientation and communication

A new model of marking behaviour was proposed from the responses of resident mice to introduced odour stimuli (Hurst 1989); the model shows how mice use urine marking both to label their physical environment and send signals to one another. Each individual mouse marks all surfaces of its home area at a low rate, so mice sharing the same area become surrounded by a familiar mixture of

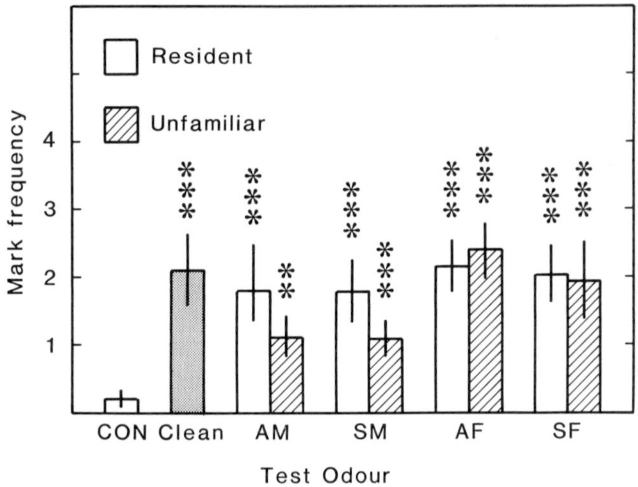


Fig. 4. Counter-marking on a patch of urine from different individuals by mice in 8 family groups during 3-min. trials (mean ± S.E., $n = 73$). Asterisks indicate significantly more marks were deposited on test tiles than on a control tile, as in Fig. 1; see Fig. 3 for notation. There were no significant differences in counter-marking on unfamiliar versus resident urine stimuli (Wilcoxon two-tailed tests). Data derived from Hurst (1989).

background marks which is dominated by marks of the dominant adult mice, especially the dominant male. Anything new coming into the area will be immediately obvious, whatever its smell, because it will not be marked by the resident mice. This stimulates investigation and also stimulates each resident to mark at a slightly higher rate, so a novel object will become incorporated into the familiar marked background. A physical edge, such as the edge of a shelf, will be highlighted where the familiar marking stops. Observations by the author under infra-red light reveal that mice, moving confidently around their familiar environment in total darkness, stop and sniff into the air over edges even if the edge location has been altered. Thus the low density of familiar resident marks covering the substrate alerts mice active in the dark to any changes in their physical environment and to the location of edges.

Mice are also alerted to social changes in their familiar substrate. The different odours of non-resident mice stand out against the resident-marked background and are investigated. More importantly, perhaps, a local increase in the frequency of marks of any familiar resident is also highlighted against the familiar mixture of marks, so a strong increase in individual mark rate acts as a signal even to other residents familiar with the odour. Like any other unexpected change in substrate marks, a local concentration of marks stimulates an increase in mark rate from all resident mice and the signal will eventually decay into the background through time. However, examination of individual responses shows that a patch of certain types of urine stimulates particular mice to counter-mark very strongly with a concentrated burst of marking. This explains how urine posts could build up. In the same way, 1 $\mu$l or more of familiar resident urine added to a resident urine post, or as little as 0.2 $\mu$l of unfamiliar urine, is immediately detected and investigated by resident mice (Hurst in preparation).

Social communication via a high frequency of urine marking could play many roles within the flexible social system found within mouse populations (see below). These potential roles have been assessed by measuring the responses of mice of differing status to specific high frequency urine marks encountered within their home area.

## Mark communication between males

The possible roles of urine marking within the territorial social system found among male mice (e.g. see Crowcroft and Rowe 1963; Reimer and Petras 1967; Mackintosh 1981; Hurst 1987*b*) fall into four main categories (Hurst, in press *a*): marks are deposited (1) to advertise territory defence against intruders; (2) to advertise dominance over other males including resident subordinates and intruders; (3) to provide information for males within the same family-based social group; or (4) males may not mark to communicate with other males, though any marks deposited to communicate with females or for their own orientation would also be available to other males in the local population.

The behavioural responses of males of differing social status towards male urine marks encountered within their home area indicate that dominant territorial males mark at a high rate to advertise their dominance over all other males within their defended territory (Hurst in press *a*); however, dominant males in isolated family groups do not respond very strongly to the urine marks of other males (Hurst 1989). Males of all ages and social status respond to local changes in their familiar marked substrate by increasing their normal mark rate, as outlined above. However, as found from arena tests, dominant males deposit significantly more marks than subordinate or young males on introduced odours and on their familiar home substrate (Hurst 1989, in press *a*). In isolated family groups, dominant males counter-mark all types of introduced male urine, but their mark rate is no greater than on a patch of clean substrate or on a patch of their own urine. Once they have to defend their territory against invasion by a neighbouring family, however, a patch of urine from another male elevates the mark rate considerably, including urine from resident subadult males; their own urine generates only the same mark rate as a patch of clean substrate. Territorially dominant males thus ensure that their own high-frequency marks always predominate over marks of all other males throughout the territory; their frequent patrolling behaviour (Crowcroft 1966; Mackintosh 1970) is likely to be an important component in this system.

Mykytowycz *et al.* (1976) proposed that dominant rabbits mark to increase their confidence in agonistic interactions with territory intruders. Gosling (1982) proposed that territory holders mark to provide a cheat-proof assessment of their dominance, allowing intruders to identify the territory owner and take appropriate action should they meet him. Both of these hypotheses can be extended to include dominance over resident subordinates and intruders, a situation more appropriate to the flexible social organization found among male house mice. Interest shown towards the high frequency marks of dominant territorial males by males of differing status indicate that males mark at a high rate to advertise their dominance to other males (Hurst in press *a*): intruders investigate substrate marks extensively during invasion; resident males of all types stop to investigate a patch of unfamiliar dominant male urine, but only resident males stop to investigate familiar dominant male urine significantly if they were clearly subordinate to the urine donor (resident or neighbour) in prior agonistic interactions. Resident dominant males, which will attack any other males invading their territory, do not stop to investigate their own urine or the familiar urine of their dominant male neighbour, while resident subadult males investigate the familiar urine of the resident dominant male intensively.

In addition to these counter-marking and investigatory responses, males of any status appear to use the high-density marks of dominant adult males to orientate to their resident territory and away from areas marked by other dominant males. Large-scale movements of the marked substrate in enclosures housing two dominant territorial males show that male mice can locate their territory boundaries using these olfactory cues (Mackintosh 1973; Harrington 1976), provided substrate marks do not conflict with any available visual cues (Mackintosh 1973). Once they have been aggressively excluded from a neighbouring territory, males of all classes tend to avoid urine of a neighbouring or unfamiliar dominant male even when encountered within their own home area (Hurst in press *a*). However dominant male urine is not simply aversive to other males (e.g. Jones and Nowell 1973; Bishop and Chevins 1987). When returned to their home area after a temporary displacement, resident dominant and subordinate males are immediately attracted to a patch of resident dominant male urine. Cox (1984) showed similar attraction to resident, and avoidance of neighbour, odours among captive feral mice. This may thus explain the apparently conflicting evidence concerning the importance of male urine marking for territorial demarcation (e.g. Mackintosh 1973) and the failure of substrate marks to prevent territory invasion (Hurst 1987*a*, 1989, in press *a*).

## Mark communication between females

Female responses towards female urine introduced into the home area of isolated (Hurst 1989) and territorial family groups (Hurst in press *b*) indicate that urine marking plays an important role in communication between female house mice, though this has received little previous attention. The social organization of female mice is more complex than the territorial dominance found among males (see Hurst 1987*b*, in press *b*, for references). Several breeding and non-breeding females usually share the same range and nest sites, breeding females sometimes helping males to defend the territory against male and female intruders. It has been suggested that females that have a low probability of raising their young within a social group benefit from delaying their reproduction (Brown 1985*b*; Vandenbergh and Coppola 1986). Correspondingly, sexual maturation and oestrous cycling of young females are inhibited by prolonged exposure to olfactory primer cues in the urine of familiar adult or grouped females under laboratory conditions (see review by Brown 1985*b*), though individuals vary in their response (e.g. Massey 1986). Little is yet known about the role of such cues within socially organized populations, but feral female mice produce puberty-inhibiting cues when populations are dense (Massey and Vandenbergh 1980), while older dominant females continue to breed even in crowded populations (e.g. Lloyd and Christian 1969).

Hurst (in press *b*) proposed four possible roles of high-frequency urine marking in inter-female communication: (1) marks play a role in the advertisement of territory defence against intruders; (2) marks advertise dominant breeding status by providing cues to influence the physiology of other females; (3) marks provide a pool of information concerning individual identity and breeding status of each group member; or (4) marks provide cues for orientation and group cohesion. Taking into account social and reproductive status and the social relationships between individual subjects and urine donors, responses towards urine marks encountered within their own home area support the hypothesis that females mark at high frequency to advertise their active breeding status to other females (Hurst in press *b*). Resident breeding females counter-mark the urine of other breeding females and neighbouring subadult females regardless of any aggressive territorial interactions between the mice; marking by non-breeding females is consistent with the low-frequency mark response of all mice to any change in their familiar olfactory background (see above). However, subadult females investigate urine from both unfamiliar and familiar resident breeding females extensively.

It is not yet known whether investigation, avoidance, and counter-marking of substrate urine marks are important components in the modulation of breeding status through reproductive primer cues (Vandenbergh and Coppola 1986), although there are differences in response to urine from familiar resident, familiar neighbour, and unfamiliar females (Hurst in press *b*). If females use substrate marks to assess local breeding conditions and then modulate their own status accordingly, they must distinguish between the marks of current resident and non-resident mice. The inhibitory effects of female primer cues require prolonged exposure to stimuli from the same individual female or females over at least several days (see Brown 1985*b*). Thus the prolonged investigation by young female mice of unfamiliar urine (Hurst 1989, in press *b*) will not inhibit their reproductive status, but may allow females to recognize future marks from the same individual; counter-marking by all residents (especially breeding females) rapidly reinstates the predominance of resident marks in the area. Familiar neighbour urine, however, could have the same inhibitory effects as urine from familiar residents. In contrast to their prolonged investigation of urine from familiar resident breeding females, young non-breeding females spend very little time investigating urine of neighbouring breeding females, and the low-frequency counter-marking response is persistent (Hurst in press *b*). Resident breeding females, initially attracted to urine of neighbouring breeding females from a distance, counter-mark very strongly and then avoid spending much time in contact with the urine (Hurst in press *b*). The dynamic nature of their marking system ensures that the marks of breeding females currently using the area predominate over all parts of the substrate.

## Mark communication between the sexes

Most studies investigating the responses of male mice to female odours or female mice to male odours have emphasized the potential roles of marking in sexual communication, through the attraction or physiological stimulation of mates (reviewed by Brown 1985*a*, *b*; also see Bronson and Coquelin 1980; Wolff and Powell 1984; Albonetti and D'Udine 1986; Barnard and Fitzsimons 1988). Sexual interactions, however, may be quite rare between mice in established populations (e.g. Hurst 1986, 1987*b*). A large proportion of females are usually pre-reproductive or non-breeding adults (e.g. Rowe *et al.* 1983; Pennycuik *et al.* 1986), breeding females will only be sexually active between pregnancies, and only dominant territorial males may succeed in mating (Wolff 1985; Hurst 1987*b*). Both territorial males and breeding females aggressively exclude unfamiliar and neighbour mice from their territory, including potential mates (e.g. Reimer and Petras 1967; Hurst 1987*b*). Cox (1984) showed that captive feral mice avoid the odours of both male and female mice from an adjacent deme.

Urine marking could thus play a number of possible roles in communication between the sexes (Hurst in press *c*): (1) marks play specific roles in sexual behaviour; (2) marks play a role in the advertisement of territory defence against intruders; (3) male marks advertise dominance by providing a cheat-proof signal that identifies the dominant male to potential mates and intruders alike; or (4) marks provide cues for orientation and group cohesion.

Strong counter-marking on female urine by dominant males living in both isolated (Hurst 1989) and territorial family groups (Hurst in press *c*) suggest that males mark to advertise their dominance to females, as well as to males in territorial groups (see above). Resident dominant males counter-mark female urine at a higher rate than a clean patch of substrate regardless of the reproductive status of the donor; urine from familiar mates, daughters, neighbours aggressively excluded from the territory, and urine from unfamiliar females are all countermarked. Differential investigation indicates that males distinguish between urine from different female donors. Substrate marks do not stop females invading another family territory (Hurst 1989, in press *b*), but like males, females also appear to use the high-frequency marks of dominant adult males to orient to their resident territory and away from areas marked by other dominant males (see Hurst in press *c*, and above). The defender of the territory will be identifiable to female intruders and to potential mates from these high-frequency marks, which may indicate the extent and quality of his territory (Hurst in press *c*).

Females living in family groups do not mark at a high rate in response to male urine, and males are not attracted to female urine, though males stop to investigate certain types of urine more than others (Hurst 1989, in press *c*). This may have been due to the pregnant or non-breeding status of the females tested, and it is still not known whether females mark to advertise their readiness to mate. Further experiments are in progress.

## Conclusions

The extensive urine-marking behaviour of house mice appears to form a highly dynamic dual system of marking which allows mice to label their physical environment both for their own orientation and to send signals to each other. All mice investigate and deposit low-density marks in response to social and non-social changes in their familiar olfactory environment. This may enhance rapid detection of changes in the environment, including edges. Dominant adult males and breeding adult females mark at a high rate to communicate with other mice. These marks do not appear to be deposited specifically to attract or repel other mice but as an advertisement of their dominance or breeding status, and are available to all members of the local population. Marks contain considerable social information about the depositor, including its individual identity. In all cases the responses of mice encountering a mark depend on whether it is from a familiar resident, familiar neighbour, or unfamiliar mouse, and the social and reproductive status of both urine donor and receiver and their previous social experiences. This results in a complex and dynamic network of communication operating in socially organized mouse populations, which only becomes apparent when mice are investigated within an appropriate social and olfactory environment.

## Acknowledgments

The paper, written and presented with financial support from the Science and Engineering Research Council, benefited from comments made by Peter Aldhous.

## References

Albonetti, M.E. and D'Udine, B. (1986). Social experience occurring during adult life: Its effects on socio-sexual olfactory preferences in inbred mice, *Mus musculus*. *Anim. Behav.*, 34, 1844–7.

Barnard, C.J. and Fitzsimons, J. (1988). Kin recognition and mate choice in mice: the effects of kinship, familiarity and social interference on intersexual interaction. *Anim. Behav.*, 36, 1078–90.

Bishop, M.J. and Chevins P.F.D. (1987). Urine odours and marking patterns in territorial laboratory mice (*Mus musculus*). *Behav. Process.*, 15, 233–48.

Boyse, E.A., Beauchamp, G.K. and Yamazaki, K. (1987). The genetics of body scent. *Trends Genet.*, 3, 97–102.

Bronson, F.H. (1976). Urine marking in mice: causes and effects. In *Mammalian olfaction, reproductive processes and behaviour* (ed. R.L. Doty), pp. 119–41. Academic Press, New York.

Bronson, F.H. and Coquelin, A. (1980). The modulation of reproduction by priming pheromones in house mice: speculations on adaptive function. In *Chemical signals: vertebrates and aquatic invertebrates* (ed. D. Müller-Schwarze and R.M. Silverstein), pp. 243–65. Plenum Press, New York.

Brown, R.E. (1985*a*). The rodents II: Suborder Myomorpha. In *Social odours in mammals* (ed. R.E. Brown and D.W. Macdonald), Vol. 1, pp. 345–457. Clarendon Press, Oxford.

Brown, R.E. (1985*b*). The rodents I: Effects of odours on reproductive physiology (primer effects). In *Social odours in mammals* (ed. R.E. Brown and D.W. Macdonald), Vol. 1, pp. 245–344. Clarendon Press, Oxford.

Cox, T.P. (1984). Ethological isolation between local populations of house mice (*Mus musculus*) based on olfaction. *Anim. Behav.*, 32, 1068–77.

Crowcroft, P. and Rowe, F.P. (1963). Social organization and territorial behaviour in the wild house mouse (*Mus musculus* L.). *Proc. Zool. Soc., Lond.*, 140, 517–31.

Crowcroft, P. (1966). *Mice all over*. Foulis, London.

Desjardins, C., Maruniak, J.A. and Bronson, F.H. (1973). Social rank in the house mouse: differentiation revealed by ultraviolet visualization of urinary marking patterns. *Science*, 182, 939–41.

Gosling, L.M. (1982). A reassessment of the function of scent marking in territories. *Z. Tierpsychol.*, 60, 89–118.

Harrington, J.E. (1976). Recognition of territorial boundaries by olfactory cues in mice (*Mus musculus* L.). *Z. Tierpsychol.*, 41, 295–306.

Hurst, J.L. (1986). Mating in free-living wild house mice (*Mus domesticus* Rutty). *J. Zool., Lond.*, 210, 623–8.

Hurst, J.L. (1987*a*). The functions of urine marking in a free-living population of house mice, *Mus domesticus* Rutty. *Anim. Behav.*, 35, 1433–42.

Hurst, J.L. (1987*b*). Behavioural variation in wild house mice (*Mus domesticus*

Rutty): a quantitative assessment of female social organization, *Anim. Behav.*, 35, 1846–57.

Hurst, J.L. (1989). The complex network of olfactory communication in populations of wild house mice, *Mus domesticus* Rutty: urine marking and investigation within family groups. *Anim. Behav.*, 37, 705–25.

Hurst, J.L. (in press *a*). Urine marking in populations of wild house mice, *Mus domesticus* Rutty. I. Communication between males. *Anim. Behav.*

Hurst, J.L. (in press *b*). Urine marking in populations of wild house mice, *Mus domesticus* Rutty. II. Communication between females. *Anim. Behav.*

Hurst, J.L. (in press *c*). Urine marking in populations of wild house mice, *Mus domesticus* Rutty. III. Communication between the sexes. *Anim. Behav.*

Jones, R.B. and Nowell, N.W. (1973). Aversive effects of the urine of a male mouse upon the investigatory behaviour of its defeated opponent. *Anim. Behav.*, 21, 707–10.

Lloyd, J.A. and Christian, J.J. (1969). Reproductive activity of individual females: three experimental freely growing populations of house mouse (*Mus musculus* L.). *J. Mammal.*, 50, 49–59

Mackintosh, J.H. (1970). Territory formation by laboratory mice. *Anim. Behav.*, 18, 177–83.

Mackintosh, J.H. (1973). Factors affecting the recognition of territory boundaries by mice (*Mus musculus* L.). *Anim. Behav.*, 21, 464–70.

Mackintosh, J.H. (1981). Behaviour of the house mouse. *Symp. Zool. Soc. Lond.*, 47, 337–65.

Maruniak, J.A., Owen, K., Bronson, F.H. and Desjardins, C. (1974). Urinary marking in male house mice: responses to novel environment and social stimuli. *Physiol. Behav.*, 12, 1035–9.

Maruniak, J.A., Darney, K.T. Jr. and Bronson, F.H. (1975→a). Olfactory perception of the nonsocial environment by male house mice. *Behav. Biol.*, 14, 237–40.

Maruniak, J.A., Owen, K., Bronson, F.H. and Desjardins, C. (1975→b). Urinary marking in female house mice: effects of ovarian steroids, sex experience, and type of stimulus. *Behav. Biol.*, 13, 211–7.

Massey, A. (1986). Variable responses to high density of female *Mus musculus*: a merger of two population regulation hypotheses. *Oecologia*, 69, 140–3.

Massey, A. and Vandenbergh, J.G. (1980). Puberty delay by a urinary cue from female house mice in feral populations. *Science*, 209, 821–2.

Mykytowycz, R., Hesterman, E.R., Gambale, S. and Dudzinski, M.L. (1976). A comparison of the effectiveness of the odours of rabbits, *Oryctolagus cuniculus*, in enhancing territorial confidence. *J. Chem. Ecol.*, 2, 13–24.

Pennycuik, P.R., Johnston, P.G., Westwood, N.H. and Reisner, A.H. (1986). Variation in numbers in a house mouse population housed in a large outdoor enclosure: seasonal fluctuations. *J. Anim. Ecol.*, 55, 371–91.

Powell, A.J. and Wolff, P.R. (1982). Sex differences in mouse urination patterns. *Anim. Behav.*, 30, 1207–11.

Reimer, J. and Petras, M.L. (1967). Breeding structure of the house mouse (*Mus musculus*) in a population cage. *J. Mammal.*, 48, 88–99.

Rowe, F.P., Swinney, T. and Quy, R.J. (1983). Reproduction of the house mouse (*Mus musculus*) in farm buildings. *J. Zool., Lond.*, 199, 259–69.

Smith, J.C. (1981). Senses and communication. *Symp. zool. Soc. Lond.*, 47, 367–93.

Vandenbergh, J.G. and Coppola, D.M. (1986). The physiology and ecology of puberty modulation by primer pheromones. *Adv. Study Behav.*, 16, 71–107.

Welch, J.F. (1953). Formation of urinating 'posts' by house mice (*Mus*) held under restricted conditions. *J. Mammal.*, 34, 502–3.

Wolff, P.R. and Powell, A.J. (1984). Urine patterns in mice: an analysis of male/female counter-marking. *Anim. Behav.*, 32, 1185–91.

Wolff, R.J. (1985). Mating behaviour and female choice: their relation to social structure in wild caught house mice (*Mus musculus*) housed in a semi-natural environment. *J. Zool., Lond.*, 207, 43–51.

# IV.12 Scent Marking Behaviour of Wild Marmosets, *Callithrix humeralifer* (Callitrichidae, Primates)

ANTHONY B. RYLANDS

## Introduction

All members of the family Callitrichidae studied so far possess specialized epidermal scent glands. These occur in a suprapubic gland pad, and in the circumgenital (scrotum and labia pudendi) and gular-sternal areas of the body. They are used in self-marking, marking of conspecifics and also items in the environment.

Epple *et al.* (1986) have described the various aspects of behaviour involved in marking with these glandular areas. The first is sit rubbing (circumgenital marking), in which the circumgenital and circumanal areas are pressed on to the substrate and rubbed back and forth and/or from side to side. A small amount of urine may also be deposited. In pull rubbing (suprapubic marking), the marmoset extends its body along a branch, and the lower abdomen and suprapubic area are pulled forward while pressed down on the substrate. Finally, in chest rubbing (sternal marking) the upper chest is applied and may simply be pushed forward along the substrate or may be rubbed back and forth.

Research on scent marking in several callitrichids is reviewed by Epple and Smith (1985) and by Epple *et al.* (1986). The scent marks from these glands can transmit information regarding the identity of the species and subspecies, sex, individual identity, hormonal condition and social status. Scent-marking behaviour has been found to play a role in aggressive, social and sexual communication within the group, as well as intergroup communication. However, studies of scent marking in this family have been almost entirely limited to captive groups, and little information is available for the contexts and functions of such behaviour in natural conditions.

In this study is reported the frequency, contexts. and localities of scent marking in a wild group of tassel-ear marmosets, *Callithrix humeralifer intermedius*.

## Methods

During May and June of 1979, observations were made on a habituated group of 13 *C. humeralifer intermedius*, at Aripuanã, in Amazonian tropical rain forest in the north of the state of Mato Grosso, Brazil (Rylands 1982). The group consisted of three adult males (ST,PT,2R), one subadult male (DT), four adult females (WT,BR,RT,GT), one subadult female (BT), two 7- to 10-month old juveniles (IP,IR), and two 2- to 3-month old infants. Observations were recorded *ad libitum*; 202 and 192 scent marks recorded during ten and eight complete observation days in May and June respectively, were outside the context of intergroup encounters. During the entire study, from September 1978 to August 1979 (115 complete observation days), 176 scent marks were recorded during intergroup encounters. Individual group members were identified by their variable tail markings, and the sex was determined by their distinctive white genitalia.

The following information was recorded for each scent mark: identity and sex of the individual, scent-mark type, context, and locality in the home range. The three scent-mark types of sit rubbing (circumgenital), pull rubbing (suprapubic), and chest rubbing (sternal) were recorded. Chest rubbing was observed on only horizontal or nearly horizontal substrates and involved only pulling the chest forward rather than to and fro movements—a pattern recorded for some other callitrichids, such as lion tamarins (see Kleiman and Mack 1980). Sternal and suprapubic rubbing were also observed in combination. This type is referred to as 'ventral rubbing' by Epple *et al.* (1986). The postures involved are illustrated in Rylands (1981). The contexts in which scent marking were observed were divided into three categories: (i) social or sexual (associated with huddling, grooming, play, or sexual behaviour); (ii) during travel; and (iii) during intergroup encounters. The marmosets frequently circumgenital scent marked in the holes that they gouged in tree trunks, branches, and vines for exudate eating, a context that was not recorded here. In order to examine the frequency of scent marking in different parts of the home range, the pattern of range use was estimated by recording at 5-min. intervals all the quadrats occupied by one or more members of the group. Quadrat use was then expressed as a percentage of the total records of quadrat occupation.

## Results

### FREQUENCY OF DIFFERENT SCENT MARK TYPES

Suprapubic marking was the pattern recorded most frequently, accounting for more than half of the total observations in both months (Fig. 1). Circumgenital marking outside of the context of tree gouging was very rare, accounting for only eight (2%) of the 394 scent marks observed. Sternal marking accounted for 31 per cent of scent marks and was more frequently observed in association with suprapubic marking.

A number of other rubbing and olfactory patterns were observed infrequently. Muzzle rubbing, described by Epple *et al.* (1986), involved wiping the cheeks on the sides of a branch. This may have been just to remove food or saliva, but, as

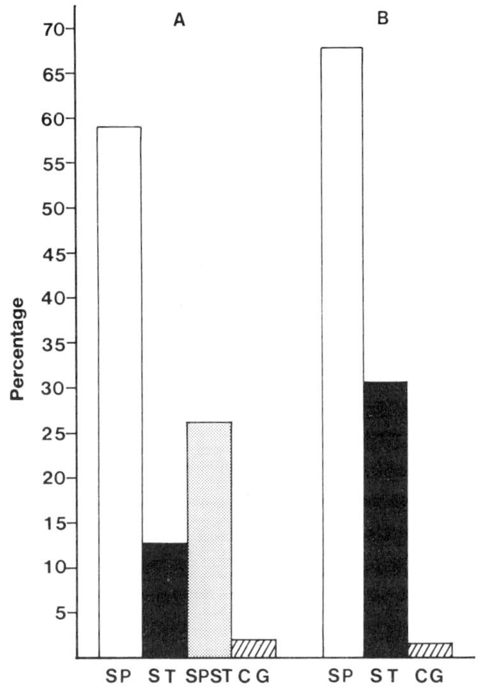


Fig. 1. The percentage frequency of different scent marking types. SP = suprapubic marking, ST = sternal marking, SPST = combined suprapubic/sternal marking, CG = circumgenital marking (except of exudate holes). A: SPST is considered as a separate scent marking type. B: the frequency of suprapubic and sternal marking irrespective of whether or not they were combined.

indicated by Epple *et al.*, the marmosets sometimes muzzle-rubbed on scent marks, which may involve the deposition of saliva or nasal discharge. Rubbing tree branches with the inside of the arms in a slow and distinct fashion was also observed on a few occasions (Rylands 1981). Marking of conspecifics usually involved anal rubbing with the tail raised, but also took the form of rubbing the entire ventral surface over the crouching body of another group member. Self-marking with urine took the form of curling the tail under the body so as to intercept the urine stream. They were not seen actively to spread urine over the body with the hands or feet.

### SCENT MARKING SEQUENCES

The marmosets were frequently observed to scent mark a site immediately after it had been marked by another group member, often sniffing or licking the previous mark. Up to seven individuals were observed marking one after the other at the

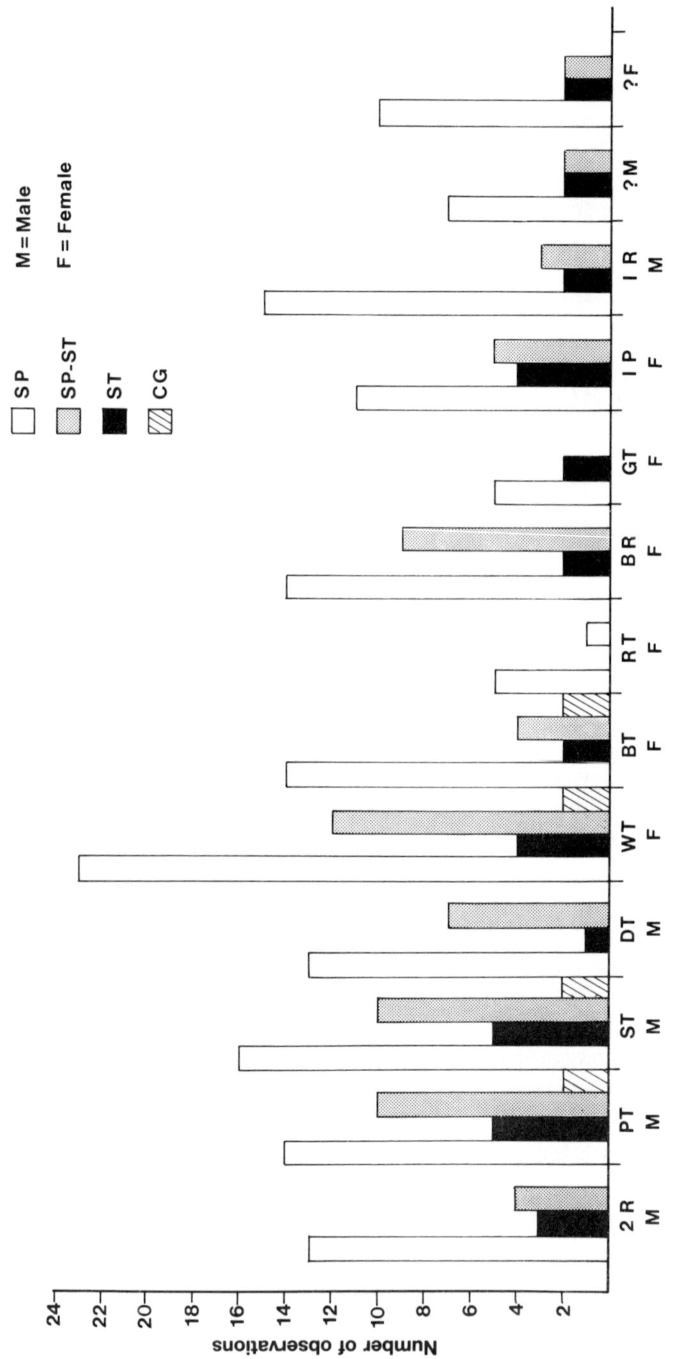


Fig. 2. The number of observations of different scent-marking types by each of the *Callithrix humeralifer* study group members.

same site. On a little less than half of the occasions when scent marking was observed at a particular site, two or more individuals scent marked in sequence.

### DISTRIBUTION OF SCENT MARKING AMONGST GROUP MEMBERS

Suprapubic marking was most frequently recorded for the breeding female WT and the two adult males ST and PT (Fig. 2); the two females, RT and GT, scent marked very little. The infants, then 2–3 months old, were not observed scent marking. The female WT showed the suprapubic/sternal combination more frequently than other group members; this scent mark pattern was not observed for the female GT and only once for the female RT. The adult male 2R, the subadult female BT, and the juveniles IP and IR also showed very little marking of this type. Circumgenital marking was observed for the adult males ST and PT, the adult female WT and the subadult female BT, but was too infrequent to permit comparisons between group members.

The scent marking frequency of the five males and five females (excluding RT) was compared, including those observations where the sex but not the individual was identified. The results showed that there was no significant difference between the sexes for the observed frequencies of scent marking (all types lumped, Mann-Whitney U test: $p > 0.20$), or for the frequencies with which they showed suprapubic, sternal, and combined suprapubic/sternal marking (U test: $p > 0.20$).

### SCENT MARKING CONTEXTS

The few circumgenital scent marks recorded outside of the context of tree gouging were in social or sexual situations, and preceded or followed grooming, huddling, or copulation. Suprapubic and sternal marking were observed during travel, when they marked especially the frequently used arboreal pathways.

Scent marking was observed more frequently during intergroup encounters than in other contexts. A rate of 2.5 scent marks per hour was recorded at times other than during intergroup encounters, whereas 11.0 marks per hour were recorded during intergroup encounters. The marmosets frequently scent marked following rather frantic huddling and grooming sessions, which punctuated collective chases of the opposing group. Suprapubic marking was the most common type recorded in this context (Fig. 3). However, relatively more sternal marking was recorded and, if considered separately, the suprapubic/sternal combination was observed more than suprapubic marking on its own. Comparing the relative frequencies of sternal marking and suprapubic marking, there was no marked difference between intergroup encounters and other contexts with respect to the frequency of suprapubic and sternal marking. The suprapubic/sternal combination, however, was recorded with a considerably higher frequency in the intergroup encounters. Circumgenital marking infrequently followed the huddling or intense but brief grooming sessions during the encounters.

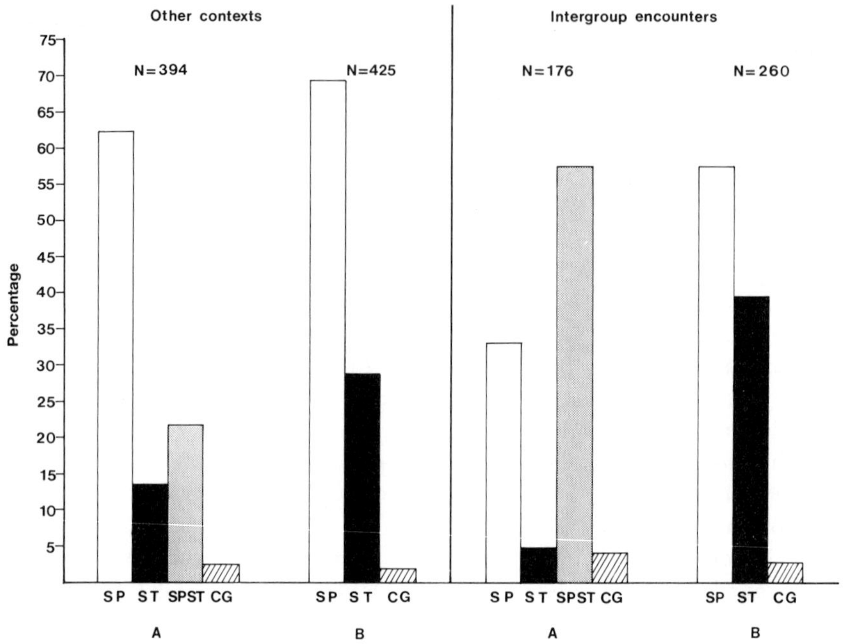


Fig. 3. The frequency of different scent-marking types in contexts other than group encounters and during intergroup encounters. A: SPST is considered as a separate scent marking type. B: the frequency of suprapubic and sternal marking irrespective of whether or not they were combined.

### DISTRIBUTION OF SCENT MARKING IN THE RANGE

The quadrats in which most scent marking occurred are displaced slightly to the east of the quadrats most used during the two months of observation (Fig. 4). However, the pattern of range use was correlated with the frequency of scent marking (Spearman rank correlation coefficient: $r = 0.81$, $p > 0.001$, two-tailed test). In order to examine the relative frequency with which the marmosets marked different parts of the range, it was, therefore, necessary to discount the effect of time spent in each quadrat. Because scent marking increased during intergroup encounters restricted to the periphery of the range, all scent marks in this context were excluded. Dividing the percentage of the total records for each scent marking type for each quadrat by the percentage of records of the occupation of the quadrat demonstrated that scent marking was not evenly distributed through the range (Figs. 4 and 5). Table 1 compares the proportion of quadrats marked and the mean rate of scent marking per quadrat in the 44 peripheral quadrats and 29 central (non-peripheral) quadrats during the two months. Circumgenital marking was too infrequent to be included in this analysis. Suprapubic, sternal, and combined suprapubic/sternal marking were each recorded in a higher percentage of the

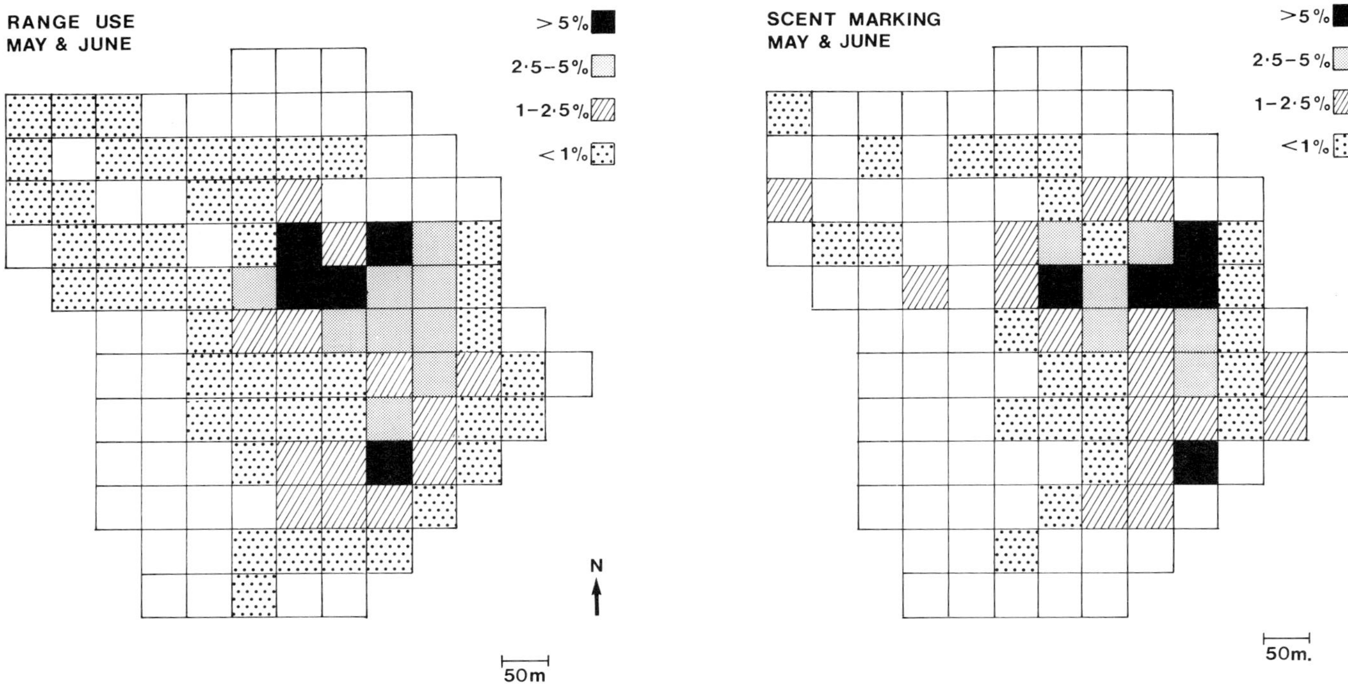


Fig. 4. The pattern of range use (left) and scent marking (right) by the *C. humeralifer* study group. Left: the percentage of records for the occupation of each quadrat. Right: the percentage of records for scent marking in each quadrat. Observation time was the same for both range use and scent marking.

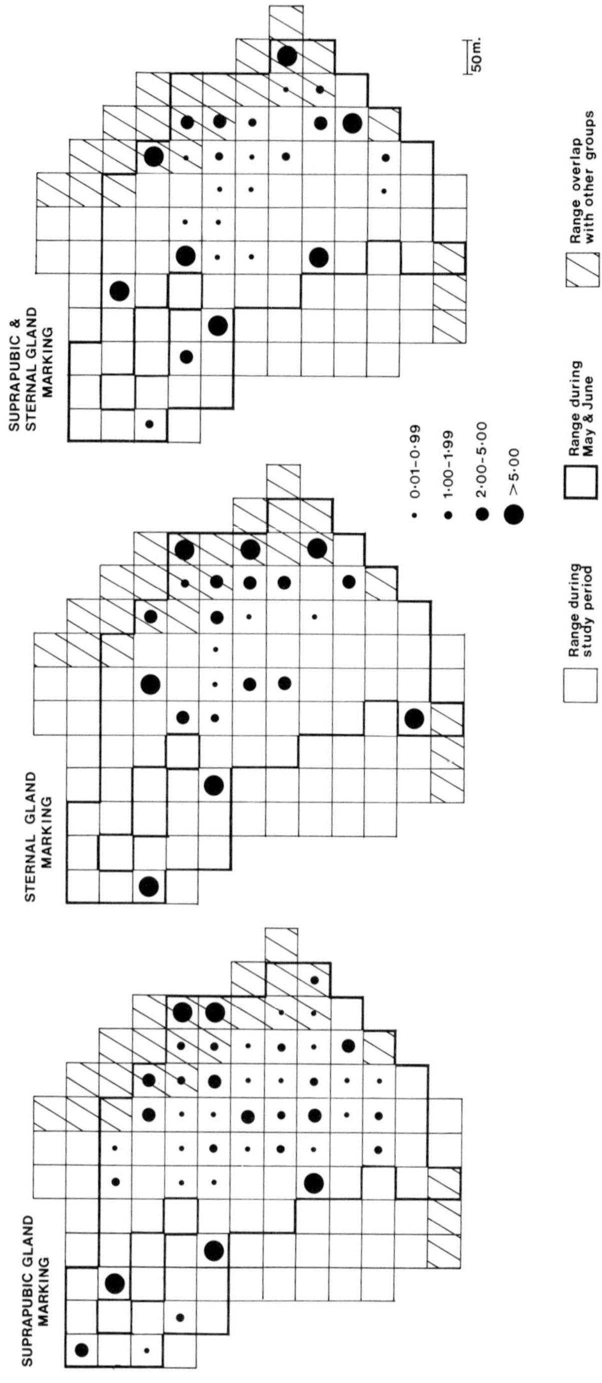


Fig. 5. The pattern of scent marking in the home range of the *C. humeralifer* study group. The larger the black dot the higher the rate of scent marking in the respective quadrat. The suprapubic/sternal combination is considered as a separate scent mark.

central quadrats than those peripheral to the home range. However, the mean rate of scent marking per quadrat was consistently higher in the peripheral than the central quadrats. Comparing the peripheral quadrats, the marmosets marked more of the quadrats on the eastern boundary than on the west (Table 2).

**Table 1.** A comparison of scent marking frequencies in the central quadrats and those of the boundary of the home range in May and June

| | Peripheral 44 quadrats | | Central 29 quadrats | |
|---|---|---|---|---|
| | No. of quads. (%) | Mean rate of marking per quad. | No. of quads. (%) | Mean rate of marking per quad. |
| Suprapubic | 20 (45%) | 1.33 | 23 (79%) | 0.87 |
| Sternal | 9 (20%) | 2.87 | 13 (45%) | 0.98 |
| Suprapubic/sternal | 12 (27%) | 1.70 | 14 (48%) | 0.79 |
| All suprapubic | 23 (52%) | 1.47 | 25 (86%) | 0.86 |
| All sternal | 16 (36%) | 2.09 | 20 (69%) | 1.17 |

The table shows the total number of central and peripheral quadrats marked and the mean rate of scent marking per quadrat (the percentage of the scent marks in each quadrat is divided by the percentage of records for the occupation of the quadrat).

**Table 2.** The number of quadrats scent marked out of the 22 peripheral quadrats to the west and the 22 peripheral quadrats to the east of the home range

| | East 22 quadrats | | West 22 quadrats | |
|---|---|---|---|---|
| | No. of quads. | % | No. of quads. | % |
| Suprapubic | 12 | 55 | 7 | 32 |
| Sternal | 6 | 27 | 3 | 14 |
| Suprapubic/sternal | 8 | 36 | 4 | 18 |
| All suprapubic | 13 | 59 | 8 | 36 |
| All sternal | 11 | 50 | 5 | 23 |

## Discussion

The suprapubic scent marking pattern shown frequently by *Callithrix humeralifer* in this study, has not been observed in captive populations of the congeneric species, *jacchus* and *argentata* (Epple 1970; Stevenson and Poole 1976; Omedes 1979; Omedes and Carroll 1980; Epple *et al.* 1986). *C. jacchus* does not have a differentiated glandular field in the suprapubic area (Sutcliffe and Poole 1978; Epple *et al.* 1986) and Epple *et al* indicate that suprapubic marking does not occur in this genus. However, this study and other observations of suprapubic marking in wild groups of *C. penicillata* (G.A.B. da Fonseca, personal communication) and *C. kuhli* (Rylands 1982) indicate an important behavioural/morphological difference amongst the species of this genus. The use of circumgenital marking is evidently common to all *Callithrix* species studied so far; and, as emphasized earlier, this scent marking pattern was frequently observed following tree gouging behaviour but was not recorded in this research. Sternal marking is reported as infrequent for *C. jacchus* (Epple 1970; Stevenson and Poole 1976; Sutcliffe and Poole 1978) and two subspecies of *argentata* (Omedes 1981).

The frequency of scent marking has been found to be closely associated with social dominance. In *C. jacchus*, dominant males and females showed more circumgenital and sternal scent marking than subordinates (Epple 1970). This is accentuated in same sex aggressive encounters and Epple concludes that both high frequency of scent marking and quality of scent marks serve to demonstrate and maintain social dominance. The importance of scent marking in dominance relations within the group has also been shown for *Saguinus fuscicollis* (Epple 1973, 1975; Epple *et al.* 1979) and *S. oedipus* (French and Cleveland 1984; Epple *et al.* 1988). Social dominance in turn relates to breeding success, and scent marking by the dominant female is likely to be involved in the reproductive inhibition of subordinate females (Abbott *et al.* 1981; Abbott 1984; Epple and Katz 1984; French *et al.* 1984; see also Abbott, this volume).

Regarding the frequency of scent marking by the different group members in this study, sternal and suprapubic marking were recorded most frequently for the adult breeding female WT and two adult males ST and PT. These animals were considered dominant in the group for a number of reasons. Apart from being most active in scent marking, the two males, along with the breeding female, were the most active infant carriers (Rylands 1986). The two males were groomers and groomees of the breeding female more than other group members and also, with the breeding female, were the most active participants in intergroup encounters in terms of the frequency of chasing, genital displaying and long calling as well as scent marking (Rylands 1982). The adult female RT showed very little scent marking. This female had entered the group in January 1978, 4 months previous to this study, and evidently still occupied a peripheral position. Although scent marking was carried out by all group members, suprapubic and sternal marking were therefore most frequent in the dominant individuals, confirming the findings for captive studies.

In *C. jacchus*, males and females showed circumgenital and suprapubic scent marking with equal frequency during encounters with same-sex strange individuals (Epple 1970). In *S. fuscicollis* females tended to scent mark more (circumgenital and suprapubic) than males in this situation (Epple 1978). Sexual dimorphism with respect to such behaviour is most marked in *S.oedipus*: females increase their suprapubic, but not circumgenital, scent marking when confronted by strangers, whereas males use overt aggression and threats (French and Snowdon 1981; French and Cleveland 1984). No sex difference was observed for any of the scent-marking types in this study due to the small sample size.

Suprapubic and sternal marking were observed throughout the home range. During group progression, single sites on frequently used arboreal pathways were marked by up to seven individuals in this way, but individuals on their own could be observed marking at any time. There was no diurnal pattern in marking activity (Rylands 1982). There was, however, a tendency for increased scent marking activity in areas where the home range overlapped with other groups. Of the three groups which overlapped the home range of the study group, only that on the north-eastern boundary was regularly involved in hostile encounters (see Fig. 5). Interactions with the groups on the south-east and southern borders were limited to vocal exchanges. There were no groups to the west, which would explain the relative lack of scent marking on this boundary (Table 2).

Suprapubic, sternal, and especially combined suprapubic/sternal scent marking increased in frequency during intergroup encounters. Suprapubic marking and circumgenital marking were observed during intergroup encounters in wild populations of *S. geoffroyi* (Dawson 1978) and *S. oedipus* (Neyman 1978). Although marking anywhere in its range (Lindsay 1979), *S. geoffroyi* also marked more frequently in areas of home range overlap (Dawson 1979). However, French and Snowdon (1981) showed that sequences of suprapubic, but not circumgenital, marking in captive female *S. oedipus* could be increased by simulating aggressive and territorial encounters; they concluded that suprapubic marking is important in intergroup spacing behaviour. This study indicates a similar conclusion for *C. humeralifer*. Scent marks are not necessarily a means of preventing invasion of an area (Epple *et al.* 1986), and the function of suprapubic marking may best fit the Competitive Assessment Model of Gosling (1982, and this volume), in which the marmosets are distributing scent to permit scent matching during intergroup encounters. Such 'scent matching' may be possible through the occasional close encounters that occur during interactions between groups, through increased scent marking activity at this time, and through genital displays that are enacted by several group members simultaneously and frequently within 1 or 2 m of the opposing group. The increased frequency of sternal marking during intergroup encounters indicates that, in addition to having a role similar to that of suprapubic marking, it may be important in active defence as an aggressive component occurring when the individual is aroused.

Circumgenital marking, outside of the context of tree gouging was, on the other hand, always recorded in highly social situations. Even during intergroup encounters it only occurred following the brief huddling and grooming bouts which punctuated chases during the peak of activity. Dawson (1979) observed circumgenital scent marking in *S. oedipus* during hostile encounters between groups, but the evidence from captivity strongly suggests a sociosexual role (French and Snowdon 1981). Circumgenital scent marking behaviour is associated with oestrus, copulation, pregnancy, and dominance relations within the group, including the reproductive inhibition of subordinate females. The association with tree gouging, first reported by Coimbra-Filho (1971), is entirely fortuitous with no identifiable role in resource or territory defence (see Lacher *et al.* 1981). Rylands (1986) considers gouge holes simply an ideal spot to place a scent mark: of all possible scent mark sites in the forest, gouge holes are the only sites guaranteed to be sniffed, gouged, and licked by other group members. Coimbra-Filho (1972) and Coimbra-Filho and Mittermeier (1976) observed that high-ranking group members marked in gouge holes more than other members, and have suggested an intragroup social function for such behaviour.

This study reinforces the arguments of French and Snowdon (1981) that scent marks of each of the glandular areas have different functions, possibly involving scents of differing chemical composition (Epple *et al.* 1988). Circumgenital marking by *C.humeralifer* plays an important role in sociosexual communication within the group. Suprapubic marking is likely to be important in communication between groups, as a possible group specific smell for scent matching in intergroup encounters. Sternal marking could be associated with high arousal situations, perhaps involving a degree of aggression not found for the other two types of scent marking.

## Acknowledgements

This work was carried out while I was employed by the Instituto Nacional de Pesquisas da Amazonia, Manaus. The study was supported by the Conselho Nacional de Desenvolvimento Cientifico e Tecnologico (CNPq) of Brazil, and formed part of a doctoral thesis for the University of Cambridge. I am most grateful to Prof. P.A. Jewell (Research Group in Mammalian Ecology and Reproduction) and Dr T.H. Clutton-Brock (Large Animal Research Group) for their encouragement and supervision. My participation in the Conference Chemical Signals in Vertebrates V at Oxford University was kindly financed by CNPq.

## References

Abbott, D.A. (1984). Behavioral and physiological suppression of fertility in subordinate marmoset monkeys. *Amer. J. Primatol.*, 6, 169–86.

Abbott, D.H., McNeilly, A.S., Lunn, S.F., Hulme, M.J. and Burden, F.J. (1981). Inhibition of ovarian function in subordinate female marmoset monkeys (*Callithrix jacchus jacchus*). *J. Reprod. Fert.*, 63, 335–45.

Coimbra-Filho, A.F. (1971). Os sagüis do gênero *Callithrix* da região oriental brasileira e um caso de duplo hibridismo entre três de suas formas (Callithricidae, Primates). *Rev. Bras. Biol.*, 31, 377–88.

Coimbra-Filho, A.F. (1972). Aspectos inéditos do comportamento do gênero *Callithrix* (Callithricidae, Primates). *Rev. Bras. Biol.*, 32, 505–12.

Coimbra-Filho, A.F. and Mittermeier, R.A. (1976). Exudate-eating and tree-gouging in marmosets. *Nature*, 262, 630.

Dawson, G.A. (1978). Composition and stability of social groups of the tamarin, *Saguinus oedipus geoffroyi*, in Panama: ecological and behavioral implications. In *The biology and conservation of the Callitrichidae* (ed. D.G. Kleiman), pp. 23–37. Smithsonian Institution, Washington, D.C.

Dawson, G.A. (1979). The use of time and space by the Panamanian tamarin, *Saguinus oedipus*. *Folia primatol.*, 31, 253–84.

Epple, G. (1970). Quantitative studies on scent marking in the marmoset (*Callithrix jacchus*). *Folia primatol.*, 13, 48–62.

Epple, G. (1973). The role of pheromones in the social communication of marmoset monkeys (Callithricidae). *J. Reprod. Fert., Suppl.*, 19, 447–54.

Epple, G. (1975). The behavior of marmoset monkeys. *Primate Behav.*, 4, 195–239.

Epple, G. (1978). Lack of effects of castration on scent marking, displays, and aggression in a South American primate (*Saguinus fuscicollis*). *Horm. Behav.*, 11, 139–50.

Epple, G., Belcher, A.M. and Smith, A.B.,III (1986). Chemical signals in callitrichid monkeys—a comparative review. In *Chemical signals in vertebrates*, 4. Ecology, evolution, comparative biology (ed. D. Duvall, D. Müller-Schwarze and R.M. Silverstein), pp. 653–72. Plenum, New York.

Epple, G., Golob, N.F. and Smith, A.B., III (1979). Odor communication in the tamarin *Saguinus fuscicollis*. Behavioral and chemical studies. In *Chemical ecology: odour communication in animals* (ed. F.J. Ritter), pp. 117–30. Elsevier North Holland, Amsterdam.

Epple, G. and Katz, Y. (1984). Social influences on estrogen excretion and ovarian cyclicity in saddle back tamarins (*Saguinus fuscicollis*). *Amer. J. Primatol.*, 6, 215–27.

Epple, G., Kuderling, I. and Belcher, A. (1988). Some communicatory functions of scent marking in the cotton-top tamarin (*Saguinus oedipus oedipus*). *J. Chem. Ecol.*, 14, 503–15.

Epple, G., and Smith, A.B., III (1985). The primates. II: a case study of the saddleback tamarin, *Saguinus fuscicollis*. In: *Social odours in mammals* (ed. R.E. Brown and D.W. Macdonald), pp. 770–803.

French, J.A. and Cleveland, J. (1984). Scent-marking in the tamarin, *Saguinus oedipus*: sex differences and ontogeny. *Anim. Behav.*, 32, 615–23.

French, J.A. and Snowdon, C.T. (1981). Sexual dimorphism in responses to unfamiliar intruders in the tamarin, *Saguinus oedipus*. *Anim. Behav.*, 29, 822–9.

French, J.A., Abbott, D.H. and Snowdon, C.T. (1984). The effect of social enviroment on estrogen excretion, scent marking, and sociosexual behavior in tamarins (*Saguinus oedipus*). *Amer. J. Primatol.*, 6, 155–67.

Gosling, L.M. (1982). A reassessment of the function of scent marking in territories. *Z. Tierpsychol.*, 60, 21–36.

Kleiman, D.G. and Mack, D.S. (1980). Effects of age, sex, and reproductive status on scent marking frequencies in the golden lion tamarin, *Leontopithecus rosalia*. *Folia primatol.*, 33, 1–14.

Lacher, T.E., Jr., Fonseca, G.A.B. da, Alves, C., Jr. and Magalhães-Castro, B. (1981). Exudate-eating, scent marking, and territoriality in wild populations of marmosets. *Anim. Behav.*, 29, 306–7.

Lindsay, N.B.D. (1979). A report on the field study of Geoffroy's tamarin *Saguinus oedipus geoffroyi*. *Dodo, J. Jersey Wildl. Preserv. Trust*, 16, 27–51.

Neyman, P.F. (1978). Aspects of the ecology and social organization of free-ranging cotton-top tamarins (*Saguinus oedipus*) and the conservation status of the species. In *The biology and conservation of the Callitrichidae* (ed. D.G. Kleiman), pp. 39–71. Smithsonian Institution, Washington, D.C.

Omedes, A. (1979). Social communication of silvery marmosets (*Callithrix argentata*), and differences between two of the three subspecies (*C.a.argentata* and *C.a.melanura*). *Dodo, J. Jersey Wildl. Preserv. Trust*, 16, 45–51.

Omedes, A. and Carroll, J.B. (1980). A comparative study of pair behaviour of four callitrichid species and the Goeldi's monkey. *Dodo, J. Jersey Wildl. Preserv. Trust*, 17, 51–62.

Rylands, A.B. (1981). Preliminary field observations on the marmoset, *Callithrix humeralifer intermedius* (Hershkovitz, 1977), at Dardanelos, Rio Aripuanã, Mato Grosso. *Primates*, 22, 46–59.

Rylands, A.B. (1982). *The behaviour and ecology of three species of marmosets and tamarins (Callitrichidae, Primates) in Brazil.* Unpublished Ph.D. thesis, University of Cambridge.

Rylands, A.B. (1985). Tree-gouging and scent-marking by marmosets. *Anim. Behav.*, 33, 1365–7.

Rylands, A.B. (1986). Infant-carrying in a wild marmoset group, *Callithrix humeralifer* evidence for a polyandrous mating system. In *A Primatologia Brasil*, II (ed. M.T. de Mello), pp. 131–44. Sociedade Brasileira de Primatologia, Brasília.

Stevenson, M.F. and Poole, T.B. (1976). An ethogram of the common marmoset

(*Callithrix jacchus jacchus*); general behavioural repertoire. *Anim. Behav.*, 24, 428–51.

Sutcliffe, A.G. and Poole, T.B. (1978). Scent marking and associated behaviour in captive Common marmosets (*Callithrix jacchus jacchus*) with a description of the histology of scent glands. *J. Zool., Lond.*, 185, 41–56.

# IV.13 Olfactory Recognition Cues and the Development of Intrasexual Kin Discrimination in Male Laboratory Mice

PETER ALDHOUS

## Introduction

It was shown by Kareem and Barnard (1982) that male laboratory mice discriminate between their unfamiliar paternal half-brothers (coefficient of relatedness $r = 0.25$) and unfamiliar, unrelated males. This is evidenced by decreased investigatory behaviour and increased passive body contact towards their relatives. The recognition that underlies such discrimination may depend on the perception and matching of olfactory cues to recognition template(s). Unfamiliar relatives are recognized by a process of stimulus generalization, since their cues overlap with those incorporated into the existing template(s). This implies that recognition cues are strongly genetically determined. In mice, genes within the highly polymorphic Major Histocompatibility Complex (MHC) are known to be particularly important in the determination of individual mouse odour (Boyse *et al.* 1987), but other loci are also known to be involved (e.g. Yamazaki *et al.* 1986).

This paper addresses several questions concerning the role of olfactory cues in kin discrimination by mice.

### 1. WHAT CUES ARE LEARNED?

Although learning of parental cues seems to have an influence on the development of intersexual kin discrimination in male mice (e.g. Kareem and Barnard 1986; Yamazaki *et al.* 1988), discrimination of unfamiliar same-sex kin seems to depend largely on learning the cues of self and/or familiar littermates. Paternal half-siblings discriminate one another despite having never encountered their common father (Kareem and Barnard 1982), and maternal influences on discrimination between males do not seem to be significant (Kareem and Barnard 1986). Is the learning of cues of self rather than those of familiar littermates more important in the development of discrimination?

2. ARE RECOGNITION CUES TRANSFERRED BETWEEN INDIVIDUALS?

Transfer of recognition cues between cohabiting mice might occur, either directly (see Taylor *et al.* 1988), or via a common housing substrate such as bedding: mouse urine is regularly deposited into the environment (e.g. Hurst 1987) and is known to be a potent source of MHC determined recognition cues (Yamaguchi *et al.* 1981). So how does recognition cue transfer affect discrimination?

In addressing these questions at the behavioural level, one can only assess discrimination and not the underlying neural processes of recognition (Waldman *et al.* 1988). Of course, it is possible that animals may recognize other individuals as kin, but not translate this into discrimination.

## Methods

The questions above were addressed using a cross-fostering design, as shown in Fig. 1. All litters were reduced to four male pups at birth. Within two days of birth, single pups were exchanged between unrelated subject litters. Would these animals discriminate (a) their own relatives (which would indicate that learning the cues of self is important except when in-utero and immediately postpartum learning of siblings' cues is involved); (b) the relatives of their foster-littermates (which would indicate that learning of littermates' cues is important; (c) both; or (d) neither?

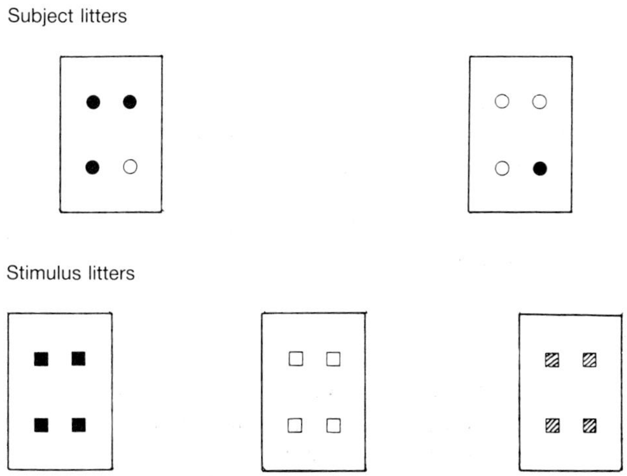


Fig. 1. Experimental design. Each symbol represents a male mouse.
Same shading, same shape: full siblings ($r = 0.5$); same shading, different shape: paternal half-siblings ($r = 0.25$); different shading: unrelated ($r < 0.01$).

Recognition cue transfer within subject litters was examined by recording the behaviour of stimulus animals with which subjects were tested. Transfer of recognition cues between unrelated foster-littermates in subject litters could result in stimulus animals failing to discriminate their own relatives. Conversely, cue transfer could result in stimulus animals treating unrelated animals as kin.

Experimental animals were outbred mice of the CFLP strain, and were separated from their mothers at 19–21 days of age. Mice were observed under dim red light in a neutral arena, which was cleaned and provided with fresh sawdust between tests. Individual subject mice were tested three times, both as juveniles (22–27 days) and as adults (68–76 days), against a member of each class of stimulus animal. For each animal, the duration of elements of social investigation (sniff, nose, investigate, follow, groom; see Mackintosh 1981), the combined total duration of these elements (social investigation), and the duration of passive body contact (touch) were recorded. Discrimination was assessed by comparing results of tests involving stimulus animals unrelated to subjects, or their foster-littermates, to results of tests involving each of the other two classes of stimulus animal. The significance of differences across the three test types was indicated by non-parametric analysis of variance. For other details of methods, see Aldhous (1989).

## Results

The results discussed below refer only to behaviours where there were significant differences across the three test types. Other analyses were not significant.

### SUBJECT ANIMALS

Juvenile cross-fostered subjects investigated their foster-littermates' relatives less (Fig. 2a) and touched them more (Fig. 2b) than unrelated animals. They also touched (Fig. 2b) and followed (Fig. 2c) their own relatives more than unrelated animals.

For adult mice, there was no evidence for discrimination among the three classes of stimulus animal by cross-fostered subjects.

### STIMULUS ANIMALS

Juvenile cross-fostered subjects were touched more by both their own relatives and the relatives of their foster-littermates, than by unrelated animals (Fig. 3a). Adult cross-fostered subjects were also discriminated by both their own and their foster-littermates' relatives, being investigated less by these two classes of stimulus animal than by unrelated animals (Fig. 3b).

The behaviour of stimulus animals towards the foster-littermates of the cross-fostered subjects was also investigated. When juvenile, these subject animals

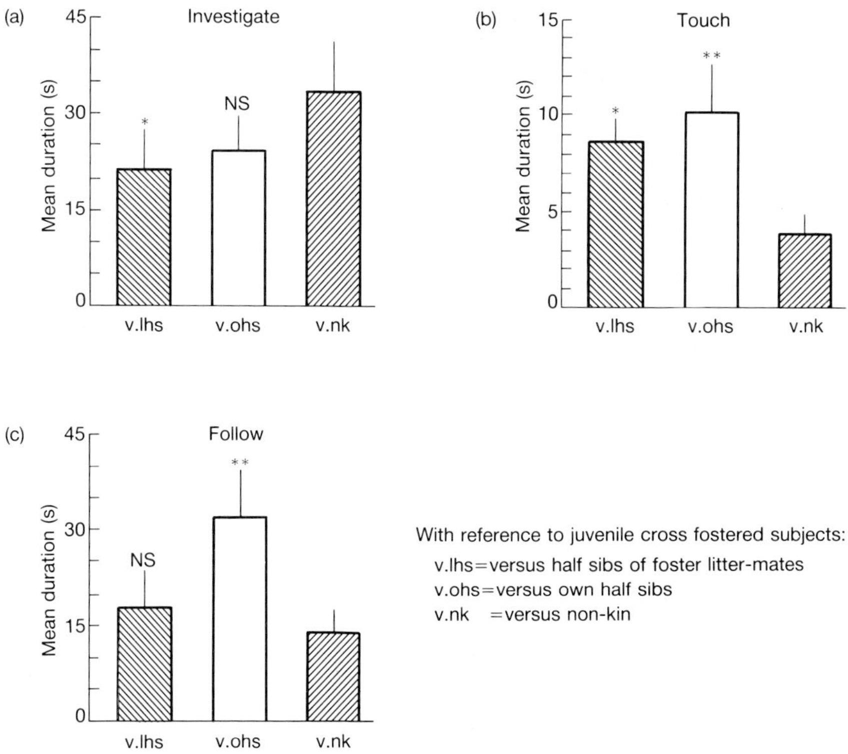


Fig. 2. Behaviour of juvenile cross-fostered subjects ($n = 12$). v.lhs, tested against the paternal half-siblings of their foster-littermates; v.ohs, tested against their own paternal half-siblings; v.nk, tested against animals unrelated to self, or their foster-littermates. Symbols refer to significance levels in Wilcoxon tests (NS, $P > 0.05$; *, $P \leqslant 0.05$; **, $P \leqslant 0.01$). The bars indicate SEM.

were investigated less by their own relatives than by unrelated animals (Fig. 3c), but were not misidentified as kin by the relatives of their single foster-littermates. When adult, however, the foster-littermates of cross-fostered subjects were discriminated by both their own relatives, and the relatives of their single foster-littermates, as evidenced by social investigation (Fig. 3d).

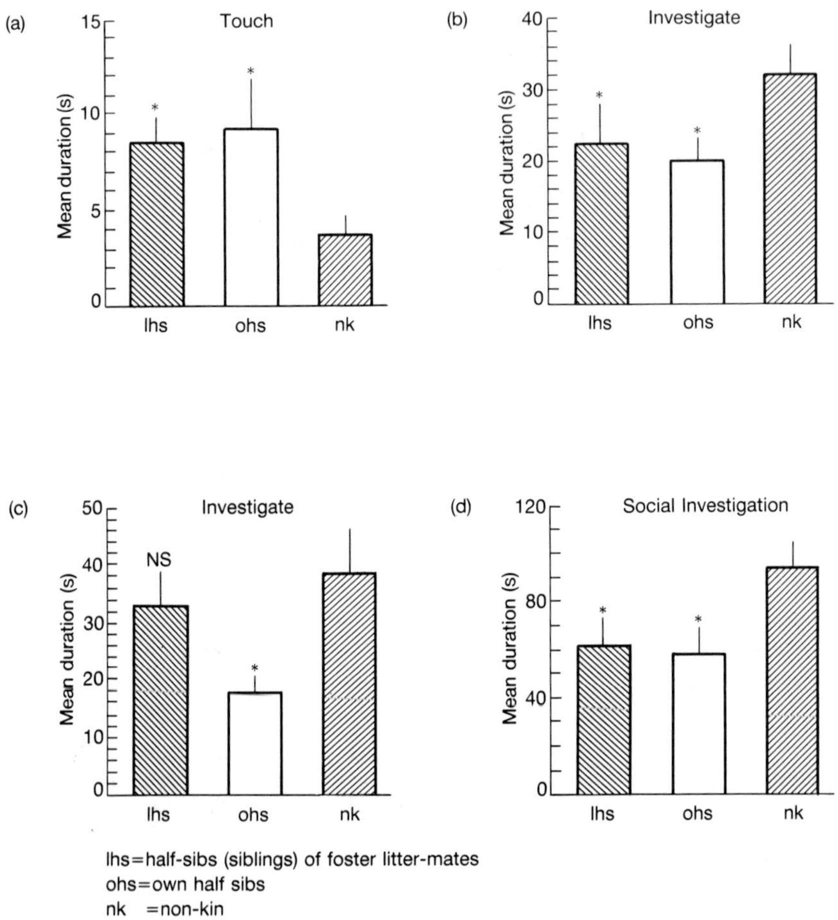


Fig. 3. Behaviour of stimulus animals. lhs, paternal half-siblings of their test opponents' foster-littermates; ohs, paternal half-siblings of their test opponents; nk, animals unrelated to test opponents, or their foster-littermates. The symbols are as for Fig. 2, but significance levels were obtained in Conover's comparisons tests.

(a) Tested against cross-fostered subjects when juvenile ($n = 12$ for each condition).
(b) Tested against cross-fostered subjects when adult ($n = 12$ for each condition).
(c) Tested against the foster-littermates of cross-fostered subjects when juvenile ($n = 9$ for each condition).
(d) Tested against the foster-littermates of cross-fostered subjects when adult ($n = 10$ for each condition).

# Discussion

## WHAT CUES ARE LEARNED?

The fact that juvenile cross-fostered subjects discriminated both their own relatives, and the relatives of their foster-littermates, suggests that both the cues of self and those of littermates (unless learning of foster-mother cues was important) were learned, and influenced discrimination. However, the result complicates this picture. Although this result does suggest discrimination on the basis of learning the cues of self, the difference was in the opposite direction from that expected for an investigatory behaviour element. This may be a consequence of being reared in the absence of relatives. Some degree of perceived similarity between the cues of self and those of littermates may be important for the development of 'normal' discrimination. N.B. When the foster-littermates of cross-fostered subjects were tested against the same three classes of juvenile stimulus animal, differences were found for investigatory behaviour consistent with previous work (see Kareem and Barnard 1982): these animals investigated both their own relatives, and the relatives of their single foster-littermates less than unrelated animals (results not shown).

Cohabitation with unrelated cagemates until adulthood seems to disrupt either the ability to recognize, or the motivation to discriminate among unfamiliar animals. The lack of discrimination shown by adult cross-fostered subjects was mirrored when their foster-littermates were tested against adult stimulus animals (results not shown).

## ARE RECOGNITION CUES TRANSFERRED BETWEEN INDIVIDUALS?

The behaviour of stimulus animals did indicate that significant recognition cue transfer occurred between unrelated foster-littermates (and possibly also from foster-mothers) in subject litters. Both juvenile and adult cross-fostered subject animals seemed, in consequence, to be misidentified as kin by the relatives of their foster-littermates.

The results obtained for stimulus animals in interaction with the foster-littermates of cross-fostered subjects (who had only been cohabiting with a single unrelated foster-littermate) indicate that recognition cue transfer between foster-littermates may have had more profound effects on discrimination in adults than juveniles. Such animals were only misidentified as kin by the relatives of their single foster-littermates as adults. This may simply reflect the longer period of cohabitation prior to adult tests, but qualitative differences between adult and juvenile odours (e.g. Taylor 1982) and increased intensity of cue transfer may also be involved.

It is possible that recognition cue transfer between unrelated animals in subject litters is a factor in the disruption of discrimination by adult subjects.

### PROBLEMS, AND PLANS FOR FUTURE WORK

The interpretation of the results presented above considers the behaviour of subject and stimulus animals separately, and assumes that discrimination, or lack of it, by one member of a test pair did not strongly affect the behaviour of the other test mouse. However, it is possible that interactions between the behaviour of subject and stimulus mice may have confounded the results. This would be a particular problem in the case of the measure of passive body contact (touch), often recorded for both members of a test pair when it could not be determined which mouse had initiated the contact. This problem could be resolved by using mice rendered temporarily anosmic by nasal lavage with $ZnSO_4$ solution (Alberts 1974) as standard test opponents (anosmic mice should not be capable of kin discrimination). I am presently applying this technique to investigate further the effects of recognition cue transfer between unrelated cohabiting mice.

## Acknowledgements

Thanks are due to Chris Barnard and Jane Hurst for commenting on the manuscript, and to the S.E.R.C. for financial support.

## References

Alberts, J.R. (1974). Producing and interpreting experimental olfactory deficits. *Physiol. Behav.*, 12, 657–70.

Aldhous, P. (1989). The effects of individual cross-fostering on the development of intrasexual kin discrimination in male laboratory mice, *Mus musculus* L. *Anim. Behav.*, 37, 741–50.

Boyse, E.A., Beauchamp, G.K. and Yamazaki, K. (1987). The genetics of body scent. *Trends Genet.*, 3, 97–102.

Hurst, J.L. (1987). The functions of urine marking in a free-living population of house mice, *Mus domesticus* Rutty. *Anim. Behav.*, 35, 1433–42.

Kareem, A.M. and Barnard, C.J. (1982). The importance of kinship and familiarity in social interactions between mice. *Anim. Behav.*, 30, 594–601.

Kareem, A.M. and Barnard, C.J. (1986). Kin recognition in mice: age, sex and maternal effects. *Anim. Behav.*, 34, 1814–24.

Mackintosh, J.H. (1981). Behaviour of the house mouse. *Symp. Zool. Soc. Lond.*, 47, 337–65.

Taylor, G.T. (1982). Urinary odors and size protect juvenile laboratory mice from adult male attack. *Dev. Psychobiol.*, 15, 171–86.

Taylor, G.T., Griffin, M. and Rupich, R. (1988). Conspecific urine marking in male rats (*Rattus norvegicus*) selected for relative aggressiveness. *J. comp. Psychol.*, 102, 72–7.

Waldman, B., Frumhoff, P.C. and Sherman, P.W. (1988). Problems of kin recognition. *Trends Ecol. Evol.*, 3, 8–13.

Yamaguchi, M., Yamazaki, K., Beauchamp, G.K., Bard, J., Thomas, L. and Boyse, E.A. (1981). Distinctive urinary odors governed by the MHC locus of the mouse. *Proc. Natl. Acad. Sci. USA.*, 78, 5817–20.

Yamazaki, K., Beauchamp, G.K., Matsuzaki, O., Bard, J., Thomas, L. and Boyse, E.A. (1986). Participation of the murine X and Y chromosomes in genetically determined chemosensory identity. *Proc. Natl. Acad. Sci. USA.*, 83, 4438–40.

Yamazaki, K., Beauchamp, G.K., Kupniewski, D., Bard, J., Thomas, L. and Boyse, E.A. (1988). Familial imprinting determines H-2 selective mating preferences. *Science*, 240, 1331–2.

# IV.14 Conspecific Odours: a Special Case for the Rat

PETER G. HEPPER

Rat pups undergo a rapid change in their preference for familiar and unfamiliar conspecific odours between 12 and 20 days of age. Given a choice between siblings or unfamiliar age-mate conspecifics, 12-day-old pups prefer their siblings whilst 20-day-old pups prefer the unfamiliar conspecifics (Hepper 1983*a*). Similarly, given a choice between their own bedding and bedding from another litter, 12-day-old pups prefer their own bedding whereas 20-day-old pups prefer the unfamiliar bedding (Carr *et al.* 1979). This preference for unfamiliar conspecifics persists into adulthood (Hepper 1983*b*). The rapid reversal in preference is due to some change in the developing pup and not to a change in the stimulus produced by the pup (Hepper 1986*a*). The change-over in preference is elicited by the odour from urine, faeces, and also by their fathers (Hepper 1986*b*). Furthermore, the change-over occurs irrespective of whether pups are tested repeatedly (days 0–22) or just once (day 12 or day 20).

In marked contrast to the rapid reversal in preference for familiar conspecific odours, acquired preference for artificial odours, i.e. not from conspecifics, such as peppermint, does not revert (e.g. Galef 1982; Galef and Kaner 1980). Furthermore this preference is long lasting; e.g. Echandia *et al.* (1982) found that after exposure to lemon from days 1–21, rats exhibited a preference for lemon at 141 days of age.

Why do pups show a reversal in their preference for familiarity to conspecific odours but not to non-conspecific odours? Change-over in preference may be due to some innate predisposition of the rat. Alternatively it may be due to a methodological difference between studies using conspecific and non-conspecific odours. First, the valence of the odours used may affect the reversal of preference. Pups aged 12 days are attracted to conspecific odour (Hepper 1983*a*, *b*). However, non-conspecific odours may be either attractive, aversive or neutral and this may determine whether pups exhibit a reversal in their preference. Second, how the pup experiences the artificial odour may affect the reversal of preference. For example, whilst a preference can be induced in a 21-day-old pup simply by introducing an artificial odour into the pup's environment, the odour needs to be painted on to the pups or mother for this preference to be subsequently maintained past this date (Galef 1982). The following experiment examined the role of odour valence and method of exposure upon the preference of pups for familiar and unfamiliar non-conspecific odours.

The response of naïve 12-day-old pups to non-conspecific odours was examined to determine the valence of the odours. Pups were tested singly for 3 mins in a circular arena, which was divided into two concentric halves. The odours were blown up through a tube in the centre of the arena and removed from the maze by suction through a central tube above the arena. It was hypothesized that pups would spend longer in the inner circle if the odours were attractive, longer in the outer circle if the odours were aversive, and show no difference in the time spent in either if the odours were neutral. Pups showed a strong preference for the odour of their siblings, and stayed significantly longer in the inner half of the arena. Of 53 non-conspecific odours—exposed at a concentration judged by human observers to be subjectively equal to that of the siblings (including food, botanical, and perfume odours) none was observed to be either attractive or aversive to the pups. It may be that valence of an odour is determined by its concentration rather than its quality. Using higher concentrations of peppermint and lemon, pups spent significantly longer in the outer circle, and the odours were classed as aversive. At lower concentrations, however, no increase in attraction to either peppermint or lemon was noted. It was decided to use both peppermint and lemon as the non-conspecific odour stimuli for exposures at either neutral or aversive concentrations.

Three types of exposure were used:

1. *Air exposure*—a cotton wool pad was placed in a wire mesh container on the top of the cage and the odour solution was placed in this twice a day.
2. *Bedding exposure*—the shavings and sawdust used as bedding material were liberally sprinkled with the odour solution and placed in the pup's cage each day.
3. *Conspecific exposure*—the odour was placed on a swab and then rubbed daily on the dorsal surface of each pup in the litter and the ventral surface of the mother.

Exposure to the odours began on day 1. Half the pups in each condition were exposed to lemon or peppermint at neutral concentration and the other half were exposed to lemon or peppermint at aversive concentration.

Pups that were not exposed to non-conspecific odours were tested for their response to conspecific odour, their littermates, or unfamiliar conspecifics of the same age. All other pups received a choice between lemon and peppermint at the same concentration to which they had been exposed, i.e. either as a neutral or aversive stimulus. Pups were tested in a rectangular arena with mesh floor, and the odours were blown into the maze at each end and withdrawn by suction under the mesh floor. Pups were tested for 2 mins at day 12 and again at day 20; the time spent on either side of the maze was noted.

The preference of pups for familiar and unfamiliar conspecific odours was analysed by ANOVA for factors of age (days 12 and 20) and preference (familiar and unfamiliar). The results indicated no significant effect of age or of preference;

there was however a significant interaction effect, pups preferring the familiar odour on day 12 but the unfamiliar odour on day 20. Table 1A indicates the mean time spent with familiar and unfamiliar odours on each day. The Fisher exact probability test revealed a significant change-over in the pup's preference from day 12 to day 20. These results support the previous work indicating a reversal in the pup's preference for familiar and unfamiliar conspecific odours.

**Table 1.** Time spent investigating familiar and unfamiliar odours

| | | Day 12 | | Day 20 | | Reversal in |
|---|---|---|---|---|---|---|
| | | Familiar | Unfamiliar | Familiar | Unfamiliar | preference |
| A | | | | | | |
| Conspecifics | | 67 (4.4) | 29 (3.9)** | 27 (3.2) | 73 (3.9)** | $p<0.0001$ |
| B | | | | | | |
| Air | *n* | 81 (5.3) | 31 (4.7)** | 71 (4.3) | 36 (4.2)** | n.s. |
| Exposure | *av* | 72 (6.1) | 33 (4.9)** | 64 (5.0) | 42 (5.2)* | n.s. |
| Bedding | *n* | 71 (5.9) | 40 (6.0)* | 77 (4.8) | 30 (3.4)** | n.s. |
| Exposure | *av* | 69 (4.4) | 37 (4.2)** | 89 (3.3) | 20 (2.5)** | n.s. |
| Conspecific | *n* | 75 (5.4) | 31 (5.0)** | 76 (4.4) | 35 (4.4)** | n.s. |
| Exposure | *av* | 75 (4.6) | 30 (4.6)** | 72 (3.7) | 36 (3.8)** | n.s. |

* $0.05<p<0.01$; ** $p<0.01$; n.s., not significant.
$n$ = neutral colour; $av$ = aversive odour.

Mean time in seconds (± S.E.) spent with either the familiar or unfamiliar odour on day 12 and day 20 for pups exposed to (A) conspecific odour and (B) artificial odours for rats exposed by air, bedding, or conspecific exposure (for both neutral and aversive odours). The significance of any changeover in preference was determined by Fisher exact probability test. The difference in time spent with familiar and unfamiliar odours on each day for each condition was assessed by matched pairs $t$-test.

The preference of pups for familiar and unfamiliar non-conspecific odours was analysed by ANOVA for between subject factors of odour valence (neutral, aversive), exposure type (air, bedding, conspecific) and within subject factors of age (days 12, 20) and preference (familiar, unfamiliar). Only one factor reached significance, that of preference. Pups spent significantly longer investigating the familiar odour than the unfamiliar odour, and this preference was unaffected by exposure type, age, or odour valence. Table 1B gives the mean time spent with familiar and unfamiliar odours for each condition. The Fisher exact probability test indicated no reversal of preference in any condition. The results indicate that neither odour valence nor method of exposure had any influence upon the pups' preference, as pups continued to prefer the familiar non-conspecific odour.

Before concluding that the changeover in preference is only elicited by conspecific odours two further factors should be considered. First, pups may commence learning about odours *in utero* (Hepper 1987*a*). Furthermore, although the non-conspecific odours were clearly perceptible to human observers, they may not have been as salient as the odours of conspecifics. Second, the non-conspecific odours may not have been as 'complex' as the conspecific odours. To examine these possibilities two further experiments were performed. In order to ensure pups were fully exposed to the odour (lemon or peppermint), litters experienced odours by all three methods of exposure from the day of birth, and were scented three times per day. Additionally the odorant was injected into the amniotic fluid on day 20 of gestation. Pups were then tested for their preference for lemon or peppermint as above. Second, to increase the complexity of the non-conspecfic odour, pups were exposed to one of two 'cocktails' consisting of five different odours (either A: orange, lemon, blackcurrant pineapple, vanilla; or B: grapefruit, strawberry, raspberry, banana, apple). The methods of exposure were the same as those used for conspecific exposure. Pups were tested for their preference between odour cocktail A or B; in both cases the pups exhibited a preference for the familiar odour at 12 and 20 days of age and showed no reversal in preference.

It would thus appear that the developing rat pup responds differently to familiar conspecific and non-conspecific odours and this is not due to differences in methodology. The fact that rat pups do respond differently raises the following questions.

## What is the difference between conspecific and non-conspecific odours?

The simple answer is that one is produced by fellow rats and the other is not. There is, however, an exception to this rule. Pups given a choice between their own mother and an unfamiliar lactating female show a preference for their own mother at 12 and 20 days of age and show no reversal of this preference (Hepper 1986*b*).

One factor that may explain the difference between conspecific and non-conspecific odours is the role played by genetic and environmental factors in mediating the production of each type of odour. Recent evidence has indicated that rats possess a genetically determined odour which forms the basis of recognition between conspecifics (Hepper 1987*b*). Whilst the mother's odour would also be influenced by genetic factors her odour is set apart from other conspecifics by production of maternal pheromone, an odour highly attractive to pups from days 12–27 (Leon and Moltz 1971). This pheromone is influenced by the mother's diet (Leon 1975), and the attraction of pups to the pheromone is based upon their prior experience with it (Leon 1975).

Thus, odours with a mainly genetic component may elicit the reversal in preference whilst odours with an environmental component do not. Such a difference may have implications for the learning and processing of such cues.

## Why is the preference for familiar and unfamiliar conspecific odours reversed?

An alternative way of posing this question is to ask why the preference for familiar non-conspecific odours is maintained. This may be explained by the different functions served by the two odour types in the rat's life. Non-conspecific odours are important in determining dietary preference. An individual must learn about its food and maintains a preference for that which has previously been encountered and is safe. Conspecific odours are important in social interactions where it will be important to investigate unfamiliar individuals fully to avoid responding in a way which reduces fitness (Hepper 1986*c*), especially as it is likely that familiar individuals will be genetically related and unfamiliar conspecifics will be genetically unrelated (Hepper 1986*c*). Similarly, inbreeding may have deleterious consequences (Hepper 1986*c*) and therefore a preference for unfamiliar (unrelated) individuals would be beneficial. Thus the differential response exhibited by rats to these two odours types may reflect different evolutionary pressures.

## Where and how the stimuli are processed?

Since rat pups respond differently to conspecific and non-conspecific stimuli, at some point the two must be processed differently. This may occur centrally in some process influencing behavioural preference or at the level of the olfactory receptor, olfactory bulb, or olfactory cortex. Alternatively since chemosensory stimulation is possible by a number of systems (main olfactory bulb, vomeronasal organ and accessory olfactory bulb, trigeminal nerve, septal organ of Masera, and nervus terminalis), these systems could be differentially activated by conspecific and non-conspecific odours. One pharmacological change that corresponds to the time of reversal in preference is an increase in acetylcholinesterase (AChE) activity in the olfactory bulb (Meisami and Firoozi 1985). The main source of AChE in the olfactory bulb appears to be from the centrifugal fibres (Meisami and Firoozi 1985) which may influence the olfactory bulb's responsiveness based on the previous experiences of the organism (Coopersmith and Leon 1988). Thus both the cholinergic system and centrifugal fibres may be influential in the reversal of preference.

Exposure to conspecific and non-conspecific odours may differentially affect the olfactory system and the consequent processing of the odours. Experience with non-conspecific odours can affect the olfactory system, e.g. the response pattern of mitral/tufted cells is altered by previous exposure to odours (Wilson *et al.* 1987), which with associated changes may influence the maintenance of preference. It remains to be seen whether conspecific odours will exert a similar effect, although such an attempt would involve considerable experimental difficulties.

## Implications for the study of olfactory function?

If the olfactory system does handle the two stimuli differently, consideration of the effects of both stimuli on the olfactory system may be important in order to obtain a true understanding of its functioning. Most studies have used non-conspecific odours to examine olfactory functioning, it may be that by restricting observations to just these odours only part of the story of how the olfactory system works will be revealed. Furthermore, and perhaps most importantly, the use of artificial odours in examining the natural behaviour of the rat must be questioned, especially with respect to social interactions.

## Acknowledgements

I thank the Association for the Study of Animal Behaviour for providing support for this research, Prof. Ken Brown for providing facilities, and Fiona Hepper for her comments on the manuscript.

## References

Carr, W.J., Marasco, E. and Landauer, M.R. (1979). Responses by rat pups to their own nest odor versus a strange conspecific nest. *Physiol. Behav.*, 23, 1149–51.

Coopersmith, R. and Leon, M. (1988). The neurobiology of early olfactory learning. In *Handbook of behavioral Neurobiology* (ed. E.M. Blass), Vol. 9, pp. 283–308. Plenum, New York.

Echandia, E.L.R., Foscolo, M. and Broitman, S.T. (1982). Preferential nesting in lemon-scented environment in rats reared on lemon-scented bedding from birth to weaning. *Physiol. Behav.*, 29, 47–9.

Galef, B.G. and Kaner, H.C. (1980). Establishment and maintenance of preference for natural and artificial olfactory stimuli in juvenile rats. *J. Comp. Physiol. Psychol.*, 94, 588–95.

Galef, B. (1982). Acquisition and waning of exposure induced attraction to a non-natural odor in rat pups. *Dev. Psychobiol.*, 15, 479–90.

Hepper, P.G. (1983*a*). Sibling recognition in the rat. *Anim. Behav.*, 31, 1177–91.

Hepper, P.G. (1983*b*). Kin recognition in the rat. Ph.D. thesis, University of Durham.

Hepper, P.G. (1986*a*). Sibling recognition in the rat: change in stimulus value or individual preference? *Anim. Behav.*, 34, 607–9.

Hepper, P.G. (1986*b*). Parental recognition in the rat. *Quart. J. Exp. Psychol.*, 38B, 151–60.

Hepper, P.G. (1986*c*). Kin recognition: functions and mechanisms. A review. *Biol. Rev.*, 61, 63–93.

Hepper, P.G. (1987*a*). The amniotic fluid. An important priming role in kin recognition. *Anim. Behav.*, 35, 1343–6.

Hepper, P.G. (1987*b*). The discrimination of different degrees of relatedness in the rat: Evidence for a genetic identifier. *Anim. Behav.*, 35, 549–54.

Leon, M. and Moltz, H. (1971). Maternal pheromone: discrimination by pre-weanling albino rats. *Physiol. Behav.*, 7, 265–7.

Leon, M. (1975). Dietary control of maternal pheromone in the lactating rat. *Physiol. Behav.*, 14, 311–9.

Meisami, E. and Firoozi, M. (1985). Acetylcholinesterase activity in the developing olfactory bulb: a biochemical study on normal maturation and the influence of peripheral and central connections. *Dev. Brain. Res.*, 21, 115–24.

Wilson, D.A., Sullivan, R.M. and Leon, M. (1987). Single-unit analysis of postnatal olfactory learning: modified bulb output response patterns to learned attractive odors. *J. Neurosci.*, 7, 3154–62.

# IV.15 Behavioural Cues to Semiochemically Important Body Regions of *Rattus norvegicus*

STEPHAN E. NATYNCZUK

## Introduction

Odour sources of possible and actual semiochemical importance in *Rattus norvegicus* include the Harderian, Meibomian, pedal, preputial, and anal glands; the urine and the faeces (Albone 1984; Brown 1985). How these odour sources feature in rat social behaviour has not been described in detail although numerous tests have been conducted to determine both male and female preferences for various odours (Brown 1985). Tests of preference and discrimination can give clues to the biological role of odours, but in the absence of natural behavioural observations the function of these odours can only be guessed. This study therefore examines the semiochemical sources sampled by rats during social behaviour.

## Methods

The rats used in this study were F1 crosses of wild trapped male *R. norvegicus* and female Wistar rats. These rats were relatively easy to handle and observe compared to the extremely timid wild parent, but were not as docile as the Wistar, and less constrained in their behaviour by persistent inbreeding (Barnett 1975).

At three to four weeks of age, rats from six litters (23 females and 21 males) were separated into single sex cages (30 × 44 × 23 cm hanging PVC with wire grid tops) but kept in their sibling groups with a normal diurnal lighting regime of 12 hours light, 8 hours darkness, and food and water available *ad libitum*. There were two rats per cage except when an odd-numbered litter size dictated that three rats be housed together. The minimum period of separation before testing was 8 weeks; the maximum 24 weeks. Using tables of random numbers, one male and one female rat were selected to provide interactions between related (R) and non-related (NR) rats, though not if that pair had been previously used.

One pair of rats was observed for 10 mins at a time in a purpose-built rattery (4.6 × 5.6 × 2.2 m) having a base of paving stones and divided, by wire mesh, into three rat-proof enclosures. The floor of each enclosure was covered with a 3:1 mixture of peat moss and sand. Abundant quantities of food and water were always available, while concrete blocks and logs provided cover for the rats.

1 = the nose
2 = the side of the face, just below the eye
3 = the neck
4 = the shoulder
5 = the flank
6 = the haunch
7 = the genital region
8 = the tail

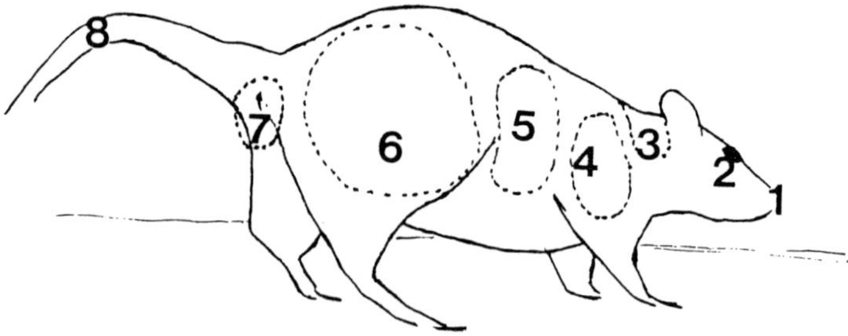


Fig. 1. Male investigation of females by proportional distribution of sniffs.

The rat's body area was divided into a number of zones (to approximate the location of potential semiochemical sources) at which sniffs could be directed (Fig. 1). When a rat placed its nose in contact with the body of another rat a sniff was recorded using an Epson HX20 microcomputer as a data logger.

After each observation session a vaginal smear was taken to determine the female's phase of oestrus (Zarrow *et al.* 1964). Scores (1 to 5) for the abundance of leucocytes (L), and nucleated (N) and cornified (C) cells were used to determine an index of oestrus (Os) calculated using the formula $Os = C/(C+N+L)$. Oestrus is indicated by the predominance of cornified cells and absence of leucocytes and nucleated cells.

For this paper I consider the sniffing behaviour of males and females for those groups of rats in which the females were oestrous (O; $Os > 0.7$) or non-oestrous (NO; $Os < 0.3$). This classification yielded four categories of male investigation of female rat: related oestrous (RO; $n = 10$); related non-oestrous (RNO; $n = 21$); non-related oestrous (NRO; $n = 9$); and non-related non-oestrous (NRNO; $n = 22$).

The mean area of each zone was calculated by tracing around the respective body area of the female ($n = 8$) rats, then cutting out and weighing the tracings. Data were transferred to a VAX minicomputer for editing and analysis using SAS (SAS Inst.) unless otherwise stated. Results for sniffs at each body zone were summed and treated as a proportion of total sniffs at all zones per rat, then transformed to their arcsine square-roots for statistical analysis (Anova) after correcting for the area of each respective zone.

## Results

Using a mixed model, three-level nested Anova (Sokal and Rohlf 1981 p. 302) no significantly different sniffing behaviour occurred dependent either on relationship ($F=1.08<F=800_{0.5[2,1]}$ two-tailed), or upon whether the females were oestrous or non-oestrous ($F=4.61<F=39.5_{0.05[28,2]}$ two-tailed). However, differences in sniffing behaviour by males at the various female body zones was found to be highly significant ($F=38.21>F=2.64_{0.005[32,28]}$ two-tailed). Variance in the sniffing behaviour of males towards females is not significant for making general remarks about the male rats' ability to distinguish either related from non-related females or oestrous from non-oestrous females. The sniffing behaviour of males within each experimental group depended upon the class of female involved. Fig. 2 shows the proportional distribution of the number of sniffs at each zone for each class of interaction.

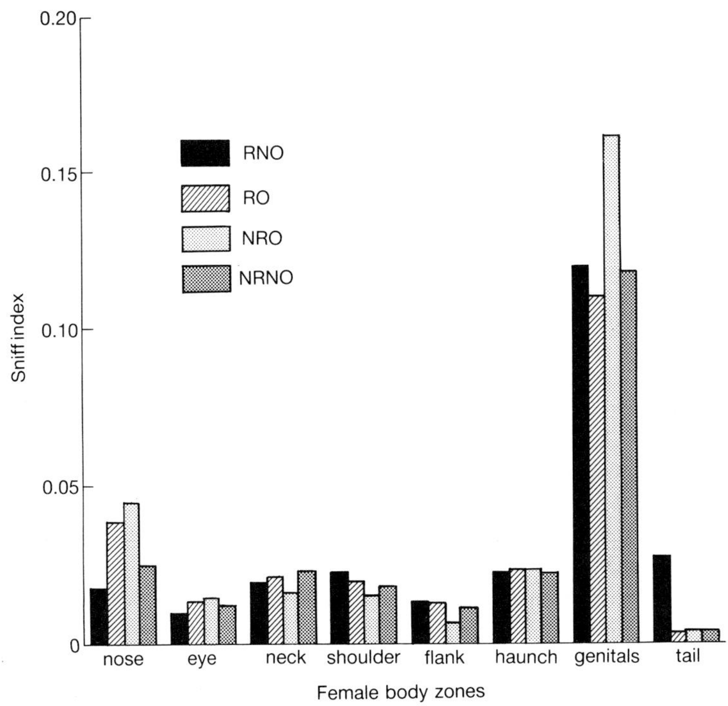


Fig. 2. Sniff index = sniffs per zone/total sniffs corrected for the area of each zone.

The number of sniffs directed at the genitals over the 10 min. observation period were compared using the Pearson Chi-squared statistic on a pairwise basis (see Table 1). From the results it is clear that the genitals of females are generally sniffed proportionally more often by males in the course of social behaviour. The genital region of non-related oestrous females was investigated significantly more frequently than the genital region of any other group.

**Table 1.** Comparison of the number of sniffs by male rats at the genitals of female rats over 10 min

| Category† | | Number of sniffs to genitals (a) | Total sniffs (b) | Area in $cm^2$ (c) | Sniff index a/(bc) |
|---|---|---|---|---|---|
| RO | 10 | 49 | 252 | 1.77 | 0.110 |
| RNO | 21 | 120 | 568 | 1.77 | 0.119 |
| NRO | 9 | 62 | 217 | 1.77 | 0.161 |
| NRNO | 22 | 83 | 398 | 1.77 | 0.118 |

Chi-squared values:

| | RNO | NRO | NRNO |
|---|---|---|---|
| RO | 0.274 | 5.268* | 0.169 |
| RNO | — | 4.806* | 0.010 |
| NRO | | — | 4.642* |

† RO = related oestrus; RNO = related non-oestrus; NRO = non-related oestrus; NRNO = non-related non-oestrus.
* $p < 0.05$, df = 1.

## Discussion

The genital region of the female rat includes several sources of odour including the anal and clitoral glands, the vagina, and the urine (Brown 1985). The urine and the vagina both contain complex mixtures of chemicals from a plethora of sources which may reveal the phase of oestrus (see Albone 1984 for review) and relatedness (see Brown *et al.* and Beauchamp *et al.* in this volume). No social functions have yet been proposed for the odours of rat anal glands.

In preference tests, males prefer the odours of clitoral glands from intact females over those of ovariectomized females (Gawienowski *et al.* 1975; Gawienowski 1977; Thody and Dijkstra 1978). Males also prefer the clitoral gland odours of oestrous and pro-oestrous females to those of dioestrous females (Gawienowski 1977; Lucas *et al.* 1982; Thody and Dijkstra 1978). Males can thus potentially use the odours from these glands to discriminate between females in various phases of the oestrous cycle. However, since whole body odours of female rats are more attractive to male rats than clitoral odours alone (see Brown 1985), it is not clear whether the male is specifically sampling the clitoral glands when nosing the female's genital area.

This study has not distinguished the provenances of odours at the genital region, nor is it necessary that male rats make such discriminations, but it is evident that signals emanating from this region enable male rats to distinguish non-related oestrous females from non-related non-oestrous females and related females irrespective of their reproductive condition. Males are generally considered to prefer oestrous over non-oestrous female odours (see Taylor and Dewsbury this vol.) but Brown (1977) reported no male preference for urine from either oestrous or ovariectomized females. However, he made no comment on the relatedness of his test subjects. The sniffing behaviour of females with respect to male rats, same-sex sniffing, and subsequent behaviour will be reported elsewhere (Natynczuk in preparation).

## Acknowledgements

I am grateful to Dr E.S. Albone, Dr D.W. Macdonald, and Dr R.E. Brown for their constructive comments on this paper.

## References

Albone, E.S. (1984). *Mammalian semiochemistry*. Wiley Interscience, Chicester.

Barnett, S.A. (1975). *The rat, a study in behaviour*. Revised edn, The University of Chicago Press.

Brown, R.E. (1977). Odor preference and urine-marking scales in male and female rats: effects of gonadectomy and sexual experience on responses to conspecific odors, *J. Comp. Physiol. Psychol.* 5, 1190–1206.

Brown, R.E. (1985). The rodents. In *Social odours in mammals* (ed. R.E. Brown and D.W. Macdonald). Clarendon Press, Oxford, pp. 245–506.

Gawienowski, A.M. (1977). Chemical attractants of the rat preputial gland. In *Chemical signals in vertebrates* (ed. D. Müller-Schwartz and M.M. Mozell), pp. 45–60. Plenum Press, New York.

Gawienowski, A.M., Orslak, P.J., Stacewicz-Sapuntzakis, M. and Joseph, B. (1975). Presence of sex pheromone in preputial gland of male rats, *J. Endocr.* 67, 283–8.

Lucas, P.D., Donohoe, S.M. and Thody, A.J. (1982). The role of oestrogen and progesterone in the control preputial gland sex odours in the female rat, *Physiol. Behav.* 28, 601-7.

Natynczuk, S.E. (in prep.) Semiochemistry and ultrasound in rat social behaviour. D.Phil. thesis, Oxford University.

Sokal, R.R., and Rohlf, F.J. (1981). *Biometry*, second edn. W.H. Freeman and Company, New York.

Stoddart, D.M. (1978). *Mammalian odours and pheromones*, *Studies in Biology*, No. 73. Arnold.

Thody, A.J. and Dijkstra, H. (1978). Effect of ovarian steroids on preputial gland odours in the female rat, *J. Endocr.*, 77, pp. 397–404.

Zarrow, M.X., Yochim, J.M. and McCarthy, J.L. (1964). *Experimental Endocrinology*. Academic Press, New York and London.

# IV.16 Social and Other Correlates of Scent Marking in Captive Virginia Opossums, *Didelphis virginiana* Kerr

DONNA J. HOLMES

## Introduction

New World opossums have been cited as models for primitive mammalian social and communication patterns (Eisenberg 1981; Lee and Cockburn 1985). The Virginia opossum, *Didelphis virginiana* Kerr, is solitary and nocturnal, suggesting that chemosignals may play a primary role in its social system (McManus 1970; Gardner 1982). In the present study, various aspects of scent marking and social behaviour were studied in captive Virginia opossums.

## Methods

### SUBJECTS AND MAINTENANCE

Subjects were members of a captive breeding colony of Virginia opossums maintained at Bowling Green State University. They comprised three groups (3 females and 2 males each) of unrelated adults born to, or trapped as, pouch young of wild mothers in January 1983 and grouped at sexual maturity (i.e. at approximately nine months of age). All had prior social experience, but not with other group members. Each group was housed in a large dog kennel ($3.66 \times 3.66 \times 2.44$ m) with an adjoining outdoor run. The 'kennels' were equipped with nestboxes and a network of wooden perches. The floor of each kennel was covered with wood shavings which were replaced regularly. Opossums were fed an appropriate diet of dry cat food supplemented with fruit, eggs, liver, and vitamins; water was provided *ad libitum*. Each animal was weighed monthly and treated for parasites, minor wounds, and infections. Vaginal smears were taken three times weekly for each female using the procedures of Jurgelski and Porter (1974), and pouches were inspected several times a week beginning in early January.

### OBSERVATIONAL PROCEDURES

A total of 288 nocturnal observation sessions (58 h total) were conducted between October 1983 and April 1984, which included the onset of the animals' first breeding season. Observation periods were divided randomly among the three groups and averaged approximately 12 min. They were performed under red light during peak activity (between dusk and midnight) from a seat outside the window of each kennel. Observations were recorded simultaneously for all group members using a behavioural inventory adapted from published ethograms (McManus 1970; Hayssen 1985). One-zero (occurrence–nonoccurrence) scores (Rhine and Linville 1980) were tabulated for each 10 s interval for each of 44 individual behaviour patterns performed by every animal in the group under observation. These included agonistic, sexual (courtship by males; submission to courtship by females), and eliminative behaviour, three marking patterns (face rubbing, licking, and anogenital dragging), an index of general activity (a composite variable calculated by summing one-zero scores for locomotory, grooming, and nest-building behaviour), and dancing (an agonistic display).

Each opossum was assigned a dominance rank (see, for example, Drickamer and Vessey 1986) based on the outcome of agonistic encounters (fights or supplants) with clear winners or losers. A rank based on the frequency of submission to courtship (sex rank) was also assigned to each female, as well as a rank (RS rank) based on the number of litters born (regardless of the subsequent fate of these litters), which represents a crude index of reproductive success.

### STATISTICAL ANALYSIS

After regression analysis on log(x + 1)-transformed data of each of the eight behavioural variables to partial out variation due to repeated measures for the same individuals, two correlation matrices (Pearson's *r*) were generated for each sex using residuals of the behavioural variables and the untransformed dominance ranks and body weights. The variables included in the first of these matrices for each sex (Fig. 2) were selected to examine relationships between social and behavioural variables and levels of scent marking overall. Variables included in the second correlation matrix (Fig. 3) were selected to explore the behavioural context of particular components of scent marking behaviour. The correlation analyses were intended to be exploratory, rather than inferential, since the likelihood of obtaining a 'significant' correlation between any two variables increases with the number of variables included in the matrix. In a separate analysis, Spearman rank correlation coefficients were calculated for all possible pairwise combinations of dominance rank, sex rank, and RS rank.

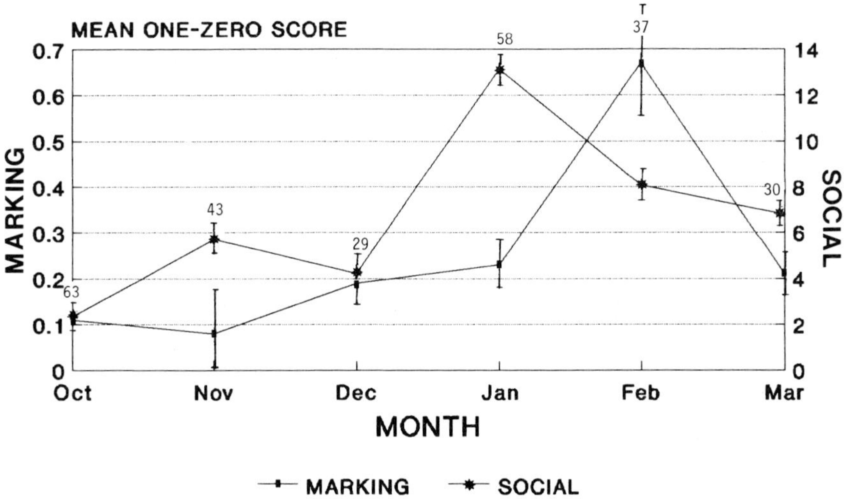


Fig. 1 Seasonal changes in frequency of social and scent marking behaviours in captive Virginia opossums. Bars represent standard errors; the number of observations sessions is shown above each month.

## Results

### SEASONAL CHANGES IN SCENT MARKING AND OTHER ASPECTS OF SOCIAL BEHAVIOUR

A sharp rise in levels of overall social behaviour corresponded to the onset of breeding in January (Fig. 1). Levels of scent marking also increased in January, but did not peak until February. A separate examination of seasonal trends in male and female marking activity (not shown here) revealed that the peak in February is attributable primarily to males. Scent marking and overall social behaviour declined gradually during the remainder of the study.

### CORRELATES OF SCENT MARKING BEHAVIOUR (ALL TYPES COMBINED)

The correlation analyses suggested an association between dominance rank and levels of sexual (courtship) behaviour in males (Fig. 2). Correlations among dominance rank, marking, and male agonistic behaviour in males were not significant. There were, however, significant associations between male weight and male agonistic behaviour directed toward females (this consisted primarily of defensive patterns). There were highly significant correlations between weight and dancing, as well as weight and activity. Marking by males was significantly correlated with weight, eliminative behaviour, dancing, and activity, but not with sexual or agonistic behaviour. There was a significant association between male dancing and activity.

| Weight M | Weight F | Sexual behaviour M | Sexual behaviour F | Scent marking M | Scent marking F | Eliminative behaviour M | Eliminative behaviour F | Agonistic toward males M | Agonistic toward males F | Agonistic toward females M | Agonistic toward females F | Dancing M | Dancing F | Activity M | Activity F | |
|---|---|---|---|---|---|---|---|---|---|---|---|---|---|---|---|---|
| −0.27 *** | −0.52 *** | −0.13 * | −0.12 ** | −0.04 | −0.21 *** | −0.02 | −0.08 * | 0.08 | −0.18 *** | 0.03 | 0.07 | 0.02 *** | −0.36 | 0.08 | −0.06 | Dominance rank |
| | | 0.14 ** | −0.12 ** | 0.11 * | 0.21 *** | 0.05 | 0.09 * | 0.08 | −0.22 *** | 0.15 ** | −0.04 | −0.18 *** | 0.34 *** | 0.11 * | 0.11 * | Weight |
| | | | | 0.06 | 0.15 *** | 0.04 | 0.11 ** | −0.07 | 0.01 | 0.10 | 0.00 | 0.02 | −0.05 | −0.02 | −0.03 | Sexual behaviour |
| | | | | | | 0.46 *** | 0.39 *** | −0.021 | 0.26 *** | 0.08 | 0.17 *** | 0.17 *** | 0.48 *** | 0.32 *** | 0.25 *** | Scent marking |
| | | | | | | | | 0.14 ** | 0.20 *** | 0.09 | 0.13 *** | 0.12 * | 0.27 *** | 0.37 *** | 0.32 *** | Eliminative behaviour |
| | | | | | | | | | | 0.11 * | 0.23 *** | 0.01 | 0.34 *** | 0.16 ** | 0.31 *** | Agonistic toward males |
| | | | | | | | | | | | | 0.09 | 0.23 *** | 0.30 *** | 0.30 *** | Agonistic toward females |
| | | | | | | | | | | | | | | 0.12 * | 0.28 *** | Dancing |

* = $p < 0.05$
** = $p < 0.01$
*** = $p < 0.001$

Fig. 2 Pearson correlations between scent marking and selected social variables. Results for males and females are shown together for ease of comparison. $N_{males} = 403$; $N_{females} = 643$.

| Weight M | Weight F | Sexual behaviour M | Sexual behaviour F | Face rubbing M | Face rubbing F | Licking M | Licking F | Anogenital dragging M | Anogenital dragging F | Urination M | Urination F | Defaecation M | Defaecation F | |
|---|---|---|---|---|---|---|---|---|---|---|---|---|---|---|
| −0.27 *** | −0.52 *** | −0.13 * | −0.12 ** | −0.02 | −0.21 *** | 0.02 | −0.10 * | 0.04 | −0.07 | −0.01 | .00 | 0.03 | −0.02 | Dominance rank |
| | | 0.14 ** | −0.12 ** | 0.10 * | −0.19 *** | 0.11 * | 0.05 | 0.10 * | 0.01 | −0.01 | 0.02 | 0.01 | −0.01 | Weight |
| | | | | 0.04 | 0.02 | 0.00 | 0.09 * | 0.04 | 0.20 *** | 0.01 | 0.00 | −0.02 | −0.01 | Sexual behaviour |
| | | | | | | 0.67 *** | 0.09 * | 0.09 * | 0.09 | 0.02 | 0.02 | 0.02 | 0.02 | Face rubbing |
| | | | | | | | | 0.12 * | −0.02 | 0.03 | −0.01 | 0.06 | 0.00 | Licking |
| | | | | | | | | | | 0.51 *** | 0.31 *** | 0.41 *** | 0.31 *** | Anogenital Dragging |
| | | | | | | | | | | | | −0.03 | 0.04 | Urination |

* = $p < 0.05$
** = $p < 0.01$
*** = $p < 0.001$

Fig. 3 Pearson correlations between individual scent marking patterns and selected social variables. Results for males and females are shown together for ease of comparison. $N_{males} = 403$; $N_{females} = 643$.

In females, on the other hand, dominance rank was highly significantly correlated with several behavioural variables and with weight. These included sexual behaviour (i.e. submission to courtship), agonistic behaviour toward males, marking overall, and dancing. Marking was highly significantly correlated with eliminative behaviour (as components of overall marking), agonistic behaviour with both male and females, dancing, and activity.

### CORRELATES OF PARTICULAR SCENT MARKING PATTERNS

There was no correlation between dominance rank and the frequency of particular scent-marking behaviours in males; weight, however, was significantly associated with both face rubbing and licking (Fig. 3). These two marking patterns were also highly significantly associated with one another. Anogenital dragging, while highly correlated with both urination and defaecation, was not associated with sexual behaviour in males. In contrast, dominance rank in females was significantly associated with face rubbing and licking. Female anogenital dragging was not significantly correlated with urination or defaecation, as in males, but was more strongly associated with sexual behaviour (i.e. submission to courtship) and face rubbing than in males.

### ASSOCIATION BETWEEN DOMINANCE, SUBMISSION TO COURTSHIP, AND NUMBER OF LITTERS BORN TO FEMALES

There was a significant association between female sex rank and RS rank (Spearman's rho $= -0.77$, d.f. 7, $p < 0.05$, two-tailed). Note that animals assigned ranks of 1 are higher ranking than those with a rank of 5, resulting in a negative correlation coefficient). Correlations between dominance rank and sex rank, and between dominance rank and RS rank, on the other hand, were not significant (Spearman's rho for dominance rank and sex rank $= -0.57$, d.f. 7, $p > 0.05$, two-tailed; Spearman's rho for dominance rank and RS rank $= 0.57$; d.f. 7, $p > 0.05$, two-tailed).

In summary, changes in the scent marking behaviour of captive opossums lagged behind increases in other aspects of social behaviour after the onset of breeding in January. While there was a general association between scent marking and the other behavioural variables examined, dominance rank, weight, and scent marking were more closely associated in females than males. There were marked sex differences of certain scent marking patterns, particularly anogenital dragging. Rank based on frequency of submission to courtship (sexual receptivity) in females was significantly associated with rank based on the number of litters born (an indicator of reproductive success), but not with rank based on outcome of agonistic encounters.

## Discussion

The function of scent marking in the social systems of non-gregarious mammals has received relatively little attention. Odours may facilitate exchange of information about reproductive status, dominance rank, and fitness in non-territorial as well as territorial mammals (Mykytowycz 1970; Thiessen and Rice 1971). Radio-telemetric data on space use patterns in opossums suggest that they probably encounter conspecific odours repeatedly in stable, overlapping home ranges (Gillette 1980; Sunquist *et al.* 1987).

In general, the results of this study suggest strong relationships among scent marking, the onset of breeding, and other behavioural variables in this species. The relationships among scent marking and other possible correlates of fitness, such as dominance, remain to be firmly established, however, and care must be taken when extrapolating from a small number of captive individuals to the natural situation. It is possible that chemocommunication plays an important role in co-ordinating social interactions in this species, and that scent marking may be related to agonistic behaviour, fitness, and priority of access to mates.

## Acknowledgements

I am grateful for the help of S. Vessey, M. Gromko, J. Rotenberry, K. Foltz, P. Brewton, K. Henderson, C. Kamen, J. Soster, and J. Harder (of the Ohio State University). The National Science Foundation (Dissertation Improvement Grant BNS-841441), the Association for Women in Science, Sigma Xi, and the Graduate College of Bowling Green State University are acknowledged for financial support.

## References

Drickamer, L.C. and Vessey, S.H. (1986). *Animal behaviour: concepts, processes, and methods*. PWS Publishers, Boston.

Eisenberg, J.F. (1981). *The mammalian radiations: an analysis of trends in evolution, adaptation, and behavior*. University of Chicago Press, Chicago.

Gardner, A.L. (1982). Virginia opossum. In *Wild mammals of North America* (ed. J.A. Chapman and G.A. Feldhamer), pp. 3–36. Johns Hopkins University Press, Baltimore.

Gillette, L.N. (1980). Movement patterns of radiotagged opossums in Wisconsin. *Amer. Midl. Nat.*, 104, 1–12.

Hayssen, V.D. (1985). A comparison of the reproductive biology of metatherian (marsupial) and eutherian (placental) mammals with special emphasis on sex differences in the behaviour of the opossum, *Didelphis virginiana*. Ph.D. thesis, Cornell University.

Jurgelski, W. Jr. and Porter, M.E. (1974). The opossum (*Didelphis virginiana*

Kerr) as a biomedical model. I. Research perspective, husbandry, and laboratory technics. *Lab. Anim. Sci.*, 24, 375–403.

Lee, A.K. and Cockburn, A. (1985). *Evolutionary ecology of marsupials.* Cambridge University Press, London.

McManus, J.J. (1970). Behavior of captive opossums, *Didelphis marsupialis virginiana*. *J. Mammal.*, 48, 486–7.

Mykytowycz, R. (1970). The role of skin glands: mammalian communication. In *Communication by chemical signals* (ed. J.W. Johnston, D.G. Moulton and A. Turk), pp. 327–360. Appleton–Century–Crofts, New York.

Rhine, R.J. and Linville, A.K. (1980). Properties of one–zero scores in observational studies of primate social behaviour: the effect of assumptions on empirical analyses. *Primates*, 21, 111–122.

Sunquist, M.E., Austad, S.N. and Sunquist, F.J. (1987). Movement patterns and home range in the common opossum (*Didelphis marsupialis*). *J. Mammal.*, 68, 173–6.

Thiessen, D. and Rice, M. (1971). Mammalian scent marking and social behavior. *Psych. Bull.*, 83, 505–39.

# IV.17 Differences Between Aggressive and Non-Aggressive Mice in Odour Signals and Marking Behaviour

N. KENNETH SANDNABBA

## Introduction

Individual variations in the production of odours affecting aggressive and investigatory behaviour in mice are related to the social status of the animal (Jones and Nowell 1975). The social status of a male mouse is partly due to inherited differences in aggressiveness and partly due to differences in fighting experience, (Lagerspetz and Sandnabba 1982). Mouse odours seem to affect aggressive encounters because a male mouse usually sniffs the opponent prior to attack (Lee and Brake 1971), and elimination of the sense of smell radically reduces the aggressiveness in mice (Brain *et al.* 1982).

Territorial behaviour seems to be connected with aggressiveness; for instance, male mice have been found to hold larger territories than females (Mackintosh 1973). There is a striking correlation between a high rate of urine marking and a high social status or dominance in mice (Ralls 1971). This study was concerned with odour signals and urine marking which influence and regulate inter-male aggressive behaviour and the investigatory behaviour of male and female mice. The purpose was to identify and integrate the roles of heredity and the environment in modulating odour communication in mice.

The questions studied were whether mice of either sex respond differently to the odour of an aggressive and a non-aggressive male and how much their response depends upon the presence of the animal producing the odour as opposed to the odour left in the environment. Experiments were also carried out to determine whether differences in odour cues, marking behaviour, and odour preferences depended upon the male being genetically non-aggressive as opposed to genetically aggressive but with its aggressive behaviour suppressed by training.

## General methods

### SUBJECTS

Four mouse strains and a total of 658 animals were used. Of these, two lines derived from mice that have been developed by selective breeding for high (TA = Turku Aggressive) and low aggressiveness (TNA = Turku Non-aggressive). Selective breeding started in 1959 from an outbred Swiss albino strain and has continued ever since (Lagerspetz 1964). The other two strains used were the parental Swiss albino and the NMRI strains, both of which are intermediately aggressive.

### PROCEDURE

*Testing the level of aggressiveness*

The level of aggressiveness of the animals was assessed by individual tests with non-aggressive standard opponents. During a 7 min. encounter the behaviour of the animals was rated according to a seven-point scale (Lagerspetz 1964, p. 42). The aggression scale was completed by recording the time of latency to the first attack. The number of bouts of tail rattling and biting were also used in a comparison between different groups of experimental animals. Males of the TA and TNA strains were not chosen randomly but were tested for aggression at the age of approximately three months. Only high scoring TA males and low-scoring TNA males were used in the experiments.

*Inhibition of aggression*

To study the effects of inhibition of aggressive behaviour, the experimental animals (TA males) were confronted daily with trained fighter mice for 21 days. The fighter mice also came from the TA line but from the previous generation of selection. They had experienced victories in earlier confrontations with other mice. The experimental animals thus suffered a defeat each day. The change in the level of aggressiveness was studied by means of individual tests (as described above).

*Collection of urine*

The urine was collected from each donor mouse separately in clean glass jars. The urine was not pooled or treated in any way. The urine was applied to two castrates by pressing the genital region to the urine. As a control, water was applied the same way. Immediately after the application of urine or water, aggression tests were begun by placing the castrated standard opponent in the glass jar where the experimental animal (a NMRI or Swiss albino male) had been placed 2 mins earlier.

*Origin of soiled sawdust*

Soiled sawdust was taken from animals that had lived on the same bedding from one to six weeks. To eliminate the possible effects of food scraps in the sawdust, the clean sawdust with mixed with grains of oats and standard laboratory food pellets. When the animals were tested for aggression on the different types of sawdust, the experimental animal was put in the glass jar followed 2 mins later by the castrated opponent. Sawdust soiled by only one animal was used for each individual testing situation.

*Recording of investigatory behaviour*

The investigatory behaviour was measured in a circular open field area (40 cm diameter) with a 20 cm-high wall. The field was divided by foam rubber strips into three main partitions, each of which was divided into six smaller partitions. The three main partitions were covered with two different types of soiled sawdust or clean (control) sawdust. The experimental animal was placed in the clean middle section of the open field and was allowed to move around for 2 mins. The investigatory behaviour of the animal on the different types of sawdust was recorded by scoring one point when it moved between partitions.

*Recording of urinary marking patterns*

The patterns of urinary marking were obtained by placing each animal in cages where the bottom was covered with a filter paper between 8 p.m. and 8 a.m. The urinary marks on the filter papers were made visible with the aid of an ultraviolet lamp.

## Results

### EFFECTS OF ODOURS FROM HIGH- AND LOW-SCORING AGGRESSIVE MALE MICE ON THE AGGRESSIVE BEHAVIOUR OF OTHER MALES (TABLE I)

The effects of male odour signals depended on whether they came from the substrate or from the animal itself. The urine of aggressive TA males, when applied to male castrates, enhanced the aggressive behaviour of the other males, whereas TA-soiled bedding depressed their aggression. In response to the TNA-soiled castrates, and on the TNA-soiled bedding, the NMRI males showed the opposite reactions.

When the aggression of the TA males was inhibited by trained fighters their level of aggressiveness declined. When applied to male castrates, TA urine depressed the aggressive behaviour of other males compared to the effects of the urine from highly aggressive controls. Soiled bedding from the TA males trained to be less aggressive lost some of its aggression inhibiting effect.

**Table 1.** Factors predisposing for high and low aggression scores in male mice

| High aggression | Low aggression |
|---|---|
| Being an offspring of the TA line (Lagerspetz 1964) | Being an offspring of the TNA line (Lagerspetz 1964) |
| Individual housing (Lagerspetz and Sandnabba 1982) | Training through individual defeats (Sandnabba 1986*a*) |
| Opponent scented with odours from a TA male (Sandnabba 1986*b*) | Opponent scented with odours from a non-aggressive male (TNA) or trained non-agressive TA male) (Sandnabba 1986*a*,*b*) |
| Exposure to non-aggressive male odours in substrate (from TNA or trained non-aggressive TA male) (Sandnaaba 1985, 1986*a*) | Exposure to TA odours in substrate (Sandnnabba 1986*b*) |

Abbreviations: TA = Turku aggressive; TNA = Turku non-aggressive.

**Table 2.** The effects of odours on investigatory behaviour and urinary marking in male mice

| Aggressive male | Non-aggressive male |
|---|---|
| Small urine spots over the whole bottom of the cage (Sandnabba 1985, 1986*a*) | Large pools of urine in few places (Sandnabba 1985, 1986*a*) |
| Repulse non-aggressive and intermediately aggressive males (Sandnabba 1985, 1986*c*) | |
| Attract highly aggressive males (Sandnnabba 1986*b*) | |
| Attract females of alien strains (Sandnabba 1985) | Repulse females of alien strains (Sandnabba 1985) |
| TA male odours repulse females of the same strain but attract TNA females (Sandnabba 1986*b*) | TA male odours repulse females of the same strain, but attract TA females (Sandnabba 1986*c*) |

Abbreviations: TA = Turku aggressive; TNA = Turku non-aggressive.

#### EFFECTS OF ODOURS FROM HIGH- AND LOW-SCORING AGGRESSIVE MALES ON THE INVESTIGATORY BEHAVIOUR OF AGGRESSIVE AND NON-AGGRESSIVE MALES (TABLE 2) AND OF FEMALES

The TA-soiled bedding discouraged investigation in males while the females preferred these areas. The males preferred areas soiled by the TA males made non-aggressive through defeats, whereas the females preferred areas soiled by the highly aggressive TA males.

The reaction of male mice to the odours differed between individuals, according to whether they were aggressive or non-aggressive. TNA males avoided the areas soiled by the TA males while the TA males were most interested in the areas soiled by other TA males.

Even in this case, learned non-aggressiveness functioned in the same way as hereditary non-aggressiveness. Losing fights also changed the odour preferences of the TA males, the soiled bedding from winner TA males now having a strong aversive effect. The TNA females preferred the odours of the TA males while the TA females preferred the odour of the TNA males.

### URINARY MARKING PATTERNS IN AGGRESSIVE AND NON-AGGRESSIVE MALE MICE

When the size and number or urine marks were examined, the TA males were found to deposit urine in small spots over the whole of the bottom of their cages while the TNA males voided in fewer places in larger pools. After inhibition of their aggressiveness, initially highly aggressive TA males started to mark in a way reminiscent of the marking behaviour of the TNA males.

## Conclusion and discussion

This study has shown that male odour cues do affect the aggressive behaviour of opponent males and that these cues are related to the behaviour of the odour donor and not to its genetic disposition for aggressiveness.

The different patterns of urine marking between aggressive males and those showing genetic or acquired non-aggression, indicates that only aggressive animals are territorial. The TA males in which aggressive behaviour has been suppressed will become territorial and aggressive again if individually housed but the TNA will not (Lagerspetz and Sandnabba 1982).

It is interesting that females, although usually attracted to males of a different genetic status to their own, do not distinguish between TNA males and TA males showing suppressed aggression.

In summary, the purpose of the different odour cues and patterns of marking in the male appears to be to reduce unnecessary fighting between potential rivals (Gosling 1982) and to inform the female of the aggressive status of the male.

## References

Brain, P.F., Goldsmith, J.F., Parmagiani, S. and Mainardi, M. (1982). Involvement of various senses in responses to individual housing in laboratory albino mice. 1. The olfactory sense. *Bolletino di Zool.*, 49, 213–22.

Gosling, L.M. (1982). A reassessment of the function of scent marking in territories. *Z. Tierpsychol.*, 60, 89–118.

Jones, R.B. and Nowell, N.W. (1975). Effects of clean and social sawdust substrates and of different urine types upon aggressive behaviour in male mice. *Aggressive Behav.*, 1, 111–21.

Lagerspetz, K.M.J. (1964). *Studies on the aggressive behaviour of mice.* Suomalainen Tiedeakatemia, Helsinki.

Lagerspetz, K.M.J. and Sandnabba, K. (1982). The decline of aggressiveness in male mice during group caging as determined by the punishment delivered by the cage mates. *Aggressive Behav.*, 8, 319–34.

Lee, C.T. and Brake, S.C. (1971). Reactions of male fighters to male and female mice, untreated or deodorized. *Psychon. Sci.*, 24, 209–11.

Mackintosh, J.H. (1973). Factors affecting the recognition of territory boundaries by house mice (*Mus musculus*). *Anim. Behav.*, 21, 464–70.

Ralls, K. (1971). Mammalian scent marking. *Science*, 171, 443–9.

Sandnabba, N.K. (1985). Differences in the capacity of male odours to affect investigatory behaviour and different urinary marking patterns in two strains of mice, selectively bred for high and low aggressiveness. *Behav. Proc.*, 11, 257–67.

Sandnabba, N.K. (1986*a*). Changes in male odours and urinary marking patterns due to inhibition of aggression in male mice. *Behav. Proc.*, 12, 349–61.

Sandnabba, N.K. (1986*b*). Differences between two strains of mice, selectively bred for high and low aggressiveness, in the capacity of male odors to affect aggressive behaviour. *Aggressive Behav.*, 12, 103–10.

Sandnabba, N.K. (1986*c*). Effects of selective breeding for high and low aggressiveness and of fighting experience on odor discrimination in mice. *Aggressive Behav.*, 12, 359–66.

# IV.18 The Role of Pheromones in the Spacing Behaviour of Snakes

NEIL B. FORD AND DEBRA HOLLAND

## Introduction

As recently as 1987, researchers have commented that there are no data to substantiate snake territoriality (Gillingham 1987). This statement was based primarily on the lack of evidence of snakes visibly defending areas. However, much of the social behaviour of snakes is controlled by pheromones (Ford 1986), and it seems likely that territoriality could also be mediated by odours, and so, difficult to observe. A number of biologists have studied aggregation tendencies in snakes and the vomeronasal system has been shown to be involved (Heller and Halpern 1982*a*, *b*). However, only one study with colubrid snakes (Porter and Czaplicki 1974) and one other with crotalines (King *et al.* 1983) have suggested any repulsion effect of conspecific odours. In neither of these studies was there convincing evidence of repulsive pheromones, but these researchers did not specifically evaluate adult males in which a territorial response would be more likely (i.e. those which exhibit intrasexual selection).

## Background

Studies on the spatial relationships of reptiles have lagged far behind those of other classes of vertebrates (Stamps 1977) and lizards are the only group in which sufficient field and laboratory data have been available to evaluate functional aspects of this behaviour. Snakes may be more mobile squamates, but field studies have given only limited information on their movements and home ranges (Macartney *et al.* 1988) and suggest that snakes exhibit little more than occasional territorial behaviour (Gregory, *et al.* 1987). The previous laboratory studies that have examined spacing in snakes typically incorporate experimental designs that make extrapolation to natural situations difficult (e.g. using small test arenas and unsexed or juvenile animals).

Social behaviour in lizards is often aggressive and sexually biased toward males defending resources (typically females or food; Stamps 1977). Territory maintenance is usually accomplished with visual displays. Territorial behaviour has not previously been documented in any snake (Porter 1972) although social aggression (combat dances) is given as evidence for territoriality (Carpenter 1984; Perry

1987). These aggressive encounters are restricted to the mating season (Schuett and Gillingham 1988), and tactile and visual cues are important (Gillingham 1987). However, pheromones are commonly important in other social contexts for snakes, i.e. reproduction (Ford 1986; Gillingham 1987), and if territories are maintained by snakes, it is likely that odours would be involved. Pheromone-mediated aggregation behaviour has been examined in the laboratory (Heller and Halpern 1982*a*, *b*) but sexual differences were not specifically examined and nor do the species utilized in those experiments (*Thamnophis* spp.) exhibit aggressive interactions. The courtship of thamnophines involves numerous males competing to mate a single female which is suggestive of epigamic sexual selection Such species would not need to defend territories for access to females. For species of snakes of which males engage in combat, the females typically will readily accept any male available during receptive periods (J. Perry, personal observation). This is suggestive of intrasexual selection (Fig. 1). If females must be prevented from mating with other males, then chemically mediated repulsion might be the most effective strategy for snakes. However, no one has examined the influence of odours on spacing in snakes that exhibit combat.

| | Family | Sexual Selection | Sensory Modality | Territoriality |
|---|---|---|---|---|
| *Thamnophis marcianus* | Colubridae | Epigamic ♀ choice | Tactile Visual | No |
| *Elaphe guttata* | Colubridae | Intrasexual ♂ combat | Tactile Visual | ? Chemically Mediated |
| *Agkistrodon contortrix* | Viperidae | Intrasexual ♂ combat | Tactile Visual | ? Chemically Mediated |

Fig. 1. A comparison of predicted spacing patterns in two families of snake.

This study examines seasonal spacing tendencies of three species of snake: both sexes of a colubrid snake that does not exhibit combat and males of a colubrid and a viperid snake known to exhibit combat dances.

## Materials and methods

### ANIMALS

The checkered garter snake (*Thamnophis marcianus*) is a colubrid snake of the south-western United States and Mexico which appears to exhibit epigamic sexual selection as several males may court a female at the same time (N. Ford, personal observation). All subjects were adults collected from the adjoining Texas counties of La Salle and McMullen, and were artificially hibernated in individual cages in a cold room at 12 °C for four months. After hibernation, snakes were maintained individually in an animal room with a natural photoperiod at 28 °C and fed weekly on *Rana catesbiana* tadpoles.

The corn snake (*Elaphe guttata*), is a colubrid snake of the south-eastern United States and males are known to exhibit combat. Although not documented, such behaviour probably decreases courtship behaviour in males that lose a combat bout (personal observation), and thus is a form of intrasexual sexual selection. All animals were raised in captivity from eggs produced by females from South Carolina. Subjects were two years of age (sexually mature as they bred at this age) at the time of the experiment. They were maintained individually under the same conditions as above but fed weekly on mice.

The southern copperhead (*Agkistrodon contortrix*), is a viperid (crotaline) snake of southern North America. It exhibits intrasexual selection by combat rituals which decrease sexual activity in losers (Schuett and Gillingham 1988). All subjects were adults collected from Smith County, Texas, USA and tested in the same year in which they were collected. Animals were maintained in an animal room at 28 °C under a photoperiod of their natural light cycles and were fed mice weekly.

## GENERAL EXPERIMENTAL METHODS

All spacing trials were conducted in a temperature controlled room in 240 cm diameter cattle stock tanks. The floor of the tank was covered either with brown wrapping paper or with corn cob animal bedding. Depending on the test either four or eight ceramic tiles (refuges) elevated 5 cm off the substrate by pieces of wood, were placed in the tank and evenly distributed in a circle 15 cm from the edge. Water was available *ad libitum* in the centre of the tank. Test animals were placed in the tank in the early evening and records started the next morning. Records were taken by gently lifting tiles and noting which refuges were occupied. After tests were completed, all cover objects were cleaned with soap and water, rinsed with 70% ethanol; all bedding or paper was emoved and the tank rinsed with 70% ethanol. The number of individuals under given cover objects was totalled for each set of experiments and the results of each test were compared to an expected distribution based on random dispersal of snakes under refuges.

## SPECIFIC EXPERIMENTS

Garter snakes were tested in tanks with eight refuges in either groups of eight males or four females and four males. Snakes were individually marked with non-toxic paint, then placed in the tank, and records were taken daily at 09.00 h. After seven days these animals were removed, and a different group of individuals of the same sex combination was placed in the tank without removing the animal bedding or tiles. At the end of the second period of seven days these snakes were removed, the cover objects and tank were cleaned, and new bedding was added. Tests were conducted during the spring breeding period, and in summer and autumn. The frequency of occurrence of snakes under each refuge during the first week was tested using a Chi-squared goodness-of-fit test against a distribution assuming such snakes to be evenly spaced under the tiles (snakes out in the open were not counted). The records from the second week were then compared to the distribution of the snakes in the first week with a Chi-squared test of independence.

Eighteen male corn snakes were tested during their spring breeding period by placing single males in tanks containing four refuges, two with substrate from cages of other males and two with fresh animal bedding under the tiles (the floor was covered with paper). Three records were taken over the next two days (morning, evening, morning). Most snakes occurred under the same type of refuge for all three records; however, for scoring preferences, if two of the three records were under refuges with substrate from another male, the animal was recorded as preferring refuges with male odours and vice versa.

Twelve male copperheads were tested in the same manner as the corn snakes, except that the tests were in the summer and autumn which is the breeding season for this species (Schuett and Gillingham 1988). In addition, since the copperheads moved only at night, records were taken only in the morning and so three days were taken to get the three records.

## Results

Results are shown in Tables 1–3. Garter snakes in mixed sex groups showed non-random distributions under the refuges (two of the three tests show aggregation) which seemed to be mediated by odours since refuges used in the first week were reoccupied during the second week (non-independence of weeks 1 and 2, Table 1). All males showed no evidence of either aggregation or avoidance behaviour (random distribution, Table 1). Male corn snakes showed no aggregation to or avoidance of other male odours (Table 2). Male copperheads showed no response to other males in the summer or autumn (Table 3).

## Discussion

In general, defence of an area or resource is necessary to consider an organism territorial (Brown and Orians 1970; Howard 1920). Although advertisement of the territory with pheromones is important, some means of expulsion of the competitor is required. Snakes in a number of families apparently utilize combat dances to determine dominance of individuals (4 species of Boidae, 21 Colubridae,

**Table 1.** The frequency of occurrence of *Thamnophis marcianus* under each of 8 refuges in an arena. The distribution of snakes in tests with fresh substrate was compared against a random distribution of those snakes under refuges using a Chi-squared goodness-of-fit test. That actual distribution was then used for the next test with the same substrate but new animals. Those distributions were compared using an 8 × 2 Contingency Table

| | **Refuge** | | | | | | | | | ***G*-value** |
|---|---|---|---|---|---|---|---|---|---|---|
| **Test** | **1** | **2** | **3** | **4** | **5** | **6** | **7** | **8** | **out** | |
| Mixed sexes: | | | | | | | | | | |
| Fresh substrate* | 1 | 19 | 1 | 7 | 0 | 19 | 0 | 4 | 5 | 20.07 |
| Old substrate | 1 | 7 | 1 | 25 | 0 | 14 | 0 | 7 | 1 | 8.96 |
| Fresh substrate* | 7 | 8 | 2 | 2 | 10 | 27 | 0 | 0 | 0 | 19.22 |
| Old substrate | 7 | 16 | 2 | 5 | 15 | 11 | 0 | 0 | 0 | 6.00 |
| Fresh substrate | 16 | 4 | 7 | 0 | 2 | 13 | 13 | 1 | 0 | 12.90 |
| Old substrate | 24 | 11 | 1 | 1 | 2 | 7 | 9 | 0 | 1 | 7.70 |
| Males: | | | | | | | | | | |
| Fresh substrate | 1 | 5 | 3 | 3 | 2 | 2 | 5 | 14 | 21 | 5.00 |
| Old substrate | 5 | 10 | 5 | 0 | 2 | 6 | 10 | 14 | 4 | 4.86 |
| Fresh substrate | 1 | 4 | 2 | 1 | 0 | 3 | 6 | 20 | 19 | 11.74 |
| Old substrate | 2 | 4 | 1 | 5 | 0 | 15 | 12 | 10 | 7 | 8.04 |
| Fresh substrate | 5 | 17 | 2 | 0 | 10 | 5 | 5 | 2 | 10 | 9.01 |
| Old substrate | 9 | 7 | 3 | 1 | 9 | 7 | 10 | 3 | 7 | 4.62 |

**Table 2.** Occurrence of male corn snakes, *Elaphe guttata* under two refuges containing substrate from the cage of a male conspecific versus two refuges containing fresh bedding. Tests were conducted in a 240-cm-diameter circular tank during the spring breeding period

| **Number with odour** | **Number without odour** | **Expected number** | $\chi^2$ | **Probability** |
|---|---|---|---|---|
| 10 | 8 | 9 | 0.22 | 0.64 |

**Table 3.** Occurrence of male southern copperheads, *Agkistrodon contortrix*, under two refuges containing substrate from the cage of a male conspecific versus two refuges with fresh animal bedding. Tests were conducted in 240-cm-diameter circular tank during the summer and the fall breeding period

| | Number with odour | Number without odour | Expected number | Chi square | Probability |
|---|---|---|---|---|---|
| Summer | 8 | 4 | 6 | 1.33 | 0.25 |
| Fall | 5 | 5 | 5 | 0.0 | 1.0 |

11 Elapidae and 25 Viperidae; S. Secor, personal communication). Because this behaviour is generally restricted to males, the competition is most likely for access to females (Gillingham *et al.* 1987). Species that do not exhibit combat tend to show prolonged courtship in which a number of males simultaneously court a single female (i.e. Thamnophines). This type of courtship does not require repulsion of competitors since selection is apparently based on female choice. Although the data are limited there is some evidence in snake species that the losers of the ritual exhibits decreased courtship activity (Gillingham *et al.* 1983). Male snakes that lose in combat also have a strong tendancy to flee the area (Carpenter 1984). Females snakes of species that exhibit combat tend to show little selection of males (J. Perry, personal communication); therefore behaviour that removes males from the area may increase the probability of a dominant male mating females present in the territory. Because snakes utilize pheromones in much of their social behaviour, it seems likely that odours should also play a role in any repulsion of competitors. If a dominant male could increase the area and time over which a competitor would be repulsed, then he could increase the area in which he would have access to females. We hypothesized that pheromonal marking of a territory should occur in males of those species that exhibit combat and these odours should repulse conspecific males.

As expected, the male snakes of species which do not combat (*T. marcianus*) showed no avoidance of other male odours even during the spring breeding period. In addition, male snakes which do combat (*A. contortrix*) showed no avoidance of other males during the non-breeding season. However, it was surprising that *A. contortrix* and *E. guttata* showed no avoidance of conspecific male odours during the breeding season. There are several possible explanations of this result:

1. As often suggested, snakes may not exhibit territorial behaviour.
2. Pheromones are not involved in the spacing of male snakes.
3. Some design flaw prevented the observation of chemically mediated avoidance behaviour in this experiment.

One additional explanation that does need to be tested is the possibility that only losers of combat bouts avoid male odours. The males in these experiments had no interaction with other males in our laboratory.

## Conclusions

This study indicates that substrate odours from male snakes are not avoided by other conspecific males regardless of the type of sexual selection exhibited by the species. It is still possible that odours are involved in maintaining territories for snakes. However, motivational factors such as participation in a combat event may need to be incorporated into tests to detect the use of pheromones to repulse competitors.

## References

Brown, J.L. and Orians, G.H. (1970). Spacing patterns in mobile animals. *Ann. Rev. Ecol. Syst.*, 1, 239–269.

Carpenter, C.C. (1984). Dominance in snakes. In *Vertebrate ecology and systematics: a tribute to Henry S. Fitch* (ed. R. Seigel, L.E. Hunt, J.L. Knight, L. Maleret and N.L. Zuschlag). *Univ. Kans. Mus. Nat. Hist. Spec. Publ.*, 10, 195–202.

Ford, N.B. (1986). The role of pheromone trails in the sociobiology of snakes. In *Advances in vertebrate chemoreception*, 4 (ed. R.M. Silverstein, D. Müller-Schwarze and D. Duvall), pp. 261–78. Plenum, New York.

Gillingham, J.C. (1987). Social Behavior. In *Snakes: Ecology and evolutionary biology* (ed. R. Seigel, J.T. Collins and S.S. Novak), pp. 184–209. Macmillan, New York.

Gillingham, J.C., Carpenter, C.C. and Murphy, J.B. (1983). Courtship, male combat and dominance in the western diamondback rattlesnake, *Crotalus atrox*. *J. Herpetol.*, 17, 265–70.

Gregory, P. T., Macartney, J.M. and Larsen, K.W. (1987). Spatial patterns and movements. In *Snakes: Ecology and evolutionary biology*, (ed. R. Seigel, J.T. Collins and S.S. Novak), pp. 366–95. Macmillan, New York.

Heller, S.B. and Halpern, M. (1982*a*). Laboratory observations of aggregative behavior of garter snakes, *Thamnophis sirtalis*. *J. Comp. Physiol. Psychol.*, 96, 967–83.

Heller, S.B. and Halpern, M. (1982*b*). Laboratory observations of aggregative behavior of garter snakes, *Thamnophis sirtalis*: Roles of the visual, olfactory, and vomeronasal senses. *J. Comp. Physiol. Psychol.*, 96, 984–99.

Howard, E. (1920). *Territory in bird life*. Dutton, New York.

King, M., McCarron, D., Duvall, D., Baxter, G. and Gern, W. (1983). Group avoidance of conspecific but not interspecific chemical cues by prairie rattlesnakes, *Crotalus viridis*. *J. Herpetol.*, 17, 196–8.

Macartney, J.M., Gregory, P.T. and Larsen, K.W. (1988). A tabular survey of data on movements and home ranges of snakes. *J. Herpetol.*, 22, 61–70.

Perry, J. (1978). An observation of 'dance' behavior in the western cottonmouth, *Agkistrodon piscivorous leucostoma* (Reptilia, Serpentes, Viperidae). *J. Herpetol.*, 12, 429–31.

Porter, H. and Czaplicki, J.A. (1974). Responses of water snakes (*Natrix r. rhombifera*) and garter snakes (*Thamnophis sirtalis*) to chemical cues. *Anim. Learn. Behav.*, 2, 129–32.

Porter, K. R. (1972). *Herpetology*. Saunders, Philadelphia.

Schuett, G.W. and Gillingham, J.C. (1988). Courtship and mating of the Copperhead, *Agkistrodon contortrix*. *Copeia*, 374–81.

Stamps, J.A. (1977). Social behavior and spacing patterns in lizards. In *Biology of the Reptilia*, 7 (ed. C. Gans and D.W. Tinkle), pp. 265–334. Academic Press, New York.

# SECTION V

## Predators and prey

# V.1 Chemically Mediated Predation in Vertebrates: Diversity, Ontogeny, and Information

GORDON M. BURGHARDT

## Introduction

There has been an enormous growth of interest in predator–prey interactions among ethologists and behavioural ecologists. Curio's (1976) seminal review would have to be double the length in but a decade. In particular, empirical studies of how animals locate and recognize prey are becoming indispensable in evaluating the relevance of elegant, but often biologically simplistic, formal models of optimal foraging, competition, social organization, and vigilance for predators. Thus, predator–prey interactions have reasserted themselves as central phenomena in evolutionary, ecological, and ethological thought.

One of the goals of this paper will be to emphasize the critical importance of two additional factors: the accumulating evidence for vast intraspecific variation in anti-predator and prey-seeking behaviour in vertebrates, and recent elegant work on the developmental plasticity and olfactory learning found in the feeding behaviour of many animals. To be sure, individual variation based on genetic, developmental, and learning processes are found in many categories of behaviour, but the extent of their role in predation has not been appreciated.

A selective brief overview of prey recognition and foraging in fish, amphibians, reptiles, birds, and mammals is considered first. However, it is striking how entangled various phenomena can become in the study of chemoreception and the ecological settings in which animals have evolved and live. Thus, using snakes as a paradigmatic group for such work, some of the complex interactions of prey cue recognition with growth, ontogeny, optimality, individual differences, experience, information transfer, and search images will be addressed.

## Vertebrate behaviour

All vertebrates share some basic anatomical and physiological characteristics, but are nonetheless extremely diverse in where and how they live. However, the gustatory, olfactory, and vomeronasal sensory mechanism in vertebrates have traceable homologies across classes which are very obscure when we try to compare them with other phyla.

An interesting facet of the various vertebrate classes is that they range from the completely aquatic to the completely terrestrial. Furthermore, members of all four of the typically terrestrial classes contain species that show this divergence. Indeed, some species of each class live important parts of their lives, if not their entire life, in the aquatic environment. One only has to think of tadpoles, turtles, sea snakes, penguins, and seals for examples of these dual environment species. Just as vision and audition in the two media differ, so too does chemoreception. There are profound consequences for chemosensory structures emanating from the reception of chemical stimuli through aqueous or gaseous media. Indeed, the division of labour between various chemosensory systems does seem more developed in terrestrial forms where, for example, gustation remains a liquid-based channel.

Basically what we are interested in are the ways in which animals perceive and respond to chemical stimuli representing or signifying prey. Other contributions in this volume cover predator recognition in detail (e.g. see Dial, Weldon, Müller-Schwarze). We should refer to these chemical stimuli as *cues* that indicate the location or presence of potential prey or enemies rather than as *signals*. Signal implies a shared code. Few prey have evolved mechanisms specifically to alert predators, and predators are unlikely to have evolved signals to warn prey that they are nearby. Signals, in the sense of communication (Burghardt 1970*b*), certainly do occur when prey or predators act to distract, startle, or deceive the other. A particularly dramatic example is the hognose snake's (*Heterodon platirhinos*) bluff attack and death feign performance (Burghardt, in press). Alarm pheromones function primarily to alert conspecifics and thus the chemical cues used do not originate from the predator itself.

The study of chemical cues in finding prey or avoiding enemies is a complex undertaking in vertebrates. Increasingly, issues of modality interaction, compound synergism and redundancy, motivation, and experience need to be considered. A serious hindrance to studying the non-social chemical ecology of vertebrates has been the paucity of unequivocal bioassays using as markers aspects of behaviour that occur only in the presence of a specific, or even not so specific, chemical cue. Often workers have had to deal with changes in base-rate response measures such as amount of locomotor activity or orientation; regardless of how they are operationally defined, these do not have the face validity of proboscis protrusion, ovoposition, or prey attack (Chiszar 1986; Cooper and Burghardt, in press).

Until recently there were few experimentally exploited aspects of behaviour in vertebrates for which specific aspects were tied to nongustatory chemoreception. Classic studies such as those of Frank Beach (1947) on the multi-sensory control in laboratory rodents of mating, maternal care, and so forth were *too* successful in convincing us that there was so much redundancy in vertebrates, read mammal, behaviour, that it was not worth attempting to tease out specific influences of the so-called minor or lower senses. Clearly, the many recent demonstrations of social behaviour, kin recognition, and so forth in this volume and elsewhere have shown the critical importance of chemical cues in many aspects of mammalian social behaviour. Is this now also true of predation?

## FISH

In the previous volume in this series, Adams and Johnsen (1986) reviewed recent work on chemical control of feeding in fish. Many studies have shown that a wide variety of fish respond with elements of feeding behaviour to either crude prey extracts, such as clam juice, or to isolated compounds. Amino acids have been found particularly attractive.

One of the most extensive research programs is that carried out by Carr (1982). He used an apparatus to evaluate various prey extracts for pigfish and pinfish in which a striker was bitten by the fish when an effective odour was introduced and the response automatically recorded. Extracts were compared by the concentration needed to elicit a given number of responses.

One of the important conclusions of the fish work is that amino acids often need to be in mixtures to be highly effective, i.e. they act synergistically. This may be related to the fact that animal tissue is composed of a limited number of amino acids and other compounds meeting the criteria for chemical prey attractants developed from Carr's work: mol. wt of $< 1000$ daltons, non-volatile, nitrogenous, amphoteric. In addition, unlike terrestrial vertebrates, gustation seems to play a much more discriminating role in fish. Weldon's (1988) studies on rejection responses of snappers (*Lutjanus*) to pieces of venomous sea snake (*Pelamis platurus*) may have involved gustation. There were indications that the snappers discriminated the sympatric sea snake from a venomous but terrestrial rattlesnake.

Ellingsen and Doeving (1986) showed that oriented bottom food-searching behaviour could be stimulated in cod by various fractions from ground shrimp and, interestingly, found evidence that the various compounds involved had both synergistic and antagonistic effects.

Such bioassays have a great potential in early identification of possible pollution effects. For example, Lemly and Smith (1987) found that in the common North American fathead minnow (*Pimephales promelas*) moderate lowering of the pH to 6.0 could lead to complete elimination of feeding responses. They suggest that the action involves inhibition of olfactory and gustatory receptor cell function (see also Royce-Malmgren and Watson 1987).

Fish have also been used to compare responses to visual and chemical cues. Rottiers and Lemm (1985) used a Y trough to demonstrate that walleyes were attracted to a variety of food cues such as glutathione, betaine, arginine, and *Daphnia* slurries and washes, but avoided a series of other potential food cues such as glycine and fish mucus. Live prey presented only visually showed that prey movement was also important.

As a final example I will mention Mearns (1986) who studied, as is rarely done with fish, responses of ingestively naïve fry of Atlantic and brown trout to various amino acids. L-proline was identified as a stimulator of the prey capture snapping response and L-alanine as a potent facilitator of locomotor prey search behaviour. Given the limited repertoire of amino acids and the diversity of prey used by these species, it is likely that prey discrimination and preferences are less clearly mediated. This is an important study in showing chemical innate recognition in naîve fish.

## AMPHIBIANS

Various studies have examined the role of chemicals in the feeding behaviour of amphibians. In this and the following vertebrate classes the reviews by Halpern (1980) and Stoddart (1980) are useful supplements to the literature. Luthardt-Lamier and Roth (1983) provide an extensive listing of references demonstrating feeding responses with controlled visual cues.

There is some evidence that urodele amphibians or salamanders give feeding responses to objects imbued with the scent of natural prey (e.g. Reese 1912). The blind cave salamander (*Proteus anguinus*) prefers water passed over prey and responds to such cues more strongly than an above-ground species (Durant *et al.* 1982).

With anurans (frogs and toads) solid evidence for the use of odours and gustation in feeding was late in coming. Risser (1914) claimed that chemical reactions to prey in toads (*Bufo americanus*) did not exist and, moreover, would be hard to demonstrate. He did find a role for chemicals in toad but not frog (*Rana*) tadpoles. Indeed, the potency of visual cues, particularly movement, have led to metamorphosed toads being considered the model animal for use in the neuroethology of visually elicited prey attack behaviour. Shinn and Dole (e.g. 1979) found that some common frogs and toads are not only attracted to odours of typical prey such as crickets, but will even attack chemical stimuli in the absence of visual cues. They also isolated from prey a lipid fraction that elicits prey attack. Dole *et al.* (1981) demonstrated that western toads (*Bufo boreas*) could learn to recognize the odour of familiar prey.

One of my points of emphasis is the interaction of chemical stimuli with other modalities in vertebrates. Salamanders are excellent examples. Lindquist and Bachman (1982) studied the role of chemical and visual cues in prey catching behaviour of the tiger salamander (*Ambystoma tigrinum*). They found that vision was most important in detecting and capturing moving prey, but chemical cues seemed most important in determining edibility.

More complex is the series of Luthardt and Roth's studies (e.g. 1983; see also Roth 1987) on prey catching in European salamanders. I will just discuss *Salamandra salamandra*. In darkness these salamanders identify prey using odour, but will ignore the same prey object in daylight. Apparently the lack of visual movement under lighted conditions inhibits snapping at prey. But experience plays a role: if you raise salamanders on stationary prey they react more in both light and darkness. Also, rearing on stationary prey led to increased use of vision but no change in response to olfactory stimuli presented in isolation under illumination.

There is little work identifying chemicals used in prey capture by amphibians. However, Mason and his coworkers have studied tiger salamanders (*Ambystoma tigrinum*) for many years and have developed a bioassay based on a conditioned avoidance paradigm which they have used successfully in several papers. In one of their most recent (Mason *et al.* 1987), they classified chemical receptors into specialist receptors, presumably pheromonal, and generalist receptors, presumably for detecting prey and predator odours. They used pure chemicals to show that a carbonyl generalist receptor does exist, although effects of natural prey and predator odours have yet to be determined.

## REPTILES

Reptiles have been the focus of much work on the chemical recognition of prey. Various reviews are available that discuss, in particular, studies on squamate reptiles (lizards and snakes) (Burghardt 1970*a*, 1977*b*, 1980; Halpern 1980, 1987; Halpern and Kubie 1984; Simon 1983). Even the recent work is voluminous and cannot be covered in detail.

All snakes have a well-developed tongue-vomeronasal organ system and have been shown to employ it in social, feeding, and reproductive behaviour almost to the exclusion of nasal olfaction (Burghardt 1980; Gillingham and Clark 1981; Halpern and Kubie 1984; Halpern 1987). It is clear that many snakes use chemicals in trailing and identifying prey, and that prey attack is often triggered primarily by chemical stimuli received via the vomeronasal organ. However, even in natricine snakes, in which the importance of chemical cues is most well established, visual cues may play an important orienting role (Burghardt and Pruitt 1975; Burghardt and Denny 1983; Chiszar *et al.* 1981; Drummond 1979), and may be sufficient to elicit attacks (Drummond 1985).

The long substantial series of papers by Chiszar and colleagues on strike-induced chemosensory searching in vipers and elapids is well-known (Chiszar *et al.* 1983). In rattlesnakes, the group they have most thoroughly studied, it appeared that visual and heat (infra-red) cues elicited the prey attack, but that after the attack and release of prey, chemical cues were used to trail and locate the envenomated prey. More recently Graves and Duvall (1985) have demonstrated that experienced rattlesnakes of at least one species fail to attack prey if the vomeronasal ducts in the roof of the mouth are sutured shut. In any event, the post-strike phenomenon has been extended to colubrid snakes (Cooper *et al.* 1989). Duvall and Chiszar (this volume) have studied the role of chemicals in locating prey patches in the field and laboratory.

In lizards the role of chemicals in various phases of predation is more variable, being far less demonstrable in the more visual iguanid and agamid lizards (Simon 1983). Nicoletto (1985) has shown the interaction of visual and chemical cues in responses of skinks (*Scincella lateralis*) to prey. This interaction is not surprising given the studies cited above which show the role of vision even in the most chemically influenced snakes. In a study of the role of vision, audition, and olfaction in two geckos, Chou *et al.* (1988) eliminated each of these senses in *Hemidactylus frenatus* and *Cosymbotus platyurus* and assessed the effect on number of fruit flies captured. In the former species no clear effect of anosmia was demonstrated. In the latter species anosmia did lead to significantly fewer insects captured, but was less detrimental than loss of sight or hearing. Current studies indicate that lizards with the most snake-like tongues discriminate prey chemically (Cooper 1989; Cooper and Burghardt, in press) and that strike-induced chemosensory searching also occurs in these species (Cooper, in press).

Considerable literature has accumulated on the importance of chemical cues in prey recognition by neonate snakes, and species differences have been demonstrated across several genera (Von Achen and Rakestraw 1984; Burghardt 1969, 1970*a*; Weldon and Schell 1984). Population differences may also be marked (Burghardt 1970*c*; Arnold 1981). Enhancement and suppression of this and other aspects of feeding may occur (Burghardt 1977*a*, 1978, and below). In most of these studies the cotton applicator method has been used, though presentation methods and scoring have varied. Cooper and Burghardt (in press) provide a review of several methods and evaluate their relative ability to discriminate among stimuli using the same data sets.

Progress has been made on chemical identification of the prey substances to which snakes respond. Sheffield *et al.* (1968) showed that garter snakes (*Thamnophis sirtalis*) responded to surface substances of earthworms (*Lumbricus terrestris*) which were non-volatile, stable, water-soluble macromolecules. Halpern and her colleagues (e.g. Halpern *et al.* 1986) extended these findings and concluded that a large molecule, perhaps of 400,000 daltons, was involved that appeared to be collagen derived from the cuticle. However, these findings have subsequently been questioned by Halpern's group (Wang *et al.* 1988). Our lab has had mixed results replicating the collagen work (Schell *et al.*, in press), but we have identified effective fractions of both low (about 1500 daltons) and high molecular weight from a fish (fathead minnow, *Pimephales promelas*) and earthworm (Burghardt *et al.* 1988). Both experienced and naîve snakes responded to our fractions.

In short, chemical cues seem to be important, although in varying ways, in the predatory behaviour of all snakes and in some lizards, especially of the families Varanidae, Geckonidae, Scinidae, and Teiidae (see Cooper, this volume). These include the locating, identification, and trailing of prey. Positive experimental demonstrations in alligators have also been reported (Scott and Weldon, in press).

## BIRDS

Some birds just seem more olfactory than others. Evidence for a clear role of chemoreception other than taste in almost every area of avian behaviour has been hard to come by, although recent advances have been made (see Mason and Clark 1986).

Carrion location by vultures has seemed to many a prime candidate for studies of the importance of olfaction and there has been a controversial, if sporadic history of research in this area. James Audubon claimed to show that the sense of smell was overrated and that the buzzard finds food visually (review in Chapman 1929). Although there was some scepticism, most secondary sources accepted the conclusion that vultures did not use smell. The total evidence seems to have been that vultures would eat fresh and putrid meat and were visually alert.

Almost 100 years later Frank Chapman carried out some experiments on vultures on Barro Colorado Island in Panama (Chapman 1929). He records how vultures were attracted to a small shack containing the carcass of a coati that had died. Vultures had not previously been seen near the building. He then did some crude tests with fish with some clear results, followed by days when a seemingly potent odour had no obvious effect. To me his conclusion seems reasonable that the vultures were sensitive to food odours, but the work, although done over 3 years, was not well-controlled.

Stager (1964) in controlled studies clearly showed that turkey vultures responded to prey odour alone, and would ignore a deer carcass decoy that looked like a deer but had no odour. The flaw of the earlier work was in not presenting visual cues isolated from chemical ones.

For other birds, there is Benham's old (1906) demonstration that kiwis (*Apteryx*) were able to find buried earthworms and meat, since confirmed with more controlled methods by Wenzel (1968). Harriman and Berger (1986) showed that ravens (*Corvus corvus*) were able to locate fish buried under as much as 2.5 cm of gravel.

The most extensive recent work on avian olfaction in natural behaviour is the the work of Grubb, Wenzel, and others on the seabirds of the avian order Procellariiformes (reviewed by Wenzel 1986). This work has shown that seabirds such as petrels, shearwaters and albatross are attracted to fish oils presented at sea on slicks or as elevated wicks using appropriate controls. Procellariiform species have enormous olfactory systems for birds, adding to the significance of such findings.

## MAMMALS

In mammals we find a veritable explosion of behavioural research which implicates the chemical senses in almost every aspect of social behaviour in every order and most families. Scent glands, flehmen, urine marking, and so forth are well-known phenomena. Even the vomeronasal organ has come into the picture. However, there is still a dearth of good studies on chemical aspects of predation in mammals; even reports on trailing and recognizing prey by dogs, bears, prosimians, and other macrosmotic mammals is not very definitive. Indeed Halpern's (1980) review did not cite a single example of chemically mediated prey detection in a mammal.

Although, unlike birds, mammals have a well-developed vomeronasal system, the use of it seems more restricted to social and reproductive behaviour and an important role in feeding behaviour has not been demonstrated. However, Wysocki *et al.* (1985), using non-volatile rhodamine dye as a marker, showed that food odours do reach the vomeronasal system in pine and meadow voles (*Microtus*).

So we are left with the olfactory system. One of the few sets of studies that seems to be comparable in method to those on reptilian prey attack is that of Langley (1983*a, b*) on the grasshopper mouse (*Onychomys leucogaster*). Using cotton swabs he found that odour is highly important in instigating prey attack but that such recognition of cricket odour develops after the first few feedings on crickets. However, audition and vision are also involved, especially in distance recognition. This was also shown in a study of responses of golden hamsters (*Mesocricetus auratus*) to live and dead crickets with olfaction, vision, and audition blocked in various combinations (Langley 1985). All three senses played a role in localization of a live cricket but with a stationary cricket vision and olfaction were used. Grasshopper mice learn quickly to recognize prey by odour and to locate buried prey (Langley 1984). Grasshopper mice also use odour when learning to suppress prey attacks to aversive prey (Langley and Knapp 1982; Whitman *et al.* 1986). These results suggest that in mammals identification of nearby prey is rarely going to be primarily chemical.

The role of olfaction and experience has also been the focus of a series of studies by Apfelbach and co-workers on ferrets (*Mustela putorius*). It was shown that prey odours are important in prey recognition (Apfelbach 1973), but that they have to be learned during a sensitive period between 60 and 90 days of age (Apfelbach 1978). Furthermore, at the time of such olfactory imprinting the granule cells in the olfactory bulb are maturing (Apfelbach 1986; Rehn *et al.* 1986).

Olfactory cues (i.e. blood) are used to detect gerbils (*Meriones unguiculatus*) by ferrets and foxes (*Vulpes fulva*) (Thiessen and Cocke 1986) and domestic cats (Cocke and Thiessen 1986). The important aspect here was that the predators were attracted to cues associated with stressed gerbils, especially to changes in the blood. On the other hand gerbils themselves avoided areas contaminated with such odours.

The finding that mammals may use chemical aspects of prey other than those with which the prey is generally associated is intriguing and so far has not been documented in reptiles. Cushing (1984) has shown that least weasels (*Mustela nivalis*) choose odours originating from urine or nesting material which are associated with oestrous versus dioestrous female prairie mice (*Peromyscus maniculatus*). Earlier (Cushing 1983) showed with both captive and field tests that polar bears (*Ursus maritimus*) prefer odour from seals, their natural prey, and odours associated with human female menstrual blood. However, they ignored non-menstrual human blood. Bears actually consumed used tampons in 42 out of 52 field trials.

## Chemoreception, variability, and experience

The demonstration of the varying role of chemoreception in predation across a wide range of vertebrates is important, as is the salience of chemical food cues to naïve animals. The studies cited above highlight the progress that has been made in various vertebrate classes. However, these demonstrations are best viewed not as an end but as a necessary starting point for the positing of evolutionary, functional, and ontogenetic questions. How does chemical prey recognition dovetail with the entire life history of a species? Here I will briefly present some findings on snakes, including unpublished data, which show the potential of this approach.

### NEONATES AND MATURATION

It has already been mentioned that neonate natricine snakes, along with other snakes and lizards, can recognize prey chemical cues at birth. It is also known that as snakes grow older or larger their prey preferences, or at least the prey selected in the field, can change (e.g. Mushinsky *et al.* 1982; Plummer and Goy 1984). This can also be shown in captive reared snakes that experience only one prey type. For example, Mushinsky and Lotz (1980), showed that *Nerodia erythrogaster* switched from a preference for fish extract to frog extract at 6 months of age regardless of the intermediary diet. Since adults of this species primarily eat frogs as adults, the authors postulate that the switch is adaptive, in part because young snakes have trouble ingesting small froglets. This implies that genetic programming of shifts in prey preference may occur. However, different species of *Thamnophis* fed large and small fish showed quantitative differences in how prey was captured, handled, and swallowed at birth and apparent improvements in learning were unequal (Halloy and Burghardt, in press). Aquatic foraging practices prior to prey capture also varied in two populations of *T. elegans* (Drummond and Burghardt 1983). Thus differences in the size or type of prey eaten may be due to maturation, experience, prey availability, or changes in foraging mode.

### INTRASPECIFIC VARIATION

Within-litter variation in the responses of previously unfed garter and water snakes to prey chemicals has been demonstrated, in addition to species and population differences (Burghardt 1975; Gove and Burghardt 1975). In the typical test where each snake is tested only once or twice on a given extract, variation is found that may reflect a genetic predisposition (e.g. Gove and Burghardt 1975; Arnold 1981) but it is not possible to affirm without other information whether individual differences are significant. However, Burghardt (1975) tested snakes repeatedly with the same fish and worm extracts and statistically significant differences in preference were shown for *individuals*. Furthermore, the correlation in responsiveness between the two prey extracts across all animals was non-significant. In all litters tested, more snakes showed significant preferences than would be expected by chance, and most, but not all, of these preferred earthworm extracts. These results held true even when different concentrations were used.

Fieldwork indicates that earthworms are the most important food for young garter snakes, *Thamnophis sirtalis* (Fitch 1965). Thus the overall worm preference does have some validity. However, in a study of geographical variation, it was found (Burghardt 1970*c*) that neonates from a gravid female captured in Wisconsin responded, as a litter, more to fish than to earthworm extracts. This result has since been replicated by Schwartz (1989). Thus innate and genetically based intraspecific variation in chemical prey preference exist (see also Arnold 1981). Do these extract results hold for actual prey ingestion as well as chemical cue prey recognition?

Studies on the long-term effects of giving garter snakes a choice between two prey from birth on were carried out in collaboration with Barbara Allen and Bruce Batts. An entire litter of neonate snakes born to two gravid *Thamnophis sirtalis* captured in either Wayne county, Michigan and Raleigh, North Carolina, were isolated at birth. Each animal was fed according to the following protocol. At each feeding the snake was gently restrained behind a piece of cardboard while two small paper medicine cups about 3 cm in diameter were placed at one end of the cage. The snake was then allowed to approach and sample or eat as many food items (equal sized pieces of nightcrawler, *Lumbricus terrestris*, or individual mosquito fish, *Gambusia affinis*). Cages were checked periodically to count prey remaining and then left undisturbed overnight, after which a final count was made. In this way the total number of each prey item ingested was recorded as well as a fairly complete record of the order in which they were eaten. Animals were not force-fed if they did not eat on their own and vitamin or mineral supplements were not given. The plan was to assess the variation found in entire litters. Naturally, in large *Thamnophis* litters, few offspring survive in the field.

In Table 1 the results of the first 20 prey ingested are recorded. Some animals ate the first time food was offered, others did not eat for some time, and others never or rarely ate at all and died. Thus we have data not only on the food types preferred but whether any food was eaten. The results are of interest in that they support the findings on within-litter variation on chemical cue preference.

In the Michigan litter 23 neonates were tested. Of the 15 animals which ate at

**Table 1.** Distribution of fish and earthworm ingestion over first 20 food items in two litters of neonate garter snakes (*Thamnophis sirtalis*)

| | Michigan | North Carolina |
|---|---|---|
| Number in litter | 23 | 41 |
| Number eating 20 prey | 15 | 22 |
| Worm preference ($p < 0.05$) | 7 | 3 |
| Fish preference ($p < 0.05$) | 5 | 8 |
| No significant preference | 3 | 11 |

least 20 food items, 7 ate significantly more worm pieces than fish. To be significant (binomial test) this means that the ratio between worm and fish had to be at least 15:5 and in fact was at least 17:3 in all 7. Five of the snakes had a comparable significant preference for fish and in all five the ratio was at least 19:1. Only three snakes did not show a significant bias for one food type.

Of the 22 animals in the North Carolina litter that ate at least 20 food items, comparable levels of bias were seen, but in this litter fish preferers were more common. Eight snakes showed a significant fish preference, three a significant worm preference, and 11 showed no preference.

Combining both litters, 37 animals ate 20 food items and 23 showed a significant ($p < 0.05$) bias divided almost equally across worm and fish, whereas by chance alone one would only expect 2. Thus the level of diversity in litters is enormous. Note that this only included two prey of the myriad available. Those animals that did not eat at all or died could very well have been predisposed to eat other kinds of prey, but often were runts or sickly from birth.

I have called this discontinuous variation within litters, chemical prey preference polymorphism (Burghardt 1975), and considered it genetically based and interpreted its function as a bet-hedging strategy on the part of females. That is, offspring would be produced that could exploit varying prey availabilities, since small vertebrate and invertebrate prey populations can vary significantly. Now we know that multiple paternity occurred in at least 50% of 32 litters we assessed in two populations in Wisconsin and Michigan (Schwartz *et al.*, in press). Thus the variation could be due, at least in part, to different paternal genotypes, a possibility we are currently assessing.

A further issue here is that fish and worm diets lead to different growth rates in *Thamnophis radix* and especially *T. sirtalis*, with fish being the more efficient diet, probably because of the higher level of calcium and phosphorus (Scudder-Davis and Burghardt 1987). The latter suggestion, based on prey and faecal analysis, is supported by our observation that neonate plains garter snakes fed on exclusive earthworm (*Lumbricus*) and fish (*Gambusia*) diets showed greater growth on the former diet when the worms were supplemented with minerals rich in calcium and phosphorus (Burghardt and Lyman, unpublished observations). There may be a tradeoff between nutrition and availability in generalist snake species such as *T. sirtalis* and *T. radix*.

### EXPERIENCE AND SEARCH IMAGES

In spite of the congenital preferences of snakes for certain prey extracts, such preferences can be altered by exposure to chemical cues through feeding (Fuchs and Burghardt 1971). Snakes fed on guppies and redworms responded increasingly more to chemicals from these prey over the first eight feedings (at 4-day intervals). When the diets were then shifted the relative differences declined. The responses of the snakes to extracts prepared from novel species of earthworm and small fish (nightcrawlers and minnows) were also tracked. While, as in Burghardt (1975), responses to different earthworm and fish species were similar, a dissociation was found *after* feeding experience, indicating that the snakes could discriminate among closely related prey when such was made meaningful. We interpreted our results as indicating that snakes may develop chemically based search images, in the sense of learning to discriminate prey chemicals in favour of the one that has paid off in the past. Other studies have shown that feeding experience does not always enhance chemical prey preference (Arnold 1978; Mushinsky and Lotz 1980); thus the effect of experience is not simply additive.

These results, in combination with those showing great intra-litter variation in preference, demonstrate that early feeding experience can override and modify the initial genetically based (Arnold 1981) preferences. That is, phenotypic convergence in prey preference among adult snakes may mask genetic variation found in neonates. Alternatively, only those with a prey preference that is most reflective of current prey availability as neonates may survive. These results indicate that this is not necessary, but field studies are needed to trace what actually happens in the field (cf. Gove and Burghardt 1975).

Subsequently Chiszar *et al.* (1985) demonstrated that cottonmouths (*Agkistrodon piscivorus*) maintain chemosensory searching for up to 70 mins after striking prey and argued that a central mechanism had to be present. More recently Melcer and Chiszar (1989) demonstrated that, during the split second of the strike, rattlesnakes learn specific chemical features of the prey struck and prefer them when encountered some minutes later. This is an important result showing that extremely rapid chemosensory (most certainly vomeronasal) learning can occur in snakes.

I once performed an experiment in which ingestively naïve neonate *Thamnophis s. sirtalis* were exposed to prey extracts of fish or earthworms. The details of the experimental design are given in Burghardt (1970*a*, p. 300). Here I report unpublished but significant retention data.

When 7 days old, two groups of 21 animals were given three 60 s tests at 20 min. intervals of earthworm (*Lumbricus terrestris*) or Baltimore minnows (non-gold *Carasius auratus*). On the following day each snake was tested once on both the fish and worm extracts, with half in each group tested on each extract first. One week later, with no intervening feeding or extract experience, the worm and fish tests were repeated.

The snakes in the two extract pre-exposed groups were subdivided into two groups, those that attacked the extract on the last of the three trials, and those who never attacked it or who habituated by the last trial. Tongue-flick attack scores (TFAS) were calculated for each snake (as in Fuchs and Burghardt 1971). The different scores for each snake were analysed with nonparametric Kruskal-Wallis tests; results are given in Table 2.

**Table 2.** Mean difference scores to previously experienced and novel prey extracts by ingestively naïve garter snakes as a function of response level at final trial on Day 1. Snakes equally divided into fish and earthworm pre-exposure groups

| | **TFAS (pre-exposed) − TFAS (novel)** | |
|---|---|---|
| **Day** | **Responders ($n=21$)** | **Non-responders ($n=20$)** |
| 2 | 13.50 | −13.48* |
| 9 | 18.82 | − 8.20* |

*$p<0.05$, Kruskal-Wallis test

The snakes that attacked the extract preferred it over the other one on the following day while those who habituated or never attacked preferred the novel prey extract ($\chi^2=5.21, p<0.025$). Even more dramatically, the difference, although declining, still persisted a week later when retested with no intervening prey or extract experience ($\chi^2=3.93$, $p<0.05$). On neither test day were there significant differences due to the kind of extract, the order of presentation, or the litter from which the snakes came. I interpreted these findings in light of the reinforcement associated with performance of the attack in association with the prey cues (Burghardt 1968). Halpern (1988) has recently reviewed evidence that vomeronasal stimulation is intrinsically rewarding. Since in this experiment the snakes that attacked the swab presumably received more vomeronasal stimulation, although non-attackers had longer exposure to the swabs, the role of attack and stimulation are confounded. More experiments are called for here, but the finding that some aspect of response to prey cues without prey ingestion was rewarding in ingestively naïve snakes is remarkable.

The development of prey search images can be approached in another way, as a decrease in stimulus threshold. Lyman and I raised a litter of 22 *Thamnophis radix*

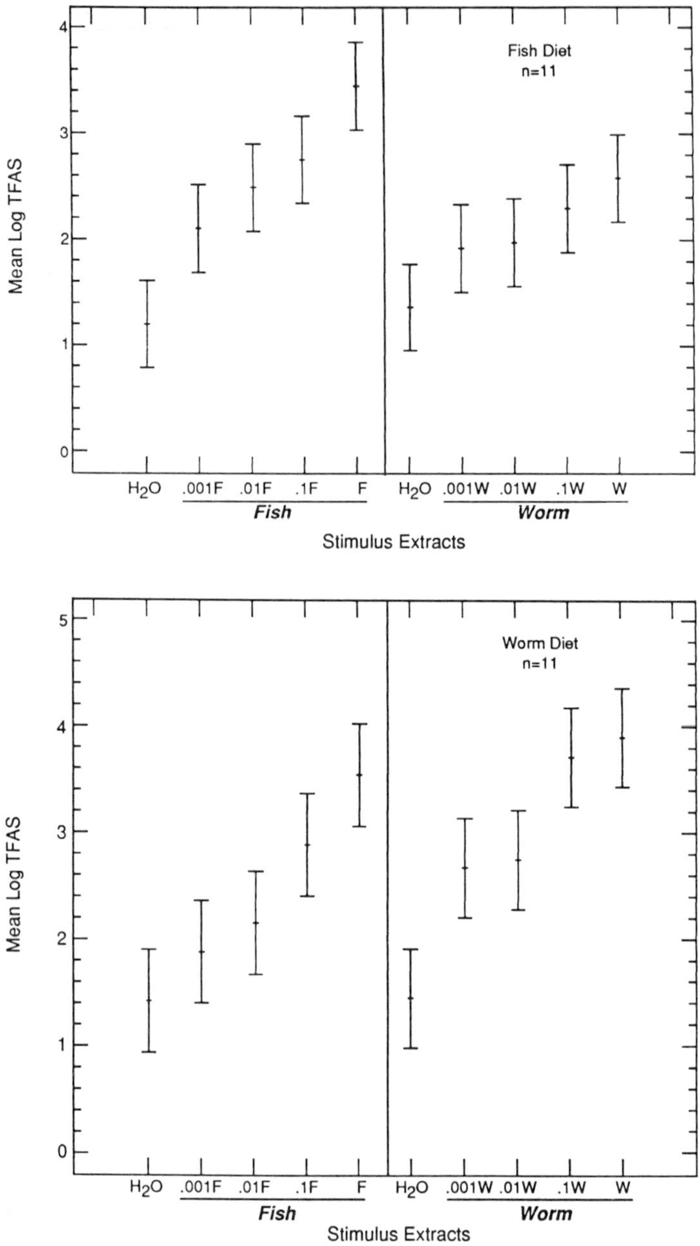


Fig. 1 Responses of *Thamnophis radix* reared on either fish or earthworm diets to control and prey extracts, (F, fish; W, worm) of various concentrations. Mean log (+1) tongue-flick attack scores (TFAS) with 95% confidence intervals.

on either mosquito fish or earthworms. After eight months we tested their responses to fish and worms and replicated the finding with *T. sirtalis*, that is, they preferred the cues from the prey they had been fed. Two months later, still maintained on the same diet, we tested the snakes on responses to extracts at four different concentrations plus a deionized water control. The concentrations were the standard raw extract (100%) and serial dilutions to 10%, 1%, and 0.1%. Because decreasing extract concentrations leads to carry-over (hysteresis) effects (Burghardt 1975), each snake was given the five test stimuli in increasing concentration beginning with water. The snakes on each diet were divided into two groups and half were given the fish extracts and half the earthworm at about 25 min. intervals. On the following day the testing was repeated with the other extract type.

The results are shown in Fig. 1; a repeated measures ANOVA (stimulus × day × subject) on log (+1) transformed TFAS scores showed a highly significant stimulus effect for both diet groups (fish: $F(9,89) = 10.26$, $p < 0.001$; worm: $F(9,89) = 16.33$, $p < 0.001$). The fish-fed animals responded most to the 100% fish extracts and vice versa for the worm-fed group. The important point to note here is that the fish-fed animals discriminated the 0.1% fish extract from water but not the 0.1% worm extract, whereas the worm-fed group discriminated the 0.1% worm extract from water but not the 0.1% fish extract. Thus feeding experience does seem to selectively lower stimulus thresholds.

## AGGREGATION AND GROWTH

Newborn (Burghardt 1983) and adult garter snakes (Noble and Clausen 1936; Heller and Halpern 1982) water snakes (*Nerodia*; Burghardt and Spottswood, unpublished), brown snakes (*Storeria*), and rattlesnakes (*Crotalus*; Graves *et al.* 1986) form aggregations. The functions of these are little understood and several hypotheses have been advanced. These include conservation of water or body heat, reduced metabolism, and reduced predation risk. One that we looked at was the effect of grouping on growth and prey assimilation. In field experiments, Burghardt and Rand (1985) showed that green iguanas (*Iguana iguana*) that slept in groups in shrubs at night (not necessarily in physical contact) had faster growth rates than those typically found in smaller groups or as singletons.

Supportive evidence for the growth hypothesis has been collected in an unpublished study in our laboratory by James Schwartz and Barbara Allen Savitsky. A litter of 16 neonate banded water snakes (*Nerodia sipedon*) born to a female captured in Highlands, North Carolina were divided at birth into three groups. The snakes were matched for initial mass. The three groups were (i) snakes isolated in individual cages and fed in isolation (II); (ii) snakes housed in groups and fed in groups (AA); and (iii) snakes housed together but removed to isolate cages for feeding (AI). The snakes were presented with prey (fish) *ad libitum* for 24 h once a week for 12 weeks, at which time final measurements were taken.

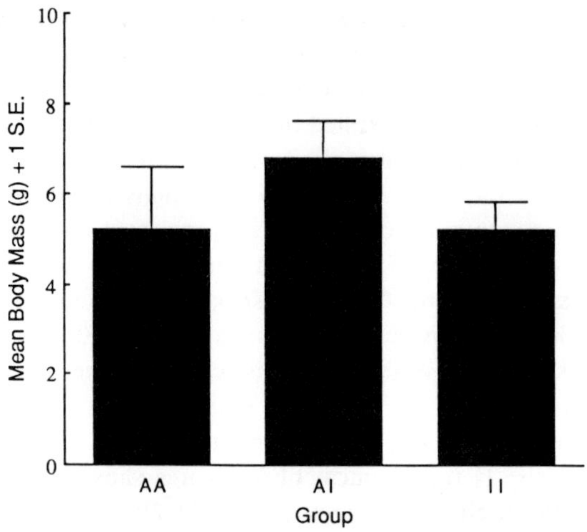


Fig. 2 Growth in *Nerodia sipedon* reared in three groups. AA, group housed and group fed; AI, group housed but fed in social isolation; II, housed and fed in isolation. There were no differences in size when the experiment began. Fish were the diet throughout.

A prediction based on social facilitation of feeding is that the AA group would grow most rapidly, since they would be stimulated to eat more by the presence of conspecifics. However, the AA and II groups did not differ in mean final weight (Fig. 2); the variance was greater in the AA group, however, indicating that social competition was involved in some way, although excess food was available. Most interesting was the finding that the AI group had the highest growth rate and the overall ANOVA was significant [$F(2,13)=11.32$, $p<0.01$]. Furthermore, the AI animals had a production/ingestion ratio almost twice as high at the II and AA groups, which were similar, indicating that the AI snakes not only ate more but converted a higher proportion of what they ate to biomass rather than to maintenance metabolism.

The mechanism of the increased growth of AI animals is unclear but could result from decreased metabolism or even reduced stress due to physical contact. The difference between the AA and the AI groups is not as paradoxical as might be first thought: although many snakes may group together, group foraging is not common. Gillingham's research (this volume) shows that even those garter snakes that retreat to cover at night in dense aggregations, leave and forage independently. This is not to say that their movements do not influence each other. One possible explanation for the lack of group foraging may be the common observation that vigorous competition occurs among snakes fed together. Garter snakes are highly attracted to prey held in a conspecific's jaws, choosing to attack such prey in preference to identical, easily available prey. This may reflect an enhanced reaction to combined movement and chemical cues (Burghardt and Denny 1983). Garter snakes capturing prey in the presence of others often attempt rapidly to carry it away using a distinctive posture in which the anterior third of the body is raised. With earthworms, in particular, two snakes often grab on to opposite ends and proceed to swallow; when they meet a prolonged conflict takes place akin to that seen in a two-headed snake (Burghardt, in press), but which may tragically end in one snake swallowing, or seriously injuring, the other.

### INFORMATION EXTRACTION

If neonate snakes are better off staying together in retreats and foraging separately, this proximity may provide opportunities for other adaptive possibilities. For example, animals in retreats defaecate and this could provide information to conspecifics as to the kind of food available in habitats other than where an individual has been foraging. This might be particularly important in neonates who, we have seen, are rather diverse in their preferences and even in the propensity to eat at all. Ward and Zahavi (1973) developed an influential information centre hypothesis with regard to certain birds. Galef's work (e.g. 1986) on rats supports this possibility for chemical cue transmission among conspecifics and it is possible that something related could occur in snakes. Lani Lyman and I are exploring this issue with garter snakes and our first study will be outlined here.

Dundee and Miller (1968) suggested that faeces function as an attractant in aggregating adult ringneck snakes (*Diadophis punctatus*). Several studies have established that adult natricine snakes can discriminate conspecific soiled from clean substrate, usually by being attracted to it (Scudder *et al.* 1980; Allen *et al.* 1984; Porter and Czaplicki 1974; Heller and Halpern 1982). With neonates, Graves and Halpern (1988) have shown that conspecific skin lipids are attractive, but no results for faeces have been reported. Unpublished studies in our laboratory have shown that dried conspecific faeces suspended in water and tested using a cotton swab procedure elicit no response either greater or less than water controls, regardless of the donor snakes' diet. The issue now is whether snakes in groups have the opportunity to use more efficiently information in the faeces or scent of other animals in locating their own food resources.

Lani Lyman and I divided 24 newborn plains garter snakes (*Thamnophis radix*) into four groups of six each balanced by initial scores to prey extract of mosquito fish and earthworms, as well as by size and sex. Two groups were raised exclusively on the fish diet and two on the worm diet. At 28 weeks the animals were placed in individual boxes containing a hiding box on either end. In this situation garter snakes will spend most of their time under cover. One cover contained a small square of cardboard substrate with faeces from unfamiliar littermates (those from another of the groups) raised on a fish diet, and the other cover contained a square with faeces from a group of unfamiliar worm-fed littermates. Stimulus positions in the cages were balanced by direction, diet, and sex. The position of the snakes was recorded at three readings per day (12 times).

After each reading the animals were gently moved to the centre by moving the shelter, so as to force the snake to make another choice.

At 35 weeks the experiment was replicated with the stimulus squares containing neither fish nor worm faeces but substances rubbed directly from the surface of either worms or fish. One difference, however, was that the worm and fish rubbings were repeated daily, while for the previous study the stimulus squares were not altered during the observations. During the repeat applications the snake was confined to a shelter and could not see the prey item.

Three weeks later all the snakes were retested for their responses to worm and fish extract using the standard swab method. As expected, diet had a major effect on the swab test scores as the familiar diet was favoured.

The measure used in the extract testing was the standard tongue-flick attack score (TFAS). To obtain a measure of relative preference the worm TFAS was subtracted from the fish TFAS. Thus here and throughout, the more positive the score the greater the fish bias, the more negative the score the more the worm bias. Absolute worm and fish scores are noted as Worm TFAS and Fish TFAS, respectively. The faeces and prey rubbing scores were derived similarly, with +12 being the maximum and −12 the minimum. If the animal was (rarely) in the centre of the cage, by the water bowl, no choice was recorded.

The snakes fed earthworms showed a significantly higher growth rate than those fed fish, as measured by either mass or snout-vent length. Scudder-Davis and Burghardt (1987) reported contrary results, but unlike them, we added calcium and phosphorus to the diets of the worm-fed snakes in this study. Scudder-Davis and Burghardt (1987) had shown that worms had lower amounts of these two minerals and even eating three times as many calories of worms barely compensated for this lack.

To summarize this complex study, Pearson correlations among the 5 measures are presented (Table 3). The F-W faeces score was positively related to the F-W prey deposit score in the habitat experiments. Surprisingly, it was also positively

**Table 3.** Product-Moment correlations among variables measurable in *Thamnophis radix* reared on fish (F) and earthworm (W) diets

| | F−W rub score | F−W TFAS | Fish TFAS | Worm TFAS | Snout-vent growth |
|---|---|---|---|---|---|
| F−W faeces score | 0.48* | −0.20 | 0.31 | 0.46* | 0.25 |
| F−W rub score | | −0.43* | −0.02 | 0.44* | 0.19 |
| F−W TFAS | | | 0.46* | −0.70*** | −0.46* |
| Fish TFAS | | | | 0.30 | 0.18 |
| Worm TFAS | | | | | 0.64** |

*$p<0.05$, **$p<0.01$, ***$p<0.001$

related to worm but not fish TFAS. What seems to be going on here is that both diet groups preferred the side marked with worm odour. Both diet groups were attracted to the side marked with fish faeces, but the worm-fed group showed the greater preference. Note that there was no significant correlation between the fish and worm TFAS scores.

What to conclude from this preliminary study? One conclusion is that different systems may underlie responses to prey cues in the environment and prey cues related to attack. It is essential for any information transfer hypothesis that snakes can discriminate available cues such as faeces deposits made by conspecifics fed on different diets. At the very least this study demonstrates that this can occur.

## Acknowledgements

This work was supported, in part, by the University of Tennessee Science Alliance, NIH training grant T32-HD07303, and NSF research grants BNS 8217569 and BNS 8709629. I thank the many present and former students who have participated in the experiments and animal maintenance, or who commented on the manuscript.

## References

Adams, M.A. and Johnsen, P.B. (1986). Chemical control of feeding in herbivorous and carnivorous fish. In *Chemical signals in vertebrates* 4 (ed. D. Duvall, D. Müller-Schwarze and R.M. Silverstein), pp. 45–61. Plenum, New York.

Allen, B.A., Burghardt, G.M. and York, D.S. (1984). Species and sex differences in substrate preference and tongue flick rate in three sympatric species of water snakes (*Nerodia*). *J. Comp. Psychol.*, 98, 358–67.

Apfelbach, R. (1973). Olfactory sign stimulus for prey selection in polecats (*Putorius putorius* L.). *Z. Tierpsychol.*, 33, 270–3.

Apfelbach, R. (1978). A sensitive phase for the development of olfactory preference in ferrets (*Mustela putorius f. furo* L.). *Z. Saugetierkd.*, 43, 289–95.

Apfelbach, R. (1986). Imprinting on prey odours in ferrets and its neural correlates. *Behav. Processes*, 12, 363–81.

Arnold, S.J. (1978). Some effects of early experience on the feeding responses in the common garter snake, *Thamnophis sirtalis*. *Anim. Behav.*, 26, 455–62.

Arnold, S.J. (1981). The microevolution of feeding behavior. In *Foraging behavior: Ecological, ethological, and psychological approaches* (ed. A.C. Kamil and T.D. Sargent), pp. 409–53. Garland STPM, New York.

Beach, F.A. (1947). Evolutionary changes in the physiological control of mating behavior in mammals. *Psychol. Rev.*, 54, 297–315.

Benham, W.B. (1906). The olfactory sense in *Apteryx*. *Nature*, 74, 222–3.

Burghardt, G.M. (1968). The reinforcement potential of the consummatory act in the modification of chemical-cue preferences. *Amer. Zool.*, 8, 745.

Burghardt, G.M. (1969). Comparative prey-attack studies in newborn snakes of the genus *Thamnophis*. *Behaviour*, 33, 77–114.

Burghardt, G.M. (1970*a*). Chemical perception in reptiles. In *Communication by chemical signals* (ed. J.W. Johnston, Jr., D.G. Moulton and A. Turk), pp. 241–308. Appleton-Century-Crofts, New York.

Burghardt, G.M. (1970*b*). Defining 'communication'. In *Communication by chemical signals* (ed. J.W. Johnston, Jr., D.G. Moulton and A. Turk), pp. 5–18. Appleton-Century-Crofts, New York.

Burghardt, G.M. (1970*c*). Intraspecific geographical variation in chemical food cue preferences of newborn garter snakes (*Thamnophis sirtalis*). *Behaviour*, 36, 246–57.

Burghardt, G.M. (1975). Chemical prey preference polymorphism in newborn garter snakes, *Thamnophis sirtalis*. *Behaviour*, 52, 202–25.

Burghardt, G.M. (1977*a*). Learning processes in reptiles. In *The biology of the Reptilia*, 7 (ecology and behavior) (ed. C. Gans and D. Tinkle), pp. 555–681. Academic Press, New York.

Burghardt, G.M. (1977*b*). The ontogeny, evolution, and stimulus control of feeding in humans and reptiles. In *The chemical senses and nutrition* (ed. M.R. Kare and O. Maller), pp. 253–75. Academic Press, New York.

Burghardt, G.M. (1978). Behavioral ontogeny in reptiles: Whence, whither, and why. In *The development of behavior: comparative and evolutionary aspects* (ed. G.M. Burghardt and M. Bekoff), pp. 149–74. Garland STPM, New York.

Burghardt, G.M. (1980). Behavioral and stimulus correlates of vomeronasal functioning in reptiles: Feeding, grouping, sex, and tongue use. In *Chemical signals: vertebrates and aquatic invertebrates* (ed. D. Müller-Schwarze and R.M. Silverstein), pp. 275–301. Plenum, New York.

Burghardt, G.M. (1983). Aggregation and species discrimination in newborn snakes. *Z. Tierpsychol.*, 61, 89–101.

Burghardt, G.M. (in press). Cognitive ethology and critical anthropomorphism: a snake with two heads and hognose snakes that play dead. In *Cognitive ethology: the minds of other animals* (ed. C.A. Ristau). Lawrence Erlbaum, Hillsdale, N.J.

Burghardt, G.M. and Pruitt, C.H. (1975). The role of the tongue and senses in feeding of newborn garter snakes. *Physiol. Behav.*, 14, 185–94.

Burghardt, G.M. and Denny, D. (1983). Effects of prey movement and prey odor on feeding in garter snakes. *Z. Tierpsychol.*, 62, 329–47.

Burghardt, G.M. and Rand, A.S. (1985). Group size and growth rate in hatchling green iguanas (*Iguana iguana*). *Behav. Ecol. Sociobiol.*, 18, 101–4.

Burghardt, G.M., Goss, S.E. and Schell, F.M. (1988). Comparison of earthworm and fish derived chemicals eliciting prey attack by garter snakes (*Thamnophis*). *J. Chem. Ecol.*, 14, 855–81.

Carr, W. (1982). Chemical stimulation of feeding behavior. In *Chemoreception in fishes* (ed. T.J. Hara), pp. 259–75. Elsevier, New York.

Chapman, E. (1929). *My tropical air castle*. D. Appleton, New York.

Chiszar, D. (1986). Motor patterns dedicated to sensory functions. In *Chemical signals in vertebrates* 4, (ed. D. Duvall, D. Müller-Schwarze and R.M. Silverstein), pp. 37–44. Plenum, New York.

Chiszar, D., Taylor, S.V., Radcliffe, C.W., Smith, H.B. and O'Connell, B. (1981). Effects of chemical and visual stimuli upon chemosensory searching by garter snakes and rattlesnakes. *J. Herpetol.*, 15, 415–24.

Chiszar, D., Radcliffe, C.W., Scudder, K.M. and Duvall, D. (1983). Strike-induced chemosensory searching by rattlesnakes: The role of envenomation-related chemical cues in the post-strike environment. In *Chemical signals in vertebrates*, 3 (ed. D. Müller-Schwarze and R.M. Silverstein), pp. 1–24. Plenum, New York.

Chiszar, D., Radcliffe, C.W., Overstreet, R., Poole, T. and Byers, T. (1985). Duration of strike-induced chemosensory searching in cottonmouths. *Can. J. Zool.*, 63, 1057–61.

Chou, L.M., Leong, C.F. and Choo, B.L. (1988). The role of optic, auditory, and olfactory senses in prey hunting by two species of geckos. *J. Herpetol.*, 22, 349–51.

Cocke, R. and Thiessen, D.D. (1986). Chemocommunication among prey and predator species. *Anim. Learn. Behav.*, 14, 90–2.

Cooper, W.E., Jr. (1989). Prey odor discrimination in the varanoid lizards *Heloderma suspectum* and *Varanus exanthematicus*. *Ethology*, 81, 250–8.

Cooper, W.E., Jr. (in press). Strike-induced chemosensory searching occurs in lizards. *Copeia*.

Cooper, W.E., Jr., McDowell, S.G. and Ruffer, J. (1989). Strike-induced chemosensory searching in the colubrid snakes *Elpahe g. guttata* and *Thamnophis s. sirtalis*. *Ethology*, 81, 19–28.

Cooper, W.E., Jr. and Burghardt, G.M. (in press). A comparative analysis of scoring methods for chemical discrimination of prey by squamate reptiles. *J. Chem. Ecol.*

Curio, E. (1976). *The ethology of predation*. Springer, New York.

Cushing, B.S. (1983). Responses of polar bears to human menstrual odors. *Int. Conf. Bear Res. Managm.*, 5, 270–4.

Cushing, B.S. (1984). A selective preference by least weasels for oestrous versus dioestrous urine of prairie deer mice. *Anim. Behav.*, 32, 1263–5.

Dole, J.W., Rose, B.B. and Tachiki, K.H. (1981). Western toads (*Bufo boreas*) learn odor of prey insects. *Herpetologica*, 37, 63–8.

Drummond, H.M. (1979). Stimulus control of amphibious predation in the northern water snake (*Nerodia s. spiedon*). *Z. Tierpsychol.*, 50, 18–44.

Drummond, H.M. (1985). The role of vision in predatory behaviour of natricine snakes. *Anim. Behav.*, 33, 206–15.

Drummond, H. and Burghardt, G.M. (1983). Geographic variation in the foraging behavior of the garter snake, *Thamnophis elegans*. *Behav. Ecol. Sociobiol.*, 12, 43–8.

Dundee, H.A. and Miller, M.C. III (1968). Aggregative behavior and habitat conditioning by the prairie ringneck snake, *Diodophis punctatus arnyi*. *Tulane Stud. Zool. Bot.*, 15, 41–58.

Durand, J.P., Parzefall, J. and Richard, B. (1982). Comparative study of chemical prey detection in the cave-living *Proteus anguinus* and its epigean relative *Necturus maculosus* (Proteidae, Urodela). *Behav. Processes*, 7, 123–34.

Ellingsen, O.F. and Doeving, K.B. (1986). Chemical fractionation of shrimp extracts inducing bottom food search behavior in cod. *J. Chem. Ecol.*, 12, 155–68.

Fitch, H.S. (1965). An ecological study of the garter snake, *Thamnophis sirtalis*. *Univ. Kansas Publ. Mus. Nat. Hist.*, 15, 493–564.

Fuchs, J. and Burghardt, G.M. (1971). Effects of early feeding experience on the responses of garter snakes to food chemicals. *Learn. Motiv.*, 2, 271–9.

Galef, B.G. (1986). Olfactory communication among rats: Information concerning distant diets. In *Chemical signals in vertebrates*, 4 (ed. D. Duvall, D. Müller-Schwarze and R.M. Silverstein), pp. 487–505. Plenum, New York.

Gillingham, J.C. and Clark, D.L. (1981). Snake tongue flicking: Transfer mechanisms of Jacobson's organ. *Can. J. Zool.*, 59, 1651–7.

Gove, D. and Burghardt, G.M. (1975). Responses of ecologically dissimilar populations of the water snake, *Natrix a. sipedon* to chemical cues from prey. *J. Chem. Ecol.*, 1, 25–40.

Graves, B.M. and Duvall, D. (1985). Avomic prairie rattlesnakes (*Crotalus viridis*) fail to attack rodent prey. *Z. Tierpsychol.*, 67, 161–6.

Graves, B.M., Duvall, D., King, M.B., Lindstedt, S.L. and Gern, W.A. (1986). Initial den location by neonatal prairie rattlesnakes: Functions, causes, and natural history in chemical ecology. In *Chemical signals in vertebrates*, 4 (ed. D. Duvall, D. Müller-Schwarze and R.M. Silverstein), pp. 285–304. Plenum, New York.

Graves, B.M. and Halpern, M. (1988). Neonate plains garter snakes (*Thamnophis radix*) are attracted to conspecific skin extracts. *J. Comp. Psychol.*, 102, 251–3.

Halloy, M. and Burghardt, G.M. (in press). Ontogeny of fish capture and ingestion in four species of garter snakes (*Thamnophis*). *Behaviour*.

Halpern, M. (1980). Chemical ecology of terrestrial vertebrates. In *Animals and environmental fitness* (ed. R. Giles), pp. 263–82. Pergamon, Oxford.

Halpern, M. (1987). The organization and function of the vomeronasal system. *Annu. Rev. Neurosci.*, 10, 325–62.

Halpern, M. (1988). Vomeronasal system functions: role in mediating the reinforcing properties of chemical stimuli. In *The forebrain of reptiles*, (ed. W.K. Schwerdtfeger and J.A.J. Smeets), pp. 142–50. Karger, Basel.

Halpern, M. and Kubie, J. (1984). The role of the ophidian vomeronasal system in species-typical behavior. *Trends Neurosci.*, 7, 472–7.

Halpern, M., Schulman, N. and Kirschenbaum, D.M. (1986). Characteristics of earthworm washings detected by the vomeronasal system of snakes. In *Chemical signals in vertebrates*, 4 (ed. D. Duvall, D. Müller-Schwarze and R.M. Silverstein), pp. 63–77. Plenum, New York.

Harriman, A.E. and Berger, R.H. (1986). Olfactory acuity in the common raven. *Physiol. Behav.*, 36, 257–62.

Heller, S.B. and Halpern, M. (1982). Laboratory observations of aggregative behavior of garter snakes, *Thamnophis sirtalis*. *J. Comp. Physiol. Psychol.*, 96, 967–83.

Langley, W.M. (1983*a*). Relative importance of distance senses in the predatory behavior of the northern grasshopper mouse (*Onychomys leucogaster*). *Anim. Behav.*, 31, 199–205.

Langley, W.M. (1983*b*). Stimulus control of feeding in the grasshopper mouse. *Z. Tierpsychol.*, 62, 291–306.

Langley, W.M. (1984). Recognition of prey species by their odors in the grasshopper mouse (*Onychomys leucogaster*). *Behav. Processes*, 9, 277–80.

Langley, W.M. (1985). Relative importance of the distance senses in hamster predatory behavior. *Behav. Processes*, 10, 229–39.

Langley, W.M. and Knapp, K. (1982). Importance of olfaction to suppression of attack response through conditioned taste aversion in the grasshopper mouse. *Behav. Neural Biol.*, 36, 368–78.

Lemly, A.D. and Smith, R.J.F. (1987). Effects of chronic exposure to acidified water on chemoreception of feeding stimuli in fathead minnows (*Pimephales promelas*): Mechanisms and ecological implications. *Environ. Toxicol. Chem.*, 6, 225–38.

Lindquist, S.B. and Bachman, M.D. (1982). The role of visual and olfactory cues in the prey catching behavior of the tiger salamander, *Ambystoma tigrinum*. *Copeia*, 1982, 81–90.

Luthardt, G. and Roth, G. (1983). The interaction of the visual and olfactory systems in guiding prey catching behaviour in *Salamandra salamandra* (L.). *Behaviour*, 83, 69–79.

Luthardt-Lamier, G. and Roth, G. (1983). Reduction of visual inhibition to stationary prey by early experience in *Salamandra salamandra* (L.). *Z. Tierpsychol.*, 63, 294–302.

Mason, J.R. and Clark, L. (1986). Chemoreception and the selection of green plants as nest fumigants by starlings. In *Chemical signals in vertebrates*, 4 (ed. D. Duvall, D. Müller-Schwarze and R.M. Silverstein), pp. 369–84. Plenum, New York.

Mason, J.R., Johri, K.K. and Morton, T.H. (1987). Generalization in olfactory detection of chemical cues containing carbonyl functions by tiger salamanders. *J. Chem. Ecol.*, 13, 1–18.

Mearns, K.J. (1986). Sensitivity of brown trout and Atlantic salmon fry to amino acids at the start of exogenous feeding. *Aquaculture*, 55, 191–200.

Melcer, T. and Chiszar, D. (1989). Striking prey creates a specific chemical search image in rattlesnakes. *Anim. Behav.*, 37, 477–86.

Mushinsky, H.R. and Lotz, K.H. (1980). Responses to two sympatric water snakes to extracts of commonly ingested prey species: ontogenetic and ecological considerations. *J. Chem. Ecol.*, 6, 523–35.

Mushinsky, H.R., Hebrard, J.J. and Vodopich, D.S. (1982). Ontogeny of water snake foraging ecology. *Ecology*, 63, 1624–9.

Nicoletto, P.F. (1985). The role of vision and the chemical senses in predatory behavior of the skink, *Scincella lateralis*. *J. Herpetol.*, 19, 487–91.

Noble, G.K. and Clausen, H.J. (1936). The aggregation behavior of *Storeria dekayi* and other snakes, with especial reference to the sense organs involved. *Ecol. Monogr.*, 6, 269–316.

Plummer, M.V. and Goy, J.M. (1984). Ontogenetic dietary shift of water snakes in a fish hatchery. *Copeia*, 1984, 550–2.

Porter, R.H. and Czaplicki, J.A. (1974). Responses of water snakes (*Natrix r. rhombifera*) and garter snakes (*Thamnophis sirtalis*) to chemical cues. *Anim. Learn. Behav.*, 2, 129–32.

Reese, A.M. (1912). Food and chemical reactions of the spotted newt, *Diemyctylus viridescens*. *J. Anim. Behav.*, 2, 190–208.

Rehn, B., Breipohl, W., Mendoza, A.S. and Apfelbach, R. (1986). Changes in granule cells of the ferret olfactory bulb associated with imprinting on prey odours. *Brain Res.*, 373, 1–2.

Risser, J. (1914). Olfactory reactions in amphibians. *J. Exp. Zool.*, 16, 617–52.

Roth, G. (1987). *Visual behavior in salamanders*. Springer, New York.

Rottier, D.V. and Lemm, C.A. (1985). Movement of underyearling walleyes in response to odor and visual cues. *Prog. Fish-Cult.*, 47, 34–41.

Royce-Malmgren, C.H. and Watson, W.H. (1987). Modification of olfactory-related behavior in juvenile Atlantic salmon by changes in pH. *J. Chem. Ecol.*, 13, 533–46.

Schell, F.M., Burghardt, G.M., Johnston, A. and Coholich, C. (in press). Analysis of chemicals from earthworms and fish that elicit prey attack by ingestively naïve garter snakes (*Thamnophis*). *J. Chem. Ecol.*

Schwartz, J.M. (1989). Multiple paternity and offspring variability in wild populations of the garter snake, *Thamnophis sirtalis* (Colubridae). Unpublished Ph.D. thesis, University of Tennessee, Knoxville.

Schwartz, J.M., MacCracken, G.F. and Burghardt, G.M. (in press). Multiple paternity in wild populations of the garter snake, *Thamnophis sirtalis*. *Behav. Ecol. Sociobiol.*

Scott, T.P. and Weldon, P.J. (in press). Chemoreception in the feeding behaviour of adult American alligators, *Alligator mississippiensis*. *Anim. Behav.*

Scudder, K.M., Stewart, N.J. and Smith, H.M. (1980). Response of neonate water snakes (*Nerodia sipedon*) to conspecific chemical cues. *J. Herpetol.*, 14, 196–8.

Scudder-Davis, R.M. and Burghardt, G.M. (1987). Diet and growth in juveniles of the garter snakes *Thamnophis sirtalis infernalis* and *Thamnophis radix radix*. *Growth*, 51, 74–85.

Sheffield, L.P., Law, J.H. and Burghardt, G.M. (1968). On the nature of chemical food sign stimuli for newborn snakes. *Comm. Behav. Biol.* [A], 2, 7–12.

Shinn, E.A. and Dole, J.W. (1979). Lipid components of prey odors elicit feeding responses in western toads (*Bufo boreas*). *Copeia*, 1979, 275–8.

Simon, C.A. (1983). A review of chemoreception in lizards. In *Lizard ecology: Studies on a model organism* (ed. R.B. Huey, E.R. Pianka and T.W. Schoener), pp. 119–33. Harvard University Press, Cambridge, Mass.

Stager, K.E. (1964). The role of olfaction in food location by the turkey vulture (*Cathartes aura*). *Los Angeles County Mus. Contrib. Sci.*, 81, 3–63.

Stoddart, D.M. (1980). *The ecology of vertebrate olfaction.* Chapman and Hall, London.

Thiessen, D.D. and Cocke, R. (1986). Alarm chemosignals in a *Meriones unguiculatus*: Prey-predator interactions. In *Chemical signals in vertebrates*, 4 (ed. D. Duvall, D. Müller-Schwarze and R.M. Silverstein), pp. 506–18. Plenum, New York.

Von Achen, P.H. and Rakestraw, J.L. (1984). The role of chemoreception in the prey detection of neonate reptiles. In *Vertebrate ecology and systematics—A tribute to Henry S. Fitch* (ed. R.A. Siegel, L.E. Hunt, J.L. Knight, L. Malaret and N.L. Zuschlag), pp. 163–72. Special Publ. Univ. Kans. Mus. Nat. Hist., Lawrence.

Wang, D., Chen, P., Jiang, X.C. and Halpern, M. (1988). Isolation from earthworms of a proteinaceous chemoattractant to garter snakes. *Arch. Biochem. Biophysics*, 267, 459–66.

Ward, P. and Zahavi, A. (1973). The importance of certain assemblages of birds as 'information centres' for food finding. *Ibis*, 115, 517–34.

Weldon, P.J. (1988). Feeding responses of Pacific snappers (genus *Lutjanus*) to the yellow-bellied sea-snake (*Pelamis platurus*). *Zool. Sci.*, 5, 443–8.

Weldon, P.J. and Schell, F.M. (1984). Responses by king snakes to chemicals from colubrid and crotaline snakes. *J. Chem. Ecol.*, 10, 1509–20.

Wenzel, B.M. (1968). Olfactory prowess of the kiwi. *Nature*, 220, 1133–4.

Wenzel, B.M. (1986). The ecological and evolutionary challenges of Procellarii-form olfaction. In *Chemical signals in vertebrates*, 4 (ed. D. Duvall, D. Müller-Schwarze and R.M. Silverstein), pp. 357–68. Plenum, New York.

Whitman, D.W., Blum, M.S. and Jones, C.G. (1986). Olfactorily mediated attack suppression in the southern grasshopper mouse toward and unpalatable prey. *Behav. Process*, 13, 77–83.

Wysocki, C.J., Beauchamp, G.K., Reidinger, R.R. and Wellington, J.L. (1985). Access of large and nonvolatile molecules to the vomeronasal organ of mammals during social and feeding behaviors. *J. Chem. Ecol.*, 11, 1147–66.

# V.2 Responses by Vertebrates to Chemicals from Predators

PAUL J. WELDON

## Introduction

This chapter describes those vertebrates for which the detection of predator chemicals has been demonstrated or is suspected. The methods by which this chemoreceptive function has been investigated are described. The observations and experiments providing the basis for this review have appeared in a variety of journals, edited volumes, and theses. Since some reports are not readily available, I have attempted to summarize and synthesize the literature on this topic.

The first three sections of this chapter describe responses to predatory fishes, amphibians, reptiles, and mammals, respectively; subsections detail reactions according to class, order, suborder, or family of responding species. Behavioural, ecological, and evolutionary aspects of chemically elicited responses to predators are discussed in the last section.

## Responses to fishes

### BY CYPRINIDS

European minnows (*Phoxinus phoxinus*) were tested for an ability to discriminate chemicals from different fishes (Göz 1941). Water from the tank of a piscivorous pike (*Esox lucius*) caused minnows to swim rapidly and leap out of water, or to halt, sink to the aquarium floor, and prop themselves aloft on their anal and ventral fins; this contrasts with their resting posture, where their belly contacts the substrate. Göz suggested that by freezing, minnows are less likely to be detected by the pike's visual and lateral line senses.

Some minnows consistently fled from pike scent. Others froze or fled in different tests; a few were unresponsive. In one test, eleven individuals that had not responded to chemicals from any stimulus fish were confined with a pike: all were attacked, and one was eaten. When retested, four survivors were still unresponsive; others showed typical reactions to the scents of pike and other fishes. Göz suggested that his subjects had developed a learned association between pike chemicals and their abrasion-released alarm pheromones (disseminated during pike attacks); the pike encounter presumably had also induced sensitivity to other heterospecific odours.

Schooling minnows failed to react to pike scent, even if the group was comprised of known responders (Göz 1941). Extreme flight, however, was observed when *P. phoxinus* alarm pheromones were presented along with pike chemicals (Frisch 1941).

Blinded minnows were tested with water conditioned by heterospecifics (Göz 1941). A pike strongly elicited flight or freezing, and a perch (*Perca fluviatilis*) also was highly effective. A trout (*Salmo fario*) elicited a moderate response. Lesser reactions were elicited by a bullhead (*Ictalurus nebulosus*), salmon (*S. salar*), stickleback (*Gasterosteus aculeatus*), and miller's thumb (*Cottus gobio*). A gudgeon (*Gobio gobio*), bitterling (*Rhodeus sericeus*), and eel (*Anguilla anguilla*) were ineffective, although the latter is notoriously piscivorous.

Visually intact minnows responded less intensely and in shorter bouts than did the blinded individuals. Strong reactions were shown to a pumpkinseed (*Lepomis gibbosus*), but neither a swordtail (*Xiphophorus helleri*) nor a gudgeon were effective.

Göz concluded that *P. phoxinus* distinguishes among chemicals from different fishes and detects dangerous ones, especially pike. The ability to respond to pike chemicals, Göz noted, may be significant given this predator's tendency to lurk in dense vegetation, where it would be difficult to detect visually.

Shiners (*Notropis texanus* and *N. venustus*) and chubs (*Hybopsis aestivalis*) were tested with human hand rinse, abrasion-released alarm pheromones, and the scents of a pike, sunfish (*Lepomis macrochirus*), bass (*Micropterus punctulatus*), and two South American cichlids, *Astronotus ocellatus* and *Cichlasoma severum* (Reed 1969). All stimulus fishes are piscivorous. Neither plain water nor human hand rinse poured into tanks had an effect, but similar reactions were exhibited to heterospecific scents and alarm pheromones. Both shiners clustered and swam to the bottom of their tanks; the chubs were less reactive.

Reed noted that piscivores need not be sympatric with test fishes to elicit escape. Even the South American cichlids were effective in many cases. The inclusion of non-piscivorous fishes among the array of stimulus species could have strengthened Reed's conclusions.

Goldfish (*Carassius auratus*) were presented with water from an aquarium containing a bullhead (*Ictalurus melas*) (Davy and Kleerekoper 1971). Most subjects avoided the area into which chemicals were introduced; others were indifferent or even attracted to the scent. However, no quantitative measures were reported.

#### BY CYPRINODONTIDS

Topminnows (*Fundulus olivaceous*) were tested by Reed (1969) with scents from the piscivores listed previously. This species occurs in habitats with some of the stimulus fishes, but it reacted weakly to only two allopatric ones, *Cichlasoma* and *Esox*.

BY ICTALURIDS

A freshly captured female bullhead (*Ictalurus natalis*) tending her fry became active when a pumpkinseed was introduced, and she attacked it vigorously (Atema 1969). Atema suggested that chemicals triggered the female's search for a predator, although no tests of this were performed.

BY POECILIDS

Mosquitofishes (*Gambusia affinis* and *G. patruelis*) normally are active swimmers. When presented with water conditioned by a pike (*Esox niger*) or pickerel (*E. americanus*), they stiffened their bodies, dampened the sweep of their tail, and remained near the water surface (George 1960). Here, the slightest noise caused these fishes to skip across the water in a zigzag pattern and to leap as much as 15 cm during their escape.

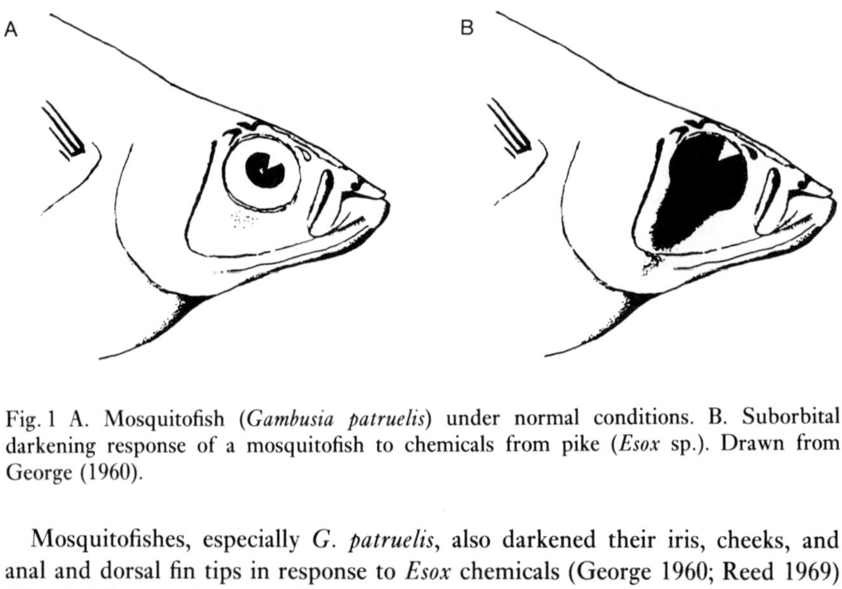


Fig. 1 A. Mosquitofish (*Gambusia patruelis*) under normal conditions. B. Suborbital darkening response of a mosquitofish to chemicals from pike (*Esox* sp.). Drawn from George (1960).

Mosquitofishes, especially *G. patruelis*, also darkened their iris, cheeks, and anal and dorsal fin tips in response to *Esox* chemicals (George 1960; Reed 1969) (Fig. 1). The significance of this response is not clear. Esocids themselves possess a dark vertical stripe along the side of their head through the eye. If *Esox* spp. hesitate to eat small conspecifics, George reasoned, a mosquitofish that resembled one may less likely be attacked. However, he favoured the idea that the suborbital darkening signals a readiness to take evasive measures. According to this declaration of awareness hypothesis, *Esox* spp. are less likely to pursue mosquitofishes prepared to flee.

*Gambusia patruelis* was tested with water conditioned by a bass (*Amblo-plites rupestris*), goldfish (*Carassius auratus*), killifish (*Fundulus heteroclitus*), mosquitofish (*G. affinis*), bullhead (*I. nebulosus*), pumpkinseed (*L. gibbosus*), shiner (*Notemigonus chrysoleucus*), trout (*Salvelinus frontialis*), bullfrog (*Rana catesbiana*), turtle (*Chrysemys picta*), and alligator (*Alligator mississippiensis*). None elicited flight or suborbital darkening. Peak flight by *G. affinis* was observed to pike scent (Reed 1969), further attesting to the extreme responses mosquitofishes display to these predators.

George determined that chemicals eliciting mosquitofishes' reactions arise from the pickerel's mucus coat and survive passage through filter paper and hours of bubbling. Activity was lost when *Esox*-conditioned water was heated or passed over activated charcoal.

#### BY SALMONIDS

Juvenile salmon (*Oncorhynchus kisutch*) in a Y maze avoided chemicals from a piscivorous squawfish (*Ptyhocheilus oregonensis*), but not those from an innocuous sucker (*Catostomus macrocheilus*) (Rehnberg and Schreck 1987). Paradoxically, chemicals from both species triggered elevations in plasma cortisol and glucose, suggesting a stress response. Rehnberg and Schreck concluded that behavioural reactions to predator chemicals need not parallel those at a physiological level.

#### BY AMPHIBIANS

Salamanders (*Eurycea bislineata*) avoided water conditioned by a predatory sunfish (*Lepomis cyanellus* ) but they showed no response to water conditioned by tadpoles (*Rana clamitans*) (Petranka *et al.* 1987). Tadpoles (*Hyla chrysoscelis*) spent more time in refuges when presented with sunfish scent than with plain water. Other amphibian larvae also avoided sunfish cues. Petranka *et al.* suggested that lingering chemicals provide erroneous information on the presence of predators, unnecessarily curtailing foraging activities of would-be prey.

Larval salamanders (*Ambystoma texanum*) increased their use of refuges when exposed to sunfish chemicals, but rinses of a turtle, snake, and crayfish had no effect (Kats 1988). Larvae with plugged nares failed to respond, suggesting that olfaction is used to detect fish predators.

### Responses to amphibians

#### BY SALAMANDERS

Salamanders (*Desmognathus monticola*) avoided filter papers conditioned by a congener, *D. quadramaculatus*, but the converse did not obtain (Roudebush and Taylor 1987). This chemosensitivity was hypothesized to lower the risk of predation to *D. monticola*, although predation has not been reported between these species.

## Responses to reptiles

### BY SALAMANDERS

Salamanders (*Plethodon richmondi* and *P. dorsalis*) presented with paper towels treated with water or conditioned by ringneck snakes (*Diadophis punctatus*) avoided the towels treated with snake odours (Cupp 1988).

### BY LIZARDS

Lizards (*Lacerta vivipara*) were tested for responses to a variety of chemicals, including those from three snakes, *Natrix natrix*, *Vipera berus*, and *Coronella austriaca*; only the latter two are saurophagous (Thoen *et al.*1986). Lizards tongue-flicked more to *Vipera* and *Coronella* scents than to other conditions. Tail vibration and foot shaking were exhibited to snakes and conspecifics; in social encounters, these aspects of behaviour were thought to act as appeasement displays. Thoen *et al.* suggested that both intraspecific interactions and encounters with predators were conflict situations to which lizards gave similar responses. Slow locomotion, often with jerking forelimbs, was observed when lizards were in cages conditioned by snakes. Thoen *et al.* stated that this is a typical response to predator cues.

Banded geckos (*Coleonyx variegatus*) exhibited tail waving in the presence of a saurophagous leaf-nosed snake (*Phyllorhyncus decurtatus*) (Dial *et al.* 1989). Chemicals from this snake's skin elicited tail-waving and defensive attacks.

### BY SNAKES

Rattlesnakes (*Crotalus* spp.) and other crotalines typically exhibit body bridging, where the mid-body is raised and undulated, when confronted by ophiophagous kingsnakes (*Lampropeltis getulus*) (Carpenter and Gillingham 1975; Weldon and Burghardt 1979; Marchisin 1980 and others). This posture may obstruct attempts to grasp the head of a crotaline and, possibly, allow a body blow to be levelled toward an advancing predator. Trunk inflation, biting, and other reactions also occur, but the vertically coiled threat posture displayed to humans, where the head is raised and the tail is vibrated, is not exhibited in encounters with snakes (Bogert 1941; Marchisin 1980).

Bogert suggested that rattlesnakes' reactions to ophiophagous snakes are chemically elicited, since subjects placed into jars conditioned by kingsnakes ceased rattling, lowered their heads, and moved erratically. Some rattlesnakes exhibited body bridging when presented with pine sticks rubbed against stimulus snakes. Moreover, air directed over a kingsnake or an extract of its skin elicited increased heart rates, as did skunk scent (Cowles and Phelan 1958).

Bogert presented rattlesnakes with sticks treated with scent gland secretions or dorsal or ventral skin substances from stimulus snakes. Dorsal skin chemicals elicited body bridging; other substances gave mixed or negative results. Marchisin (1980) confirmed Bogert's contention that dorsal skin chemicals, and not those from the scent glands or ventral skin, are effective by presenting materials from kingsnakes and indigo snakes (*Drymarchon corais*).

Members of several crotaline genera, including *Agkistrodon*, *Bothrops*, *Crotalus*, and *Sistrurus*, respond defensively to ophiophagous snakes. In one study, more than 20 South, Central, and North American species were observed body bridging to cotton balls rubbed against the dorsal skins of known or suspected ophiophagous snakes (Weldon and Burghardt 1979). Tests of litters of naïve crotalines indicated that defensive reactions occur when ophiophage chemicals are first presented (Weldon and Burghardt 1979; Marchisin 1980). Not all individuals in the litters, however, gave identical responses, and some failed to respond.

Kingsnakes have been the focal stimulus predator in a number of studies of crotaline defences. Other snakes also elicit reactions. Many, such as indigo snakes, are ophiophagous; others apparently are not. A rattlesnake (*C. atricaudatus*), for example, exhibited body bridging to a water snake (*Nerodia fasciata*), a predator of amphibians and fishes (Marchisin 1980). Skin rubbings of *Pseustes sulfureus*, a South American bird- and lizard-eating species, elicited reactions in several North American rattlesnakes (Bogert 1941).

A number of ophiophages have been documented to elicit reactions in allopatric crotalines (Bogert 1941; Weldon and Burghardt 1979; Marchisin 1980). However, fresh skin rubbings from boid, crotaline, and colubrid snakes that are not habitually ophiophagous were ineffective (Bogert 1941).

Rattlesnakes and other crotalines tongue-flick infrequently during encounters with snake predators (Carpenter and Gillingham 1975; Marchisin 1980), a feature that presumably reduces from attention being drawn to their head. Chemicals elicited this response since water moccasins (*Agkistrodon piscivorus*) exhibited less tongue-flicking in a cage previously occupied by a kingsnake than in one occupied by a (rodent-eating) hognose snake (*Heterodon nasicus*) or in an open field (Chiszar *et al.* 1978). A similar trend was observed in a rattlesnake (*C. viridis*).

Some colubrid snakes also detect chemicals from snake predators. Garter snakes (*Thamnophis* spp.) exhibited more tongue-flicks during airborne and swab presentations of chemicals from an ophiophagous racer (*Coluber constrictor*) and a kingsnake than to those from some generally non-ophiophagous species or to blank conditions (Weldon 1982). Corn snakes (*Elaphe guttata*) exhibited more tongue-flicks to the skin extracts of heterospecific snakes, but they apparently did not discriminate between ophiophagous and non-ophiophagous species in all cases (Weldon *et al.* 1990).

Pine snakes (*Pituophis melanoleucus*) followed trails of conspecifics, but not those deposited by kingsnakes (Burger 1989). The results of trail-following tests may provide results related more readily than tongue-flicking to survivorship *vis-à-vis* ophiophagous snakes. Experiments comparing responses to trails from ophiophagous and non-ophiophagous snakes are needed.

### BY RODENTS

Laboratory mice (*Mus musculus*) were placed into an arena with a floor lined, on one half, with plain paper sprayed with water and, on the other half, with paper that had lined the cages of a rat snake (*Elaphe obsoleta*) and an (earthworm-eating) earth snake (*Virgina striatulata*) (Weldon *et al.* 1987). Male mice failed to respond, but females deposited more faecal boli on the side of the arena treated with rat snake scent. Females also ate less of food pellets treated with rat snake skin extract than of those treated with solvent alone.

Wood rats (*Neotoma albigula*) responded to a variety of snakes by vibrating their tail and kicking their hindfeet on to the substrate (Richardson 1942). The involvement of chemical releasers was tested by presenting rats with snakes in glass jars with or without lids. Rats oriented slowly to the sealed jar and gave brief, low key alarm. Snakes in uncapped jars elicited alarm typical of that given during actual encounters, leading Richardson to conclude that snake chemicals are important in eliciting a defensive reaction. The source of relevant chemicals was not determined, but the scent gland secretions of a (non-rodent eating) garter snake (*Thamnophis* sp.) had no effect and a shed snake skin elicited mild reactions.

Unfortunately, Richardson's study lacked quantitative behavioural measures. Moreover, it is not clear that visual cues were eliminated in his tests of chemical elicitation.

Kangaroo rats (*Dipodomys merriami*) with surgically induced sensory deprivations were placed in encounters with a sidewinder (*Crotalus cerastes*) (Webster 1973). Rats subjected to olfactory bulbectomy, unlike intact, blinded, or deafened individuals, did not avoid the snake, and even walked on it. Webster concluded that olfactory bulbs—and presumably olfaction—are necessary for kangaroo rats to identify sidewinders. Another kangaroo rat, *D. spectabilis*, was observed sniffing snake chemical deposits (Randall and Stevens 1987).

The involvement of chemical releasers in the snake-directed defences of California ground squirrels (*Spermophilus beecheyi*) was suggested by the observation that a squirrel flagged its tail and kicked sand at an artifical burrow in which a conspecific had encountered a snake (Hennessy and Owings 1979). To test for chemical effects, squirrels were presented with gopher snakes (*Pituophis melanoleucus*) or rattlesnakes (*Crotalus viridis*) in perforated or intact, clear plastic bags. Squirrels kicked sand and approached more frequently when snakes were confined in perforated bags, indicating that they perceived snake chemicals. Rattlesnakes, which are suspected of posing a greater danger to squirrels, elicited more extreme reactions. These results agree with the observation that squirrels plugged their burrows more frequently when rattlesnakes intruded than when gopher snakes did so (Coss and Owings 1978).

Fig. 2 A. Siberian chipmunk (*Eutamias sibiricus asiaticus*) gnaws on the skin of a rat snake (*Elaphe* sp.) carcass. B. Chipmunk applies snake-derived substances to its coat.

Siberian chipmunks (*Eutamias sibiricus*) encountering live snakes vibrated their tail and thumped their feet (Kobayashi 1987). Although chipmunks reacted to chemicals from several vertebrates, none elicited these reactions as strongly as did snakes. Chipmunks encountering a dead snake, bit off pieces of its skin and rubbed these saliva laden materials on to their fur (Kobayashi and Watanabe 1986) (Fig. 2). The carcass and glandular exudates of other vertebrates were presented, but only snake chemicals elicited scent application behaviour.

Kobayashi and Watanabe hypothesized that snake chemicals applied a chipmunk's fur render it less vulnerable to chemosensory oriented predators. Rat snakes abstained from eating mice treated with snake-derived substances, thus supporting this hypothesis. Similar effects possibly are achieved by other animals using endogenously produced substances that mimic the scent of their predators.

## Responses to mammals

### BY SALMONIDS

Salmon (*Oncorhynchus kisutch* and *O. tshawytscha*) rapidly retreated to lower levels of a fish ladder when rinses of human hand, bear paw, deer feet, and dog and sea-lion meat were introduced upstream (Brett and MacKinnon 1954). Human hand rinse also repelled *O. keta*, *O. gorbuscha*, and *O. nerka*, with successive hand rinses decreasingly effective (Idler *et al.* 1956). Extracts of seal and sea-lion skin elicited similar reactions. Commercially available phosphate esters, fatty acids, purines, and other compounds were ineffective.

Electrophoretic and chromatographic separations of human hand rinse indicated that L-serine is a salmon repellent (Idler *et al.* 1956; cf. Jones and Hara 1982). This amino acid reportedly was effective in concentrations of the same order of magnitude as the hand rinses. The reactions elicited by L-serine were less dramatic and of shorter duration than those elicited by human scent. Idler *et al.*, therefore, suggested that other chemicals are also involved.

Rehnberg and Schreck (1986) agreed that *O. kisutch* avoids L-serine, but they expressed doubt that this reaction is related to predator avoidance. They suggested that salmonids avoided certain amino acids because they are novel or unpleasant when presented in concentrations above those normally encountered in nature.

### BY INSECTIVORES

Tenrecs were tested for responses to urine, faeces, and scent gland secretions of the viverrids *Fossa fossa*, *Galidia elegans*, and *Viverricula indica* (Eisenberg and Gould 1970). *Microgale dobsoni* gave an open mouth threat display to *Galidia* scent. *Hemicentetes semispinosus* erected its quills and bucked while running toward cloth impregnated with the odours of *Galidia* or *Fossa*. *Tenrec ecaudatus* hissed and erected its hair when presented with viverrid chemicals, and responded with slashing bites when touched. Some of these reactions also were shown to dogs and humans. The extent, therefore, to which they were elicited by chemicals needs further study.

### BY RODENTS

The literature on mammal's responses to predator chemicals is dominated by reports of rodents' reactions to carnivore scents. Laboratory rats have been popular subjects in these studies.

Rats (*Rattus norvegicus*), three weeks old or older, huddled in corners, froze, and crouched when cats were placed near their cages (Griffith 1919, 1920). Trembling, back muscle twitches, and whining often occurred, and feeding, nursing, and other activities ceased. Griffith believed that chemicals elicited fear since rats reacted to a damp cloth or a hand rubbed against a cat, and when placed into an arena laden with cat scent. Anosmic rats and those presented with a cat in a sealed glass jar were unresponsive.

Rats were presented with cat faeces, urine, heart, and other tissues in unsuccessful attempts to elicit reactions (Griffith 1920). Other attempts to arouse rats with domestic cat faeces (Cattarelli and Chanel 1979; Cattarelli *et al.* 1974; Vernet-Maury *et al.* 1968) or urine (Reiff 1956) have failed. Griffith suspected that chemicals from cat hair or skin, especially those from around the nose, were effective.

Curti (1935) observed rats freeze in the presence of cats, but she attributed little importance to chemical releasers since her subjects showed no fear in a cat-scented box or when separated from a cat by a cloth mesh. She concluded that movement cues (visually perceived) were important in eliciting fear. Blinded rats apparently responded to auditory stimuli, but again little importance was attributed to chemicals (Curti 1942). The conflicting results of Griffith and Curti are difficult to reconcile given their differing methods and their failure to quantify the behaviour of their subjects.

Wistar rats were presented with a live cat, taped cat vocalizations, and other cues (Blanchard *et al.* 1975). An anaesthetized cat shrouded in cheesecloth (used to present chemicals) elicited marginal responses. Freezing and avoidance were most pronounced when a moving cat or dog was visually perceived. Chemical effects were not explicitly discounted, but Blanchard *et al.* state that movement is the primary fear-eliciting stimulus.

Rats were trained to traverse a runway for water, and the effects of substances applied to the apparatus were observed (Courtney *et al.* 1968). The scent of a cat, deposited by walking on the runway, suppressed the speed with which rats completed their runs more than did the scent of shocked rats or a commercial deodorant.

Rats' perception of predator chemicals may be surgically impaired. Naïve intact rats placed into a chamber laden with cat scent exhibited freezing and micturition more frequently than did rats whose olfactory bulbs had been removed (Sieck *et al.* 1974). Mollenauer *et al.* (1974) observed intact and bulbectomized rats separated from a cat by a wire mesh. Intact rats showed signs of fear, defaecating more and exploring less. Mollenauer *et al.* suggested that the olfactory bulbs contribute to emotionality. The possibility that their results were due to a chemosensory impairment was not discussed.

Cattarelli, Vernet-Maury, and their colleagues have shown that the scent of fox (*Vulpes vulpes*) faeces is a potent elicitor of fear in Wistar rats. An array of chemicals has been presented and diverse behavioural and physiological measures have been scored in these investigations.

In one study, plain air or air passed over conspecifics or fox and lion faeces were presented to two- and four-month-old rats (Vernet Maury *et al.* 1968). The older rats explored least and defaecated most to fox scent. The two-month-olds gave less pronounced reactions. The heart rates of rats presented with fox and lion odours did not appear to differ significantly, but lymphocyte counts of anaesthetized rats were highest when exposed to the fox.

Carnivore faeces, synthetic chemicals, and scents of stressed, rewarded, unrewarded, grouped, and isolated conspecifics have been presented in a number of studies. Fox scent inhibited exploration, grooming (Cattarelli *et al.* 1975), ultrasonic vocalizations (except in three- to four-week-old individuals) (Vernet-Maury *et al.* 1982), and operant conditioning (Cattarelli and Vigouroux 1981); stimulated defaecation and micturition (Cattarelli *et al.* 1974); and roused rats from sleep (Cattarelli and Chanel 1979; Cattarelli 1982*a*, *b*). No significant effect on respiratory rate was observed (Cattarelli et al. 1977*a*). Lion and cat faecal scents, on the other hand, were classified as neutral or even reassuring on the basis of the responses they elicited (Cattarelli *et al.* 1974, 1975).

Neurophysiological studies indicate that olfactory bulb mitral cells are active when cat and other reassuring scents are presented (Cattarelli *et al.* 1977*b*). Fox and other alarming chemicals, in contrast, inhibited mitral cell firing. Higher brain centres apparently influence these cells since inhibition was abolished in deafferentiated rats.

Autoradiographic mappings of the olfactory bulbs (Astic and Cattarelli 1982) and piriform cortex (Cattarelli *et al.* 1988) were obtained after rats were presented with a neutral odour or fox faecal scent. Considerable overlap was observed in the brain sites exhibiting heightened metabolic activity. Although some regions of the olfactory bulbs consistently were active with only one condition, no differences were observed in the piriform cortex. Alarm reactions and mitral cell inhibition to fox odours disappeared after a bilateral transection of the lateral olfactory tract, but the awakening influence of this carnivore's scent remained unchanged (Cattarelli, 1982*a*).

2,5-Dihydro-2,4,5-trimethyl thiazoline from fox droppings were comparable to whole faecal extracts in eliciting emotional responses and elevating plasma corticosterone in rats (Vernet-Maury 1980). The presentation of other sulphur-containing compounds indicated that another faecal product, 4-mercapto-4-methyl penta-2-one, also was avoided (Vernet-Maury *et al.* 1984; see also Fomban and Polak 1987).

Scent sources from carnivores other than faeces may be recognized. A variety of rodents were presented with the urine of domestic and exotic mammals, including canid, felid, hyaenid, and mustelid carnivores (Reiff 1956). Fresh urine from martens (*Martes* spp.) was highly aversive, but lost its activity after a week or two. Fox and lion scents also were shunned. Rats were attracted to cattle, horse, and, curiously, puma urine; domestic dog and cat urine were ineffective.

Hamsters (*Mesocricetus auratus*) were presented with air from jars containing carnivore bedding materials (Dieterlen 1959). Ferret and polecat scents elicited defaecation, micturition, screeching, and a threat posture involving rearing and teeth chattering in subjects three weeks old or older. Dog scent and, to a lesser extent, that of cats elicited freezing or flight. Each response seemed a fitting line of defence against these predators. Flight from dogs, for example, is appropriate since they do not pursue hamsters into burrows.

Adult hamsters raised in isolation were less responsive than were group-reared individuals. Since neither dog and polecat urine nor polecat faeces were effective, Dieterlen suggested that hamsters responded to chemicals from the fur or skin. Preserved mustelid pelts, however, had no effect.

Voles (*Microtus agrestris*) were tested with traps daubed with an extract of weasel (*Mustela nivalis*) anal sac secretions and with unscented traps set out in old field grassland in England (Stoddart 1976). They avoided traps treated with weasel chemicals; other tests suggested that jaguar (*Panthera onca*) and tiger (*P. tigris*) urine (Stoddard 1980, 1982) also were aversive. No significant responses, however, were detected in woodmice (*Apodemus sylvaticus*). Despite the threat of predation, woodmice either do not perceive chemicals from weasels or they extract no information from them.

Similar results were obtained with Scottish rodents (Gorman 1984). Clean traps or ones treated with stoat (*Mustela erminea*) or guinea-pig anal sac secretions were used. *Microtus agrestis* and *M. arvalis* avoided traps treated with stoat scent, and in the laboratory they shunned a side of a chamber treated with stoat extract. These results are worth noting since the voles were from populations isolated from stoats for at least 5,000 years. Woodmice (*A. sylvaticus*) showed no responses in laboratory tests and, as in Stoddart's study, entered traps without regard to treatment.

Sullivan and his colleagues have conducted extensive studies on the use of carnivore exudates for rodent and lagomorph control (see Sullivan *et al.* 1988, *c*). 2-Propylthietane and 3-propyl-1,2-dithiolane from stoat anal glands and 2,5-dihydro-2,4,5-trimethylthiazoline from fox faeces repelled voles (*M. montanus* and *M. pennsylvanicus*) (Sullivan *et al.* 1988*a*; Sullivan *et al.* in press *a*). These compounds also repelled gophers (*Thomomys talpoides*) in laboratory tests, but chemicals from mustelid anal glands generally were ineffective

(Sullivan and Crump 1986*a*; Sullivan *et al.* 1988*b*). Recently, however, 3,3-dimethyl-1,2-dithiolane and other components of stoat anal glands produced a significant avoidance in *T. talpoides* during field tests (Sullivan *et al.*, in press *b*)

Voles (*M. pennsylvanicus*) were tested in outdoor pens for responses to chemicals from shrews (*Blarina brevicauda*) (Fulk 1969). They avoided boxes containing shrew droppings more than clean boxes. In the laboratory, voles avoided a chamber occupied by a shrew or one containing its droppings more than one conditioned by domestic mice. These results agree with observations in the field indicating that bait stations used heavily by shrews were used only infrequently by voles.

Deer mice (*Peromyscus leucopus*) were tested with traps containing cotton balls treated with conspecific odours or weasel (*M. frenata*) faeces (Madzer *et al.* 1976). No avoidance of weasel scent was indicated (see also Sullivan *et al.* 1988*b* on *P. maniculatus*). Since weasels were rarely sighted in the study area, Madzer *et al.* thought that they were an insignificant predator of mice.

### BY LAGOMORPHS

Hares (*Lepus americanus*) were tested with extracts of carnivore faeces, urine, skin odours, and mustelid anal gland secretions as feeding deterrents (Sullivan *et al.* 1985*a*). Scented materials in plates or vials were tied to seedlings, and browsing was monitored. Bobcat (*Lynx rufus*) and lynx (*L. canadensis*) faeces most consistently suppressed feeding. Coyote (*Canis latrans*), wolf (*C. lupus*), and wolverine (*Gulo gulo*) faeces had no effect.

The urine of all carnivores sampled significantly reduced browsing. Neither domestic dog urine nor a synthetic substance (2-methylbutyric acid) suppressed feeding, but coyote scent, obtained from bedding materials and by rubbing a wet cloth against the fur, was effective.

Feeding by hares was inhibited by the anal gland secretions of weasel (*Mustela erminea*), mink (*M. vison*), and other mustelids. The scent of the wolverine inhibited feeding for the first day, but insufficient quantities may have been present for lasting effects. In another study, wolverine urine inhibited browsing by hares, but deer (*Odocoileus hemionus*) scent had no effect (Sullivan 1986).

Tests of pure compounds indicated that 3-propyl-1,2-dithiolane from the stoat (*M. erminea*) suppressed feeding (Sullivan and Crump 1984). 2,2-Dimethylthietane, from both stoat and ferret, and 3-methyl-3-butenyl methyl sulfide from fox urine also were effective; other sulphur-containing compounds were not (Sullivan and Crump 1986*b*).

### BY UNGULATES

Anecdotal accounts suggest that some ungulates were alarmed by carnivore scent or avoided sites recently visited by them (e.g. Mech 1970; Schaller 1967). Experimental studies also have been performed.

Predator naïve deer (*Odocoileus hemionus*) were tested with faecal scents of a coyote (*C. latrans*), mountain lion (*Felis concolor*), African lion (*Panthera leo*), leopard (*P. unica*), and tiger (*P. tigris*) (Müller-Schwarze 1972). The deers' responses to bowls with unscented food or food with vials of carnivore faeces were observed. Feeding inhibition and avoidance occurred with the faeces of coyote and mountain lion, the only two stimulus species with which deer are sympatric. These scents, however, did not differ significantly in their effects from those of the leopard and lion. Even the least effective stimulus—tiger faeces—significantly reduced deers' tendency to feed compared with unscented food. Aqueous faecal extracts of an array of sympatric canids and felids had similar effects (Melchiors and Leslie 1985).

Deer (*O. virginianus*) were presented with polythene glycol, ethanol, and ethanol extracts of wolf (*Canis lupus*) dung, coyote (*C. latrans*) dung and urine, domestic dog (*C. familiaris*) urine, muskrat (*Odonata canadensis*) castor, deer metatarsal secretions (which contain alarm pheromones), and onion (Müller-Schwarze 1983). Their responses were measured by noting the days during which they visited gardens containing these materials. Not all sites gave reliable results, but wolf scats appeared most aversive. Metatarsal secretions, muskrat castor, and coyote scats were intermediate in their effects.

Carnivore faecal and urine odours were tested as feeding deterrents of *O. hemionus* (Sullivan *et al.* 1985*b*). Cougar, coyote, wolf, lynx, and bobcat scents suppressed feeding, even more than a commercial repellent. Coyote, jaguar, and wolf faeces in a second trial suppressed feeding for several days, but their effectiveness declined dramatically after a heavy rain. The urine of all carnivores, especially that of coyotes, reduced feeding. Human urine had no effect.

#### BY CARNIVORES

Anecdotal reports indicate that meerkats (*Suricata suricata*) growled at the scent of dog droppings (Ewer 1968), and that hyaenas (*Crocuta crocuta*) were alarmed by lion and leopard scents (Kruuk 1972).

#### BY PRIMATES

Tamarins (*Saguinus labiatus*) were presented with faecal extracts of predatory and non-predatory neotropical mammals (Caine and Weldon 1989). The jaguar (*Panthera onca*), jaguarundi (*Herpailurus yagouaroundi*), and margay (*Felis wiedi*) elicited more sniffing and avoidance than did the scents of non-predators, and the margay elicited occasional alarm calls. It is not known, however, whether any of the stimulus species prey on this tamarin.

## Behavioural, ecological, and evolutionary considerations

Studies of predator odour responses have revealed previously undescribed action patterns and thus have extended the known behavioural repertoires of some species. The display of topographically unique responses to chemicals from particular predators, as some investigations demonstrate, should caution against attempts to characterize a species-typical defence on the basis of observations of reactions to one or a few predators, or to isolated cues from them.

Tests of the function of responses to predators have provided insight on the ecology and evolution of anti-predator adaptations, but empirical survivorship studies have seldom been performed for chemically elicited responses. George (1960) attempted to evaluate mosquitofishes' reactions to pickerels by staging encounters. The observation that mosquitofish occupying the upper region of an aquarium were captured less frequently than those below them suggested to George that these fishes were less vulnerable when near the water surface (the involvement of chemoreception in this case is unclear, however, since the cues to which subjects were exposed were not restricted).

Staged encounters are not practical in every case. Indeed, whether they can even provide useful results depends on when defences are enacted in the chain of events leading to predator-prey confrontations. Schaller (1972), for instance, found that lions stalking Thompson's gazelles (*Gazella thompsoni*) took no obvious account of wind direction, but were most successful if they approached their quarry from downwind. Gazelles and other species may benefit merely by attending to windborne chemicals or by avoiding areas laden with the scent of a predator—without physically encountering one.

The involvement of chemoreception in anti-predator defences may be demonstrated by comparing the survivorship of intact subjects with that of individuals whose chemical senses have been surgically impaired. Laboratory studies on rodents provide incentive for this approach, which could be adopted in field studies.

The extent to which predators pose a threat to some species has been inferred from their ability chemically to elicit responses (e.g. Boonstra *et al.* 1982). The demonstrated effectiveness of chemicals from allopatric predators and non-predators in some studies should serve to caution against such an approach. Atavistic responses also may occur, as with the voles in Gorman's (1984) study.

Where initial experiments have failed to indicate responses to predator scents, sex, rearing conditions, age, individual variation, and other factors need to be considered before the involvement of chemical elicitors is dismissed. Lyons and Banks (1982), for example, found that mustelid scents failed to elicit distress calls in nine-day-old Sprague-Dawley rats, and they suggested that artificial breeding had rendered their subjects unresponsive. Older rodents, however, may give more pronounced reactions to predator chemicals, as shown in some studies. Distress calling, moreover, is just one parameter to be measured; other behavioural and physiological responses may be more sensitive indicators.

As a number of studies make clear, substances from different anatomical regions of predators may differ in their defence-releasing properties. Tests, therefore, should be performed with samples obtained from a variety of anatomical sites. Dorsal skin chemicals from ophiophagous snakes, for example, elicit body bridging in crotaline snakes, but scent gland secretions and even substances from the ventral skin are ineffective. Since scent gland secretions, at least, are discharged when snakes themselves are molested, potential prey may attend more reliably to skin chemicals to identify snake predators. The significance of responses to chemicals from other sources may be more difficult to interpret. Field tests with hares, for instance, indicated the coyote and wolf faeces did not deter feeding, but their urine did (Sullivan *et al.* 1985*a*). Hamsters were unreactive to dog and polecat urine, but showed extreme reactions to carnivore bedding scents (Dieterlen 1959). Studies of foraging patterns, scent marking habits, and details of the microhabitats in which predators and prey co-occur may contribute to hypotheses on the significance of responses to particular body scents.

Negative results in a number of studies have been indicated in tests with chemicals from predators. Space limitations have not permitted a discussion of them all, but these findings are of potential value in indicating conditions that select against or constrain the use of chemoreception in predator recognition. McMahon and Tash (1979), for example, suggested that shad (*Dorosoma pentenense*) are unresponsive to piscivorous fish scents since they and their predators swim more rapidly than chemicals are transmitted in an aquatic medium. The results of natural history and stimulus control investigations may indicate more specific hypotheses on where and when chemically elicited defences are enacted.

## Acknowledgements

J.F. Eisenberg, R.D. Estes, J.J. Lepri, D.M.Stoddart, P.J. Sutherland, and F. Walther provided valuable correspondence. D.A. Baines, G.M. Burghardt, M. Cattarelli, D. Duvall, H.W. Greene, H. Drummond, D. Müller-Schwarze, R.J.F. Smith, T.P. Sullivan, and E. Vernet-Maury commented on the manuscript. T. Kobayashi provided photographs and gave permission for their use. M.J. Tanner helped prepare the final version of the manuscript. This paper was written while the author was supported by grants from the Smithsonian Institution, Sea Grant, H.E.A.R.T., Kempner Fund, and National Geographic Society.

## References

Astic, L. and Cattarelli, M. (1982). Metabolic mapping of functional activity in the rat olfactory system after a bilateral transection of the lateral olfactory tract. *Brain Res.*, 245, 17–25.

Atema, J. (1969). The chemical senses in feeding and social behavior of the catfish, *Ictalurus natalis*. Ph.D. dissertation, University of Michigan.

Blanchard, R.J., Mast, M. and Blanchard, D.C. (1975). Stimulus control of defensive reactions in the albino rat. *J. Comp. Physiol. Psychol.*, 88, 81–8.

Bogert, C.M. (1941). Sensory cues used by rattlesnakes in their recognition of ophidian enemies. *Ann. N.Y. Acad. Sci.*, 41, 329–43.

Boonstra, R., Rodd, F.H. and Carlton, D.J. (1982). Effect of *Blarina brevicauda* on trap response of *Microtus pennsylvanicus*. *Can. J. Zool.*, 60, 438–42.

Brett, J.R. and MacKinnon, D. (1954). Some aspects of olfactory perception in migrating adult coho and spring salmon. *J. Fish. Res. Bd. Canada*, 11, 310–8.

Burger, J. (1989). Following of conspecific and avoidance of predator chemical cues by pine snakes (*Pituophis melanoleucus*). *J. Chem. Ecol.*, 15, 799–806.

Caine, N.G., and Weldon, P.J. Responses by red-bellied tamarins (*Saguinus labiatus*) to fecal scents of predatory and non-predatory neotropical mammals. *Biotropica*, 21, 186–9.

Carpenter, C.C. and Gillingham, J.C. (1975). Postural responses to kingsnakes by crotaline snakes. *Herpetologica*, 31, 293–302.

Cattarelli, M., Vernet-Maury, E. and Chanel, J. (1974). Influences de différentes odeurs biologiques sur le comportement émotif du Rat placé dans un 'espace vide d'informations'. *C.R. Acad. Sci. Paris*, 278, 2653–65.

Cattarelli, M., Vernet Maury, E., Chanel, J., MacLeod, P. and Brandon, A.M. (1975). Olfactory bulb and integration of some odorous signals in the rat—behavioural and electrophysiological study. In *Olfaction and taste*, V (ed. D.A. Denton and J.P. Coghlan), pp. 235–7. Academic Press, New York.

Cattarelli, M., Pager, J. and Chanel, J. (1977*a*). Modulation des responses multiunitaires du bulbe olfactif et de l'activité respiratoire en fonction de la signification des odeurs chez le rat non contraint. *J. Physiol.* (*Paris*), 73, 963–84.

Cattarelli, M., Vernet-Maury, E. and Chanel, J. (1977*b*). Modulation de l'activité du bulbe olfactif en fonction de la signification des odeurs chez le rat. *Physiol. Behav.*, 19, 381–7.

Cattarelli, M. and Chanel, J. (1979). Influence of some biologically meaningful olfactory odorants on the vigilance states of the rat. *Physiol. Behav.*, 23, 831–8.

Cattarelli, M. and Vigouroux, M. (1981). Evaluation des reactions émotionelles déclenchées par des odeurs a signification biologique chez le rat au cours d'un conditionnement operant. *Physiol. Behav.*, 27, 445–55.

Cattarelli, M. (1982*a*). Transmission and integration of biologically meaningful olfactory information after bilateral transection of the lateral olfactory tract in the rat. *Behav. Brain Res.*, 6, 313–37.

Cattarelli, M. (1982*b*). The role of the medial olfactory pathways in olfaction: behavioral and electrophysiological data. *Behav. Brain Res.* 6, 339–64.

Cattarelli, M., Astic, L. and Kauer, J.S. (1988). Metabolic mapping of 2-deoxyglucose uptake in the rat piriform cortex using computerized image processing. *Brain Res.*, 442, 180–4.

Chiszar, D., Scudder, K., Knight, L. and Smith, H.M. (1978). Exploratory behavior in prairie rattlesnakes (*Crotalus viridis*) and water moccasins (*Agkistrodon piscivorus*). *Psychol. Rec.*, 28, 363–8.

Coss, R.G. and Owings, D.H. (1978). Snake-directed behavior by naïve and experienced California ground squirrels in a simulated burrow. *Z. Tierpsychol.*, 48, 421–35.

Courtney, R.J., Jr., Reid, L.D. and Wasden, R.E. (1968). Suppression of running times by olfactory stimuli. *Psychon. Sci.*, 12, 315–6.

Cowles, R.B. and Phelan, R.L. (1958). Olfaction in rattlesnakes. *Copeia*, 1958, 77–83.

Cupp, P.V., Jr. (1988). Avoidance of predators by salamanders through the detection of chemical odors. *Amer. Zool.*, 28, 156A.

Curti, M.W. (1935). Native fear responses of white rats in the presence of cats. *Psychol. Rev. Monogr.*, 46, 78–98.

Curti, M.W. (1942). A further report on fear responses of white rats in the presence of cats. *J. Comp. Psychol.*, 34, 51–3.

Davy, F.B. and Kleerekoper, H. (1971). Observations of the effects of the odour of predatory fishes on the locomotor behavior of a prey (*Carassius auratus*). *Texas J. Sci.*, 22, 299–300.

Dial, B.E., Weldon, P.J. and Curtis, B. (1989). Chemosensory identification of snake predators (*Phyllorhyncus decurtatus*) by banded geckos (*Coleonyx variegatus*). *J. Herpetol.*, 23, 224–9.

Dieterlen, F. (1959). Das Verhalten des Syrischen Goldhamsters (*Mesocricetus auratus*). *Z. Tierpsychol.*, 16, 47–103.

Eisenberg, J.F. and Gould, E. (1970). *The tenrecs: A study in mammalian behavior and evolution*. Smithsonian Institution Press, Washington, D.C.

Ewer, R.F. (1968). *Ethology of mammals*. Plenum, New York.

Fomban, A.M. and Polak, E.H. (1987). Odor similarity between stress-inducing odorants in Wistar rats. *J. Chem. Ecol.*, 13, 153–66.

Frisch, K. von. (1941). Die Bedeutung des Geruchsinnes im Leben der Fische. *Naturwissen.*, 29, 321–33.

Fulk, G.W. (1969). The effect of shrews on the space utilization of voles. *J. Mammal.*, 53, 461–78.

George, C.J.W. (1960). Behavioral interaction of the pickerel (*Esox niger* LeSueur and *Esox americanus* LeSueur) and the mosquitofish (*Gambusia patruelis* Baird and Girard). Ph.D. dissertation, Harvard University.

Gorman, M.L. (1984). The response of prey to stoat (*Mustela erminea*) scent. *J. Zool. London*, 202, 419–23.

Göz, H. (1941). Über den Art- und Individualgeruch bei Fischen. *Z. Vergl. Physiol.*, 29, 1–45.

Griffith, C.R. (1919). A possible case of instinctive behavior in the white rat. *Science*, 50, 166–7.

Griffith, C.R. (1920). The behavior of white rats in the presence of cats.

*Psychobiology*, 2, 19–28.

Hennessy, F.F. and Owings, D.H. (1979). Snake species discrimination and the role of olfactory cues in the snake-directed behavior of the California ground squirrel. *Behaviour*, 65, 115–24.

Idler, D.R., Fagerlund, U.H.M. and Mayoh, H. (1956). Olfactory perception in migrating salmon. II.L-serine, a salmon repellent in mammalian skin. *J. Gen. Physiol.*, 39, 889–92.

Jones, K.A. and Hara, T.J. (1982). Behavioral response by arctic charr (*Salvelinus alpius*) to taurocholic acid and L-serine, two putative semiochemicals. *Amer. Zool.*, 22, 295A.

Kats, L.B. (1988). The detection of certain predators via olfaction by small-mouthed salamander larvae (*Ambystoma texanum*). *Behav. Neural Biol.*, 50, 126–31.

Kobayashi, T. and Watanabe, M. (1986). Analysis of snake-scent application behaviour in Siberian chipmunks (*Eutamias sibiricus asiaticus*). *Ethology*, 72, 40–52.

Kobayashi, T. (1987). Does the Siberian chipmunk respond to the snake by identifying it? *J. Ethol.*, 5, 137–44.

Kruuk, H. (1972). *The spotted hyena: a study of predation and social behavior.* University of Chicago Press, Chicago.

Lyons, D.M. and Banks, E.M. (1982). Ultrasounds in neonatal rats: novel, predator and conspecific odor cues. *Develop. Psychobiol.*, 15, 455–60.

Madzer, E., Capone, M.R. and Drickamer, L.C. (1976). Conspecific odors and trappability of deer mice (*Peromyscus leucopus noveboracensis*). *J. Mammal.*, 67, 607–9.

Marchisin, A., (1980). Predator-prey interactions between snake-eating snakes and pit vipers. Ph.D. dissertation, Rutgers University.

McMahon, T.E. and Tash, J.C. (1979). The use of chemosenses by thread fin shad, *Dorosoma petenense*, to detect conspecific, predators and food. *J. Fish. Biol.*, 14, 289–96.

Mech, L.D. (1970). *The wolf: the ecology and behaviour of an endangered species.* Natural History Press, New York.

Melchiors, M.A. and Leslie, C.A. (1985). Effectiveness of predator fecal odors as black-tailed deer repellents. *J. Wildl. Manage.*, 49, 358–62.

Mollenauer, S., Plotnik, R. and Snyder, E. (1974). Effects of olfactory bulb removal on fear responses and passive avoidance in the rat. *Physiol. Behav.*, 12, 141–4.

Müller-Schwarze, D. (1972). Responses of young black-tailed deer to predator odors. *J. Mammal.*, 53, 393–4.

Müller-Schwarze, D. (1983). Experimental modulation of behavior of free-ranging mammals by semiochemicals. In *Chemical signals in vertebrates*, 3 (ed. D. Müller-Schwarze and R.M. Silverstein), pp. 235–44. Plenum, New York.

Petranka, J.W., Kats, L.B. and Sih, A. (1987). Predator-prey interactions among fish and larval amphibians: use of chemical cues to detect predatory fish. *Anim. Behav.*, 35, 420–5.

Randall, J.A. and Stevens, C.M. (1987). Footdrumming and other anti-predator responses in the bannertail kangaroo rat (*Dipodomys spectabilis*). *Behav. Ecol. Sociobiol.*, 20, 187–94.

Reed, J.R. (1969). Alarm substances and fright reaction in some fishes from the southeastern United States. *Trans. Am. Fish Soc.*, 98, 664–8.

Rehnberg, B.G. and Schreck, C.B. (1986). The olfactory L-serine receptor in coho salmon: biochemical specifity and behavioral response. *J. Comp. Physiol.*, 159A, 61–7.

Rehnberg, B.G. and Schreck, C.B. (1987). Chemosensory detection of predators by coho salmon (*Oncorhynchus kisutch*): behavioral reaction and the physiological stress response. *Can. J. Zool.*, 65, 481–5.

Reiff, M. (1956). Untersuchungen über natürliche und synthetische Geruchstoffe, die bei Ratten und Maüsen eine stimulierende Wirkung auslosen. *Acta Tropica*, 13, 289–318.

Richardson, W.B. (1942). Reaction toward snakes as shown by the wood rat, *Neotoma albigula*. *J. Comp. Psychol.*, 34, 1–10.

Roudebush, R.E. and Taylor, D.H. (1987). Chemical communication between two species of desmognathine salamanders. *Copeia*, 1987, 744–8.

Schaller, G. (1967). *The deer and the tiger*. University of Chicago Press, Chicago.

Schaller, G.B. (1972). *The Serengeti lion: a study of predator-prey relations*. University of Chicago Press, Chicago.

Sieck, M.H., Baumbach, H.C., Gordon, B.L. and Turner, J.F. (1974). Changes in spontaneous, odor modulated and shock induced behavior patterns following discrete olfactory system lesions. *Physiol. Behav.*, 13, 427–39.

Stoddart, D.M. (1976). Effect of the odour of weasels (*Mustela nivalis* L.) on trapped samples of their prey. *Oecologia*, 22, 439–41.

Stoddart, D.M. (1980). Some responses of a free living community of rodents to the odours of predators. In *Chemical signals in vertebrates*, 3 (ed. D. Müller-Schwarze and R.M. Silverstein), pp. 1–10. Plenum, New York.

Stoddart, D.M. (1982). Does trap odour influence estimation of population size of the short-tailed vole, *Microtus agrestis*? *J. Anim. Ecol.*, 51, 375–86.

Sullivan, T.P. (1986). Influence of wolverine (*Gulo gulo*) odor on feeding behavior of snowshoe hares (*Lepus americanus*). *J. Mammal.*, 67, 385–8.

Sullivan, T.P. and Crump, D.R. (1984). Influence of mustelid scent-gland compounds on suppression of feeding by snowshoe hares (*Lepus americanus*). *J. Chem. Ecol.*, 10, 1809–21.

Sullivan, T.P., Nordstrom, L.O. and Sullivan, D.S. (1985*a*). Use of predator odors as repellents to reduce feeding damage by herbivores. I. Snowshoes hares (*Lepus americanus*). *J. Chem. Ecol.*, 11, 903–19.

Sullivan, T.P., Nordstrom, L.O. and Sullivan, D.S. (1985*b*). Use of predator odors as repellents to reduce feeding damage by herbivores. II. Black-tailed deer (*Odocoileus hemionus columbianus*). *J. Chem. Ecol.*, 11, 921–35.

Sullivan, T.P. and Crump, D.R. (1986*a*). Avoidance response of pocket gophers (*Thomomys talpoides*) to mustelid anal gland compounds. In *Chemical signals in vertebrates*, 4 (ed. D. Duvall, D. Müller-Schwarze and R.M. Silverstein), pp. 519–31. Plenum, New York.

Sullivan, T.P. and Crump, D.R. (1986*b*). Feeding responses of snowshoe hares (*Lepus americanus*) to volatile constituents of red fox (*Vulpes vulpes*) urine. *J. Chem. Ecol.*, 12, 729–39.

Sullivan, T.P., Crump, D.R. and Sullivan, D.S. (1988*a*). Use of predator odors as repellents to reduce feeding damage by herbivores. III. Montane and meadow voles (*Microtus montanus* and *Microtus pennsylvanicus*). *J. Chem. Ecol.*, 14, 363–77.

Sullivan, T.P., Crump, D.R. and Sullivan, D.S. (1988*b*). Use of predator odors as repellents to reduce feeding damage by herbivores. IV. Northern pocket gophers (*Thomomys talpoides*). *J. Chem. Ecol.*, 14, 379–89.

Sullivan, T.P, Sullivan, D.S., Crump, D.R., Wieser, H. and Dixon, E.A. (1988*c*). Predator odors and their potential role in managing pest rodents and rabbits. In *Proc. 13th Vertebrate Pest Conference* (ed. A.C. Crabbe and R.E. Marsh), pp. 145–50. Univerisity of California, Davis.

Sullivan, T.P, Crump, D.R., Wieser, H. and Dixon, E.A. (in press *a*). Comparison of release devices for stoat (*Mustela erminea*) semiochemicals used as montane vole repellents. *J. Chem. Ecol.*

Sullivan, T.P, Crump, D.R., Wieser, H. and Dixon, E.A. (in press *b*). Responses of pocket gophers (*Thomomys talpoides*) to an operational application of synthetic semiochemicals of stoat (*Mustela erminea*). *J. Chem Ecol.*

Thoen, C., Bauwens, D. and Verheyen, R.F. (1986). Chemoreceptive and behavioural responses of the common lizard *Lacerta vivipara* to snake chemical deposits. *Anim. Behav.*, 34, 1805–13.

Vernet-Maury, E., Le Magnen, J. and Chanel, J. (1968). Comportement émotif chez le rat; influence de l'odeur d'un prédateur et d'un non-prédateur. *C.R. Acad. Sci. Paris*, 267, 331–4.

Vernet-Maury, E. (1980). Trimethyl-thiazoline in fox feces: a natural alarming substance for the rat. In *Olfaction and taste*, VIII (ed. H. van der Starre), p. 407. IRL Press, Washington.

Vernet-Maury, E., Duveau, A. and Chanel, J. (1982). Ontogeny of ultrasonic vocalizations in the rat after their alteration by fox odor. Abstract 5th ECRO Congress, Regensberg.

Vernet-Maury, E., Polak, E.H. and Damael, A. (1984). Structure-activity relationship of stress-inducing odorants in the rat. *J. Chem. Ecol.*, 10, 1007–18.

Webster, D.B. (1973). Audition, vision, and olfaction in kangaroo rat predator avoidance. *Am. Zool.*, 13, 1346A.

Weldon, P.J. and Burghardt, G.M. (1979). The ophiophage defensive response in crotaline snakes: extension to new taxa. *J. Chem. Ecol.*, 5, 141–51.

Weldon, P.J. (1982). Responses to ophiophagous snakes by snakes of the genus *Thamnophis*. *Copeia*, 1982, 788–94.

Weldon, P.J., Divita, F.M. and Middendorf, G. A., III (1987). Responses to snake odors by laboratory mice. *Behav. Processes*, 14, 137–46.

Weldon, P.J., Ford, N. and Perry-Richardson, J. (1990). Responses by corn snakes (*Elaphe guttata*) to chemicals from heterospecific snakes. *J. Chem. Ecol.*, 16, 37–44.

# V.3 Chemosensory Orientation and Earthworm Location by Foraging Eastern Garter Snakes, *Thamnophis S. Sirtalis*

JAMES C. GILLINGHAM, JOHN ROWE AND MARY ANN WEINS

## Introduction

The Eastern garter snake, *Thamnophis s. sirtalis*, is probably the most widely distributed snake in North America (Fitch 1965) and may be one of the most successful (Garstka and Crews 1986). Part of the reason for this success is the wide range of food items taken by this species (Carpenter 1952). However, throughout their geographical range garter snakes eat earthworms which may be the most commonly taken prey (Gillingham and Rowe 1984). The mechanisms used by garter snakes to forage on earthworms are of interest because it is likely they reflect on the snake's evolutionary success.

Under laboratory conditions, garter snakes exhibit great interest in, and even attack, stimuli derived from the surface wash from earthworms (Burghardt 1966, Burghardt and Pruitt 1975). More recently, attempts have been made to isolate the potent chemoattractant of living earthworms (Halpern *et al* 1984; Halpern *et al.* 1986; Kirschenbaum *et al.* 1986); the attractant compound has been found to be a relatively low molecular weight sulphydryl-containing protein (M. Halpern, personal communication). However, knowledge that garter snakes feed on earthworms and that earthworms produce a protein attractive to garter snakes does not elucidate the predator–prey interface between these two organisms.

In this chapter, we will demonstrate that Eastern garter snakes make daily forays into habitats containing earthworms, on which they feed in large numbers. Further, we present evidence that to locate earthworms garter snakes use chemical cues from faecal castings deposited at the soil's surface.

## Study site

A natural population of *Thamnophis s. sirtalis*, was studied on Beaver Island, Charlevoix County, Michigan, over the spring and summer months of 1984–6. More than 150 individual snakes were measured and permanently scale-clipped

(Blanchard and Finster 1933) in a 4.7-ha grassy meadow crossed diagonally by an intermittent stream. The predominant vegetation consisted of apple (*Pyrus malus*), ground juniper (*Juniperus communis*), bracken fern (*Pteridium aquilineum*), milkweed (*Asclepias* spp.), and various grasses including *Poa compressa, Phleum pratense*, and *Agropyron repens*. The meadow is classified as 'cleared land' (Darlington 1940) and contains two old cement foundations and a large variety of wooden and metal debris, evidence of an active farmstead 40 years previous to this study.

## Foraging pattern

Preliminary observations on the meadow indicated that garter snakes frequently used the cracks and fissures within the foundations to enter cavities beneath the concrete slabs. In order to document the daily movements of garter snakes, a wooden observation platform (1.2 × 1.2, 2.5 m in height) was erected adjacent to one of these slabs. Over 50 snakes which used this slab were paint-marked for remote identification. A walkway 4 m high and 35 m long was constructed diagonally across the meadow from which foraging snakes could be observed.

Each day garter snakes emerged from the slab under observation between 08.00 and 10.00 hours. Individual snakes would first extend their heads and necks from the fissures in the cement and gradually glide forward until their bodies were entirely exposed on the slab surface or on the soil next to the foundation margin. The amount of time spent resting on or next to the slab was variable (probably basking time), following which the snakes would glide off into the meadow.

Paths taken by individual garter snakes away from the slab were recorded and plotted daily; the paths were found to be almost identical day after day. Moreover each individual was found to spend most of the remaining daylight hours foraging in the same general area in which it had foraged previously (Fig. 1). After daily foraging, snakes would return to the concrete slab along the same path taken at emergence. This emergence–foraging–return pattern was monitored for 56 snakes between 22 June and 27 July 1984; individual snakes used the same paths at emergence and return, and for foraging within a confined portion of the meadow. In July 1984, snakes were scored for their tendency to emerge and return from and to the same site each day. A snake was scored 'different site' if on its emergence or return it entered or exited from a crack or fissure different from that at which it was last seen, and/or if its pathway within 3 m of the slab deviated more than 0.5 m from that it used previously on emergence or return. In a typical 24-h period, of 50 snakes, 46 emerged to use the same site (Chi $\chi^2 = 35.3$, df $= 1$, $p < .01$); of 20 returning snakes, 17 used the same path as previously (Chi $\chi^2 = 10.80$, df $= 1$, $p < .01$).

Considerable spatial overlap occurred among individuals within these foraging areas. Such overlap occurred out of necessity when one considers that at least 56 individuals foraged daily in a meadow of less than 5 ha. No aggressive interactions were observed even though individuals made frequent contact with one another.

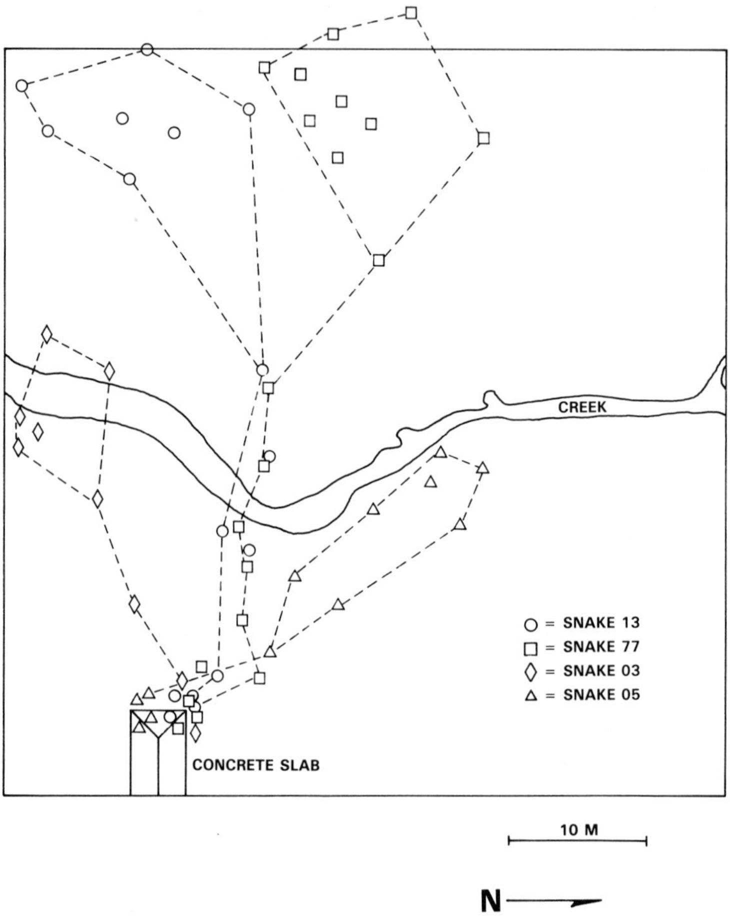


Figure 1. Diagram of the study site on Beaver Island, Charlevoix Co., Michigan,showing a one-week location record for four Eastern garter snakes, *Thamnophis s. sirtalis*, as they emerged from the cement slab, moved to their foraging site and returned to the cement slab. (See text for details.)

Adjacent to the grassland meadow in which garter snakes foraged was a sandy field containing sparce scrub vegetation and ground juniper (*J. communis*). No snakes were ever observed foraging in this sandy area. Four 1-m$^2$ quadrats were selected randomly along a 50-m transect through the meadow and the sandy field. Each quadrat was carefully sampled by soil layer for earthworms. Plots 1–3 (taken in the meadow) have much higher earthworm densities than plot 4 (taken in the sandy field) (Table 1). Also, more worms were located near the surface in the meadow than in the sandy field.

**Table 1.** Results of earthworm survey made in four 1-$m^2$ quadrats randomly located along a 50-m transect on Beaver Island, Charlevoix, Michigan, in 1985. Earthworms were counted at four levels within the quadrat, as indicated

| Stratum | Plot 1 | Plot 2 | Plot 3 | Plot 4 |
|---|---|---|---|---|
| Surface/litter | 24 | 27 | 29 | 0 |
| 1–2 cm | 35 | 71 | 48 | 5 |
| 2–5 cm | 54 | 65 | 43 | 26 |
| Below 5 cm | 18 | 14 | 14 | 8 |
| *Total* (worms/$m^2$) | 131 | 177 | 134 | 39 |

## Foraging success and orientation

To assess the ability of garter snakes to locate earthworms, individual snakes leaving the foundation at morning emergence were intercepted during July 1985. These snakes were palpated for the presence of food in their gut; this information, along with time of emergence, was recorded. Later on the same day, the meadow was searched for foraging garter snakes each of which was captured, identified, palpated and if food was present it was forcibly regurgitated.

**Table 2.** Earthworm foraging rate for 11 garter snakes tested on Beaver Island, Charlevoix, Michigan, during July 1985 (see text for details)

| Time of emergence | Time of capture | Total worms | Rate (worms/h) |
|---|---|---|---|
| 08 43 | 11 06 | 17 | 7.1 |
| 08 29 | 11 42 | 24 | 7.4 |
| 09 13 | 10 24 | 16 | 13.6 |
| 08 59 | 11 21 | 22 | 9.3 |
| 09 32 | 11 43 | 19 | 8.7 |
| 09 16 | 10 58 | 20 | 11.8 |
| 10 03 | 12 22 | 17 | 7.6 |
| 09 56 | 11 31 | 15 | 9.5 |
| 08 50 | 12 05 | 17 | 7.6 |
| 08 21 | 10 31 | 21 | 9.5 |
| 09 00 | 12 33 | 18 | 5.1 |

Results obtained on 12 July 1985 are shown in Table 2; the data are for 11 snakes which left the slab with empty stomachs and were later recaptured. These snakes had an average of 19 worms in their stomachs on capture. Further, it was calculated that these garter snakes located between 5.1 and 13.6 worms per hour. The maximum number of worms found in any snake was 24; this may be a conservative indication of capacity because the snakes' foraging day was interrupted. To determine the relative importance of vision and chemoreception in the ability of garter snakes to orient themselves within their habitat, a temporary blinding experiment was conducted during July 1986. Six paint-marked snakes were captured at 12.00 hours in the meadow and their spectacles (clear scales covering the eyes) were painted with two coats of black enamel paint. When the paint was dry, the animals were released. Simultaneously, six different marked individuals (controls) were also captured and handled for a time comparable with that needed for the painting and drying procedure, and then released. The concrete slab was then monitored for evening returns.

Of the six temporarily blinded snakes, three returned to the slab by nightfall. Of the sighted controls, five out of six returned to the slab. There is no statistical difference in return performance between the blinded and the control animals ($p = 0.41$; Fisher Exact Probability Test). The blinded animals who returned were captured and inspected; no sign of wear on the paint over the spectacles was visible. The stomach of one garter snake was found to contain earthworms, which had been taken since the midday painting.

## Earthworm location

Our observations of foraging garter snakes in the field suggest that the cues mediating earthworm location are located at the soil surface. Garter snakes in this study were observed to forage by probing their snouts into available microhabitat spaces; for example under leaves and debris, and into tufts of grass. When they encountered an area with appropriate cues, tongue-flicking increased and probing became more frenzied. When capturing an earthworm, the snake rapidly bent its head earthwards (at an angle of about 90° to its neck) and thrust its snout into the soil, thereby extracting the prey. A surface cue was undoubtedly responsible for the feeding thrust and we hypothesized that this cue could be the earthworm's casting.

To test this hypothesis, soil from the meadow was placed in the bottom of a steel cattle watering tank (0.75 × 1.5 × 1.0 m) to a depth of 10–12 cm. The soil was moistened and about 80 earthworms from the meadow were added to the tank. After the earthworms had completely burrowed into the soil, the surface was moistened again and smoothed by hand. Smoothing created a surface on which earthworm castings could easily be seen following their deposit and could then be harvested for testing with garter snakes.

In the laboratory, an aquarium (30 × 60 × 30 cm) was used as a test chamber (Fig. 2). All four glass sides were covered from the outside with opaque 'butcher paper' and a mirror (60 × 30 cm) was mounted above the chamber at an angle of 45° to facilitate a view of the entire chamber bottom. The chamber floor was lined

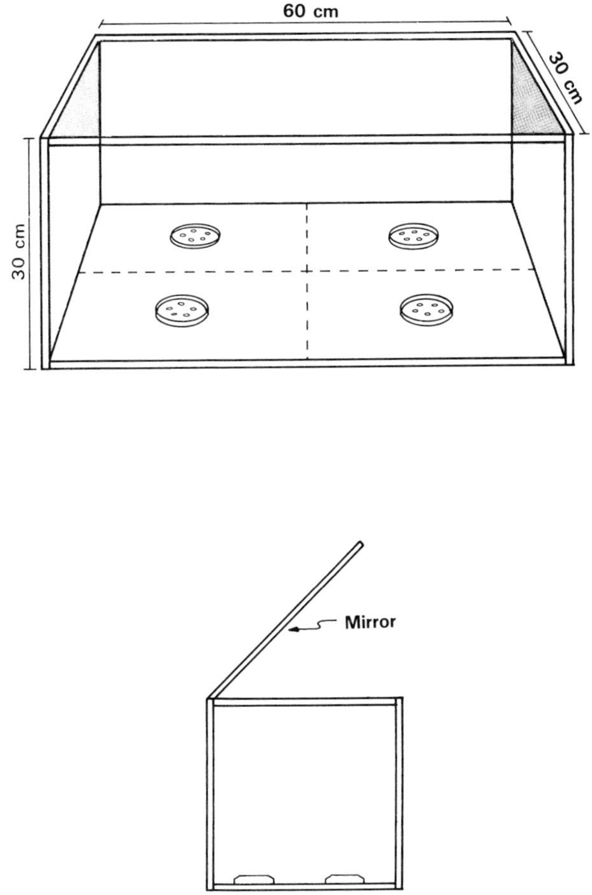


Figure 2. Diagram of experimental arena used to test garter snakes for their attraction to earthworm castings of various ages as well as control substances. Top: front view; bottom: side view. (See text for procedural details.)

with 'butcher paper' (30 × 60 cm) which was divided by pencil lines into four rectangles of equal-size. Syracuse dishes with perforated Plexiglas tops were used as test odour receptacles, following the method of Gillingham and Clark (1981).

Earthworm castings of various ages beyond harvest (from fresh up to 120 h), distilled water, blank receptacles and soil from the study site were all used to test garter snakes. Using a random number table, castings from various age categories and/or controls were placed in the odour receptacles, each of which was centrally located in a grid rectangle. A single fresh-caught garter snake, held in captivity for no more than 24 h, was placed in the chamber with the four test odour combinations and allowed to adjust for 5 mins. Subsequently, the garter snake was observed for 5 mins and its head position recorded in one of the four test areas at 10-s intervals. Each snake generated 30 records; 180 records were obtained for the six snakes tested in each combination of test categories. The aquarium was washed with soap, rinsed with tap water and ethanol following each snake's testing and fresh paper was placed in the bottom.

**Table 3.** Results of five casting trials, each testing six different Eastern Garter snakes for their interest in castings of various age as well as control substances (water, soil). See text for procedural details

| Test | Cumulative Score | |
|---|---|---|
| Casting trial 1 | | |
| 3-day Casting | 17 | |
| 0–5 hr casting | 144 | |
| Water | 14 | |
| Blank | 5 | |
| | | N = 6, df = 3, $\chi^2 = 292.2$, $p < 0.01$ |
| Casting trial 2 | | |
| 5-day Casting | 30 | |
| 0–5 hr casting | 122 | |
| soil | 22 | |
| blank | 6 | |
| | | N = 6, df = 3, $\chi^2 = 182.2$, $p < 0.01$ |
| Casting trial 3 | | |
| 5-day casting | 22 | |
| 3-day casting | 31 | |
| 10–15 hr casting | 58 | |
| 5–10 hr casting | 69 | |
| | | N = 6, df = 3, $\chi^2 = 32.7$, $p < 0.01$ |
| Casting trial 4 | | |
| 0–5 hr casting | 56 | |
| 5–10 hr casting | 27 | |
| 10–15 hr casting | 45 | |
| 15–20 hr casting | 52 | |
| | | N = 6, df = 3, $\chi^2 = 11.0$, $p < 0.05$ |
| Casting trial 5 | | |
| 10–15 hr casting | 64 | |
| 15–20 hr casting | 26 | |
| 20–24 hr casting | 36 | |
| 24–30 hr casting | 18 | |
| | | N = 6, df = 3, $\chi^2 = 32.4$, $p < 0.01$ |

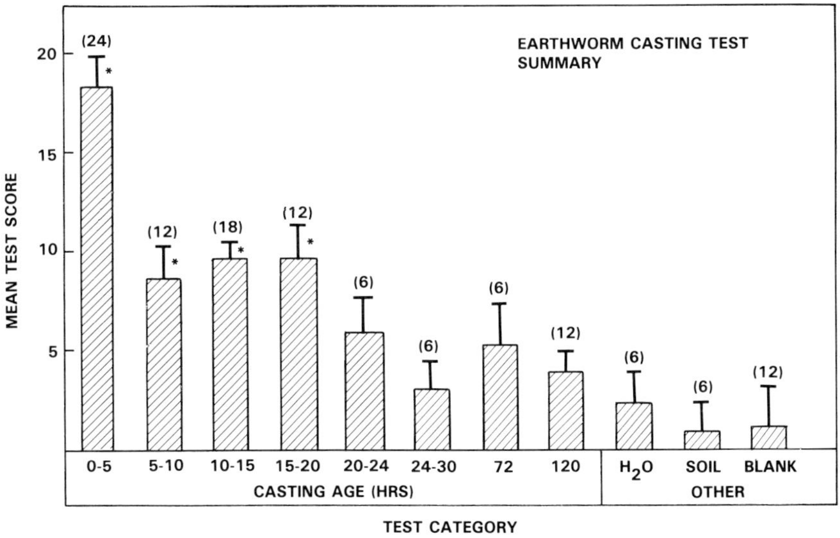


Figure 3. Mean test scores for Eastern garter snakes tested with variously aged earthworm castings and control substances. Asterisk indicates that mean score was significantly different from controls (water, soil, blank) at the .05 confidence level using the Mann–Whitney U-test. Sample size is indicated parenthetically.

The results of five trials, each testing six different snakes, are shown in Table 3. It is clear that earthworm castings elicit significantly higher test scores when compared with controls (water, soil, blank). Further, a comparison between mean test scores across categories (Figure 3) shows that fresher castings are more chemoattractive than those aged 24 h or more. Although not reflected in the numerical data, yet no less significant was the fact that test snakes consistently made obvious attempts at pushing their snouts into test receptacles containing fresh (0–20 h castings) and showed rapid tongue-flicking and searching movements when in the vicinity of these receptacles.

A garter snake's interest in an earthworm casting, and its subsequent snout-thrust into the earth at that point, suggests that there is some probability greater than zero that an earthworm lies beneath the casting. To test this, casting removal observations were conducted to determine the frequency with which earthworms observed in tunnels just beneath the castings. On 12 consecutive mornings earthworm castings were harvested and as each was removed using a forceps, it was noted whether or not a worm could be seen below the casting in the tunnel. The earth's surface in the tank was smoothed at midnight each night before, so the age of the castings was never more than 8 h. Table 4 shows that in only one instance were no earthworms exposed following casting harvest. On all other days, at least one earthworm was observed, and in one instance, earthworms were observed below 25% of the castings removed.

**Table 4.** Results of 12 days of observations made on earthworm burrows beneath castings immediately following casting removal. For each day's observation, by column, the number of castings harvested, the frequency of burrows containing visible earthworms and the percent of total castings harvested with earthworms visible beneath, are indicated

| Day | Number castings harvested | Frequency worms observed | Percentage |
|---|---|---|---|
| 1 | 15 | 3 | 20 |
| 2 | 7 | 1 | 14 |
| 3 | 19 | 0 | 0 |
| 4 | 12 | 2 | 17 |
| 5 | 20 | 3 | 15 |
| 6 | 18 | 1 | 6 |
| 7 | 17 | 3 | 18 |
| 8 | 21 | 2 | 10 |
| 9 | 18 | 2 | 11 |
| 10 | 12 | 3 | 25 |
| 11 | 17 | 1 | 6 |
| 12 | 21 | 1 | 5 |

## Discussion

Garter snake movements between their daily refugia and foraging areas are regular and can be accomplished without vision. The importance of these snakes' chemosensory systems to such daily movements is borne out by the fact that temporarily blinded animals can return unerringly to their daily refugia and may even successfully forage for earthworms in the process. Although other workers have shown that snake movements may depend on visual cues such as landmarks and celestial information (Landreth 1973; Parker and Brown 1980; Duvall, King and Gutzwiller 1985). Such cues apparently are redundant in this context. The most impressive aspect of these snakes' daily movements is their close adherence to the previous day's paths to and from the concrete slab under which they spent the night.

This population of garter snakes exploits the abundant earthworms within its habitat, and seems to avoid foraging in areas where either earthworm density is reduced or where worms are found at greater depths. Although no data are currently available for foraging rates for this or any other snake (but see Greene 1983), it seems impressive that *T. s. sirtalis* can locate and consume up to 14 worms per hour. Garter snakes are not the only vertebrates able to exploit earthworms as a food source. The American robin (*Turdus migratorius*) is notorious for its diurnal foraging for earthworms on lawns. These birds are capable of locating earthworms visually by noting movement near the earth's surface (Heppner 1965). The European badger (*Meles meles*) is capable of locating 'patches' of earthworms and concentrates its foraging efforts in such areas (Kruuk 1978). However, this nocturnal mammal locates its prey by making direct contact with them on the surface and is presumably using either chemosensory or tactile channels. Our data reduce the importance of garter snake use of these channels for the location of subterranean earthworms.

The importance of the garter snake chemosensory system in feeding is accepted (Halpern 1983); laboratory studies have shown it to be a reliable indicator of the food preferences of this snake (Burghardt 1980). The identification of a chemical produced by earthworms and shown to be attractive to garter snakes (Halpern *et al.* 1986; Kirschenbaum *et al.* 1986; Halpern Personal Communication) has added significantly to our understanding of how a garter snake's chemosensory system functions. Our field study provides strong supporting evidence for the hypothesis that earthworm castings alone play a role in prey location by garter snakes. We have further shown that a garter snake's interest in freshly deposited castings is reinforced by the probability that below that casting lies an earthworm. Our data begin to complete the picture of garter snake foraging success, success that has led to its current position as one of the most abundant snakes in North America.

## Acknowledgements

We would like to thank Central Michigan University Biological Station on Beaver Island for its logistical support and generous use of facilities. Our thanks are also extended to Skip McDonough for permission to use the field study site. Field-work was assisted at various times by Jacque LaFreniere and Tracy Miller. We also thank Brent Graves for many helpful discussions and criticisms of earlier drafts. This research was supported in part by a grant from Central Michigan University.

## References

Blanchard, R.N. and Finster, E.B. (1933). A method of marking living snakes for future recognition, with a discussion of some problems and results. *Ecology*, 14, 334–7.

Burghardt, G.M. (1966). Stimulus control of the prey attack response in naive garter snakes. *Psychon. Science*, 4, 37–8.

Burghardt, G.M. (1980). Behavioral and stimulus correlates of vomeronasal functioning in reptiles: feeding, grouping, sex, and tongue use. In *Chemical signals in vertebrates and invertebrates* (ed. D. Müller-Schwarze and R. Silverstein). Plenum Press, New York.

Burghardt, G.M. and Pruitt, C.H. (1975). Role of the tongue and senses in feeding of naive and experienced garter snakes. *Physiol. Behav.*, 14, 185–94.

Carpenter, C.C. (1952). Comparative ecology of the common garter snake (*Thamnophis sirtalis*), the ribbon snake (*Thamnophis s. sauritus*), and Butler's garter snake (*Thamnophis butleri*) in mixed populations. *Ecol. Monog.*, 22, 235–58.

Darlington, H. T. (1940). Some vegetational aspects of Beaver Island, Lake Michigan. *Pap. Michigan Academy of Science Arts and Letters*, 25, 31–7.

Duvall, D., King, M.B. and Gutzwiller, K.J. (1985). Behavioral ecology and ethology of the prairie rattlesnake. *Natl. Geog. Res.*, 1, 80–111.

Fitch, H.S. (1965). An ecological study of the garter snake, *Thamnophis sirtalis. Univ. Kansas Publ. Mus. Nat. Hist.*, 15, 493–564.

Garstka, W.R. and Crews, D. (1986). Pheromones and reproduction in garter snakes. In *Chemical signals in vertebrates*, Vol. 4 (ed. D. Duvall, D. Müller-Schwarze and R.M. Silverstein), pp. 243–60. Plenum Press, New York.

Gillingham, J.C. and Clark, D. (1981). Snake tongue-flicking: transfer mechanics to Jacobson's organ. *Can. J. Zool.*, 59, 1651–7.

Gillingham, J.C. and Rowe, J. (1984). Daily foraging behavior of the Eastern garter snake *Thamnophis sirtalis. Am. Zool.*, 24, 17A.

Greene, H. (1983). Field studies of hunting behavior by bushmasters. *Am. Zool.*, 23, 897.

Halpern, M. (1983). Nasal chemical senses in snakes. In *Advances in vertebrate neuroethology*, Series A, Vol. 56 (ed. J. Ewert, R.R. Capranica and D.J. Ingle), pp. 141–6. Plenum Press, New York.

Halpern, M., Schulman, N. and Kirschenbaum, D.M. (1986). Characteristics of earthworm washings detected by the vomeronasal system of snakes. In *Chemical signals in vertebrates*, Vol. 4 (ed. D. Duvall, D. Müller-Schwarze and R.M. Silverstein), pp. 63–77. Plenum, New York.

Halpern, M., Schulman, N., Scribani, L. and Kirschenbaum, D.M. (1984). Characterization of vomeronasally-mediated response-eliciting components of earthworm wash-II. *Pharmacol. Biochem. Behav.*, 21, 655–62.

Heppner, F. (1965). Sensory mechanisms and environmental clues used by the American Robin in locating earthworms. *Condor*, 67, 247–56.

Kirschenbaum, D.M., Schulman, N. and Halpern, M. (1986). Earthworms produce a collagen-like substance detected by the garter snake vomeronasal system. *Proc. Natl. Acad. Sci.*, 83, 1213–6.

Kruuk, H. (1978). Foraging and spatial organisation of the European badger, *Meles meles* L. *Behav. Ecol. Sociobiol.*, 4, 75–80.

Landreth, H. (1973). Orientation and behavior of the rattlesnake, *Crotalus atrox. Copeia*, 26–31.

Parker, W.S. and Brown, W.S. (1980). Comparative ecology of two colubrid snakes, *Masticophis t. taeniatus* and *Pituophis melanoleucus deserticola*, in Northern Utah. *Publ. Milwaukee Pub. Mus. Biol. Geol.*, 7, 1–104.

# V.4 Prey Odour Discrimination by Lizards and Snakes

WILLIAM E. COOPER, JR.

## Introduction

It has long been suspected that snakes and lizards detect chemical cues from prey by tongue-flicking (reviewed by Burghardt 1970; Simon 1983), but systematic studies of detection and discrimination of prey odours from other chemical stimuli have not been done for many families of squamates. In this paper I give a preliminary report on the distribution of use of chemical cues in feeding among squamate families.

## Method of study

Present knowledge of prey odour detection and discrimination, i.e. differential responsiveness to two or more odours, has been obtained in lizards largely via a single experimental paradigm; more varied experimental approaches have been used in studies of snakes. In the commonest experimental approach, prey odours and control stimuli are presented to reptiles on cotton-tipped applicators. The number of tongue-flicks and latency to attack the cotton swab are typically combined into a single measure of response to the stimulus, the tongue-flick attack score, TFAS (Burghardt 1967, 1970; Cooper and Burghardt 1990). I have conducted studies of response to prey odour in several lizard families (i.e. Iguanidae, Agamidae, Scincidae, Lacertidae, Teiidae, Anguidae, Helodermatidae, and Varanidae) using this technique (see Cooper and Burghardt 1990). Responses of reptiles were tested to prey odour (mouse or cricket depending on diet), deionized or distilled water, and cologne, which was used as a pungency control. Inclusion of the pungency control allowed assessment of responses to a readily detectable but non-food chemical stimulus, and was used in lieu of chemical stimuli from non-prey species, to which various snakes and lizards do not respond strongly (e. g. Burghardt 1969; Von Achen and Rakestraw 1984). Cologne also had the advantage of being a standardized stimulus across predator species.

## Detection and discrimination in squamate reptiles

The known distribution of prey odour detection and discrimination among squamate families based on experimental demonstrations is presented in Table 1.

**Table 1.** Distribution of experimentally demonstrated prey odour detection and discrimination in squamate reptiles

| Taxon | Detection | Discrimination | Reference |
|---|---|---|---|
| **Lacertilia:** | | | |
| Iguanidae | absent | absent | Cooper (1989a) |
| Agamidae | absent | absent | Cooper (1989a) |
| Chamaeleontidae | unknown | unknown | |
| Gekkonidae | present | present | Dial (1978, and unpublished data) |
| Pygopodidae | unknown | unknown | |
| Xantusiidae | unknown | unknown | |
| Scincidae | present | present | Burghardt (1973); Loop and Scoville (1972); Nicoletto (1985); Von Achen and Rakestraw (1984); Cooper and Vitt (1989) |
| Lacertidae | present | present | Kahmann (1939); Cooper (1990) |
| Teiidae | present | present | Kahmann (1939); Cooper (1990) |
| Dibamidae | unknown | unknown | |
| Anguidae | present | present | Kahmann (1939); Von Achen and Rakestraw (1984); Cooper (unpublished) |
| Xenosauridae | unknown | unknown | |
| Cordylidae | unknown | unknown | |
| Helodermatidae | present | present | Cooper (1989b) |
| Varanidae | present | present | Cooper (1989b) |
| **Ophidia:** | | | |
| Typhlopidae | unknown | unknown | |
| Leptotyphlopidae | present | unknown | Watkins *et al.* (1967) |
| Boidae | unknown | unknown | |
| Pythonidae | unknown | unknown | |
| Tropidophiidae | unknown | unknown | |
| Uropeltidae | unknown | unknown | |
| Aniliidae | unknown | unknown | |
| Acrochordidae | unknown | unknown | |
| Viperidae | present | present | Naulleau (1965); Chiszar and Scudder (1980); Chiszar *et al.* (1986a,b) |
| Elapidae | present | present | Stimac *et al.* (1982); Radcliffe *et al.* (1986) |
| Colubridae | present | present | e. g. Burghardt (1969, 1970); Arnold (1981); Von Achen and Rakestraw (1984) |

For the data on lizard families and Burghardt's studies of colubrid snakes, prey odour detection is listed as present if TFAS was significantly higher for prey odour than water, and prey odour discrimination is listed as present if TFAS additionally was significantly higher for prey odour than for the pungency control.

## Taxonomic relationships and ecological correlates

Use of the tongue during foraging by lizards was noted by Evans (1961) to be widespread among members of Autarchoglossa and virtually absent in Ascalabota (Camp 1923). In Camp's (1923) outdated classification Ascalabota included Agamidae, Chamaeleontidae, Gekkonidae, and Iguanidae; all other lizard families were placed in Autarchoglossa. Because most early studies of possible chemosensory function in lizards focused on scincids and iguanids, Evan's (1961) generalization was supported. The discovery that the gecko *Coleonyx brevis* can detect and discriminate prey odours (Dial 1978, and unpublished data) contradicts this, but is consistent with the more recent inclusion of geckos in Scleroglossa with all former autarchoglossans (Estes *et al.* 1988).

Reliance on chemical cues in feeding behaviour in lizards appears to be associated with foraging mode, diet activity, and perhaps searching for hidden prey. Insectivorous iguanids and agamids are sit-and-wait (ambush) foragers which attack moving prey objects detected by visual cues (e.g. Burghardt 1964). Chamaeleontids are slow-moving visual hunters having a highly modified tongue that cannot be used for sampling chemical cues of potential prey items before attack. The other lacertilian families in which prey odour discrimination has been studied are primarily active foragers. Tongue structure varies dramatically among these families; nevertheless, members of those families that have been studied do detect prey odours and discriminate them from control odours.

Another factor that might affect use of chemical cues in feeding behaviour is availability of light. Diurnal, widely foraging lizards that eat exposed prey may use visual and/or chemical cues during foraging; examples are the skinks *Eumeces fasciatus* (Burghardt 1964, 1973), *E. laticeps* (Cooper 1981; Cooper and Vitt 1989), and *Scincella lateralis* (Nicoletto 1985). Many diurnal active foragers sometimes locate prey hidden in dark places under or within objects or earth, and, hence, may use chemical cues to do so. Among these are scincids (Vitt and Cooper 1986), teiids (Noble and Kumpf 1936), lacertids (Simms 1970), anguids (Smith 1946), helodermatids (Bogert and del Campo 1956), and varanids (Auffenberg 1984). Lizards that forage nocturnally can make only limited use of visual cues. Chemical cues might be important to them in identifying food.

Snakes make extensive use of the tongue-vomeronasal system in foraging; all species studied in detail are capable of detecting prey odours and discriminating them from other odours. However, responses to chemical prey cues differ among taxa. Among colubrids, the natricine garter snakes (*Thamnophis* sp.) show high TFAS to prey odours and frequently attack applicators bearing prey odours (e.g.

Burghardt 1969), whereas the colubrine corn snakes (*Elaphe g. guttata*) show high TFAS to prey odours but attack a much lower proportion of swabs (Burghardt and Abeshaheen 1971; Cooper unpublished data). Among viperids, an elevated rate of tongue-flicking is released by striking (strike-induced chemosensory searching; SICS). Viperids are primarily ambush foragers that release potentially dangerous prey after envenomation and locate the prey by scent-trailing after the venom has had time to take effect (Radcliffe *et al.* 1986). Although striking may be guided by visual or thermal cues in rattlesnakes, its release appears to require vomeronasal stimulation (Graves and Duvall 1985). The cobra *Naja mossambica pallida*, an actively foraging elapid, also exhibits SICS (Radcliffe *et al.* 1986).

## Areas for future study

Prey odour detection and discrimination are now known from all of the speciose scleroglossan lizard families other than Cordylidae (Table 1), but these abilities have been studied in only a few groups of snakes. However, more detailed data on responses to prey odours are available for several colubrids and viperids than for any lizard. Future study might profitably extend the taxonomic range of basic knowledge and examine relationships with ecological factors and foraging mode.

## Acknowledgments

This work was partially supported by contract DE-AC09-76SR00819 between the U.S. Department of Energy and the University of Georgia through its Savannah River Ecology Laboratory and by a grant-in-aid from Auburn University at Montgomery.

## References

Arnold, S.J. (1981). Behavioral variation in natural populations. I. Phenotypic, genetic, and environmental correlations between chemoreceptive responses to prey in the garter snake, *Thamnophis elegans*. *Evolution*, 35, 489–509.

Auffenberg, W. (1984). Notes on feeding behaviour of *Varanus bengalensis* (Sauria: Varanidae). *J. Bombay Nat. Hist. Soc.*, 80, 286–302.

Bogert, C.M. and del Campo, R.M. (1956). The gila monster and allies: the relationships, habits, and behavior of the lizards of the family Helodermatidae. *Bull. Amer. Mus. Nat. Hist.*, 109, 1–238.

Burghardt, G.M. (1964). Effects of prey size and movement on the feeding behavior of the lizards *Anolis carolinensis* and *Eumeces fasciatus*. *Copeia*, 576–8.

Burghardt, G.M. (1967). Chemical-cue preferences of inexperienced snakes: comparative aspects. *Science*, 157, 718–21.

Burghardt, G.M. (1969). Comparative prey-attack studies in newborn snakes of the genus *Thamnophis*. *Behaviour*, 33, 77–114.

Burghardt, G.M. (1970). Chemical perception in reptiles. In *Advances in chemoreception*, Vol. 1, Communication by chemical signals (ed. J.W. Johnston, D.G. Moulton and A. Turk), pp. 241–308. Appleton-Century-Crofts, New York.

Burghardt, G.M. (1973). Chemical release of prey attack: Extension to naive newly hatched lizards, *Eumeces fasciatus*. *Copeia*, 1973, 178–81.

Burghardt, G.M. and Abeshaheen, J.P. (1971). Responses to chemical stimuli of prey in newly hatched snakes of the genus *Elaphe*. *Anim. Behav.*, 19, 486–9.

Camp, C.L. (1923). Classification of the lizards. *Bull. Amer. Mus. Nat. Hist.*, 48, 290–491.

Chiszar, D. and Scudder, K.N. (1980). Chemosensory searching by rattlesnakes during predatory episodes. In *Chemical signals: vertebrates and aquatic invertebrates* (ed. D. Müller-Schwarze and R.M. Silverstein), pp. 125–39. Plenum, New York.

Chiszar, D., Radcliffe, C., Boyd, R., Radcliffe, A., Yun, H., Smith, H.M., Boyer, T., Atkins, B. and Feiler, F. (1986*a*). Trailing behavior in cottonmouths (*Agkistrodon piscivorus*). *J. Herpetol.*, 20, 269–72.

Chiszar, D., Radcliffe, C. and Feiler, F. (1986*b*). Trailing behavior in banded rock rattlesnakes (*Crotalus lepidus klauberi*) and prairie rattlesnakes (*C. viridis viridis*). *J. Comp. Psychol.*, 100, 368–71.

Cooper, W.E., Jr. (1981). Visual guidance of predatory attack by a scincid lizard, *Eumeces laticeps*. *Anim. Behav.*, 29, 1127–36.

Cooper, W., Jr. (1989*a*). Absence of prey odor discrimination by iguanid and agamid lizards in applicator tests. *Copeia*, 1989, 472–8.

Cooper, W.E., Jr. (1989*b*). Prey odor discrimination in the varanoid lizards *Heloderma suspectum* and *Varanus exanthematicus*. *Ethology*, 81, 250–8.

Cooper, W.E., Jr. (1990). Prey odor detection by teiid and lacertid lizards and the relationship of prey odor detection to foraging mode in lizard families. *Copeia* (in press).

Cooper, W.E., Jr. and Burghardt, G.M. (1990). A comparative analysis of scoring methods for chemical discrimination of prey by squamate reptiles. *J. Chem. Ecol.* (in press).

Cooper, W.E., Jr. and Vitt, L.J. (1989). Prey odor disciminination by the broad-headed skink (*Eumeces laticeps*). *J. Exp. Zool.*, 249, 11–16.

Dial, B.E. (1978). Aspects of the behavioral ecology of two Chihuahuan desert geckos (Reptilia, Lacertilia, Gekkonidae). *J. Herpetol.*, 12, 209–16.

Estes, R., de Queroz, K. and Gauthier, J. (1988). Phylogenetic relationships within Squamata. In *Phylogenetic relationships of the lizard families* (ed. R. Estes and G. Pregill), pp. 119–281. Stanford University Press, Stanford.

Evans, L.T. (1961). Structure as related to behavior in the organization of populations of reptiles. In *Vertebrate speciation* (ed. W.F. Blair), pp. 148–178. University of Texas Press, Houston.

Graves, B.L. and Duvall, D. (1985). Avomic prairie rattlesnakes *Crotalus viridis* fail to attack prey. *Z. Tierpsychol.*, 67, 161–6.

Kahmann, H. (1939). Uber das Jacobsonsche Organ der Echsen. *Z. Verrgl. Physiol.*, 26, 669–95.

Loop, M.S. and Scoville, S.A. (1972). Response of newborn *Eumeces inexpectatus* to prey-object extracts. *Herpetologica*, 28, 254–6.

Naulleau, G. (1965). La biologie et le comportement prédateur de *Vipera aspis* au laboratoire et dans la nature. *Bull. Biol. Fr. Belg.*, 49, 395–524.

Nicoletto, P.F. (1985). The roles of vision and the chemical senses in predatory behavior of the skink, *Scincella lateralis*. *J. Herpetol.*, 19, 487–91.

Noble, G.K. and Kumpf, K.P. (1936). The function of Jacobson's organ in lizards. *J. Genet. Psychol.*, 48, 371–82.

Radcliffe, C.W., Estep, K., Boyer, T. and Chiszar, D. (1986). Stimulus control of predatory behaviour in red spitting cobras (*Naja mossambica pallida*) and prairie rattlesnakes (*Crotalus v. viridis*). *Anim. Behav.*, 34, 804–14.

Simms, C. (1970). *Lives of British lizards.* Goose and Son, Norwich.

Simon, C.A. (1983). A review of lizard chemoreception. In *Lizard ecology: studies of a model organism* (ed. R.B. Huey, E.R. Pianka and T.W. Schoener), pp. 119–133. Harvard University Press, Cambridge, Mass.

Smith, H.M. (1946). *Handbook of lizards.* Cornell University Press, Ithaca, New York.

Stimac, K., Radcliffe, C.W. and Chiszar, D. (1982). Prey recognition learning by red spitting cobras, *Naja mossambica palllida*. *Bull. Psychon. Soc.*. 19, 187–8.

Vitt, L.J. and Cooper, W.E., Jr. (1986). Foraging and diet of a diurnal predator (*Eumeces laticeps*) feeding on hidden prey. *J. Herpetol.*, 20, 408–15.

Von Achen, P.H. and Rakestraw, J.L. (1984). The role of chemoreception in the prey selection of neonate reptiles. In *Vertebrate ecology and systematics: a tribute to Henry S. Fitch* (ed. R.A. Seigel, L.E. Hunt, J.L. Knight, L. Malaret and N.L. Zuschlag), pp. 163–172. Special Publication, University of Kansas Museum of Nat. Hist., Lawrence.

Watkins, J.F., II, Gehlbach, F.R. and Baldridge, R.S. (1967). Ability of the blind snake, *Leptotyphlops dulcis*, to follow pheromone trails of army ants, *Neivamyrmex nigrescens* and *N. opacithorax*. *Southw. Natur.*, 12, 455–462.

# V.5 Behavioural and Chemical Ecology of Vernal Migration and Pre- and Post-Strike Predatory Activity in Prairie Rattlesnakes: Field and Laboratory Experiments

DAVID DUVALL AND DAVID CHISZAR

## Introduction

Studies of the chemical ecology and semiochemistry of a broad range of vertebrates (Duvall *et al.* 1986) and reptiles, especially snakes and lizards (Burghardt 1970, 1980, this volume; Halpern and Kubie 1984) seem to have accelerated in pace and increased in depth in recent years. In this paper we discuss the chemical ecology of two key events in the predatory ecology and ethology of prairie rattlesnakes (*Crotalus viridis viridis*): (1) the search for prey and the areas in which they reside (e.g. King and Duvall 1990); and (2) prey trailing, once live rodents have been bitten, envenomated, and then released (e.g. Chiszar *et al.* 1983, 1990; Golan *et al.* 1982). In the present context, these two phases of predatory activity may be termed pre- and post-strike, respectively. It is our view that dual track or parallel field and laboratory analyses of prairie rattlesnake behavioural ecology and ethology leads to more complete and robust prediction and explanation (Duvall *et al.* 1985, 1990; Graves and Duvall 1984, 1987; Graves *et al.* 1986).

## Vernal movements and foraging/ambush site selection: the pre-strike phase

Prairie rattlesnakes occurring on the high, seasonal deserts of southern Wyoming exhibit lengthy, fixed-angle (i.e. straight-line) vernal migrations (Duvall *et al.* 1985; King 1987; King and Duvall 1990). Such movements can take animals several kilometres away from their overwintering den; males typically travel further and more efficiently than females (Duvall *et al.* 1985; King and Duvall 1990). Other snake taxa are known to exhibit similar habits (e.g. Brown *et al.* 1982; Fitch and Shirer 1971; Gregory *et al.* 1987; Landreth 1973; Macartney *et al.* 1988; Reinert and Zappalorti 1990).

Duvall *et al.* (1985), King (1987), and King and Duvall (1990) have obtained quantitative field data suggesting that the primary function of the vernal phase of the larger season-long migration among prairie rattlesnakes is to find food (cf. Duvall *et al.*, submitted). In the high desert communities along the Continental Divide of south-central Wyoming, prairie rattlesnakes prey principally on deer mice (*Peromyscus maniculatus*), the most common small mammal in the area. Locally active patches of these rodents are spatially unpredictable from year to year, populations are small in size, and they are widely and patchily (= clustered; see Ludwig and Reynolds 1988) distributed (Duvall *et al.* 1985; King 1987). This poses a significant foraging problem for the rattlesnakes, especially given their brief active or growing season (ca. 90 days/year; Martner 1986). As the active season progresses, males must also search for receptive or non-pregnant females (King and Duvall 1990), and this compounds the importance of their finding food early in the season. The survival value of lengthy, straight-line vernal movements lies in the search for spatially unpredictable and sparse small mammal prey. Once the few rodent patches that exist are located, migrating snakes temporarily reduce rates of locomotion or cease movement altogether. They then forage within a patch for up to 3–4 weeks by active searching or sit-and-wait ambushing (Duvall *et al.* 1985; King 1987; King and Duvall 1990).

Additionally, males face the dual problems of finding food *and* mates (King and Duvall 1990), and their reproductive success depends on relatively greater attention to the former (e.g. Trivers 1972). Accordingly, male prairie rattlesnakes should be especially responsive to deer mouse patches located in the spring, and relatively more responsive to these rodents and stimuli derived from them than females. Females can continue to eat intermittently across the entire season, whereas males become relatively unresponsive to prey during the midsummer to fall breeding season, concentrating instead on locating mates (King and Duvall 1990). Only a small proportion of males locates receptive females each season at the Wyoming site (Duvall *et al.* unpubl.).

Given the especial importance of the tongue-vomernasal organ (VNO) system to the ethology of squamate reptiles such as prairie rattlesnakes (Bellairs 1984; Burghardt 1980; Chiszar 1986; Duvall 1986; Ford 1986; Graves and Duvall 1984; Halpern and Kubie 1984; Weldon and Burghardt 1979) there are two additional questions that merit attention: (i) would migrating prairie rattlesnakes respond to prey-derived semiochemicals as cues reflecting the presence of rodents in a patch (e.g. urine, integumentary materials, etc.) (Chiszar and Radcliffe 1986)?; and (ii) would males be more sensitive in this regard than females? Certainly both thermal and visual cues play major roles in prey location and attack (e.g. de Cock Buning 1983; Dickman *et al.* 1987; Hennessy and Owings 1989; Kardong 1986). Nevertheless, it is appropriate to ask whether semiochemicals alone might play a role in prey location, in related decisions about when to cease moving, and when to explore slowly a prey patch or stop altogether and sit-and-wait in ambush.

In the course of a detailed, seven-year study of ethology and behavioural ecology of radio-tagged, free-ranging prairie rattlesnakes in Wyoming, Duval *et al.* (submitted) observed opportunistic foraging in these snakes, ranging from active searching and prey capture, to carrion feeding, to sit-and-wait ambushing (cf. Diller 1989; Gillingham and Baker 1981; Gillingham and Clark 1981; Klauber 1972; Reinert *et al.* 1984). Assuming that chemical cues do play a role in signalling the presence or absence of rodents in a patch, it is meaningful to expect a reduction in activity, and not just its complete cessation.

Snakes drawn from different populations may exhibit differing behaviour according to local habitat. Work by Arnold (1981) and Drummond and Burghardt (1983) has shown that different populations of garter snakes exhibit qualitatively different patterns of behavioural adaptation. It is therefore possible that the behaviour of prairie rattlesnakes captured in the Pawnee grasslands of north-eastern Colorado and in the high deserts of south-central Wyoming might differ in certain meaningful ways in our system. There are differences in habitat parameters between the two sites: for example, the active seasons are at least two months longer on the Pawnee grasslands (Duvall and Chiszar unpubl.; cf. Martner 1986); also the uniform distribution of dominant, native grasses could in turn lead to a corresponding uniform distribution of rodents. Such differences may have resulted in the derivation of differing patterns of foraging behaviour, including functional reasons for vernal migrations and responses to prey-derived semiochemicals.

The following hypotheses address the roles of prey and prey-derived cues, including semiochemical ones, in the search for foraging and/or ambushing sites:

1. One function of vernal movements of prairie rattlesnakes studied in the laboratory is to find food, as evidenced by reduced activity in, and preferences for, regions of test arenas containing live prey or certain prey-derived stimuli.
2. Vernal migrations of animals tested in the field will likewise decrease in magnitude or cease altogether when subjects contact experimentally augmented patches containing live deer mice or deer mouse-derived stimuli.
3. Prey-derived semiochemicals alone can signal the presence of rodent prey in an area to migrating or locomoting snakes, in the field or laboratory, respectively.
4. Males from the Wyoming site are more responsive to prey and prey-derived stimuli than females.
5. Subjects drawn from two allopatric populations, occurring in regions with very different habitats and communities, respond differently to semiochemicals derived from their rodent prey.

## Field experiments

### EXPERIMENT I

*Methods and procedures*

In May 1987, as male and non-pregnant female snakes emerged from hibernation, they were captured and given surgical implants of radio-telemetry tracking tags (AVM Instrument Co., Livermore, CA). Radiotag mass amounted to an average of 4% of snakes' body masses. Pregnant females were not used in this study, since they do not eat in seasons in which they are pregnant (Graves *et al.*, 1986); non-pregnant females are referred to simply as females hereafter. On recovery from surgery in late May, snakes were assigned to one of two treatment groups and released at the den from which they emerged (i.e. Blind Canyon or Coal Creek Canyon dens, Haystack Mountains, Wyoming). The experimental group (6 males and 4 females) was exposed to cages containing 2–4 deer mice that were freshly trapped on site and placed in snakes' paths as they migrated. The control group (4 males and 2 females) were exposed to cages prepared identically except that they did not contain live deer mice.

Once individual migrating snakes had either travelled over 200 m from the den and been away from the den for two days from the time of release, they were randomly assigned to one of the two treatment groups and the experiment commenced. Advantage was taken of migrating rattlesnakes' tendency to hole-up each evening in rocky cracks and crevices, abandoned rodent burrows, or other temporary refuges, to set treatment conditions in place. As individual snakes ceased to move for the day and retreated to refuges, treatment conditions were placed within 1 m of the opening to the cavity that the snake had entered. Treatment conditions consisted of: (1) a plastic rodent cage (29.5 × 18.5 × 13 cm) outfitted with a hardware cloth top, sawdust, a daily ration of wild bird seed, a full water bottle, and one large strip of duct tape to hold the assembly in place; (2) four small cheesecloth scent bags (8 × 5 × 3 cm), which were filled with clean or deer-mouse-soiled sawdust and scattered within a 1 m radius of the cage; (3) an A-frame, plywood lean-to (40 × 40 cm) to protect the cages and their contents from sun, wind, and rain. Cages contained 2–4 live deer mice and the bags were filled with deer-mouse soiled sawdust. Control cages contained no live deer mice and the scent bags were filled with unsoiled or clean sawdust. Once treatment conditions were put into place, the cages were kept ca 1 m of the snakes for no longer than 21 days (mean = 19.4 days). On days when snakes moved, treatments were relocated to the new locale of the animal that evening.

*Results*

Selected data obtained from Experiment 1 are presented in Table 1. Based upon a Condition Site Fidelity Index (CSFI) calculated for each snake (i.e. a standardized

**Table 1.** Per cent staying scores ($\bar{X} \pm SE$) of free-ranging prairie rattlesnakes exposed to live deer mice and a control treatment[a]

| Sex[b,c] | Conditions[c] Deer mice | Control |
|---|---|---|
| Males | $81.0 \pm 3.9$ ($n=6$) | $47.5 \pm 10.1$ ($n=4$) |
| Females | $32.1 \pm 6.0$ ($n=4$) | $57.2 \pm 9.6$ ($n=2$) |

[a] Values are CSFI; see text for explanation. All proportional data are Arcsin transformed prior to ANOVA. Sample sizes are given in parentheses.
[b] Significant main effect; see text for discussion.
[c] Significant interaction effect; see text for discussion.

staying score, or percentage, determined for each subject and based upon the number of evenings individuals stayed within a 5 m radius of conditions over the total number of days that the experiment was run), male but not female prairie rattlesnakes ceased migration and remained significantly longer than control animals [$F_{interaction}$ $(1,12) = 10.95$, $p<0.01$]. Males were significantly more responsive to both conditions than females [$F$ $(1,12) = 5.28$, $p<0.05$]. Males were also occasionally found coiled atop cages contining deer mice, but never atop clean cages.

EXPERIMENT 2

A goal not accomplished in Experiment 1 was to include a deer mouse odour condition. More than the 16 subjects actually radio-tagged and tested would have been necessary, but as it turned out in 1987, they were not available. Accordingly, a within subjects pilot study was conducted (Experiment 2): 3 males and 1 female out of the control animal sample that continued to migrate following exposure to clean cages, were exposed to cages containing soiled deer mouse bedding and four scent bags filled with this material. Staying scores (CSFI values, as above) were determined over a 21-day period. It was found that snakes exposed to clean versus soiled conditions generated CSFI or staying values ($\bar{X} \pm SE$) of $44.1 \pm 9.0$ and $66.1 \pm 6.3$, respectively; the difference was not significant.

## Laboratory experiments

*General methods of Wyoming experiments*

Three experiments (Experiments 3, 4, and 5) were conducted in the spring and early summer, using prairie rattlesnakes freshly captured in the field as they emerged from hibernation and then returned to our laboratory at the University of Wyoming, in Laramie (cf. Duvall *et al.* 1990). We wished to determine the extent to which live deer mice or deer mouse-derived chemical signals would affect snake movement and certain predatory actions in laboratory enclosures. Behavioural tests were done in large, rectangular arenas (76 × 240 × 90 cm), divided into two sections; a mouse cage containing one of the experimental conditions (see below) was placed in one section, and the second (control) condition was placed in the other. Nine trials, each of 10 mins, were spread evenly over a period of 3 days. Dependent variables measured were: (1) proportion of accumulated time spent on either side of the arena (TSide); (2) proportion of accumulated time spent within a rectangular area (34.5 × 45 cm) (TRectangle), drawn in the centre of the floor of respective halves of the arena, and within which cages containing the treatments were placed; (3) proportion of accumulated time spent in ambush postures within either of the two rectangles (TAmbush); (4) the number of peers (Peers) in which snakes surveyed the contents of cages at close range (8 cm or less).

The latter action usually precedes a predatory strike or attack. All proportional data obtained were Arcsin transformed prior to analysis of variance. Deer mice used in experiments were freshly captured on site and returned to Laramie. Snakes were maintained and tested under 14 h light/10 h dark photoperiod at ambient arena temperature of 23.6 °C (range = 19.4–27.8 °C).

### EXPERIMENT 3

In this study, the two conditions were deer mouse odours (cages containing soiled sawdust bedding) or clean cages containing fresh, unsoiled sawdust. Of the 8 males and 7 females tested, no significant F ratios were detected. This suggests that snakes freshly emerged from natural hibernation and presumably hungry did not employ deer mouse-derived chemical signals to determine if rodents were present in respective arena zones.

### EXPERIMENT 4

False bottoms were fitted into the mouse cages containing soiled, odorous sawdust bedding or those containing clean, non-odorous bedding. This was done to raise the substrate closer to the screen lid so that the snakes could readily tongue flick directly into the sawdust, and detect any non-volatile chemical signals that might be present. Presumably this would facilitate tongue-VNO detection of relatively non-volatile chemical signals (Halpern and Kubie 1980; Kubie and Halpern 1978). Of the 12 males and 9 females tested, both sexes exhibited significantly more TAmbush when near the soiled as opposed to the clean cage [$\bar{X} \pm SE = 19.4 \pm 4.1$ versus $5.6 \pm 2.4$, respectively; $F(1,19) = 4.321, p < 0.049$].

### EXPERIMENT 5

In this study, cages containing two live deer mice were paired with cages containing soiled bedding (i.e. everything but live deer mice). Selected descriptive statistics for the 11 males and 9 females tested are presented in Table 2. Rattlesnakes of both sexes were equally and highly responsive to cages containing live deer mice. Some important and significant differences detected included ones for (i) TSide [$F(1,18) = 10.03, p < 0.005$], (ii) TRectangle [$F(1,18) = 15.87, p < 0.001$], (iii) TAmbush [$F(1,18) = 13.14, p < 0.002$], and (iv) Peers [$F(1,18) = 5.95, p < 0.024$].

**Table 2.** Selected responses ($\bar{X} \pm SE$) of prairie rattlesnakes in the laboratory to live deer mice versus deer mouse-soiled bedding[a]

| | | **Dependent measures[b]** | | | |
|---|---|---|---|---|---|
| **Condition** | **Sex** | **TSide[c]** | **TRectangle[c]** | **TAmbush[c]** | **Peers[c]** |
| Live deer mice | Males | $64.5 \pm 4.7$ | $45.1 \pm 4.8$ | $34.2 \pm 4.7$ | $0.6 \pm 0.3$ |
| | Females | $67.7 \pm 4.6$ | $44.9 \pm 5.4$ | $37.6 \pm 5.3$ | $0.3 \pm 0.1$ |
| Deer mouse bedding | Males | $35.3 \pm 4.7$ | $11.5 \pm 3.0$ | $9.0 \pm 2.9$ | $0.1 \pm 0.1$ |
| | Females | $32.4 \pm 4.6$ | $12.9 \pm 3.7$ | $3.7 \pm 2.1$ | 0.0 |

[a] All proportional data Arcsin transformed prior to ANOVA. Selected data from Duvall *et al.* (1990).
[b] TSide = per cent duration spent on respective sides of the arena mid-line; TRectangle = per cent duration spent in rectangular areas surrounding respective cages/conditions, corrected (divided) by the per cent time spent on a particular side; TAmbush = per cent duration spent in ambush postures within respective angular areas, corrected (divided) by respective TRectangle; Peers = the number of times snakes surveyed the contents of cages at close range.
[c] Significant main effect; see text for discussion.

### EXPERIMENT 6

Evidence suggests that snakes vary in patterns of activity in nature (e.g. Gregory *et al.* 1987; Macartney *et al.* 1988). We have argued that the length, duration, and direction of vernal and seasonal movements exhibited by prairie rattlesnakes may vary systematically between allopatric populations as a function of (selective) variables reflecting local community and habitat factors (Duvall *et al.* 1985, 1989; King and Duvall 1990). Accordingly, prairie rattlesnakes residing in the grasslands of north-eastern Colorado might be more likely to exhibit a random walk while foraging than the lengthy straight-line movements of snakes living in the patchy and unpredictable prey communities characteristic of the Wyoming site. Quantitative simulation models (Duvall, Chiszar and Mintzer, unpubl.) of search paths do indeed suggest that the maximal route for foraging in a patchy array (in terms of probability of actual prey location) is a straight line, whereas that in a community with spatially uniform prey is a random walk. The latter behaviour is more likely to be characteristic of a true sit-and-wait forager, particularly one with no prior information on the location of potentially successful ambush sites.

This means that Colorado snakes should be more responsive than Wyoming snakes to chemical cues derived from potential rodent prey, because waiting in ambush for rodents, as opposed to actively seeking them out, is more likely to pay off in a community where prey resources are uniformly distributed in space. Thus, when tested in the lab, rattlesnakes from the grasslands of north-eastern Colorado would be especially sensitive to, and prefer to occupy arena zones marked with, semiochemicals derived from rodent prey.

*Methods and procedures*

The experimental subjects were five long-term resident prairie rattlesnakes (*C. v. viridis*), captured in the Pawnee grasslands of north-eastern Colorado. The study was conducted at the University of Colorado, Boulder, where snakes were maintained on a regular diet of house mice (*Mus musculus*).

Sex was not a variable; this experiment was designed solely to manipulate the presence or absence of house mouse-derived semiochemicals in a large ($3 \times 5$ m, floor area), semi-naturalistic indoor arena. The five snakes studied were successively and individually placed into the enclosure for two weeks, throughout which no prey odours were presented. Ambient temperature was maintained at ca. 26 °C by an electrical heater and a 12 h light/12 h dark photoperiod was employed. The floor of the arena was marked so that 15 1-$m^2$ zones could be distinguished.

On days 15–20, two steel mouse cages with screen tops were placed in the enclosure, with each cage in a randomly selected square. One cage contained clean sawdust bedding, the other bedding that had been soiled by house mice for one week. No mice were present in either container when placed into the arena. The squares occupied by snakes were recorded 10 times/day for six days.

During the first two weeks snakes moved more or less randomly in the arena. There was a slight preference for corner squares, but the snakes did move throughout the arena. Accordingly, since corner squares were preferred, cages containing bedding were never placed there. This was the only restriction on the otherwise random selection of sites for cages.

When the clean and soiled containers were in place (days 15–20), it was observed that snakes initially positioned themselves next to or on top of the container with soiled shavings, although this behaviour disappeared over succeeding days. Hence, data for days 15–17 and 18–20 were analysed separately.

Single sample *t*-tests were used to analyse mean per cent spatial differences of snakes during days 15–17 and for days 18–20.

*Results*

During days 15–17, snakes were found more frequently than expected near the soiled cage [$t(4) = 11.30$, $p < 0.05$], and less frequently than expected near the clean cage [$t(4) = -10.00$, $p < 0.05$] and all other squares [$t(4) = -9.76$, $p < 0.05$]. All of these effects disappeared during days 18–20. A paired *t*-test revealed a significant drop in the mean per cent attendance at the soiled container between days 15–17 and 18–20 [$t(4) = 4.58$, $p < 0.05$] (see Table 3).

**Table 3.** Per cent ($\bar{X} \pm SE$) spatial preferences of prairie rattlesnakes in the laboratory for house mouse-soiled versus unsoiled bedding[a]

| | **Preference for:** | | |
|---|---|---|---|
| **Days** | **Soiled Squares** | **Unsoiled squares** | **All other squares** |
| 15–17 | 70.0 ± 5.6[b] (6.7) | 0.7 ± 0.6[b] (6.7) | 29.1 ± 5.9[b] (86.7) |
| 18–20 | 12.9 ± 7.0 (6.7) | 4.1 ± 1.5 (6.7) | 83.0 ± 12.2 (86.7) |

[a] Expected per cent spatial preference scores given in parentheses. Data from Duvall *et al.* (1990).
[b] Deviates significantly from expected. See text for discussions of statistical significance.

*Discussion*

Experiments 1–6 support the hypothesis that a major function of prairie rattlesnake vernal migrations, particularly among males, is to find food. Though sample sizes in the field need to be boosted, especially for females, at least some of this picture is confirmed in both laboratory and field. It is also significant that successful field experiments could actually be conducted on subjects as secretive and cryptic as prairie rattlesnakes. While developments in behavioural ecology and sociobiology have quickly expanded to include matters of quantitative strategy (e.g. Stephens and Krebs 1986) and evolution (e.g. Arnold 1981), they have not kept pace with experimental studies of the ecological, or survival value function of behaviour in free-ranging and laboratory populations of animals. Experiments 1–6 help to fill this gap.

The role of rodent prey-derived semiochemicals in foraging and/or ambush site selection, or the pre-strike phase, is much less clear. Concerning the snakes tested in the field in Wyoming—first, more work must be done; we need an improved experimental design for deer mouse odour presentation. Such an effort is currently in progress. Second, snakes drawn from two allopatric populations, residing in different habitats and occurring in different communities, responded very differently in the laboratory to chemical cues derived from their rodent prey; those from Colorado were very responsive while those from Wyoming were almost completely unresponsive. Though this disparity may be due to any number of methodological differences in the Laramie versus Boulder work, it is also possible that microevolutionary history may have contributed to the observed variation in patterns of responsiveness. Duvall *et al.* (1990, submitted) consider these problems in more detail.

Other field studies on our Wyoming site have shown clear sex differences in certain key patterns of behaviour (Duvall *et al.* 1985; King 1987; King and Duvall 1990). The sample of females employed in field experiments must be increased before any preliminary conclusions can be drawn and, again, efforts to accomplish this are in progress.

## Trailing envenomated prey: the post-strike phase

Prairie rattlesnakes studied in the laboratory are known to release adult rodents after the envenomating strike or bite, allowing the victim to wander freely until the venom takes its lethal effect (e.g. Kardong 1986; for reviews see Chiszar and Scudder 1980; Chiszar *et al.* 1983). Snakes then locate the carcasses by following the chemical trail that is deposited by the rodent as it wanders away from the site of attack (Chiszar *et al.* 1983, 1986, 1990; Golan *et al.* 1982). Rodents urinate copiously after envenomation, and it seemed likely that rodent urine was the material detected by rattlesnakes as they follow trails (cf. Burghardt 1970). Experiment 7 was designed to test this hypothesis.

### EXPERIMENT 7

#### *Methods and procedures*

The procedures of Golan *et al.* (1982) were used. The apparatus was a wooden pen (180 × 66 × 81 cm), and for a week a snake was confined to one side by a partition containing a guillotine door. On the paper floor cover of the large compartment a meandering line (150 cm in length) was drawn, beginning at the centre of the partition separating the two compartments and ending behind a rock at the opposite end of the large compartment. This line was paralleled on each side by a dashed line, 2 cm from the central one. Chemical trails were deposited along the central line as described below, and we considered the snake to be following the trail if it moved within the area bounded by the dashed lines (see below). This area constituted 12% of the total area in the large compartment.

Three types of trails were made; the first was a urine trail. House mouse (*Mus musculus*) urine was collected in a metabolism cage and refrigerated at 13 °C for up to 48 h until needed, but not more than 48 h. To make a trail, the urine was warmed to 35 °C and applied to the solid line by pipetting one drop at 5-cm intervals, starting at the guillotine door and ending behind the rock. Urine drops usually expanded to form a circle about 1 cm in diameter. The second (control) trail was made with tap water by pipetting drops at 5-cm intervals. The third trail was made by dragging a recently killed house mouse (in dorsal recumbency and with the inguinal area covered with cotton) along the solid line. No visible urine spots resulted from this procedure, and we presume that only materials from the integument were deposited (e.g. sebaceous secretions, Harderian exudates, and possible inguinal expressions that the rodent had earlier spread over its fur while grooming). These three types of trails are referred to as urine (U), water (W), and integumentary (I) trails, respectively.

Each of five prairie rattlesnakes was exposed to each type of trail in random order at the rate of one trial per week. A snake remained in the apparatus until it completed three trials. A trial began by preparing the trail, placing a non-envenomated house mouse carcass behind the rock, and then suspending a house mouse into the snake's compartment and allowing the snake to strike. This rodent was removed and the guillotine door was opened. We recorded: (1) latency for the snake's head to make initial contact with the trail area; (2) the total number of tongue flicks emitted; (3) the number of tongue flicks that were emitted while the snake's head was within the trail area; (4) the time in seconds that the snake's head was within the trail area; (5) the time required for the snake to contact the house mouse carcass behind the rock.

A maximum of 20 mins was allowed; if a snake failed to locate the carcass during this period, a score of 1200 s was assigned. In addition, snakes were observed for 10 min prior to the start of each trial. Tongue flicks were recorded during this baseline period for comparison with subsequent post-strike rates.

It is known from previous experimental work that trail following usually does not occur in laboratory tests if snakes are not permitted to strike prey just prior to the trial (Chiszar *et al.* 1983, 1986; Golan *et al.* 1982). The present experiment included five snakes that were exposed to the three types of trails after no-strike presentations of house mice (for 5 s) just before opening the guillotine door. These snakes consistently failed to follow trails of any sort, and these data will not be discussed further.

### *Results and discussion*

Rate of tongue flicking (per minute) during the baseline period was 2.7, indicating that the snakes were quiescent. Average rate of tongue flicking during the first 10 min after striking prey was 40.4, a significant increase over the baseline value [$F(1,4) = 134.57, p < 0.01$]. The post-strike rate of tongue flicking did not differ as a function of the type of trail ($F < 1$). All other data and effects are summarized in Table 4.

**Table 4.** Mean responses of prairie rattlesnakes to trail types[a]

| Row[b] | Dependent measure | Type of trail: Water | Urine | Integu-mentary | F(2,8)[b] |
|---|---|---|---|---|---|
| A | Trail contact latency (s) | 282.2* | 251.1* | 275.1* | 1.04 |
| B | Tongue flicks on trail (%) (−0.12) | 11.7* (0.87) | 17.6* (3.88[c]) | 44.9† | 4.96[c] |
| C | Time on trail (%) (−0.38) | 9.4* (0.61) | 16.1* (3.38[c]) | 45.8† | 5.88[c] |
| D | Carcass contact latency (s) | 941.2* | 646.6*† | 388.1† | 5.17[c] |

[a] Numbers within parentheses are single sample t-test values (df = 4) comparing the tabled mean to the chance value of 12%. See text for discussion.
*† Within rows, means bearing the same superscript do not differ significantly by Duncan's new multiple range test (alpha = 0.05).
[b] F ratios from repeated measures analyses of variance applied to respective dependent measures. See text for discussion.
[c] $p < 0.05$

Mean latency to contact the trail area did not differ as a function of the type of trail; together with the tongue flicking data this suggests that striking activates the searching pattern and that chemical cues in the environment make no additional contribution to the initial part of the post-strike performance (Golan *et al.* 1982). However, once the trail area has been contacted, snakes moved in relation to the trail only when integumentary materials had been applied. This can be seen in several ways in Table 4. First, in rows B and C the means are significantly higher for I trails than for the other types of trail. Second, when the means in rows B and C are compared to the chance value of 12%, only the means for I trails are significantly higher than this value. Finally, the latency to contact the hidden carcass is numerically lowest for snakes given an I trail (although this mean is not significantly different from that for snakes given a U trail).

The *t*-tests reported in rows B and C of Table 4 are convincing, and we conclude that only trails made with integumentary materials were followed by prairie rattlesnakes. Thus, we reject the hypothesis that rodent urine is the key ingredient used by these predators to follow rodent trails. Rodent urine has great information value to other rodents (e.g. Drickamer 1986; Jemiolo *et al.* 1986; Vandenbergh 1980), but it appears to be irrelevant to prairie rattlesnakes. Concerning the active components of the I trail, it is likely that under natural conditions the trail is comprised of integumentary exudates and expressions of certain other exocrine glands (e.g. Harderian glands; see Blake *et al.* 1983) that have been spread over the fur during bouts of grooming. It is also noteworthy that desert and prairie rodents urinate much less than laboratory strains of house mice (Duval *et al.* 1990). Perhaps prairie rattlesnakes have been selected to focus attention during prey trailing on non-urinary rodent semiochemicals because urinary cues are less available and less thermostable under natural conditions.

## Acknowledgements

We thank C. Radcliffe, M. Goode, W. Hayes, J. Leonhardt, and A. Redder for their many contributions to this research. The editors and anonymous reviewers provided comments that improved the quality of the manuscript. Experiments 1–5 were supported by NSF Grant No. RII-8610680.

## References

Arnold, S.J. (1981). The microevolution of feeding behavior. In *Foraging behavior*, (ed. A. Kamil and T. Sargent), pp. 409–53. Garland STPM Press, New York.

Bellairs, A.d'A. (1984). Closing address. In *The structure, development and evolution of reptiles* (ed. M.W.J. Ferguson), pp. 665–83. Zool. Soc. London Symp. N0. 52, Academic Press, New York.

Blake, R., Duvall, D., Gern, W.A. and Chiszar, D. (1983). Mouse Harderian glands: A source of chemical signals mediating head-first swallowing by prairie rattlesnakes. *J. Colo. Wyo. Acad. Sci.*, 15, 46–7.

Brown, W.S., Pyle, D.W., Greene, K.R. and Friedlaender, J.B. (1982). Movements and temperature relationships of timber rattlesnakes (*Crotalus horridus*) in northeastern New York. *J. Herpetol.*, 16, 151–61.

Burghardt, G.M. (1970). Chemical perception in reptiles. In *Communication by chemical signals* (ed. J.W. Johnston, Jr., D.G. Moulton and A. Turk), pp. 241–308. Appleton-Century-Crofts, New York.

Burghardt, G.M. (1980). Behavioral and stimulus correlates of vomeronasal functioning in reptiles: Feeding, grouping, sex, and tongue use. In *Chemical signals* (ed. D. Müller-Schwarze and R.M. Silverstein), pp. 275–301. Plenum, New York.

Chiszar, D. and Scudder, K.M. (1980). Chemosensory searching by rattlesnakes during predatory episodes. In *Chemical signals* (ed. D. Müller-Schwarze and R.M. Silverstein), pp. 125–39. Plenum, New York.

Chiszar, D., Radcliffe, C.W., Scudder, K. and Duvall, D. (1983). Strike-induced chemosensory searching by rattlesnakes: The role of envenomation-related cues in the post-strike environment. In *Chemical signals in vertebrates* 3, (ed. D. Müller-Schwarze and R.M. Silverstein), pp. 25–44. Plenum, New York.

Chiszar, D. (1986). Motor patterns dedicated to sensory functions. In *Chemical signals in vertebrates*, 4 (ed. D. Duvall, D. Müller-Schwarze and R.M. Silverstein), pp. 37–44. Plenum, New York.

Chiszar, D. and Radcliffe, C.W. (1986). Responses to prey-derived stimuli bucolubrid, elapid, and viparid snakes: Proximate mechanisms mediating prey behavior. ASIH Symposium on Snake Ecology and Behavior, Victoria, B.C., Canada.

Chiszar, D., Radcliffe, C.W. and Feiler, F. (1986). Trailing behavior in banded rock rattlesnakes (*Crotalus lepidus klauberi*) and prairie rattlesnakes (*Crotalus viridis viridis*). *J. Comp. Psychol.*, 100, 368–71.

Chiszar, D., Melcer, T., Lee, R., Radcliffe, C.W. and Duvall, D. (1990). Chemical cues used by prairie rattlesnakes (*Crotalus viridis*) as they follow trails of rodent prey. *J. Chem. Ecol.*, in press.

de Cock Buning, T. (1983). Thermal sensitivity as a specialization for prey capture and feeding in snakes. *Amer. Zool.*, 23, 363–75.

Dickman, J.D., Colton, J.S., Chiszar, D. and Colton, C.A. (1987). Trigeminal responses to thermal stimulation of the oral cavity in rattlesnakes (*Crotalus viridis*) before and after bilateral anaesthetization of the facial pit organs. *Brain Res.*, 400, 365–70.

Diller, L.V. (1989). A field observation on the feeding behavior of *Crotalus viridis lutosus*. *J. Herpetol.*, in press.

Drickamer, L.C. (1986). Puberty-influencing chemosignals in house mice: Ecological and evolutionary considerations. In *Chemical signals in vertebrates*, 4 (ed. D. Duvall, D. Müller-Schwarze and R.M. Silverstein), pp. 441–55. Plenum, New York.

Drummond, H. and Burghardt, G.M. (1983). Geographic variation in the foraging behavior of the garter snake, *Thamnophis elegans*. *Behav. Ecol. Sociobiol.*, 12, 43–8.

Duvall, D., King, M.B. and Gutzwiller, K.J. (1985). Behavioral ecology and ethology of the prairie rattlesnake. *Nat. Geog. Res.*, 1, 80–111.

Duvall, D. (1986). A new question of pheromones: Aspects of possible chemical signaling and reception in the mammal-like reptiles. In *The ecology and biology of the mammal-like reptiles* (ed. N. Hotton III, P.D. MacLean, J.J. Roth and E.C. Roth), pp. 219–38, Smithsonian Institution, Washington, D.C.

Duvall, D., Müller-Schwarze, D. and Silverstein, R.M. (ed.). (1986). *Chemical signals in vertebrates*, 4. Plenum, New York.

Duvall, D., Chiszar, D., Hayes, W.K., Leonhardt, J.K. and Goode, M.J. (1990). Chemical and behavioral ecology of foraging activity in prairie rattlesnakes. *J. Chem. Ecol.*, in press.

Duvall, D., Goode, M.J., Hayes, W.K., Leonhardt, J.K. and Brown, D., (submitted), Why prairie rattlesnakes migrate in the spring.

Fitch, H.S. and Shirer, H.W. (1971). A radiotelemetric study of spatial relationships in some common snakes. *Copeia*, 1971, 118–28.

Ford, N.B. (1986). The role of pheromone trails in the sociobiology of snakes. In *Chemical signals in vertebrates*, 4 (ed. D. Duvall, D. Müller-Schwarze and R.M. Silverstein), pp. 261–78. Plenum, New York.

Gillingham, J.C. and Baker, R.E. (1981). Evidence for scavenging behavior in the western diamondback rattlesnake (*Crotalus atrox*). *Z. Tierpsychol.*, 55, 217–27.

Gillingham, J.C. and Clark, D.L. (1981). An analysis of prey searching behavior in the western diamondback rattlesnake, *Crotalus atrox*. *Behav. Neural Biol.*, 32, 235–40.

Golan, L., Radcliffe, C., Miller, T., O'Connell, B. and Chiszar, D. (1982). Trailing behavior in prairie rattlesnakes (*Crotalus viridis*). *J. Herpetol.*, 16, 287–93.

Graves, B.M. and Duvall, D. (1984). Avomic prairie rattlesnakes (*Crotalus viridis*) fail to attack rodent prey. *Z. Tierpsychol.*, 67, 161–6.

Graves, B.M., Duvall, D., King, M.B., Linstedt, S.L. and Gern, W.A. (1986). Initial den location by neonatal prairie rattlesnakes: Functions, causes, and natural history in chemical ecology. In *Chemical signals in vertebrates*, 4 (ed. D. Duvall, D. Müller-Schwarze and R.M. Silverstein), pp. 285–304. Plenum, New York.

Graves, B.M. and Duvall, D. (1987). An experimental study of aggregation and thermoregulation in prairie rattlesnakes (*Crotalus viridis viridis*). *Herpetologica*, 43, 259–64.

Gregory, P.T., Macartney, J.M. and Larsen, K.W. (1987). Spatial patterns and movements. In *Snakes* (ed. R.A. Seigal, J.T. Collins and S.S. Novak), pp. 366–95. Macmillan, New York.

Halpern M. and Kubie, J.L. (1980). Chemical access to the vomeronasal organs of garter snakes. *Physiol. Behav.*, 24, 367–71.

Halpern, M. and Kubie, J.L. (1984). The role of the ophidian vomeronasal system in species-typical behavior. *Trends Neurosci.*, 7, 472–7.

Hennessy, D.F. and Owings, D.W. (1988). Rattlesnakes create a context for localizing their search for potential prey. *Ethology*, 77, 317–29.

Jemiolo, B., Andreolini, F. and Novotny, M. (1986). Chemical and biological investigations of female mouse pheromones. In *Chemical signals in vertebrates*, 4 (ed. D. Duvall, D. Müller-Schwarze and R.M. Silverstein), pp. 79–85. Plenum, New York.

Kardong, K. (1986). The strike behavior of the rattlesnake, *Crotalus viridis oreganus*. *J. Comp. Psychol.*, 100, 304–14.

King, M.B. (1987). Movements and natural history of prairie rattlesnakes, *Crotalus viridis viridis*, in Wyoming. Thesis, University of Wyoming, Laramie.

King, M.B. and Duvall, D. (1990). Prairie rattlesnake seasonal migrations: episodes of movement, vernal foraging and sex differences. *Anim Behav.*, in press.

Klauber, L.M. (1972). *Rattlesnakes*. University of California Press, Berkeley and Los Angeles.

Kubie, J.L. and Halpern, M. (1978). Garter snake trailing behavior: Effects of varying prey extract concentration and mode of prey-extract presentation. *J. Comp. Psychol.*, 92, 362–73.

Landreth, H.F. (1973). Orientation and behavior of the rattlesnake, *Crotalus atrox*. *Copeia*, 1973, 26–31.

Ludwig, J.A. and Reynolds (1988). *Statistical ecology*. John Wiley and Sons, New York.

Macartney, J.M., Gregory, P.T. and Larsen, K.W. (1988). A tabular survey of data on movements and home ranges of snakes. *J. Herpetol.*, 22, 61–73.

Martner, B.E. (1986). *Wyoming climate atlas*. University of Nebraska Press, Lincoln.

Reinert, H.K., Cundall, D. and Busher, L.M. (1984). Foraging behavior of the timber rattlesnake, *Crotalus horridus*. *Copeia*, 976–81.

Reinert, H.K. and Zappalorti, R.T. (1988). Timber rattlesnakes (*Crotalus horridus*) of the Pine Barrens: Their movement patterns and habitat preference. *Copeia*, 964–78.

Stephens, D.W. and Krebs, J.R. (1986). *Foraging theory*. Princeton University Press, Princeton.

Trivers, R.L. (1972). Parental investment and sexual selection. In *Sexual selection and the descent of man, 1871–1971* (ed. B. Campbell), pp. 136–79. Aldine, Chicago.

Vandenbergh, J.G. (1980). The influence of pheromones on puberty in rodents. In *Chemical signals* (ed. D. Müller-Schwarze and R.M. Silverstein), pp. 229–41. Plenum, New York.

Weldon, P.J. and Burghardt, G.M. (1979). The ophiophage defensive response in crotaline snakes: Extension to new taxa. *J. Chem. Ecol.*, 5, 141–51.

# V.6 Predator–prey Signals: Chemosensory Identification of Snake Predators by Eublepharid Lizards and its Ecological Consequences

BENJAMIN E. DIAL

## Introduction

Predation is a major selective force acting within natural populations of vertebrates; in response to high rates of predation, vertebrates have evolved a suite of antipredator tactics (Harvey and Greenwood 1978). Behavioural, morphological, and physiological characteristics of these tactics work in concert to increase a prey's probability of escape during direct encounters with predators (e.g. Dial and Fitzpatrick 1983). However, avoidance of these encounters altogether would have greater selective advantage to prey. Thus, multiple sensory mechanisms that facilitate predator avoidance, including chemosensory identification of predators, should represent important components in the evolution of antipredator tactics (Endler 1986).

Lizards have well-developed chemosensory systems (see Stebbins 1948; Simon 1983) and the potential for chemosensory-based behaviour appears to be considerable. Although several aspects have been investigated (e.g. Greenberg 1943; Duvall *et al.* 1980; 1981; Burghardt 1980; Pederson 1988; Cooper and Vitt 1986*a*, *b*), one ecologically-important behaviour that has received little experimental investigation is the chemical identification of predators by lizard prey. Snakes are major predators of lizards, particularly in desert ecosystems (e.g. Pianka 1986). In addition, several chemical products are produced by the integument of snakes (Landmann 1979, 1980). Thus, it seems possible that lizards possessing a well-developed chemosensory apparatus might use snake-integumentary chemicals to their advantage.

In this paper, the few studies that have addressed chemical identification of predators by various types of prey are briefly reviewed. The use of an antipredator tactic of some lizards (tail display) is then discussed as a behavioral bioassay of chemosensory identification of predators. Lastly, the results are reviewed of some recent studies of predatory-snake identification by *Coleonyx variegatus*, a eublepharid gecko (Order Sauria: Family Eublepharidae; Grismer 1988) with well-developed chemosensory abilities.

## Detection of snake-integumentary chemicals

Snake-integumentary chemicals are important to both snake physiology and behaviour (e.g. Roberts and Lillywhite 1980, 1983; Gartska and Crews 1981; Mason and Crews 1985; Ford 1986). Several preliminary studies also indicate a function in the predator–prey context; for example, Hennessey and Owings (1978) demonstrated that ground squirrels (genus *Spermophilus*) could detect skin chemicals from predatory snakes, while Randall and Stevens (1987) suggested that kangaroo rats (genus *Dipodomys*) may also be capable of such chemosensory detection. Weldon and colleagues have shown that laboratory mice as well as snakes themselves can detect chemicals from their own snake predators (see Weldon *et al.* 1987 and citations therein).

Chemical identification of snakes by lizards has received little attention. Berry (1974) suggested than an iguanid lizard, *Sauromalus obesus,* could possible detect snake-skin chemicals; however, no experimental data are available to support this hypothesis. Simon *et al.* (1981) found no evidence that another iguanid, *Sceloporus jarrovi,* could chemically detect a kingsnake predator, *Lampropeltis pyromelana.* Members of more chemically-oriented lizard families (e.g. Scincidae, Lacertidae, Gekkonidae, Eublepharidae) appear to be more likely candidates. Cooper and Vitt (1984*a*, *b*) predicted that the well-developed chemosensory apparatus of the scincid *Eumeces laticeps* might function in this context. Thoen *et al.* (1986) observed that the lacertid *Lacerta vivipara* responded by tongue-flicking, tail vibrations, and leg shaking to snake-skin chemicals. However, the ecological significance of these responses is not clear.

The major problem in testing chemosensory identification of predators is how to distinguish between the prey's ability to simply detect a novel stimulus and its capacity to identify the source of the stimulus as predator. In this context, it is important to use a predator-specific, defensive aspect of behaviour as a bioassay.

## Tail display: a behavioural bioassay

Two criteria must be met when using a behavioural pattern as a bioassay of chemical identification (O'Connell 1977). First, the functional basis of the behaviour during naturally occurring events must be thoroughly understood; and second, the response should be as specific as possible to the particular stimulus.

Several species of eublepharid geckos exhibit a predator-specific behaviour, tail display, accompanied by some morphological and physiological adaptations, all of which are designed to facilitate escape from predators. Thus, these lizards offer a model system with which to examine chemosensory identification of predators.

Fig. 1. Tail display by an adult *Coleonyx variegatus* (mean snout-vent length = 70 mm). During the display, the tail was arched from the horizontal, locomotory position to a vertical position, and then slowly undulated to the left and right. The display functions to attract a predator's attack to the tail, which is then autotomized, resulting in the escape of the prey.

For example, when a snake is visually detected, a gecko positions its tail towards the approaching predator, elevates it from the horizontal position (the normal, locomotory position) to a verticle arch, and then slowly undulates it to the left and right (Fig. 1; Dial and Fitzpatrick 1981). Experimental studies (Congdon *et al.* 1974; Dial 1978) have shown that this display functions to attract the predator's attack to the tail, resulting in tail autotomy and the lizard's escape (Dial and Fitzpatrick 1983). Tails are then regenerated within four to six weeks (Dial and Fitzpatrick 1981).

In addition, geckos possess specialized plates (caudal fracture planes) within their caudal vertebrae. These plates develop postembryologically, and pass transversely through the vertebral centrum (Fig. 2; Bellairs and Bryant 1985). Caudal fracture planes have an important structural arrangement with the surrounding myosepta: each caudal myoseptum begins and ends at a fracture plane, rather than at an intervertebral joint (Fig. 2), and this has important functional consequences. When a snake grasps an autotomous tail, the myosepta anterior and posterior to the point of attack contract, using the snake's grip as a fulcrum for leverage (Dial unpublished data). This results in autotomy occurring very easily and rapidly at a fracture plane, rather than with more difficulty at an intervertebral joint.

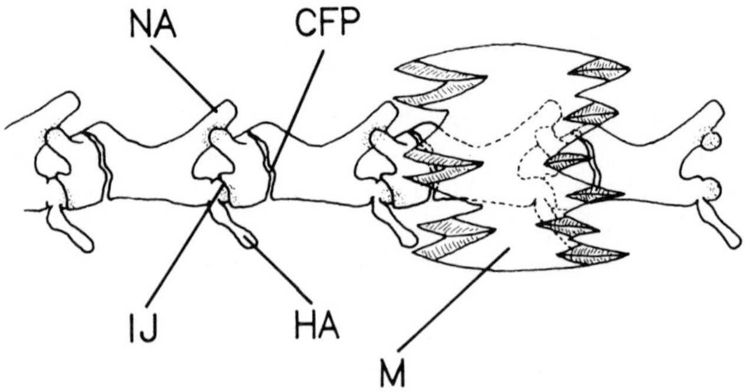


Fig. 2. Relationship between several caudal vertebrae and a myoseptum. Note that the myoseptum begins and ends adjacent to a caudal fracture plane, rather than adjacent to an intervertebral joint. Drawn from a juvenile specimen of *Ctenosaura similus* (Texas Cooperative Wildlife Collection 20813, Texas A&M University). CFP = caudal fracture plane, HA = hemal arch, IJ = intervertebral joint, M = myoseptum, NA = neural arch.

Finally, tails of these lizards exhibit very rapid and forceful axial-muscle contractions *after* autotomy, usually lasting 3–4 mins. Experimental studies (Dial and Fitzpatrick 1983) have demonstrated that this functions to increase prey (autotomized tail) handling time to a predator, thereby increasing escape time for lizards. These behavioural, morphological, and physiological aspects of the tail display/tail autotomy complex attest to its importance during encounters with predators, and support the use of tail display as a bioassay of chemical identification of predators.

## Chemosensory identification of predators by eublepharid geckos

Some of my recent studies involving eublepharid geckos have examined several aspects of predator-prey relations in these lizards, and the following questions are among the topics addressed here. (a) What is the chemosensory ability of eublepharid geckos to detect integumentary chemicals from snakes? (b) What is their capacity to identify the source of these chemicals as being from a predator? And (c) what is their capacity to distinguish between integumentary chemicals of a saurophagous (lizard-eating) and a non-saurophagous snake species?

Detailed accounts of the experimental procedures and results of studies involving chemosensory behaviour (and antipredator tactics in general) can be found in the original publications, particularly in Dial (1978), Dial and Fitzpatrick (1981), and Dial *et al.* (1989).

## The prey and predators

*Coleonyx variegatus* is a terrestrial, eublepharid gecko (mean snout-vent = length SVL = 71 mm) which occurs in several desert ecosystems of south-western North America. These lizards are nocturnal and insectivorous, and are important prey items of several species of desert snakes (e.g. *Phyllorhynchus decurtatus*, *Hypsiglena torquata*). They have evolved a complex antipredator mechanism that includes elaborate tail-display behaviour (Fig. 1), as well as tail autotomy and extensive postautotomy tail-muscle contractions. Tail autotomy occurs frequently within natural populations; in a recent study, 67% of the adults in a population in San Diego Co., California exhibited regenerated tails (Dial, unpublished).

*Phyllorhynchus decurtatus* is a nocturnal, terrestrial snake (mean SVL = 408 mm) which occurs sympatrically with *C. variegatus* over approximately 80% of the distribution range of geckos. This is a saurophagous species, and examination of alcohol-preserved museum specimens has revealed that *C. variegatus* constitutes a considerable portion of its diet. In addition, autotomized gecko tails are frequently observed in the stomachs of these snakes, attesting to the success of tail display and tail autotomy as an antipredator tactic.

*Chionactis occipitalis* is a nocturnal, fossorial snake (mean SVL = 244 mm) which occurs sympatrically with *C. variegatus* over approximately 57% of the distribution range of geckos. These snakes prey exclusively on arthropods.

## Experimental procedures

### VISUAL STIMULI

Although field and laboratory observations suggested that tail display in *C. variegatus* is used largely during encounters with predators, many lizard species also use tail display in other contexts, such as stalking of prey (Dial, unpublished) and long-distance signalling to predators (Dial 1986). To establish the specific context in which tail display is used by *C. variegatus*, and thus to establish its validity as a bioassay, encounters were staged between geckos and live prey (crickets) and live predators (*P. decurtatus*). To ensure that only visual cues were perceived, and to prevent chemical cues from being detected, I placed either a cricket or a snake in a sealed, 1000 ml glass jar, and presented this to geckos in their resident terraria. Stimuli were presented in random order, and only one stimulus was presented to an individual per day. Gecko responses were scored for either presense or absence of tail display.

### CHEMICAL STIMULI

The procedure in these experiments was to present, in random order, distilled water as a neutral control, a commercial cologne as a pungency control, and skin chemicals from both snake species on cotton-tipped swabs to geckos, and to score for a response. Swabs were held from 1 to 2 cm in front of a gecko's snout for 60 s. Control liquids were applied to swab tips by dipping tips into liquids. Snake-skin chemicals were applied to tips by rolling dry tips along the dorsal, lateral, and ventral skin surfaces of snakes, from the nape to approximately two-thirds the length of the snake. Chemicals were not collected from the skin on the cloacal region of snakes. Thus, the chemicals deposited on swab tips represented only integumentary chemicals, and were not contaminated by cloacal exudates. Trials were scored for several responses, including tongue-flick, defensive attack (see Dial *et al.* 1989), tail display, and flight.

## Results

### TAIL DISPLAY TO VISUAL STIMULI

Experiments with live crickets and snakes revealed that tail display in *C. variegatus* is used almost exclusively during encounters with snakes, and rarely when these geckos are stalking insect prey (Table 1; Dial *et al.* 1989). These data, together with other experimental data (e.g. Dial 1978), indicate that the functional basis of tail display is predator defence, and that it is used specifically during encounters with potential predators. Therefore, I suggest that this behaviour represents an accurate bioassay of predator identification (see O'Connell 1977).

**Table 1.** Reponses of *Coleonyx variegatus* to visual stimuli: live cricket prey (*Gryllus domesticus*) and live snake predators (*Phyllorhynchus decurtatus*). To prevent chemical cues from being detected, crickets and snakes were presented in 1000 ml, air-tight glass containers

| | **Visual stimulus** | | | | | | |
|---|---|---|---|---|---|---|---|
| | **Cricket prey** | | **Snake predator** | | | | |
| **Behaviour** | **Number** *n* | % | **Number** *n* | % | d.f. | G[a] | *p*-value |
| Tail display | 1 | 3 | 12 | 80 | 1 | 28.44 | <0.001 |
| No tail display | 26 | 97 | 3 | 20 | | | |

[a] G-test of independence

CHEMOSENSORY IDENTIFICATION OF SNAKE PREDATORS

The results of experiments with chemical stimuli revealed that *C. variegatus* is capable of detecting integumentary chemicals from snakes. Geckos presented with snake chemicals exhibited the most diverse array of responses compared to all other stimuli. For example, none of the trials with cologne as pungency control resulted in a response, and the only response exhibited to water controls was tongue-flicking (Table 2). In contrast, there were several responses to snake-skin chemicals (see Table 2).

**Table 2.** Summary of responses of *Coleonyx variegatus* to neutral and pungency control chemicals and skin chemicals from a saurophagous and a non-saurophagous snake species

| Chemical stimulus | Response | Number *n* | % |
|---|---|---|---|
| Water | | | |
| | no response | 16 | 62 |
| | tongue flick | 10 | 38 |
| Cologne | | | |
| | no response | 20 | 100 |
| *Phyllorhynchus decurtatus* | | | |
| | no response | 1 | 4 |
| | tongue-flick | 1 | 4 |
| | defensive attack | 3 | 13 |
| | tail display[a] | 12 | 50 |
| | tail display followed by flight | 7 | 29 |
| | distress call | 4 | 17 |
| *Chionactis occipitalis* | | | |
| | no reponse | 7 | 70 |
| | flight | 3 | 30 |

[a] Total trials with tail display = 79%. *P. decurtatus* vs. all other chemicals: G = 69.92, $p < 0.001$; *P. decurtatus* vs. *C. occipitalis*: G = 22.12, $p < 0.0001$.

More importantly, these experiments suggest that geckos can identify the source of the chemicals as predator. Seventy-nine per cent of the geckos tested with skin chemicals from *P. decurtatus* exhibited tail display, and this behaviour occurred *only* in response to these chemicals (Table 2; Dial *et al.* 1989). Tail displays to chemicals were identical in posture and timing to displays given to visual snake-predator stimuli. Twenty-nine per cent of the geckos that displayed exhibited rapid flight behaviour immediately after concluding the display, and 17% of these gave a distress call at the initiation of flight behaviour. In addition, the results suggest that the primary sensory modality used to detect snake chemicals may have been nasal olfaction rather than the vomeronasal system (see Simon 1983; Graves and Halpern 1988); only 4% of the responses to snake chemicals were preceded by tongue-flicking, whereas 96% of the trials resulted in an active response.

The test of the hypothesis that geckos can distinguish between chemicals from a saurophagous and a non-saurophagous species is confirmed. Tail display was *not* exhibited to chemicals from *Chionactis occipitalis* (Table 2); however, 30% of these trials resulted in flight behaviour. It is possible that geckos exhibiting flight were affected by previous exposure to *Phyllorhynchus* chemicals, but this remains to be determined.

## Ecological consequences of predator identification

Chemical identification of predators may affect several aspects of the behaviour and ecology of eublepharid lizards. Recent studies have demonstrated that some snakes leave chemical trails composed of integumentary chemicals as they move across the substrate (reviewed by Ford 1986). Trails are partially composed of non-volatile, lipid-based pheromones which are released from the integument (Gartska and Crews 1981; Burken *et al.* 1985; Schell and Weldon 1985), and probably have considerable longevity.

Two closely related species, *Coleonyx brevis* and *Coleonyx reticulatus,* utilize chemoreception extensively while foraging at night (Dial 1978). Predator identification might occur when foraging *Coleonyx* come into contact with trails left by passing snakes, and if so, geckos might avoid these areas. This would place short-term time/energy constraints on gecko foraging, but the benefits in terms of predator avoidance would outweigh these costs (see Endler 1986). Thus, chemosensory-based predator identification might have important consequences which affect the relationship between optimal foraging and predation in eublepharid geckos. A similar phenomenon has been described in the responses of larval amphibians to chemicals from their fish predators (Petranka *et al.* 1987).

*Coleonyx variegatus* and *P. decurtatus* are nocturnal species; both remain secluded in burrows and retreat sites under rocks during the day. Current studies (Dial and Grismer, in preparation) indicate that several species of *Coleonyx* have very narrow and specific ranges of microhabitat requirements, particularly in terms of temperature and moisture content. However, the microhabitat characteristics of burrows and other retreat sites in desert ecosystems are often quite heterogeneous. For example, at 07.30 h in July 1988, near Study Butte, Brewster Co., Texas, the temperature difference between the surface of a rock shelter (32.6 cm in length, 7.4 cm in depth) and various locations within a sand burrow under that rock ranged from 6 to 18 °C (Dial, unpublished observation; see Huey *et al.* 1988). Consequently, within desert ecosystems, retreats with species-specific microhabitat characteristics may be a very limited resource. If integumentary chemicals remain in retreats that are used by snakes, their presence may limit the selection of retreats that are acceptable to geckos. This may have important consequences to nest selection and the reproductive success of eublepharid geckos, and possibly to other aspects of their life history.

## Acknowledgements

It is my pleasure to thank H. Erle Rawlins Jr. for his encouragement and support, as well as his assistance both in the field and in writing this paper. I thank Harry W. Greene, Laurie J. Vitt, Paul Weldon, David Duvall, and Baunnie Sea for their discussions of gecko biology and chemical ecology. Financial support was provided in part by a Faculty Summer-Research Grant and the Department of Biology, Chapman College, Orange, California.

## References

Bellairs, A.d'A. and Bryant, S.V. (1985). Autotomy and regeneration in reptiles. In *Biology of the Reptilia* (ed. C. Gans and F. Billett), Vol. 15, Development B, pp. 301–410. Wiley, New York.

Berry, K.H. (1974). The ecology and social behavior of the chuckwalla, *Sauromalus obesus obesus* Baird. *Univ. California Publ. Zool.*, 101, 1–60.

Burghardt, G.M. (1980). Behavioral and stimulus correlates of vomeronasal functioning in reptiles: feeding, grouping, sex, and tongue use. In *Chemical signals: Vertebrates and aquatic invertebrates* (ed. D. Müller-Schwarze and R.M. Silverstein), pp. 295–301. Plenum, New York.

Burken, R.R., Wertz, P.W. and Downing, D.T. (1985). A survey of polar and non-polar lipids extracted from snake skin. *Comp. Biochem. Physiol.*, 81B, 315–8.

Congdon, J.D., Vitt, L.J. and King, W.W. (1974). Geckos: adaptive significance and energetics of tail autotomy. *Science*, 184,1379–80.

Cooper, W.E., Jr. and Vitt, L.J. (1984*a*). Detection of conspecific odors by the female broad-headed skink, *Eumeces laticeps*. *J. Exp. Zool.* 229,49–54.

Cooper, W.E., Jr. and Vitt, L.J. (1984*b*). Conspecific odor detection by the male broad-headed skink, *Eumeces laticeps*: effects of sex and site of odor source and of male reproductive condition. *J. Exp. Zool.*, 230, 199–209.

Cooper, W.E., Jr. and Vitt, L.J.. (1986*a*). Interspecific odor discrimination among syntopic congeners in scincid lizards (Genus *Eumeces*). *Behavior*, 97, 1–9.

Cooper. W.E., Jr. and Vitt, L.J. (1986*b*). Tracking of female conspecific odor trails by male broad-headed skinks (*Eumeces laticeps*). *Ethology*, 71, 242–8.

Dial, B.E. (1978). Aspects of the behavioral ecology of two Chihuahuan Desert geckos (Reptilia, Lacertilia, Gekkonidae). *J. Herpetol.*, 12, 209–16.

Dial, B.E. and Fitzpatrick, L.C. (1981). The energetic costs of tail autotomy to reproduction in the lizard *Coleonyx brevis* (Sauria: Gekkonidae). *Oecologia*, 51, 310–7.

Dial, B.E. and Fitzpatrick, L.C. (1983). Lizard tail autotomy: function and energetics of postautotomy tail movement in *Scincella lateralis*. *Science*, 219, 391–3.

Dial, B.E. (1986). Tail display in two species of iguanid lizards: a test of the 'predator signal' hypothesis. *Amer. Nat.*, 127, 103–11.

Dial, B.E., Weldon, P.J. and Curtis, B. (1989). Chemosensory identification of snake predators (*Phyllorhynchus decurtatus*) by banded geckos (*Coleonyx variegatus*). *J. Herpetol.*, 23, 224–9.

Duvall, D. (1979). Western fence lizard (*Sceloporus occidentalis*) chemical signals. 1. Conspecific discrimination and release of a species-typical visual display. *J. Exp. Zool.*, 210, 321–6.

Duvall, D. (1981). Western fence lizard (*Sceloporus occidentalis*) chemical signals. 11. A replication with naturally breeding adults and a test of the Cowles and Phelan hypothesis of rattlesnake olfaction. *J. Exp. Zool.*, 218, 351–61.

Endler, J.A. (1986). Defense against predators. In *Predator–prey relationships. Perspectives and approaches from the study of lower vertebrates* (ed. M.E. Feder and G.V. Lauder), pp. 109–34. University of Chicago Press, Chicago.

Ford, N.B. (1986). The role of pheromone trails in the sociobiology of snakes. In *Chemical signals in vertebrates*, 4; *ecology, evolution, and comparative biology* (ed. D. Duvall, D. Müller-Schwarze and R.M. Silverstein), pp. 261–78. Plenum, New York.

Garstka, W.R. and Crews, D. (1981). Female sex pheromone in the skin and circulation of a garter snake. *Science*, 214, 681–3.

Graves, B.M. and Halpern, M. (1988). The role of lingual transport in vomeronasal organ chemoreception of a lizard. Program and abstracts, American Society of Ichthyologists and Herpetologists, p. 101.

Greenberg, B. (1943). Social behavior of the western banded gecko, *Coleonyx variegatus* Baird. *Physiol. Zool.*, 16, 110–22.

Grismer, L.L. (1988). Phylogeny, taxonomy, classification, and biogeography of eublepharid geckos. In *Phylogenetic relationships of the lizard families* (ed. R. Estes and G. Pregill), pp. 369–470. Stanford University Press, Stanford, California.

Harvey, P.H. and Greenwood, P.J. (1978). Anti-predator defense strategies: some evolutionary problems. In *Behavioral ecology: an evolutionary approach* (ed. J.R. Krebs and N.B. Davies), pp. 129–151. Blackwell, Oxford.

Hennessey, D. F. and Owings, D.H. (1978). Snake species discrimination and the role of olfactory cues in the snake-directed behavior of the California ground squirrel. *Behavior*, 65, 115–24.

Huey, R.B., Peterson, C.R., Arnold, S.J. and Porter, W.P. (1989). Hot rocks and not-so-hot rocks: retreat site selection by garter snakes and its thermal consequences. *Ecology*, 70, 931–44.

Landmann, L. (1979). Keratin formation and barrier mechanisms in the epidermis of *Natrix natrix* (Reptilia: Serpentes). An ultrastructural study.

*J. Morphol.*, 162, 93–126.

Landmann, L. (1980). Lamellar granules in mammalian, avian and reptilian epidermis. *J. Ultrastruct. Res.*, 72, 245–63.

Mason, R.T. and Crews, D. (1985). Female mimicry in garter snakes. *Nature*, 316, 59–60.

O'Connell, R.J. (1977). From insect to mammal: complications of the bioassay. In *Chemical signals in vertebrates* (ed. D. Müller-Schwarze and M.M. Mozell), pp. 377–89. Plenum, New York.

Pedersen, J.M. (1989). Laboratory observations on the function of tongue extrusion in the desert iguana, *Dipsosaurus dorsalis*. *J. Comp. Psychol.*, 102, 193–6.

Petranka, J.W., Kats, L.B. and Sih, A. (1987). Predator–prey interactions among fish and larval amphibians: use of chemical cues to detect predatory fish. *Anim. Behav.*, 35, 420–5.

Pianka, E.R. (1986). Ecology and natural history of desert lizards. Princeton University Press, Princeton, NJ.

Randall, J.A. and Stevens, C.M. (1987). Foot drumming and other anti-predator responses in the bannertail kangaroo rat (*Dipodomys spectabilis*). *Behav. Ecol. Sociobiol.*, 20, 187–94.

Roberts, J.B. and Lillywhite, H.B. (1980). Lipid barrier to water exchange in reptile epidermis. *Science*, 207, 1077–9.

Roberts, J.B. and Lillywhite, H.B. (1983). Lipids and permeability of epidermis from snakes. *J. Exp. Zool.*, 228, 1–9.

Schell, F.M. and Weldon, P.J. (1985). C-NMR analysis of snake skin lipids. *Agric. Biol. Chem.*, 49, 3579–600.

Simon, C.A., Gravelle, K., Bissinger, B.E., Eiss, I. and Ruibal, R. (1981). The role of chemoreception in the iguanid lizard *Sceloporus jarrovi*. *Anim. Behav.*, 29, 46–54.

Simon, C.A. (1983). A review of lizard chemoreception. In *Lizard ecology: studies of a model organism* (ed. B. Huey, E.R. Pianka and T.W. Schoener), pp. 119–33. Harvard University Press, Cambridge, Mass.

Stebbins, R.C. (1948). Nasal structure in lizards with reference to olfaction and conditioning of inspired air. *Amer. J. Anat.*, 83, 183–222.

Thoen, C., Bauwens, D. and Nerheyen, R.F. (1986). Chemoreceptive and behavioral responses of the common lizard *Lacerta vivipara* to snake chemical deposits. *Anim. Behav.* 34, 1805–13.

Weldon, P.W., Divita, F.M. and Middendorf, III, G.A. (1987). Responses to snake odors by laboratory mice. *Behav. Proc.*, 14, 137–46.

# V.7 The Effect of Mink Odour on Rabbits and Small Mammals

IAN ROBINSON

## Introduction

Recently there has been much interest in the use of predator odours as biological repellents to reduce feeding damage by herbivores. Stoddart (1976), Dickman and Doncaster (1984), and Gorman (1984) have shown that small mammals tend to avoid traps tainted with predator odours. Sullivan and Crump (1984) showed that 3-propyl-1,2,dithiolane from the stoat (*Mustela erminea*) and 2,2-dimethylthietane from the mink (*M. vison*) diminished the feeding behaviour of snowshoe hares (*Lepus americanus*). Further work suggested that faeces, urine, or glandular secretions from some predators were effective in reducing feeding damage by snowshoe hares, black-tailed deer (*Odocoileus hemionus columbianus*), montane voles (*Microtus montanus*) and meadow voles (*M. pennsylvanicus*) (Sullivan *et al.* 1988; Sullivan 1985, 1986).

The major volatile component of mink anal sac secretion, 2,2-dimethylthietane (DMT) was synthesized as part of a larger study of olfactory communication in mink (Robinson 1987). In the present work the effect of this compound on potential prey of mink was determined by comparing the trapping efficiency of Longworth traps tainted with DMT with those tainted with control odours, and by determining whether the odour of DMT suppressed feeding behaviour in rabbits (*Oryctolagus cuniculus*).

## Methods

### TRAPPING RESPONSE OF SMALL MAMMALS

The number of small mammals caught in Longworth traps containing experimental odour was compared with those caught in traps containing control odours. The experimental odour consisted of 0.5 ml of DMT mixed with 10 g petroleum jelly and the control odour consisted of 1 ml of butyric acid mixed with 10 g of petroleum jelly. Butyric acid provided a control for any attraction or repulsion that may have been caused by any strong odour. The petroleum jelly was used as an odour carrier to prevent the odour sources evaporating too rapidly. Traps containing petroleum jelly alone were therefore used as a second control.

Approximately 0.5 g of the petroleum jelly was smeared on to a 2 $cm^2$ piece of paper, which was then positioned half-way along the tunnel of a Longworth trap. All traps contained barley and carrot for food, and hay for bedding.

Two groups of traps were set in woodland habitat at Culterty Field Station, Aberdeenshire, Scotland, in the spring of 1986. The first group of 15 traps was laid out in a 5 by 3 grid with traps positioned approximately 10 m apart. Each row of three traps contained one of each type of odour source, but the position of the traps was varied so that no two rows contained the same order of odours. To control for spatial biases on each experimental day, the position of the odours within a row of three was altered by moving each trap to the next point in the row, and bringing the third trap to the first position. The second set of 15 traps were set in groups of three traps at each of five points, chosen as areas that appeared favorable for small mammals. At each point, three traps containing different odours were positioned side by side so that their entrances were close together. The position of each trap (i.e. left, right, or centre) was randomly determined and was varied daily.

All traps were checked daily in the morning; the sex and species of individuals caught were recorded before they were released. Traps that had caught animals were replaced with clean traps containing fresh food, bedding and odour, and the old trap was removed for cleaning. If no animal was caught in a trap the odour was replaced every second day and the bedding was replaced if it became damp.

### FEEDING BEHAVIOUR OF RABBITS

From May to August 1986, twelve sites (feeding stations) were selected along a transect on the Sands of Forvie National Nature Reserve, Aberdeenshire, Scotland. The feeding stations were located in areas where rabbits were known to be present, and were positioned on areas of open sand so that any deer feeding at a site could be detected by their tracks. Feeding stations were spaced 60–150 m apart with no station being visible from any other station. Carrots were placed at each feeding station three times per week for one month before the experiments commenced to allow rabbits to become accustomed to finding food at those locations.

The three odours (including two controls) used for the small mammal trapping were used again in these experiments. Each odour source was prepared by placing 10 g of petroleum jelly into a petri dish and heating it in an oven until the jelly melted and spread over the base of the dish. After cooling 0.5 ml DMT or 1.0 ml butyric acid was added and mixed with the petroleum jelly, except for dishes which contained only petroleum jelly.

In experimental trials one petri dish was positioned at each feeding station with a small quantity of dry sand sprinkled into them so that the petroleum jelly was just covered. Five pieces of carrots were then positioned on the sand in each petri dish and an additional five pieces placed around the outside of the dish about 2 cm from the edge. Experiments were set up approximately 2 h before sunset and checked after 24 and 48 h. The amount of carrots eaten was recorded, and the sand around each feeding station examined for deer tracks.

Trials were conducted at weekly intervals and the odour presented at any particular feeding station was varied in a systematic way so that each feeding station contained each odour on an equal number of occasions. Odour sources were removed between trials but a few pieces of carrot were left behind so that rabbits would continue to find food at the feeding stations.

## Results

The number of occasions when *Apodemus sylvaticus* or *Clethrionomys glareolus* were caught in Longworth traps tainted with DMT, butyric acid, or petroleum jelly alone are presented in Table 1. The data from both sexes were pooled as no significant differences were found between the numbers of each sex caught in each trap. For each species the number of captures in traps containing each type of odour was compared with the number expected if animals were equally likely to be caught in each trap. When the traps were positioned together, the expected number of captures was calculated by dividing the total number of captures by the number of different treatments.

For traps positioned together, there was a significant difference between the observed and expected numbers of *Apodemus* and *Clethrionomys* caught in each trap. Fewer animals were caught in traps tainted with DMT than expected by chance, but more *Apodemus* than expected were caught in traps tainted with butyric acid. Slightly more *Apodemus* were caught in the traps tainted only with petroleum jelly than those tainted with DMT, but this difference was more marked for *Clethrionomys*. For traps in the grid, significantly fewer *Apodemus* were caught in traps tainted with DMT than in those tainted with butyric acid or containing only petroleum jelly. Significantly fewer *Clethrionomys* were caught in traps baited with DMT than in traps containing petroleum jelly alone.

When the number of each species caught in grid traps tainted with a particular odour was compared with the corresponding value for the traps positioned together, no significant differences were found in any comparison (for *Apodemus*, $\chi\}2=2.78$, $p>0.05$; for *Clethrionomys*, $\chi\}2= 0.27$, $p>0.05$).

The number of occasions when rabbits had eaten all, some, or no carrots both in and around the dishes containing each of the odours are presented in Table 2. The numbers of carrots eaten after 24 h were examined by log-linear model contingency tests (Zar 1974) and a significant difference was found in the amount of carrots eaten both in ($G=28.5$, 4 d.f.; $p<0.001$) and around ($G=15.3$, 4 d.f.; $p<0.01$) the various odour sources (Table 2). All of the carrots were eaten more often when associated with butyric acid or petroleum jelly alone, than with DMT. Carrots were rejected more often when in association with DMT than with the other odours. The numbers of carrots either in or around the dish were eaten

**Table 1.** Number of *Apodemus sylvaticus* and *Clethrionomys glareolus* caught in Longworth traps tainted with 2,2-dimethylthietane (M), butyric acid (C), or control odour (O)

**Traps together**

| | Apodemus M | C | O | Total | Clethrionomys M | C | O | Total |
|---|---|---|---|---|---|---|---|---|
| Number caught[a] | 18 (25.3) | 36 (25.3) | 22 (25.3) | 76 | 2 ( 6.7) | 6 ( 6.7) | 12 ( 6.7) | 20 |
| Total no trap | 80 | 80 | 80 | | 80 | 80 | 80 | |

$\chi 2 = 7.05$ (2 d.f.); $p < 0.05$
[a]Values given in parentheses are those that would be expected by chance.

**Traps in grid**

| | Apodemus M | C | O | Clethrionomys M | C | O |
|---|---|---|---|---|---|---|
| Number caught[a] | 4 | 16 | 14 | 2 | 4 | 11 |
| | $p < 0.05$ | | | $p =$ N.S. | | |
| | | $p < 0.02$ | | | $p < 0.02$ | |
| Total trap nights | 40 | 40 | 40 | 40 | 40 | 40 |

equally often when tainted with DMT ($G=4.51$, 2 d.f.; $p>0.05$). After 48 h all carrots in and around all dishes had been eaten. No deer tracks were recorded close to any of the dishes suggesting that deer did not eat any of the carrots.

**Table 2.** The number of occasions when all, some or no carrots, in or outside the dish, were eaten by rabbits. Dishes contained either 2,2 -dimethylthietane in petroleum jelly (M), butyric acid in petroleum jelly (C), or petroleum jelly alone (O). Data are for carrots placed in and around the dishes and checked after 24 hours

| | Odour | Not eaten | Some eaten | All eaten | Total presented |
|---|---|---|---|---|---|
| Carrots in dish | M | 12 | 3 | 9 | 24 |
| | C | 1 | 2 | 21 | 24 |
| | O | 0 | 2 | 22 | 24 |
| Carrots around dish | M | 5 | 4 | 15 | 24 |
| | C | 1 | 0 | 23 | 24 |
| | O | 0 | 2 | 22 | 24 |

## Discussion

Fewer *Apodemus* and *Clethrionomys* were captured in traps tainted with DMT than in traps containing butyric acid or petroleum jelly alone (Table 1). When traps were together, the butyric acid apparently had an attractive effect on *Apodemus*, as most animals were caught in traps tainted with that odour. The responses of *Clethrionomys* to traps tainted with mink odour are similar to those of other microtine rodents responding to odours from potential predators (Stoddart 1976, Dickman and Doncaster 1984, Gorman 1984, Sullivan *et al.* 1988). However, the response of *Apodemus* in this study is in contrast to its response to odours from some other Mustelines (Stoddart 1976, Gorman 1984). Stoddart (1976) found that the number of *Apodemus* caught in traps containing weasel (*Mustela nivalis*) odour was not significantly different to the number caught in control traps. Gorman (1984) showed that whereas *Microtus agrestis* and *M. arvalis* (Orkney vole) avoided the half of an arena that was tainted with stoat odour, *Apodemus* did not avoid the stoat odour. Little (1985) showed that *Apodemus* did not avoid Longworth traps tainted with the odour of mink faeces, but Dickman and Doncaster (1984) found that *Apodemus* did avoid traps tainted with fox (*Vulpes vulpes*) faecal odour.

The experiments showed that rabbits ate carrots associated with DMT less often than those associated with other odours. However, total feeding suppression did not occur, as carrots associated with DMT were eaten on some occasions

(Table 2). The reduction in feeding occurred only over a 24 h period, and all carrots were eaten after 48 h. This may be because the rabbits became habituated to the odour of the DMT after 48 h or because much of the DMT had evaporated after this period.

Sullivan *et al.* (1985) found that carnivore odours were effective in reducing feeding damage on tree seedlings by snowshoe hares but that novel odours such as 2-methylbutyric acid or domestic dog (*Canis familiaris*) urine did not reduce feeding. Of those mustelid odours that have been tested in field trials, 3-propyl-1,2-dithiolane from stoats and 2,2-dimethylthietane (present in the anal sac secretion of mink and stoats) produced a very negative effect on feeding behaviour for up to 38 days (Sullivan and Crump 1984). This suggests that snowshoe hares do not habituate to the predator odours. In addition, Sullivan *et al.* (1988) showed that a 1:1 mixture of 2-propylthietane (the major component of stoat anal sac secretion) and 3-propyl-1,2-dithiolane suppressed feeding by voles on trees for 4–5 months.

Avoidance of predator odours by potential prey may be a mechanism by which they avoid areas where predators are present. As mink are a recent introduction to Britain, it is possible that the avoidance of mink odour by rodents and rabbits may be an innate response to general predator odours. However, as 2,2-dimethylthietane is also present in the anal sac secretion of stoats (Brinck *et al.* 1983), the observed responses may really have been a response to an odour of stoats, which were occasionally present in both study areas. The suggestion that avoidance of predator odours is innate and generalized rather than a learned response to a specific predator is supported by several earlier studies. Stoddart (1982*a*, *b*) observed that British *Microtus agrestis* avoid traps tainted with urine from tigers (*Panthera tigris*); Dickman and Doncaster (1984) found that rodents avoid badger (*Meles meles*) odour although badgers had never been recorded in their study area; and Gorman (1984) showed that Orkney voles avoid stoat odours although they have been isolated from stoats for several millennia. In contrast, however, are the responses of *Apodemus* to weasel and stoat odour, and mink faeces (Stoddart 1976; Gorman 1984; Little 1985), and the lack of response by snowshoe hares to 2-propylthietane (the major component of stoat anal sac secretion) or to domestic dog urine (Sullivan and Crump 1984, Sullivan *et al.* 1985). Further research is required to determine why *Apodemus* should avoid the odour of the major component of mink anal sac secretion but not that of mink faeces, or weasel or stoat odour, and why snowshoe hares are unaffected by 2-propylthietane.

The use of fear or avoidance inducing odours as repellents to reduce damage by herbivores to crops is an important line for continuing research. Sullivan *et al.* (1988) have shown the efficacy of several synthetic predator odours as area repellents rather than merely as protection for individual plants. This illustrates the future potential of using predator odours to provide crop and forest protection.

## References

Brinck, C., Erlinge, S. and Sandell, M. (1983). Anal sac secretion in mustelids: a comparison. *J. Chem. Ecol.*, 9, 727–45.

Dickman, C.R. and Doncaster, C.P. (1984). Responses of small mammals to Red fox (*Vulpes vulpes*) odour. *J. Zool., Lond.*, 204, 521–31.

Gorman, M.L. (1984). The response of prey to stoat (*Mustela erminea*) scent. *J. Zool., Lond.*, 202, 419–23.

Little, D. (1985). The effect of mink, *Mustela vison*, faeces on three species of rodents. B.Sc. Honours thesis, Department of Zoology, University of Aberdeen.

Robinson, I.H. (1987). Olfactory communication and social behaviour in the mink (*Mustela vison*). Ph.D. thesis, University of Aberdeen.

Stoddart, D.M. (1976). Effect of the odour of weasels (*Mustela nivalis*) on trapped samples of their prey. *Oecologia*, 22, 439–41.

Stoddart, D.M. (1982*a*). Demonstration of olfactory discrimination by the short-tailed vole, *Microtus agrestis* L. *Anim. Behav.*, 30, 293–4.

Stoddart, D.M. (1982*b*). Does trap odour influence estimation of population size of short-tailed vole, *Microtus agrestis* L.? *J. Anim. Ecol.*, 51, 375–86.

Sullivan, T.P. (1985). Use of predator odours as repellents to reduce feeding damage by herbivores. ii. Black-Tailed Deer (*Odocoileus hemionus columbianus*). *J. Chem. Ecol.*, 11, 921–35.

Sullivan, T.P. (1986). Influence of wolverine (*Gulo gulo*) odor on feeding behavior of snowshoe hares (*Lepus americanus*). *J. Mammal.*, 67, 385–8.

Sullivan, T.P. and Crump, D.R. (1984). Influence of mustelid scent-gland compounds on the suppression of feeding by snowshoe hares (*Lepus americanus*). *J. Chem. Ecol.*, 10, 1809–21.

Sullivan, T.P., Nordstrom, L.O. and Sullivan, D.S. (1985). Use of predator odours as repellents to reduce feeding damage by herbivores. i. Snowshoe Hares (*Lepus americanus*). *J. Chem. Ecol.*, 11, 903–9.

Sullivan, T.P., Crump, D.R. and Sullivan, D.S. (1988). Use of predator odours as repellents to reduce feeding damage by herbivores. iii. Montane and Meadow Voles (*Microtus montanus* and *Microtus pennsylvanicus*). *J. Chem. Ecol.*, 14, 363–77.

Zar, J.H. (1974). *Biostatistical analysis*. Prentice Hall, Englewood Cliffs, N.J.

# V.8 Olfactory Cues in Salmonid Migration

KJELL B. DØVING

## Introduction

Migrating salmon are known to use their olfactory sense to return to their place of origin after a sojourn in the sea (Wisby and Hasler 1954; for reviews see Hasler 1966; Hasler and Scholz 1983; Stabell 1984; Døving 1986). Less well known, however, is how the salmon tracks an odour trail at sea. The question becomes more pertinent when we realize that salmon have no visual contact with the sea-bed even when it is as little as 10 or 20 m away. Currents will carry the salmon along, thus a reference system or directional cue is needed by which to navigate or orientate (Arnold 1974).

It has been suggested that salmon can follow an odour trail by comparing odour concentrations between different locations; to find a home-stream, the salmon should be able to swim up a concentration gradient. There are several contraindications to such a hypothesis. Firstly, sea gradients are much larger in the vertical than the horizontal direction. Secondly, there is no convincing evidence that salmon actually perceive a gradient: the olfactory system is well suited to detect and discriminate odorants, but is less precise at detecting or measuring the concentrations. One way the salmon could use its olfactory sense for orientation would be by comparing the impressions made at the left and right nostrils, a process known as tropotaxis. However, as will be discussed below, our evidence suggests that the salmon does not use this method.

Westerberg (1982*b*, 1983, 1984) has investigated the behaviour of salmon in open water, using an ultrasound tag which continuously monitors the salmons' swimming depth (Westerberg 1982*a*). By simultaneously monitoring of fish movement and measurement of the hydrographic structures present, he demonstrated that the salmon in Lake Vänern, Sweden, kept to certain layers of the water masses. The salmon in fact preferred the interphase between the homogeneous layers. Typically they swam up and down from one layer to the next at amplitudes of about 1 to 2 m.

At the time of salmon migration, lakes and oceans are built up of a series of horizontal layers. These layers range from decimeters to meters thick, and typically the horizontal dimensions are much larger than the vertical. The layers exist for days or even weeks, and comprise homogeneous layers separated by interphase layers with gradients that can be very steep. The water layers arise because of differences in density; in fresh water, the main factor is temperature, but in salt water salinity is also important.

The water layers move in relation to the shore and in relation to one another. Why this is so can best be illustrated in the context of salmon migration. In a fjord system fresh water and salt water mix to form brackish water. These various water masses find their level in the layered structure according to their density. In the spring the seawater is usually more salty and colder than the brackish water, and will be found at greater depths. Thus we have brackish layers near the surface flowing out and salt water flowing into the fjord. Depending upon the wind, surface temperature, and topography of the system, the currents flowing in opposite directions will be situated in layers at different levels.

Salmon have to navigate in these complex water masses. Students of salmon migration have suggested that sun, moon, and stars may provide cues for orientation (Hasler 1966). While not rejecting the possibility that salmon make use of the earth's magnetic field or standing infrasonic waves (Sand and Karlsen 1986), the present work will focus on the immediate environment of the salmon. Our experiments of migrating salmon have convinced us that the layered structure of the water is decisive for successful navigation, and that olfaction is crucial in the selection of one of the many layers. Only when this selection is made, will the salmon be able to make a decision as to the direction it will swim.

## Methods

The experiments were carried out in the Stavanger fjord system, east and south-east of Stavanger, South-west Norway, in autumn 1983. The inter-connected fjords have a total length of more than 100 km, a typical width of 3 km, and depth of typically 200–300 m. Several mostly shallow sounds connect the fjords to the open sea. The deepest sill depth is 56 m. Hydrographically the upper layer of the fjord, approximately down to the sill depth, is strongly stratified, with weak estuarine circulation dominated by transient wind-driven currents. The basin below the sill depth is almost stagnant.

The salmon used in this study (Table 1) were mature adults of the river Imsa stock, except fish H which was of the Sandvik river stock, of eastern Norway. This fish was reared from the egg stage at Imsa and was released when two years old in May 1982 together with fish E and F. All fish were captured in the fjord near the river Imsa or in a fish trap at the river mouth. Their identity was established by tags and marks. Fish H returned to the river Imsa following the experiment.

Fish that were part of an ablation experiment were anaesthetized with chlorobutanol and allowed to recover for at least 6 h before release. Experimental fish (D, E) were rendered anosmic by transection of the olfactory nerve mid-way between the eyes and the nares (See Fig. 1). The sham operation (F) was made by a transverse cut through the skin only. The unilateral ablation sham (H) was made by severing the right olfactory nerve with a sham cut made on the left side. The

**Table 1.** Data for salmon tracked in the fjord near Ims, Norway, in September 1983. Only salmon H was recaptured in the fish trap of Imsa River. The trajectory length, ground speed parameters, and Straightness Index (SI) were calculated excluding the first 2 h of tracking. SI is defined as the distance from the start to the end of the track as a percentage of the trajectory length. The ground speeds (cm/s) are average measures over 30-min time intervals. The treatments were: I, intact; A, anosmic; S, sham operated; U, unilateral section of the olfactory nerve (from Døving *et al.* 1985)

| | | | Tracking | | Ground speed | | |
|---|---|---|---|---|---|---|---|
| **Fish** | **Length (cm)** | **Treatment** | **Duration (h)** | **Length (km)** | **maximum cm/s** | **% of time >15 cm/s** | **SI (%)** |
| C | 83 | I | 25 | 24.4 | 63 | 78 | 65 |
| D | 80 | A | 11 | 3.3 | 20 | 26 | 15 |
| E | 60 | A | 14 | 5.3 | 35 | 25 | 86 |
| F | 66 | S | 22 | 12.6 | 56 | 53 | 47 |
| H | 72 | U | 16 | 4.5 | 35 | 25 | 39 |

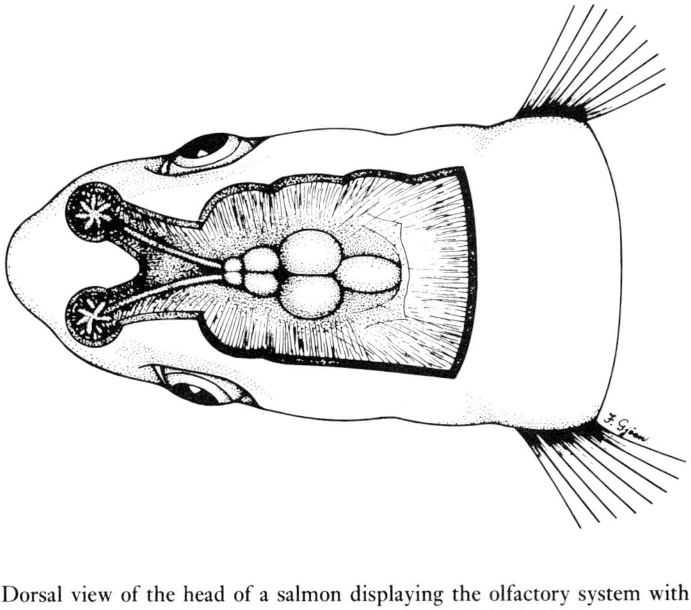

Fig 1. Dorsal view of the head of a salmon displaying the olfactory system with receptor organs, olfactory nerves, and olfactory bulbs. The size of the brain is exaggerated.

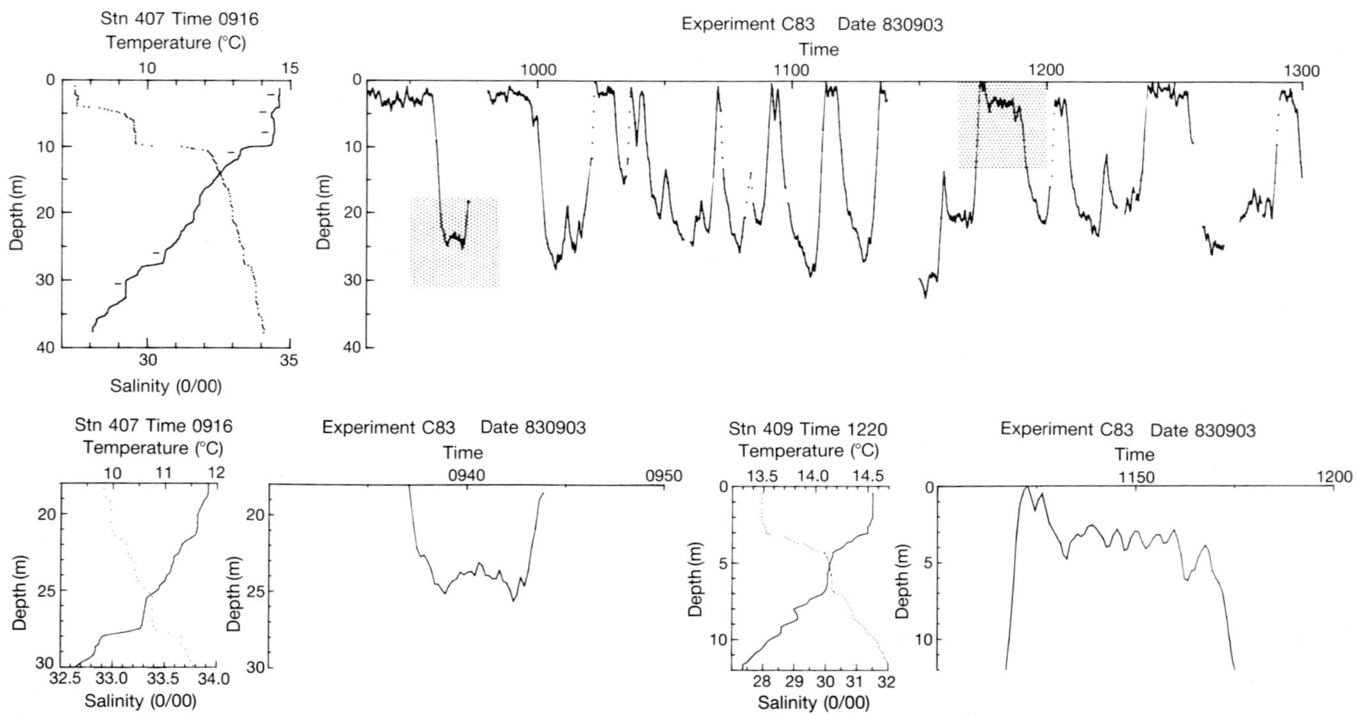


Fig. 2. Swimming depth record chart of salmon C. Simultaneous temperature and salinity profiles are shown at the left. The shaded parts of the upper diagram are shown expanded in the lower diagrams. In this and Fig. 4 the continuous line shows successive 10 s averages, the dotted portions indicate periods of low signal-to-noise ratio and data drop-out. The time-scale shows local time. The horizontal bars by the upper left temperature profile indicate the depths where water samples were collected. (From Døving *et al.* 1985).

success of this treatment was confirmed at *post-mortem* following recapture of fish H in the river Imsa on 31 October 1983.

A pressure-sensitive ultrasonic transmitter was placed into the fish's stomach via the oesophagus. The transmitter measured 60 mm in length, 15 mm in diameter and weighed 9.5 g in water. It operated at 49.5 kHz, producing a 30 ms pulse repetition period varying from 0.5 to 1.5 s over the full pressure range. Each transmitter was calibrated in the sea immediately before tracking. Details of the transmitter, calibration procedure, and data analysis can be found in Westerberg (1983).

Tracking was done from the 31 m R/V Svanic of the University of Gothenburg, Sweden. Positions were obtained by radar ranging. Since the distance to the shore was in the order of one nautical mile the positions were accurate to 50 m or less. Ship's sonar (Simrad SJ 3) was used in the passive mode as the tracking receiver recorded continuously on a Sony TCD-5 tape recorder for later analysis. During the tracking an online plot of the fish's swimming depth was displayed on a Hewlett-Packard 85 computer connected to a timer/counter (Hewlett-Packard 5316 A) to measure the pulse periods. This information was used to plan the hydrographic sampling. The same equipment was used for later decoding of the audio tapes. Each pulse period was meaured and converted to a depth value. Swimming depths were then averaged over successive 10-sec intervals, giving 360 data points per hour of tracking. The time series were used for the time-depth plots. The relative resolution of the depth data is approximately 0.2 m.

## Results

The behaviour of Atlantic salmon observed in the Stavanger fjord was very similar to that observed among Baltic salmon (*Salmo salar*) in the Gulf of Bothnia and among land-locked salmon (*Salmo salar*) in Lake Vänern in Sweden (Westerberg 1982*b*). Typically, the salmon made continuous small-scale vertical movements in the gradient layer, with occasional larger excursions in the water column (Fig. 2). However, the high-amplitude movements never went below the sill depth, which marked the division between the stratified upper layers and the stagnant lower basin.

The behaviour of anosmic salmon differed in several respects. They did not select specific depths and often made high-amplitude movements up and down through the water column. These movements took them much deeper into the fjord, often well into the lower basin. Unlike the intact fish, which remained in mid-water, the anosmic fish were also observed to follow bottom contours. The various behaviours of anosmic fish observed in this study were similar to those described by Westerberg (1982*b*) for an anosmic salmon tracked in Lake Vänern.

The behaviour of the intact and anosmic salmon, described above, indicates that the olfactory sense is used to select layers and to use the information as a reference for purposeful orientation. This would suggest that the salmon must have the ability to discriminate the odour or temperature of water layers in the ocean. To verify this assumption water samples were taken from different depths along the trajectory of the salmon (Fig. 2). In later physiological experiments, impulse activity of single neurones of the olfactory bulb were recorded on exposure to the various water samples. For details of the methods used see Døving *et al.* (1985).

An example of a series of recordings from a single neurone is shown in Fig. 3. Two-thirds of the neurones investigated displayed differential changes in discharge rates in reponse to stimulation with the water samples. These results indicate the capacity of the olfactory system in salmon to discriminate the odorant qualities of fine-scale hydrographic features.

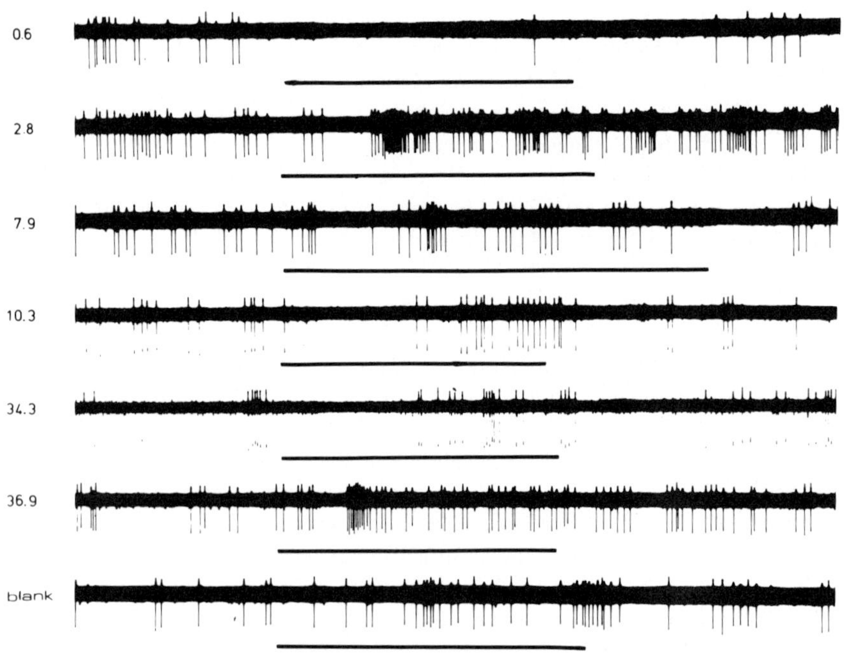


Fig. 3. The responses of a single olfactory bulb neurone in salmon to blank and collected water samples. The numbers to the left of each trace indicate the depth in the fjord where the water samples were taken (see Fig. 2). Stimulation period is indicated by the horizontal line under each trace. Total trace length is 20 s. (From Døving *et al.* 1985).

## Discussion

Critics of sensory impairment studies (Brett and Groot 1963; Harden-Jones 1968; Peters 1971; Ramsey 1961) argue that fish suffering such treatment may home with less precision due to generalized traumatic or inhibitory effects rather than loss of the olfactory sense. In spite of the possible trauma caused by the ablation procedure, the anosmic fish demonstrated large-scale vertical movements although net horizontal progress was slow. Thus the surgical procedure does not seem to affect motivation for active swimming. The high-amplitude swimming behaviour of anosmic fish may be described as an active, but unsuccessful, search for water layers with suitable odour characteristics (Døving *et al.* 1985). This is further supported by the behaviour of a salmon with unilateral ablation, which ultimately entered the river of release.

An explanatory model of salmon olfactory orientation in open water must account for: (i) large-scale horizontal turbulence and (ii) lack of visual points of reference, which renders rheotactic orientation impossible (Arnold 1974). The following hypotheses for salmon orientation in the coastal zone is suggested, based on the results of tracking experiments by Westerberg (1982*b*) and Døving, Westerberg and Johnsen (1985). The basic concept is that information for orientation is found locally in the vertical distribution of odorants rather than in a complicated, large-scale horizontal distribution. The salmon can gain information about the waters' qualities by traversing the water layers. The fish will be able to scan through the water masses more effectively by moving up and down than by moving in the horizontal direction.

The homogeneous layers of stratified bodies of water contain home-stream odorants, and the home-stream layers are assumed to be the layers of interest for the salmon. These layers contain specific odorants, and the adjacent layers, establish a reference system for oriented movement. In order to maintain its position in relation to the home-stream layer, the salmon moves from the layer of interest to the adjacent one with low-amplitude zigzag movements, and may on occasion pass through the interface layer of a strong vertical density gradient (Fig. 4).

An interesting similarity to this vertical zigzag movement can be found in the horizontal zigzag pattern of salmon in a stream. Upstream movements of migrating coho salmon (*Oncorhynchus kisutch*) were shown to be rheotactic responses to the artificial homestream chemical signal to which the fish were exposed during smoltification (Johnsen and Hasler 1980). Regions of confluence between two tributaries are characterized by discrete masses of water bearing different odours and separated by a common interface boundary. Detailed observations of migratory behaviour indicate that the fish move along the interface boundary in a zigzag pattern, which permits the animal to monitor the edges of the two water masses of differing chemical composition (Johnsen 1982). Zigzag behaviour in the horizontal plane allows the fish to follow the chemical trail into

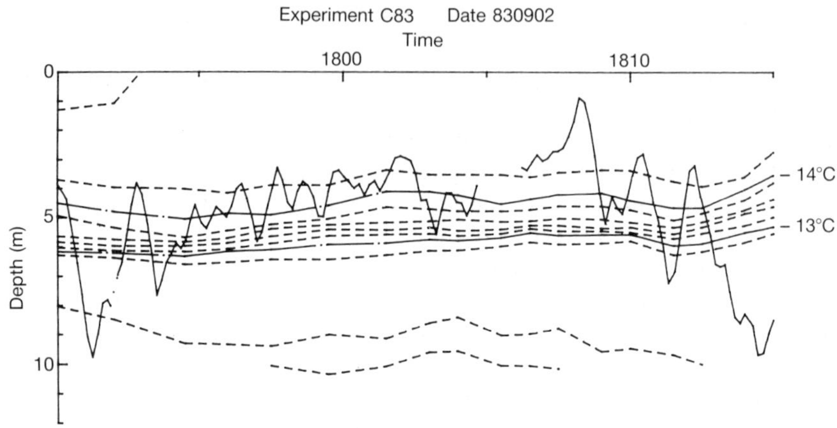


Fig. 4. Detail of the swimming depth of intact salmon C in relation to thermal structure. The temperature soundings were made with approximately 2-min. intervals and the isotherms are drawn with a spacing of 0.2 °C. (From Døving *et al.* 1985.)

the correct tributary. In this case the directional cue is visual contact with the bottom. If the interface of adjacent layers in the ocean becomes indistinct, the fish will begin high-amplitude, up-and-down movements to locate another layer of interest. Salmon without an olfactory sense will continue to swim up and down in search of a layer of interest. In the absence of odour information, the vertical extent of the search is limited only by physical boundaries—the surface, the bottom, or a strong thermocline—and such unsuccessful search behaviour has been observed in anosmic fish.

What directional cues do the salmon use to choose and maintain horizontal swimming direction? A likely cue is the local direction of the current shear, i.e. the direction of the relative motion of the layer of interest and the adjacent layer. Statistically this will be related to the direction of the source of the odorants, but if and how this shear can be detected is not known. One proposal (Westerberg 1984) is that the fish are capable of sensing any extension of micro-scale fluctuations in the gradient layer. If the fluctuations are detected by the olfactory system this should imply a tropotactic orientation. However, the successful homing of a salmon with a unilateral ablation of the olfactory nerve seems to contradict this hypothesis (Døving *et al.* 1985).

At present, the tracking technique, including occasional hydrographic measurements in the vicinity of the fish, is too imprecise to provide more than additional support for an interpretation of the observed behaviour. Physiologically, it has been shown that the salmon is able to discriminate water in adjacent fine-structure layers by olfaction. Behaviourally, the loss of the olfactory sense caused high-amplitude vertical movements that can be interpreted as olfactory search. Future studies will investigate the external directional cues used, and whether sensory systems other than the olfactory system are required for successful homing of salmon.

The Norwegian zoologist Hans Nordeng suggested that the homeward migration of salmon is initiated and directed by odour trails emanating from descending smolt of the same stock (Nordeng 1971, 1977). The smolts establish a trail of odours leading from their respective fresh-water home localities to the salmon out at sea. The maturing salmon respond innately to the smolt odours and start to swim homewards.

The attractant substances are found in the intestinal content of the smolt and are possibly of bile origin (Selset 1980; Selset and Døving 1980). Physiological experiments have shown that bile is a potent source of odours for fish such as salmon and sea char (Døving *et al.* 1980, Døving 1986). The neurones of the olfactory bulb in char react differentially to the odorants from different char populations (Døving *et al.* 1974). The results of various experiments carried out subsequent to Nordeng's hypothesis provide supporting evidence that population-specific odours constitute a trail which the salmon follows on its way home (see Døving 1989).

## References

Arnold, G.P. (1974). Rheotropism in fishes. *Biol. Rev.*, 49, 515–76.

Brett, J.R. and Groot, C. (1963). Some aspects of olfactory and visual responses in Pacific salmon. *J. Fish. Res. Board Can.*, 20, 287–303.

Døving, K. B. (1986). Functional properties of the fish olfactory system. *Prog. Sensory Physiol.*, 6, 39–104.

Døving, K.B. (1989). Molecular cues in salmonid migration. In Molecules in Physics, Chemistry and Biology. (ed. J. Muruani). *Topics Mol. Organiz. Engin.*, 4, 299–329.

Døving, K.B., Nordeng, H., and Oakley, B. (1974). Single unit discrimination of fish odors released by char (*Salmo alpinus* L.) populations. *Comp. Biochem. Physiol.*, 47A, 1051–63.

Døving, K.B., Selset, R. and Thommesen, G. (1980). Olfactory sensitivity to bile acids in Salmonid fishes. *Acta Physiol. Scand.*, 108, 123–31.

Døving, K.B., Westerberg, H. and Johnsen, P.B. (1985). Role of olfaction in the behavioral and neural responses of Atlantic Salmon, *Salmo salar*, to Hydrographic stratification ch. I. *Can. J. Fish. Aquat. Sci.*, 42, 1658–67.

Harden-Jones, F.-R. (1968). *Fish migration*. Edward Arnold, London.

Hasler, A.D. (1966). *Underwater guideposts.* Homing in salmon. University of Wisconsin Press, Madison, Milwaukee and London.

Hasler, A.D. and Scholz, A.T. (1983). Olfactory imprinting and homing in salmon. *Zoophysiology*, 14, 134.

Hawkins, A.D. and Sand, O. (1977) Directional hearing in the medial vertical plane by the cod. *J. Comp. Physiol.*, 122, 1–8.

Johnsen, P.B. (1982). A behavioural control model for homestream selection in salmonids. In *Proceedings of the salmon and trout migratory behaviour symposium.* (ed. E.L. Brannon and E.O. Salo), pp. 266–73. School of Fisheries, University of Washington, Seattle Wash.

Johnsen, P.B. and Hasler, A.D. (1980). The use of chemical cues in the upstream migration of coho salmon, *Oncorhynchus kisutch* Walbaum. *J. Fish Biol.*, 17, 67–73.

Nordeng, H. (1971). Is the local orientation of anadromous fishes determined by pheromones? *Nature*, 233, 411–3.

Nordeng, H. (1977). A pheromone hypothesis for homeward migration in anadromous salmonids. *Oikos*, 28, 155–9.

Peters, M. (1971). Sensory mechanisms of homing in salmonids: a comment. *Behavior*, 39, 18–9.

Ramsey, D.A. (1961). Olfactory cues in migrating salmon. *Science*, 133, 56–7.

Sand, O. and Karlsen, H.E. (1986). Detection of infrasound by the atlantic cod. *J. Exp. Biol.*, 125, 197–204.

Selset, R. (1980). Chemical methods for fractionation of odorant produced by char smolts and tentative suggestions for pheromone origins. *Acta Physiol. Scand.*, 108, 97–103.

Selset, R. and Døving, K.B. (1980). Behaviour of mature anadromous Char (*Salmo alpinus* L.) towards ordorants produced by smolts of their own population. *Acta Physiol. Scand.*, 108, 113–22.

Stabell, O.B. (1984). Homing and olfaction in salmonids: a critical review with special reference to the Atlantic salmon. *Biol. Rev.*, 59, 333–88.

Westerberg, H. (1982*a*). Ultrasonic tracking of Atlantic salmon (*Salmo salar* L.)—I. Movements in coastal regions. *Rep. Inst. Freshw. Res., Drottningholm*, 60, 81–101.

Westerberg, H. (1982*b*). Ultrasonic tracking of Atlantic salmon (*Salmo salar* L.)—II. Swimming depth and temperature stratification. *Rep. Inst. Freshw. Res., Drottningholm*, 60, 102–20.

Westerberg, H. (1983). Monitoring fish behaviour in relation to hydrographic fine structure. In *Proceedings of the 4th International Conference on Wildlife Biotelemetry* (ed. D.G. Pincock), pp. 154–66.

Westerberg, H. (1984). The orientation of fish and the vertical stratification at fine- and microstructure scales. In *Mechanisms of migration in fishes* (ed. J.D. Cleave, W.H. Neill, J.J. Dodson and G.P. Arnold), pp. 179–204. Plenum, New York.

Wisby, W.J. and Hasler, A.D. (1954). The effect of olfactory occlusion on migrating river silver salmon (*O. kisutch*). *J. Fish. Res. Board Can.*, 11, 472–8.

# SECTION VI

## Applications

# VI.1 Leading them by Their Noses: Animal and Plant Odours for Managing Vertebrates

D. MÜLLER-SCHWARZE

## Introduction

To use odour signals for attracting or repelling animals and to manipulate their feeding and mating behaviour, are age-old goals of hunters, farmers, foresters, gardeners, pet owners, animal handlers, and many others. Today, odours have been applied or are candidates for application to animal husbandry, biological pest control, wildlife management, and population monitoring. In addition, increasingly we will have to cope with perturbances to chemical communication by environmental pollution. In this review I try to identify the current research front that lies between practitioners' time-proven, established procedures, and scientific enthusiasts' high, but often naïve, hopes for high-tech solutions.

## Animal husbandry

In livestock management chemical signals are important for *Reproduction* and *Feeding behavior*, and these two areas of animal husbandry have recently received much attention. Zoos, animal parks, and captive breeding programs for endangered species are paying particular attention to the use of animal odours.

### FEEDING BEHAVIOUR

It has long been known that domestic herbivores use their chemical senses in selecting food (e.g. Arnold 1966; Arnold *et al.* 1980; Sheehy and Winward 1981). However, the precise relative contributions of smell and taste to food choices and the plant chemicals responsible for such discrimination are still being explored. Sheep with impaired senses of smell, taste, and touch still show food preferences (Arnold 1966); taste was important for discriminating between fine- and broadleaved varieties of *Dactylis glomerata* from different localities, while brome grasses were always highly acceptable, and recognized by their touch.

Sheep can be manipulated by odours to vary their food intake; more of a hay ration was eaten if butyric acid or amyl acetate were added, and less if mixed with coumarin or glycine. Anosmic sheep were not affected in their food consumption (Arnold *et al.* 1980). Of known plant compounds, coumarin, gramine, tannic acid, malonic acid, and glycine reduced hay intake, and none enhanced feeding. Sheep rendered anosmic and agustatory were not affected, and Arnold *et al.* (1980) conclude that sensory stimuli are not involved in food consumption. Since *in vitro* digestibility is decreased by tannic acid, gramine, and somewhat by coumarin, it was suggested that these compounds have their effect in the rumen, resulting in reduced food intake.

Reducing the grazing availability of certain toxic range plants remains an important goal. For instance, many cattle and, to a lesser degree, sheep on mountain rangeland in the western United States and Canada are lost to poisoning by diterpenoid alkaloids from larkspur (*Delphinium*). This is the most serious poisonous plant problem for that region (Ralphs *et al.* 1988). In a search for feeding repellents for mountain pastures with larkspur, faecal odours from predators were placed near food in enclosure experiments. Faecal odours from coyote (*Canis latrans*), fox (*Vulpes vulpes*), and mountain lion (*Felis concolor*) reduced food consumption by cattle and sheep, while bear (*Ursus americanus*) odours did not. These results suggested that whilst livestock were not kept from entering a faecal odour-treated area, they did spend less time grazing there (Pfister *et al.* 1990).

Knowles and Tahan (1979) investigating different repellents to protect *Pinus radiata* seedlings from browsing by sheep, found that fermented egg with acrylic adhesive and red dye were effective.

## REPRODUCTION

The potential use of sex odours for manipulation of reproductive behaviour in domestic animals has been long recognized and should also apply to wild species in captivity. We can distinguish *priming* from *signalling* effects.

### *Priming*

Of the various effects of priming pheromones, acceleration or delay of puberty in both sexes, and regulation of the oestrous cycle by delay, acceleration, or synchronization may aid animal husbandry. In swine (*Sus scrofa*), the puberty of gilts was accelerated by one month by the presence of the boar and his odour (Izard 1983), while development of female beef cattle was accelerated by two weeks by exposure to bull urine (Izard 1983).

Bull urine sprayed daily into the nasal cavity accelerated onset of ovarian activity, thus hastening both puberty and calving dates (Izard and Vandenbergh 1982*a*). Oestrus was synchronized by cervical mucus odours from oestrous cows which increases the efficiency of artificial insemination and calf rearing (Izard and Vandenbergh 1982*b*).

*Signalling*

Perhaps the most important information transmitted between the sexes to influence reproduction is the female's oestrous status. Numerous studies have examined the ability of males to distinguish odours of oestrus versus non-receptive females, and some in cattle; these are covered in detail by Signoret (this volume) and Klemm (this volume). The sex odours of female cattle can be used in practice to test bulls for libido, and to enhance libido as chemosensory aphrodisiacs. Klemm *et al.* (1987) isolated and identified a number of compounds in extracts of cow cervico-vaginal mucus. Of these, six compounds, and particularly four ketones, were found in extracts, which even after 1 year of storage were able to effect positive responses of bulls. Bulls can be used to detect 'silent', i.e. undetected oestrus occurring in nearly 40% of oestrous cycles of dairy cows and in 25–30% of ovulating cows, which causes delays in successful artificial insemination. In contrast to non-cycling cows, those with 'silent' oestrus still have the sex pheromone in the cervico-vaginal mucus that stimulates the bull (Kiddy and Mitchell 1981).

## Biological pest control

For long reliant on odours and tastants, the baiting and trapping of rodents and other mammals are still being refined by a better understanding of chemoreception and whether compounds are biologically active.

Relatively high olfactory sensitivity of a pest species should be exploited for selective pest control. For example, feral pigs (*Sus scrofa*) have an acute sense of smell; captive feral pigs were able to locate and open experimental bait packages which were not olfactorily noticed by raptors or granivorous birds and could not be opened by small mammals or birds. Thus, together with design features of bait application, most non-target species will be naturally excluded from the baiting (O'Brien 1986). To attract and to induce rodent pests to consume bait with chemosterilants, sex attractant odours can be added to the bait; an example is the use of 3-chloro-1,2-propanediol (Field 1972).

## Wildlife management

### ATTRACTANTS

*Trapping*

Trappers have traditionally used a great variety of odour lures. As long as the mammal *is* attracted to the odour, it does not matter to the trapper *why* it is attracted. Chemical analysis of intra- and interspecific odour signals and better understanding of stimulus-response relationships will result in more scientifically based attractants. Discussing coyote attractants, Bullard *et al.* (1983) noted that 'virtually nothing is known about what kinds of compounds elicit what kinds of responses'.

*Predator control*

The elimination of carnivores such as coyotes in North America or jackals in South Africa, which prey on domestic livestock, has remained a controversial issue. New approaches are continually being tried, such as the use of volatile compounds that are sufficiently specific to attract only the target species to a trap or poison bait, thus sparing other wildlife. Bullard *et al.* (1983) experimented with synthetic fermented egg lures to attract coyotes, and found the behaviour responses to vary with odour intensity and quality. After finding that coyotes (*Canis latrans*) are attracted to sheep liver extracts (Barnum *et al.* 1982) a series of volatile aldehydes from sheep liver extracts were isolated by Lorenz *et al.* (1983), and Scrivner *et al.* (1984) then tested the attractiveness to captive coyotes of the aldehyde series from octanal to hexadecanal. Octanal, nonanal, decanal, and undecanal released as much sniffing and 'rub-rolling' as trimethylammonium decanoate (TMAD), a traditional coyote attractant. The aldehydes triggered very little lick-chewing or biting—behaviour observed more frequently in response to TMAD. Therefore, the investigators concluded that the four aldehydes may be suitable as lures for traps, although they would be of limited use with toxicant-delivering systems that are triggered by licking or biting.

An effective coyote attractant (synthetic fermented egg; designated DRC-6500) is at the same time also a deer repellent. It consists of 41.8% caproic acid, 35.1% butyric acid, hexylamine, trimethylamine, dimethyl disulfide, 2-mercapto ethanol, and ethyl caproate (Turkowski *et al.*, 1983).

*Aiding colonization: attracting to habitat and opposite sex*

It has been suggested repeatedly that scenting the habitat is likely to attract members of a particular species to a designated area and entice them to remain there. This approach may be promising for a large species if an individual's own odour or that of the opposite sex is used, although it has not been tried experimentally.

## REPELLENTS

*Wildlife damage control*

Chemical wildlife repellents that prevent or reduce damage to vegetation can be based on plant defence compounds or predator odours, but other materials, such as fermented egg or fish have been used successfully.

An example of a plant-derived repellent is pine oil, a mixture of terpene alcohols and monoterpenes. When mixed with laboratory rations, pine oil reduced feeding in snowshoe hare, *Lepus americanus*, and voles, *Microtus townsendii* (Bell and Harestad 1987). Red deer calves (*Cervus elaphus*), consumed less of their standard pelleted ration if it was contaminated with odours of crushed lodgepole pine (*Pinus contorta*) and the monoterpenes borneol, limonene, terpinolene, or alpha-pinene (Elliot and Loudon 1987). Likewise, mule deer *Odocoileus hemionus*, prefer the one species of juniper (*Juniper deppeana*), which has the lowest concentration of oxygenated monoterpenes, over two others tested (*J. osteosperma* and *J. scopularum*). Food consumption is inversely related to the volatile oil content of the respective species (Schwartz *et al.* 1980). Sagebrush (*Artemisia*) is an important winter diet for mule deer, pronghorn, and sage grouse, and concentrations of volatiles vary between the plant taxa. One-fifth or less of variability in utilization by mule deer is attributable to the relative percentages of volatiles in leaves. Oxygenated compounds, particularly methacrolein and arthole, are more highly (negatively) correlated with sagebrush utilization than hydrocarbons (Scholl *et al.* 1977). However, Welch *et al.* (1983) found no clear correlation between mule deer browsing of sagebrush and its monoterpenoid contents. Likewise, White *et al.* (1982) found no clear preference by pygmy rabbits *Brachylagus idahoensis* for different sagebrush (*A. tridentata*) populations.

Predator-derived compounds have been used successfully to repel rodents. Sulfur compounds from red fox (*Vulpes vulpes*) faeces 'stress' Norway rats, *Rattus norvegicus*, (Vernet-Maury *et al.* 1984), and sulfur compounds from stoat (*Mustela erminea*) and red fox protect apple trees from damage by the apple orchard vole, *Microtus montanus* (Sullivan *et al.* 1988).

Faecal odours from bobcat (*Felis rufus*) reduced browsing on salal (*Gaultheria shallon*) by black-tailed deer (*Odocoileus hemionus columbianus*) by 51% and those from mountain lion (*F. concolor*), wolf (*Canis lupus*) and coyote (*C. latrans*) had lesser effects ranging from 27% to 8% feeding reduction (Melchiors and Leslie 1985). White-tailed deer (*O. virginianus*) also avoided feeding near the odour of wolf droppings applied to gardens (Müller-Schwarze 1983). Naive young black-tailed deer avoid the odours of mountain lion and coyote droppings (Müller-Schwarze 1972).

Sullivan and co-workers have examined in detail the effects of predator odours on feeding by mammalian herbivores. Urine, faeces and anal gland secretion of various North American mustelids, canids, and felids suppressed feeding by snowshoe hares (Sullivan *et al.* 1985*a*). Black-tailed deer fed less on salal, red cedar (*Thuja plicata*), and Douglas fir (*Pseudotsuga menziesii*) if treated with urine or faeces of large cats and canids (Sullivan *et al.* 1985*b*). Snowshoe hares fed less on lodgepole pine (*Pinus contorta*) seedlings if they were treated with wolverine (*Gulo gulo*) urine (Sullivan 1986) or the anal gland sulfur compound 3-propyl-1,2 -dithiolane from stoat, *Mustela erminea* and 2,2- dimethylthietane from mink, *M. vison* (Sullivan and Crump 1984). The sulfur compounds 3-methyl-3-butenyl methyl sulfide, 2-phenylethyl methyl sulfide, and 3-methylbutyl methyl sulfide also reduced feeding by snowshoe hare on lodgepole pine seedlings, with the first being most effective (Sullivan and Crump 1986). Tables 1 and 2 list the identified sulfur compounds, and their sources and effects.

**Table 1.** Some odour components of mustelid anal gland with effects on prey species

| Predator | Compound | Prey | Effect | Reference |
|---|---|---|---|---|
| Stoat (*Mustela erminea*) | 3-propyl-1,2- dithiolane | Snowshoe Hare (*Lepus americanus*) | Feeding suppression | Sullivan and Crump 1984 |
| | | Meadow Vole (*Microtus pennsylvanicus*) | Area avoidance | Sullivan *et al*. 1988 |
| | 2-propylthietane | Pocket Gopher (*Thomomys talpoides*) | Area avoidance | Sullivan and Crump 1986*b* |
| | | Montane Vole (*Microtus montanus*) | Area avoidance | Sullivan *et al*. 1988 |
| | 3-propyl-1,2-dithiolane | Montane Vole | Colonization deterrent | Sullivan *et al*. 1988 |
| Mink (*M. vison*) | 2,2-dimethylthietane | Snowshoe Hare | Feeding supression | Sullivan and Crump 1984 |
| Ferret (*M. putorius*) | 3,3-dimethyl-1, 2-dithiolane | Pocket Gopher | Colonization deterrent | Sullivan and Crump 1986*b* |

**Table 2.** Some red fox (*V. vulpes*) odour components with effects on prey species

| Compound | Prey species | Effect | Reference |
|---|---|---|---|
| *Faeces*: | | | |
| 2,5-dihydro-2,4,5,trimethyl thiazole | Laboratory Rat | 'Stress' | Vernet-Maury *et al*. 1984 |
| | Montane and Meadow Voles | Area avoidance | Sullivan *et al*. 1988 |
| 3,3-dimethyl-1,2-dithiolane | Laboratory Rat | 'stress' | Vernet-Maury *et al*. 1984 |
| *Urine*: | | | |
| 3-methyl-3-butenyl methyl sulfide | Snowshoe Hare | Feeding supression | Sullivan and Crump 1986*a* |

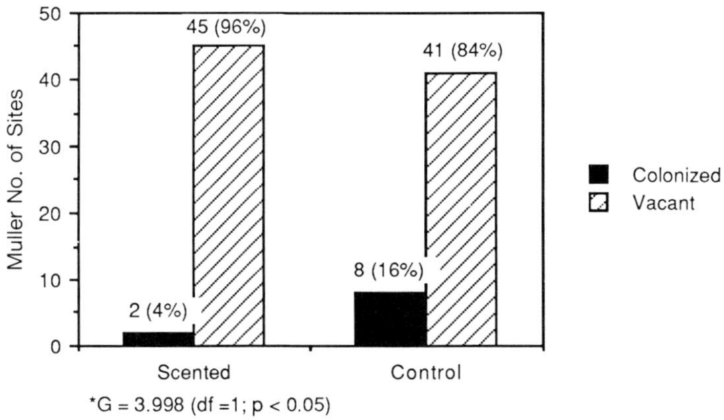


Fig. 1. Colonization of vacant beaver sites that were experimentally scented with castoreum, compared with unscented controls. Control sites (on right) were four times as likely to become colonized (16%) as scented sites (4%). The vast majority of both types of sites (84% and 96% respectively) remained vacant, because population pressure was low, with only 30% of all potential sites occupied. This was a result of intense management by trapping. After data in Welsh and Muller-Schwartz (1989).

*Preventing colonization by conspecific territorial odours*

Few studies have addressed the question of whether scenting the habitat with 'area repellents' derived from typically avoided odours of predators or rivals will deter newcomers from settling in a section of otherwise suitable habitat. North American beaver (*Castor canadensis*) can be prevented from colonizing vacant habitat by experimental application of castoreum from the castor sacs that this species uses for territorial scent marking (Müller-Schwarze and Heckman 1980; Welsh and Müller-Schwarze 1989). Figure 1 summarizes the results of a field experiment described by Welsh and Müller-Schwarze (1989).

For such treatment to succeed, certain conditions must be met. Above all, the population density should be moderate so that vacant beaver habitat is still available for colonization. If given a chance to choose between artificially scented and unscented sites, beaver are more likely to colonize the latter. If, on the other hand, there is a shortage of vacant sites, beaver will be forced to settle in scented sites. In our study area, only about 30% of the potential beaver sites were occupied, providing ample choice between scented and unscented sites. By contrast, a saturated population in the western part of New York State did colonize all scented sites, presumably because there was little choice.

Anal gland secretion of yellow voles (*Lagurus luteus*) may have potential as a repellent. In field experiments, previously occupied burrows were treated with anal gland secretion from strange individuals. For one day, voles avoided many of these scented burrows, but re-entered them later on (Fan *et al.* 1981; Fan 1983). Adult pocket gophers (*Thomomys talpoides*) are deterred from colonizing an area treated with sulfur compounds (3-propyl-1,2-dithiolane and 3,3-dimethyl-1,2-dithiolane) from the anal glands of the stoat, *Mustela erminea* (Sullivan and Crump 1986*b*).

*Aversive conditioning*

In addition to *primary food aversions* (without learning, the animal avoids an often bitter food from the outset), much has been expected of *conditioned aversions.* Aversive conditioning of mammalian predators and scavengers has been tested with emetics, especially lithium chloride (LiCl). Coyotes that prey on sheep or goats have been conditioned with mixed results. Coyotes that experienced LiCl on a collar worn by a sheep were later not deterred from attacking sheep. Instead, they learned to avoid the specific collar area and aimed their attacks at the throat and rear end of lambs (Burns *et al.* 1984). Similarly, coyotes did not develop an aversion to rabbits after three different types of LiCl induced illness. These were intraperitoneal injections after consuming rabbit meat, eating LiCl-laced rabbit meat packaged in rabbit fur, and eating on LiCl-laced carcasses (Horn 1983). However, in a different study, cinnamic aldehyde on collars of lambs reduced coyote predation from 8.2% to 4.4% (Faller 1975).

Aversions are generally bait-specific. For this reason, aversive conditioning of black and polar bears with emetics resulted in aversions that were limited to particular food (Harting 1987). Lithium Chloride-treated sheep flesh wrapped in sheep skin caused bait aversion in coyotes, but did not stop predation on live lambs (Burns 1983).

Primary aversion to compounds that taste bitter to humans has been found in many animal taxa from protozoans (*Stentor*) to coelenterates, molluscs, echinoderms, *Limulus*, earthworms, insects, fish, birds, and mammals. This universal aversion to anthropocentrically bitter compounds is thought to result from natural selection for avoidance of toxic alkaloids and glycosides that are bitter to humans (Garcia and Hankins 1975).

When searching for repellents, it is important to remember that the same compound may be avoided by one group of vertebrates, but can be attractive to another. For example, capsaicin from hot red peppers is avoided by mammals (Kare 1965) but eaten by birds, and is actually recommended as parrot food (Reidinger and Mason 1983). Similarly, dimethylanthranilate is used as artificial grape flavouring in human food, while birds reject it (Kare and Pick 1960).

## Population monitoring

Odours, especially those introduced inadvertently during field experiments, may influence the results of small rodent censuses. Stoddart (1982*a*) live-trapped short-tailed voles (*Microtus agrestis*) until only few unmarked, i.e. 'new' animals were caught; this occurred after 13 nights. Then the traps were cleaned with hot water and detergent, and trapping continued. During the three subsequent nights the number of new, i.e. unmarked voles caught in the traps increased, indicating that voles had been deterred from entering the soiled traps. The population estimate by the Lincoln Index increased by 27% after the traps had been cleaned, and by 62.4% after they had been cleaned a second time after three nights' use. Conversely, the population estimate dropped by 14.6% after the traps had been treated with the urine of a male tiger (Stoddart 1982*b*).

The density of scent marks in a territory or home range can be indicative of the population density of that species. Visible scent marks such as the faecal piles of muskrats (*Ondatra zibethicus*), scent mounds of beaver (*Castor canadensis*), carnivore droppings on conspicuous features or along trails, and fox or wolf urine marks in the snow can be more easily counted. Not only do more individuals produce more of such signs, but any given colony or family mark more when contact with neighbours and transients is more frequent (Müller-Schwarze and Heckman 1980).

The responses of rodents to scented traps may also be used to infer the status and density of the population. Adult muskrats (*Ondatra zibethicus*) avoided musk-scented traps in a saturated breeding population (van den Berk and Müller-Schwarze 1984), but were attracted to such traps in a Dutch immigrating pioneer population (Ritter *et al.* 1982).

## Environmental pollution: interference with chemical communication

Very little is known about disturbances by environmental pollution to chemical communication or chemoselection of food, which in turn would call for remedial action. A better investigated example is the response of fish to amino acids that are feeding attractants. Juvenile Atlantic salmon (*Salmo salar*) change their orientation with respect to waterborne amino acids if the pH of the water is lowered. At a pH of 7.6 they are attracted to glycine and avoid L- alanine. When the pH is experimentally dropped to 5.1, the fish become indifferent to glycine and are now attracted by L-alanine (Royce- Malmgren and Watson 1987). If the home stream odour contains amino acids, the fish could be disoriented by lowered pH; thus water acidification could impair migration, and subsequently reproduction, with severe consequences for the entire population. The precise mechanism of the chemoreception impairment is not known. It is currently being debated whether changes in water pH affect stimulus molecules (Hara 1976) or receptor proteins (Tierney and Atema 1988).

## Discussion

Where are the opportunities and limitations for applying natural odours to manipulate wild vertebrates?

The available data justify the generalization that consistent effects have been demonstrated, but that most of these are not sufficiently robust to be of interest for serious large- scale field applications. Either the response is not very strong—as on colonization by beaver, or not durable enough—as in avoidance of predator odours by mammalian herbivores. It is now abundantly clear that the mammalogist's expectations have to be vastly different from those of specialists of other disciplines, such as the entomologist.

During the experimental stage several biological and chemical considerations are important. Biologically, one has to determine the precise effects of a stimulus, understand the role of the various forms of learning, be aware of multisensory control of responses, and offer choices between treated and untreated resources. The same stimulus may have different effects, depending on the status of the population. As in the muskrat studies cited above, a male odour may act as repellent to other adults in a territorial breeding population, while at low population densities it may attract the opposite or even the same sex. The effect may also change over time with the phases of the breeding cycle.

Chemically, we need to identify more active compounds. For field application, an appropriate slow-release formulation, which is water-resistant, easy to apply, and cost-effective, has to be developed. It is important to remember that odours are species- and context-specific. The greater the biological sophistication required for successful use of odours in a specific situation, the smaller the market for a potential product, hence there is likely to be little interest by industry.

For odours and tastants that require conditioning, the main disadvantage is the specificity of the response. Coyotes, rats, or bears avoid only those foods, locales, or body parts of prey, that had been treated with the aversive chemical stimulus.

We have learned much since earlier reviews (see, for example, Shumake 1977), especially in the areas of biological determinants of odour responses and chemical identification of active compounds, while progress in actual practical applications has been slow. Odours will, however, have their place as one measure among several, in integrated management of both domestic livestock and wildlife species.

## References

Arnold, G.W. (1966). The special senses in grazing animals. II. Smell, taste, and touch and dietary habits in sheep. *Aust. J. Agric. Res.*, 17, 531–42.

Arnold, G.W., de Boer, E.S. and Boundy, C.A. (1980). The influence of odour and taste on the food preferences and food intake of sheep. *Aust. J. Agric. Res.*, 31, 571–87.

Barnum, D.A., Fagre, D.B. and Howard, W.E. (1982). Liver extracts as potential lures. In *Min. Proc. Annu. Meeting Western Regional Coord. Committee*, 26, Waco, Texas, Aug. 10–11, pp. 9–10.

Bell, C.M. and Harestad, A.S. (1987). Efficacy of pine oil as repellent to wildlife. *J. Chem. Ecol.*, 13, 1409–17.

Bullard, R.W., Turkowski, F.J. and Kilburn, S.R. (1983). Responses of free-ranging coyotes to lures and their modifications. *J. Chem. Ecol.*, 9, 877–88.

Burns, R.J. (1983). Microencapsulated Lithium chloride bait aversion did not stop coyote predation on sheep. *J. Wildl. Managm.*, 47, 1010–7.

Burns, R.J., Conolly, G.E. and Griffiths, R.E. Jr. (1984). Repellent or aversive chemicals in sheep neck collars did not deter coyote attack. *Proc. Vertebrate Pest Conf.*, II, 146–53.

Elliot, S. and Loudon, A. (1987). Effects of monoterpene odors on food selection by red deer calves (*Cervus elaphus*). *J. Chem. Ecol.*, 13, 1343–9.

Faller, T.C. (1975). *Field evaluation of a collared repellent to reduce coyote predation on sheep.* Final report to: U.S. Fish and Wildlife Service, Contract No. 14-16-0008-835, North Dakota State Univ. Expt. Station, Hettinger, N.D., 47 pp.

Fan, Zhigin, Hao Shoushem, Liang Yingnan, Wang Mingyue and Zhao Yong. (1981). Chemical signals and their role in social groups of two volves (*Lagurus luteus* and *Lagurus lagurus*). *Acta Ecol. Sin.*, 1, 66.

Fan, Zhigin (1983). The ecological importance of the anal gland secretion of yellow voles (*Lagurus luteus*). In *Chemical signals in vertebrates*, 3 (ed. D. Müller-Schwarze and R.M. Silverstein), pp. 211–22. Plenum, New York.

Field, R.J. (1972). The use of sexual attractant pheromones to increase the acceptability of the male chemosterilant 3-chloro-1,2-propanediol by wild Norway rats (*Rattus norvegicus*). *Diss. Abstr. Int. [B] Sci. Eng.*, 32, 5533.

Garcia, J. and Hankins, W.G. (1975). The evolution of bitter and the aquisition of toxiphobia. In *Olfaction and taste*, V (ed. D.A. Denton and J.P. Coghlan), pp. 39–45. Academic Press, New York.

Hara, T.J. (1976). Effects of pH on the olfactory responses to amino acids in rainbow trout, *Salmo gairdneri*. *Comp. Biochem. Physiol.*, 54, 37–9.

Harting, A.L. (1987). Aversive conditioning, deferrents and attractants. In *Grizzly bear compendium*, pp. 89–94. Interagency Grizzly Bear Committee, National Wildlife Federation, Washington, D.C.

Horn, S.W. (1983). An evaluation of predatory suppression in coyotes using lithium chloride induced illness. *J. Wildl. Managm.*, 47, 999–1009.

Izard, M.K. and Vandenbergh, J.G. (1982*a*). The effects of bull urine on puberty and calving date in crossbred beef heifers. *J. Anim. Sci.*, 55, 1160–8.

Izard, M.K. and Vandenbergh, J.G. (1982*b*). Priming pheromones from oestrous cows increase synchronization of oestrus in dairy heifers after PGF-2 injection. *J. Reprod. Fert.*, 66, 189–96.

Izard, M.K. (1983). Pheromones and reproduction in domestic animals. In *Pheromones and reproduction in mammals* (ed. J.G. Vandenbergh), pp. 253–85. Academic Press, New York.

Kare, M.R. (1965). The special senses. In *Avian physiology* (ed. P.D. Sturke), pp. 406–46. Cornell University Press, Ithaca, New York.

Kare, M.R. and Pick, H.L. (1960). The influence of the sense of taste on feed and fluid consumption. *Poult. Sci.*, 39, 697–706.

Kiddy, .A. and Mitchell, D.S. (1981). Estrus-related odors in cows: time of occurrence. *J. Dairy Sci., 64*, 267–271.

Klemm, W.R., Hawkins, G.N. and De Los Santos, E. (1987). Identification of compounds in bovine cervico-vaginal mucus extracts that evoke male sexual behavior. *Chem. Senses*, 12, 77–87.

Knowles, R.I. and Tahan, F. (1979). A repellent to protect radiata pine seedlings from browsing by sheep. *New Zealand J. For. Sci.*, 9, 3–9.

Lorenz, G., Stern, D.J., Flath, R.A., Haddon, W.F., Tillan, S.J. and Teranishi, R. (1983). Identification of sheep liver volatiles. *J. Agric. Food Chem., 31*, 1052–7.

Melchiors, M.A. and Leslie, C.A. (1985). Effectiveness of predator fecal odors as black-tailed deer repellents. *J. Wildl. Managm.*, 49, 358–62.

Müller-Schwarze, D. (1972). Responses of young black-tailed deer to predator odors. *J. Mammal.*, 53, 393–4.

Müller-Schwarze, D. (1983). Experimental modulation of behavior of free-ranging mammals by semiochemicals. In *Chemical signals in vertebrates*, 3 (ed. D. Müller-Schwarze and R.M. Silverstein), pp. 235–44. Plenum, New York.

Müller-Schwarze, D. and Heckman, S. (1980). The social role of scent marking in beaver (*Castor canadensis*). *J. Chem. Ecol.*, 6, 81–95.

O'Brien, P.H. (1986). An approach to the design of target-specific vertebrate pest control systems. In *Proc. 12th Vertebrate Pest Conf.* (ed. T.P. Salmon), pp. 247–52. University of California, Davis, California.

Pfister, J.A., Müller-Schwarze, D. and Balph, O.F. (1990). Effects of predator fecal odors on feed selection by sheep and cattle. *J. Chem. Ecol.*, 16.

Ralphs, M.H., Olsen, J.D., Pfister, J.A. and Manners, G.D. (1988). Plant–animal interactions in larkspur poisoning in cattle. *J. Anim. Sci.*, 66, 2334–42.

Reidinger, R.F. and Mason, J.R. (1983). Exploitable characteristics of neophobia and food aversions for improvements in rodent and bird control. In *Vertebrate pest control and management materials*, Fourth Symp. 1982 (ed. D.E. Kaukeinen), pp. 20–36. ASTM Publ. 04-817000-481, Philadelphia.

Ritter, F.J., Bruggemann, I.E.M., Gut, J., Persoons, C.J. and Verwiel, P.E.J. (1982). Chemical stimuli of the muskrat. In *Determination of behaviour by chemical stimuli* (ed. J.E. Steiner and J.R. Ganchrow), pp. 77–89. IRL Press, London.

Royce-Malmgren, C.H. and Watson, W.H. (1987). Modification of olfactory-related behavior in juvenile Atlantic salmon by changes in pH. *J. Chem. Ecol.*, 13, 533–46.

Scholl, J.P., Kelsey, R.G. and Shafizadeh, F. (1977). Involvement of volatile compounds of Artemisia in browse preference by mule deer. *Bio. Chem. Syst. Ecol.*, 5, 291–5.

Schwartz, C.C., Regelin, W.L. and Nagy, J.G. (1980). Deer preference for juniper forage and volatile oil treated foods. *J. Wildl. Managm.*, *44*, 114–20.

Scrivner, J.H., Howard, W.E. and Teranishi, R. (1984). Aldehyde volatiles for use as coyote attractants. In *Proc. 11th Vertebrate Pest Conf.*, pp. 157–60. University of California, Davis, California

Sheehy, D.P. and Winward, A.H. (1981). Relative palatability of seven *Artemisia* taxa to mule deer and sheep. *J. Range Managm.*, 34, 397–9.

Shumake, S.A. (1977). The search for applications of chemical signals in wildlife management. In *Chemical signals in vertebrates*, I (ed. D. Müller-Schwarze and M.M. Mozell), pp. 357–76. Plenum, New York.

Stoddart, D.M. (1982*a*). Does trap odour influence estimation of population size of the short-tailed vole, *Microtus agrestis*? *J. Anim. Ecol.*, 51, 375–86.

Stoddart, D.M. (1982*b*). Demonstration of olfactory discrimination by the short-tailed vole, *Microtus agrestis*. *Anim. Behav.*, 30, 293–4.

Sullivan, T.P. and Crump, R.D. (1984). Influence of mustelid scent gland compounds on suppression of feeding by snowshoe hares (*Lepus americanus*). *J. Chem. Ecol.*, 10, 1809–21.

Sullivan, T.P., Nordstrom, L.O. and Sullivan, D.S. (1985*a*). The use of predator odors as repellents to reduce feeding damage by herbivores. I. Snowshoe hares (*Lepus americanus*). *J. Chem. Ecol.*, 11, 903–19.

Sullivan, T.P., Nordstrom, L.O. and Sullivan, D.S. (1985*b*). The use of predator odors as repellents to reduce feeding damage by herbivores. II. Black-tailed deer (*Odocoileus hemionus columbianus*). *J. Chem. Ecol.*, 11, 921–35.

Sullivan, T.P. (1986). Influence of wolverine (*Gulo gulo*) odor on feeding behavior of snowshoe hares (*Lepus americanus*). *J. Mammal.*, 67, 385–8.

Sullivan, T.P. and Crump, D.R. (1986*a*). Feeding responses of snow-shoe hares (*Lepus americanus*) to volatile constituents of red fox (*Vulpes vulpes*) urine. *J. Chem. Ecol.*, 12, 729–39.

Sullivan, T.P. and Crump, D.R. (1986*b*). Avoidance responses of pocket gophers (*Thomomys talpoides*) to mustelid anal gland compounds. In *Chemical signals in vertebrates*, IV (ed. D. Duvall, D. Müller-Schwarze and R.M. Silverstein), pp. 519–31. Plenum, New York.

Sullivan, T.P., Crump, D.R. and Sullivan, D.S. (1988). Use of predator odors as repellents to reduce feeding damage by herbivores. III. Montane and meadow voles (*Microtus montanus* and *Microtus pennsylvanicus*). *J. Chem. Ecol.*, 14, 363–77

Tierney, A.J. and Atema, J. (1988). Amino acid chemoreception. Effects of pH on receptors and stimuli. *J. Chem. Ecol.*, 14, 135–141.

Turkowski, F.J., Popelka, M.L. and Bullard, R.W. (1983). Efficacy of odor lures and baits for coyotes. *Wildl. Soc. Bull.*, 11, 136–145.

van den Berk, J. and Müller-Schwarze, D. (1984). Responses of wild muskrats (*Ondatra zibethicus* L.) to scented traps. *J. Chem. Ecol.*, 10, 1411–5.

Vernet-Maury, E. (1980). Trimethyl-thiazoline in fox feces: a natural alarming substance for the rat. *Olfaction and Taste, VII*: 407.

Vernet-Maury, E., Polak, E.H. and Demal, A. (1984). Structure-activity relationship of stress-inducing odorants in the rat. *J. Chem. Ecol.*, 10, 1007–18.

Welch, B.L., McArthur, E.D. and Davis, J.N. (1983). Mule deer preference and monoterpenoids (essential oils). *J. Range Managm.*, 36, 485–7.

Welsh, R.G. and Müller-Schwarze, D. (1989). Experimental habitat scenting inhibits colonization by beaver, *Castor canadensis*. *J. Chem. Ecol.*, 15, 887–93.

White, S.M., Flinders, J.T. and Welch, B.L. (1982). Preference of pygmy rabbits (*Brachylagus idahoensis*) for various populations of big sagebrush (*Artemisia tridentata*). *J. Range Managm.*, 35, 724–6.

# VI.2 A Natural Deer Repellent: Chemistry and Behaviour

D.H. ABBOTT, D.A. BAINES, C.G. FAULKES, D.C. JENNENS, P.C.Y.K. NING AND A. J. TOMLINSON

## Introduction

The olfactory detection of predators by their prey is an important aspect of mammalian chemical communication (Sullivan and Crump 1986). Some species of ungulate respond with equal aversion to predator odours as to the actual sighting of predators (Schaller 1967). This effect of predator odour on ungulates has been employed to repel game from agricultural, horticultural, and forestry areas, but has met with limited success (Ives 1960; Van Haaften 1963).

The most effective preventative measures against deer damage in the UK presently involve high fencing of plantations or fitting protective collars to saplings. Both measures are expensive and time-consuming and the former can also result in injury to deer. A repellent based on a predator odour might avoid these disadvantages. Preliminary developments include anthraquinone and ziram repellents (Royal Society of Chemistry 1983) and faecal odours produced by decomposition of proteinaceous and fatty materials (Rochelle *et al.* 1974; Bullard *et al.* 1978; Harris *et al.* 1983).

Faeces from predator species suppressed feeding by European roe deer (*Capreolus capreolus*) and red deer (*Cervus elaphus*) (Van Haaften 1963). Müller-Schwarze (1972) demonstrated that faeces from predators such as the mountain lion (*Felis concolor*) and coyote (*Canis latrans*), which are sympatric with North American black-tailed deer (*Odocoileus hemionus columbianus*), resulted in consistent suppression of feeding by these deer. Similar results with black-tailed deer were obtained by Melchiors and Leslie (1985) and Sullivan *et al.* (1985). Interestingly, faeces from the African lion (*Panthero leo*) were equally as effective as the native predators in deterring North American black-tailed deer from feeding (Müller-Schwarze 1972). Furthermore, lion dung was one of the faecal odours used successfully to deter roe and red deer from feeding in farmland (Van Haaften 1963). The use of lion faeces *per se* as a universal deer repellent is impractical and uneconomical. Nevertheless, a long-acting synthetic replica of such predator faeces would be an extremely useful and viable deterrent.

Consequently, this paper will describe the first steps taken towards this goal of developing a viable deer repellent from lion dung. For the first time, analytical, chemical, and behavioural techniques were combined in a multi-disciplinary approach to: (i) describe the chemistry of compounds found in lion dung; (ii) show that the volatile compounds were an effective feeding deterrent for red deer in comparison to other sources of similar compounds; and (iii) briefly show that synthetic formulations based on the volatiles identified in lion dung, also acted as effective repellents. This project was not a comprehensive study of faecal odours as deer repellents and concentrated solely on applications related to lion dung. Details relating to the synthetic formulations and their development have been published elsewhere (Baines *et al.* 1988).

## Materials and methods

The lion faeces used in this study were freshly obtained from two adult male and two adult female African lions housed at London Zoo, Regent's Park, between October and November, 1985. The faeces were stored at $-20°C$ until required for analysis or behavioural studies. Each lion was fed with a 10 lb joint of beef for 5 days of each week. This was supplemented daily with a feline mineral powder preparation and once weekly with two herrings.

### ANALYTICAL METHODS

#### *Headspace extractions*

Lion faeces (27 g) were mixed with 70 ml distilled water and the volatiles extracted on to an activated carbon filter over a period of 1 h at 40°C using the Grob headspace stripping technique (Grob and Zurcher 1976). The carbon filter was washed with dichloromethane ($3 \times 20$ $\mu$l). Hexadecane was added at the rate of 0.05 mg/ml to the dichloromethane to act as an internal standard. Portions (1 $\mu$l) of this extract were analysed by flame ionization (FID-GC) and specific sulphur (SSD-GC) gas chromatography and by combined gas chromatography-mass spectrometry (GC-MS).

The GC-MS analyses were performed with the aid of a Hewlett Packard 5890 gas chromatograph, which was linked via a direct interface to a VG 70-250 SE mass spectrometer. The chromatography was performed on a DB-1 fused silica capillary column (60 m x 0.32 mm i.d., 1 $\mu$m film thickness, ex J and W Scientific) with helium employed as the carrier gas. The oven temperature programme was initially 40°C which was maintained for 5 mins then increased at 4°C/min. to a maximum temperature of 250°C. This final temperature was maintained for 30 mins.

#### *Soxhlet extraction*

A total extract was prepared by soxhlet extraction of 400 g of lion faeces with 500 ml 1,1,2-trichlorotrifluoroethane (freon; b.p. 45–47°C) over 4–5 h, as previously described by Teranishi *et al.* (1980). Of this extract 300 ml was concentrated to 10 ml using the Kadurna-Danish method (Wauchope 1975) and subjected to FID-GC, SSD-GC, and GC-MS employing gas chromatography phases of various polarities. The remaining 200 ml of the total extract was fractioned into acidic, basic, and neutral fractions as described by Teranishi *et al* (1980).

### DEVELOPMENT OF THE SYNTHETIC FORMULATION

Initial behavioural trials indicated the greatest repellency for lion dung and the total extract, with the acidic, basic, and neutral fractions having progressively less repellency (D.H. Abbott *et al.*, unpublished results). A six-day behavioural trial was then carried out to test the relative repellent properties of lion dung, the total extract, the acid fraction of the total extract, an artificial blend (ABI) based on the acid fraction, and dung from an Asiatic elephant (*Elaphus maximus*) as a herbivore control. The elephant dung was freshly collected from two young females housed at London Zoo, Regent's Park and was stored at $-20°C$ until required for behavioural trials.

Subsequently, two further synthetic formulations were tested versus controls and lion dung in a second 6-day behavioural trial.

### BEHAVIOURAL TRIALS

#### *Animals*

Two groups of five red deer hinds were maintained at the Institute of Zoology research facility at Whipsnade Zoo, Bedfordshire. Group A were all hand-reared at Whipsnade, while all of Group B were dam-reared at either the Hill Farming Research Organization, Edinburgh ($n=1$) or the Rowett Research Institute, Aberdeen ($n=4$). During the behavioural trials, group A were 1–2 years old and Group B were adults between 4 and 9 years old. All the deer were well accustomed to the presence and scent of humans. Group A was used in the first six-day trial and Group B was used in the second.

#### *Trial procedure*

All behavioural trials were conducted at the Institute of Zoology research facility at Whipsnade Zoo. The trials were held in an outdoor enclosure (8.7 x 13.0 m), cleared of vegetation, and containing a regular rectangular pattern (3.8 x 6.5 m) of six food buckets. Each bucket contained 2 kg of pelleted food (Dengie dried lucerne nuts). The animals were brought in from pasture 24 h before each trial and deprived of food until the trial, but water was always available *ad libitum*. Trials usually occurred at weekly intervals, lasted for 8 h, and were conducted between 08.00 and 19.00 h.

The deer were only allowed access to the testing enclosure at the start of each trial and were removed at the end of a trial. During the trials, each of the six food buckets was weighed hourly to establish the amount of food pellets consumed. An identical bucket of food pellets was kept outside the testing enclosure to permit adjustments to the bucket weights in respect of rainfall or evaporation of solvent in the scent treatment containers.

Two six-day behavioural trials were carried out from 30 April to 2 July 1987, and 4 March to 7 April 1988, respectively. During each of the six-day trials, treatment and control scents were randomly assigned to a bucket position for the first day of the trial and then rotated clockwise one position at the start of each of the next five trial days. The scent treatments were freshly prepared for each day. In the first six-day trial, scent treatments were placed inside a perforated plastic cylinder (180 $cm^3$) which was placed on top of the food in each bucket. Liquid treatments were absorbed on to a pad of cotton wool inside the cylinder. In the second six-day trial, the synthetic formulations were each sprayed on to the food pellets as a 50 ml aqueous solution and the pellets were then air-dried. The control food was not treated. Lion dung (100 g) was presented in an identical container to that of the first six-day trial and the other buckets contained empty containers as controls. Details relating to the quantified behavioural measurements made during both six-day trials will be published separately.

### *Statistical analysis*

The food intake data were subjected to an analysis of variance (ANOVA) based on a Latin square design. In the first six-day trial, comparisons of means were made *post hoc* using Duncan's multiple range test. In the second six-day trial, comparisons between individual means were made *post hoc* using either a two-sample or an approximate t-test, depending on the outcome of the F-statistic for comparing the equality of the variances of the means involved (SAS; Freund and Littell 1981).

## Results

### *Analysis of lion dung*

The mass spectrometric total ion trace of the total extract of lion dung and the analysis of the Grob headspace extract resulted in the identification of 125 compounds (Table 1). The acidic fraction provided the greatest contribution to the total extract in terms of numbers of compounds and its effect versus the human-perceived aroma of the total extract itself. This fraction contained acids ranging in chain length from acetic to docosanoic acid and included various branched and unsaturated derivatives. Steroids (mainly cholesterol) and nitrogen -containing compounds (mainly indole, 2-piperidone, oxindole, and simple amines) formed the next largest group of contributory chemicals. Five sulphur compounds were identified as follows: dimethyl disulphide, dimethyl trisulphide, dimethyl tetrasulphide, 3-mercaptopropan-1-ol and 3-mercaptothiopropan-1-ol.

**Table 1.** Peak identifications of components detected in lion faeces

| Scan number | Component identity | Scan number | Component identity |
|---|---|---|---|
| 31 | Solvent impurity | 873 | Nonanoic acid |
| 34 | 2-Methyl-3-buten-2-ol | 880 | 2,2,2-Trifluoro-*N*- |
| 36 | Unknown ($M^+88$) | | 2(phenyl ethyl) |
| 38 | Solvent impurity | | acetamide? |
| 40 | Dimethyl disulphide | 898 | Tetradecanol |
| 47 | Butanoic acid propyl | 925 | 3-Ethyl-4-methyl |
| | ester | | pyrrole-2,5-dione |
| 48 | Ethyl benzene | 939 | Unknown |
| 61 | 3-Methyl butanol | 954 | Decanoic acid |
| 74 | n-Pentanol | 978 | Pentadecanol |
| 128 | n-Hexanol | 1007 | Laurolactone |
| 143 | Dimethyl trisulphide | 1019 | Dihydro-5-(2-octenyl) |
| 209 | Acetic acid | | furan-2-one |
| 264 | 2-Ethyl hexan-1-ol | 1032 | Undecanoic acid |
| 279 | 3-Mercaptopropan-1-ol | 1042 | Indole |
| 297 | Propanoic acid | 1055 | Hexadecanol |
| 329 | 2-Methyl propanoic acid | 1109 | Dodecanoic acid |
| 288 | Butanoic acid | 1136 | Phenyl acetic acid |
| 430 | 3-Methyl butanoic acid | 1141 | *N*-(2-phenyl ethyl) |
| 470 | 3-(Methylthio)propan- | | acetamide |
| | 1-ol | 1149 | *N*-(2-phenyl propyl) |
| 476 | Dimethyl tetrasulphide | | acetamide |
| 500 | Pentanoic acid | 1175 | Phenyl propanoic acid |
| 518 | Unknown | 1219 | Phthalate |
| 529 | 2-Methyl pentanoic acid | 1231 | Tetradecanoic acid |
| 561 | 4-Methyl pentanoic acid | | (branched chain?) |
| 600 | Hexanoic acid | 1256 | Tetradecanoic acid |
| 614 | Benzyl alcohol | | (straight chain?) |
| 628 | Phenyl propanoic acid | 1289 | Pentadecanoic acid |
| | ethyl ester | | (branched chain?) |
| 629 | *N*-Butyl-*N*-ethyl | 1297 | Oxindole |
| | acetamide | 1321 | Pentadecanoic acid |
| 623 | 2-Methyl hexanoic acid | 1355 | Unknown ($M^+238$) |
| 644 | Phenyl ethanol | 1374 | Hexadecanoic acid |
| 673 | BHT | | (branched chain?) |
| 696 | Heptanoic acid | 1395 | Palmitic acid |
| 704 | Phenyl propanoic acid | 1416 | Hexadecanoic acid |
| | propyl ester | 1423 | Heptadecanoic acid |
| 724 | Phenol | | (branched chain?) |
| 759 | Phenyl propanol | 1445 | Heptadecanoic acid |
| 793 | 4-Methyl phenol | 1452 | Heptadecanoic acid |
| 794 | Octanoic acid | 1511 | Octadecanoic acid |
| 803 | 2-Piperidone | | (branched chain) |
| 831 | 7-Octenoic acid | 1529 | Stearic acid |
| 875 | Imidazole | 1534 | Oleic acid |

**Table 1.** cont.

| Scan number | Component identity | Scan number | Component identity |
|---|---|---|---|
| 1563 | Linoleic acid | 1774 | Docosanoic acid |
| 1569 | Octadecadienoic acid | 1810 | Docosanoic acid |
| 1596 | Octadecatrienoic acid | 1910 | Hydroxynonadecanoic acid |
| 1603 | Octadecadienoic acid | | |
| 1632 | Eicosanoic acid | 1935 | Unknown |
| 1637 | Cholestra-3,5-diene | 2048 | 3-Alpha cholestan-3-ol |
| 1644 | Eisosenoic acid | 2187 | Cholestan-3-ol |
| 1652 | Eicosadienoic acid | 2280 | Cholesterol |
| 1667 | Eicosadienoic acid | 2290 | Vitamin E? (Tocopherol) |
| 1677 | Eicosatrienoic acid | 2432 | Cholest-5-ene-3-one |
| 1685 | Eicosatrienoic acid | 2453 | Unknown steroid |
| 1726 | Unknown | 2570 | Lanosterol |
| 1749 | Eicosapentenoic acid | | |

*The effectiveness of volatile components in lion dung as a deer repellent*

During the first six-day trial, there were significant differences between the amounts of food consumed after 8 h by the five red deer hinds, depending on treatment (Fig. 1a; $F(5,20) = 13.42, p < 0.001$). Whilst buckets containing the odours of

lion dung or the total extract from lion dung (total extract) were virtually untouched, all the other buckets contained significantly less food at the end of the trial. The control bucket had the least food remaining, while buckets containing the odours of the acid fraction of the total extract, elephant dung, and the synthetic blend of the acid fraction had statistically similar and intermediate amounts. Lion dung and the total extract were considered as the only effective repellents.

To check whether or not the repellency found in the first six-day trial was due to a non)specific, unpleasant, odour effect a variety of animal and unpleasant odours were tested. All produced poor repellency in comparison to lion dung and the total extract (results not shown). The odours included cattle dung, 2,2-dimethylthietane (present in the anal sac secretions of the stoat; Crump 1980) and the five sulphur-containing compounds identified in lion dung.

*Effect of synthetic formulations as deer repellents*

During the second six-day trial, there was again a significantly different amount of food consumed from the scented buckets after 8 h (Fig. 1b; $F(5,20) = 23.10, p < 0.001$). The two buckets containing the odours of each of the two synthetic formulations (composition not shown for commercial reasons) were untouched. A small, but insignificant amount of food was consumed from the bucket with the lion dung odour and the food in all three control buckets was almost completely consumed. Thus, the synthetic formulations were considered as effective as lion dung itself.

In all the behavioural trials it was obvious from both quantitative behavioural

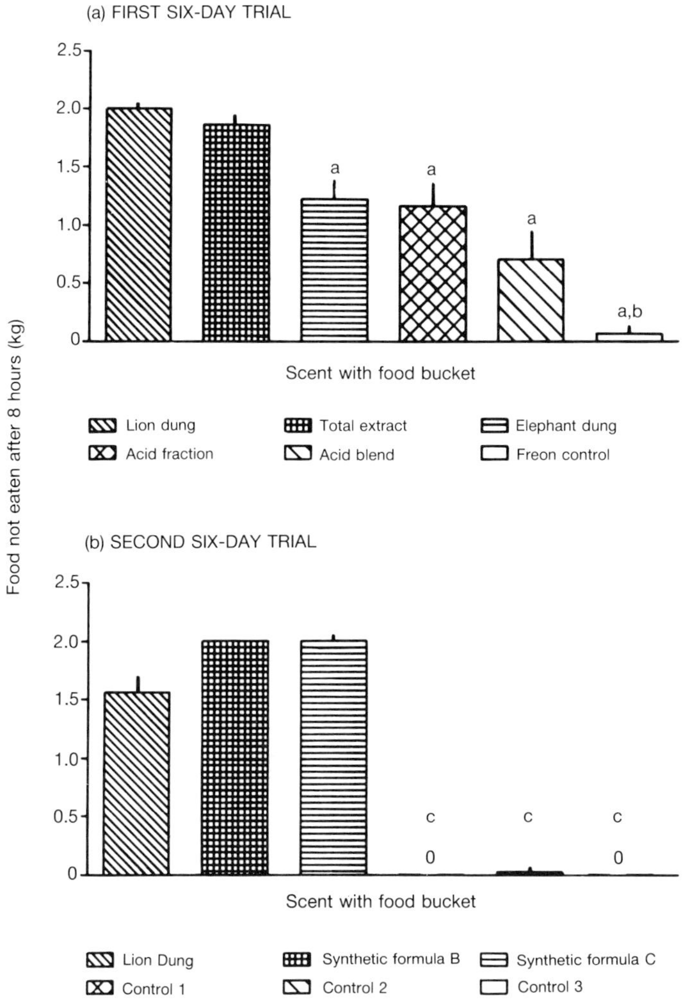


Fig. 1. Food remaining in scented buckets (mean ± SEM) after six 8-h test periods for (a) the first 6-day trial and (b) the second 6-day trial. All buckets contained 2 kg of food at 0 h.
Acid fraction = acid fraction of the total extract of lion dung.
Acid blend = synthetic blend of the acid fraction.
Freon control = equivalent volume of freon used in 'total extract' and 'acid fraction'.
Controls 1–3 = untreated buckets.
a: $p < 0.05$ vs. lion dung and total extract;
b: $p < 0.05$ vs. all others. (Duncan's multiple range test following ANOVA).
c: $p < 0.05$ vs. lion dung and synthetic formulations b and c (Two-sample or approximate t-test following ANOVA).
o: Food always totally consumed.

measurements (results not shown) and measurement of the food consumed, that a repellent in one bucket did not influence the deers' responses to neighbouring buckets. This is best illustrated in Fig. 1b where there was no influence on the food consumed in control buckets from surrounding buckets containing repellents. Repellency, in these trials, was apparently only effective over a short distance.

## Discussion

This paper describes for the first time a multi-disciplinary approach, combining chemical and behavioural techniques, to develop a viable, odour-based deer repellent from lion dung. A comprehensive chemical analysis of lion dung was produced for the first time. Many of the compounds identified were similar to those identified in the anal sac secretions of the lion (Albone 1984; Albone and Gronneberg 1977) and of other carnivores (Albone and Perry 1975).

The deer repellent properties of lion dung previously demonstrated by Van Haaften (1963) and Müller-Schwarze (1972) were also confirmed. In addition, solvent extracts of lion dung and synthetic formulations based on lion dung were found to be as effective as lion dung in repelling deer. The precise odour cues that the deer responded to have not yet been completely determined. Certainly, vapour pressure has been shown to be an important factor in deterring deer from feeding when monoterpene odours were involved (Elliott and Loudon 1987). However, the vapour pressures of tested odours were not measured in the present work.

The repellency of lion dung may be due to a general carnivore odour signal produced by the products of meat and animal fat digestion, and their oxidative and microbiological breakdown. This hypothesis would be in keeping with other observations of predator odour repellency (e.g. Müller-Schwarze 1972) and may help to explain why European red deer react to the smell of the African lion. They may not be recognizing a lion odour *per se*, but rather, a common carnivore signal. If this is not the case, then these deer must have an inherent memory of the smell of lion as predator, which has been selectively retained through many generations. This may not be such an unlikely suggestion because lions were widespread in Europe at least 15,000 years ago and were still found in Greece as recently as 200–300 BC (Day 1981; Neff 1986; Van Orsdol 1984). As Europe was an area of dense forest prior to the Roman expansion, red deer and fallow deer (*Dama dama*) would have been the prime prey for larger species of lions. Certainly, there is increasing evidence that specific predator–prey olfactory communication may be well established between herbivorous small mammals and their major predators (Sullivan and Crump 1986).

It was interesting that faeces from a herbivore, in this study a pair of Asian elephants, also showed significant properties as a deer repellent during the behavioural trials. While the repellency was not nearly as marked as that produced by lion dung, it may explain why dung from another herbivore, the domestic cow, is used as a component in various 'home-made' deer repellents (Ueckermann 1969, cited by Novellie *et al* 1982). The reasons for these apparently non-specific effects are not known, but dung from herbivores may contain some of the same volatile chemical components as lion dung. Certainly, the high degree of repellency produced by lion lung in this study did not appear to be due to an unpleasant odour effect because a variety of unpleasant odours were tested, including 2,2-dimethylthietane (a sulphur-containing compound present in the anal sac secretions of the stoat; Crump 1980), and all displayed poor repellency (D.H. Abbott *et al.* unpublished results). However, deer may simply avoid feeding near dung deposits; horses and cattle have indeed been shown to avoid grazing near dung from their own species (Odberg and Francis-Smith 1977). The latter behaviour might aid avoidance of parasitic re-infestation (Taylor 1954). Nevertheless, since cattle did not avoid grazing beside horse and sheep dung (Odberg and Francis-Smith 1977), it is difficult to simply attribute the limited deer repellency of elephant dung to a general avoidance of feeding close to dung.

There are many questions of a chemical nature relating to the repellency of lion dung which still remain unanswered. The potential usefulness of such a naturally occurring odour as a repellent in protecting forestry, agriculture, and horticulture from extensive deer damage should provide the impetus to generate such work. If long-lasting repellency of a synthetic formulation of lion dung is proven in field trials, this may revolutionize the management and conservation of forests world-wide. It would also fulfil Van Haaften's prediction of 26 years ago that a synthetic 'dispensive substance' based on the chemical analysis of lion dung would provide 'an excellent repellent'.

## Acknowledgements

We thank Professors A.P.F. Flint and N.W.R Daniels and Dr P.M. Summers for criticism of this manuscript; A. Beasey and B. Eves for valuable technical assistance and maintenance of the animals; H. Pepper of the Forestry Commission, Dr A.S.I. Loudon and D.M. Richardson for their help and advice during the study; E. Swain, D.M. Richardson and the keepers on the Lion Terraces of London Zoo for providing the lion dung for analysis and use in the behavioural trials, and the keepers in the Elephant House of London Zoo for providing the elephant dung; G. Norman and Dr J. Barrett for assistance during the behavioural trials; N. Sutcliffe for technical assistance in preparing the synthetic formulations; and T. Grose and A. King for typing the manuscript.

## References

Albone, E.S. (1984). *Mammalian semiochemistry*. Wiley, Chichester.

Albone, E.S. and Gronneberg, T.O. (1977). Lipids of the anal sac secretions of the red fox, *Vulpes vulpes*, and of the lion, *Panthero leo*. *J. Lipid Res.*, *18*, 474–9.

Albone, E.S. and Perry, G.C. (1975). Anal sac secretion of the red fox, *Vulpes vulpes*; volatile fatty acids and diamines: implications for a fermentation hypothesis of chemical recognition. *J. Chem. Ecol.*, 2, 101–11.

Baines, D.A., Faulkes, C.G., Tomlinson, A.J. and Ning, P.C.Y.K. (1988). Improvements in or relating to repellent compositions. *European patent publication*, No. 0 280 443. European Patent Office.

Bullard, R.W., Shumake, S.A., Campbell, D.L. and Turkowski, F.J. (1978). Preparation and evaluation of a synthetic fermented egg coyote attractant and deer repellent. *J. Agric. Food Chem.*, 26, 160–3.

Crump, D.R. (1980). Thietanes and dithiolanes from the anal gland of the stoat, *Mustela erminea*. *J. Chem. Ecol.*, 6, 341–7.

Day, D. (1981). *The Doomsday book of animals*. Ebury Press, London.

Elliott, S. and Loudon, A.S.I. (1987). Effects of monoterpene odours on food selection by red deer calves (*Cervus elaphus*). *J. Chem. Ecol.*, 13, 1343–9.

Freund, R.J. and Littell, R.C. (1981). *SAS for linear models. a guide to the ANOVA and GLM procedures*. SAS Institute, Cary, North Carolina.

Grob, K. and Zurcher, F. (1976). Stripping of trace organic substances from water. Equipment and procedure. *J. Chromatog.*, 117, 285–94.

Harris, M.T., Palmer, W.T. and George, J.L. (1983). Preliminary screening of white-tailed deer repellents. *J. Wildl. Managm.*, 47, 516–9.

Ives, F.F. (1960). Deer repellents. *Annu. Rep. Oreg. State Hort. Soc.*, 52–9.

Melchiors, M.A. and Leslie, C.A. (1985). Effectiveness of predator faecal odours as black-tailed deer repellents. *J. Wildl. Managm.*, 49, 358–62.

Müller–Schwarze, D. (1972). Responses of young black-tailed deer to predator odours. *J. Mammal.*, 53, 393–4.

Neff, N.A. (1986). *The big cats: the paintings of Guy Coheliach*. Harry N. Abrams, New York.

Novellie, P., Bigalke, R.C. and Pepler, D. (1982). Can predator urine be used as a buck or rodent repellent? *S. Afr. For. J.*, 123, 51–5.

Odberg, F.O. and Francis-Smith, K. (1977). Studies of the formation of ungrazed eliminative areas in fields used by horses. *Appl. Anim. Ethol.*, 3, 27–34.

Rochelle, J.A., Gauditz, I., Oita, K. and Oh, J.H.K. (1974). New developments in big-game repellents. In *Wildlife and forest management in the Pacific Northwest*, (ed. H.C. Black), pp. 103-12. Oregon State University, Corvallis, Ore.

Royal Society of Chemistry (1983). The Agrochemical Handbook. Unwin Bros.

Schaller, G.B. (1967). *The deer and the tiger*. University of Chicago Press, Chicago, Illinois.

Sullivan, T.P. and Crump, D.R. (1986) Avoidance response of pocket gophers (*Thomomys talapoides*) to mustelid anal gland compounds. In *Chemical signals in vertebrates*, 4 (ed. D. Duvall, D. Müller-Schwarze and R.M. Silverstein), pp. 514–31. Plenum, New York.

Sullivan, T.P., Nordstrom, L.O. and Sullivan, D.S. (1985). Use of predator odours as repellents to reduce feeding damage by herbivores. II. Black-tailed deer (*Odocoileus hemionuss Columbianus*). *J. Chem. Ecol.*, 11, 921–35.

Taylor, E.L. (1954). Grazing behaviour and helmintic disease. *Br. J. Anim. Behav.*, 2, 61–2.

Teranishi, R., Murphy, E.L., Stern, D.J., Howard, W.E. and Fagre, D.F. (1980). Chemicals useful as attractants and repellents for coyotes. In *Worldwide Furbearers' Conference proceedings*, Vol 3 (ed. J.A. Chapman and D. Pursley), pp. 1839–51.

Van Haaften, J.L. (1963). A natural repellent. In *International Union of Game Biologists, transactions VI Congress*, 7–12 Oct., pp. 389–92. The Nature Conservancy, London.

Van Orsdol, K.G. (1984). Lion. In *The encyclopedia of mammals: 1* (ed. D.W. MacDonald). pp. 28–33. Allen and Unwin, London.

Wauchope, R.D. (1975). Solvent recovery and reuse with the Kuderna-Danish evaporator. *Anal. Chem.*, 47, 1879.

# VI.3 Chemical Signals in Domestic Ungulates

J.P. SIGNORET

## Introduction

In domestic ungulates, the frequency of mutual sniffing and existence of the specialized display or 'flehmen' emphasizes the importance of chemical communication. Anatomically, the size of the olfactory bulbs and the development of the vomeronasal system are similar to macrosmatic rodents or carnivores. Chemical communication appears to play important roles in reproductive and social behaviours, as reviewed in this report.

## Chemical signals and sexual behaviour

### CHEMICAL SIGNALS INVOLVED IN THE ORGANIZATION OF SEXUAL BEHAVIOUR: 'SIGNALLING PHEROMONES'

*Signalling pheromones from the male acting on the receptive female*

In all mammalian species, the female contribution to the organization of sexual behaviour is twofold: first, she actively searches for the male (proceptivity), and, when in contact, displays the posture of receptivity that allows copulation to take place.

Sex-specific odours have been identified in the boar (16-androstenes; Patterson 1968), and the male goat (4-ethyloctanoic acid; Sugiyama *et al*. 1981). Sex-specific scent glands comprise head and skin glands of the male goat, preputial pouch and submaxillary gland of the boar, and anteorbital gland of the ram. In some cases, discrete behavioural patterns achieve chemosensory communication, such as self-enurination of the male goat (Price *et al.* 1986), and urine and faeces marking by the stallion.

*Chemical communication and proceptive behaviour of the female.* In the ruminants, the social organization of matriarchal female units and male bachelor groups is disrupted at the time of reproduction. In feral Soay sheep, the rams move to special rutting areas where they are visited by females (Jewell and Grubb 1974). A similar organization has been described in domestic sheep and cattle: the receptive females move to the male, gathering around him in temporary harems or sexually active groups. In large paddocks, ewes in oestrus approach and mate with tethered rams as efficiently as when the males are moving freely to the females (Inkster 1957; Lindsay and Robinson 1961; Lindsay and Fletcher 1972). In the pig, the proportion of females taking the initiative in olfactory contact with a male is elevated during oestrus.

The proceptivity of the sow can be measured in preference tests using a T maze. Neither visual nor auditory cues from the boar are necessary for attracting the oestrous female (Signoret 1974). In the absence of visual cues, she exhibits a clear preference for normal over sialectomized males. This operation—the surgical removal of the submaxillary gland—prevents the release of frothy saliva containing androstenes and reduces the attractive value of the boar, suggesting a role of the salivary androstenes as attracting pheromone (Pearce and Hughes 1987) (Fig. 1a). The attraction of the oestrous sow by the boar is not influenced by experience, and operates in females isolated from birth from any contact with males (Signoret 1970).

*Chemical communication and the release of the copulatory posture of the female.* The copulatory posture of the sow—immobilization—can be released by tactile stimulation of the female's back. The presence of olfactory stimuli from the male (Signoret *et al.* 1961), or of androstenes (Melrose *et al.* 1971; Signoret 1974) increases the reaction rate. However, as for attraction by the male, visual and auditory cues have a cumulative effect on the female's response (Fig. 1b).

*Signalling pheromones from the female acting on the male*

*Attracting pheromones* do not appear to be necessary for the detection of oestrous females by the male. In pens, male pigs presented to a group of females approach them with an equal frequency whether or not they are in oestrus (Signoret 1972). When freely interacting in paddocks, control and anosmic rams are equally successful in detecting and mating with oestrous ewes. However, intact males preferentially approach oestrous ewes whereas anosmic ones do not discriminate (Lindsay 1965).

In a T maze, boars spend more time near oestrous rather than anoestrous sow (Signoret 1972; McGlone and Morrow 1987).

The existence of a *pheromone eliciting mounting by the male*, when in contact with an oestrous female, is controversial: several authors report no difference in the sexual reactions of males presented to immobilized oestrous versus anoestrous females in cattle (Chenoweth *et al.* 1979; Blockey 1981), and in sheep (Signoret 1975; Fig. 2b). Cervico-vaginal mucus of oestrous cows has been reported to stimulate mounting in the bull (Hart *et al.* 1946; Sambraus and Waring 1975; Paleologou 1977; Garcia *et al.* 1986), whereas several other authors have not oberved any effect (Hale 1966).

An increase in olfactory investigation of the genital area (Perry 1985) and of flehmen (Hradecky *et al.* 1983) occurs before oestrus in cows. However, a detailed

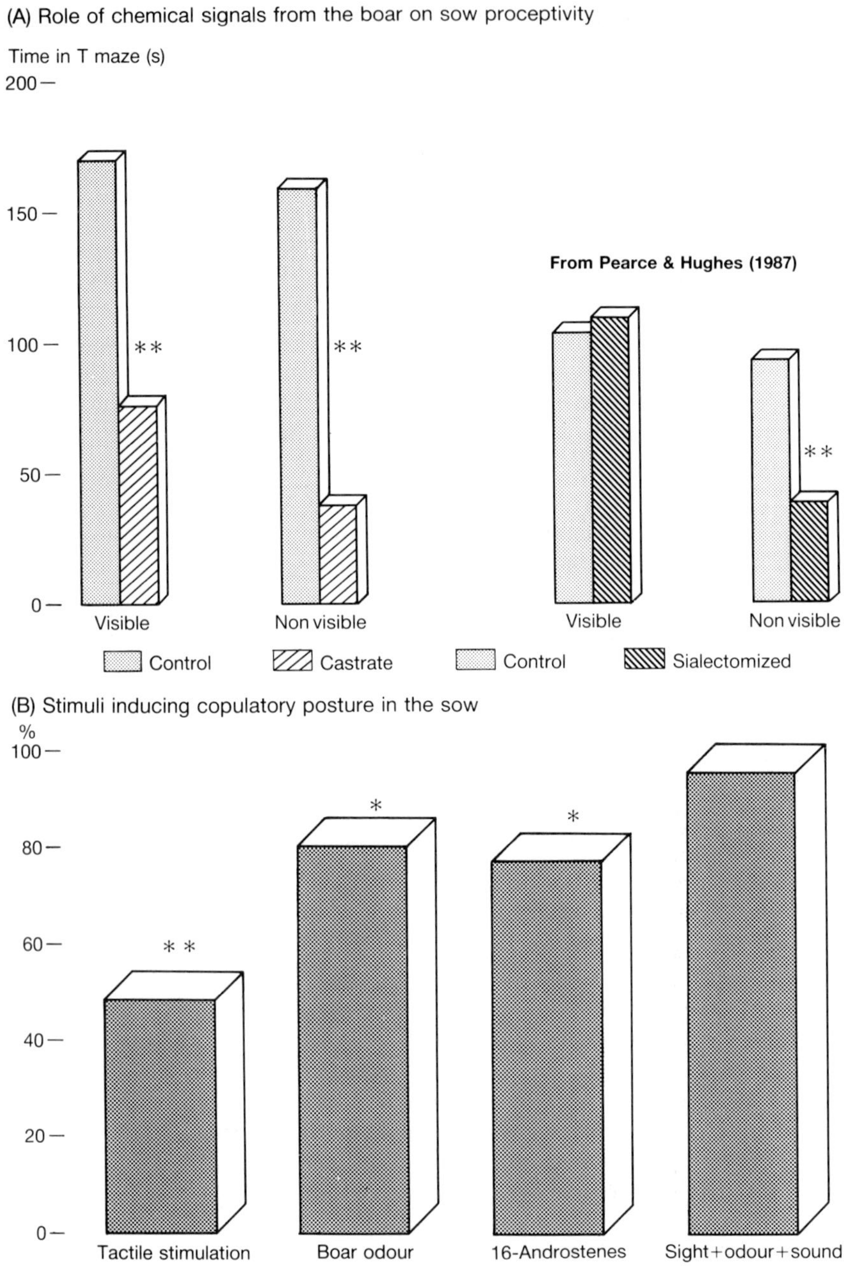


Fig. 1. Chemical signals from the male pig attract the oestrous female (a) and induce the copulatory posture (b) (from Signoret 1972). In both cases, the 16-androstenes appear to be the major pheromones involved (* $p<0.05$; * $p<0.01$).

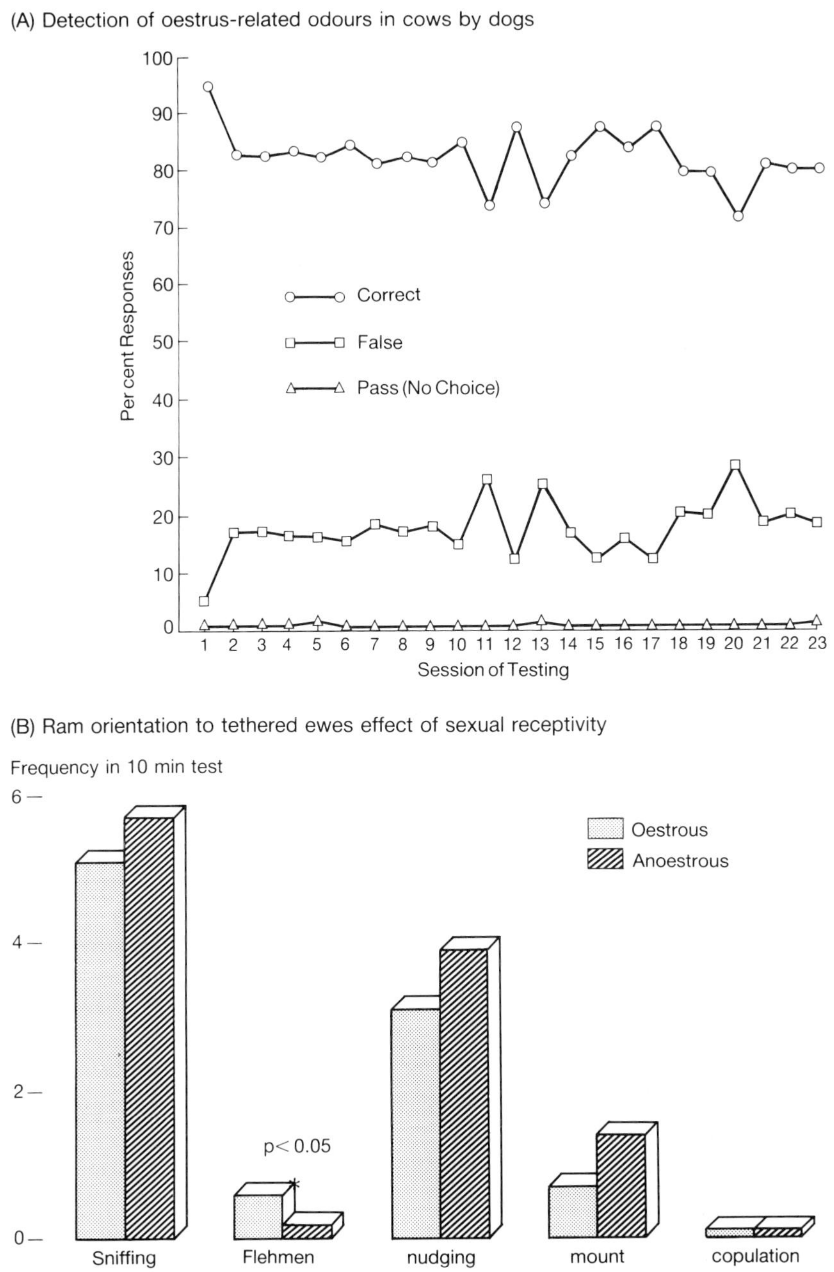


Fig. 2. The oestrous female cow produces pheromones that could be detected by dogs (a) (from Kiddy *et al.* 1970). However, when a ewe is immobilized—the copulatory posture—the ram does not discriminate (b) (from Signoret 1975) (* $p<0.05$).

analysis in goats shows that the flehmen reaction occurs more frequently in response to urine from dioestrous than oestrous females (Ladewig *et al.* 1980).

The bull (Klemm *et al.* 1987), but not the stallion (Marinier *et al.* 1988), reacts differentially to urine and vaginal secretion of oestrous versus anoestrous female.

Vaginal fluids, urine, milk and blood plasma from oestrous versus anoestrous cows are distinguished by dogs (Fig. 2a; Kiddy *et al.* 1978; Kiddy and Mitchell 1981) and rats (Ladewig and Hart 1981; Dehnhard and Claus 1988). A first attempt to identify the pheromone has been carried out in cattle (Klemm *et al.* 1987).

## CHEMICAL SIGNALS REGULATING THE PHYSIOLOGICAL PROCESSES OF REPRODUCTION: 'PRIMER PHEROMONES'

### *Pheromonal acceleration of puberty*

The onset of puberty in gilts is synchronized after transport and presentation to the boar (du Mesnil du Buisson and Signoret 1962) and occurs earlier when the females are kept in the presence of the male (Brooks and Cole 1970; Hughes 1982). This effect is eliminated by removal of the olfactory bulbs (Kirkwood *et al.* 1981). There is an indication that 16-androstenes have a primer effect on puberty attainment, but they probably act in association with other non-olfactory stimulation from the boar (Kirkwood *et al.* 1983; Booth 1984).

In cattle, weekly spraying with bull urine accelerates puberty (Izard and Vandenbergh 1982), supporting the hypothesis of a primer pheromone effect. Reaction to stimulation by the male appears to depend on the heifers' body weight and the level of nutrition (M. Roberson, personal communication).

### *Pheromonal induction of ovulation*

Domestic sheep and goats are seasonal breeders, mating taking place in summer/autumn. At the end of seasonal anoestrus (spring), mating occurs in a highly specific pattern involving a double peak on days 17–19 and 24–26 after introduction of the rams (Eyal 1958; Prud'hon and Denoy 1969). Watson and Radford (1960) have shown that contact and visual isolation do not prevent this effect which appears to be mediated by the sense of smell. A similar phenomenon has been described in goats (Chemineau 1983; Over 1988), and anosmia prevents such a response (Chemineau *et al.* 1986).

Despite the long delay in the occurrence of overt oestrus, the physiological response of anoestrous ewes to the male's presence takes place rapidly: most of the females present a silent ovulation within 48 h. This is the consequence of a change in secretion of luteinizing hormone (LH) evident within 10 mins of presentation of the ram (Martin *et al.* 1980; Poindron *et al.* 1980).

The existence and origin of the pheromonal effect have been studied. The odour of the rams' wool induces LH release (Signoret and Lindsay 1982) and

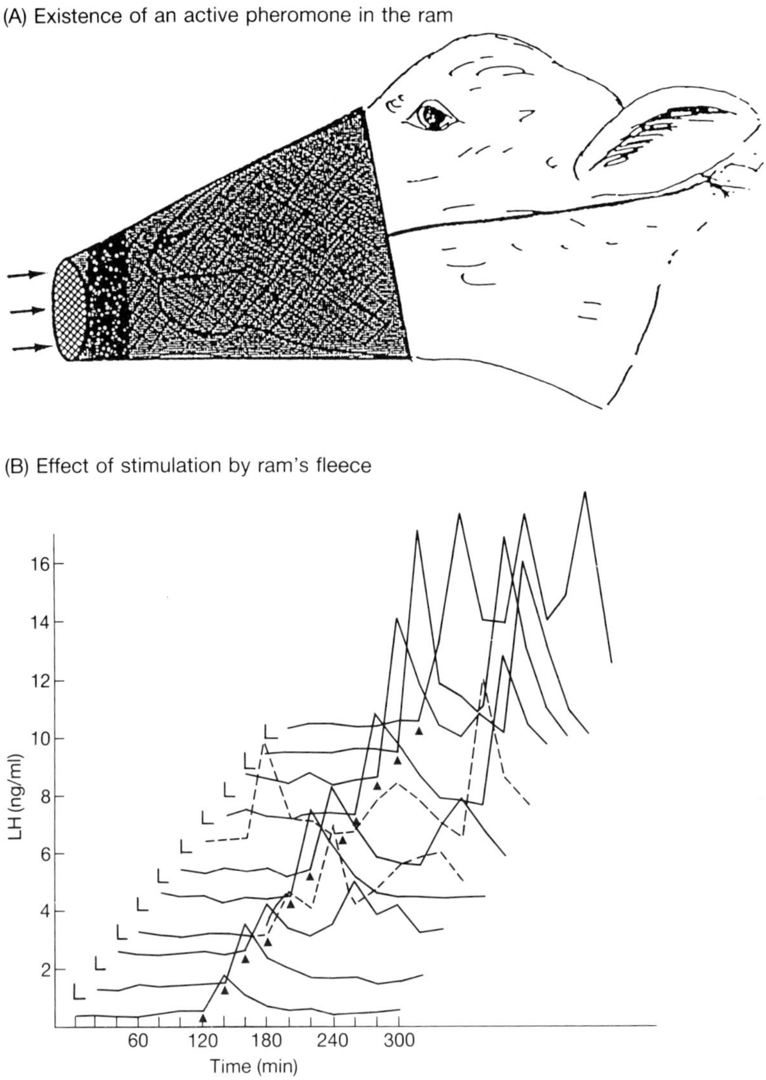


Fig. 3. Primer pheromone in rams' fleece induces LH release in the anoestrous ewe (from Cohen-Tannoudji 1988). The circle indicates the presentation of the odour.

ovulation (Knight and Lynch 1980). The secretion of the anteorbital glands has a similar effect, unlike either the urine or the secretion of the inguinal glands of the male (Cohen-Tannoudji 1988). The extract of males' fleece stimulates LH secretion in female sheep (Cohen-Tannoudji 1988) (Fig. 3) and goats (Over 1988). A cross-species stimulation is possible: the odour of the male goat induces ovulation (Knight and Birch 1983) and LH secretion (Cohen-Tannoudji 1988) in ewes.

The pheromone appears to be a mixture of several components. In sheep, the diols 1,2-C16 and -C18 are involved, but they have to be associated with unidentified compound(s). The chemically identified male pheromones, 4-ethyloctanoic acid (goat) and the 16-androstenes (boar), are uneffective in both goats and ewes (Cohen-Tannoudji 1988; Over 1988).

In contrast to rodents, pheromonal stimulation of the ewe is not necessarily mediated by the vomeronasal system: destruction of the Jacobson's organ or sectioning of the vomeronasal nerves do not suppress the LH secretion induced by rams' fleece extract (Cohen-Tannoudji 1988). Furthermore, chemical signals alone do not solely account for the effect of rams on ewes: surgical removal of both main and accessory olfactory bulbs does not eliminate the secretion of LH induced by the presence of the ram (Cohen-Tannoudji *et al.* 1986).

The effect of the male's presence in inducing ovulation has been cited during post-partum anoestrus in cows (Zalesky *et al.* 1984) and in sows (Rowlinson and Bryant 1974). During lactation anoestrus of the ewe, the ram changes the rhythm of LH secretion (Poindron *et al.* 1980).

After puberty, female pigs undergo regular oestrous cycles throughout the year. Olfactory bulbectomy disturbs this cyclicity (Signoret and Mauléon 1962), which becomes seasonal (Booth and Baldwin 1986). Unidentified olfactory inputs appear to maintain the regular oestrous cyclicity in this species.

*Stimulation of the endocrine secretions of the male by female pheromones*

The female stimulation of the male's endocrine secretions has long been controversial. The permanence of the process of spermatogenesis and the high frequency of LH and androgen secretion make this effect difficult to establish. However, sexual stimulation has been shown to induce LH and androgen secretion in the bull (Katongole *et al.* 1971), the boar (Liptrap and Raeside 1978), and the ram (Sanford *et al.* 1974; Illius *et al.* 1976; Moore *et al.* 1978). In fact, in the ram such an effect is more obvious during the period corresponding to seasonal anoestrus in the female (Yarney and Sanford 1983; Schanbacher *et al.* 1987); however, its existence has been established throughout the year (Gonzalez *et al.* 1988*a, b*). Direct contact of the male with an oestrous female is necessary for the stimulation of LH and androgen secretion to take place (R. Gonzalez, personal communication), but there is no direct relationship between the sexual behaviour and the response (Gonzalez *et al.* 1988*b*). In the bull, nasogenital contact with an oestrous heifer induces LH and testosterone secretion. However, urine from oestrous and anoestrous cows induces flehmen reaction without influencing endocrine secretions (Godfrey *et al.* 1988).

## Chemical signals and maternal behaviour

Chemical signals are involved in each step of the mother–young relations, neonatal care, and development of the mutual recognition of the dam and the young. Most experimental evidence has been obtained from sheep.

In ruminants, the most striking feature of parturient female behaviour is the strong attraction to any object coated with foetal fluids: not only the newborn, but the placenta as well as straw or any other object (Collias 1956). Such attraction is due to the odour of the amniotic fluid, but is limited to a few hours around parturition and turns to repulsion at any other time (Levy *et al.* 1983). It appears to be associated with the occurrence of uterine contractions and the resulting sensory inputs from the genital tract.

The immediate attraction to the amniotic fluids ensures the first contact and licking of the neonate by the dam. Elimination of amniotic fluids in washing of the newborn lamb disturbs the normal development of maternal behaviour in experienced ewes, and totally disorganizes it in primiparous ones (Fig. 4; Levy and Poindron 1987).

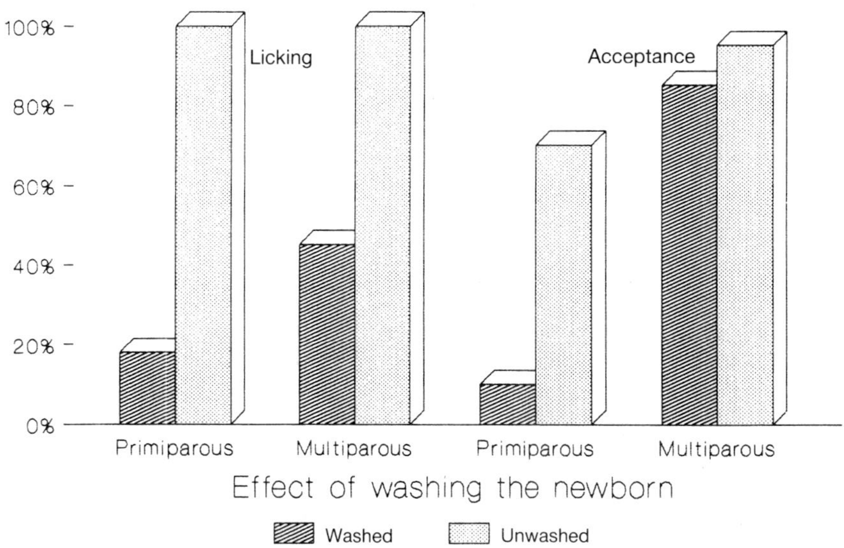


Fig. 4. The odour of amniotic fluid attracts the parturient ewe to the newborn lamb, induces licking, and facilitates development of the maternal bond, especially in primipara. The reactions of the ewe are disorganized when the lamb has been washed (* $p<0.05$; ** $p<0.01$) (from Levy and Poindron 1987).

A similar attraction for amniotic fluids is observed in cattle and goats. In contrast, the reactions of sows are entirely different as the parturient sow appears disinterested in the placenta, amniotic fluids, and even the neonates.

*The maternal bond* takes place rapidly: immediately after birth, the parturient ewe is unable to discriminate her lamb from an alien newborn, whereas 2 to 3 h later she actively refuses any other lamb. The development of this exclusive bond relies on the individual odour of the lamb and requires close proximity (Fig. 5; Poindron and Leneindre 1980). Deprivation of the sense of smell in the dam prevents the development of the bond without disturbing suckling behaviour (Bouissou 1968; Baldwin and Shillito 1974; Poindron 1976).

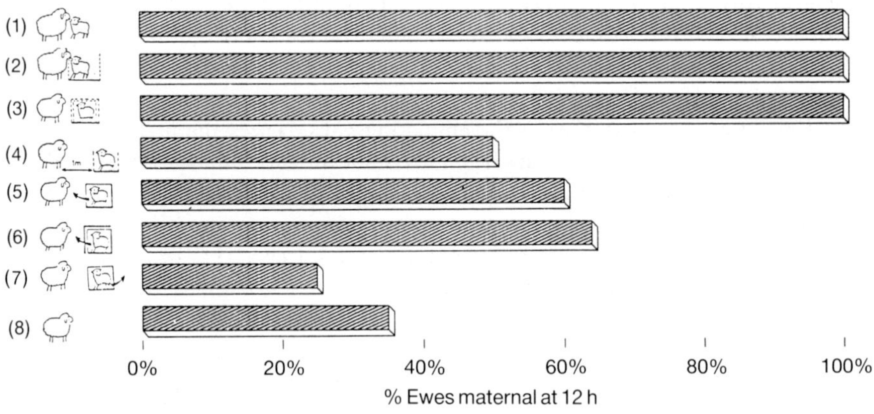


Fig. 5. Development of the maternal bond requires close proximity and especially the possibility of olfactory contact. The treatments during the first 12 h post-partum were:

1. Full contact;
2. contact and licking possible, suckling prevented;
3. lamb placed in a wiremesh cage, and could be seen and smelled, but no direct contact possible;
4. lamb maintained at the distance of 1 m;
5. lamb placed in a plexiglas cage: it could be seen, and its odour is sent by ventilation in the direction of the ewe;
6. same as 5, but the cage has opaque walls;
7. same as 5, but the ventilation is directed away from the ewe;
8. the ewe is isolated from the lamb.

(* $p<0.05$) (from Poindron and Leneindre 1980).

This aspect of chemical communication has been used for practical fostering in sheep and cattle: a cloth placed on the newborn and transferred to the alien young (Alexander and Stevens 1985*a*; Dunn *et al.* 1987), or an unguent odorant (Leneindre *et al.* 1978; Alexander and Stevens 1985*b*), facilitate fostering.

*Beyond the post-partum period*, sniffing by the ewe appears to be critical for individual recognition of the lamb (Poindron 1974). Post-partum anosmia eliminates individual recognition, and anosmic female sheep and goats suckle any young (Bouissou 1968; Klopfer and Gamble 1966; Poindron 1976). However, mutual recognition at a distance involves auditory and visual cues (Tschanz 1962, Alexander and Shillito 1977*a*, *b*; Poindron and Carrick 1976).

### CHEMICAL SIGNALS AND BEHAVIOUR OF THE YOUNG

The newborn lamb succeeds in suckling within 30 mins of birth. Odour, together with tactile and thermal stimulation, accounts for orientation to the udder (Vince *et al.* 1987). The wax produced by the inguinal gland of the dam not only orientates the lamb to the udder but even allows individual recognition of the dam (Vince and Ward 1984).

As early as 12 h after birth, lambs discriminate their dam from other ewes, but only at close quarters (Nowak *et al.* 1987). During a sucking attempt, the lamb first moves to the dam's head, which facilitates sniffing by the ewe (Poindron 1974). Such an organization of the behavioural sequence is disturbed when the ewe is anosmic: the frequency of the lambs approaching the head of the dam is then considerably reduced. This suggests that the lamb learns how to be recognized by its dam (Poindron and Leneindre 1980).

Piglets are rapidly fixed to one teat of their dam, which they actively defend against litter mates or alien young. The location of their usual teat appears to be given by the odour and/or taste. When suckling an 'artificial sow' comprised of a row of rubber teats, the piglets establish fixed positions as normally. Their positions are completely disturbed by thorough washing of the rubber teats (Jeppesen 1982).

### CHEMICAL SIGNALS AND SOCIAL RELATIONS

Wild and feral ungulates are associated in social groups. The importance of chemical signals in social relations is emphasized by the frequency of olfactory control occurring at first contact, as observed in cattle (Bouissou 1985). The possibility for the domestic ungulates to use olfactory cues for *individual recognition* has been experimentally demonstrated. Operant conditioning techniques show that sheep are able to distinguish between odours of a variety of excretory and secretory products from two different individuals: urine, wool, faeces, saliva, secretion of interdigital pouch, anteorbital gland, and inguinal wax (Baldwin and Meese 1977). With the same techniques, pigs (Meese *et al.* 1975), goats, cattle (Baldwin 1976), and stallions (Marinier *et al.* 1988) also successfully discriminate between individual conspecific urine samples.

The display and maintenance of dominance submission relationships in adult cows do not require visual communication. Since generally no acoustic signals are emitted, a role for individual recognition based on olfactory signals has been hypothesized (Bouissou 1971). Surprisingly, the establishment of such relationships among animals at their first contact takes place in the absence of visual communication (Bouissou 1985), but also between anosmic animals (Mansard and Bouissou 1980).

There is evidence for *aggression-modulating pheromones* in pigs: the intensity of fighting is reduced in anosmic pigs (Meese and Baldwin 1975), and the existence of a pheromone signalling submission has been mentioned by McGlone (1985). Spraying urine and blood plasma from pigs recently engaged in fighting reduced the duration of aggression when compared with subjects sprayed with urine of non-fighting pigs, or water. The 16-androstenes have a similar action (McGlone *et al.* 1987).

## Conclusions

Chemical signals play a major role in different aspects of inter-individual relations in domestic ungulates. However, the type of message, the complexity of the chemical signal, and the use of the information by the animal vary between different species and situations.

The simplest type of signal concerns a stable and permanent trait, such as the characteristics of the male in a given species. The male pheromones have been identified in goat and pig.

The situation is slightly more complex when the signal characterizes a physiological stage such as the oestrous female or the newborn. In some cases such signals are immediately efficient, whereas in others they have to be learned from previous experience, and/or have a cumulative action or could be replaced by other sensory cues. The chemical identification of such compounds remains a reasonable aim for research. Another promising area is the analysis of nervous mechanisms that control gonadotrophin release as a response to olfactory stimulation.

A higher level of complexity is reached when individual odours are involved. A 'chemical fingerprint' would allow individual recognition. Such a signal could be learned with extreme efficiency and reliability, as in post-partum development of the maternal bond. In this case, the complexity of both the signal and learning processes offer long-term future research objectives.

## References

Alexander, G. and Shillito, E.E. (1977*a*). The importance of odour, appearance and voice in maternal recognition of the young in merino sheep (*Ovis aries*). *Appl. Anim. Ethol.*, 3, 127–35.

Alexander, G. and Shillito, E.E. (1977*b*). Importance of visual clues from various body regions in maternal recognition of the young in merino sheep (*Ovis aries*). *Appl. Anim. Ethol.*, 3, 137–43.

Alexander, G. and Stevens, D. (1985*a*). Fostering in sheep: II Use of hessian coat to foster an additional lamb on to ewes with single lambs. *Appl. Anim. Behav. Sci.*, 14, 335–44.

Alexander, G. and Stevens, D. (1985*b*). Fostering in sheep: III Facilitation by the use of odorants. *Appl. Anim. Behav. Sci.*, 14, 345–54.

Baldwin, B.A. and Shillito, E.E. (1974). Effects of the ablation of the olfactory bulbs on parturition and maternal behaviour in Soay sheep. *Anim. Behav.*, 22, 220–3.

Baldwin, B.A. (1976). Ability of goats and calves to distinguish between conspecific urine samples using olfaction. *Appl. Anim. Ethol.*, 3, 145–50.

Baldwin, B.A. and Meese, G.B. (1977). The ability of sheep to distinguish between conspecifics by means of olfaction. *Physiol. Behav.*, 15, 803–8.

Blockey, M.A. de B. (1981). Modification of a serving capacity test for beef bulls. *Appl. Anim. Ethol.*, 7, 321–36.

Booth, W.D. (1984). Boar odour and puberty attainment in gilts. In *The male in farm animal reproduction* (ed. M. Courot), pp. 305–9. M. Nijhoff publ. Dordrecht (Netherlands).

Booth, W.D. and Baldwin, B.A. (1986). Evidence for a seasonal component in the delay of first oestrus following olfactory bulbectomy in prepubertal gilts. *Appl. Anim. Behav. Sci.*, 15, 196–7.

Bouissou, M.F. (1968). Effet de l'ablation des bulbes olfactifs sur la reconnaissance du jeune par sa mère chez les ovins. *Rev. Comp. Anim.*, 2, 77-83.

Bouissou, M.F. (1971). Effet de l'absence d'informations optiques et de contact physique sur la manifestation des relations hiérarchiques chez les bovins domestiques. *Ann. Biol. Anim. Bioch. Biophys.*, 11, 191–8.

Bouissou, M.F. (1985). Contribution á l'étude des relations interindividuelles chez les bovins domestiques femelles (*Bos taurus*, L.). Ph.D. thesis, University of Paris VI.

Brooks, P.H. and Cole, D.J.A. (1970). Effect of the presence of a boar on attainment of puberty in gilts. *J. Reprod. Fert.*, 23, 435–40.

Chemineau, P. (1983). Effect on oestrus and ovulation of exposing creole goats to the male at three times of the year. *J. Reprod. Fert.*, 67, 65–72.

Chemineau, P., Levy, F. and Thimonnier, J. (1986). Effects of anosmia on LH-secretion, ovulation and oestrus behaviour. *Anim. Reprod. Sci.*, 10, 125–32.

Chenoweth, P.J., Brinks, J.S. and Nett, T.M. (1979). A comparison of three methods of assessing sex-drive in yearling beef bulls and relationships with testosterone and LH levels. *Theriogenology*, 12, 223–33.

Cohen-Tannoudji, J. (1988). Effet du mâle sur la secretion gonadotrope (LH) de la brebis (*Ovis aries* L.). Ph.D. thesis, University of Paris VI.

Cohen-Tannoudji, J., Locatelli, A. and Signoret, J.P. (1986). Non-pheromonal stimulation by the male of LH release in the anoestrous ewe. *Physiol. Behav.*, 36, 921–4.

Collias, N.E. (1956). The analysis of socialization in sheep and goats. *Ecology*, 37, 228–39.

Dehnhard, M. and Claus, R. (1988). Reliability criteria of a bioassay with rats, trained to detect the estrus-specific odor in cow urine. *Theriogenology*, 30, 1127–38.

Dunn, G.C., Price, E.O. and Katz L.S. (1987). Fostering calves by odor transfer. *Appl. Anim. Behav. Sci.*, 17, 33–9.

Eyal, E. (1958). The introduction of rams as a factor influencing sexual activity of ewes. *Israel J. Agric. Res. Katvim*, 8, 359–66.

Garcia, M.C., McDonnell, S.M. and Kenney, R.M. (1986). Bull sexual behavior tests: stimulus cow affects performance. *Appl. Anim. Behav. Sci.*, 16, 1–10.

Godfrey, R.W., Lunstra, D.D. and Schanbacher, B.D. (1988). Influence of nasogenital contact or urine from estrus females on serum luteinizing hormone (LH) and testosterone (T) concentrations and flehmen in beef bulls. Proc. 20th Ann. Conf. Reprod. Behav., Omaha (Nebraska), June 1988.

Gonzalez, R., Poindron, P. and Signoret, J.P. (1988*a*). Temporal variation in LH and testosterone responses of rams to the introduction of females during the breeding season. *J. Reprod. Fert.*, 83, 201–8.

Gonzalez, R., Orgeur, P. and Signoret, J.P. (1988*b*). Luteinizing hormone, testosterone and cortisol responses in rams upon presentation of oestrous females during the non-breeding season. *Theriogenology*, 30, 1075–86.

Hale, E.B. (1966). Visual stimuli and reproductive behavior in bulls. *J. Anim. Sci.*, 25 (Suppl.), 36–44.

Hart, G.H., Mead, S.W. and Regan, W.M. (1946). Stimulating the sex-drive of bovine males in artificial insemination. *Endocrinology*, 39, 221–3.

Hradecky, P., Sis, R.F. and Klemm, W.R. (1983). Distribution of flehmen reactions of the bull throughout the oestrous cycle. *Theriogenology*, 20, 197–203.

Hughes, P.E. (1982). Factors affecting natural attainment of puberty. In *Control of pigs' reproduction* (ed. D.J.A. Cole and G.R. Foxcroft), pp. 117–138. Butterworth, London.

Illius, A.W., Haynes, N.B. and Lamming, G.E. (1976). Effects of ewe proximity on peripheral plasma testosterone levels and behaviour in the ram. *J. Reprod. Fert.*, 48, 25–32.

Inkster, I.J. (1957). The mating behaviour of sheep. *New Zealand Sheep Fmg A.*, pp.163–169

Izard, M.K. and Vandenbergh, J.G. (1982). The effects of bull urine on puberty and calving date in crossbred beef heifers. *J. Anim. Sci.*, 55, 1160–8.

Jeppesen, L.E. (1982). Teat-order in groups of piglets reared on an artificial sow. II Maintenance of teat-order with some evidence for the use of odour cues. *Appl. Anim. Ethol.*, 8, 347–55.

Jewell, P.A. and Grubb, P. (1974). In *Island survivors* (ed. P.A. Jewell, C. Milner and J. Morton Boyd), pp. 224–41. Oxford University Press, Oxford.

Katongole, C.B., Naftolin, F. and Short, R.V. (1971). Relationship between blood levels of luteinizing hormone and testosterone in bulls and the effects of sexual stimulation. *J. Endocr.*, 50, 457–66.

Kiddy, C.A. and Mitchell, D.S. (1981). Estrus-related odors in cows: time of occurrence. *J. Dairy Sci.*, 64, 267–71.

Kiddy, C.A., Mitchell, D.S., Bolt, D.J. and Hawk, H.W. (1978). Detection of oestrus-related odors in cows by trained dogs. *Biol. Reprod.* 19, 389–95.

Kirkwood, R.N., Forbes, J.M. and Hughes, P.E. (1981). Influence of boar contact on attainment of puberty in gilts after removal of the olfactory bulbs. *J. Reprod. Fert.*, 61, 193–6.

Kirkwood, R.N., Hughes, P.E. and Booth, W.D. (1983). The influence of boar-related odours on puberty attainment in gilts. *Anim. Prod.*, 36, 131–6.

Klemm, W.R., Hawkins, G.N. and De Los Santos, E. (1987). Identification of compounds in bovine cervico-vaginal mucus extracts that evoke male sexual behavior. *Chem. senses*, 12, 77–87.

Klopfer, P.H. and Gamble, J. (1966). Maternal imprinting in goats: the role of chemical senses. *Z. Tierpsychol.*, 23, 588–92.

Knight, T.W. and Birch, E.J. (1983). Identification of pheromones which will stimulate anovular ewes to ovulation. *Proc. Endocr. Soc. Aust.* 26, suppl. 2:3 (Abstr.).

Knight, T.W. and Lynch, P.R. (1980). Source of ram pheromones that stimulate ovulation in the ewe. *Anim. Reprod. Sci.*, 13, 74–6.

Ladewig, J. and Hart, B.L. (1981). Demonstration of estrus-related odors in cow urine by operant conditioning of rats. *Biol. Reprod.*, 24, 1165–9.

Ladewig, J., Price, E.O. and Hart, B.L. (1980). Flehmen in male goats: role in sexual behavior. *Behav. Neural Biol.*, 30, 312–22.

Leneindre, P., Petit, M. and Garel, J.P. (1978). Suckling of two calves by Salers cows. II Adoption of calves. *Ann. Zootech.*, 27, 553–69.

Levy, F. and Poindron, P. (1987). Importance of amniotic fluids for the establishment of maternal behaviour in relation with maternal experience in sheep. *Anim. Behav.*, 35, 1188–92.

Levy, F., Poindron, P. and Leneindre, P. (1983). Attraction and repulsion by amniotic fluid and their olfactory controls in the ewe around parturition. *Physiol. Behav.* 31, 687–92.

Lindsay, D.R. (1965). The importance of olfactory stimuli in the mating behaviour of rams. *Anim. Behav.*, 13, 75–8.

Lindsay, D.R. and Fletcher, I.C. (1972). Ram-seeking activity associated with oestrous behaviour in ewes. *Anim. Behav.*, 20, 452–6.

Lindsay, D.R. and Robinson, T.J. (1961). Studies of the efficiency of mating in the sheep. II. The effect of freedom of rams, paddock size, and age of ewes. *J. Agric. Sci.*, 57, 141–5.

Liptrap, R.M. and Raeside, J.I. (1978). A relationship between plasma concentrations of testosterone and corticosteroids during sexual and aggressive behaviour in the boar. *J. Endocr.*, 76, 75–85.

Mansard, C. and Bouissou, M.F. (1980). Effect of olfactory bulbs removal on the establishment of the dominance-submission relationships in domestic cattle. *Biol. Behav.*, 5, 169–78.

Marinier, S.L., Alexander, A.J. and Waring, G.H. (1988). Flehmen behaviour in the domestic horse: discrimination of conspecific odours. *Appl. Anim. Behav. Sci.*, 19, 227–37.

Martin, G.B., Oldham, C.M. and Lindsay, D.R. (1980). Increased plasma LH levels in seasonally anovular merino ewes following the introduction of the rams. *Anim. Reprod. Sci.*, 3, 125–32.

McGlone, J.J. (1985). Olfactory cues and pig aggressive behavior: evidence for a submissive pheromone. *Physiol. Behav.*, 34, 195–8.

McGlone, J.J. and Morrow, J.L. (1987). Individual differences among mature boars in T-maze preference for oestrous or non-oestrous sows. *Appl. Anim. Behav. Sci.*, 17, 77–82.

McGlone, J.J., Curtis, S.E. and Banks, E.M. (1987). Evidence for aggression-modulating pheromones in prepubertal pigs. *Behav. Neural Biol.*, 47, 27–39.

Meese, G.B. and Baldwin, B.A. (1975). Effects of ablation of the olfactory bulbs on aggressive behaviour in pigs. *Appl. Anim. Ethol.*, 1, 251–62.

Meese, G.B., Conner, D.J. and Baldwin, B.A. (1975). Ability of the pig to distinguish between conspecific urine samples using olfaction. *Physiol. Behav.*, 15, 121–5.

Melrose, D.R., Reed, H.C.B. and Patterson, R.L.S. (1971). Androgen steroids associated with boar odour as an aid to the detection of oestrus in pig artificial insemination. *Brit. Vet. J.*, 127, 497–502.

Mesnil du Buisson, F. du and Signoret, J.P. (1962). Influences externes sur le déclenchement de la puberté chez la truie. *Ann. Zootech.*, 11, 53–9.

Moore, R.W., Whyman, D. and Wilson, P.R. (1978). Effects of sexual stimulation on plasma levels of LH and testosterone in rams from high- and low-fertility flocks. *J. Reprod. Fert.*, 53, 67–70.

Nowak, R., Poindron, P., Leneindre, P. and Putu, I.G. (1987). Ability of 12-hours-old merino and crossbred lambs to recognize their mothers. *Appl. Anim. Behav. Sci.*, 17, 263–71.

Over, R. (1988). Wirkung von pheromonen des Ziegenbockes auf die LH-freiseztung bei weiblichen saisonal azyklischen Ziegen des Rasse bunte deutsche Edelziege. Thesis, University of Stuttgart-Hohenheim.

Paleologou, A.M. (1977). Detecting oestrus in cows by a method based on bovine sex pheromones. *Vet. Rec.*, 100, 319–20

Patterson, R.L.S. (1968). 5α-Androst-16-ene-3-one: compound responsible for taint in boar fat. *J Food Sci. Agric.*, 19, 31–8.

Pearce, G.P. and Hughes, P.E. (1987). An investigation of the roles of boar-component stimuli in the expression of proceptivity in the female pig. *Appl. Anim. Behav. Sci.*, 18, 287–99.

Perry, G.C. (1985). Mating behaviour. In *World animal science. A-5 Ethology of farm animals* (ed. A.F. Fraser), pp.325–33. Elsevier, Amsterdam.

Poindron, P. (1974). Etude de la relation mère-jeune chez des brebis (Ovis aries), lors de l'allaitement. *C.R. Acad. Sci.*, 282, 2691–4.

Poindron, P. (1976). Mother–young relationships in intact or anosmic ewes at the time of suckling. *Biol. Behav.*, 2, 161–77.

Poindron, P. and Carrick, M.J. (1976). Hearing recognition of the lamb by its mother. *Anim. Behav.*, 24, 600–2.

Poindron, P. and Leneindre, P. (1980). Endocrine and sensory regulation of maternal behavior in the ewe. *Adv. Study Behav.*, 11, 75–119.

Poindron, P., Cognie, Y., Gayerie, F., Orgeur, P., Oldham, C.M. and Ravault, J.P. (1980). Changes in gonadotropins and prolactin levels in isolated (seasonally or lactationally) anovular ewes associated with ovulation caused by the introduction of rams. *Physiol. Behav.*, 25, 227-36.

Price, E.O., Smith, V.M. and Katz, L.S. (1986). Stimulus conditions influencing self-enurination, genital grooming and flehmen in male goats. *Appl. Anim. Behav. Sci.*, 16, 371–81.

Prud'Hon, M. and Denoy, I. (1969). Effet de l'introduction de béliers vasectomisés dans un troupeau de Mérinos d'Arles, quinze jours avant le début de la lutte de printemps, sur l'apparition des oestrus, la fréquence des erreurs de détection des ruts et la fertilité des brebis. *Ann. Zootech.*, 18, 95–106.

Rowlinson, R. and Bryant, M.J. (1974). Un système permettant de maîtriser l'anoestrus de lactation: l'effet du mâle. *Congr. Int. Med. Vet. Porcine (Lyon)*, C5 (Abstr.).

Sambraus, H.H. and Waring, G.H. (1975). Der Einfluss des Harns brünstiger Kühe auf die Geschlechstlust von Stieren. *Z. S ugertierk.*, 40, 49–54.

Sanford, L.M., Palmer, W.M. and Howland, B.E. (1974). Influence of sexual activity on serum levels of LH and testosterone in the ram. *Canad. J. Anim. Sci.*, 54, 579–85.

Schanbacher, B.D., Orgeur, P., Pelletier, J. and Signoret, J.P. (1987). Behavioural and hormonal responses of sexually experienced Ile de France rams to oestrous females. *Anim. Reprod. Sci.* 14, 293–300.

Signoret, J.P. (1970). Sexual behaviour patterns in female domestic pigs (*Sus scrofa*) reared in isolation from males. *Anim. Behav.* 18, 165–168.

Signoret, J.P. (1972). The mating behaviour of the sow. In *Pig production* (ed. D.J.A. Cole), pp. 295–314. Butterworth, London.

Signoret, J.P. (1974). Role des différentes informations sensorielles dans l'attraction de la femelles en oestrus chez les Porcins. *Ann. Biol. Anim. Bioch. Bioph.* 14, 747–55.

Signoret, J.P. (1975). Influence of the sexual receptivity of a teaser ewe on the mating preferences in the ram. *Appl. Anim. Ethol.* 1, 229–32.

Signoret, J.P. and Lindsay, D.R. (1982). The male effect in domestic mammals: effect on LH secretion and ovulation. Importance of olfactory cues. In *Olfaction and endocrine regulation* (ed. W. Breiphol), pp. 63–72. IRL Press, London.

Signoret, J.P. and Mauléon, P. (1962). Influence de l'ablation des bulbes olfactifs sur les mécanismes de la reproduction chez la truie. *Ann. Biol. Anim. Biochem. Bioph. Biophys.*, 2, 167–74.

Signoret, J.P. and Mesnil du Buisson, F. du (1961). Etude du comportement de la truie en oestrus. *Congr. Int. Reprod. Anim. La Haye*, pp. 171–5.

Sugiyama, T., Sasada, H., Masaki, F. and Yamashita, K. (1981). Unusual fatty acids with specific odor from mature male goats. *Agric. Biol. Chem.* 45, 2655–58.

Tschanz, B. (1962). Über die Beziehungen zwischen Muttertier und Jungen beim Mufflon (*Ovis aries musimon*, Pall.). *Experientia* 18, 187–90.

Vince, M.A. and Ward, T.M. (1984). The responsiveness of newly born clun forest lambs to odour sources in the ewe. *Behaviour*, 89, 117–27.

Vince, M.A., Lynch, J.J., Mottershead, B.E., Green, G.C. and Elwin, R.L. (1987). Interactions between normal ewes and newly born lambs deprived of visual, olfactory and tactile sensory information. *Appl. Anim. Behav. Sci.* 19, 119–36.

Watson, R.H. and Radford, H.M. (1960). The influence of rams on the onset of oestrus in merino ewes in spring. *Aust. J. Agric.Res.* 11, 65–71.

Yarney, T.A. and Sanford, L.M. (1983). The reproductive-endocrine response of adult rams to sexual encounters with estrual ewes is season dependent. *Horm. Behav.* 17, 169–82.

Zalesky, D.D., Day, M.L., Garcia-Winder, M., Imakawa, K., Kittok, R.J., D'Occhio, M.J. and Kinder, J.E. (1984). Influence of exposure to bulls on resumption of estrous cycles following parturition in beef cows. *J. Anim. Sci.* 59, 1135–9.

# VI.4 Sample Contact Required for Complete Bull Response to Oestrous Pheromone in Cattle

G. RIVARD AND W.R. KLEMM

## Introduction

The role and importance of chemical signals in reproductive behaviour has been clearly established in several species of mammals (Halpin 1986). In ungulates, for example, during the courtship ritual the male presents a unique oro-facial behavioural pattern, the flehmen, by which he seems to identify some specific chemical message about the state of oestrus.

Some literature suggests that the vomeronasal organ (VNO) mediates direct-contact chemoreception. Beauchamp *et al.* (1983) studied male guinea pigs (*Cavia porcellus*) that were prevented from contacting female conspecific urine; subsequent male interest in investigating these samples declined following repeated testing. In contrast, contact with the female urine sample evoked more sexual behaviour, and it was concluded that non-volatile molecules provide sensory input to the VNO. In ruminants, we know from the movement of fluorescent dye that flehmen is associated with the introduction of liquids into the VNO (Ladewig and Hart 1980; Melese-d'Hospital and Hart 1985).The neural pathways for VNO detection in vertebrates are separate and distinct from olfactory pathways (Estes 1972; Wysocki 1979; Johns 1980; Halpern 1987).

The ultimate criterion for effective pheromonal communication of the sexual status by a female is a clear message emitted from her body fluids. The design feature of the molecule(s) should optimize reproductive behaviour by being able to attract a male from a distance and then to permit confirmation at close range. The pheromone must be volatile and perhaps of low molecular weight for air diffusibility, in order to announce imminent ovulation. We postulate the existence of a second chemical message that would confirm, at close range, the message implied at long range, stimulating the male receiver to generate the appropriate mating behaviour. Such a second messenger need not be volatile or of low molecular weight, as long as the male can gain direct access to it by taste or via the vomeronasal senses.

Our laboratory has shown that bull (*Bos taurus*) flehmen is specifically related to oestrus detection (Jacobs *et al.* 1980, 1981; Hradecky *et al.* 1983; Klemm *et al.* 1987). This species seems best suited to evaluate the role of contact in pheromone detection because it is well studied, commonly available, and readily housed. Bulls sniff and touch the vulva of oestrous cows and then display two to six tongue compression strokes of the hard palate, presumably transferring vaginal materials to the vomeronasal organ, followed by the flehmen response (Jacobs *et al.* 1981). During the preparatory phase of mating behaviour, the oestrous pheromone causes a neurological response that relaxes the retractor penis muscles and releases blood flow into the corpus cavernosum penis in preparation for erection (Walker 1984).

We previously defined the bull behaviour in response to serum from oestrous cows (Rivard and Klemm 1989). Here, we test the hypothesis that initiation of mating behaviours requires licking, and not just sniffing, of oestrous samples.

## Methods

### SAMPLE PREPARATION

We already know that the best sources of pheromone are the serum or the vulval skin glands collected from day 3 to day 1 prior to ovulation (Rivard and Klemm 1989). Therefore, we collected blood just before oestrus from 12 mature, mixed-breed cows. In one group, oestrus was induced with 6 ml of prostaglandin F2 alpha (dinoprost), Upjohn's Lutalyse (cows 206, 210, 253, and 474). Using a second method, natural oestrus was identified at or near ovulation time by physical examination of both ovaries via rectal palpation (cows 10, 13, 14, 217, 404, 606, 808, and 1152). About 100 ml of blood was collected per cow in vacutainers each day at the time of oestrus but for no more than three consecutive days. Then the 15–20 ml serum was extracted by centrifugation and stored at −70 °C until the bull behavioural assay.

Serum samples were divided randomly into two groups of nine samples, but with alternate distribution of samples coming from artificial or natural oestrus. Deionized water (20 ml) was used as control presentation and was poured into the container from the bottle just before the bull assay. Water was deemed a suitable control because we had previously determined that no responses were elicited from serum samples from four anoestrous cows or from two bulls.

### BEHAVIOURAL ASSAY

We performed the assay on unrestrained, freely behaving bulls under conditions that simulated the natural environment in which bovine reproductive behaviour occurs.

Samples were presented in (i) a large, 25 cm diameter, flat stainless-steel plate that allowed bulls to touch and to lick the serum sample or the control sample; and (ii) a narrow, 12.5 cm diameter, deep stainless-steel bowl that prevented any direct contact with serum sample or control. The same sample could not be tested in both the bowl and the plate because much of the volatile pheromone evaporated during a given test.

Before presentation either the serum sample or the water control was heated to 37°C using an electronic, battery-powered heating unit. The plate or the bowl was placed on the ground at an extremity of a 450 $m^2$ pen where five mature bulls were housed. There were no cows in this pen.

Testing was conducted only on days when there was no rain or strong wind. In order to reduce conditioning, the tests were conducted irregularly, from November 1987 to April 1988, with no more than two sessions per week. Also, we never presented more than three serum samples and three water controls in either form of presentation. Both plate and bowl presentation were used alternately, with the water control presented in the same form immediately afterwards. The bulls were observed for a minimum of 15 mins at each presentation.

### DATA RECORDING

The data were recorded as simple event sequences as soon as the serum-borne odour stimulus was presented to a given bull, which became the focal individual. The presence or absence (+/-) of a given event was tallied for serum samples from each cow. The water control never elicited attraction behaviour.

Chi-squared goodness-of-fit tests with pooled data were performed separately for each type of stimulus presentation. Finally, we determined one-way state transition probabilities for both plate and bowl stimuli.

## Results and discussion

### BULL REPERTOIRE

The behaviour of bulls in response to serum presentation was assessed as a chain of stereotyped behaviors, which we describe as follows:

1. Head directed ($H_d$): The bull turned or oriented his head in the direction of the sample. The nostrils were apparently enlarged. The muzzle was wet with secretion.
2. Distance inspection ($D_i$): Without body displacement, the bull breathed deeply with apparent movements of the muzzle.
3. Attraction (A): The bull walked carefully toward the sample, increasing his olfactory investigation.
4. Close inspection ($C_i$): The bull extended his head over the sample, took deep breaths (2 to 3), and stayed near the odour source.
5. Salivation (S): From the muzzle and the lips, the saliva was apparent

while the bull wandered around the sample for a maximum of 10 feet.

6. Urination (U): Either standing or walking, the bull urinated in a stream-like fashion, followed by jets of urine.
7. Licking of sample ($L_s$): Coming back to the sample, the bull touched the sample with his tongue, one to three times.
8. Tongue manipulations ($T_m$): Wiping of the muzzle and tongue-compression strokes were apparent.
9. Hypersalivation (H): Strings of saliva dropped from the bull's mouth.
10. Yawning (Y): Contracting the muscles of his neck and lower jaw, the bull opened his mouth and made some lateral movements of the lower jaw and tongue.
11. Flehmen (F): With the head extended horizontally, the bull had his upper lip curled and retracted for some seconds to minutes.
12. Vocalization (V): Two to three bellowings were emitted at short intervals.
13. Penis protrusion ($P_p$): The penis was out of the preputial skin fold.
14. Penis dribbling ($P_d$): Liquids dropped from the penis. Contractions of the penis were sometimes apparent.
15. Head butting ($H_b$): Males show agonistic manifestations, such as threatening, butting, chasing, fighting, and withdrawing (Reinhardt *et al.* 1986).
16. Mounting (M): The same male who touched the sample mounted a nearby male. The observer occasionally saw ejaculation from a standing position.

## BOWL VS. PLATE PRESENTATION DATA

From the bowl data, we observed ten different action patterns of which the yawning behaviour was a unique response to smell, and in most cases (77%) terminated the sequence of events. With the plate presentation, the first six events of the sequence were similar to those observed after bowl presentation; but they were followed by four events (testing responses) in which licking triggered the flehmen instead of the yawning response. Then, five sex-related events completed the full sequence of action patterns (Table 1 a, b).

When we tallied the incidence of each behaviour (df 15) over 18 samples from all cows, the frequency of the various behaviour patterns was significantly different than expected by chance with the bowl data (Chi square $= 48.25$, $p < 0.001$). The cell values in the Freeman-Tukey Z scores that were less than expected by chance were for licking, flehmen, vocalization, penis protrusion, penis dribbling, head butting, and mounting. When both bowl and plate presentations were compared, licking of the sample appeared as the only behaviour that occurred less than expected by chance.

The state-transition diagram (Fig. 1) revealed a high degree of serial

**Table 1.** Incidence of various behaviour patterns

a. No Touching the sample (bowl presentation)

| Cow No.: | 210 | 210 | 206 | 404 | 253 | 253 | 217 | 1152 | 808 | Total |
|---|---|---|---|---|---|---|---|---|---|---|
| Head directed | + | + | + | + | + | + | + | + | + | 9 |
| Distance inspection | + | + | + | + | + | + | + | + | + | 9 |
| Attraction | + | + | + | + | - | + | + | + | + | 8 |
| Close inspection | + | + | + | + | - | + | + | + | + | 8 |
| Salivation | + | - | + | + | - | + | + | + | + | 7 |
| Urination | - | - | - | - | - | + | - | + | + | 3 |
| Tongue manipulation | + | + | + | + | - | - | - | + | + | 6 |
| Hypersalivation | - | + | + | + | - | - | - | - | - | 3 |
| Yawning | + | + | + | + | - | - | - | - | - | 4 |
| Penis protrusion | - | - | - | - | - | - | - | + | + | 2 |

b. Touching the sample (plate presentation)

| Cow No. | 606 | 217 | 253 | 014 | 013 | 010 | 206 | 474 | 474 | Total |
|---|---|---|---|---|---|---|---|---|---|---|
| Head directed | + | + | + | + | + | + | + | + | + | 9 |
| Distance inspection | + | + | + | + | + | + | + | + | + | 9 |
| Attraction | + | + | + | + | + | + | + | + | + | 9 |
| Close inspection | + | + | + | + | + | + | + | + | + | 9 |
| Salivation | + | + | + | + | + | + | + | + | + | 9 |
| Urination | + | + | + | + | + | + | + | + | + | 9 |
| Licking | + | + | + | + | + | + | + | + | + | 9 |
| Tongue manipulation | + | + | + | + | + | + | + | + | + | 9 |
| Hypersalivation | + | - | + | + | + | + | + | + | + | 8 |
| Flehmen | + | + | + | + | + | + | + | + | + | 9 |
| Vocalizations | - | - | - | + | + | + | - | + | - | 4 |
| Penis protrusion | - | - | + | + | + | + | - | + | + | 6 |
| Penis dribbling | - | - | + | + | + | + | - | + | + | 6 |
| Head butting | - | - | + | + | + | + | - | + | + | 6 |
| Mounting | - | - | + | + | + | + | - | + | + | 6 |

dependency in the sequence of behaviour patterns, particularly when the sample could be tasted (plate data). The bowl data demonstrated that the chain of actions was interrupted after salivation. The serum samples also elicited agonistic behaviour prior to the sexual mating attempts. The difference between the two chained sequences showed that the yawning behaviour was a response to smell, and no further events were usually recorded. Licking sample was an important component of the testing and mounting ritual. The data support the view that when chemosensory input to the VNO is prevented, as in bowl presentation,

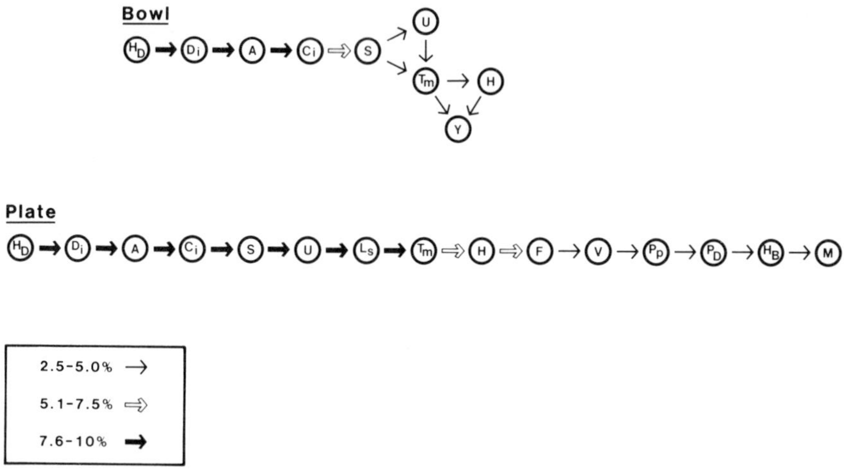


Fig. 1. State-transition diagrams (with transition probabilities).

responsiveness to conspecific oestrous pheromone diminishes. The completion of one phase in this complex behavioural sequence provided the stimulus that initiated the next phase.

We conclude that the sequence has two parts: (1) a set of orienting and approaching reactions triggered by an olfactory signal; (2) a pattern of overt reproductive behaviour that is triggered by the same or a different pheromone that is tasted and/or sampled via the VNO.

## Acknowledgements

This research was supported by a grant from the U.S. Department of Agriculture, No. 87-CRCR-1-2579.

## References

Beauchamp, G.K., Martin, G.I., Wellington, J.L. and Wysocki, C.J. (1983). The accessory olfactory system: role in maintenance of chemoinvestigatory behaviour. In *Chemical Signals in Vertebrates* 3, (ed. D. Müller-Schwarze and R.M. Silverstein), pp. 73–86. Plenum, New York.

Estes, R.D. (1972). The role of the vomeronasal organ in mammalian reproduction. *Mammalia*, 36, 317–30.

Halpern, M. (1987). The organization and function of the vomeronasal system. *Annu. Rev. Neurosci.*, 10, 325–62.

Halpin, Z.T. (1986). Individual odors among mammals: Origins and functions.

*Adv. Study Behav.*, 16, 39–70.

Hradecky, P., Sis, R.F. and Klemm, W.R. (1983). Distribution of flehmen reactions of the bull throughout the bovine estrous cycle. *Theriogenology*, 20, 197–204.

Jacobs, V.L., Sis, R.F., Chenoweth, P.J., Klemm, W.R., Sherry, C.J., and Coppock, E.E. (1980). Tongue manipulation of the palate assists estrous detection in the bovine. *Theriogenology*, 13, 353–6.

Jacobs, V.L., Sis, R.F., Chenoweth, P.J., Klemm, W.R. and Sherry, C.J. (1981). Structures of the bovine vomeronasal complex and its relationships to the palate: tongue manipulation. *Acta Anat.*, 110, 48–58.

Johns, M.A. (1980). The role of the vomeronasal system in mammalian reproductive physiology. In *Chemical Signals in Vertebrates*, 3 (ed. D. Müller-Schwarze and R.M. Silverstein), pp. 341–63. Plenum, N.Y.

Klemm, W.R., Hawkins, G.N. and De Los Santos, E. (1987). Identification of compounds in bovine cervico-vaginal mucus extracts that evoke male sexual behavior. *Chem. Senses*, 12, 77–87.

Ladewig, J. and Hart, B.L. (1980). Flehmen and vomeronasal organ function in male goats. *Physiol. Behav.*, 24, 1067–71.

Melese-d'Hospital, P.Y. and Hart, B.L. (1985). Vomeronasal organ cannulation in male goats: evidence for transport of fluid from oral cavity to vomeronasal organ during flehmen. *Physiol. Behav.*, 35, 941–4.

Reinhardt, C., Reinhardt, A. and Reinhardt, V. (1986). Social behaviour and reproductive performance in semi-wild Scottish Highland cattle. *Appl. Anim. Behav. Sci.*, 15, 125–36.

Rivard, G. and Klemm, W.R. (1989). Two body fluids containing bovine estrous pheromone(s). *Chem. Senses*, 14, 273–9.

Walker, D.F. (1984). Causes of copulatory failure in beef bulls. *Mod. Vet. Pract.*, 65, 252–6.

Wysocki, C.J. (1979). Neurobehavioral evidence for the involvement of the vomeronasal system in mammalian reproduction. *Neurosci. Biobehav. Rev.*, 3, 301–41.

# VI.5 Using Chromatography and a Dog to Identify Some of the Compounds in Human Sweat which are Under Genetic Influence

BARBARA A. SOMMERVILLE, MICHARD A. GREEN AND DAVID J. GEE

## Introduction

Chemical messages have long been known to play an important part in the lives of most mammals, and it is becoming apparent that the information transmitted may not only describe the social and reproductive status of the producer but also its genotype. Beauchamp *et al.* (1985) have shown that mice can detect a difference of only one gene by monitoring the urine chemistry of a companion and Hepper (1987) has made a similar observation in the rat. Although humans are regarded as being more visually motivated and less influenced by smell than most mammals, they can show a surprising acuity in discriminating each other on the basis of body odour (BO) (Russell 1976; Porter and Moore 1981). Dogs find it easy to distinguish individual humans by their scent (Sommerville and Gee 1987) unless they are presented with identical twins, but Kalmus (1955) found that they could distinguish twins if the scent of both was available for comparison. Hepper (1988) has extended this observation and shown that T-shirts worn by infant identical twins on the same breast-fed diet, in a similar environment proved indistinguishable but, after weaning, sufficient differences occurred to enable a dog to distinguish between them.

The Hungarian police actually use dogs to match scent samples collected at the scene of a crime with the BO of criminal suspects (Szinak 1985). A similar procedure is carried out in Holland. The British police restrict their use of dogs to tracking a known individual or locating hidden bodies, drugs, and explosives; one dog is programmed for one particular task, rather than several, owing to the communication barrier between dog and man. Our research project was conceived as an attempt to bypass this barrier, by building a mechanical dog capable of distinguishing one person from another on the sole basis of BO and able to supply a hard copy of its decision.

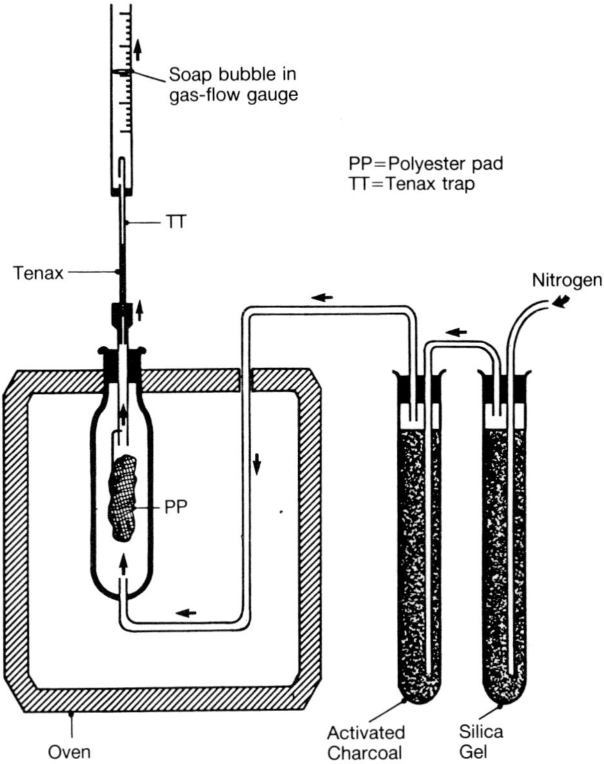


Fig 1. Apparatus for loading a Tenax trap used for concentrating headspace samples of human axillary sweat.

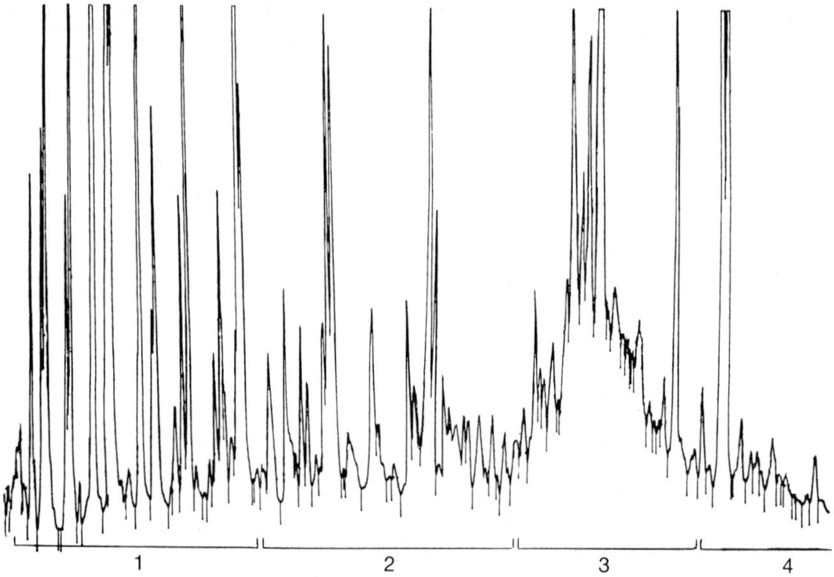


Fig 2. Typical gas chromatogram of human axillary sweat run on a BP10 column, frations 1–4 are indicated.

## Methods and results

We chose underarm sweat as the starting point for analysis since it is a notorious source of BO and also carries a pheromonal component (Cutler *et al.* 1986; Gower *et al.* 1985). Sweat donors were issued with a collecting kit containing unperfumed soap, with which to wash the armpits during the 24 h prior to and during sampling. The kit also contained nine polyester squares, to pin in pairs under the two sleeves of a T-shirt. The donors were requested to wear the squares for a 12 h daytime period and fresh ones for a 12 h night period, then to place them in labelled polythene bags and seal these in aluminium foil. One spare square, handled and packed as the others but not pinned under the arms was used as a blank to assess the amount of environmental contamination.

The apparatus shown in Fig. 1 was used to remove the volatiles from the skin-side polyester square (pp) and concentrate them in a Tenax trap (TT). Each sample was then desorbed from the Tenax at 320°C in a suitably modified gas chromatograph injector, freeze focused, and run down a capillary column of either BP1, BP10 or BP20 polarity, to a flame ionization detector (FID).

It was found that:

1. There were of the order of 200 peaks visible on the chromatogram (Fig. 2).
2. Most of those of lower boiling point (Fig. 2, region 1) were present in the blank sample.

**Table 1.** Results of the computerized pattern-matching system carried out on two different sets of sweat chromatograms: those from a pair of identical twins and those from a pair of unrelated people

| A and B[a] | Twin pair $n=6$ | | Unrelated people $n=10$ |
|---|---|---|---|
| Common peaks | $\frac{17}{27}$ | | $\frac{15}{26}$ |
| (ED) Euclidian Distance | $0.66 \pm 0.07$ | * | $0.94 \pm 0.06$ |
| (BCD) Box car distance | $1.03 \pm 0.17$ | * | $2.26 \pm 0.19$ |

Values are means ± SEM.
* $p<0.05$ determined by Student's t-test.
[a] If A and B are identical, the ED and BCD will be 0.00.

3. The region of the chromatogram marked 2 in Fig. 2 showed certain individual variations in pattern, which remained fairly constant for several weeks and these were more similar in identical twins than in unrelated people. Table 1 shows the results of a computerized pattern-matching analysis carried out on this region. The chromatograms were reconstructed from the ion peaks obtained by mass spectrometry, and the print-outs were digitized and transferred to an ATARI personal computer. The general alignment of the pair of chromatograms (A and B) was carried out on the basis of a small number of compounds common to both A and B but not all of the 17 (or 15) peaks recognized as being aligned by the analysis system were identical compounds. Nevertheless, the Euclidian and Box Car Distances (derived by adding together the discrepancies in height between each aligned peak in the two chromatograms) did show a statistically significant difference between the value for twins and that for unrelated people.
4. No conspicuous difference which might be attributed to age or sex was noted in this region.

The dog discrimination tests were set up by laying the fabric squares containing the sample over the top of a disposable plastic cup containing warm water. The cup was covered with an upturned plastic pot perforated by holes so that convection currents carried the volatiles up to the dog's nose. A trained German Shepherd dog was given a duplicate of one of the samples to sniff and then asked to retrieve a good match which it did by knocking over the appropriate pot and picking up the square. When matching unrelated people, three samples were presented at a time and when matching identical twins, two samples.

The dog made 13 correct matches out of 17 tests for unrelated people (a random score would have given about 6 out of 18), and 14 correct matches out of 21 for identical twins (random score, 10 out of 20).

The next stage was to find out whether the region of the chromatogram, in which individual differences had been noted, provided the dog with sufficient information to carry out human discrimination. This was done by fitting an outlet splitter to the distal end of the capillary column, so that fractions of the chromatogram could be captured in a moist polyester square while a portion of the outflow was still monitored by the F.I.D. Dog discrimination tests, using the fraction of the chromatogram marked 2 in Fig. 2, resulted in a correct scoring rate of 11 out of 14 for unrelated people (random score about 5 out of 15) but the identical twin samples were retrieved indiscriminately. Using the successive fraction of the chromatogram (3 in Fig. 2), both unrelated and identical twins could be distinguished with a score rating slightly better than that of the whole sample. (This may reflect the obscuring effect of environmental noise in a whole sample.)

Work is now in progress to split down fraction 2 still further to identify compounds that are the same in identical twins.

## Discussion

The dog discrimination tests have provided a valuable bioassay to narrow down the region of the sweat chromatogram that warrants detailed analysis. It is commonly found in pheromone research that the compounds of biological importance are represented by exceedingly small peaks in a gas chromatography run, the larger ones being either environmental contaminants or metabolic products common to all individuals. Because of this tendency, it is encouraging that the pattern recognition system has shown definite evidence of genetic variability among the peaks that are clearly visible. Nevertheless, it is a formidable challenge to develop an artificial system that can even approach the efficiency of an integrated biological system in the living animal.

## References

Beauchamp, G.K., Yamazaki, K. and Boyse, (1985). The chemosensory recognition of genetic individuality. *Sci. Amer.*, July, 66–72.

Cutler, W.N., Preti, C., Krieker, A., Huggins, G.R., Garcia, C.R. and Lawley, H.J. (1986). Human axillary secretions influence women's menstrual cycles: the role of donor extract from men. *Horm. Behav.*, 20, 443–73.

Gower, D.B., Bird, S., Sharma, P. and House, F.R. (1985). Axillary 5 $\alpha$ androst-16-en-3-one in men and women: relationships with olfactory acuity of odorous 16-androstenes. *Experientia*, 41, 1134–6.

Hepper, P.G. (1987). The discrimination of different degrees of relatedness in the rat: evidence for a genetic identifier. *Anim. Behav.*, 35, 549–54.

Hepper, P.G. (1988). The discrimination of human odour by the dog. *Perception*, 17 (4), 549–54.

Kalmus, H. (1955). The discrimination by the nose of the dog of individual human odours and in particular the odour of twins. *Anim. Behav.*, 3, 25–31.

Porter, R.H. and Moore, J.D. (1981). Human kin recognition by olfactory cues. *Physiol. Behav.*, 27, 493–5.

Russell, M.J. (1976). Human olfactory communication. *Nature*, 260, 520–3.

Sommerville, B.A. and Gee, D.J. (1987). Research on body odours: new prospects for combatting crime? *Int. Crmin. Police Rev.*, 407, 18–22.

Szinak, J. (1985). Identification of odours. A new and effective aid to law enforcement: the identification of offenders by comparing their personal odours with odours in stock. *Int. Crimin. Police Rev.*, March.

# VI.6 Potential for Applications of Anal Sac Volatiles from Domestic Dogs

J.W.S. BRADSHAW, S.E. NATYNCZUK AND D.W. MACDONALD

## Introduction

Despite the high densities of faecal and urinary scent marks of *Canis familiaris* in most centres of human habitation, remarkably little is known of their functions, either pre-domestication or current, and there is little information on their chemical composition. Urinary scent marks of domestic dogs appear to carry individual-specific information (Dunbar and Carmichael 1981; Brown and Johnston 1983), and may function as territorial markers in males (Stoddart 1980) and contain a sexual attractant in females (Doty and Dunbar 1974; Schultz *et al.* 1985). It has been suggested that the odour of domestic dogs' anal sac secretions could carry information relating to sexual attraction, territory marking, or alarm (Fox 1971; Doty and Dunbar 1974; Preti *et al.* 1976; Dunbar 1977).

The anal sac of canids may be discharged during, or independently of, defaecation apparently under voluntary control, although much more is known in this respect about the wolf *Canis lupus* than the domestic dog (Asa *et al.* 1985). If the anal sac secretions of domestic dogs were to contain information relating to dominance or territory, it should, at least theoretically, be possible to manipulate the signal and thereby control their defaecation patterns. This might be achieved either by attracting dogs to defaecate in chemically marked areas or by repelling dogs from areas such as footpaths and children's playgrounds.

The feasibility of whether such applications ultimately become possible will depend upon the characteristics of the signals from the anal sacs, and the reactions of both naïve and experienced dogs to those signals. The concept, borrowed from insect semiochemistry, of simple releaser compounds which reliably induce stereotyped behaviour patterns, has not proved useful in examining mammalian chemical signals. However, it is not yet firmly established that all mammalian responses to chemical signals are entirely learned. It was therefore thought worthwhile to examine the anal sacs of domestic dogs to see whether, beneath the expected pattern of individual variation, any universal, perhaps sex- or species-specific features could be detected. Published analyses of the components of the secretion (Preti *et al.* 1976) suggest that its composition varies little within *C. familiaris*, even between males and females or between females in different stages of oestrus. If any semiochemical signal could be decoded, it might contain attractant and/or repellent compounds which could be used in the ways mentioned above. A detailed chemical analysis was made of samples of anal sac secretion from domestic dogs of both sexes and two breeds, in an attempt to separate species-specific from individual-specific and randomly variable components. In this paper we summarize our findings on the physical composition of the sac contents, their pH, and their volatile components.

## Materials and methods

Details of collection of samples from the sacs are given in Natynczuk *et al.* (1989). Individual samples were collected from 12 male and 11 female English beagles at approximately four-weekly intervals over a 10-month period. Each was centrifuged for 10 mins at 2000–3000 x *g*. On removal from the centrifuge the size and colour of each of the layers was recorded, and the pH of the aqueous supernatant layer measured using narrow range pH paper (B.D.H.). Further samples were collected from 10 of the same female and 6 of the same male beagles, two months after the last sample in the series above, and from 10 female Labrador retrievers. Secretions from each breed/sex combination were pooled and analysed by headspace gas chromatography/mass spectrometry.

## Results

### APPEARANCE

The appearance of a typical centrifuged sample is shown in Fig. 1. The sac contents appear to consist of lipids, an aqueous layer, and particulate materials of several sizes. This concords with the role of the sac as fermentation vessel, in which bacteria generate and modify odour volatiles from precursors produced in the associated glands and epithelial tissue (Albone 1984). It was difficult to quantify the proportions of the different layers in each sample, which may in any case have been affected by the proportion of sac contents expressed on each occasion. However, the variation in colour and proportions was similar in extent to that reported by Doty and Dunbar (1974), and it was evident that there was considerable variation both between and within individuals.

### PH

Despite the rather imprecise method used for measuring pH, consistent differences were found between individual dogs (Fig. 2) over the entire 10-month period. There were no consistent differences between males and females ($F_{(1,18)} = 0.02$). The mean pH increased from 6.18 at the first sample to 6.79 at the fourth, after which its stayed relatively constant; this may have been an artefact of repeated sampling from the same animals.

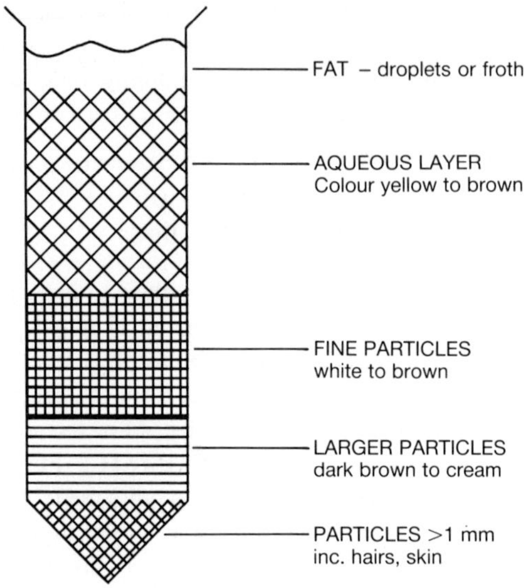


Fig. 1. Description of the appearance of sample after centrifugation.

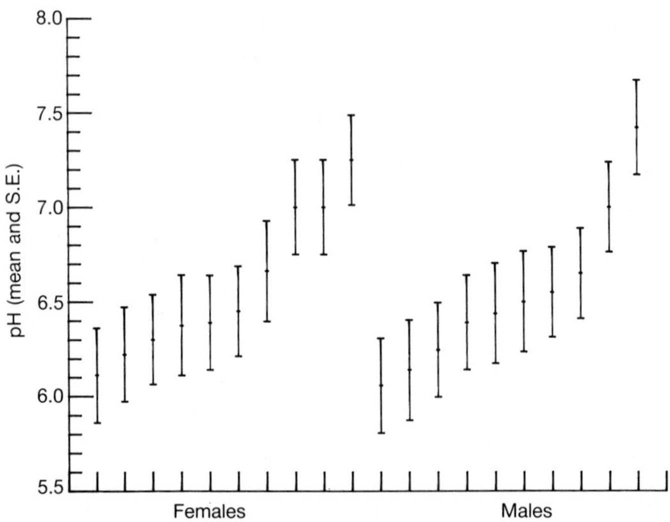


Fig. 2. Variation of pH between individual dogs.

HEADSPACE VOLATILES

Considerable differences were found between the three samples (Natynczuk *et al.* 1989). Out of 37 compounds identified, 19 were detected in only one sample, five in two, and only 13 in all three. While some of these differences may have been caused by threshold effects of the instrumentation employed, there were also major differences in the relative proportions of major components. For example, trimethylamine was found only in the male beagle sample, while acetic acid was found only in the female beagles; both of these compounds were major components in the analysis by Preti *et al.* (1976) of American beagles. Our analyses, although being from pooled samples, do indicate some of the diversity of the volatile signals which may be expected from individual dogs.

## Discussion

The volatile chemical signal is potentially controlled or modified at several distinct stages. These include the synthesis and release of volatile and non-volatile compounds into the sac; their modification by micro-organisms in the sac flora; changes in the activity and composition of that flora; the relative solubilities of the odour molecules in lipid and aqueous phases; for significantly ionizable molecules—the pH of the aqueous phase; and the substrate on to which the secretion is released. We have shown that the degree of control exerted by individual dogs appears to differ at two of these stages. Individuals hold the pH of their sac contents relatively constant, in contrast to the variation observed between individuals, but the proportion of lipid appears to vary from one sample to the next. The pH will have profound effects on the volatility of, for example, acetic acid and trimethylamine (Albone 1984), and the proportion of lipid will affect the relative volatilities of lipophilic and lipophobic compounds (Regnier and Goodwin 1977). Our headspace studies, reflecting the end-product of all but the last of the processes listed above, also indicate a high degree of individual variation between odours.

The marked individual and inter-class differences in secretion properties argue against any species-specific compounds or ratios of compounds which might have stereotyped effects on domestic dog behaviour. Rather, an individual-specific signal is suggested, which would have value in territorial or other marking, by means of a learned association between the scent mark and its depositor or by odour matching between the two (Gosling, this volume). The changing nature of the secretion would mean that, to be useful, a responding dog's odour 'picture' of its neighbours would have to be continually modified; dogs can remember individual urine scent signals from one day to the next (Brown and Johnston 1983), and may therefore be able to do the same for anal sac odours.

## References

Albone, E.S. (1984). *Mammalian Semiochemistry*. John Wiley, Chichester.

Asa, C.S., Peterson, E.K., Mech, L.D. and Seal, U.S. (1985). The deposition of anal sac secretions by captive wolves, *Canis lupus*. *J. Mammal.*, 66, 89–93.

Brown, D.S. and Johnston, R.E. (1983). Individual discrimination on the basis of urine in dogs and wolves. In *Chemical Signals in Vertebrates*, 3 (ed. D. Müller-Schwarze and R.M. Silverstein), pp. 343–6. Plenum, New York.

Doty, R.L. and Dunbar, I. (1974). Attraction of beagles to conspecific urine, vaginal and anal sac secretion odours. *Physiol. Behav.*, 12, 825–33.

Dunbar, I.F. (1977). Olfactory preference in dogs: the response of male and female beagles to conspecific odours. *Behav. Biol.*, 20, 471–81.

Dunbar, I. and Carmichael, M. (1981). The response of male dogs to urine from other males. *Behav. Neural Biol.*, 31, 465–70.

Fox, M.W. (1971). *Behaviour of Wolves, Dogs and Related Canids*. Jonathan Cape, London.

Natynczuk, S., Bradshaw, J.W.S. and Macdonald, D.W. (1989). Chemical constituents of the anal sacs of domestic dogs. *Biochem. Syst. Ecol.*, 17, 83–7.

Preti, G., Muetterties, E.L., Furman, J.M., Kennelly, J.J. and Johns, B.E. (1976). Volatile constituents of dog (*Canis familiaris*) and coyote (*Canis latrans*) anal sacs. *J. Chem. Ecol.*, 2, 177–86.

Regnier, F.E. and Goodwin, M. (1977). On the chemical and environmental modulation of pheromone release from vertebrate scent marks. In *Chemical Signals in Vertebrates* (ed. D. Müller-Schwarze and M.M. Mozell), pp. 115–33. Plenum, New York.

Schultz, T.H., Kruse, S.McK. and Flath, R.A. (1985). Some volatile constituents of dog urine. *J. Chem. Ecol.*, 11, 169–75.

Stoddart, D.M. (1980). *The Ecology of Vertebrate Olfaction*. Chapman and Hall, London.

# INDEX